Nicole Oresme, *Questiones in Meteorologica de ultima lectura, recensio parisiensis*

Medieval and Early Modern Philosophy and Science

VOLUME 32

The titles published in this series are listed at *brill.com/memps*

Nicole Oresme, *Questiones in Meteorologica de ultima lectura, recensio parisiensis*

Study of the Manuscript Tradition and Critical Edition of Books I–II.10

Edited by

Aurora Panzica

BRILL

LEIDEN | BOSTON

Cover illustration: Detail of Uppsala, Universitetsbibliotek, Ms. C 596, f. 2r.

SWISS NATIONAL SCIENCE FOUNDATION

The Open Access publication of this volume was financed by the Swiss National Science Foundation.

The Library of Congress Cataloging-in-Publication Data is available online at http://catalog.loc.gov
LC record available at http://lccn.loc.gov/2021016049

Typeface for the Latin, Greek, and Cyrillic scripts: "Brill". See and download: brill.com/brill-typeface.

ISSN 2468-6808
ISBN 978-90-04-46140-6 (hardback)
ISBN 978-90-04-46310-3 (e-book)

This book is printed on acid-free paper and produced in a sustainable manner.

Contents

Nicole Oresme, Questiones in Meteorologica de ultima lectura, recensio parisiensis

Acknowledgements

This critical edition was prepared as part of my doctoral dissertation at the University of Fribourg, Switzerland. I would like to express my gratitude to my supervisor, Tiziana Suarez-Nani, to my co-supervisor, Nicolas Weill-Parot, as well as to William O. Duba.

I am very thankful to the staff of the libraries that own copies of Oresme's commentary, from whom I frequently requested materials and information.

I am also very grateful to the Swiss National Science Foundation, which financed my doctoral project and the Open Acess publication of this volume, as well as to Paul Bakker and Christoph Lüthy, who agreed to publish this edition in the series *Medieval and Early Modern Philosophy and Science*. Paul's careful proofreading of the manuscript has proven invaluable to avoid a considerable number of mistakes and inconsistencies. Any remaining errors or oversights are my own.

Introduction: The Strange Case of the Second Redaction of Oresme's *Questions on Meteorology*

The redaction of Oresme's *Questions on Meteorology* that is partially edited in this volume was until recently considered to be Oresme's only commentary on Aristotle's *Meteorology*.[1] However, recent investigations have revealed the existence of two other commentaries on this Aristotelian text that should also be ascribed to Oresme: a literal commentary (*Sententia*) and a question commentary (*Questiones*).[2]

1 As an example, Weijers's and Lohr's inventories mention only this redaction of Oresme's *Questions on Meteorology*: O. Weijers and M. Calma, *Le travail intellectuel à la Faculté des arts de Paris: textes et maîtres (ca. 1200–1500). Répertoire des noms commençant par L-M-N-O*, Turnhout 2005 (Studia Artistarum, 13), 175; Ch. Lohr, *Latin Aristotle Commentaries*, I.2. *Medieval Authors M–Z*, Florence 2010 (Corpus Philosophorum Medii Aevi. Subsidia, 18), 34–35.

2 Both texts are transmitted in a Parisian manuscript which bears the signature Darmstadt, Universitäts- und Landesbibliothek, Hs. 2197. The importance of this manuscript has been pointed out by Stefano Caroti in his introduction to the edition of Oresme's *Questions on De generatione et corruptione*: Nicolaus Oresme, *Questiones super De generatione et corruptione*, ed. S. Caroti, München 1996 (Veröffentlichungen der Kommission für die Herausgabe ungedruckter Texte aus der mittelalterlichen Geisteswelt, 20), 35*–46*. The literal commentary is transmitted on ff. 100ra–123ra; the question commentary, on ff. 58ra–82rb and 125ra–127vb. Both texts are ascribed to Nicole Oresme in the manuscript: "Expliciunt questiones primi *Metheororum* compilate ante magistrum Nicholaum de Oresme Normannum Deo gratias. Incipiunt questiones secundi eiusdem ab eodem" (f. 81rb); "Explicit sententia primi *Metheororum* reportata ante magistrum Nicholaum Oresme nationis Normannorum. Incipit sententia secundi eiusdem reportata ab eodem" (f. 106ra). My research enabled me to identify five other copies of this redaction of Oresme's *Questions on Meteorology*: München, Bayerische Staatsbibliothek, Clm 4375, ff. 19ra–46rb; Kraków, Biblioteka Jagiellońska, cod. 753, ff. 51ra–83vb; cod. 635, pp. 194^{a}–209^{a} (only questions I.19–31); cod. 686, ff. 110vb–120ra (only questions I.19–31 and II.1–6); cod. 686, ff. 81ra–97va (only questions I.3 and I.12–32). For more information on this redaction of Oresme's commentary, see A. Panzica, "Une nouvelle rédaction des *Questions sur les Météorologiques* de Nicole Oresme", *Bulletin de philosophie médiévale* 57 (2015), 257–264; Ead., "Nicole Oresme à la Faculté des Arts de Paris: les *Questions sur les Météorologiques*", *Archives d'histoire doctrinale et littéraire du Moyen Âge* 84 (2017), 7–89, esp. 13–21; Ead., "Albert of Saxony's *Questions on Meteorology*: Introduction, Study of the Manuscript Tradition, and Edition of Book I–II.2", *Archives d'histoire doctrinale et littéraire du Moyen Âge* 86 (2019), 231–356, esp. 232–239; Ead., "L'hypothèse de la cessation des mouvements célestes au XIVe siècle: Nicole Oresme, Jean Buridan et Albert de Saxe", *Vivarium* 56 (2018), 83–125; Ead., *De la Lune à la Terre: les débats sur le premier livre des Météorologiques d'Aristote au Moyen Âge latin (XIIe–XVe siècles)*, forthcoming. An edition of this commentary will soon be published in the Brill series *Medieval and Early Modern Philosophy and Science*.

There is textual and doctrinal evidence that these two newly discovered texts predate the question commentary on Aristotle's *Meteorology* by Oresme that was previously known in the literature. Both texts are transmitted in a Parisian codex which dates back to 1346.[3] This important manuscript, which was copied by two students writing under Oresme's direct supervision,[4] is the oldest identified witness of an Oresmian text. A doctrinal comparison of the two redactions of Oresme's *Questions on Meteorology* suggests that the newly discovered commentary is the earlier of the two: in this text, Oresme's arguments are rich and profound, but sometimes inconsistent, and not as clearly formulated as in the previously known set of questions. Moreover, at some points, it seems that in the first redaction of his *Questions on Meteorology*, Oresme is searching for his own philosophical path, exploring solutions that he would subsequently abandon. For instance, he defends the astrological theses that astral influence acts not only on primary, but also on secondary qualities (question I.3); that planetary conjunctions affect social and political events (question I.5); and that the observation of the stars allows us to predict atmospheric phenomena (question II.5). By contrast, in the second redaction of his *Questions on Meteorology*, which was discovered first, Oresme tends to belittle the importance of celestial causes and to explain natural phenomena mainly on the basis of sublunary principles—an attitude that he maintains and further develops in his later writings.[5] In light of these considerations, I shall refer to the commentary edited in this volume as the *ultima lectura*, and to the recently discovered one as the *prima lectura*.[6]

There is neither internal nor external evidence that allows us to establish a precise dating for the second redaction of Oresme's *Questions on Meteorology*. Although the oldest manuscripts transmitting this text date back to the second half of the 1360s, we can reasonably suppose that Oresme prepared this commentary in the context of his teaching career at the Parisian Faculty of Arts. This

3 This date appears in the colophon at f. 192vb: "Expliciunt questiones super librum *De anima* reportate ante magistrum Johannem de Wesalia in vico straminum Parisius per manus Johannis Margan de Yvia, anno domini m ccc 46°".

4 The students in question are Iohannes Margan de Yvia and Henricus de Danderiche, both native of the diocese of Liège. For more information on these students, see Nicole Oresme, *Questiones super de generatione et corruptione*, ed. S. Caroti, 46*, as well as the introduction to the newly discovered redaction of Oresme's *Questions on Meteorology*, in this series (forthcoming).

5 Oresme's differing positions on astral influence in the two redactions of his *Questions on Meteorology* are addressed in Panzica, *De la Lune à la Terre*.

6 In some of the publications cited in footnote 2, I used the labels *redactio antiqua* (for the *prima lectura*) and *redactio nova* (for the *ultima lectura*).

period extends from the beginning of the 1340s[7] to the first half of the 1350s. The date commonly assumed to mark the end of Oresme's teaching activity at the Arts Faculty is 1356, the year in which he became Great Master of Theology at the College of Navarre.[8] Oresme's self-references provide us with some information about the relative chronology of the second redaction of his *Questions on Meteorology*.[9] At the end of question I.8, Oresme refers to his commentary on Aristotle's *Physics*.[10] According to the modern editors of his *Questions on the Physics*, the *terminus ante quem* of this text is 1347, as this was the year in which a theory of accidents that Oresme used extensively in his commentary was officially condemned.[11] Question I.8 also contains another self-reference, in which Oresme mentions the theory of *impetus* discussed in the first book of his commentary on Aristotle's *De celo*.[12] Oresme's *Questions on De celo* has been edited

7 Courtenay has shown that Oresme was already Master of Arts by 1342: W.J. Courtenay, "The Early Career of Nicole Oresme", *Isis* 91 (2000), 542–548. For biographical information about Oresme, see E. Grant's edition of Oresme's *De proportionibus proportionum* and *Ad pauca respicientes*, Madison 1966 (University of Wisconsin Publications in Medieval Science), 3–10; S.M. Babbitt, *Oresme's Livre de Politiques and the France of Charles V*, Philadelphia 1985 (Transactions of the American Philosophical Society, 75), 1–12; F. Neveux, "Nicole Oresme et le clergé normand du XIVe siècle", in J. Quillet (ed.), *Autour de Nicole Oresme. Actes du colloque Oresme organisé à l'Université de Paris XII*, Paris 1990 (Bibliothèque d'histoire de la philosophie), 9–36; M. Lejbowicz, "Nicole Oresme dans la lumière de l'urbanité", in P.J.J.M. Bakker, E. Faye, and C. Grellard (eds.), *Chemins de la pensée médiévale: Études offertes à Zénon Kaluza*, Turnhout 2002 (Textes et Etudes du Moyen Âge, 20), 675–708; Id., "Nicole Oresme, spectateur engagé", in J. Celeyrette and C. Grellard (eds.), *Nicole Oresme philosophe*, Turnhout 2014 (Studia Artistarum, 39), 21–61.

8 On this institution, founded by the King within the University of Paris, see N. Gorochov, *Le Collège de Navarre: de sa fondation (1305) au début du XVème siècle (1418): histoire de l'institution, de sa vie intellectuelle et de son recrutement*, Paris 1997 (Études d'histoire médiévale, 1) and W.J. Courtenay, "The University of Paris at the Time of Jean Buridan and Nicole Oresme", *Vivarium* 42 (2004), 3–17, esp. 12–14.

9 On the relative chronology of Oresme's commentaries used for teaching at the Arts Faculty see J. Celeyrette, "Les *Questions sur la Physique* dans l'œuvre de Nicole Oresme", in Celeyrette and Grellard (eds.), *Nicole Oresme philosophe*, 63–82, esp. 64–66 and Panzica, *Nicole Oresme à la Faculté des Arts de Paris*, 27–33.

10 This reference appears in the answer to the fifth argument *contra*: "dico quod motus calefacit accipiendo 'motum' pro 'mobili taliter se habere'; sed utrum mobile sic se habere sit ipsum mobile, vel aliud, vel que res sit, dictum est in libro *Physicorum*". Oresme deals with this problem in questions III.2–III.7 of his commentary on the *Physics*.

11 S. Caroti, J. Celeyrette, S. Kirschner, and E. Mazet (eds.), *Nicole Oresme. Questiones super Physicam (Books I–VII)*, Leiden/Boston 2013, XXV. See also S. Caroti, "*Modi rerum* and Materialism: a Note on a Quotation of a Condemned Articulus in Some Fourteenth-Century Parisian *De anima* commentaries", *Traditio* 55 (2000), 211–234.

12 This reference appears in the answer to the third argument *contra*: "concedo quod lapis in descendendo calefit. Et quando dicebatur: 'igitur leve fieret', concedo—hoc est, fit minus

by Kren, who considers it to be a very early work to be dated to the beginning of Oresme's teaching at the Arts Faculty—that is to say, according to Kren, to the second half of the 1340s.[13] In the text edited by Kren, Oresme refers to his *Questions on the Physics* and to his *Questions on Meteorology*. The fact that, in the second redaction of his *Questions on Meteorology*, he refers to the *Questions on De celo* can be explained in two different ways: either he is referring to the first redaction of his *Questions on Meteorology* in the *Questions on De celo*, or he is pointing to another version of his *Questions on De celo* in the second redaction of his *Questions on Meteorology*. Both solutions are possible. If we look more closely at the self-references in the *Questions on De celo*, we find that parallel passages can be found in both redactions of Oresme's *Questions on Meteorology*, yet these topics are dealt in more detail in the first redaction.[14] But it is

---

gravis. Et cum dicebatur: 'igitur in fine moveretur tardius quam in principio', concedo nisi aliud obesset; sed modo in eius descensu acquirit quemdam impetum de quo dicebatur super primum *Celi*".

13 C. Kren, *The Questiones super de Celo of Nicole Oresme*, unpublished Ph.D. Dissertation, University of Wisconsin 1965, X–XII.

14 The self-references at issue occur in question II.13, *Utrum tota terra semper quiescat*, ed. Kren 681[144–147]: "respondetur quod non oportet quia sicud dictum est super librum *Meteororum*, motus calefacit propter confricationem corporum que non est in proposito quia aer usque prope speram ignis movetur hoc modo". A corresponding passage can be found in question I.7 (*Utrum motus localis sit causa caloris*) of the first redaction of Oresme's *Questions on Meteorology*: "Secunda conclusio est quod motus turbidus calefacit. Patet, quia ex tali motu partes distrahuntur et rarefiunt, per primam suppositionem, et ad rarefactionem sequitur calefactio, per secundam [...]. Ultima conclusio: quod motus tranquillus calefacit, non tamen solus, sed quia ex confricatione cum exteriori continente fit motus turbidus, ex quo sequitur distractio, deinde rarefactio et calefactio consequentes. Patet ex exemplo de sagitta proiecta, et etiam de ferro confricato lapidi". Oresme addresses the same topic in question I.8 (*Utrum motus celi sit causa calefactionis ignis in sua spera et etiam aeris superioris*) of the second redaction of his commentary, but with greater concision: "Sic ergo motus qui fit cum confricatione est causa caloris. Unde ymaginandum est quod ex tali confricatione fit quedam rarefactio et quedam partium distractio, quam consequitur caliditas". The second reference to the commentary on *Meteorology* contained in the *Questions on De celo* is the following: "Ad rationes in oppositum ad primam, sicud est de igne, et cetera, dico quod non est simile quia ignis est propinquior celo quam terra et propter hoc movetur ut patuit primo *Meteororum*" (II.13, Kren, 695[266–269]). Oresme is probably referring to question I.4 (*Utrum aliquis motus localis in istis inferioribus sit effective a celo*) of the first redaction of his *Questions on Meteorology*: "Quinto movet celum inferiora mediante solu motu locali; sic dicitur quod movet speram ignis vel ignem in spera, quia propter nimiam velocitatem trahit secum ignem tali motu". The second redaction of his commentary does not contain any specific question devoted to the motion of the sphere of fire, but this topic is briefly addressed in question I.8 (*Utrum motus celi sit causa calefactionis ignis in sua spera et etiam aeris superioris*): "Secunda conclusio: ignis sic movetur, videlicet circulariter una cum celo, per virtutem sibi impressam a celo, eo modo

equally possible that, in the second redaction of his *Questions on Meteorology*, Oresme is referring to a different redaction of his *Questions on De celo* than the one edited by Kren. In fact, while Oresme refers to the first book of his *Questions on De celo* for a discussion of *impetus* theory, in the text edited by Kren this topic is discussed in the second book.[15] This might be an additional argument for ascribing to Oresme the anonymous set of questions on Aristotle's *De celo* transmitted in the manuscript München, Bayerische Staatsbibliothek, Clm 4375, ff. 47ra–76ra, whose Oresmian paternity was suggested by Kirschner.[16] In this commentary, *impetus* theory is discussed in the first book.[17]

The second redaction of Oresme's *Questions on Meteorology* is transmitted in nineteen manuscripts, one of which was identified during the preparation of this edition (Poznań, Archivum Archidiecezjalne, Cms 53, ff. 1^{r}–95^{v}). Almost all of these manuscripts originated in Central and Eastern Europe, particularly in Prague. This situation has precise historical reasons, since Oresme's commentary was used at Prague University for teaching of Aristotle's *Meteorology*. Oresme's *Questions* had a great impact on the medieval reception of this Aristotelian text, starting with other Parisian masters of the fourteenth century, namely Albert of Saxony and Themo Iudeus, who based their commentaries on that of Oresme.[18] While Albert's *Questions* had only a limited circulation,[19]

quo ferrum movetur insequendo magnetem. Patet, quia ex quo non movetur motu raptus, sicut dicebat prima conclusio, non videtur esse alius modus dicendi nisi dicatur quod sic movetur per virtutem sibi impressam a celo".

15 Nicole Oresme, *Questions on De celo* II.7 (*Utrum motus naturalis sit velocior in fine quam in principio*), ed. Kren, 559$^{343-346}$: "Quarto modo quod est ad propositum, ex velocitatione motus per quam acquiritur quedam habilitas vel impetus et quedam fortificatio accidentalis ad velocius movendum".

16 S. Kirschner, "Eine weitere Fassung eines lateinischen 'De caelo-Kommentars' von Nicolaus Oresme?", in B. Fritscher, G. Brey (eds.), *Cosmographica et Geographica. Festschrift für Heribert M. Nobis zum 70. Geburtstag*, 1, München 1994 (Algorismus 13), 209–222.

17 München, Bayerische Staatsbibliothek, Clm 4375, I.22 (*Utrum motus naturalis sit velocior in fine quam in principio*), f. 61va: "omne motum naturaliter ab intrinseco in velocitando motum acquirit fortitudinem et habilitatem coadiuvantem ipsum motum, quod potest dici impetus, vel inclinatio accidentalis".

18 For the list of questions discussed in Albert's and Themo's commentaries, see respectively Panzica, "Nicole Oresme à la Faculté des Arts de Paris", 57–63 and 64–72. For some examples of the dependence of Albert of Saxony's *Questions on Meteorology* on that of Oresme's, see Panzica, "Albert of Saxony's Questions on Meteorology", 232–241. This article contains an edition of books I–II.2 from Albert's commentary (268–356): a comparison between this text and the corresponding part of Oresme's commentary, edited in the present volume, clearly shows that Albert relies on Oresme.

19 Only two complete manuscripts of Albert of Saxony's *Questions on Meteorology* are known: Erfurt, Universitäts- und Forschungsbibliothek, Dep. Erf., CA 4° 299, ff. 53^{r}–103^{v}

Themo's *Questions* was printed several times between the end of the fifteenth and the beginning of the sixteenth century.[20] Probably through the intermediary of Themo, the influence of Oresme's commentary spread to England and Italy: many of Oresme's theses were adopted by the Italian master Blasius of Parma[21] and by an anonymous Scotist master at the end of the fourteenth century. Ascribed to Duns Scotus, this anonymous question commentary on Aristotle's *Meteorology* was printed by Wadding in Scotus's *Opera omnia* in the

---

and Berlin, Staatsbibliothek—Preußischer Kulturbesitz, lat. fol. 387, ff. 63^{r}–102^{v}. Two Krakow manuscripts contain a compilation consisting of questions I.19–31 from the first redaction of Oresme's commentary on *Meteorology* and questions I.1–I.14, II.7–17, as well as books III and IV, from Albert's commentary: Kraków, Biblioteka Jagiellońska, BJ 635, pp. 177–236 and 686, ff. 101ra–134va. A Parisian manuscript also contains a compilation from Albert (II.11–III.9) and Oresme (in this case, the second redaction of Oresme's commentary: I.1–II.10): Paris, Bibliothèque Nationale de France, lat. 15156, ff. 226^{r}–288^{v}. Two other manuscripts contain fragments from the first question of Albert's commentary: Wien, Österreichische Nationalbibliothek, Cod. 5453, f. 48vb and Kraków, Biblioteka Jagiellońska, cod. 751, f. 2^{r-v}. On these manuscripts see Panzica, "Albert of Saxony's *Questions on Meteorology*", 248–266.

20 Themo's *Questions on Meteorology* is transmitted in a considerable number of manuscripts, the majority of which are of Italian origin: Città del Vaticano, Biblioteca Apostolica Vaticana, Vat. lat. 2177, ff. 1^{r}–92^{v}; Padova, Seminario vescovile, cod. 24, ff. 1^{r}–50^{v}; Ferrara, Biblioteca comunale Ariostea, Ms. Classe II, n. 380, ff. 1^{r}–60^{v}; München, Bayerische Staatsbibliothek, Clm 6962, ff. 93^{r}–146^{r}; Venezia, Biblioteca Nazionale Marciana, lat. XIV, 129, ff. 77^{r}–122^{v} (formerly: Venezia, Biblioteca del Monastero di S. Michele, 136); Paris, Bibliothèque Nationale de France, lat. 6547, ff. 1^{r}–33^{r} (incomplete copy). The first edition was printed in Pavia around 1480; other editions were printed in Venice by Ottavianus Scotus in 1496, 1507, 1515, 1522 (along with Gaetanus of Thiene's commentary on *Meteorology*). Georges Lockert printed Themo's *Questions* in Paris in 1516 and 1518 (along with other commentaries on Aristotle by Parisian masters of the fourteenth century: Albert of Saxony's *Questions on Physics* and on *De celo*; John Buridan's *Questions on De anima* and on the *Parva naturalia*).

21 Blasius of Parma's *Questions on Meteorology* is transmitted in five manuscripts: Città del Vaticano, Biblioteca Apostolica Vaticana, Chigi O IV 41, ff. 59^{r}–105^{r}; Città del Vaticano, Biblioteca Apostolica Vaticana, Vat. lat. 2160, ff. 62^{r}–138^{v}; Firenze, Biblioteca Medicea Laurenziana, Ashburnham 185, ff. 1^{r}–59^{v}; Chicago, University Library, ms. 10, ff. 1ra–37va (incomplete copy); Città del Vaticano, Biblioteca Apostolica Vaticana, Vat. lat. 4082, ff. 82vb–85va (only qq. I.8–9). For the recent identification of the last witness, see A. Panzica, "Les *Questions sur les Météorologiques* du manuscrit Vat. Lat. 4082: Blaise de Parme, Nicole Oresme et l'*Inter omnes impressiones*", *Bulletin de philosophie médiévale* 61 (2019), 153–182. At question III.6 of his commentary, Blasius refers explicitely to Oresme's theory of the rainbow: "Prime difficultati respondet Nicholaus Horen et facit hanc ymaginationem ut quod in apparitione iridis sunt ymaginandi plures ordines guttarum aque, quarum alique sunt propinquiores oculo", ms. Città del Vaticano, Biblioteca Apostolica Vaticana, Vat. lat. 2160, f. 123va.

seventeenth century.[22] In Central and Eastern Europe, the influence of the second redaction of Oresme's *Questions on Meteorology* was direct and profound. This is particularly evident in the question commentaries on Aristotle's *Meteorology* by three Polish masters who taught in the 1420s, namely Benedictus Hesse de Cracovia, Paul of Worczyn and Peter of Sienno, who rely heavily on Oresme's text and explicitly quote it.[23]

Rediscovered at the end of the nineteenth century, the second redaction of Oresme's *Questions on Meteorology* attracted the attention of the historian of science Alexander Birkenmajer, who provided a first list of manuscripts and studied their interrelations.[24] Birkenmajer also pointed out a puzzling fact about this set of questions, namely that the greater part of the third book of Oresme's commentary is almost identical to the corresponding book in Themo Iudeus' *Questions*. In this portion of the text, Oresme's questions reveal many philosophical and textual inconsistencies. For example, they contain theses rejected elsewhere by Oresme and internal references that cannot be found in Oresme's text, but only in Themo's. Since all of the manuscripts to which he had access stem from Central and Eastern Europe, Birkenmajer concluded that the form of Oresme's commentary as transmitted cannot be the original one, but rather shows evidence of a contamination from Oresme's and Themo's *Questions*. The origin of the oldest manuscripts suggested that the compilation was made at Prague University. The methods used to teach Aristotle's philosophy at the Prague Arts Faculty make this hypothesis very likely: as documented by the Statutes, bachelors and masters at this university used to base their lectures on commentaries by famous masters from Paris and Oxford. According to this hypothesis, Oresme's text was therefore reworked in Prague for use in lectures and, in the process, contaminated with Themo's *Questions*.[25]

Birkenmajer's reconstruction was rejected by McCluskey, who edited the questions concerning light and vision from the third book of the commentary

---

22 Iohannes Duns Scotus, *Opera omnia*, ed. L. Wadding, Lyon 1639, reprint Hildesheim 1968 (12 vols.), 3: 1–125. On this text see L. Petrescu, "The Threefold Object of the Scientific Knowledge. Pseudo-Scotus and the Literature on the *Meteorologica* in Fourteenth-century Paris", *Franciscan Studies* 72/1 (2014), 465–502 and Panzica, *De la Lune à la Terre*.

23 See A. Panzica, "Commenter les *Météorologiques* à l'Université de Cracovie: de l'assimilation des modèles parisiens à la naissance d'une tradition polonaise", *Recherches de Théologie et Philosophie Médiévales* 87/1 (2020), 77–166, esp. 108–140, and Panzica, *De la Lune à la Terre*.

24 A. Birkenmajer, *Études d'histoire des sciences en Pologne*, choix d'articles par J.B. Korolec, A.M. Birkenmajer, textes polonais trad. par C. Brendel [et al.], revus par J. Wolf, Wrocław/Warsaw/Krakow 1972 (Studia Copernicana, 4), 178–239.

25 Birkenmajer, *Études d'histoire des sciences en Pologne*, 181–198.

ascribed to Oresme.[26] According to McCluskey, it is not necessary to postulate the intervention of a third master in order to explain the current form of Oresme's commentary: McCluskey suggests that while preparing his lectures on Aristotle's *Meteorology*, Oresme himself took long passages from Themo's *Questions* without fully harmonising the two texts.[27] Interestingly, McCluskey argues that if a Prague master had compiled this set of questions on Aristotle's *Meteorology* from two different sources, he would have made his lectures consistent, which is not the case.[28] One wonders why McCluskey refuses to accept that Oresme himself might also have been concerned about consistency. As I have shown elsewhere,[29] and as I will argue in more detail in the preface to the edition of the third book, I consider McCluskey's explanation to be unsatisfactory.

It is clear that the solution to this puzzle could only come from a Parisian copy of Oresme's text. Such a copy was discovered by Thorndike in 1954 in the manuscript Paris, Bibliothèque Nationale de France, lat. 15156.[30] Unfortunately, this manuscript cannot help us to evaluate the authenticity of the third book of Oresme's commentary that is transmitted in the Central European family of manuscripts, since the text stops abruptly at question II.10, breaking off in the middle of a sentence. After a blank leaf, the text restarts with question II.11 of Albert of Saxony's *Questions on Meteorology*. Yet this witness is not without importance for the reconstruction of the tradition of Oresme's *Questions*. While Birkenmajer and McCluskey focused on the third book, I collated the whole of the first and part of the second book transmitted in the Parisian manuscript with the other witnesses. As I will prove later in more detail, this comparison reveals many important differences between the Parisian manuscript and the other copies of Oresme's *Questions*. These differences, which are subtler than in the third book, reveal, on the one hand, the existence of two different traditions of Oresme's text and, on the other, the superiority of the Parisian tradition.

---

26 S.C. McCluskey, *Nicole Oresme on Light, Color, and the Rainbow: an Edition and Translation, with Introduction and Critical Notes, of Part of Book III of his "Questiones super IV libros Meteororum"*, unpublished Ph.D. dissertation, University of Wisconsin 1974.

27 McCluskey, *Nicole Oresme on Light, Color, and the Rainbow*, 30–63.

28 McCluskey, *Nicole Oresme on Light, Color, and the Rainbow*, 61–62, fn. 51.

29 Panzica, "Nicole Oresme à la Faculté des Arts de Paris", 12; Ead., "Albert of Saxony's *Questions on Meteorology*: Introduction, Study of the Manuscript Tradition and Edition of Book I–II.2", *Archives d'histoire doctrinale et littéraire du Moyen Âge* 86 (2019), 231–356, esp. 231–241.

30 L. Thorndike, "Oresme and Fourteenth Century Commentaries on the *Meteorologica*", *Isis* 45 (1954), 145–152.

The particular situation of the manuscript tradition of Oresme's commentary makes necessary the adoption of different editorial principles for the reconstruction of the text. In light of the differences between the Parisian manuscript and the Central European copies, I refrained from producing a composite text that switches from the extant part of the Parisian manuscript to the other copies. Rather, I deemed it reasonable to produce an edition of Oresme's commentary based on the Parisian manuscript first: this edition, which adopts the variants of the Central European family only when the Parisian witness is faulty, covers the portion of the text transmitted in the Parisian manuscript, i.e. the first book and the first ten questions of the second book. The rest of the text will be presented in a separate edition following different editorial principles.

In so doing, my intention is not to avoid addressing the problem of the authenticity of the third book. On the contrary, I am convinced that this partial edition will provide us with important evidence that will facilitate the evaluation of the problems posed by the third book in the broader context of Oresme's commentary. Other important data for the solution of this enigma will be provided by the edition of the first redaction of Oresme's *Questions on Meteorology*, on which I am working in parallel. A comparison of the list of questions from the third book discussed in the two redactions of Oresme's *Questions* reveals an interesting fact: most of the questions found in the second redaction that contain inconsistencies and that were thought to be spurious by Birkenmajer are missing from the first redaction. I am confident that a parallel edition and study of the two redactions will help us understand the strange case of contamination in the second redaction of Oresme's *Questions on Meteorology*.

The Manuscript Tradition of the Second Redaction of Nicole Oresme's *Questions on Meteorology*: Manuscript Descriptions and a Study of Their Relationships

The consultation of regional inventories and studies on Nicole Oresme, as well as the direct inspection of the manuscript sources, enabled me to identify nineteen manuscripts transmitting the second redaction of Nicole Oresme's *Questions on Meteorology* and one manuscript that contains a partial table of questions of this commentary.[1]

1) Basel, Universitätsbibliothek, lat. F I 11, ff. 4^r–85^r = *Ba*
2) Basel, Universitätsbibliothek, lat. F V 2, ff. 2^r–63^v = *Ba¹*
3) Berlin, Staatsbibliothek—Preußischer Kulturbesitz, lat. fol. 631, ff. 39^r–114^r = *Be*
4) Erfurt, Universitäts- und Forschungsbibliothek, Dep. Erf., CA 2° 334, ff. 158^v–167^r (IV) = *E*
5) Kassel, Universitätsbibliothek—Landesbibliothek und Murhardsche Bibliothek, 2° Phys. et hist. nat. 12, ff. 1^v–107^r = *Ka*
6) Klagenfurt, Bischöfliche Bibliothek, XXXI b 5, ff. 1^r–124^r = *Kl*
7) Kraków, Biblioteka Jagiellońska, cod. 749, ff. 59^v–110^v = *Kr*
8) Kraków, Biblioteka Jagiellońska, cod. 751, ff. 3^r–53^r = Kr_1
9) Kraków, Biblioteka Jagiellońska, cod. 2095, ff. 245^r–307^r = Kr_2
10) Kraków, Biblioteka Jagiellońska, cod. 2117, ff. 195^r–322^r [pp. 389–643] = Kr_3
11) Leipzig, Universitätsbibliothek, ms. 1387, ff. 181^r–275^r = *L*
12) München, Bayerische Staatsbibliothek, Clm 4376, ff. 1^r–64^v = *M*
13) München, Bayerische Staatsbibliothek, Clm 17226, ff. 1^r–140^v (I.2, 4–6, 10, 12–13, 17, II.1–7, III.7–10, IV.1–11) = *M¹*
14) Paris, Bibliothèque Nationale de France, lat. 15156, ff. 226^r–265^v (I–II.10) = *P*
15) Poznań, Archivum Archidiecezjalne, Cms 53, ff. 1^r–95^v = *Po*
16) St. Gallen, Stiftsbibliothek, Cod. Sang. 839, ff. 1^r–177^r = *Sg*

1 L. Thorndike, "Oresme and Fourteenth Century Commentaries on the Meteorologica", 145–152; Birkenmajer, *Études d'histoire des sciences en Pologne*, 178–239; McCluskey, *Nicole Oresme on Light, Color, and the Rainbow*, 81–98; Weijers, *Le travail intellectuel à la Faculté des arts de Paris*, 6: 175; Lohr, *Latin Aristotle Commentaries, I.2. Medieval Authors M–Z*, 34–35; M. Markowski, "Aristotelica Poznaniensia", *Acta mediaevalia* 17 (2004), 281–293.

17) Uppsala, Universitetsbibliotek, Ms. C 596, ff. 2^r–97^r = *U*
18) Wien, Österreichische Nationalbibliothek, Cod. 5453, ff. 49^r–109^v = *Wi*
19) Wrocław, Biblioteka Uniwersytecka, ms. IV Q 27, ff. 1^r–163^r = *Wr*

A partial list of questions of Oresme's commentary can be found in ms. Città del Vaticano, Biblioteca Apostolica Vaticana, Pal. lat. 1045, f. 118^v.

Manuscript Descriptions

1 *Basel, Universitätsbibliothek, lat. F I 11, ff. 4^r–85^r* (Ba)

Paper; ff. II+209+I; mm 291×204; 1369

Date and origin. The colophons of the texts transmitted in this manuscript inform us about the date and the origin of this codex, which was copied in 1369 at Prague University: "Expliciunt questiones super librum *De anima* per reverendum magistrum Byridanum, Parysius compilate, Prage reportate, etc." (f. 118ra); "Expliciunt questiones totius libri *De celo* per reverendum magistrum Albertum compilate et finite sub anno domini 1369°" (f. 209ra). The term *reportate* indicates that the text derives from a *pronuntiatio*, which was a common practice at Prague University. It consisted of a sort of collective dictation, which the Prague masters and bachelors used to teach the Aristotelian texts, based on the commentaries of famous masters from Paris and Oxford.[2] The manuscript was bought by Peter of Ulma, physician and master, at Heidelberg; Peter sold it in 1559 to the University of Basel.[3]

Composition. Wooden binding partially covered with white leather. Clasps on the front cover. The strap and catch plate are missing. The binding is identical to that of manuscript F II 5 (see below). Two guard-leaves are numbered by a modern hand "1" and "2". On the first leaf, a medieval hand wrote: "Librum hunc emebat universitas † de Olma et continentur primo questiones super libris *Metheororum* Aristotelis; secundo continet questiones Biridani super libris *De anima*; tertio questiones super libris *De generatione et corruptione*; quarto questiones Alberti super libris *De celo et mundo*. [*alia manu*] Ex libris bibliotece Basileiensis Academie, 1559". The guard-leaves are followed

2 On this practice see L. Moonan, "The Scientific Writings of Lawrence of Lindores (d. 1437)", *Classica et Mediaevalia* 39 (1988), 273–317, at 283–284 and J. Kejř, "Pronuntiatio", *Studie o rukopisech* 36 (2005–2006), 57–65.

3 For biographical information about Peter of Ulma, see E. Wickersheimer, *Dictionnaire biographique des médecins en France au Moyen Âge*, Geneva 1979 (Hautes études médiévales et modernes 34/1–2) (2 vols.), 2: 665a; D. Jacquart, *Supplément à E. Wickersheimer, Dictionnaire biographique des médecins en France au Moyen Âge*, Geneva 1979 (Hautes études médiévales et modernes, 35), 244.

by 209 leaves grouped according to this formula:[4] I (3–19)$^{8+8}$; II (19–30)$^{7+7}$; III (31–46)$^{8+7}$; IV (46–55)$^{(8-1)+(7-4)}$; V (56–69)$^{7+7}$; VI (70–83)$^{7+7}$; VII (84–85)$^{1+1}$; VIII (86–99)$^{7+7}$; IX (100–113)$^{7+7}$; X (114–121)$^{4+4}$; XI (122–129)$^{4+4}$; XII (130–143)$^{7+7}$; XIII (144–153)$^{5+5}$; XIV (154–155)$^{1+1}$; XV (156–167)$^{6+6}$; XVI (168–179)$^{6+6}$; XV (180–191)$^{6+6}$; XVII (192–201)$^{5+5}$; XVIII (202–210)$^{5+(5-1)}$. At the end of the volume, we find another guard-leaf, numbered 211. Catchwords: f. 17^{v}, f. 45^{v}, f. 55^{v}, f. 69^{v}, f. 99^{v}, f. 113^{v}. Signatures: f. 156^{r} (pus); f. 168^{r} (2); 180^{r} (3); f. 192^{r} (4); f. 202^{r} (5).

The volume consists of two main parts: ff. 1–119 and ff. 120–211, which present two different medieval foliations. A modern hand has numbered the leaves from the beginning to the end of the volume.[5] As a consequence, the medieval and the modern foliations are shifted by 119 units in the second part of the volume. In the first part, the gap between the two foliations is three units wide, since the medieval foliation starts with Oresme's *Questions on Meteorology* (numbered 4 according to the modern foliation), while the modern one also includes the guard-leaves (numbered 1 and 2) and the table of contents (numbered 3).

Content. This manuscript contains a collection of commentaries on natural philosophy by Parisian masters of the fourteenth century: Nicole Oresme's *Questions on Meteorology* (ff. 3ra–85vb) is followed by a short version of the *redactio A* of Buridan's *Questions on De anima* (ff. 86ra–118ra),[6] by the *redactio B* of Buridan's *Questions on De generatione et corruptione* (ff. 120ra–155vb)[7] and

---

4 To represent the structure of the fascicules, I have adopted the method used in the catalogue of medieval Latin manuscripts of the Jagiellonian Library in Krakow: *Catalogus codicum manuscriptorum medii aevi latinorum qui in Bibliotheca Jagellonica Cracoviae Asservantur*, Wrocław 1980–. The Roman numeral before the brackets indicates the order of the fascicule; the two Arabic numbers between brackets indicate the first and the last leaf of the fascicule according to the foliation adopted; the formula in the exponent expresses the number of leaves that compose the fascicule. The numbers before and after the + sign indicate, respectively, how many leaves there are before and after the thread. A quaternion is thus expressed by the formula 4+4; a quinion, by the formula 5+5; a senion, by the formula 6+6, and so on. The addition or subtraction of some leaves is expressed by operands + and -, followed by the quantity of the added or missing leaves. For example, the formula 5+(5–1) represents a fascicule composed by two quinions, the second of which has a missing leaf. I have preferred this codicological formula to the more widely used German one because it allows us to express clearly which part of the fascicule is concerned by the addition or the subtraction of leaves.

5 In this description, I refer to the modern numbering.

6 B. Michael, *Johannes Buridan. Studien zu seinem Leben, seinen Werken und zur Rezeption seiner Theorien im Europa des späten Mittelalters*, Berlin 1985 (2 vols.), 2: 684, nr. 20.21; Lohr, *Latin Aristotle Commentaries, i.2. Medieval Authors M–Z*, 257–257, nr. 24.

7 Michael, *Johannes Buridan*, 2: 635–641.

by Albert of Saxony's *Questions on De celo* (ff. 156ra–209ra). This codex also contains some texts on logic: a fragment from the *Libellus de terminis naturalibus*, a dictionary of basic terminology used by the Oxford Calculators (ff. 118^{r-v})[8] and a fragment of Albert of Saxony's *Expositio* on Aristotle's *Posterior Analytics* (ff. 209rb–210va).

The *Questions on Meteorology*. The colophon attributes the commentary to Nicole Oresme: "Expliciunt questiones libri *Meteororum* vel *Metheororum* seu *Methaurorum* reverendi magistri Nycolai Orem, viri multum experti, ut in istis dictis suis poterit apparere" (f. 85vb). The text is copied in two columns of about fifty lines each. The content of ff. 52^{v} and 53^{r} has been crossed out with a big X that covers the entire page. On f. 53rb, the copyist wrote: "verte folium et invenies ultra quantum ad secundum, etc. Et hic nichil deficit, ut videbis, sed indirecte et incorrecte nimis scriptum erat, ideoque absasum [*sic*] est et secundario scriptum est, et hoc magis correcte et bene. Deo gratias". On f. 75vb, l. 6, the hand changes, as happens on f. 83ra. The incipits are written in textualis but the decoration has not been completed, since the white space set aside for the initials is at times left blank (for example at f. 39vb). Moreover, the headings of the questions only reach question I.6, on f. 9rb, after which the space set aside for them remains empty. The pilcrows stop at f. 10rb. At the beginning of the text, a big Q extends over the space of eight lines. Pointing hands (f. 5va); some rare marginal notes related to the structure of the text. Some *nota bene* in a medieval hand (for example at f. 49rb: "nota bene quod templa et monasteria percutiantur a fulmine sepius quam alie structure"). On f. 3, we find a table of contents.

Incipit and explicit of the books

I, ff. 4ra–26va: Queritur circa librum *Metheororum* primo ista questio: utrum possibile sit de impressionibus metheorologicis habere scientiam et opinionem. Et arguitur quod non, quia de eis non contingit habere scientiam nec opinionem; igitur questio falsa ...X... Ad quartam: 'si galaxia esset de natura celi', etiam concedo. Sed tamen Aristoteles determinavit de ea in isto libro propter hoc quia plures antiquorum tradiderunt galaxiam esse de natura elementari. Et sic est finis questionum huius libri primi. Explicit primus liber *Metheororum*.

II, ff. 26vb–39vb: ⟨C⟩irca secundum librum *Metheororum* queritur utrum locus generationis pluvie sit media regio aeris. Arguitur quod non: nullum elementum debet generari extra locum sibi naturalem ...X... Ad rationes in oppositum patet ex secundo articulo.

8 *Inc.*: "Natura est principium motus et quietis eius in quo est. Istam diffinitionem ponit Aris-

III, ff. 39vb–78rb: ⟨Q⟩ueritur consequenter circa tertium librum *Methaurorum* utrum ventus sit exalatio calida et sicca. Et arguitur quod non sit calida, quia sic sequeretur quod in temporibus ventosis aer deberet esse calidus ...X... Ad ultimam dico quod non in vanum laborant, ex quo aliam artem ignorant. Et sic patent questiones tertii *Metheororum*.

IV, ff. 78rb–85vb: Circa quartum librum *Metheororum* queritur primo ista questio: utrum sunt quatuor qualitates prime, scilicet caliditas, humiditas, frigiditas et siccitas. Et arguitur primo quod non, quia quod per superhabundantiam dicitur, uni soli convenit ...X... Ad sextam dico quod in salamandra non dominatur ignis in quantitate nec etiam nutritur ab igne; sed tale animal non est ustibile ab igne; et ideo, quia non uritur ab igne, dicimus quod tale animal habitet in igne. Et sic patet ultima questio istius. Expliciunt questiones *Metheororum* magistri Nycolai Orem. [85vb] Expliciunt questiones *Meteororum* vel *Metheororum* seu *Methaurorum* reverendi magistri Nycolai Orem, viri multum experti, ut in istis dictis suis poterit apparere.

Bibliography

Catalogues: B.M. von Scarpatetti, R. Gamper, and M. Stähli, *Katalog der datierten Handschriften in der Schweiz in lateinischer Schrift vom Anfang des Mittelalters bis 1550*, 1: *Die Handschriften der Bibliotheken von Aarau, Appenzell und Basel*, Dietikon/Zurich 1977, 171; Lohr, *Aristotelica Helvetica*, 34–36; Ch. Lohr, *Latin Aristotle Commentaries, I.2. Medieval Authors M–Z*: 40, 256, 258; 2: 34 [here the manuscript is incorrectly dated to the 15th century].

---

toteles in secundo *Physicorum* et etiam Commentator". This text is ascribed to Iohannes Garisdale by Ch. Lohr (*Aristotelica Helvetica*, Fribourg (Switzerland) 1994, 34–36), but this attribution is still a matter of debate, as texts with similar incipits have been ascribed also to William Heytesbury, Iohannes Gardisdale and Thomas Netter. A.R. Verboon studied and edited a version of this text ascribed to a certain "master Albertus". See A.R. Verboon, *Lines of Thought. Diagrammatic Representation and the Scientific Texts of the Arts Faculty, 1200–1500*, unpublished Ph.D. dissertation, Leiden University 2010, Appendix A, 239–250, available online at https://openaccess.leidenuniv.nl/handle/1887/16029; revised edition: A.R. Verboon and E.P. Bos, *Master Albert: Termini naturales*, in C. Angotti, M. Calma, and M. Teeuwen (eds.), *Portraits de Maîtres offerts à Olga Weijers*, Turnhout 2012 (Textes et Etudes du Moyen Âge, 65), 335–360. This attribution can be found in the two manuscripts used by Verboon for her edition: Paris, Bibliothèque Nationale de France. lat. 566, ff. 49^{r}–58^{r} and Assisi, Biblioteca del Sacro Convento di S. Francesco, fondo antico, ms. 690, ff. 314^{r}–319^{r}. Our manuscript was unknown to A.R. Verboon. Lohr (*Aristotelica Helvetica*, 110) mentions another text with a similar incipit kept at Basel University library: F VI 58, ff. 301^{r}–306^{v}. I wish to express my gratitude to Miroslav Hanke (Czech Academy of Sciences) for information on this text.

Literature: M. Burckardt, "Aus dem Umkreis der ersten Basler Universitätsbibliothek", *Basler Zeitschrift für Geschichte und Alterumskunde*, 58/59 (1959), 155–191, nr. 164; B. Michael, *Johannes Buridan*, 2: 636, 686 [here the manuscript is incorrectly dated to the 15th century]; McCluskey, *Nicole Oresme on Light, Color, and the Rainbow*, 83–84; M. Streijger, P.J.J.M. Bakker, and J.M.M.H. Thijssen (eds.), *John Buridan, Quaestiones super libros De generatione et corruptione Aristotelis*, Leiden/Boston 2010 (Medieval and Early Modern Philosophy and Science, 14) [here the manuscript is incorrectly dated to the 15th century].

2 ***Basel, Universitätsbibliothek, lat. F v 2, ff. 2ra–63va* (Ba^1)**

Paper; mm 288 × 213; 140 ff.; 1369

Date and origin. Just like the manuscript F I 11, this codex comes from the University of Prague. The first part, which contains Nicole Oresme's *Questions on Meteorology*, dates to 1369, as is evident from the colophon at f. 63rb: "Anno domini millesimo CCC° sexagesimo nono finite sunt questiones *Metheororum*". But the similarities between this codex and manuscript F I 11 do not stop there. Like the latter, this manuscript was also bought by Peter of Ulma, who resold it to the University of Basel. On the first inside cover and on the first leaf we find a possession mark: "Ex libris Bibliothece Academie Basileiensis, 1559"; and at f. 1^r: "Hunc librum emebat universitas de libris magistri Petri de Olma". On f. 63ra, between Oresme's *Questions on Meteorology* and the table of contents, we find an epigram that is considered the oldest document in Old Prussian: "Kayle rechyse / Thoneaw labonache thewelyse /eg koyte poyte nykoyte pennega doyte".[9]

Composition. Wooden binding partially covered with white leather. Two clasps on the front cover. The strap and the catch-plate are missing. The binding is identical to that of manuscript F I 11. A paper guard-leaf, numbered 1, contains the following statement: "Hunc librum emebat universitas de libris magistri Petri de Olma et continetur in eo vide signa. Item primo continentur in eo questiones super libris *Metheororum* magistri N. Orem; secundo questiones Biridani super libris *De generatione et corruptione*; tertio questiones super libris *De anima*; quarto questiones super libris *De celo et mundo*; quinto quedam puncta circa computum manualem". The guard-leaf is followed by 139 paper

9 See A. Bammesberger, *Anmerkungen zum Baseler Epigramm*, in Id., *Baltistik: Aufgaben und Methoden. Akten der Tagung vom 10. bis 11. November 1995 in Eichstätt*, Heidelberg 1968, 121–126; D. Petit, *Untersuchungen zu den baltischen Sprachen*, Leiden/Boston 2010 (Brill's Studies in Indo-European Languages and Linguistics, 4), 22.

leaves grouped mainly in senions, according to the following formula: I (2–13)$^{6+6}$; II (14–25)$^{6+6}$; III (26–37)$^{6+6}$; IV (38–49)$^{6+6}$; V (50–63)$^{7+7}$; VI (64–74)$^{6+6}$;[10] VII (75–86)$^{6+6}$; VIII (87–98)$^{6+6}$; IX (99–111)$^{6+6}$; X (112–121)$^{5+5}$; XI (122–135)$^{7+7}$; XII (136–140)$^{(4-1)+(4-2)}$.

The manuscript consists of two main parts: the first one covers fascicules I–V (ff. 2–63); the second one, senions VI–XII (ff. 64–140). Fascicules I–V contain signatures that prove that the volume was arranged differently during the Middle Ages. These senions are numbered: "nonus", "decimus", "undecimus", "duodecimus", "tredecimus". Another arrangement is suggested by signatures in fascicules IX (2us), X (3us) and in fascicules XI (1us) and XII (2us). Fascicules IX and X are in fact the second and the third fascicule of the *Questions on De anima*, while fascicules XI and XII are the first and the second quire of the *Questions on De celo*. Other signatures in the lower margins of senions IV and V, between the columns, reflect the current arrangement of the volume: "quartus" and "quintus".

The original division of the volume into two parts is confirmed by the medieval foliations: the first one covers fascicules I–V, numbered 2–63 by a medieval hand; the second one covers fascicules VI–XII, numbered 1–78 by a medieval hand too. A third foliation, written by a modern hand, stretches across the whole volume.[11] An error in the modern foliation occurs between f. 72 and f. 73 (9 and 11, according to the second medieval foliation): f. 72 is followed by another 72, which was therefore numbered 72°. In the first part of the volume, the medieval and the modern foliation correspond to each other. In the second part of the volume, they are shifted by 63 units until f. 72 according to the modern foliation, and by 62 units from the subsequent folio, numbered 72° according to the modern foliation.

Content. The content of this codex is very similar to that of ms. F I 11. As in the latter manuscript, Oresme's *Questions on Meteorology* (ff. 2ra–63va) is followed by the *redactio B* of Buridan's *Questions on De generatione et corruptione* (ff. 64ra–86rb), by a short version of the *redactio A* of Buridan's *Questions on De anima* (ff. 87ra–120ra) and by Albert of Saxony's *Questions on De celo*, this time in a shortened form (ff. 122ra–139rb).[12] The codex also contains an anonymous

---

10 The leaf between f. 72 and f. 73 is numbered 72°.

11 In this description I always refer to the modern foliation.

12 *Inc.*: "Circa librum *De celo et mundo* Aristotelis queritur primo questio talis: utrum cuilibet corpori simplici naturaliter insit tantum unum motus simplex. Et arguitur quod non"; q. I.2: "Secundo queritur utrum in mundo sint quinque corpora distincta secundum speciem, scilicet quatuor elementa et celum, scilicet quinta essentia"; q. I.3: "Tertio, utrum preter elementa et celum sit ponendum sextum corpus simplex"; q. I.4, "Quarto,

question on logic (120va–121ra)[13] and some *Puncta* on the *Computum manuale* (ff. 139rb–140va).[14]

The *Questions on Meteorology*. The colophon and the table of contents ascribe the commentary to Nicole Oresme: "Et sic est finis questionum Oren super *Metheororum*", f. 63ra; "et tantum de questionibus *Metheororum* magistri N. Orem", f. 63va. The text is copied by different hands using different inks (some examples are: the change of hand and ink on f. 9vb, l. 14; the change of ink on f. 20vb; hands repeatedly changing on ff. 22^{v}–23^{r}; abrupt change of hand on f. 30^{v}) and laid out in two columns. Brown-inked pilcrows are highlighted with a vertical red stroke. Some marginal notes refer to the structure of the text. Corrections in black ink (particularly recognisable between ff. 2ra–9vb in contrast with the pale brown ink of the main text). The first words of the questions between ff. 2ra–10va are in textualis. From f. 11va onwards, the questions are separated by a simple blank space. On f. 35^{v} and f. 43^{r}, we find drawings related to the text. The commentary is followed by a table of contents (ff. 63^{r}–63^{v}) written by a different hand from those that copied the text.

Incipit and explicit of the books

I, ff. 2ra–17ra: ⟨C⟩irca primum librum *Metheororum* utrum possibile sit de impressionibus metheoroloycis haberi scientia simul et opinio. Et arguitur quod non, quia de eis non contingit habere scientiam nec opinionem; igitur questio falsa ...X... Ad quartam: 'si esset [17ra] de natura celi', etc., dico: tamen Aristoteles determinavit de ea propter hoc quia plures antiquorum crediderunt galaxiam esse de natura elementari. Et sic est primi libri finis.

II, ff. 17ra–25vb: Utrum locus generationis pluvie sit media regio aeris. Arguitur quod non: nullum elementum debet generari extra locum sibi naturalem ...X... Ad rationes in oppositum patet ex secundo articulo.

III, ff. 25vb–57ra: Utrum ventus sit exalatio calida et sicca. Et arguitur quod non, quia non sit exalatio calida, quia si sic, sequeretur quod in temporibus

utrum talis mundus aggregatus ex quatuor corporibus simplicis sit unum ens continuum"; *expl.*: "Sed tamen non determinant sibi istam ex sua natura propria, sicut quod ignis est rotundus sive spericus, hoc est ratione continentis spere lune. Et sic est finis. Expliciunt questiones breves Alberti *De celo et mundo*".

13 *Inc.*: "Utrum sic esse in re sicut per propositionem significatur aliquid dictum ab ipsa re et eius propositione. Et arguitur primo quod sic, quia sic esse in re sicut per propositionem significatur potest esse quando nec res est nec eius propositio"; *expl.*: "quia omne ens quod non est chimera est de facto, sequitur quod res significata per istam propositionem nulla chimera est, etiam est de facto. Et sic patet questio".

14 *Inc.*: "Notandum est primo quod duplex est cyclus, quorum unus est solaris et alter lunaris. Cyclus lunaris est iste qui complet cursum suum in 19 annis"; *expl.*: "et hoc idem faciendum

ventosis aer deberet esse calidus ...X... Ad ultimam dico quod non in vanum laborat ex quo aliam artem ignorat. Et sic est finis questionum tertii.

IV, ff. 57ra–63ra: Incipiunt quarti *Metheororum* questiones. Utrum tantum sint quatuor qualitates prime, scilicet caliditas, humiditas, frigiditas et siccitas. Et arguitur quod non, quia ⟨quod⟩ per superhabundantiam dicitur, uni soli convenit ...X... Ad sextam dico quod in salamandra non dominatur ignis in quantitate nec etiam nutritur ab igne; sed tale animal non est ustibile ab igne; et ideo, quia non uritur ab igne, dicimus quod tale animal habitet in igne. Et sic est finis questionum Oren super *Metheororum*. Ad honorem Dei gloriosi. Amen. Deo gratias. Expliciunt questiones *Metheororum*. [*alia manu*]: anno Domini millesimo CCC sexagesimo nono finite sunt questiones *Metheororum* per manus illius qui scripsit eas, etc., et in vigilia Epiphanie per manus illius qui scripsit eas. Omnibus omnia non mea sompnia dicere possumus. Amen.

Bibliography

Catalogues: Von Scarpatetti, Gamper, and Stähli, *Katalog der datierten Handschriften in der Schweiz*: 1, 190; Lohr, *Aristotelica Helvetica*, 78–79; Id., *Latin Aristotle Commentaries. Medieval Authors* 1: 256, 258; 2: 34 [here the manuscript is incorrectly dated 1469].

Literature: Burckardt, "Aus dem Umkreis der ersten Basler Universitätsbibliothek", 155–191, nr. 165; Michael, *Johannes Buridan*, 2: 636–637, 686; Streijger, Bakker, and Thijssen (eds.), *John Buridan Quaestiones super libros De generatione et corruptione Aristotelis*, 13.

3 *Berlin, Staatsbibliothek—Preußischer Kulturbesitz, lat. fol. 631, ff. 39^r–114^r* (Be)

Paper; mm 250 × 190; ff. III+118+II; 1470

Date and origin. The colophons of the texts transmitted in this codex inform us that they were copied between 1470 and 1471: "Rescripte sunt hee questiones venerabilis magistri Nicolai Orem super libros *Metheororum* Aristotelis. Anno Domini 1470" (f. 114^r); "finis anno 1471. He sunt questiones, puto magistri Christiani de Ackoy, super *Sphera materialis*" (f. 37^v). The origin of the manuscript is unknown, but various elements prove that it circulated in Central Europe. On f. 38^v, we find a medieval note that has later been scraped. It refers to the apparition of a halo around the moon that could be seen from Basel: "anno 63 [...] apparuit Basilea halo circa lunam diu durans [...]". On the second guard-

est de primis duabus litteris duarum sillabarum primarum cuiuslibet dictionis illorum versorum, et hoc non computetur pro littera".

leaf, we find a modern possession mark: "Eduardo Machado, Hamburgo, Mayo 29, 1868". In 1903, the Prussian state library bought the manuscript from the antiquary J. Halle of Munich (see the note on the front cover: "acc. 1903, 177").

Codicology. Modern pasteboard binding covered with brown leather. Three modern paper guard-leaves numbered I–III at the beginning of the volume; at the end of the volume, we find three medieval paper guard-leaves numbered 115–117, followed by two modern guard-leaves numbered IV and V. The leaves are grouped primarily into quaternions: I (III–3)$^{2+2}$; II (4–11)$^{4+4}$; III (12–19)$^{4+4}$; IV (20–27)$^{4+4}$; V (28–35)$^{4+4}$; VI (36–43)$^{4+4}$; VII (44–51)$^{4+4}$; VIII (52–59)$^{4+4}$; IX (60–66)$^{4+4}$; X (67–74)$^{4+4}$; XI (75–82)$^{4+4}$; XII (83–90)$^{4+4}$; XIII (91–98)$^{4+4}$; XIV (99–106)$^{4+4}$; XV (107–114)$^{4+4}$; XI (115–IV)$^{2+2}$. Modern numbering in pencil in the upper outer margin. Two subsequent folios were numbered 62 and 62bis. Watermark P similar to Piccard XII 641 (Freiburg im Breisgau, 1472).[15]

Content. The manuscript contains two texts: at ff. 1^{r}–37^{v}, some *Questions* on Sacrobosco's *De spera*, which the colophon hesitantly ascribes to Christianus de Ackoy;[16] at ff. 39^{r}–114^{r}, the second redaction of Oresme's *Questions on Meteorology*.

The *Questions on Meteorology*. The text, which is ascribed to Nicole Oresme in the colophon (f. 144^{r}), was copied by a German scribe in the fifteenth century. The handwriting, a bastarda, is regular and very readable. The text is laid out in one column with fifty-three lines per page. The first words of the questions are written in textualis. Some marginal notes by a different hand from that of the copyist are written alongside the last question of the third book, which discusses the artificial transformation of metals (for example, on the lower margin of f. 39^{r} we read: "Nihil agit ultra gradum suum proprium / Nihil agit ultra suam speciem / Activa non producunt formas substantiales / Nihil producit perfectius se ipso").

---

15 G. Piccard, *Die Wasserzeichenkartei im Hauptstaatsarchiv Stuttgart*, Stuttgart 1961–1997 (17 vols.).

16 *Inc.*: "Queritur circa initium *Spere materialis* primo utrum astronomia sit scientia. Et arguitur primo quod non, nam nulla scientia est prohibita; astronomia est prohibita; igitur astronomia non est scientia"; *expl.*: "sed potius debet dici obscuratio quam eclypsis Solis. Auctoritas post oppositum est pro conclusione responsali. Et hec de questione, et per consequens de omnibus questionibus *Spere materialis*". *Colophon*: "Finis anno 1471. He sunt questiones, puto magistri Christiani de Ackoy, supra *Sphera materiali*". Christianus de Ackoy, native from the diocese of Utrecht, obtained the degree of Master of Arts in Paris and subsequently enrolled in the Universities of Heidelberg (in 1387) and Cologne (in 1394). See Lohr, *Latin Aristotle Commentaries, I.2. Medieval Authors M–Z*: 97–98. Lohr only mentions Christianus of Ackoy's *Questions* on the *Posterior Analytics* and his treatise on the natural terms, but not the commentary on the *Sphere* transmitted in the Berlin manuscript.

Incipit and explicit of the books

I, ff. 39r–58r: Circa primum *Metheororum* queritur primo utrum possibile sit de impressionibus metheorologicis habere simul scientiam et opinionem. Et arguitur quod non, quia de impressionibus metheorologicis non contingit habere scientiam et opinionem; igitur questio falsa …X… Propter hoc quia plures antiquorum crediderunt gallaxiam esse de natura elementari, propter eorum oppinionem ipse talem materiam hic tangit, etc. Explicit primus *Metheororum*.

II, ff. 59r–68v: ⟨C⟩irca secundum *Metheororum* queritur primo utrum locus generationis pluviarum sit media regio aeris. Et arguitur quod non, nam nullum elementum debet generari extra suum locum naturalem …X… Rationes patent in secundo articulo huius questionis, etc. Explicit secundum *Metheororum*, etc.

III, ff. 69r–105r: Circa tertium *Metheororum* queritur, secundum distinctionem istius totius in principio positam ab Alberto, utrum ventus sit exalatio calida et sicca. Et arguitur primo quod non sit exalatio calida et sicca …X… Ad ultimam dico quod non in vanum laborant, ex quo aliam artem ignorant.

IV, ff. 105v–114r: ⟨C⟩irca quartum [*corr. sup. lin. ex*: secundum] *Metheororum* queritur primo utrum tantum quatuor sint qualitates prime, scilicet caliditas, frigiditas, humiditas, siccitas. Et arguitur primo quod non, quia quod per superhabundantiam dicitur, uni soli convenit …X… Sed tale animal non est ustibile ab igne; et ideo, quia non uritur ab igne, dicimus quod tale animal habitet in igne, et non quod nutriatur ab igne nec in eo dominetur ignis, sed illo modo quo dictum est, scilicet quod sit inustibile ab ipso igne. Rescripte sunt hee questiones venerabilis magistri Nicolai Orem super libros *Metheororum* Aristotelis. Anno Domini 1470.

Bibliography

Catalogues: Ch. Lohr, "Aristotelica Berolinensia", *Traditio* 54 (1999), 353–423, at 378; Id., *Latin Aristotle Commentaries. Medieval Authors*, 1: 97–98; 2: 34.

Literature: Birkenmajer, *Études d'histoire des sciences en Pologne*, 182–183; Michael, *Johannes Buridan*, 2: 661, nr. 6; McCluskey, *Nicole Oresme on Light, Color and the Rainbow*, 81–82.

4 *Erfurt, Universitäts- und Forschungsbibliothek, Dep. Erf.* CA *2° 334, ff. 158v–167r* (E)

Paper; mm 305×105; ff. III+208+II; 1421

Date and origin. The colophon of the *Questions on Meteorology* informs us that the text was copied in 1421 (f. 167v: "Expliciunt questiones *Metheororum* reverendi magistri Biridani finite ac peracte in festo sancti Mathie per me, per Petrum filium Francisci de Grinstad. Quarum conservator est reverendus magister Petrus dictus Culmach; Paulus vero de Tr. harum est possessor. Domo Domini 1421"). This part of the codex was copied in Göttingen, as stated in the colophon of f. 179^{v}: "Explicit liber *De sensu et senssato* [*sic*] et dicta per me Henricum Wetter [Wecter or Voecter] in Gottingen". The volume belonged to the library of the Collegium Porta Celi, founded at Erfurt University by Amplonius Rating de Berka in 1433, and was registered under the signature "7" in Amplonius's catalogue.[17]

Codicology. Wooden binding covered with brown leather, restored in 1974. Two intact clasps attached to the back cover. Two modern paper guard-leaves, followed by a medieval parchment guard-leaf. The first guard-leaf is glued to the inside cover. On the recto of the parchment guard-leaf, a medieval hand has written in black ink: "liber librarie collegii porte celi" and "septimus naturalis in novis". Another medieval hand has written in red ink: "liber collegii porte celi". On the verso of the same guard-leaf, a third medieval hand has written: "septimus philosophie naturalis in novis Collegii Porte Celi". There is also a table of contents by the same hand: "In isto volumine continentur: Questiones *Parvorum Naturalium* reverendi magistri Marsilii de Inghen; Questiones *Meteororum* reverendi magistri Biridani; Commentum *Parvorum Naturalium* reverendi magistri Marsilii; *Problemata* Aristotelis principis philosophorum; item quod quatuor modis febris cotidiana solet evenire". The volume consists of three parts, of different sizes with different kinds of paper: I (ff. 1–63); II (ff. 64–167); III (168–208). The leaves are bound as follows: I $(1–13)^{6+(6+1)}$; II $(14–25)^{6+6}$; III $(26–37)^{6+6}$; IV $(38–49)^{6+6}$; V $(50–63)^{7+7}$; VIII $(64–75)^{6+6}$; $(76–87)^{6+6}$; $(88–99)^{6+6}$; $(100–111)^{6+6}$; $(112–123)^{6+6}$; $(124–135)^{6+6}$; $(136–147)^{6+6}$; $(148–159)^{6+6}$; $(160–167)^{6+(6-4)}$; $(168–177)^{5+5}$; $(178–189)^{6+6}$; $(190–205)^{8+8}$; $(206–208)^{(3-2)+(3-1)}$. At the end of the volume, we find two modern paper guard-leaves; the second one is glued to the inside back cover, in the same manner as at the beginning of the volume. The first four fascicules of the second part contain signatures in red ink, in the central lower margin: f. 64^{r} (i); f. 76^{r} (ii); f. 88^{r} (iii); f. 100^{r} (iiii). Ff. 61^{v}–63^{r} are blank. Modern foliation in pencil in the upper outer margin. Two consecutive leaves were numbered 196; for this reason, an additional foliation starts with the second leaf numbered 196, to correct the previous one.

Content. This manuscript contains Marsilius of Inghen's commentaries on the *Parva naturalia*;[18] a set of "problems" concerning the body of animals,

---

17 G. Oergel, "Das Collegium zur Himmelspforte während des Mittelalters", *Mitteilungen des Vereins für Geschichte und Altertumskunde von Erfurt* 19 (1898), 19–114.

18 Ff. 1ra–33rb: *Questiones super librum De sensu et sensato*; ff. 33va–38va: *Questiones super librum De memoria et reminiscentia*; ff. 38vb–46va: *Questiones super librum De somno et vigilia*; ff. 46vb–55va: *Questiones super librum De longitudine et brevitate vite*; ff. 55va–61rb:

which is partially taken from Aristotle's *Problemata*,[19] a note on four different kinds of fever[20] and a set of questions on Aristotle's *Meteorology* (ff. 64ra–167rb), which the colophon ascribes to John Buridan.

The *Questions on Meteorology*. Although the colophon ascribes the whole commentary to Buridan, this only applies to the first three books, the fourth being by Oresme.[21] The text is laid out in two columns of about forty-five lines each. The ink is brown and the decoration has not been completed, as the rubrics stop on f. 68va and the incipits of the questions in textualis only reach question 1.2, on f. 65vb. Only the incipit of the first book has been copied: in some other cases, the space intended for this purpose has been left blank (see for example f. 121va).

Questiones super librum De inspiratione et respiratione (unfinished); ff. 168ra–179va: *Expositio libri De sensu et sensato*; ff. 179vb–183ra: *Expositio libri De memoria et reminiscentia*; ff. 183rb–189va: *Expositio libri De somno et vigilia*; ff. 189va–193va: *Expositio libri De longitudine et brevitate vite*; ff. 191va–193ra: *Expositio libri De iuventute et senectute*; ff. 193ra–194rb: *Expositio libri De respiratione et inspiratione*; ff. 194rb–194vb: *Expositio libri De morte et vite*; ff. 194vb–195rb: *Expositio libri de motu cordis*; ff. 195rb–195vb: *Expositio libri De motibus animalium*. Colophon: "Explicit liber *De sensu et senssato* [*sic*] et dicta per me, Henricum Wetter in Gottingen". On Marsilius of Inghen's commentaries on Aristotle's *Parva naturalia* see S. Ebbesen, C. Thomsen-Thörnqvist, and V. Decaix, "Questiones on De sensu et sensato, De memoria, and De somno et vigilia. A Catalogue", *Bulletin de Philosophie Médiévale* 57 (2015), 59–116, esp. 85–86 (*De sensu et sensato*), 95 (*De memoria*), 113–114 (*De somno et vigilia*).

19 *Inc.*: "Omnes homines naturaliter scire desiderant, scribit philosophorum princeps"; *expl.*: "cognitis et solutis propleumatibus de partibus corporis humani ex variis Aristotelis et aliorum philosophorum et pariter medicorum codicibus lucide collectis et extractis, hic iam consequenter aliqua propleumata generalia, attamen valentia et utilia, sunt subiungenda". L. Thorndike, and P. Kibre, *A Catalogue of Incipits of Scientific Writings in Latin*, Cambridge (MA), 1963 (The Medieval Academy of America, 29), col. 986.

20 *Inc.*: "febris cottidiana quattuor modis solet evenire"; *expl.*: "sudorem emittat de mane". See Thorndike and Kibre, *A Catalogue of Incipits of Scientific Writings in Latin*, col. 551.

21 As I have pointed out elsewhere (Panzica, "Nicole Oresme à la Faculté des Arts de Paris", 38), this ascription caused some confusion in catalogues and modern inventories, which often ascribe the whole text to Buridan: W. Schum, *Beschreibendes Verzeichnis der Amplonianischen Handschriften-Sammlung zu Erfurt*, Berlin 1887, 230–231; E. Faral, "Jean Buridan: notes sur les manuscrits, les éditions et le contenu de ses ouvrages", *Archives d'histoire doctrinale et littéraire du Moyen Âge* 35 (1960), 1–53, esp. 21, Ch. Lohr, "Medieval Latin Aristotle Commentaries. Authors: Jacobus-Johannes Juff", *Traditio* 26 (1970), 135–216, at 172, M. Markowski, *Repertorium commentariorum medii aevi in Aristotelem Latinorum quae in Bibliotheca Amploniana Erffordiae asservantur*, Wrocław 1987, 182. Birkenmajer (*Études d'histoire des sciences en Pologne*, 181–182) and McCluskey (*Nicole Oresme on Light, Color and the Rainbow*, 96) correctly distinguish the first three books, which should be ascribed to Buridan, from the fourth book, which is by Oresme.

Incipit and explicit of the books

Prol.: Ut habetur in prohemio *De anima*, constat quod scientiam libri *Metheororum* valde reputare debemus aliis nobiliorem et super cetera appetibilem; opera namque metheoroloyca inter cetera nobis apparentia sunt mirabiliora.

I, ff. 64rb–104rb: Queritur primo utrum de impressionibus metheoroloicis sit tradenda scientia naturalis distincta a scientia *De celo et mundo* et *De generatione et corruptione*, *De mineralibus* et *De anima et animatis*. Arguitur primo quod non sit scientia de eis, quia non est scientia de singularibus, contingentibus et potentibus aliter se habere ...X... Veniunt enim alia mineralia de alio loco ad illum. Et sic sit finis huius questionis et primi libri.

II, ff. 104rb–121va: Circa secundum librum *Metheororum* queritur primo de perpetuitate vel generatione maris, scilicet utrum mare fuit perpetuum vel aliquando factum. Arguitur quod non fuit perpetuum ...X... Solutiones aliarum rationum satis haberi possunt ex predeterminatis. Et sic patet questio, et per consequens secundus liber.

III, ff. 121va–158va: Circa principium tertii libri primo notandum, et verum est, quod ab antiquo solet istum totum librum *Metheororum* dividi in quatuor partes ...X... Sive lapidibus, sive plantis, sive avibus, sive earum plumis et metallis. Et sic finitum tertii libri [*sic*] *Metheororum*, date a reverendo magistro Biridano.

IV, ff. 158va–167rb: Circa principium quarti libri *Metheororum* queritur utrum tantum sint quatuor qualitates prime, scilicet caliditas, humiditas, frigiditas et siccitas. Arguitur quod non ...X... Ad sextam dico quod in salamandra non duratur ignis in quantitate nec ibi nutritur ab igne; sed tale animal non est ustibile ab igne; et ratio quia non uritur ab igne, dicimus quod tale animal non est ustibile ab igne. Et sic patet questio, etc.

Bibliography

Catalogues: W. Schum, *Beschreibendes Verzeichnis der Amplonianischen Handschriften Sammlung zu Erfurt*, Berlin 1887, 230–231; Bénédictins du Bouveret, *Colophons de manuscrits occidentaux des origines au XVIe siècle*, Fribourg (Switzerland) 1979 (Spicilegii Friburgensis Subsidia) (6 vols.), 5: 87; M. Markowski, *Repertorium commentariorum medii aevi in Aristotelem Latinorum quae in Bibliotheca Amploniana Erffordiae asservantur*, Wrocław 1987, 70, 182; Lohr, *Latin Aristotle Commentaries, I.2. Medieval Authors M–Z*: 257; 2: 34.

Literature: Birkenmajer, *Études d'histoire des sciences en Pologne*, 181–182; McCluskey, *Nicole Oresme on Light, Color and the Rainbow*, 96; S. Bages, *Les Questiones super tres libros Metheororum Aristotelis de Jean Buridan: étude suivie de l'édition du livre I*, Thèse de doctorat de l'École des Chartes, 1986 (2 vols.), 1: 148–150; Panzica, "Nicole Oresme à la Faculté des Arts de Paris", 35.

5 *Kassel, Universitätsbibliothek—Landesbibliothek und Murhardsche Bibliothek, 2° Phys. et hist. nat. 12, ff. 1ᵛ–107ʳ* (Ka)

Paper; mm 220×331; ff. 203; 1424

Date and origin. This manuscript, which is dated 1424 (f. 107^{r}: "Expliciunt questiones Orem, reverendi magistri Wilhelmi super *Metheororum*. Anno domini millesimo quadragentesimo vicesimo quarto, sabbato post festum sancti Egidii"), was owned by the library of the Augustinian Canons of Fritzlar and came into the possession of the Murhardsche Bibliothek of Kassel in 1804.

Codicology. Wooden binding covered with leather. The manuscript contains 203 paper leaves grouped as follows: I $(1–12)^{6+6}$; II $(13–24)^{6+6}$; III $(25–36)^{6+6}$; IV $(37–48)^{6+6}$; V $(48–60)^{6+6}$; VI $(61–72)^{6+6}$; VII $(73–84)^{6+6}$; IX $(85–96)^{6+6}$; X $(97–108)^{6+6}$; XI $(109–120)^{6+6}$; XII $(121–132)^{6+6}$; XIII $(133–144)^{6+6}$; XIV $(145–156)^{6+6}$; XV $(157–166)^{5+5}$; XVI $(167–179)^{(1+6)+6}$; XVII $(180–191)^{6+6}$; XVIII $(192–203)^{6+6}$. Modern page numbering in pencil in the outer margin. Signatures in the upper margin of the recto leaves: f. 61 (6us); f. 73 (7us); f. 85 (octavus); f. 97 (nonus).

Content. In addition to Oresme's *Questions on Meteorology* (ff. 2ra–107rb), this manuscript contains an incomplete set of anonymous questions on Aristotle's *Physics* (109ra–198rb). These questions are preceded by a table of contents of the questions for the first three books (ff. 107vb–108ra).[22]

The *Questions on Meteorology.* The commentary is ascribed to Guillaume Oresme in the colophon (f. 107^{r}).[23] The text, copied in a single hand, is laid out

22 The first book contains 24 questions; the second book, 12 questions; the third book, 19 questions; the fourth, contains 15 questions; the fifth book, 10 questions; the sixth book, 8 questions; the seventh book, 6 questions; the eighth book, 5 questions. Some of the questions listed in the table are: I.1, *Utrum phylosophia naturalis sit considerativa de omnibus rebus mundi*; I.2, *Utrum totali scientie naturali assignandum est unum subiectum*; I.3, *Utrum ens mobile sit subiectum scientie libri primi* […]; II.1, *Utrum res artificialis sit distincta a re naturali*; II.2, *Utrum differentia quam ponit Philosophus inter naturalia ex una parte et artificialia ex altera sit bona*; II.3, *Utrum figura sit distincta a re figurata* […]; III.1, *Utrum, ignorato motu, necesse sit ignorare naturam*; III.2, *Utrum in alteratione, preter alterabile et qualitatem que acquiritur, sit ponendus fluxus distinctus*; III.3, *Utrum qualitates contrarie possunt se simul compati in aliquo subiecto*.

23 Very little is known about Guillaume Oresme, who was a theologian, member of the Collège of Navarre from 1352: J. de Launoy, *Regii Navarre Gymnasii Parisiensis Historia*, Paris 1677 (2 vols.), 1: 93; F. Meunier, *Essai sur la vie et les ouvrages de Nicole Oresme*, Paris 1857, 10. He is mentioned in a bull of 1367, which informs us that at that time he held a bachelor in Theology and was canon of Bayeux: H. Denifle and E. Chatelain, *Chartularium Universitatis Parisiensis*, 1200–1452, Paris, 1889–1897 (4 vols.), 2 (1891), 641, n. 3 ("Guillelmo Oresme, bacallario in theologia, canonico Baiocensi"). Guillaume Oresme is the author of a translation of Ptolemy's *Quadripartitum* prepared at the request of Charles V and transmitted in ms. Paris, Bibliothèque Nationale de France, franç. 1348.

in two columns of about fifty lines each. On f. 1^{r}, there is a sort of draft of the first question, which stops after the words: "ex parte ipsius opinantis". The handwriting is not as polished as that of the following text. On f. 1^{r}, we find many pen testing marks ("amice benignissime contrariam doctrinam exquiris fraternitatem; ave maria, ave maria gratia"). After the table of contents on f. 1^{v}, the text starts from scratch on f. 2^{r}. The words that mark the beginning of the books are in textualis and are placed so as to form a square. Rubrics and pilcrows in red. Marginal notes in red or brown ink. Some discontinuous notes refer to the content of the text (see for example f. 38^{r}, *marg. sup.*: "nota distinctionem marium et quare unum sit salsum et aliud non"). The initials of the words that introduce new paragraphs are marked by a red stroke. The decoration is incomplete: on f. 2vb, for instance, the space left for the title of the question has been left blank; similarly, on f. 92ra, the space for the initial words of the fourth book has been left blank. On f. 1^{v}, we find a table of contents for the first three books; the table of contents of the fourth book can be found on f. 107va.

Incipit and explicit of the books

I, ff. 2ra–24vb: Circa initium primi libri *Metheororum* queritur primo utrum possibile sit de impressionibus metheorologicis simul habere scientiam et opinionem. Et arguitur primo quod non, quia de impressionibus metheorologicis non contingit habere scientiam et opinionem, ergo questio falsa ...X... Ad quartam: 'si galaxia non esset de natura celi, tunc de ea non esset determinandum in isto libro', concedo. Sed tamen Aristoteles determinavit de ea hic propter hoc quia plures antiquorum crediderunt quod galaxia esset de natura elementari. Et sic est finis questionum primi libri. Et sic est finis primi libri *Metheororum*. Sequuntur questiones *Metheororum* libri secundi.

II, ff. 25^{ra-va}: ⟨C⟩irca initium secundi libri *Metheororum* queritur primo utrum locus generationis pluvie sit media regio aeris. Et arguitur quod non: nul-

---

On the ascription of this translation see M. Lejbowicz, "Guillaume Oresme, traducteur de la *Tétrabible* de Claude Ptolémée", *Pallas* 30 (1983), 107–133. The edition of this translation was prepared by J.W. Gossner, *Le Quadripartit Ptholomee*, unpublished Ph.D. dissertation, Syracuse University 1951. Menut has written that "well authenticated works of Nicole have been occasionally attributed to Guillaume both in MS. and in printed editions" (A.D. Menut (ed.), *Maistre Nicole Oresme: Le livre de Éthiques d'Aristote, Published from the Text of MS. 2902, Bibliotheque Royale de Belgique, with a Critical Introduction and Notes*, New York 1940, 11, fn. 8). Two manuscripts transmitting Nicole Oresme's *De communicatione ydiomatum* ascribe the text to Guillaume Oresme: Erfurt, Universitäts- und Forschungsbibliothek, Dep. Erf. CA 4° 150, ff. 83^{v}–92^{v} (Colophon: "Explicit tractatus magistri Guillemi Horem *De communicatione ydiomatum*") and Toulouse, Bibliothèque d'Étude et du Patrimoine, ms. 246, ff. 219^{v}–225^{v} (Initium: "Incipit tractatus *De communicatione ydiomatum* editus a magistro Guillelmo Oreme, doctore in theologia Parisiensi").

lum elementum debet generari extra locum sibi naturalem ...X... Ad rationes ante oppositum patet ex secundo articulo. Et sic est finis questionum secundi *Metheororum*.

III, ff. 39vb–91vb: ⟨C⟩irca tertium librum *Metheororum* queritur utrum ventus sit exalatio calida et sicca. Et arguitur primo quod non sit exalatio calida, nam si sic, sequeretur quod temporibus ventosis aer debet esse calidus ...X... Ad ultimam dico quod non in vanum laborant, ex quo aliam artem ignorant. Et sic patet questio. Et sic est finis questionum tertii libri *Metheororum*, de quo laudetur Sancta Trinitas in secula seculorum. Amen.

IV, ff. 92ra–107rb: Utrum tantum quatuor sunt qualitates prime, scilicet caliditas, humiditas, siccitas et frigiditas. Et arguitur primo quod non, quia quod per habundantiam dicitur, uni soli rei convenit ...X... Et quia non uritur ab igne, dicimus quod ipsum habitet in igne. Et sic est finis.

Bibliography

Catalogues: Lohr, *Latin Aristotle Commentaries, I.2. Medieval Authors M–Z*: 34.

Literature: McCluskey, *Nicole Oresme on Light, Color and the Rainbow*, 89–90.

6 *Klagenfurt, Bischöfliche Bibliothek, XXXI b 5, ff. 1^r–124^r* (Kl)

Paper; mm 290 × 200; ff. 168; ca. 1395–1405

Date and origin. Very little is known about this manuscript. According to Michael, it comes from southern Germany or Carinthia.[24] The codex is mentioned in a manuscript catalogue of Gurk's diocesan library from about 1700 under the signature 138.[25] The watermark, a long bell with three parallel clips, which appears in fascicules X–XII, was particularly widespread between the last decade of the fourteenth and the first years of the fifteenth century.[26]

Codicology. Wooden binding covered with leather. A parchment sheet glued to the inside front cover contains the table of contents of Oresme's *Questions*. The leaves are grouped mainly in senions, according to the following formula: I $(1–12)^{6+6}$; II $(13–24)^{6+6}$; III $(25–36)^{6+6}$; IV $(37–48)^{6+6}$; V $(49–60)^{6+6}$; VI $(61–72)^{6+6}$; VII $(73–84)^{6+6}$; VIII $(85–96)^{6+6}$; IX $(97–106)^{5+5}$; X $(107–118)^{6+6}$; XI $(119–124)^{3+3}$; XII $(125–136)^{6+6}$; XIII $(137–148)^{6+6}$; XIV $(149–161)^{6+6}$; XV $(162–168)^{4+4}$.

24 Michael, *Johannes Buridan*, 2: 741.

25 *Catalogus super libros bibliothecae episcopatus Gurcensis cum praevia instructione ad inveniendum quemvis librum in promptum*. This catalogue is now preserved under the signature XXX b 11 in the Klagenfurt Diocesan Library. See Menhardt, *Handschriftenverzeichnis der Kärntner Bibliotheken 1: Klagenfurt*, Vienna 1927 (Handschriftenverzeichnisse Österreichischer Bibliotheken 1: Kärnten), 33–34.

26 See the database *Wasserzeichen des Mittelalters*, http://www.wzma.at/2901.

Another parchment leaf is glued to the inside back cover. This leaf contains a fragment from Alanus of Lille's *De planctu nature*.[27] Signatures: f. 25^{r} (3us *Metheororum*), f. 36^{v} (3us), f. 48^{v} (4us), f. 60^{v} (5us), f. 72^{v} (6us), f. 83^{v} (7us), f. 96^{v} (8us), f. 106^{v} (9us), f. 118^{v} (10us), f. 123^{v} (11us), f. 136^{v} (primus). Catchwords: f. 36^{v}, f. 48^{v}, f. 72^{v}, f. 83^{v}, f. 96^{v}, f. 106^{v}. Medieval foliation in the outer upper margin of the recto leaves. The foliation skips from 159 to 161

Content. The codex contains Oresme's *Questions on Meteorology* (ff. 1^{v}–124^{r}) and John Buridan's *Questions on the Parva naturalia* (ff. 125rb–168vb).[28]

The *Questions on Meteorology*. The commentary, which is anonymous, is laid out in two columns of about forty lines each. The text was copied by two scribes. The first one copied ff. 1^{v}–41ra (l. 28); 107ra–108ra (l. 2); ff. 112ra–124rb; the second one, which has a curvier handwriting, copied ff. 41ra (l. 29)–106vb; ff. 108ra (l. 3)–111vb.

A table of contents of the *Questions* is copied on a parchment sheet glued to the inner front cover. This sheet is numbered 119 in the upper margin. On f. 119, we find an alphabetical index that presents the content of the questions in alphabetical order (for example: "aqua an naturaliter ascendat ad orificia fontium", "animalia habitantia in locis calidis, an sint"). It is interesting to notice that the subjects listed do not always match the titles of the questions. In compiling this alphabetical list, the copyist therefore wished to provide the reader with as detailed an index as possible.

In the upper margin, we find a medieval numbering in brown ink. In some cases, the first words and the structural parts of the questions are copied in textualis (see for instance f. 7rb and f. 28vb). More often, however, the space set aside for these words has been left blank (f. 7^{v} and f. 8^{r}). The first words of the text (⟨C⟩irca initium libri *Metheororum* Aristotelis queritur questio talis) are placed so as to form a square of a width of about seventeen lines. The decoration is incomplete: the space intended for the initials of the books has been left blank (see, for example, the incipit of the first book on f. 1ra and the explicit of the second book on f. 33rb).

Incipit and explicit of the books

I, ff. 1va–33rb: ⟨C⟩irca initium libri *Metheororum* Aristotelis queritur questio talis: utrum possibile sit de impressionibus metheroloicis [*sic*] simul habere scientiam et opinionem. Et arguitur primo quod non, quia de impressionibus

27 The fragment, written in two columns, begins with the words: "melius esse producitur per me homo procreatur ad mortem per ipsum recreatur" and ends with the words: "falsitate vesperascit occasum".

28 Ff. 125rb–143vb: *Questiones in librum De sensu et sensato*; ff. 143vb–146va: *Questiones in librum De memoria et reminiscientia*; ff. 145va–153rb: *Questiones in librum De somno et vigilia*;

metholoycis [*sic*] non contingit habere scientiam nec opinionem; ergo questio falsa …X… Ad quartam: 'si galaxia esset de natura celi, tunc de ea non esset determinandum in isto libro', concedo tamen quod Aristoteles determinat de ea propter quod quia plures antiquorum crediderunt galaxiam esse de natura elementari. Et sic patet questio. Hoc de isto primo libro. Deo laus.

II, ff. 33rb–40va: ⟨C⟩irca librum secundum *Metheororum* Aristotelis queritur questio prima, circa secundum huius, secundum distinctionem in principio positam ab Alberto, et est questio ista: utrum locus generationis pluvie sit media regio aeris. Et arguitur primo quod non …X… Ad rationes in oppositum patet in secundo articulo.

III, ff. 40va–109va: ⟨C⟩irca tertium, secundum distinctionem illius libri totius in principio positam ab Alberto, sit questio prima: utrum ventus sit exalatio calida et sicca. Et arguitur primo quod non sit exalatio calida …X… Ad ultimam dico quod non in vanum laborant, ex quo aliam artem ignorant.

IV, ff. 109va–124^{r}: ⟨C⟩irca principium istius quarti queritur primo utrum tantum sint quatuor qualitates prime, scilicet caliditas, humiditas, frigiditas, siccitas. Et arguitur quod non, quia quod per superhabundantiam dicitur, uni soli convenit …X… Ad quartam dico quod in salamandra non dominatur ignis in quantitate; sed tale animal non est ustibile ab igne nec etiam nutritur ab igne; ratio huius quia non uritur ab igne, dicimus quod tale animal habitat in igne. Hoc de questione et per consequens de omnibus. Ecce finis.

Bibliography

Catalogues: H. Menhardt, "Die Handschriftensammlung der bischöflichen Bibliothek in Klagenfurt", *Zentralblatt für Bibliothekswesen* 39 (1922), 363–381, esp. 375; Id., *Handschriftenverzeichnis der Kärntner Bibliotheken 1: Klagenfurt*, 68 [here Oresme's *Questions* are attributed with doubt to Albert the Great]; M. Markowski, *Repertorium commentariorum medii aevi in Aristotelem Latinorum qui in bibliothecis austriacis: Admont, Furt bei Göttweig, Graz, Heiligenkreuz, Klagenfurt, Klosterneuburg, Kremsmünster, Linz, Melk, Salzburg, Sankt Florian, Vorau, Zwettl asservantur*, Krakow 2008, 65, 196; Lohr, *Latin Aristotle Commentaries, I.2. Medieval Authors M–Z*: 260–262; 2: 35.

Literature: McCluskey, *Nicole Oresme on Light, Color and the Rainbow*, 90–91; Michael, *Johannes Buridan*, 2: 741, 749, 755, 763, 770.

ff. 153rb–157va: *Questiones in librum De longitudine et brevitate vite*; ff. 157va–168vb: *Questiones in libros De morte et vita, De respiratione, De iuventute et senectute*. See Ebbesen, Thomsen-Thörnqvist, Decaix, "Questiones", 82, 94, 112.

7 ***Kraków, Biblioteka Jagiellońska, cod. 749, ff. 59^{v}–110^{v}*** **(Kr)**

Paper; mm 300×215; ff. 110 + III; ca. 1360–1370

Date and origin. This manuscript comes from the University of Prague, as stated by the colophon on f. 59^{r}: "Expliciunt questiones *De generatione et corruptione* Byrydani magistri Parariensis [*sic*] Prage reportate". The majority of the texts contained in this manuscript were copied by a certain Iohannes (see below the colophon of Oresme's *Questions on Meteorology*). The watermarks allow us to date the manuscript to the 1370s.

Composition. See the modern catalogue, 5, 293.[29]

Content. The manuscript contains some commentaries on the Aristotelian natural books by Parisian masters of the fourteenth century: Albert of Saxony's *Questions on De celo* (ff. 1^{r}–42^{v}); John Buridan's *redactio B* of the *Questions on De generatione et corruptione* (ff. 43^{r}–59^{r}), as well as the second redaction of Oresme's *Questions on Meteorology* (ff. 59^{r}–110^{r}).

The Questions on Meteorology. The colophon ascribes the text to Nicole Oresme: "Expliciunt questiones *Metheororum* magistri Nicholai dicti cognomine Orem, finite per manus Iohannis et reportate ante festum Sancti Galli" (f. 110va). The text is laid out in two columns of about sixty-five lines each. The handwriting is tiny and compact; the ink is brown. From f. 60rb onwards, we notice an abrupt change in ink. The copyist, who stopped writing at f. 60ra and restarted on f. 60rb, completely skipped the main body of question I.2 (*Utrum impressiones metheorologice fiant secundum naturam inordinatiorem quam sit natura celi*) jumping directly from the contrary arguments (f. 60ra) to the answer. On f. 59va, some space has been left for the initial that should have introduced the text. The titles of the questions are in textualis. The folios have been numbered with a pencil by a modern hand. Another modern hand has numbered the questions with a pencil.

Incipit and explicit of the books

I, ff. 59va–69vb: Queritur circa librum *Methorum* [*sic*] primo utrum possibile sit de impressionibus metlogicis [*sic*] habere scientiam et opinionem. Et arguitur quod non, quia de eis non convenit habere scientiam nec opinionem; ergo questio falsa …X… Ad quartam: 'si galaxia esset de natura celi', etc., concedo, tamen Aristoteles determinat de ea propter hoc quia [69vb] plures antiqui crediderunt galaxiam esse de natura elementari. Et sic patent questiones primi libri.

II, ff. 69vb–76va: ⟨T⟩unc restat tractare questiones secundi *Metheororum*. Utrum locus generationis pluvie sit media regio aeris. Et arguitur quod non:

29 M. Kowalczyk, A. Kozłowska, M. Markowski, S. Włodek, and M. Zwiercan, *Catalogus*

nullum elementum debet generari extra locum sibi naturalem ...X... Oppositum patet ex secundo articulo. Et sic est finis secundi libri *Metheororum*.

III, ff. 76va–101rb: ⟨T⟩unc secuuntur questiones tertii libri *Metheororum*. Utrum ventus sit exalatio calida et sicca. Et arguitur quod non sit exalatio calida, quia si sic, sequitur quod in temporibus ventosis aer deberet esse calidus ...X... Ad ultimam dico quod non in vanum laborant, ex quo aliam artem ignorant.

IV, ff. 101rb–110^{v}: ⟨Q⟩ueritur circa quartum librum et ultimum *Metheororum*. Queritur circa quartum librum utrum sunt tantum quatuor qualitates prime, scilicet caliditas, humiditas, frigiditas et siccitas. Et arguitur primo quod non, quia quod per superhabundantiam dicitur, uni soli convenit ...X... Sed tale animal non est combustibile ab igne et ideo, quia non uritur ab igne, dicimus quod ipsum habitat in igne. Et sic est finis, etc. Expliciunt questiones *Metheororum* magistri Nicolai dicti cognomine Orem. Finite per manus Iohannis et reportate ante festum sancti Galli [16th October]. Laus Christo sit etc.

Bibliography

Catalogues: W. Wisłocki, *Catalogus codicum manuscriptorum Bibliothecae Universitatis Jagellonicae Cracoviensis*, Krakow 1877 (2 vols.), 1: 222; M. Kowalczyk [et al.], 292–294; M. Markowski, and Z. Włodek, *Repertorium commentariorum medii aevi in Aristotelem Latinorum quae in Bibliotheca Jagellonica Cracoviae asservantur*, Wrocław 1974, 59; Lohr, *Latin Aristotle Commentaries, I.2. Medieval Authors M–Z*: 40, 256; 2: 35.

Literature: Birkenmajer, *Études d'histoire des sciences en Pologne*, 188–189; Michael, *Johannes Buridan*, 2: 638–639; Streijger, Bakker, and Thijssen (eds.), *John Buridan Quaestiones super libros De generatione et corruptione Aristotelis*, 15–16; M. Markowski, *Burydanizm w Polsce w okresie przedkopernikańskim: studium z historii filozofii nauk ścisłych na Uniwersytecie Krakowskim w XV wieku*, Wrocław 1971, 404; A. Panzica, "Commenter les *Météorologiques* à l'Université de Cracovie", 80–82.

8 *Kraków, Biblioteka Jagiellońska, cod. 751, ff. 3^r–53^r* (Kr_1)

Paper; mm 275 × 220; 150 ff.; ca. 1360–1370

Date and origin. The watermarks allow us to date this manuscript to around 1360–1370. Some notes in Czech prove its Bohemian origins (see f. 114^v). The

codicum manuscriptorum medii aevi latinorum qui in Bibliotheca Jagellonica Cracoviae asservantur, 5, Numeros continens inde a 668 usque ad 771, Wrocław/Warsaw 1992, 292–294, at 293.

manuscript was brought to Krakow in the second quarter of the fifteenth century by a Polish scholar. Starting from 1462, it belonged to master Iacobus Boksicze, as witnessed by the mark of possession on the inside cover: "Liber magistri Boxicze".[30]

Composition. See Kowalczyk [et al.], *Catalogus*, 5, 299.

Content. The manuscript mainly contains commentaries on the Aristotelian corpus by Parisian masters of the fourteenth century: a fragment from Albert of Saxony's *Questions on Meteorology* (f. 2ra–^{vb});[31] the second redaction of Oresme's *Questions on Meteorology* (ff. 3ra–53rb); the short version of Buridan's *Questions on De anima* (ff. 54ra–73vb); Albert of Saxony's *Questions on De celo* (ff. 74ra–124ra) and the *redactio B* of Buridan's *Questions on De generatione et corruptione* (ff. 124ra–147rb).

The *Questions on Meteorology*. The commentary is ascribed to Nicole Oresme in the colophon and in the table of contents that precedes the text ("Expliciunt tabule questionum libri *Metheororum* a magistro Horem confectarum", f. 1^{v}; "Et sic est finis *Metheororum* Horem", f. 53rb). The text is laid out in two columns of about fifty-five lines each. The handwriting is compact and regular, in brown ink. Corrections (possibly by the copyist himself) above the line and in the margins (e.g. on f. 1^{r}, l. 4 and l. 9). Some marginal notes in another hand refer to the structure of the text and its contents (e.g. on f. 3ra: "prima"; "oppositum"; "que sit scientia"). The questions are numbered in red by a medieval hand. There is an error in question III.25, *Utrum yris solum dupliciter et non multipliciter possit apparere*: the number 25 has erroneously been replaced by 21. The wrong numbering is consistently used until the end of the third book. A modern hand has corrected this numbering by writing the correct number in pencil and putting the incorrect one in parentheses. There are two foliations, both by modern hands: the first one, in grey pencil, in the upper margin of each recto; the second one, in red pencil, in the lower margin. The former is not always present or clearly readable: this might explain why a second foliation was added. The decoration has not been completed: the lines intended for the initial and the incipit of the third book, on f. 22va, have been left blank, just as the space intended for

30 Iacobus de Boxicze is mentioned in the register of students of Krakow University for the year 1456: *Album studiosorum Universitatis Cracoviensis*, Tomus I (ab Anno 1400 ad Annum 1489), Krakow 1887, 147. He obtained the degree of Master of Arts in 1462 and that of Doctor in Medicine in 1476 (*Polski słownik biograficzny*, 2, 244–245). Two other manuscripts from the Jagellonian Library belonged to him: ms. 499 (*Catalogus*, 3, 130–132) and ms. 565 (*Catalogus*, 4, 7–9).

31 "⟨C⟩irca librum *Metheororum* queritur primo utrum de impressionibus metheoroloycis sit scientia. Et arguitur quod non, quia huiusmodi impressiones, sicut nubes, tonitrua, non

the titles of books and questions. The pilcrows are marked with a stroke in red ink. The numbering of the book in the upper margin of each verso leaf is also in red, sometimes using Roman numerals, sometimes the Arabic equivalent.

Incipit and explicit of the books

I, ff. 3ra–14vb: Utrum de impressionibus metheorologicis possibile sit habere scientiam simul et opinionem. Arguitur quod non, quia de eis non contingit habere scientiam nec opinionem; igitur questio falsa ...X... Ad quartam: 'si galaxia esset de natura celi', etc., concedo, tamen Aristoteles determinat de ea hic propter hoc quia plures antiquorum crediderunt galaxiam esse de natura elementari. Et sic est primi libri finis.

II, ff. 14vb–22rb: Utrum locus generationis pluvie sit media regio aeris. Arguitur quod non: nullum elementum debet generari extra locum sibi naturalem ...X... Ad rationes in oppositum patet ex secundo articulo. Et sic patet questio. Incipiunt questiones tertii libri *Metheororum.*

III, ff. 23ra–47va: Utrum ventus sit exalatio calida et sicca. Et arguitur quod non, quia quod non sit exalatio calida ...X... Ad ultimam dico quod non in vanum laborant, ex quo aliam artem ignorant. Et sic est finis questionum tertii libri *Metheororum.*

IV, ff. 47va–53rb: Utrum tantum sint quatuor qualitates prime, scilicet caliditas, humiditas, frigiditas et siccitas. Et arguitur primo quod non, quia ...X... Ad sextam dico quod in salamandra non dominatur ignis in quantitate nec etiam nutritur ab igne; sed tale animal non est ustibile ab igne; et ideo, quia non uritur ab igne, dicimus quod tale animal habitet in igne. Et sic est finis *Metheororum* Horem ad honorem Dei etc.

Bibliography

Catalogues: Wisłocki, *Catalogus*, 1: 223; Kowalczyk [et al.], *Catalogus*, 5: 296–300; Markowski and Włodek, *Repertorium*, 160; Lohr, *Latin Aristotle Commentaries, I.2. Medieval Authors M–Z*: 40, 256; 2: 35.

Literature: Birkenmajer, *Études d'histoire des sciences en Pologne*, 1: 189–191; Markowski, *Burydanizm w Polsce*, 403–404; Michael, *Johannes Buridan*, 2: 639, 687; Streijger, Bakker, and Thijssen (eds.), *John Buridan Quaestiones super libros De generatione et corruptione Aristotelis*, 16; Panzica, "Albert of Saxony's *Questions on Meteorology*", 254; Panzica, "Commenter les *Météorologiques* à l'Université de Cracovie", 82.

---

sunt perpetue ...X... tunc sunt quatuor conclusiones. Prima est generalis, quod universum et omnia que in eo sunt bene sunt ordinata et post pone illas etc.". On this fragment see A. Panzica, "Albert of Saxony's *Questions on Meteorology*", 254.

9 *Kraków, Biblioteka Jagiellońska, cod. 2095, ff. 245^r–307^r* (Kr_2)

Paper; mm 210×150; ff. I+308; 1406

Date and origin. The colophons of the texts contained in this manuscript offer plenty of information about its history. The manuscript was copied at the University of Prague by a Polish student, Iohannes Stolle de Glogovia, in 1406.[32] Iohannes attended the course on Aristotle's *Physics* held by Master Lindore of Scotland (f. 221rb: "Et sic terminantur questiones totius libri *Physicorum* Aristotilis summi philosophorum principis, compilate per honorabilem magistrum Laurentium Londoriensem de Scotia, reportate vero per Johannem Stolle de Glogovia Maiori in Studio alme Universitatis Pragensis, anno Domini M° quadragintesimo sexto") and listened to the *Questions on Meteorology* by Nicole Oresme, which was also used at the University of Prague (f. 307rb: "Et sic patent questiones totius libri *Metheororum* compilate per reverendum magistrum Orem Parisiensem et comparate per Johannem Stolle in Studio alme Universitatis Pragensis, a. D. 1406"). The possession marks on the second guard-leaf and on the first leaf confirm the Prague origin of this codex and the fact that it belonged to Iohannes Stolle: "Liber Johannis Stolle de Glogovia, scriptus ab eodem in Studio Pragensi" (this possession mark has been written from right to left by inverting the orientation of each letter); "liber comitatis Maioris Collegii Artistarum. [*alia manu*] Nicolai a Schadek Prokopiadis liber vetustissimus Pragensis Studii".

Composition. Wooden binding covered with brown pasteboard. Traces of the central boss and of pegs on both covers. The front cover is detached. The first guard-leaf has been torn away. On the second guard-leaf, numbered II by a modern hand, a medieval hand, probably Iohannes Stolle himself, has sketched out a table of contents.[33] This guard-leaf is followed by 308 leaves grouped mainly in senions according to the following formula: I (1–12)$^{6+6}$; II (13–24)$^{6+6}$;

---

32 Iohannes Stolle is mentioned as *notarius*, namely the official *reportator*, in another manuscript of the Jagellonian Library, ms. 2116, f. 1^r: "Super tractatus P⟨etri⟩ Hyspani datus per notarium, id est Joannem Stolle de Glogovia Maiori" (see Markowski, *Repertorium*, 115). After having obtained the degree of Master of Arts, he subsequently taught at Krakow and Vienna. On Iohannes Stolle and his collection of philosophical books, see J. Zathey, A. Lewicka-Kamińska, and L. Hajdukiewicz, *Historia Biblioteki Jagiellonskiej*, 1, 1364–1775, Krakow, 1966, 82–84 and L. Moonan, "The Scientific Writings", 273–317.

33 "In hoc libro continentur primo questiones *Phisicorum* compilate per magistrum Laurentium Londoriensem de Scotia; secundo problemata collecta de *Problematibus* Aristotilis; tertio questiones *Metheororum* Magistri [*add. sup. lin.*: Nicolai] Orem, quorum librorum autor est Aristotiles, Nycomaci filius, Grecorum sapientissimus, qui scientias invenit et complevit, quia quidquid scripti invenitur ab antiquis scientiis, non est dignum ut sit pars huius scientie vel principium. Nullus eorum qui secuti sunt hunc usque ad hoc tempus, quod est mille et quigentorum annorum, aliquid addidit vel invenit in dictis eius errorem

III (25–36)$^{6+6}$; IV (37–48)$^{6+6}$; V (49–60)$^{6+6}$; VI (61–72)$^{6+6}$; VII (73–84)$^{6+6}$; VIII (85–96)$^{6+6}$; IX (97–108)$^{6+6}$; X (109–120)$^{6+6}$; XI (121–132)$^{6+6}$; XII (133–144)$^{6+6}$; XIII (145–156)$^{6+6}$; XIV (157–169)$^{6+(6+1)}$; XV (170–181)$^{6+6}$; XVI (182–193)$^{6+6}$; XVII (194–205)$^{6+6}$; XVIII (206–217)$^{6+6}$; XIX (218–230)$^{(6+1)+6}$; XX (231–244)$^{7+7}$; XXI (245–256)$^{5+5}$, f. 255 is a piece of paper bound to the fascicule; XXII (256–267)$^{6+6}$; XXIII (268–279)$^{6+6}$; XXIV (280–292)$^{6+6}$, f. 292 is a piece of paper bound to the fascicule; XXV (293–297)$^{4+(4-3)}$; XXVI (298–308)$^{6+(6-1)}$.

Catchwords on ff. 156^{v} and 169^{v}. Signatures: f. 157^{r} (14); f. 170^{r} (15us); f. 182^{r} (16); f. 194^{r} (17us); f. 206^{r} (18); f. 231^{r} (20). A medieval foliation on the right side of the upper margin. The numbering skips one leaf between 107 and 108. This is why, from that point on, a modern foliation accompanies the medieval one: the two foliations are shifted by one unit up to f. 199 (200 according to the modern foliation). The leaf after 199 has been numbered 110 by the medieval hand: the numbering thus continues with 111, 112, 113, and so on. The modern hand numbers these leaves 209, 210, 211, and so on. The leaf that follows 120 has been numbered 200 by the medieval hand. The medieval foliation stops at f. 231 (225 according to the modern numbering).

Content. The manuscript contains Lawrence of Lindores' *Questions on Aristotle's Physics* (ff. 1ra–221rb); some anonymous questions on matter and form (ff. 221va–226vb);[34] a set of *problemata* collected from Aristotle's *Problemata* (ff. 227ra–244bis^{r}), and the second redaction of Oresme's *Questions on Meteorology* (ff. 245ra–307rb).

The *Questions on Meteorology*. The initium and the colophon ascribe the commentary to Nicole Oresme: "Incipiunt questiones *Metheororum* reverendi magistri Orem, Parisius date ac ab ipso collecte" (f. 245ra); "Hoc de questione et per consequens de omnibus questionibus librorum *Metheororum*, compilate per reverendum magistrum Orem Parisiensem et comparate per Johannem Stolle in Studio alme Universitatis Pragensis A.D.1406, pro quarum complectione Dominus noster Iesus Christus" (f. 307rb). The text is laid out in two columns of about fifty lines each. The handwriting is very readable while being rather rushed. In the margins we find some notes in cursive handwriting, probably by a different hand than that of the copyist (see f. 246^{v}, 247^{r}). A modern foliation in blue pencil in the upper outer margin. The initial words of the text

alicuius quantitatis, et talem esse virtutem in uno individuo miraculosum et extraneum existit. Et quia illa dispositio fuit in illo homine, dignus est potius dici divus quam homo". This is drawn from the prologue of Averroes's *Commentary on the Physics*, ed. Venetiis apud Iunctas, 1552–1564, 4: f. 4va-5ra.

34 "Prima questio: Dubitatur utrum appetitus materie sit ipsa materia naturale vel sit quasi distinctum ab ipsa materia. Et videtur primo quod sit res distincta ab ipsa materia. Secunda questio: Queritur hic utrum forme rerum naturalium sunt distincte a rebus naturalibus".

("Incipiunt questiones *Metheororum* reverendi magistri Orem"), in textualis, are placed so as to form a square with a height of about eighteen lines. Pointing hands (see f. 246rb), rubrics and pilcrows in red. The words that refer to the structure of the text are sometimes underlined in red (see f. 245^{v}). Questions I.10 and III.16 are copied on separate leaves of reduced size that were then bound to the manuscript and included in the foliation. On the verso of one of them, the copyist has drawn a polygon around whose perimeter he has written: "hic nichil desinit nisi aurum et argentum, oves et boves et alia multa quorum non est numerus" (f. 292^{v}). On f. 297^{r}, after question III.19, we find a table of contents corresponding to the questions on *Physics*, which continues until the following verso leaf. Many leaves have been torn out, and the remaining questions of the third book (III.20–35) are missing. The fourth book begins at f. 298^{r}.

Incipit and explicit of the books

I, ff. 245ra–265ra: Incipiunt questiones *Metheororum* reverendi magistri Orem, Parisius date ac ab ipso collecte, quarum prima est: utrum possibile sit de impressionibus metheorologicis habere scientiam et opinionem. Arguitur quod non, quia de eis non contingit habere scientiam nec opinionem, ergo questio falsa ...X... propter hoc quod plures antiquorum crediderunt galaxiam esse de natura elementari.

II, ff. 265ra–276vb: Circa secundum librum *Metheororum*: utrum ⟨locus⟩ generationis pluvie sit media regio aeris. Arguitur quod non, quia nullum elementum debet generari extra locum sibi naturalem ...X... Ad rationes in oppositum patet ex secundo articulo.

III, ff. 276vb–297ra: Circa tertium: utrum ventus sit exalatio calida et sicca. Et arguitur quod non sit exalatio calida, quia, si sic, sequeretur ...X... in hoc seculo nichil scimus. Hoc de questione.

IV, ff. 289ra–307rb: Queritur circa principium ipsius quarti primo utrum tantum sint quatuor qualitates prime, scilicet caliditas, humiditas, frigiditas et siccitas. Et arguitur primo quod non, quia ...X... sed tale animal non est ustibile ab igne, nec etiam nutritur ab igne. Ratio huius, quia non uritur ab igne, dicimus quod tale animal habitat in igne. Hoc de questione et per consequens de omnibus questionibus librorum *Metheororum*, compilate per reverendum magistrum Orem Parisiensem et comparate per Johannem Stolle in Studio alme Universitatis Pragensis a. D. 1406.

Bibliography

Catalogues: Wisłocki, *Catalogus*, 2: 507; Markowski and Włodek, *Repertorium*, 109; Lohr, *Latin Aristotle Commentaries, I.2. Medieval Authors M–Z*: 365–366, 2: 35.

Literature: Birkenmajer, *Études d'histoire des sciences en Pologne*, 183–186; Markowski, *Burydanizm w Polsce*, 404–405; McCluskey, *Nicole Oresme on Light,*

Color and the Rainbow, 86; L. Moonan, "The Scientific Writings of Lawrence of Lindores (d. 1437)", *Classica et Mediaevalia* 39 (1988), 273–317, at 278–283; Panzica, "Commenter les *Météorologiques* à l'Université de Cracovie", 83.

10 *Kraków, Biblioteka Jagiellońska, cod. 2117, ff. 195^{r}–322^{v} [pp. 389–643]* (Kr$_3$)

Paper; mm 215×155; I+326; 1444–1450

Date and origin. This manuscript consists of two parts. The first one (ff. 1–74) was copied in 1444 in Głogów (Lower Silesia), as emerges from the colophon on f. 74ra: "Finitus 6 feria sub sermone ante festum Michaelis per Vincentium Kenczel in Glogouia Maiori, sub anno Domini 1444". The second part (ff. 75–322) was copied in 1450 by Paulus de Zorawia, a student in Frankfurt (Oder), as proved by the colophon on f. 322^{v}: "Et finito libro post festum Pasce, quarta feria in die Viti anno Domini 1450 per Paulum de Zorawia imbuens scientys gymnasy Frankenfordensis".

Composition. A parchment guard-leaf, followed by 326 paper leaves grouped in senions: I (1–14)$^{6+6}$;[35] II (15–26)$^{6+6}$; III (27–38)$^{6+6}$; IV (39–50)$^{6+6}$; V (51–62)$^{6+6}$; VI (63–74)$^{6+6}$; VII (75–86)$^{6+6}$; VIII (87–98)$^{6+6}$; IX (99–110)$^{6+6}$; X (111–122)$^{6+6}$; XI (123–134)$^{6+6}$; XII (135–146)$^{6+6}$; XIII (147–158)$^{6+6}$; XIV (159–170)$^{6+6}$; XV (171–182)$^{6+6}$; XVI (183–194)$^{6+6}$; XVII (195–206)$^{6+6}$; XVIII (207–218)$^{6+6}$; XIX (219–230)$^{6+6}$; XX (231–242)$^{6+6}$; XXI (243–254)$^{6+6}$; XXII (255–266)$^{6+6}$; XXIII (267–278)$^{6+6}$; XXIV (279–290)$^{6+6}$; XXV (291–302)$^{6+6}$; XXVI (303–315)$^{6+6}$;[36] XXVII (316–326)$^{6+6}$, the last leaf has been torn out. In the outer margin, we find a pagination in black ink and a foliation in grey pencil, both by modern hands.

Content. This manuscript contains some texts on natural philosophy: a commentary on Peter of Dresden's *Compendium of natural philosophy* (ff. 1^{r}–74^{v});[37] Marsilius of Inghen's *Questions on the Parva naturalia* (ff. 75^{r}–194^{v}); Nicole Oresme's *Questions on Meteorology* (ff. 195^{r}–322^{v}). These texts are followed by the corresponding table of contents: that of Oresme's commentary on ff. 325^{r}–336^{v}, and that of Marsilius's *Questions* on ff. 323^{r}–324^{v}.

The *Questions on Meteorology*. The colophon ascribes the commentary to Guillaume Oresme: "Et sic est finis *Metheororum* reverendi magistri Wilhelmi de Orem", f. 322^{v}. The text has been assembled from two copies by different copyists. The first part (ff. 195–206) stops in the middle of question I.7. The page

35 Ff. 4–5 are pieces of paper bound to the manuscript.

36 Folio 304 is a piece of paper bound to the manuscript.

37 Inc.: "'omnis gloria eius filie regis ab intus in fimbriis aureis' [*Ps.* 44, 14]. Hec verba, quamvis prophetarum non minimus David in laudem et honorem"; f. 2vb: "⟨N⟩atura est principium. Iste libellus introductorius in philosophiam naturalem principali sui divi-

layout of the two parts is different. In the first part, the text is displayed in two columns of about fifty lines each. The words that refer to the internal structure of the questions are in textualis. The second part (ff. 207–322), copied by Paulus de Zorawia, is written *longis lineis*. The space intended for the internal divisions of the questions has remained blank. A modern hand has numbered the questions in grey pencil, restarting the numbering at the beginning of each book. Signatures in the upper margin, between the two columns (ai, aii, aiii …).

The last question of the third book ends on the recto of a piece of leaf that has been bound to the manuscript, as in the case of ms. 2095. The fourth book begins on the verso of the same leaf, which is numbered 304. The decoration has not been completed. On f. 195^{r}, the first words of the text ("⟨C⟩irca materiam Orem super *Metheorum* queritur primo utrum") are written in textualis and placed so as to form a square with a width of nineteen lines; the space intended for the initial "C" has also been left blank. The pilcrows start on f. 195va and stop shortly thereafter on f. 196rb. On ff. 195^{v}–196^{r}, the words that refer to the structure of the questions are written in large characters and surrounded by a red frame. We do not find any rubrics in the rest of the text.

The manuscript was misbound before it was foliated. On the bottom margin of f. 218^{v}, a medieval hand wrote "quere in quarto sexternio" and a modern hand added: "= 509". This anomaly is also indicated by a pointing hand on f. 218^{v}, next to which we find a small *a*. The pointing hand, just like the *a*, reappears on f. 255^{r}. The text stops again at f. 266^{v}: another pointing hand tied with a thin string to the letter *a* shows up, this time without any reference to a specific folio. In the upper margin of f. 243^{r}, we find another pointing hand next to the letter *b*. The text then proceeds until f. 254^{v}, where another interruption occurs, and we find another letter *b* in the upper margin. In the upper margin of f. 219^{r}, we find the letter *c*; in the upper margin of f. 242^{v}, the letter *d*, and in the upper margin of f. 267^{r} the letter *e*. The correct order of the folios would thus be: ff. 195–218; 255–266; 243–254; 219–242; 267–322.

Incipit and explicit of the books

1, f. 195ra–257^{v}: ⟨C⟩irca materiam Orem super *Metheorum* [*sic*] queritur primo utrum possibile sit de impressionibus metroloycis [*sic*] simul habere scientiam et opinionem. Et arguitur quod non, quia de impressionibus metroloycis [*sic*] non contingit habere scientiam nec opinionem; ergo etc. …X… Ad quartam: 'si galaxia esset de natura celi, tunc de ea non esset determinandum in isto libro', concedo. Et cum dicitur: 'tamen Aristoteles determinavit de ea hic',

---

sione dividitur in tres tractatus"; expl.: "in hac vita in suis effectibus, sed perfecte cognosci etiam in vita eterna, que donare dignetur Iesus Christus benedictus in secula seculorum. Amen".

respondetur quod hoc fecit propter hoc quod plures antiquorum crediderunt quod galaxia est de natura elementari, et propter eorum opinionem ipse talem materiam hic tetigit. Et sic est finis questionum primi libri *Metheororum*.

II, ff. 257v–250r: Circa secundum librum *Metheororum* queritur primo utrum locus generationis pluvie sit media regio aeris. Et arguitur quod non: nullum elementum debet generari extra locum sibi naturalem …X… Ad rationes in oppositum patet ex secundo articulo. Et sic est finis secundi libri *Me* …

III, ff. 250r–304r: Queritur primo utrum ventus sit exalatio calida et sicca. Et arguitur primo quod non sit exalatio calida, nam sic sequeretur quod in temporibus ventosis …X… Ad ultimam dico quod in vanum laborant [*sic*], ex quo aliam artem ignorant. Sequitur quartus.

IV, ff. 304v–322v: Circa initium quarti libri *Metheororum* queritur primo utrum tantum quatuor sunt qualitates prime, scilicet caliditas, humiditas, siccitas et frigiditas. Et arguitur primo quod non, quia quod per superhabundantiam dicitur, uni soli rei convenit …X… Ad sextam dico quod in salamandra non dominatur ignis in quantitate ⟨nec⟩ etiam nutritur ab igne; sed tale animal non est ustibile ab igne; et ideo, quia non uritur ab igne, dicimus quod ipsum habitat in igne. Et sic est finis *Metheororum* reverendi magistri Wilhelmi de Orem, pro quo sit benedictus Deus in secula seculorum. Et finito libro post festum Pasce, quarta feria in die Viti anno Domini 1450 per Paulum de Zorawia imbuens sciencys gymnasy Frankenfordensis.

Bibliography

Catalogues: Wisłocki, *Catalogus*, 2: 511; Markowski and Włodek, *Repertorium*, 115–116; Lohr, *Latin Aristotle Commentaries, I.2. Medieval Authors M–Z*: 35.

Literature: Birkenmajer, *Études d'histoire des sciences en Pologne*, 186–189; Markowski, *Burydanizm w Polsce*, 405; McCluskey, *Nicole Oresme on Light, Color and the Rainbow*, 86–87; Panzica, "Commenter les *Météorologiques* à l'Université de Cracovie", 83.

11 *Leipzig, Universitätsbibliothek, ms. 1387, ff. 181r–275r* (L)

Paper; mm 300×210; ff. I+276+I; 1429–1438

Date and origin. The first two texts contained in this manuscript, Marsilius of Inghen's *Questions on Metaphysics* and *Questions on De generatione et corruptione*, were copied in Frankfurt (Oder) by Iacobus Thysenaw of Prettin (see below, codicology).[38] The colophon of the second text informs us of the date

38 G. Erler, *Die Matrikel der Universität Leipzig 1409–1599*, 1 (Leipzig 1985): 116; 2 (Leipzig 1987): 120, 125.

when the copy was made: “Et sic est finis huius libri *De generatione et corruptione* venerandi magistri Marsilii anno Domini m° cccc xxxvii, quarta feria ante festum Viti, reportatum per quemdam Iacobum de Prettin et ab eodem pronuntiatum” (f. 176vb). The third text transmitted in this manuscript, Nicole Oresme’s *Questions on Meteorology*, was copied in Leipzig in 1429, as stated by the colophon on f. 274rb: “Expliciunt questiones magistri Orem super libros *Metheororum* scripte in studio Lipzii et finite anno domini millesimo quadragintesimo vicesimo nono”. Ff. 275^{r}–276^{v} contain some texts related to the Council of Basel of the years 1437–1438 and that were written in Frankfurt and Leipzig. The manuscript is mentioned in the catalogue of the Leipzig Faculty of Philosophy of 1480 under the signature “H inferior n. 150” and in a catalogue of the same library compiled around 1560 under the signature 209.

Codicology. Wooden binding. Traces of pegs and of the central boss. On the inside front cover, we find a paper guard-leaf with a medieval inscription: “Questiones Marsilii de Inghen super *Methaphysicam* et super *De generatione et corruptione*. Item questiones Wilhelmi Orem super *Metheororum*”. On a parchment guard-leaf we read: “Liber magistri Jacobi Thysenaw de Prettin, quem ipse propriis manibus conscripserat et ipsum estimat in valore vi f. renensium in auro. *Metaphisicam* ipse exemplar collegit Frankfenfordis ibidem pro tunc locatus existens. Marsilium Lipczk pronunciavit et pronunciando sic conscripsit Orem super *Metheororum*, intendita pro pecunia comparavit. [*alia manu*]: super *De generatione et corruptione*”. This guard-leaf is followed by 275 paper leaves: I (2–13)$^{6+6}$; II (14–25)$^{6+6}$; III (26–37)$^{6+6}$; IV (38–49)$^{6+6}$; V (50–61)$^{6+6}$; VI (62–73)$^{6+6}$; VII (74–85)$^{6+6}$; VI (86–98)$^{6+6}$; IX (99–100) two cut folios + 2; X (101–112)$^{6+6}$; XI (113–124)$^{6+6}$; XII (125–133)$^{5+5-1}$; XIII (134–144)$^{6+6}$; XIV (145–156)$^{6+6}$; XV (157–168)$^{6+6}$; XVI (169–180)$^{6+6}$; XVII (181–192)$^{6+6}$; XVIII (193–204)$^{6+6}$; XIX (205–216)$^{6+6}$; XX (217–228)$^{6+6}$; XXI (229–240)$^{6+6}$; XXII (241–252)$^{6+6}$; XXIII (253–264)$^{6+6}$; XXIV (265–276)$^{6+6}$. At the end of the volume, we find a parchment guard-leaf numbered 277 by a modern hand. Signatures on the recto leaves: f. 38 (traces of a crossed out signature); f. 74 (7); f. 86 (8? cut off); signature 9 on the remains of a cut off leaf after f. 98; f. 113 (10); f. 125 (11); f. 134 (12? cut); f. 145 (13); f. 157 (14); f. 169 (15).

The manuscript has two foliations: a medieval one in brown ink, in the upper margin between the two columns, and a modern one in pencil, in the outer margin. The medieval foliation begins with the second leaf of the first text, the *Questions on Metaphysics*, numbered 2, while the modern foliation assigns this number to the preceding leaf. This discrepancy of one unit between the two foliations continues up until f. 89 (88 according to the medieval foliation). Leaf 42 is numbered 43 by the medieval hand, but this mistake has no effect on the overall numbering, as the following leaf is also numbered 43. Starting with f. 89, the difference between the medieval and the modern numbering increases to two units due to the modern one jumping from 89 to 91. Another error occurs in the modern numbering, which numbers the leaf between 141 and 142 as 141a, thus reducing again the gap between the two numberings. Leaves 168–169 (169–170 according to the modern numbering) have not been numbered by the medieval hand, which however resumes the numbering correctly with f. 170 (171 according to the modern numbering). In the same way, leaves 178–179 (179–180 according to the modern numbering), which are blank, have not been numbered by the medieval hand, which restarts with f. 180 (181 according to the modern numbering).

Content. This manuscript contains Marsilius of Inghen's *Questions on the Metaphysics* (ff. 2ra–112rb: books I–II, IV–X, XII),[39] followed by a corresponding table of contents (ff. 112^{rb-vb}); Marsilius of Inghen's *Questions on De generatione et corruptione* (ff. 113ra–176vb);[40] an anonymous question about universals (ff. 177va–178ra);[41] the second redaction of Oresme's *Questions on Meteorology* (ff. 181ra–274rb), followed by the corresponding table of contents (ff. 274va–275ra), and some juridical texts concerning the University of Leipzig.[42]

The *Questions on Meteorology*. The colophon ascribes the commentary to Oresme (though without mentioning his first name), while the references on the front and on the back plates ascribe it to Guillaume Oresme. The text is copied in two columns of about forty lines each. The handwriting is widely spaced and easy to read; the ink is brown. We find two foliations: the first, a medieval one, in the central upper margin; the second, a modern one, in the

39 *Inc.*: "Circa initium *Methaphysice*. Utrum methaphysica sit sapientia. Et arguitur primo quod non, quia sapientia est scientia principalium; sed methaphysica non est huiusmodi"; *expl.*: "auctoritates post oppositum sunt pro dictis in tertio articulo. Hic est finis *Methaphysice*".

40 *Inc.*: "Circa librum *De generatione et corruptione queritur utrum ens mobile ad formam sit subiectum scientie libri De generatione et corruptione*"; *expl.*: "ratio post oppositum est pro ultima conclusione de questione ultima huius libri secundi, et consequenter de omnibus. Et sic est finis huius libri *De generatione et corruptione* venerandi magistri Marsilii. Anno domini m° cccc xxxvii, quarta feria ante festum Viti. Reportatum per quemdam Iacobum De Prettin et ab eodem pronuntiatum. Sit laus Deo". On f. 140rb, we find a table of contents for the first book; at the end of the text, on f. 177ra, there is a table of contents for the second book.

41 *Inc.*: "Utrum universale reale in essendo secundum viam Platonis et Socratis sit ponendum. Et probatur multipliciter sic. Primo: illa est concedenda: 'bonum est'"; *expl.*: "omnibus singularibus non distinctum ab eis".

42 "Copia littere dominorum electorum" (colophon: "Datus Frankfordie die xvii, mensis Marcy, Anno domini Mo cccc xxxviii", f. 275^{v}); "Copia bulle lecte in sessione generali" (f. 276^{r}); "Responsium domini Eugenii super scripta Universitatis studii Lipzii" (f. 276^{r}); "Missiva nuncii horum" (f. 276^{r}).

outer upper margin. Marginal notes by the copyist refer to the content of the text. At the beginning of the books, we find large initials. In the first and second books, the initials are bichrome, red and blue. The incipit of the fourth book is written in the shape of a square around a big initial C in red. Rubrics and pilcrows in red. Some structuring words are sometimes surrounded by a squared box in red (as for example on f. 181vb). The first words of the questions, in textualis, are also squared in red.

An error in the numbering of the questions occurs at the end of the third book. The question that in other manuscripts is numbered III.33 (*Utrum, visa yride et Sole vel astro elevato notabiliter super horizontem, apparet minor portio maioris circuli quam appareat Sole existente in fine horizontis*) is referenced as III.34, without III.33 appearing anywhere else. Consequently, the subsequent questions, which in the other manuscripts are numbered III.34 and III.35, are numbered III.35 and III.36. Another error occurs at the end of the fourth book. The question that is numbered IV.7 in the other manuscripts (*Utrum balnea conferant ad digestionem*), is numbered IV.6 here, just like the previous one. As a result, question IV.8 (*Utrum sint tantum tres digestiones*) becomes question IV.7. This discrepancy continues until the end of the fourth book. It is important to notice that this difference in the Leipzig manuscript does not affect the questions themselves, but only their numbering, as no question has been added or removed.

Incipit and explicit of the books

I, ff. 181ra–203va: Circa initium *Metheororum*, utrum possibile sit de impressionibus meteorologicis habere scientiam simul et opinionem. Et arguitur primo quod non, quia de impressionibus metroloycis non contingit habere scientiam nec opinionem, ergo questio falsa ...X... Ad quartam: 'si galaxia esset de natura celi, tunc de ea non esset determinandum in isto libro', concedo. Et cum dicitur: 'tamen Aristoteles determinavit de ea hic', respondetur quod fecit propter hoc quia plures antiquorum crediderunt galaxiam esse de natura elementari, et propter eorum opinionem ipse talem materiam hic tetigit. Et sic est finis libri primi. Explicit liber primus *Metheororum*.

II, ff. 203vb–216vb: Circa initium libri secundi *Metheororum* primo queritur utrum locus generationis pluvie sit media regio aeris. Et arguitur primo quod non: nullum elementum debet generari extra locum sibi naturalem ...X... Rationes patent in secundo articulo huius questionis. Et tantum de ista questione, et per consequens de toto libro secundo huius *Metheororum*.

III, ff. 216vb–261rb: Liber tertius *Metheororum*. Circa initium tertii libri *Metheororum* primo queritur utrum ventus sit exalatio calida et sicca. Et arguitur quod non sit exalatio calida ...X... Ad ultimum: quod non in vanum laborant, ex quo aliam artem ignorant, etc. [*rubr.*: Sequitur liber quartus, etc. Ave Gertrudis virgo grata ex regali stirpe nata leprosos mundare non sprevisti ob amorem Iesu Christi, cecis viam demonstrasti, etc. Sponsa mea].

IV, ff. 261va–274rb: Circa primum quarti *Metheororum* primo movetur ista questio, utrum tantum sint quatuor qualitates prime, scilicet caliditas, humiditas, frigiditas et siccitas. Et arguitur quod non, quia quod per superhabundantiam dicitur, uni soli convenit ...X... Ad ultimam dico quod in salamandra non dominatur ignis in quantitate, sed tale animal non est ustibile ab igne nec uritur ab igne; et ideo, quia non uritur ab igne, dicimus quod tale animal habitat in igne, etc. Expliciunt questiones magistri Orem super libros *Metheororum*, scripte in Studio Lipczensi et finite anno domini millesimo quadragintesimo vicesimo nono.

Bibliography

Catalogues: L.J. Feller, *Catalogus codicum manuscriptorum Bibliothecae Paulinae in Academia Lipsiensi*, Lipsiae 1686, 362, nr. 71 [in this catalogue the *Questions on Meteorology* are attributed to Guillaume Oresme, as in the colophon of the manuscript]; Lohr, *Latin Aristotle Commentaries, 1.2. Medieval Authors M–Z*: 35; M. Markowski, *Repertorium commentariorum medii aevi in Aristotelem Latinorum, qui in Bibliotheca Universitatis Lipsiensis asservantur*, Krakow 2012, 76 and 234.

Literature: Birkenmajer, *Études d'histoire des sciences en Pologne*, 191–192; D. Döring, *Die Bestandsentwicklung der Bibliothek der Philosophischen Fakultät der Universität zu Leipzig von ihren Anfängen bis zur Mitte des 16. Jahrhunderts: ein Beitrag zur Wissenschaftsgeschichte der Leipziger Universität in ihrer vorreformatorischen Zeit*, Leipzig 1990 (*Beiheft 99 zum Zentralblatt für Bibliothekswesen*), 60, 70; H.-U. Wöhler, "Die erste philosophische Fakultät in Sachsen bis zum Beginn der Reformation in localen, regionalen und überregionalen Kontext", *Bochumer Philosophisches Jahrbuch für Antike und Mittelalter* 13 (2008), 217–240.

12 *München, Bayerische Staatsbibliothek, Clm 4376, ff. 1^{r}–64^{r}* (M)

Paper; mm 285×215; ff. II + 152; 1365–1386

Date and origin. The texts contained in this manuscript were copied at Prague University between 1365 and 1386, as emerges from the initia and the colophons: "Expliciunt questiones primi [*corr. sup. lin. ex* secundi] libri *Metheororum*, finite die beati Erhardi martiris atque pontificis, anno Domini 1366" (f. 15vb); "expliciunt questiones quatuor librorum *Metheororum* Byridani, finite Prage anno Domini 13°66to, in vigilia beate Dorothee virginis, per pedes Johannis Krichpaumi de Ingol⟨stat⟩" (f. 64rb); "incipit declaratio magistrorum sancte theologie sancte Universitatis studii Pragensis [...], anno Domini 1386" (f. 66va);

"expliciunt questiones super totalem librum *De senso et sensato* collecte Parysius per reverendum magistrum Albertum de Rychmersdorf, pronunciate Prage in quadam bursa tunc temporis, anno MCCCLXV, feria quarta proxima post assumptionem virginis gloriose per Johannem Krichpaumum de Ingolstat, finite in die Sancti Bernhardi" (f. 86rb); "expliciunt questiones *De sompno et vigilia*, anno Domini 1367mo" (f. 100vb); "hic sit finis questionum *De longitudine et vite brevitate* in Praga anno Domini millesimo CCC°LXVII, in vigilia gloriosissime annunciationis virginis Marie per Johannem Krichpaumum de Ingol⟨stat⟩" (f. 104va); "et sic patent questiones primi *De anima* finite sabato ante festum palmarum in Praga, anno Domini 1365to" (f. 107va).

Most of the texts have been copied by Iohannes Krichpaum from Ingolstadt, a student at the Prague Arts Faculty who identifies himself in the colophons (see f. 64rb, f. 86rb and f. 107va). Similarly to the ms. Clm 4375 of the Bavarian State Library, which transmits a copy of the first redaction of Oresme's *Questions on Meteorology*, this manuscript came into the possession of the abbey of SS. Ulrich and Afra in Augsburg. In the upper margin of f. 1^{r}, we read: "Monasterii S. Udalrici Aug".

Composition. Wooden binding; traces of clasps on the front cover. On the lower part of the front cover, we find a votive image of the abbey of SS. Ulrich and Afra in Augsburg, which is accompanied by the words: "P.P. Benedictinorum, Lib. et Imp. Monasterii S.S. Udalrici et Affrae Augustae". Two parchment guard-leaves. On the first one, a medieval hand wrote: "Buridani". These leaves are followed by 152 paper leaves, grouped mainly in quinions: I $(1–10)^{(6-1)+5}$; II $(11–20)^{5+5}$; III $(21–30)^{5+5}$; IV $(31–41)^{6+(6-1)}$; V $(42–51)^{5+5}$; VI $(52–61)^{5+5}$; VII $(62–67d)^{5+5}$; VIII $(68–77)^{5+5}$; IX $(78–87)^{5+5}$; X $(88–97)^{5+5}$; XI $(98–104c)^{5+5}$; XII $(105–114)^{5+5}$; XIII $(115–124)^{5+5}$; XIV $(125–134)^{5+5}$; XV $(134–144)^{5+5}$; XVI $(145–152)^{4+4}$.

Signatures in the first leaf of each fascicule (from *a* to *q*). The first fascicule begins with the signature "a ii", as the first leaf has been ripped.

Two modern foliations in the upper margin: the first one in black ink; the second one in pencil. Since the leaves that should have been numbered 68, 69, 70 and 71 are blank, they have been omitted from the numbering by the black-ink hand. The pencil hand has numbered them 67a, 67b, 67c, and 67d. The same circumstances apply to ff. 105, 106 and 107, which are numbered in pencil 104a, 104b, and 104c. Ff. 17–19, 26–29, 31, 33–34, 36–39, 42, 47, 48–53, 57–59, 61–63, 68–69, 71–74, 76, 77, 78, 79, 81, 82, 83, 84, 127, 128, 129, 131, 132, 133, 134, 136, 137, 138, 139, 143, 144, 146, 147, 148, 149, 152 have not been numbered.

Content. The academic origin of this manuscript is reflected in its content, which consists mostly of commentaries on Aristotle's natural philosophy by Parisian masters of the second half of the fourteenth century: the second redaction of Oresme's *Questions on Meteorology* (ff. 1ra–64rb); an anonymous question about the plague (ff. 64va–66rb);[43] a declaration of the masters of Prague University (ff. 66va–67vb);[44] the *Questions on De sensu et sensato* ascribed to Albert to Saxony or Nicole Oresme (ff. 68ra–86rb);[45] some questions on *De memoria et reminiscentia* (ff. 86va–91ra) and on *De somno et vigilia* (91ra–100vb), which should probably be ascribed to Albert of Saxony; two anonymous questions on *De somno et vigilia* (100vb–101vb);[46] a set of questions on *De longitudine et brevitate vite* (ff. 102va–104va), which should also probably be ascribed to Albert of Saxony; the second redaction of Buridan's *Questions on De anima* (ff. 105ra–124vb; ff. 150va–151va);[47] the *redactio B* of Buridan's *Questions on De generatione et corruptione* (ff. 125ra–150rb).

The *Questions on Meteorology*. The colophon ascribes the commentary to Buridan: "Expliciunt questiones quatuor librorum *Metheororum* Byridani" (f. 64rb). The text is copied in two columns of about fifty lines each. The handwriting is regular; the ink is brown. Annotations in the margin by the copyist. The incipits of the books, just like the first words of each question, are copied in textualis and have a rubricated initial. The numbering of the questions is inconsistent. A first mistake occurs after question III.11 (*Utrum typho et ecnephya et incensio sint exalationes calide et sicce*), correctly introduced by the words "undecimo queritur". The subsequent question (*Utrum visus refrangatur*

---

43 *Inc.*: "Quia secundum Aristotelem in prohemio *Metaphysice*: 'propter admirationem antiqui inceperunt phylosophari', et quia omnes miramur de pestilentia [...], ergo circa libellum *Secretorum* Aristotelis queritur ista questio: utrum morbus pestilentie effective proveniat ab influentia corporum supercelestium"; *expl.*: "ad ultimam oportet si Saturnus et Mars etc., sed parvam et transitoriam; ergo non est efficax ad ista inferiora".

44 *Inc.*: "Incipit declaratio magistrorum facultatis theologie sancte Universitatis studii Pragensis".

45 The text is attributed to Albert in the colophon of this manuscript ("Expliciunt questiones super totalem librum *De sensu et sensato* collecte Parisius per reverendum magistrum Albertum de Richmersdorf, pronunciate Prage in quadam bursa tunc temporis anno MCCCLXV per Johannem Krichpaumum de Ingolstat finite in die sancti Bernardi") and to Oresme in manuscript Erfurt, Universitäts- und Forschungsbibliothek, Dep. Erf. 4° 299 (f. 128^{r}: "Questiones *Parvorum naturalium* Orem"). This commentary was edited by J. Agrimi, *Le Questiones de sensu attribuite a Oresme e Alberto di Sassonia*, Florence 1983 (Pubblicazioni della Facoltà di lettere e filosofia dell'Università di Pavia, 29), who did not manage to solve the problem of its attribution (see particularly 9–33 of the introduction).

46 *Inc.*: "Queritur utrum sompnus sit causa sanitatis. Arguitur quod non, quia si sic, sequitur quod vigilia est"; *expl*: "nec etiam generatur disproportio humorum nec qualitatum. Et sic patet questio". II. *Inc.*: "Queritur utrum oculus mulieris menstruose inficiat speculum. Et videtur quod non: omne agens est simile cum passo"; *expl.*: "quod eodem modo mulier menstruosa ..." [incomplete].

47 In his inventory of medieval commentaries on Aristotle in the Bavarian State Library, Markowski fails to notice that this text does not end at f. 124vb, but continues on f. 150va.

in occursu medii densioris et rarioris), which is the twelfth of the third book, is, however, introduced by the words "tredecimo queritur". This mistake persists until the question *Utrum yris sit forma dyaphanalis vel specularis, vel utrum fiat per reflexionem vel per refractionem*, which in this manuscript is numbered III.23 and in the others III.22. The following question (*Utrum, supposito quod yris fiat per reflexionem radiorum, utrum tunc fiat in nube vel super guttulas roridas vel super stillicidia*), at f. 43vb, is preceded by the words "queritur vicesimo secundo", even if it is numbered III.24 (III.23 in the other manuscripts). As a result, the following two questions (*Utrum omnes yris debeat esse tricolor* and *Utrum yris solum dupliciter et non multipliciter potest apparere*) are numbered III.23 and III.24. Surprisingly, the subsequent question (*Utrum semper, apparentibus duabus yridibus, superior yris debet habere colores conversim positos*) is introduced by the words "queritur vicesimo octavo". This new numbering continues until the question *Utrum, visa yride et Sole vel astro elevato notabiliter super horizontem, apparet minor portio maioris circuli quam appareat Sole existente in fine horizontis* (f. 52va), which is numbered III.35 in our codex and III.33 in the other manuscripts. At that point, the copyist, who probably became aware of the mistake, numbered the following questions III.34 and III.35 (and not III.36 and III.37, as he should have done on the basis of his erroneous numbering) in agreement with the numbering of the other manuscripts.

Incipit and explicit of the books

I, ff. 1ra–15vb: Primo queritur circa initium primi libri *Metheororum* utrum possibile sit de impressionibus metroloycis [*sic*] habere simul scientiam et opinionem. Et arguitur quod non, quia de impressionibus metheorologicis non contingit habere scientiam nec opinionem; ergo questio falsa ...X... Ad quartam: 'si galaxia non esset de natura celi, tunc de ea non esset determinandum in isto libro', condedo. Tamen Aristoteles determinavit de ea hic propter hoc quia plures antiquorum crediderunt galaxiam esse de natura elementari. Expliciunt questiones primi libri *Metheororum*, finite die beati Erhardi martiris atque pontificis, anno Domini 1366.

II, ff. 15vb–24va: Circa secundum *Metheororum* queritur secundum distinctionem in principio huius libri positam ab Alberto, et est questio ista: utrum locus generationis pluvie sit media regio aeris. Et arguitur quod nullum elementum debet primo generari extra locum non naturalem ...X... Ad rationes in oppositum patet ex secundo articulo. Et sic est finis questionum secundi libri *Metheororum*, finite in die sancti Vincentii in civitate Pragensi tunc temporis anno Domini millesimo 1366-to hora quasi crepusculi.

III, ff. 24vb–56rb: Circa tertium *Metheororum* queritur primo secundum distinctionem totius libri istius in principio positam ab Alberto. Ista sit questio prima, utrum ventus sit exalatio calida et sicca. Et arguitur primo quod non sit exalatio calida, quia si sic, sequeretur quod in temporibus ventosis aer deberet esse calidus ...X... Ad ultimam dico quod non in vanum laborant, ex quo aliam artem ignorant. Et sic est finis questionum tertii libri *Metheororum*. Finitus Prage anno Domini 1366 sabbato, die proximo ante festum purificationis Virginis gloriose.

IV, ff. 56rb–64rb: Circa quartum *Metheororum* queritur utrum tantum sunt quatuor qualitates prime, scilicet caliditas, humiditas, frigiditas et siccitas. Et arguitur quod non, quia quod per superhabundantiam dicitur, uni soli convenit ...X... Ad sextam dico quod in salamandra non dominatur ignis in quantitate; sed tale animal non est ustibile ab igne; et ideo, quia non uritur ab igne, dicimus quod tale animal habitet in igne, etc. Deo gratias. Expliciunt questiones quatuor librorum *Metheororum* Byridani, finite Prage anno Domini 13°66-to in vigilia beate Dorothee virginis per pedes Johannis Krichpaumi de Ingol⟨stat⟩.

Bibliography

Catalogues: K. Halm, G. v. Laubmann, and W. Meyer, *Catalogus codicum latinorum Bibliothecae Regiae Monacensis. I.2, Codices num. 2501–5250 complectens*, Munich 1894 (Catalogus codicum manu scriptorum Bibliothecae Regiae Monacensis III.2), 181–182 [here Oresme's *Questions on Meteorology* are ascribed to Buridan, as in the colophon of the manuscript]; M. Markowski, *Buridanica quae in codicibus manu scriptis bibliothecarum Monacensium asservantur*, Wrocław/Warsaw 1981, 70, 73–74, 94, 133; Lohr, *Latin Aristotle Commentaries, I.2. Medieval Authors M–Z*: 43, 258; 2: 35.

Literature: P. Duhem, *Le système du monde: histoire des doctrines cosmologiques de Platon à Copernic*, Paris 1913–1959 (10 vols.), 4: 129ss.; Agrimi, *Le "Quaestiones de sensu"*; Michael, *Johannes Buridan*, 2: 639–640, 661, fn. 10, 736; J. Agrimi, "Les *Questions de sensu* attribuées à Albert de Saxe. Quelques remarques sur les rapports entre philosophie naturelle et médecine chez Buridan, Oresme et Albert", in J. Biard (ed.), *Itinéraires d'Albert de Saxe. Paris-Vienne au XIVe siècle. Actes du colloque organisé le 19–22 juin 1990 dans le cadre des activités de l'URA 1085 du CNRS à l'occasion du 600e anniversaire de la mort d'Albert de Saxe*, Paris 1991 (Études de philosophie médiévale, 69), 191–204, esp. 193; Streijger, Bakker, and Thijssen (eds.), *John Buridan Quaestiones super libros De generatione et corruptione Aristotelis*, 17.

13 *München, Bayerische Staatsbibliothek, Clm 17226, ff. 1ra–140rb* (M¹)

Paper; mm 310 × 215; ff. 140 + III; 1413

Date and origin. The colophon informs us that the text was copied in 1413 by a certain Lanzman: "Et sic est finis huius operis. Deo gratias, etc. Et questiones ille scripte sunt per Lanzman ... suo domino sub anno incarnationis domini Iesu Christi M CCCC XIII° etc." (f. 140rb).

Composition. Pasteboard binding covered with leather. On the front cover a medieval hand has written: "Questiones *Metheororum* Burydani in primo, in secundo et in quarto libro. Questiones *Metheororum* Orem(?) in primo, in tertio et in quarto libro". The volume consists of 140 paper leaves grouped in senions according to the following formula: I (1–12)$^{6+6}$; II (13–24)$^{6+6}$; III (25–36)$^{6+6}$; IV (37–48)$^{6+6}$; V (49–60)$^{6+6}$; VI (61–72)$^{6+6}$; VII (73–84)$^{6+6}$; VIII (85–96)$^{6+6}$; IX (97–108)$^{6+6}$; X (109–120)$^{6+6}$; XI (121–132)$^{6+6}$; XII (133–140c)$^{6+(6-1)}$. Catchwords on the verso leaves at the end of the fascicules: f. 12, f. 24, f. 36, f. 48, f. 60, f. 72, f. 96, f. 108, f. 120. Signatures on the recto leaves at the beginning of the fascicules: f. 13 (secundus); f. 25 (tertius); f. 37 (quartus). Modern foliation in black pen and pencil. The last three leaves are blank and have not been numbered.

Content. This manuscript contains a compilation of questions on Aristotle's *Meteorology* based on the commentaries of John Buridan, Nicole Oresme and Themo Iudeus. In the first two books, we find an alternation of Oresme's and Buridan's questions; starting from the third book, the compilator also introduces some excerpts from Themo. In his catalogue of commentaries on Aristotle at the Bavarian State Library, Markowski attributes the whole text to John Buridan, probably having been misled by the incipit, which does indeed correspond to that of Buridan.[48] (f. 1ra: "⟨S⟩i scientiam oppinamur honorabiliorem ex eo ceteris paribus quod est de magis mirabilibus ...").[49] However, as has been pointed out by Birkenmajer, the explicit does not correspond to the other manuscripts of Buridan's text,[50] but rather to the second redaction of Oresme's *Questions* (f. 140ra: "Ad sextam dico quod in salamandra non dominatur ignis in quantitate nec etiam nutritur igne, sed tale animal non est ustibile ab igne; et ideo, quia non uritur ab igne, dicitur quod tale animal habitat in igne").[51] The compilator does not make an effort for consistency, as he does not omit the frequent references in Buridan's questions that point to passages in the

48 Markowski, *Buridanica*, 72–73, 140.

49 The prologue and the first book of Buridan's *Questions on Meteorology* have been edited by S. Bages in her doctoral dissertation: S. Bages, *Les Questiones super tres libros Metheororum Aristotelis de Jean Buridan*. The prologue can be consulted in vol. 2: 1–2.

50 The list of these manuscripts can be consulted in E. Faral, "Jean Buridan: notes sur les manuscrits, les éditions et le contenu de ses ouvrages", *Archives d'histoire doctrinale et littéraire du Moyen Âge* 21 (1946), 1–53, esp. 22–24, and in Bages, *Les Questiones super tres libros Metheororum Aristotelis de Jean Buridan*, 1: 185–188.

51 The explicit of Buridan's *Questions* is as follows: "sive lapidibus sive plantis sive animalibus sive earum principiis et metallis et aliis". For the different forms in which this explicit is

original text, rendering them completely meaningless after the process of interpolation. For example, in question I.11 we read: "Sicut de motu quesitum est, ita quinto querimus de lumine, videlicet utrum omne lumen sit calefactivum", f. 18rb.

The text is copied in a single hand in two columns of about forty lines per page. The handwriting is regular and easily readable. The decoration has not been completed: the space intended for the initials of the first (f. 1ra) and of the third (f. 91vb) book has been left blank. The incipit of the second (f. 63ra) and of the fourth (f. 130ra) book have not been copied at all. The incipits of the questions in textualis stop at f. 9ra. From then on, a plain blank space separate the questions. Only ff. 83va–84vb and ff. 87ra–91va contain rubrics. Corrections by the copyist himself.

Incipit and explicit of the books

Prohemium, ff. 1^{ra-rb}: ⟨S⟩i scientiam oppinamur honorabiliorem ex eo ceteris paribus quod est de magis mirabilibus et appetibiliorem, prout habetur in prohemio *De anima*, constat quod scientiam libri *Metheororum* valde reputare debemus ...X... Ex omnibus autem coarticulatis fieri magnitudine aliqua. Sequitur nunc questio prima.

I, ff. 1rb–63ra: ⟨C⟩irca materiam, utrum de impressionibus metheorologicis sit tradenda scientia naturalis distincta a scientia *De celo et mundo, De generatione et corruptione, De mineralibus* et *De animalibus et animatis*. Arguitur primo quod non sit scientia de eis ...X... De alio loquendo ad illum. Et hec sufficiant de ista materia. Et sic est finis primi libri *Methaurorum*. Sequitur.

II, ff. 63ra–91va: Iam de mari quesita fuerunt multa super primum librum, sed adhuc super secundum queritur de perpetuitate vel generatione maris, scilicet utrum fuit perpetuum vel aliquando factum. Et arguitur quod non ...X... Inflamabantur et ardebant consequenter grossiora. Hec ergo sint dicta de hac materia. Sequitur questio. Et sic est finis illarum questionum predictarum, etc.

III, ff. 91vb–130ra: ⟨C⟩irca tertium librum *Metheororum* queritur utrum omnis radius visualis reffrangatur in occursu medii densioris vel radioris [*sic*]. Arguitur primo quod non ...X... Ad ultimam dicendum quod omnino in vanum laborant, ex quo aliam artem ignorant. Expliciunt questiones libri *Methaurorum*, primi, secundi et tertii; sequuntur quarti.

IV, ff. 130ra–140rb: Utrum tantum quatuor sint qualitates prime, scilicet caliditas, humiditas, frigiditas et siccitas. Et arguitur primo quod non: quod per

transmitted in the manuscripts, see Bages, *Les Questiones super tres libros Metheororum Aristotelis de Jean Buridan*, 1: 148–177.

superhabundantiam dicitur, uni soli convenit ...X... Ad sextam dico quod in salamandra non dominatur ignis in quantitate nec etiam nutritur igne; sed tale animal non est ustibile ab igne; et ideo, quia non uritur ab igne, dicimus quod tale animal habitat in igne, etc. Et sic est finis huius operis. Deo gratias, etc. Et questiones ille scripte sunt per Lanzman ... suo domino sub anno incarnationis Domini nostri Jesu Christi M°CCCC°XIII, etc.

Bibliography

Catalogues: K. Halm, G. von Laubmann, and W. Meyer, *Catalogus codicum latinorum Bibliothecae Regiae Monacensis. IV.3, Codices num. 15121–21313 complectens*, Munich 1878 (Catalogus codicum manu scriptorum Bibliothecae Regiae Monacensis IV.3), 89; Markowski, *Buridanica*, 72–73, 140; Lohr, *Latin Aristotle Commentaries, I.2. Medieval Authors M–Z*: 35.

Literature: Birkenmajer, *Études d'histoire des sciences en Pologne*, 193; Bages, *Les Questiones super tres libros Metheororum Aristotelis de Jean Buridan*, 1: 176–177; Michael, *Johannes Buridan*, 2: 656–657; Panzica, "Nicole Oresme à la Faculté des Arts de Paris", 35 and 73–78 [list of questions with indication of their sources].

14 ***Paris, Bibliothèque Nationale de France, lat. 15156, ff. 226^{r}–265^{v}* (P)**

Paper; mm 225×150; ff. II+227+II; ca. 1369

Date and origin. The manuscript belonged to the Library of Saint Victor at the University of Paris (f. 34^{r}, *mg. inf.*: "Iste liber est Sancti Victoris parisiensis. Cuicumque eum furatus fuerit vel tolaverit vel titulum istum deleverit anathema sit. Amen"). On the verso of the second parchment guard leaf, we find an old signature: "n.n.20". In the upper margin of the first paper leaf (numbered 26), we find three numbers (probably old signatures) written by modern hands: "288" (crossed out), "729" and "c.2.20". On the same line, a modern hand has written: "S. Victor 514". One of the texts contained in this manuscript is dated 1369 (f. 34^{v}: "anno Domini 1369").

Composition. Wooden binding covered with leather. Traces of two pegs on the front cover. On the inside cover, we find a paper rectangle with the following modern inscription: "Volume de 288 feuillets. Moins les feuillets 3–26, 28–31, 194–226. Les feuillets 32, 33, 56, 59, 181–266 sont blancs. Le feuillet 1 est mutilé. 10 Mai 1869". Two parchment guard-leaves. On the verso of the second leaf, a medieval hand has written: "Tabulam hic contentorum reperies folio 297". The folio number was thereafter erased by a modern hand, which wrote on the same line "31". On the same parchment leaf, we find an old signature: "n.n.20". The guard-leaves are followed by 227 paper leaves grouped mainly in senions: I (3–33)$^{(3-1)+3}$; II (34–45)$^{6+6}$; III (46–57)$^{6+6}$; IV (58–69)$^{6+6}$; V (70–81)$^{6+6}$; VI

(82–93)$^{6+6}$; VII (94–105)$^{6+6}$; VIII (106–117)$^{6+6}$; IX (118–129)$^{6+6}$; X (130–141)$^{6+6}$; XI (142–153)$^{6+6}$; XII (154–165)$^{6+6}$; XIII (166–177)$^{6+6}$; XIV (178–181)$^{2+2}$; XX (182–193)$^{6+6}$; XXI (226–237)$^{6+6}$; XXII (238–247)$^{6+6}$; XXIII (248–259)$^{6+6}$; XXIV (260–271)$^{6+6}$; XXV (272–285)$^{7+7}$; XXVI three folios. At the end of the volume some leaves have been cut off (nine, as we can infer from the numbering of the following leaf). Two parchment guard-leaves, numbered by a medieval hand 298 and 299. Signatures in the upper margin of each recto leaf: f. 34 (primus 6); f. 46 (2us); f. 58 (3us); f. 70 (4us); f. 82 (5us); f. 94 (6us); f. 106 (7us); f. 118 (8us); f. 130 (9us); f. 142 (10us); f. 154 (11us); f. 166^{r} (12us); f. 178 (13us). Catchwords in the inner lower margin of the verso leaves: f. 45; f. 69; f. 81; f. 93; f. 105; f. 117; f. 129; f. 141; f. 153; f. 165; f. 177; f. 259; f. 237; f. 247; f. 271; f. 285.

Blank leaves: ff. 31^{v}–33^{v}; ff. 46^{r}–47^{v}; f. 146^{v}; f. 149^{r}; f. 170^{v}; ff. 176^{v}–177^{r}; ff. 180^{v}–181^{v}; f. 193^{r}. Medieval foliation in the upper outer margin. The first paper leaf is numbered 26, as leaves 1–25 have been torn out. A modern hand added "3–" before "26", in order to make clear that the preceding leaves were missing. The same hand numbered the two parchment guard-leaves 1 and 2. Folio 27 is followed by f. 31. The modern hand wrote "28–" before 31, to make clear that folios 28–30 were missing. Ff. 34–181 must have circulated independently in the Middle Ages, as they are consecutively numbered with another foliation. This foliation has been crossed out. F. 193 is followed by f. 226. A modern hand has written "194–" before 226, to make clear that ff. 194–225 were missing.

Content. This manuscript gathers various heterogeneous texts. On f. 31^{r}, a modern hand has compiled a table of contents: "Tabula contentorum / Metra quedam fol. 26 / Plura theologicalia a fol. 44 usque 180 / tractatus quidam de consequentiis 183 / Questiones super librum *Metheororum* 226". The table is followed by a list of missing texts: "Desiderata in hoc codice / Tractatus quidam de Rhetorica fol. 2 / Epistola Luciferi ad prelatos Ecclesie 28 / Aristotelis *Secreta secretorum* sive *De regimine principum* 194". I have identified the verses on ff. 26^{r}–27^{v} as Geoffroi of Vinesauf's *Poetria nova*.[52] The second text contained in this codex (the *Plura theologicalia*) is a commentary on the *Sentences*. A fifteenth-century hand, different from that of the copyist, ascribes it to John Gerson (f. 178^{v}: "Sententie magistri Joannis Gerson"). This ascription is accepted by A. Combes[53] and contested by Glorieux, who ascribes this commentary

52 J.V. Le Clerc, B. Hauréau, and P. Meyer (eds.), *Histoire littéraire de France, XIIIe siècle*, 18, Paris 1835, 305–311.

53 A. Combes, "Note sur les *Sententiae Magistri Joannis Gerson* du manuscrit B.N.LAT. 15156", *Archives d'histoire doctrinale et littéraire du Moyen Âge* 14 (1939), 365–385.

to Jean Régis, a Franciscan who lectured on the *Sentences* in the academic year 1369–1370.[54] The third text, on ff. 182r–192v, is a treatise on logic in ten chapters.[55]

The *Questions on Meteorology*. The text, which is transmitted anonymously, is copied in a single column of about forty lines per page. The handwriting, typical of mid-fourteenth-century Paris, is regular and easy to read. The large initials that should appear at the beginning of each question are missing. The first words of the books and of the questions are copied in textualis, as are the words that introduce a new part of the question ("in oppositum, quantum ad primum, ad rationes"). The text stops, unfinished, on f. 288v, at question III.20 (*Utrum omni hora diei yris posset apparere*), with the words: "quod semper semidyameter iridis est equalis, ideo dicendum est quod hoc est". A first interruption occurs on f. 265v, at question II.10 (*Utrum mare debeat fluere et refluere*), with the words: "eius a Sole augetur eius virtus a Sole"; the commentary resumes on f. 267r, with the question *Utrum aqua maris debeat esse salsa*, f. 266 having been left blank. McCluskey has noticed that, starting from this question, the text no longer follows Oresme's *Questions*, but rather Albert of Saxony's. The collation of the Parisian commentary with the other witnesses of Albert's commentary enabled me to confirm this statement,[56] which nonetheless requires some clarification. While the third book of Albert's *Questions* only contains nine questions, the Parisian manuscript consists of twenty. This can easily be explained by the different division of the books adopted in the Parisian manuscript: while in Albert's manuscripts the question *Utrum ventus sit exalatio calida et sicca* is the seventh of the second book, in the Parisian manuscript it is placed at the beginning of the third book. The *Questions on Meteorology* contained in this manuscript is therefore a compilation consisting of questions I.1–II.10 by Nicole Oresme and questions II.11–III.9 by Albert of Saxony.

Incipit and explicit of the books

I, ff. 226r–253r: ⟨C⟩irca librum *Metheororum*, qui est quartus naturalium, queritur primo utrum possibile sit de impressionibus metheorologicis habere simul scientiam et opinionem. Et arguitur primo quod non: de impressionibus metheorologicis non contingit habere scientiam nec etiam opinionem; ergo

54 P. Glorieux, "Le *Commentaire sur les Sentences* attribué à Jean Gerson", *Recherches de théologie ancienne et médiévale* 18 (1951), 128–139. Glorieux has shown that the *principia*, the commentary on the *Sentences*, and the table of contents, which are written in the same hand, form a textual unity (134–135). He also identified the masters referred to in the *principia* of the commentary (133).

55 *Inc.*: "Benivolentia theologorum studentium me compulit".

56 For the other manuscripts of Albert's *Questions on Meteorology* see Panzica, "Albert of Saxony's Questions on Meteorology", 248–256.

questio falsa ...X... Ad quartam: 'si esset de natura celi, de ea non esset determinandum', concedo. Sed tamen Aristoteles determinavit de ea hic propter hoc quod plures antiquorum crediderunt galachiam esse de natura elementari. Et sic patet ad questionem.

II, ff. 253ᵛ–269ʳ: ⟨C⟩irca secundum librum *Metheororum* queritur utrum locus generationis pluvie sit media regio aeris. Et arguitur primo quod non: nullum elementum debet generari extra locum sibi naturalem ...X... Ratio in oppositum soluta est, quia considerabatur quod radii Solis sunt causa salsedinis, et intelligebatur de salsedine maris. Hoc de questione.

III, 269ʳ–288ᵛ: ⟨C⟩irca tertium librum *Metheororum* queritur utrum ventus sit exalatio calida et sicca. Et videtur primo quod sit aer motus, quia aliquis motus est ventus, ut patet in follibus ...X... Semper semidiameter yridis est equalis, ideo dicendum est quod hoc est ...

Bibliography

Catalogues: L. Delisle, *Inventaire des manuscrits de l'Abbaye de Sainct-Victor conservés à la Bibliothèque Impériale, sous les numéros 14232–15175 du fonds latin*, Paris 1869, 76–77; W. Seńko, *Repertorium commentariorum Medii Aevi in Aristotelem Latinorum quae in bibliothecis publicis Parisiis asservantur* (*Bibliothèque Nationale, Arsenal, Mazarine, Sorbonne, Ste Geneviève*), Warsaw 1982, 192–193; G. Ouy, *Les manuscrits de l'Abbaye de Saint-Victor: catalogue établi sur la base du répertoire de Claude de Grandrue* (*1514*), Turnhout 1999 (Bibliotheca Victorina, 10) (2 vols.), 2: 341; Lohr, *Latin Aristotle Commentaries, I.2. Medieval Authors M–Z*: 35.

Literature: A. Combes, "Note sur les Sententiae Magistri Joannis Gerson du manuscrit B. N. LAT. 15156", *Archives d'histoire doctrinale et littéraire du Moyen Âge* 14 (1939), 365–385; P. Glorieux, "Le Commentaire sur les Sentences attribué à Jean Gerson", *Recherches de théologie ancienne et médiévale* 18 (1951), 128–139; L. Thorndike, "Oresme and Fourteenth Century Commentaries on the *Meteorologica*", *Isis* 45 (1954), 145–152 [list of questions at pp. 150–152]; McCluskey, *Nicole Oresme on Light, Color and the Rainbow*, 34, fn. 3, 97–98; Panzica, "Nicole Oresme à la Faculté des Arts de Paris", 34.

15 *Poznań, Archivum Archidiecezjalne, Cms 53, ff. 1ʳ–95ᵛ* (Po)

Paper; mm 205×150; I+368+I; 15th C

Date and origin. Nothing is known about the date or provenance of this codex, except that it belonged to the Cistercensian friars, as is evident from a note on f. 1ʳᵃ: "Ex libris fratrum B.M.V. de [...] Ord. Cisterc".

Codicology. Medieval wooden binding. The front cover is missing. A clasp on the back cover. A parchment guard-leaf followed by 368 paper leaves grouped mainly in senions, according to the following scheme: i (1–12)$^{6+6}$; ii (13–24)$^{6+6}$; iii (25–36)$^{6+6}$; iv (37–48)$^{6+6}$; v (49–60)$^{6+6}$; vi (61–70)$^{5+5}$; vii (71–82)$^{6+6}$; viii (83–94)$^{6+6}$; ix (95–107)$^{7+(7-1)}$; x (108–119)$^{6+6}$; xi (120–131)$^{5+5}$; xii (132–143)$^{6+6}$; xiii (144–155)$^{6+6}$; xiv (156–167)$^{6+6}$; xv (168–179)$^{6+6}$; xvi (180–191)$^{6+6}$; xvii (192–207)$^{8+8}$; xviii (208–219)$^{6+6}$; xix (220–231)$^{6+6}$; xx (232–243)$^{6+6}$; xxi (244–250)$^{4+(4-1)}$; xxii (251–262)$^{6+6}$; xxiii (263–274)$^{6+6}$; xxiv (275–286)$^{6+6}$; xxv (287–298)$^{6+6}$; xxvi (299–310)$^{6+6}$; xxvii (311–322)$^{6+6}$; xxviii (323–334)$^{6+6}$; xxix (335–346)$^{6+6}$; xx (347–358)$^{6+6}$; xxi (359–368)$^{6+(6-2)}$. At the end of the volume we find another parchment guard-leaf.

Three main codicological units can be distinguished: i (ff. 1–95); ii (ff. 96–248); iii (ff. 251–367). In the first unit, we find signatures in the inner margin of the first leaf of each fascicule: f. 13 (2us); f. 25 (3us); f. 37 (4us); f. 49 (5us); f. 61 (6us); f. 71 (7us); f. 83 (8us). In the second codicological unit, the signatures can be found sometimes in the inner, sometimes in the upper margin, between the two columns: f. 110 (2); f. 120 (3); f. 132 (4us); f. 144 (5us); f. 156 (6us); f. 168 (7us); f. 180 (8us); f. 192 (nonus); f. 208 (decimus); f. 220 (undecimus, upper margin between the two columns); f. 232 (12us, upper margin between the two columns); f. 244 (13us). Finally, in the third codicological unit, the signatures are always in the upper central margin: f. 263 (2us); f. 275 (3us); f. 287 (4us); f. 299 (5us); f. 311 (6us); f. 323 (7us); f. 335 (8us); f. 347 (9us); f. 359 (10us). Modern foliation in pencil in the upper outer margin, every ten leaves. Ff. 148^v–150^v and ff. 267^v–268^v are blank.

Content. Three textual units correspond to the three codicological units of this manuscript. The first part (ff. 1^r–95^v) contains the second redaction of Oresme's *Questions on Meteorology*; the second part (ff. 96^r–248^r), a set of anonymous *Questions on the Physics*; the third part (ff. 251^r–367^r), Lawrence of Lindores's *Questions on De anima*.[57]

The *Questions on Meteorology*. The text is anonymous. In his catalogue of commentaries on Aristotle in Poznan's Libraries, Markowski ascribes it to John Buridan, without justifying this choice. Yet the incipits and the explicits of the books do not correspond to Buridan's text, but to the second redaction of Oresme's *Questions on Meteorology*. Moreover, Buridan did not comment on the fourth book of Aristotle's *Meteorology*, while Oresme did. Following a direct study of the text, we can now confirm that it is a copy of the second redaction of Oresme's *Questions on Meteorology*.

57 For the incipit and the explicit of these texts, see M. Markowski, "Aristotelica Poznaniensia", 281–293, at 290.

The text is written in different hands and laid out in two columns of about forty-eight lines each. There is no decoration. The words that introduce the structural parts of the questions are sometimes written in bigger character; in other cases, these words are missing and replaced by a blank space. Nine questions from the third book are missing: III.24–III.32. At the end of question III.32 (f. 78va), the copyist has written: "Item sequuntur adhuc decem questiones, scilicet inter illam 'supposito quod yris fiet secundum reflexionem', et illam 'utrum, visa yride et Sole' ". This note is followed by a blank space of eleven lines. The scribe then resume the copy with question III.33 (*Utrum, visa yride et Sole vel astro elevato notabiliter super horizontem, apparet minor portio maioris circuli quam appareat Sole existente in fine horizontis*).

Incipit and explicit of the books

I, ff. 1ra–29vb: Queritur primo utrum possibile sit de impressionibus meteoroloycis [*sic*] simul habere scientiam et opinionem. Et arguitur primo quod non, quia de impressionibus metheorologicis non contingit habere scientiam nec opinionem; ergo etc. ...X... Ad quartam: 'si galaxia non esset de natura celi, tunc de ea non esset determinandum in isto libro', concedo. Et cum dicitur: 'tamen Aristoteles determinavit de ea hic', respondetur quod hoc fecit propter hoc quod plures antiquorum crediderunt galaxiam esse de natura elementari, et propter eorum opiniones ipse talem materiam hic tegit. Et sic est finis questionum primi libri *Metheororum*.

II, ff. 29vb–44vb: Circa secundum librum *Meteororum* queritur primo utrum locus generationis pluvie sit media regio aeris. Et arguitur primo quod non: nullum elementum debet generari extra locum sibi naturalem ...X... Ad rationes ante oppositum patet ex secundo articulo. Et sic est finis secundi libri, etc. Sequitur tertius liber.

III, ff. 45ra–85ra: Queritur primo utrum ventus ⟨sit⟩ exallatio calida et sicca. Et arguitur primo quod non sit exallatio calida, nam sic sequeretur quod in temporibus ventosis aer deberet esse calidus ...X... Ad ultimam dico quod non in vanum laborant, ex quo aliam artem ignorant. Sequitur quartus.

IV, ff. 85ra–95va: Circa initium quarti libri *Methaurorum* queritur primo utrum tantum quatuor sunt qualitates prime, scilicet caliditas, humiditas, siccitas et frigiditas. Et arguitur primo quod non, quia quod ⟨per⟩ superhabundantiam dicitur, uni soli convenit ...X... Ad sextam dico quod in salamandra non dominatur ignis in quantitate nec nutritur ab igne; sed tale animal non est combustibile ab igne; et ideo dicimus quod ipsum habitat in igne. Et sic est finis.

Bibliography

Catalogues: A. Lisiecki, *Katalog rękopisów bibliotecki seminarynej w Poznaniu*, Poznań 1905, 30; Markowski, "Aristotelica Poznaniensia", 289–290. On the manuscript collection of the Diocesan Archiv of Poznan, see T. Makowski, *Rękopisy w zbiorach kościelnych*, Warsaw 2014 (Zbiory rękopisów w Polsce, 2), 279–293.

Literature: Panzica, "Commenter les *Météorologiques* à l'Université de Cracovie", 84, 103, 160; Ead., "Henricus Totting de Oyta's and Nicole Oresme's Commentaries on *Meteorology*: Some New Identifications in Eastern Europe", *Bulletin de philosophie médiévale* 62 (2020), 195–211, at 200, n. 7.

16 *St. Gallen, Stiftsbibliothek, Cod. Sang. 839, ff. 1ʳ–177ʳ* (Sg)

Paper; mm 220 × 165; ff. 180; 1459

Date and origin. This manuscript was copied in 1459, as stated by the colophon: "Rescripte sunt hee questiones venerabilis magistri Orem *Metheororum* Aristotelis paripatetici [*sic*] anno domini 1459 pridie idus mensis septembris indictione septima" (f. 175ᵛ). A scribe's marginal note in Middle High German proves that the codex was copied in a German speaking region (f. 54ʳ: "Is Schiff ist gut"). The same scribe copied another manuscript conserved in St. Gallen: ms. 840, which contains Nicholaus Theoderici of Amsterdam's *Questions* on Aristotle's *Metaphysics*. The manuscript belonged to the library of the monastic community of St. Gallen already in the sixteenth century, as witnessed by the coat of arms of Prince-Abbot of St. Gallen Diethelm Blarer (1530–1564), on f. 175ᵛ.

Composition. The binding, restored in the eighteenth century, is made of leather and pasteboard. The first leaf, numbered A in pencil by a modern hand, was probably added afterwards, as it has not been numbered by the medieval hand and as the subsequent leaf is rather worn, which indicates that it was previously the first leaf of the volume. Pius Kolb, an eighteenth-century librarian, has written on this leaf: "Tractatus egregius seu questiones in IV libros *Metheororum* Aristotelis". In the upper margin of the same leaf, Kolb entered the signature "s.n.308", under which the manuscript was registered in an old catalogue of 1755. The old signature is followed by the modern one, "839", written in red ink.

The volume consists of fifteen senion quires: I (A–11)$^{6+6}$; II (12–23)$^{6+6}$; III (24–35)$^{6+6}$; IV (36–47)$^{6+6}$; V (48–59)$^{6+6}$; VI (60–71)$^{6+6}$; VII (72–84)$^{6+6}$; VIII (85–96)$^{6+6}$; IX (97–108)$^{6+6}$; X (109–120)$^{6+6}$; XI (121–132)$^{6+6}$; XII (133–144)$^{6+6}$; XIII (145–156)$^{6+6}$; XIV (157–168)$^{6+6}$; XV (169–179B)$^{6+6}$.

Starting with the second fascicule, we find signatures in the recto lower margin of every first leaf: f. 12 (2us); f. 24 (3us); f. 36 (4us); f. 48 (5us); f. 60 (6us); f. 72 (7us); f. 85 (8us); f. 97 (9us); f. 109 (10us); f. 121 (11us); f. 133 (12us); f. 145 (13us); f. 157 (14us); f. 169 (15us). Medieval foliation in the upper outer margin. Ff. 177ᵛ–179ᵛ are blank. The numbering skips from 76 to 78.

The *Questions on Meteorology*. This manuscript only contains the second redaction of Oresme's *Questions on Meteorology* (ff. 1^r–175^r), followed by a table of contents (ff. 176^r–177^r). The text is copied in a single column with about forty lines per page. The handwriting is a fifteenth-century bastarda; the ink is brown. The incipits of the questions are boxed in red. The decoration is unfinished: the space intended for the initials of the books has been left blank (f. 1^r, f. 42^v, f. 67^r, f. 152^v).

Incipit and explicit of the books

I, ff. 1^r–42^r: ⟨C⟩irca primum *Metheororum* queritur primo utrum possibile sit de impressionibus metheorologicis habere simul scientiam et oppinionem. Et arguitur quod non, quia de impressionibus metheoroloycis non contingit habere scientiam et oppinionem; igitur questio falsa …X… Ad quartam dico: 'si gallaxia esset de natura celi, tunc de ea non esset determinandum in isto libro', concedo. Et cum dicitur: 'tamen Aristoteles determinavit de ea', respondetur quod hoc facit propter hoc quia plures antiquorum crediderunt gallaxiam esse de natura elementari, propter eorum oppinionem ipse talem materiam hic tangit. Explicit primus *Metheororum*.

II, ff. 42^v–66^v: ⟨C⟩irca secundum *Metheororum* queritur primo utrum locus generationis pluviarum sit media regio aeris. Et arguitur primo quod non, nam nullum elementum debet generari extra suum locum naturalem …X… Rationes patent in secundo articulo huius questionis.

III, ff. 67^r–152^v: ⟨C⟩irca tertium *Metheororum* queritur, secundum distinctionem istius totius in principio positam ab Albertho, utrum ventus sit exalatio calida et sicca. Et arguitur quod non sit exalatio calida et sicca, nam sic sequeretur quod in temporibus ventosis aer deberet esse calidus …X… Ad ultimam dico quod non in vanum laborant, ex quo aliam artem ignorant. Et sic finis est questionum tertii *Metheororum* Aristotelis.

IV, ff. 152^v–175^v: ⟨C⟩irca quartum *Metheororum* queritur primo utrum tantum quatuor sint qualitates prime, scilicet caliditas, humiditas, frigiditas, siccitas. Et arguitur primo quod non, quia quod per superhabundantiam dicitur, uni soli convenit …X… Ad sextam dico quod in salamandra non dominatur ignis in quantitate nec etiam nutritur ab igne; sed tale animal non est ustibile ab igne; et ideo, quia non uritur ab igne, dicimus quod tale animal habitet in igne, et non quod nutriatur ab igne vel quod in eo dominetur [175^v], sed illo modo quo dictum est, scilicet quod sit inustibile ab ipso igne. Et sic finis est huius libri, de quo fine laudetur Deus gloriosus in secula benedictus. Amen. Rescripte sunt hee questiones venerabilis magistri Orem supra libris *Metheororum* Aristotelis paripotetici [*sic*] anno domini 1459, pridie ydiis mensis septembris, indictione septima.

Bibliography

Catalogues: G. Scherrer, *Verzeichniss der Handschriften der Stiftsbibliothek von St. Gallen*, Halle 1875, 285; B.M. von Scarpatetti, R. Gamper, and M. Stähli, *Katalog der datierten Handschriften in der Schweiz in lateinischer Schrift vom Anfang des Mittelalters bis 1550, 3: Die Handschriften der Bibliotheken St. Gallen-Zürich*, Dietikon/Zurich 1991, nr. 195, 71; Ch. Lohr, *Aristotelica Helvetica*, 259–260; Lohr, *Latin Aristotle Commentaries. Medieval Authors*, 2: 35.

Literature: H. Suter, "Eine bis jetzt unbekannte Schrift des Nic. Oresme", *Zeitschrift für Mathematik und Physik*, 27 (1882), Hist.-Lit. Abteilung, 121–125; L. Thorndike, "More Questions on the *Meteorologica*", *Isis* 46 (1955), 357–360.

17 ***Uppsala, Universitetsbibliotek, Ms. C 596, ff. 2ʳ–97ʳ* (U)**

Paper; mm 310×220; ff. 313; 1424

Date and origin. The colophon of the *Questions on Meteorology* informs us that the text was copied in 1424 (f. 96ᵛ: "Anno MCCCC 24°"). The manuscript may have originated in Germany. It belonged to the monastery of Vadstena in Swabia until the latter was abandoned in 1595. The manuscript was then bought by the King of Sweden. In 1620, king Gustavus Adolphus of Sweden donated it to the University library of Uppsala.

Composition. See M. Andersson-Schmitt, H. Hallberg, and M. Hedlund, *Mittelalterliche Handschriften der Universitätsbibliothek Uppsala, Katalog über die C-Sammlung*, 6: Handschriften C 551–935, Stockholm 1995 (Acta Bibliothecae Regiae Universitatis Upsaliensis, 26), 80–81.

Content. The manuscript contains two texts: the second redaction of Oresme's *Questions on Meteorology* (ff. 2ʳ–97ʳ), followed by Marsilius of Inghen's *Questions on the Metaphysics* (ff. 98ʳ–308ᵛ).

The *Questions on Meteorology.* The colophon ascribes the commentary to Guillaume Oresme: "Et sic est finis quarti libri *Metheorologicorum* reverendi magistri Wilhelmi de Orem" (f. 96ᵛ). The text is laid out on two columns of about forty-five lines each. The ink is brown; the handwriting regular. Rubrics and rubricated pilcrows. The first words of the questions are copied in textualis. The words that refer to the argumentative structure of the text are sometimes highlighted in red, as are other parts of the text that were probably regarded as important (see for example f. 3ᵛ). In the margins, we find brief annotations in red, which refer to the structure of the text. Large initials in red at the beginning of the books. On f. 97ʳ⁻ᵛ, we find a table of contents written by the hand of the copyist.

Incipit and explicit of the books

I, ff. 2ʳᵃ–23ᵛᵇ: *Metrorum* [*sic*] questiones. Incipiendo ita queritur primo utrum possibile sit de impressionibus metroloycis [*sic*] simul habere scientiam et opinionem. Et arguitur primo quod non, quia de impressionibus metroloycis [*sic*] non contingit habere scientiam …X… Ad quartam: 'si galaxia esset de natura celi, tunc de ea non esset in illo libro determinandum', concedo. Sed tamen Aristoteles declarat de ea hic propter hoc quia plures antiquorum crediderunt quod galaxia esset de natura elementari. Et sic est ⟨finis⟩ questionum primi libri.

II, ff. 24ra–36ra: Apud initium secundi libri *Metheororum*, secundum distinctionem in principio illius positam ab Alberto, sit prima questio illa: utrum locus generationis pluvie sit media regio aeris. Et arguitur quod non: nullum elementum debet generari extra locum sibi naturalem …X… Ad rationes ante oppositum patet ex secundo articulo. Et sic est finis questionum secundi *Metheororum*.

III, ff. 36ra–81va: Circa tertium librum *Metheororum* sit prima questio illa, utrum ventus sit exalatio calida et sicca. Et arguitur primo quod non sit exalatio calida, nam sic sequeretur quod in temporibus ventosis aer deberet esse calidus …X… Ad ultimam dico quod non in vanum laborant, ex quo alia arte ignorant, etc. Et patet questio. Et sic est finis questionum tertii libri *Metheororum*.

IV, ff. 81vb–96vb: Nunc circa quartum *Metheororum* primo queritur utrum tantum quatuor sunt qualitates prime, scilicet caliditas, humiditas, siccitas et frigiditas. Et arguitur quod non, quia quod per habundantiam dicitur, uni soli rei convenit …X… Ad sextam dico quod in salamandra non dominatur ignis in quantitate nec etiam nutritur ab igne; sed tale animal non est combustibile ab igne; et ideo, quia non comburitur ab igne, dicimus quod ipsum habitet in igne, etc. Et sic est finis quarti libri *Metheorologicorum* reverendi magistri Wilhelmi de Orem.

Bibliography

Catalogues: Andersson-Schmitt, Hallberg, and Hedlund, *Mittelalterliche Handschriften der Universitätsbibliothek Uppsala, Katalog über die C-Sammlung*, 6, 80–81; P.O. Kristeller, *Iter Italicum, 5: Alia itinera III and Italy III, Sweden to Yugoslavia, Utopia, Supplement to Italy (A–F)*, London 1963, 20; Lohr, *Latin Aristotle Commentaries. Medieval Authors*, 2: 35.

Literature: Birkenmajer, *Études d'histoire des sciences en Pologne*, 193; McCluskey, *Nicole Oresme on Light, Color, and the Rainbow*, 95.

18 *Wien, Österreichische Nationalbibliothek, Cod. 5453, ff. 49^r^–109^v^* (Wi)

Paper; mm 330×225; ff. 109; 1370

Date and origin. The colophon of Buridan's *Questions on De generatione et corruptione* informs us that the text was written in Prague in 1370: "Expliciunt questiones magistri Biridani super *De generatione et corruptione* reportate Prage sub anno millesimo C°C°C°LXX° in die sancti Petri ad vincula" (f. 48^{v}). Like the texts in manuscript BJ 749, the commentaries contained in this codex derive from a *pronuntiatio*, as is witnessed by the use of the expression *reportate*. The material aspects of the texts confirm this hypothesis: the handwriting is hasty and there are many interruptions.

Codicology. Parchment binding. The volume contains 109 paper leaves grouped as follows: I $(1–9)^{4+(4+1)}$; II $(10–22)^{6+(6+1)}$; III $(23–29)^{5+(5-3)}$; IV $(30–41)^{7+(7-2)}$; V $(42–49)^{4+4}$; VI $(50–61)^{6+6}$; VII $(62–73)^{6+6}$; VIII $(74–85)^{6+6}$; IX $(86–97)^{6+6}$; X $(98–109)^{6+6}$. Signatures in the central upper margin of the first leaf of each fascicule: f. 10 (2); f. 23 (3); f. 30 (4); f. 42 (5); f. 50 (6); f. 62 (7); f. 74 (8); f. 86 (9); f. 98 (10). Modern foliation in the upper outer margin. Blank leaves: ff. 9–10, f. 13, ff. 28rb–29vb, f. 45^{v}.

Content. The manuscript contains texts on natural philosophy by Parisian masters of the fourteenth century: some questions abbreviated from the first two books of Albert of Saxony's *Questions on De celo* (ff. 1ra–8vb);[58] an incomplete copy of Buridan's *Questions on De anima, redactio A* (ff. 11ra–28ra); the *redactio B* of Buridan's *Questions on De generatione et corruptione* (ff. 30ra–48va); a fragment from Albert of Saxony's *Questions on Meteorology* (f. 48vb),[59] and the second redaction of Oresme's *Questions on Meteorology* (ff. 49ra–109vb).

The *Questions on Meteorology*. The text, which is anonymous, is laid out in two columns of about sixty lines each. The handwriting is hasty and tight, sometimes almost unreadable (see for instance ff. 82^{v}–83^{r}). Long marginal notes have also been written very hastily (see for instance f. 52^{r}, f. 64^{r}, f. 106$^{r–v}$). The incipits of the first and the second book are copied in textualis; the other incipits are missing. The incipits of the questions are written in textualis only until f. 67^{r}, after which a simple blank space separates the questions. In the margins of f. 79^{v} and f. 80^{r}, we find some rough figures drawn by the scribe in order to illustrate the geometrical demonstrations contained in the text.

58 *Inc.*: "Utrum cuilibet corpori simplici naturaliter insit unus motus. Arguitur quod non, quia celo non inest naturaliter unus simplex motus; igitur questio falsa"; I.2, *Utrum in mundo sint etiam* [*sic*] *corpora distincta secundum speciem, scilicet elementa et celum, seu quinta essentia* (f. 1rb); I.3, *Utrum preter elementa et celum sit pondendum sextum corpus simplex* (f. 1va); I.4, *Utrum totalis mundus aggregatus ex quatuor corporibus simplicibus sit unum continuum* (f. 1vb); I.5, *Utrum sursum et deorsum debeant capi in celo secundum polos, sic quod unus sit sursum et aliter deorsum* (f. 2ra); I.6, *Utrum polus articus nobis apparens sit sursum et aliter deorsum* (f. 2rb). The text is interrupted at the very beginning of the third book, with the words: "queritur primo circa tertium librum *De celo et mundo* utrum aliquod elementum", and so on. Ff. 9^{r}–10^{v}, which were intended to host the rest of the text, have been left blank.

59 Panzica, "Albert of Saxony's Questions on Meteorology", 255–257.

Incipit and explicit of the books

I, ff. 49ra–63rb: ⟨Q⟩ueritur circa primum *Metheororum* utrum de impressionibus meteoroloycis possibile sit habere scientiam simul et opinionem. Et arguitur quod non, quia de eis nec contingit habere scientiam nec opinionem; igitur questio falsa ...X... Ad quartam: 'si [63rb] galaxia esset de natura celi', concedo. Sed tamen Aristoteles determinavit de ea in isto libro quia plures antiquorum crediderunt galaxiam esse de natura elementari.

II, ff. 63rb–71vb: ⟨Q⟩ueritur primo circa secundum librum *Metheororum* utrum locus generationis pluvie sit media regio aeris. Arguitur quod non debet generari extra locum suum naturalem ...X... Ad rationes in oppositum patet ex secundo articulo. Et sic sequitur tertius liber.

III, ff. 71vb–103ra: Utrum ventus sit exalatio calida et sicca. Et arguitur quod non sit exalatio calido [*sic*], quia sic sequitur quod in temporibus ventosis aer deberet esse calidus ...X... Ad ultimam dico quod non in vanum laborant, ex quo aliam artem ignorant. Et sic est finis questionum tertii libri huius.

IV, ff. 103ra–109vb: Utrum tantum sint quatuor qualitates prime, scilicet caliditas, humiditas, frigiditas, siccitas. Et arguitur primo quod non, quia quod per superhabundantiam dicitur, uni soli convenit ...X... Ad sextam dico quod in salamandra non dominatur ignis in quantitate nec etiam nutritur ab igne; sed tale animal non est ustibile ab igne; et ideo, quia non uritur ab igne, dicimus quod tale animal habitat in igne, etc. Deo gratias.

Bibliography

Catalogues: F. Unterkircher, *Die datierten Handschriften der Österreichischen Nationalbibliothek von 1501 bis 1600*, Vienna 1976 (Katalog der datierten Handschriften in lateinischer Schrift in Österreich 4), 208; M. Markowski, *Repertorium commentariorum medii aevi in Aristotelem Latinorum quae in Bibliothecis Wiennae asservantur*, Wrocław 1985, 73, 82, 83, 89, 269; Lohr, *Latin Aristotle Commentaries. Medieval Authors*, 1: 258; 2: 35.

Literature: Michael, *Johannes Buridan*, 2: 641, 689; Streijger, Bakker, and Thijssen (eds.), *John Buridan Quaestiones super libros De generatione et corruptione Aristotelis*, 18–19; Panzica, "Nicole Oresme à la Faculté des Arts de Paris", 36; Ead., "Albert of Saxony's *Questions on Meteorology*", 255–257.

19 *Wrocław, Biblioteka Uniwersytecka, IV Q 27, ff. 1r–163r* (Wr)

Paper; mm 220×150; ff. II+165; 14th century

Date and origin. On f. 165^{v}, we find a mark of possession that has subsequently been erased: "Liber Nicolai de scola † artium baccalaurei". The date and provenance of this manuscript are unknown. On the back inside-cover, we read: "Anno millesimo trecentesimo nonagesimo nono obyt serenissima princeps regina

Hedwigis in Cracovia in die sancti Galli". This note proves that the manuscript circulated at Krakow University. It came into the possession of the Wrocław University Library in 1810.

Codicology. Wooden binding. Traces of pegs on both covers; traces of the central boss on the front cover. A paper guard-leaf, on the verso of which we find a table of contents of the first book of Oresme's *Questions on Meteorology*. At the bottom of this leaf, numbered I on the verso by a modern hand, we find a piece of modern paper, upon which is written: "Aus der Bibliothek des Kollegiatstifts zu Glogau". A second guard-leaf, this one in parchment, has been numbered II on the recto by a modern hand. On the verso of this leaf, we read: "Dominus Petrus Nachancze dedit". The guard-leaves are followed by 165 paper leaves grouped mainly in senions: I (1–12)$^{6+6}$; II (13–26)$^{7+7}$; III (27–38)$^{6+6}$; IV (39–50)$^{6+6}$; V (51–62)$^{6+6}$; VI (63–74)$^{6+6}$; VII (75–86)$^{6+6}$; VIII (87–98)$^{6+6}$; IX (99–110)$^{6+6}$; X (111–122)$^{6+6}$; XI (123–134)$^{6+6}$; XII (135–146)$^{6+6}$; XIII (147–158)$^{6+6}$; XIV (159–163)$^{(3-1)+3}$; XV (164–165)$^{1+1}$. Signatures on the recto of the following leaves: f. 1 (1us), f. 13 (2us), f. 27 (3us), f. 39 (4us), f. 51 (5us), f. 63 (6us), f. 75 (7us), f. 87 (8us), f. 99 (9us), f. 111 (10), f. 123 (XI), f. 135 (XIIus), f. 147 (XIIIus), f. 159 (XIIII). Modern foliation in black ink in the upper outer margin.

Content. This manuscript only contains Oresme's *Questions on Meteorology*, (ff. 1^r–163^r), followed by a table of contents of the four books (ff. 164^r–165^r). Another table of contents, concerning only the first book, can be found on the verso of the first guard-leaf.

The *Questions on Meteorology*. A rubricated initium by the hand of the copyist ascribes the text to Nicole Oresme: "Hee sunt questiones magistri Orem super libris *Metheororum*" (f. 1^r). This ascription is repeated in the colophons of the third and of the fourth book (see below). The text, which was copied in a single hand in brown ink, is written in a single column of about thirty-five lines per page. The handwriting is regular and easily readable. The incipits of the books and of the questions are written in textualis and framed in red. Short marginal notes referring to the structure of the text, written by the copyist. Rubricated pilcrows; pointing hands (for example on f. 2^r).

Incipit and explicit of the books

I, ff. 1^r–41^v: Circa librum *Metheororum* Aristotelis, peritissimi philosophi, queritur utrum possibile sit de impressionibus habere metroloycis [*sic*] simul scientiam et opinionem. Et arguitur primo quod non, quia de impressionibus metroloycis non contingit habere scientiam nec opinionem; igitur questio falsa ...X... Propter quod quia plures antiquorum crediderunt gallaxiam esse de natura elementari. Et sic patent questiones primi libri.

II, ff. 41^v–64^v: Questio prima circa secundum, secundum distinctionem in principio positam ab Alberto, et est questio ista: utrum locus generationis pluvie sit media regio aeris. Et arguitur primo quod non ...X... Ad rationes in oppositum patet in secundo articulo. Explicit secundus liber libri *Metheororum*. Sequitur tertius.

III, ff. 64v–144v: Circa tertium, secundum distinctionem illius libri totius in principio positam ab Alberto, sit questio prima: utrum ventus sit exalatio calida et sicca. Et arguitur primo quod non sit exalatio calida ...X... Ad rationem ultimam dico quod non in vanum laborant, ex quo aliam artem ignorant. Hec sunt dicta de questionibus tertii. Expliciunt questiones tertii libri *Metheororum* reverendi magistri Orem.

IV, ff. 145r–163r: ⟨C⟩irca principium istius quarti queritur primo utrum tantum sint quatuor qualitates prime, scilicet caliditas, frigiditas, humiditas et siccitas. Et arguitur quod non, quia quod per superhabundantiam ...X... ratio huius, quia non uritur ab igne, dicimus quod tale animal habitat in igne. Hec de questione. Amen. Expliciunt questiones *Metheororum* seu *Metherorum* magistri Orem.

Bibliography

Catalogues: W. Göber, and J. Klapper, *Katalog der Handschriften der ehemaligen Universitätsbibliothek Breslau*, ca. 1920–1944 [manuscript catalogue], 19 (IV Q 1–50); Lohr, *Latin Aristotle Commentaries, I.2. Medieval Authors M–Z*: 35.

Literature: A.G.E.Th. Henschel, *Catalogus codicum Medii Aevi medicorum ac physicorum qui manuscripti in bibliothecis Vratislaviensibus asservantur*, Wrocław 1847 (2 vols.), 2: 30; Birkenmajer, *Études d'histoire des sciences en Pologne*, 193–194; McCluskey, *Nicole Oresme on Light, Color and the Rainbow*, 95; Panzica, "Commenter les *Météorologiques* à l'Université de Cracovie", 78, 157.

20 ***Città del Vaticano, Biblioteca Apostolica Vaticana, Pal. lat. 1045, f. 118v***

Paper; mm 275 × 225; II+120+I; second half of the fourteenth century

Date and origin. The codex consists of two parts: the first one (ff. 1–90) was copied in Prague in the second half of the fourteenth century; the second part (ff. 91–120) was probably copied in Paris in the same period. The manuscript belonged to Marsilius of Inghen, who left a short annotation on f. 120v.

Content. This manuscript contains Albert of Saxony's *Questions on De celo* (ff. 1ra–61rb), Albert of Saxony's *Sophismata* (ff. 65ra–90rb); John Buridan's *Questions on De anima* (91ra–118ra) followed by its table of contents; a partial table of contents of the second redaction of Oresme's *Questions on Meteorology* (118v), and a note on the magnets by the hand of Marsilius of Inghen (120v).

The table of contents of Oresme's *Questions on Meteorology*. The table of contents is preceded by the following words, which ascribe the text to Nicole Oresme: "Hii sunt tituli questionum quas format magister Nicolaus Orem super librum *Metheororum*". The table only covers the first two books of Oresme's commentary. As far as the first part of the text is concerned, the table is quite selective and omits questions I.1–2, I.4, I.6–7. Starting with question I.8, the table corresponds to the actual content of Oresme's commentary. The titles of the questions are followed by their answer, in an extremely concise form. In the following I provide a transcription of this table.

I.3 Primum, utrum iste mundus inferior sit contiguus lationibus superioribus ut eius virtus inde gubernetur, et dicit cum Aristotele quod sic.

I.5 Utrum eedem opiniones infinities reiterentur. Dicit cum Aristotele quod sic.

I.8 Utrum motus celi sit causa calefactionis ignis in spera sua et etiam supreme regionis aeris. Dicit cum Philosopho quod sic.

I.9 Utrum lumen sit productivum caloris. Dicit cum Philosopho quod sic.

I.10 Utrum contrarium circumstans suum contrarium fortificet ipsum per anthiparistasim. Dicit cum Philosopho quod sic.

I.11 Utrum media regio aeris sit semper frigida. Dicit cum Philosopho quod sic.

I.12 Utrum omnium impressionum metheorologicarum exalatio et vapor sunt materia. Dicit cum Philosopho quod sic.

I.13 Utrum impressiones ignite que fiunt per inflammationem exalationis fiunt naturaliter in aere. Dicit quod sic.

I.14 Utrum de nocte, serenitate existente, debeant apparere hyatus et voragines et sanguinei colores. Dicit cum Philosopho quod sic.

I.15 Utrum cometa sit de natura celesti vel de natura elementari. Dicit cum Philosopho quod sit de natura elementari.

I.16 Utrum cometa sit exalatio callida [*sic*] inflamata [*sic*]. Dicit cum Philosopho quod sic.

I.17 Utrum motus comete sit naturalis vel violentus. Dicit quod sit naturalis.

I.18 Utrum comete significent mortem principum, siccitates, ventos et terre motus. Dicit quod sic.

I.19 Utrum galaxia sit de natura elementari vel celesti. Dicit quod sit de natura celesti. Sequitur de secundo libro.

II.1 Utrum locus generationis pluvie sit media regio aeris. Dicit cum Philosopho quod sic.

II.2 Utrum pluvia, nix et pruina sint eiusdem speciei. Dicit cum Philosopho quod sic.

II.3 Utrum grandines magis debeant generari in hyeme et estate quam in vere et autumpno. Dicit cum Philosopho quod magis in vere et autumpno.

II.4 Utrum aqua calida applicata frigori congelanti citius congeletur quam aqua frigida. Dicit cum Philosopho quod sic.
II.5 Utrum rubedo matutina sit signum pluvie. Dicit quod sic.
II.6 Utrum caligo sit signum future pluvie. Dicit cum Philosopho quod non.
II.7 Utrum aqua naturaliter ascendat ad orificia fontium. Dicit quod non.
II.8 Utrum aque fontium generentur in terra. Dicit cum Philosopho quod sic.
II.9 Utrum mare sit perpetuum vel aliquando fuerit factum. Dicit cum Philosopho quod sic.
II.10 Utrum mare debeat fluere et refluere. Dicit quod sic.
II.11 Utrum aqua maris debeat esse salsa. Dicit quod sic.
II.12 Utrum aque puteales debeant esse salse. Dicit quod sic.

Catalogues

D. Walz, *Die historischen und philosophischen Handschriften der Codices Palatini Latini in der Vatikanischen Bibliothek* (Cod. Pal. Lat. 921–1078), Wiesbaden 1999 (Kataloge der Universitätsbibliothek Heidelberg, 3), 200–202.

Overview of the Manuscripts

Signature	Siglum	Date	Origin
Basel, UB, lat. F I 11	Ba	1369	Prague
Basel, UB, lat. F V 2	Ba	1369	Prague
Berlin, SB–PK, lat. fol. 631	Be	1470	Central Europe
Erfurt, CA, 2° 334	E	1421	Göttingen
Kassel, LB, 2° Phys. et hist. nat. 12	Ka	1424	Central Europe
Klagenfurt, Bisch. Bibl., XXXI b 5	Kl	ca. 1395–1405	Central Europe
Kraków, BJ, 749	Kr	1360–1370	Prague
Kraków, BJ, 751	Kr_1	1360–1370	Prague
Kraków, BJ, 2095	Kr_2	1406	Prague
Kraków, BJ, 2117	Kr_3	1444–1450	Frankfurt (Oder)
Leipzig, UB, 1387	L	1429	Leipzig
München, BSB, Clm 4376	M	1365–1386	Prague
München, BSB, Clm 17226	M^1	1413	Central Europe
Paris, BnF, lat. 15156	P	ca. 1369	Paris
St. Gallen, Stibi, Cod. Sang. 839	Sg	1459	Germany?
Uppsala, UB, C 596	U	1424	Germany?

(*cont.*)

Signature	Siglum	Date	Origin
Wien, ÖNB, Cod. 5453	Wi	1370	Prague
Wrocław, BU, IV Q 27	Wr	14th C expl.	Central Europe
Vaticano, BAV, Pal. lat. 1045	Va	14th	Prague/Paris?

Relationships between the Manuscripts: A Twofold Tradition

The collation clearly shows that the manuscript tradition of the second redaction of Nicole Oresme's *Questions on Meteorology* is divided into two branches: on the one side family α, of which only one member has so far been identified, namely *P*; on the other side family β, which comprises all other witnesses (*Ba*, *Ba*^{1}, *Be*, *Ka*, *Kl*, *Kr*, *Kr*^{1}, *Kr*^{2}, *Kr*^{3}, *L*, *M*, *M*^{1}, *Po*, *Sg*, *U*, *Wi*, *Wr*).

I shall discuss the particularities of *P* as the sole representative of family α later in more detail. For the moment, I will present the common accidents of family β and its subfamilies. I consider an accident as typical for one family if it is shared by half of the members of this family + 1. If the accident is also shared by other manuscripts, I do not consider it as typical for this family. The following lists present the results of a sample portion of the collation, namely the beginning of question I.1 (lines 1–50 in our edition) and of question II.2 (lines 1–44 in our edition). The text before the brackets is the text of *P*.

Additions of family β

1) I.1, 1 non] quia *add. Ba Ba*^{1} *Be Ka Kl Kr Kr*^{1} *Kr*^{2} *Kr*^{3} *L M Po Sg U Wi Wr*
2) I.1, 3 posteriorum] sed *add. Ba Ba*^{1} *Be Ka Kl Kr*^{1} *Kr*^{2} *Kr*^{3} *L M Po Wi Sg U Wr*
3) I.1, 5 scientia] illa *add. Ba*^{1} *Be Ka Kr*^{1} *Kr*^{2} *Kr*^{3} *L M Po Sg U Wi*
4) I.1, 6 formidine] ad oppositum *add. Ba*^{1} *Be Ka Kl Kr Kr*^{1} *Kr*^{2} *Kr*^{3} *L M Po Sg U Wi Wr*
5) I.1, 6 formidine] ad oppositum *add. Ba Ba*^{1} *Ka Kr Kr*^{1} *Kr*^{2} *Kr*^{3} *Po U Wi*
6) I.1, 6 sicut] prius *add. Ba Ba*^{1} *Be Ka Kl Kr Kr*^{1} *Kr*^{2} *Kr*^{3} *M Po Sg U Wi Wr*
7) I.1, 6 non] est *add. Ba Ba*^{1} *Ka Kr*^{1} *Kr*^{2} *Kl U Wi Wr*
8) I.1, 6 subiectum] istius *add. Ba*^{1} *Kl Kr Kr*^{1} *Kr*^{2} *M Po Sg U Wi Wr*
9) II.1, 1 nec aqua] nec pluvia nec aqua *Ba*^{1} *Ka K*^{1} *L M Po M*^{1} *Wi Wr*] nec aqua nec pluvia *Ba Kl K*^{1} *Kr*^{2} *Kr*^{3} *Sg U*
10) II.1, 3 regionem] aeris *add. Ba Be Ka Kl Kr Kr*^{1} *Kr*^{2} *L M M*^{1} *Po Sg U Wr*
11) II.1, 3 illius] in *add. Ba Be Ka L Kl Kr Kr*^{1} *Kr*^{2} *Kr*^{3} *U Sg Wi Wr*

12) II.1, 4 regionem] aeris *add. Be Ka L M Kl Kr*^{1} *Kr*^{2} *Kr*^{3} *M*^{1} *Po Sg U Wi Wr*
13) II.1, 7 probatur] quia *add. Be Kl Kr*^{3} *L M*^{1} *Po Sg U K Wr*
14) II.1, 7 vapores] sufficientes *add. Ba*^{1} *Be Ka Kl L M Kr Kr*^{1} *Kr*^{2} *Kr*^{3} *M*^{1} *Po Sg U Wi Wr*
15) II.1, 7 regionis] aer *add. Ba Ka Kl Kr*^{3} *L Wi Wr*] aeris *add. M*] aeris aer *add. Ba*^{1} *Be Kr Kr*^{1} *Kr*^{2} *M*^{1} *Po Sg W*
16) II.1, 7 sufficientes] ibi *add. Ba Ba*^{1} *Be Ka L M Kr Kr*^{1} *Kr*^{2} *M*^{1} *Po Sg U Wi Wr*

Omissions of family β

1) I.1, 1 primo *om. Ba Ba*^{1} *Be Ka Kr Kr*^{1} *Kr*^{2} *Kr*^{3} *M Po Sg Wi*
2) I.1, 1 etiam *om. Ba Ba*^{1} *Be Ka Kl Kr*^{1} *Kr*^{2} *Kr*^{3} *M Po Sg U Wi Wr*
3) I.1, 2 de se *om. Ba Ba*^{1} *Be Ka Kl M Po Sg Kr Kr*^{1} *Kr*^{2} *Kr*^{3} *Wi Wr*
4) I.1, 2 in *om. Ba Ba*^{1} *Be Ka Kl Kr*^{1} *Kr*^{2} *Kr*^{3} *L M Sg U Wi Wr*
5) I.1, 3 et *om. Ba Ba*^{1} *Be Ka Kl Kr*^{1} *Kr*^{2} *Kr*^{3} *L M Po Sg U Wi Wr*
6) I.1, 6 qui tradiderunt nobis scientiam de impressionibus metheorologicis et sic patet auctoritatibus quod de impressionibus metheorologicis potest esse scientiam *om. Ba*^{1} *Be Ka Kl Kr Kr*^{1} *Kr*^{2} *Kr*^{3} *L M* Po *Sg U Wi Wr*
7) II.1, 1 primo *om. Ba*^{1} *Be Ka Kl Kr Kr*^{1} *Kr*^{2} *Kr*^{3} *U Wi*
8) II.1, 3 usque *om. Ba Ba*^{1} *Be Ka Kl Kr Kr*^{1} *Kr*^{2} *Kr*^{3} *L M M*^{1} *Po Sg U Wi Wr*
9) II.1, 4 in vallibus *om. Ba Ba*^{1} *Be Ka Kl Kr Kr*^{1} *Kr*^{2} *Kr*^{3} *L M M*^{1} *Po Sg U Wi Wr*
10) II.1, 4 nubes et *om. Ba*^{1} *Be Ka Kl Kr Kr*^{1} *Kr*^{2} *Kr*^{3} *L M M*^{1} *Po Sg U Wi Wr*
11) II.1, 5 alique *om. Be Ka Kl Kr*^{3} *L M*^{1} *Po U Sg Wr*
12) II.1, 6 videndum est *om. Ba Ba*^{1} *Be Ka Kl Kr Kr*^{1} *Kr*^{2} *Kr*^{3} *L M M*^{1} *Sg Po U Wi*
13) II.1, 6 videndum est^{2} *om. Ba Ba*^{1} *Be Ka Kl Kr Kr*^{1} *Kr*^{2} *Kr*^{3} *L M M*^{1} *Po Sg U Wi Wr*
14) II.1, 6 videndum est *om. Ba Ba*^{1} *Be Ka Kl Kr Kr*^{1} *Kr*^{2} *Kr*^{3} *L M M*^{1} *Po Sg U Wi Wr*
15) II.1, 7 vel *om. Ba Ba*^{1} *Be Ka Kl Kr Kr*^{1} *Kr*^{2} *Kr*^{3} *L M M*^{1} *Po Sg U Wi Wr*
16) II.1, 7 si *om. Ba Ba*^{1} *Be Ka Kl Kr Kr*^{1} *Kr*^{2} *Kr*^{3} *L M M*^{1} *Po Sg U Wi Wr*

Variants of family β

1) I.1, 1 impressiones autem metheorologice] sed impressiones metheorologice *Ba Ba*^{1} *Kr Kr*^{1} *Kr*^{2} *L U Wi*
2) I.1, 3 nota est] tenet *Be Ka Kl Kr L M Po Sg U Wi Wr*] patet *Ba Ba*^{1} *Kr*^{1} *Kr*^{2}
3) I.1, 3 quod non est verum non potest sciri] nichil contingit sciri nisi verum *Ba Ba*^{1} *Be Ka Kl L Kr*^{1} *Kr*^{2} *Kr*^{3} *M Po Sg U Wi Wr*[60]

---

60 We can consider this variant to be shared by all the members of family β: in fact, the only witness that transmits a different text, *Kr*, reads: "nichil convenit sciri nisi verum", which

4) I.1, 4 quia] ex eo quod *Ba Kr Kr¹ Kr³ L M U Wi Wr*] eo quod *Kr²*] ex eo quia *Ka Kl*] de eo quod *Ba¹*
5) I.1, 4 de illo de quo] de illis de quibus *Ba Ba¹ Be Kr Kr¹ Kr² L M Sg U Wi Wr*] ex hiis de quibus *Kl Kr³*
6) I.1, 4 corruptibili] corruptibilibus *Ba Ba¹ Be Ka Kl Kr¹ Kr² Kr³ L M Po Sg U Wi Wr*
7) I.1, 4 ut videtur non possunt] *Ba Ba¹ Kr Kr¹ Kr² L M Wi Wr*] non videntur posse *Kl*
8) I.1, 5 fiunt] sunt *Ba Ba¹ Kl Kr¹ Kr² Wi Wr*
9) I.1, 5 debet esse] habetur *Ba Ba¹ Be Ka Kl M Po L Kr Kr¹ Kr² Kr³ Sg U Wi Wr*
10) I.1, 5 in littera] in primo illius/huius libri *Be Ka Kr³ L M Po Sg U*] in primo huius *Ba Kl Kr¹ Kr² Wi Wr*
11) I.1, 5 dicit enim] ubi dicit *Ba Ba¹ Be Ka Kl Kr¹ Kr² Kr³ M Po Sg U Wi Wr*] ut dicit *L*
12) I.1, 5 accidunt] contingunt *Ba Ba¹ Be Ka Kl Kr¹ Kr² Kr³ L M Po Sg U Wi Wr*
13) I.1, 6 in oppositum est aristoteles et albertus et alii] oppositum vult aristoteles albertus et alii *Ba Ba¹ Kl Kr¹ Kr² L M Sg Wi Wr*] oppositum vult albertus et alii *Be*
14) I.1, 6 ipsis] impressionibus *Ba¹ Be Kr Kr¹ Kr² M Sg Wi*
15) I.1, 6 quia] nam *Ba Ba¹ Be Ka Kl Kr¹ Kr² Kr³ L M Po Sg U Wi Wr*
16) I.1, 6 quia²] nam *Ba Ba¹ Be Ka Kl Kr¹ Kr² Kr³ L M Po Sg U Wi Wr*
17) I.1, 6 quia de eadem re possibile est] immo possibile est de eadem re *Ba Ba¹ Be Kl Kr¹ Kr² Kr³ L M Po Sg U Wi Wr*
18) I.1, 6 et iret] ambulans *Ba Ba¹ Kl Kr Kr¹ Kr² Wi Wr*] et ambularet *Be L M Sg*] ambulat *Kr³*] ambulet *Ka U*
19) I.1, 6 ista] tunc hac *Ba Ba¹ Be Ka Kl Kr¹ Kr² Kr³ Sg U Wi Wr*] hac *L M*
20) I.1, 6 demonstrando patrem tuum] et sic *Ba Ba¹ Be Ka Kl L M Po Kr¹ Kr² Kr³ Sg U Wi Wr*
21) I.1, 6 est] esset *Ba Be Ka Kr Kr³ L Po U Sg Wi*
22) I.1, 6 ipse enim est] quia pater tuus est *Ba Ba¹ Be Ka Kl Kr Kr¹ Kr² Kr³ L Po Sg U Wi Wr* (quia pater est *M*)
23) I.1, 6 istius propositionis] talis conclusionis *Ba Ba¹ Ka Kl Kr* (*ante correctionem*) *Kr¹ L M Sg U Wi Wr*] istius conclusionis *Kr² Kr³*] illius conclusionis *Be*
24) I.1, 6 similiter eandem conclusionem quam sor opinatur plato potest scire] similiter eandem conclusionem quam sor scit plato potest opinari

---

is nothing other than a somewhat corrupted form of the variant shared by family β due to a faulty reading of the abbreviation used for the word "contingit".

Ba Ba^1 Be Ka Kr^1 Kr^2 Kr^3 L M Po Sg U Wi] similiter eandem conclusionem quam sor scit plato potest opinare *Wr Kl*] similiter eandem conclusionem quam scit sor plato potest opinari *Kr*

25) I.1, 6 potest esse] esset *Ba Ba^1 Be Ka Kl Kr Kr^1 Kr^2 L M Sg U Wi Wr*] sit *Kr^3*

26) II.1, 2 videtur tenere] tenet *Ba^1 Ka Kl Kr^1 Kr^3 M M^1 Po U Wi Wr*] nota est L] est nota *Be Sg*] patet *Ba K^1 Kr^2*

27) II.1, 2 idem non est] idem non videtur esse *Ba Ba^1 Be Kl Kr Kr^1 Kr^2 Kr^3 M^1 Po U Sg Wi Wr*] non videtur esse idem *M*

28) II.1, 2 fulgura] fulmina *Ba Ba^1 Be Ka Kr^1 Kr^2 Kr^3 L M M^1 Po Sg U Wi Wr*] ultima *Kl*

29) II.1, 2 per] propter *Ba Ba^1 Ka Kr Kr^1 Kr^2 Kr^3 L M^1 Po Sg Wi Wr*

30) II.1, 3 inde] deinde *Ba Ba^1 Be Ka Kr Kr^1 Kr^2 Kr^3 L M Po Sg U Wi*

31) II.1, 3 converteretur in aquam] sequeretur pluvia *Be Ka Kl Kr^3 L Po Sg Wr*] sequitur pluvia *Ba Kr Kr^1 Kr^2 M M^1 U Wi*

32) II.1, 3 deberet fieri] fieret *Ba Ba^1 Be Ka Kl Kr Kr^1 Kr^2 Kr^3 L M M^1 Po U Sg Wi Wr*

33) II.1, 3 ita] sic *Ba Ba^1 Be Ka Kl Kr Kr^1 Kr^2 Kr^3 L M M^1 Po U Wr om. Sg*

34) II.1, 4 existens] stans *Ba^1 Ka Kl Kr^1 Kr^2 Kr^3 M M^1 Po U Wi Wr*] steteat *L*] stetit *Ba*] steterat *Be Sg om. K*

35) II.1, 4 montes enim illi non sunt] quia montes non sunt *Ba Ba^1 Be Ka Kr Kr^1 Kr^2 Kr^3 Po Sg U Wi*] quia sunt *Kl Wr*] quia montes sunt *M M^1*] quia aliqui montes sunt *L*

36) II.1, 4 ita] tam *Ba Be Ka Kl Kr Kr^1 L M Sg Wi Wr om. Kr^2*

37) II.1, 4 attingerent] attingunt *Ba^1 Ka Kl Kr^1 Kr^2 M M^1 Po Wi Wr*] attingant *Ba K*

38) II.1, 5 in oppositum est] oppositum dicit *Ba Ba^1 Be Kl Kr Kr^1 Kr^2 L Sg Wi Wr* (in oppositum est *U*)

39) II.1, 6 exponende] ponende *Be Ka Kl Kr^3 L M M^1 Po Sg U Wr*

40) II.1, 6 pluvie] pluviarum *Ba^1 Be Ka Kr^1 Kr^2 Kr^3 L M M^1 Po Sg U Wi Wr* (pluvie *Kl*)

41) II.1, 6 pluviarum quoad quantitatem] guttarum ipsius pluvie *Ka Kl Kr^3 M M^1 U Po Wr*] guttarum *Be Sg*] guttarum earum *L* (de quantitate pluviarum *Ba Ba^1 Kr Kr^1 Kr^2 Wi*)

42) II.1, 7 declaratur] probatur *Ba Ba^1 Be Ka Kl Kr Kr^1 Kr^2 Kr^3 L M^1 Po Sg U Wi Wr* (*M lect. dub.*)

43) II.1, 7 elevatum in nubem et convertere in aquam] et convertere in nubem et sic sequitur pluvia *Ba Ba^1 Kl Kr^1 Kr^3 L Po Wi*] et convertere in nubem et ut sic sequitur pluvia *Kr*] et convertitur in nubem et sic sequitur pluvia *Kr^2*] et convertere in nubem et sequitur pluvia *Ka M M^1 U*] et vertere in pluvias et sequitur inde pluvia *Be Sg*

44) II.1, 7 ecce causas sufficientes in media regione aeris] ergo sunt in media regione aeris cause sufficientes *Ba Ba*^{1} *Kl Kr M Wi*] igitur in media regione aeris sunt cause sufficientes *Be Kr Kr*^{1} *Kr*^{2} *Sg*] ergo in media regione aeris cause sufficientes *Wr*] ideo sunt cause sufficientes in media regione aeris *Ka Kr*^{3} *M M*^{1} *Po U*

45) II.1, 7 et fieret] quod fieret *Ba Ba*^{1} *Ka Kl Kr Kr*^{1} *Kr*^{2} *Kr*^{3} *L M M*^{1} *Po U Wi Wr*] quod fiat *Be Sg*

46) II.1, 7 reliquum] alterum *Ba*^{1} *Ka Kl Kr Kr*^{1} *Kr*^{3} *L M M*^{1} *Po U Wi Wr*] aliud *Ba Kr*^{2}

Inversions and transpositions of family β

1) I.1, 6 aliquis assentit] assentit aliquis *Ba Ba*^{1} *Be Ka Kl M Kr Kr*^{1} *Kr*^{2} *Kr*^{3} *M Sg U Wi Wr* (consentit aliquis *Po*)

2) I.1, 6 super manus et pedes] super pedes et manus *Ba Ba*^{1} *Ka Kl Kr Kr*^{1} *Kr*^{3} *L Po Sg U Wi Wr*] super pedes et super manus *M*] per pedes et manus *Be*

3) I.1, 6 a te opinata] opinate a te *Ba Ba*^{1} *Be Ka Kl Kr Kr*^{1} *Kr*^{2} *Kr*^{3} *L M Po Sg U Wi Wr*

Question I.5 provides a good example of the differences between *P* and manuscripts of family β. In the following table, I present an example taken from the solution of this question, transcribed from manuscript *P* and from manuscript *Kl*, chosen as representative for family β:

Text of *P* 233ʳ	**Text of family β (*Kl*)**
Quantum ad tertium, sit ista conclusio: huiusmodi variatio fit effective a corpore celesti tamquam ab agente universali et generali. Patet ex precedentibus duabus questionibus, nam ista inferiora reguntur ab illis superioribus, sicut prius dicebatur; ergo, causa variationis in istis inferioribus est influentia celestis. Istud etiam patet per astrologos, ponentes quod aliquando fiunt alique magne coniunctiones que sunt cause inceptionis quarundam sectarum et opinionum que diu durant, si non sint mediocres; si autem sint parve, tunc sunt cause inceptionis sectarum et.	Ad tertium sit conclusio hec: quod huiusmodi variatio opinionum fit effective a corpore celesti tamquam a causa effective universali et generali. Patet ex duabus questionibus precedentibus, nam inferiora reguntur a superioribus, sicut prius dicebatur; ergo causa variationis in istis inferioribus est influentia celestis. Hoc etiam patet per astrologos, volentes quod aliquando fiunt alique magne coniunctiones que sunt cause inceptionis quarundam sectarum et quarundam opinionum, et quedam coniunctiones, si sunt magne, tunc sunt cause inceptionis quarundam

(*cont.*)

Text of *P* 233r	Text of family β (*Kl*)
opinionum que per modicum tempus durant [...]	sectarum et quarundam opnionum que diu durant; si autem tales coniunctiones sunt mediocres, tunc essent cause inceptionis opinionum que mediocriter durant, sed si sunt parve, tunc etiam cause talis durationis sunt parve. Hoc de tertio [...].
Similiter variatio opinionum in aliquo regno potest fieri per guerras in illo regno; unde, cum sunt guerre in aliquo regno, homines non possunt insistere studio, sed oportet eos insistere aliis, et sic potest accidere variatio opinionum in regno. Etiam potest fieri propter sterilitatem; unde, acquisitis necessariis, Antiqui inceperunt philosophari.	Etiam variatio opinionum in aliqua regione potest fieri per guerras, nam in regionibus homines propter guerras non coexistunt studio. Potest etiam fieri propter sterilitatem, quia, habitis necessariis, antiqui ceperunt phylosophari, sicut recitat Aristoteles in prohemio *Metaphysice*, ubi dicit quod antiqui sacerdotes in Egipto iam habentes omnia vite necessaria invenerunt scientias speculativas.
Similiter potest accidere variatio opinionum propter variationem complexionum et inclinationum hominum; unde aliquando homines magis inclinantur ad praxim quam ad speculandum, et hoc aliquando accidit propter malitiam hominum. Cum enim speculativi non honorentur, qui bene insisterent speculationibus ponunt se ad practicam, sicut tempore hodierno experimur. Plures enim se dant legibus quam artibus, et isto modo opiniones variantur in ignorantiam et non in opinionem contrariam.	Similiter, etiam fieri potest variatio opinionum propter naturalem inclinationem hominum ad opiniones, unde aliqui homines magis inclinantur ad praxim, id est ad scientias practicas, quam ad speculandum, et hoc aliquando fit propter malitiam hominum. Cum nunc viri speculativi non honorantur, ponunt se ad leges, et in isto fit variatio opinionum in ignorantiam et non in scientiam, ut si artiste student leges. Hoc de quarto.
Quantum ad quintum, potest dici quod causa finalis huiusmodi variationis et reiterationis opinionum est decor universi, unde propter talem variationem, universum est pulchrius. Unde propter aliquam rem in se turpem una tota multitudo potest fieri pulchrior, sicut potest	Quantum ad quintum, potest dici quod causa finalis huiusmodi variationis et reiterationis opinionum est decor universi, unde propter talem variationem opinionum universum est pulchrius. Nam etiam propter aliquam rem in se turpem una tota multitudo potest fieri pulchrior, sicut

(*cont.*)

Text of *P* 233r	Text of family β (*Kl*)
apparere picturis. Alia causa similiter huiusmodi variationis potest esse exercitium et intellectus hominum delectatio, quia delectatur in novitatibus et variationibus.	patet in picturis, ut si aliquis paries est depictus albo colore et viridi et nigro, qui color tamen in se turpis est, et sic de alijs coloribus, tamen ex hoc ille color redderetur pulchrior. Secundo dico quod causa finalis istius variationis est exercitium et delectatio intellectui humani, nam intellectus humanus delectatur in varietatibus et novitatibus. Hoc de articulo quinto.
Ad rationes. Ad primam dico quod secundum Philosophum conceditur quod mundus est eternus, et isto sumpto infinities eedem opiniones reiterantur; cuius causa dicta est in questione, quia infinities consimiles fiunt coniunctiones planetarum que aliquando erant, que sunt causa huiusmodi variationis et reiterationis opinionum.	Ad primam dico: si conceditur secundum Aristotelem quod mundus esset eternus, tunc eedem opiniones infinities reiterentur, et huiusmodi causa dicta est.

Manuscripts of family β alter the arguments *contra* of question I.11:

Text of *P* 241v	Text of family β
Et arguitur primo quod non: semper media regio aeris est calida; igitur non frigida. Consequentia tenet; antecedens patet ex eo quod aer per se et naturaliter est calidus et humidus.	Et arguitur primo quod non, quia semper media regio aeris est calida; ergo non frigida. Consequentia tenet et antecedens patet ex eo quod aer per se et naturaliter est calidus et humidus.
Secundo: si semper media regio aeris esset frigida, vel ergo illa frigiditas inesset sibi naturaliter, vel violente. Non potest dici quod naturaliter, cum contrarium frigiditatis insit sibi naturaliter, puta caliditas; nec potest dici quod violente, quia tunc	Secundo: si semper media regio aeris esset frigida, sequeretur quod esset multum densa, sed hoc est falsum. Consequentia tenet ex eo quod frigus est causa condensationis. Falsitas consequentis patet, quia tunc nos impediret ne videmus celum et

(*cont.*)

Text of *P* 241^v^	Text of family β
non semper media regio aeris esset frigida, cum nullum violentum semper duret.	astra, nam propter eius densitatem species visibiles celi et species astrorum non possent multiplicari usque ad visum nostrum.
Tertio: si sic, sequeretur quod media regio aeris sit semper densa; sed hoc est falsum. Consequentia tenet ex quo frigiditas est causa condensationis. Falsitas consequentis patet, nam tunc impediret ne videremus celum et astra, nam propter eius densitatem species visibiles celi et etiam astrorum non possent multiplicari usque ad visum nostrum.	Tertio: si sic, sequeretur quod aer qui est in media regione aeris esset gravior aere qui est hic inferius. Consequentia tenet ex eo quod magis frigida minus frigidis sunt graviora: aer autem hic inferius est minus frigidus aere existente in media regione aeris. Falsitas consequentis patet, quia tunc aer ille qui est in media regione aeris descenderet et aer inferius ascenderet, cum ordo gravium et levium sit graviora esse sub levioribus et leviora super graviora.
Quarto sequeretur quod aer qui est in media regione aeris esset gravior quam aer qui est hic inferius. Hoc est falsum. Consequentia tenet ex eo quod frigida minus frigidis sunt graviora; aer autem hic inferius esset minus frigidus aere existente in media regione, si semper media regio aeris esset frigida. Falsitas consequentis patet, quia tunc aer qui est in media regione descenderet et aer qui est hic inferius ascenderet, cum ordo gravium et levium sit graviora esse sub levioribus et leviora esse super graviora.	Quarto: si sic, sequitur quod in media regione aeris non generarentur impressiones ignee; sed hoc est falsum. Et consequentia tenet propter nimiam frigiditatem aeris. Falsitas consequentis patet, quia ibi generantur corruscationes et fulgura; illa autem sunt impressiones ignee.
Quinto: si sic, sequeretur quod in media regione non generantur impressiones in aere; sed hoc est falsum. Consequentia tenet propter nimiam frigiditatem medie regionis ipsius aeris. Falsitas consequentis patet, nam ibi generantur corruscationes	Quinto: si semper media regio aeris esset frigida, vel ergo ista frigiditas inesset sibi naturaliter vel violente. Non potest dici quod naturaliter, quia tunc contrarium inesset sibi naturaliter, puta caliditas. Nec etiam potest dici quod insit sibi violente,

(*cont.*)

Text of *P* 241v	Text of family β
et fulgura. Ista autem sunt impressiones ignee.	quia tunc non semper media regio aeris esset frigida, cum nullum violentum semper duret.

The first sentence of the second argument in manuscripts of family β is identical to the corresponding argument in *P*: afterwards, manuscripts of family β mistakenly switch to the third argument in *P*. The content of the second argument in *P* can be found in the fifth argument of family β. Moving the second argument to the fifth place creates a difference in the numbering of the arguments which persists until the sixth argument. We find the same difference in the answer to the arguments *contra*.

The same principle applies to the arguments *contra* in question I.15. In this case, the order of arguments 5 and 6 is inverted and argument 4 is formulated slightly differently in *P* and in the manuscripts of family β. However, as in the other cases, this modification does not affect the content of the text:

Text of *P*	Text of family β
Quarto ad idem rationibus Senece, nam, si esset de alia natura quam de natura celi, tunc fieret in aere, sicut alie impressiones; sed quecumque impressiones fiunt in aere sunt parve durationis, sicut videmus de nubibus et consimilibus; sed comete non sunt parve durationis, immo magne; igitur etc.	Quarto confirmatur rationibus Senece, nam, si esset de alia natura quam de natura celi, tunc tales impressiones non diu durarent, et hoc si fieret in aere; sed comete diu durant.
Quinto: impressiones que fiunt in aere non durant nec manent sub eodem astro nec moventur ad motum illius astri, et hoc si fiunt per inflammationem; sed visum est quod cometa movebatur ad modum astri.	Quinto: comete sunt magne significationis et significant aliquos effectus futuros qui non statim eveniunt; sed impressiones alie non significant aliquos effectus futuros, cum illi cito eveniunt, sicut patet de rubedine matutina ad quam statim sequitur pluvia, ex quo videtur quod

(*cont.*)

Text of *P*	Text of family β
	cometa non sit de natura elementari, sed de natura celi.
Sexto: comete sunt magne significationis et significant aliquos effectus futuros qui non statim eveniunt; sed impressiones alie, si significant aliquos effectus proprios, illi statim eveniunt, sicut patet de rubedine matutina, postquam statim videtur sequi pluvia.	Sexto: impressiones que fiunt in aere non diu manent nec moventur sub aliquo astro erratico nec ad motum istius astri, et hoc si fiunt per inflammationem; sed visum est quod cometa movebatur ad motum astri, ex quo sequitur quod non est de natura elementari de qua sunt alie impressiones.

In the second article of question I.14, manuscripts of family β omit a long passage:

Text of *P* 246v	Text of family β
Quantum ad secundum, sit prima suppositio: quando aliqua duo equaliter distant a visu, quorum unum est coloratum vel saltem apparens sub colore magis claro, reliquum vero sub colore minus claro, tunc istud quod apparet sub colore clariori apparet esse propinquius quam istud quod est coloratum colore minus claro. Et istud sciunt pictores: partem enim que debet apparere supereminere colorant colore magis claro et illam que debet apparere densa colore minus claro, sicut colore nigro vel alio tendente ad nigredinem. Et propter istam causam potest contingere quod aliqua superficies plana potest apparere concava vel convexa. Si enim continue versus medium coloretur colore claro et versus extremum colore	Quantum ad secundum, sit prima suppositio: quando aliqua duo equaliter distant a visu quorum unum est coloratum vel saltem apparens sub colore magis claro, reliquum vero sub colore minus claro, tunc illud quod apparet sub colore clariori apparet esse propinquius quam istud quod est magis coloratum. Et hoc bene sciunt pictores: quando enim volunt aliquam partem facere apparere magis prope, tunc colorant ipsam colorem magis claro quam illam partem quam volunt apparere minus. Et propter istam causam potest contingere quod aliqua superficies plana potest apparere concava et profunda.

(*cont.*)

Text of *P* 246^{v}	Text of family β
obscuro, iudicaretur convexa, quia partes medie apparerent propinquiores, quia essent colorate colore clariori. Si autem partes versus medium alicuius superficiei plane essent colorate colore obscuro et versus extremum colore claro, tunc illa superficies apparet concava, quamvis esset plana, quia partes extreme apparent propinquiores quam partes medie; nam essent colorate colore clariori. Sic similiter superficies convexa posset apparere plana, sicut si partes magis versus nos essent colorate colore obscuro et remotiores a nobis colore claro, tunc posset fieri recompensatio claritatis et distantie.	
Secunda suppositio: cum lux vel album videtur per nigrum, apparet rubeum. Ista suppositio patet cum videmus Solem per fumum, et propter hoc ignis etiam in carbone apparet coloris rubei: carbo enim est nigri coloris; ignis autem est lucidus.	Secunda suppositio: cum lux vel album videtur per nigrum, tunc apparet rubeum. Ista suppositio patet, quia, cum Solem videmus per fumum, tunc apparet nobis rubeus, et propter hoc etiam ignis in carbone apparet coloris rubei: carbo enim est nigri coloris; ignis autem est lucidus.

All these examples confirm that the manuscript tradition of the second redaction of Oresme's *Questions on Meteorology* is divided in two branches: on the one side, an isolated manuscript, namely *P*; on the other side, all other copies.

Manuscript P

Birkenmajer—who, as we have already seen in the introduction, did not know of manuscript *P*—noticed that all the manuscripts of the second redaction of Oresme's *Questions on Meteorology* he had identified could be related to Central Europe. According to him, "the father of all German copies" did not transmit the original text of Oresme's commentary, but a reworking of it by an anonymous Prague master.[61] McCluskey, for his part, mentioned other arguments in favour of Birkenmajer's thesis. He pointed out that seven witnesses of Oresme's *Questions*—including some of the oldest ones—came from the University of Prague. There bachelors and masters of the Arts Faculty commonly based their lectures on the commentaries of famous masters from Paris and Oxford.[62] The discovery of a Parisian manuscript provides possible confirmation of Birkenmajer's assumption, as this codex transmits a slightly different text than the other witnesses of Oresme's *Questions*.

Given the differences between manuscript *P* and all other copies, one might consider it to be corrupted, but following a complete collation of the text, it seems to me that in the majority of cases where *P* and manuscripts of family β disagree, it is *P* that provides the better variants. In what follows, I shall give some examples to demonstrate this.

In question I.2, Oresme states that nature is ordered. To corroborate this principle, manuscript *P* refers to the third book of Aristotle's *De celo* and to the eighth book of Aristotle's *Physics*: "natura non est inordinata. Tertio enim *Celi* et octavo *Physicorum* dicitur: 'nichil est inordinatum eorum que secundum naturam fiunt' ". The literal quotation can indeed be found in the Latin translation of

61 Birkenmajer, *Études d'histoire des sciences en Pologne*, 239.

62 This practice is mentioned by the statutes of the University. The *Liber decanorum* of the Prague Arts Faculty contains a decision, dated 1367, which regulates the practice of the *pronuntiatio*. See *Statuta et Acta rectorum Universitatis Carolinae Pragensis: 1360–1614*, eds. F. Šmahel and G. Silagi, Prague 2018 (Documenta Historica Universitatis Carolinae Pragensis, 1), 235–236: "Item eodem anno domini 1367, 20. die aprilis in plena congregacione facultatis magistri considerantes, quod multe inordinaciones et deformitates fierent per pronunciatores, et multi errores, ex quibus studentibus in artibus magnum dispendium, et toti facultati, imo toti universitati grave scandalum posset exoriri; nam quivis scholaris pronuntiavit, quidquid et quandocunque voluit, propria temeritate et scripta incorrecta et ignota, multos errores continentia dabant ad pennam, ipsa reverendis magistris false adscribendo, ut tanto plures sibi allicerent reportantes: facultas igitur cupiens illorum pronunciatorum temerariam abusionem supprimere et utilitati studentium fideliter providere, matura deliberacione prehabita et unanimi consensu statuerunt, quod quivis magistrorum poterit, super quolibet libro de facultate artium propria dicta dare, per se vel per alium pronunciando, poteritque scripta et dicta aliorum per se vel per alium pronunciare, dummodo sint ab aliquo vel aliquibus famoso vel famosis de universitate Pragensi, Parisiensi, vel Oxoniensi magistro vel magistris compilata, et dummodo ista antea fideliter correxerit, et pronunciatorem assumpserit idoneum et valentem. Baccalarii super libros Aristotelis et alios libros difficiles propria dicta dare, vel pronunciare non debebunt, dicta tamen aliorum magistrorum de ista universitate, vel aliorum de universitatibus, Parisiensi scilicet vel Oxoniensi, famosorum dare potuerunt, non quidem per alios, sed per semetipsos pronunciando, dummodo tamen decanum, qui pro tempore fuerit, prerequirant, ut ista committat alicui magistrorum, qui examinet, si sint dicta istius, cui adscribuntur, aut etiam sint correcta".

the *Physics* by Iacobus Venetus and in the revision by William of Moerbeke.[63] Interestingly, manuscripts of family β do not refer either to *De celo* or to the *Physics*, but rather to the third book of the *Meteorology* ("ut dicitur in tertio huius"), in which this quotation cannot be found.

In question I.3, Oresme deals with the concepts of continuity and contiguity, defined by Aristotle in the fifth book of the *Physics*. Oresme argues against the view according to which the parts of continuous bodies have to move homogeneously. The Seine, for instance, consists of a continuous mass of water, but its parts can move in different directions: "Secana est una massa continua, et tamen una eius pars potest moveri versus unam partem et alia versus aliam". The Latin word for the Seine is "Secana". In most of the manuscripts of family β this word is replaced by the expression "lancea mota ad latus". While this modification radically changes the meaning of the text, the reasoning is still applicable with some effort. A long body rotated around its core will have some of its parts moving in the opposite direction to the others. Yet it seems evident that, in contrast to what happens with a river, such an object moves homogeneously. Moreover, strictly speaking, its parts do not move in opposite directions, as they all have the same angular difference. In addition to this, the characterisation of continuous mass ("Secana est una massa continua") is more suitable to a river than to a long body such as a spear. The example found in the manuscripts of family β is therefore not as pertinent as the one we find in *P*. Three manuscripts of family β have other variants, semantically different from each other, but graphically quite similar: *Be*, *Ka*, *Sg* and *Po* read "locus", *L* reads "lapis", *M* and *U* read "lacus", which restores, though in a slightly modified way, the original meaning of *P*.

The same question provides us with further examples of ways in which the text of family β can be considered as a corruption of the text of *P*. An interesting case concerns the following passage: "Et cum arguebatur: 'materia Sortis et materia aeris circumstantis Sortem sunt continue; ergo Sortes et aer sunt continui', negetur antecedens". Manuscripts of family β replace the term "consequentia" with the term "antecedens". In fact, Oresme would not have admitted the antecedent of his conclusion, namely that the matter of Socrates and that of the air around him are continuous. Based on Aristotle's *Physics* (V, 4, 228a20–22), Oresme maintains that in order for two bodies to be continuous,

---

63 Aristotle, *Physica*, VIII, 1, 252 a 12–13, *Translatio Iacobi Veneti* (translatio vetus), eds. F. Bossier and J. Brams, Leiden/New York 1990 (Aristoteles Latinus VII 1–2) (2 vols.), 1, 282[11]: "At vero nichil inordinatum est eorum que natura et secundum naturam sunt"; *Translatio Guillelmi de Morbeka* (translatio noua), *AL* VII 3, 373[10]: "At vero nihil inordinatum eorum quae natura et secundum naturam sunt".

it is not enough for them to be juxtaposed: they must form a substantial unity (I.3, 14: "ad hoc quod aliqua sint continua, requiritur quod ultima eorum sint unum per se"). This is far from being the case with Socrates' matter and that of the air surrounding him, which possess different forms. The substitution of the term "consequentia" for the term "antecedens" should thus be considered as an error in manuscripts of family β.

The following paragraph of question I.3 contains a substitution that is easily explainable from a paleographical point of view. Manuscripts of family β read "taliter" instead of "totaliter" in the following passage: "sciendum est quod illud dicitur fieri a motibus supercelestibus quod fit ab intelligentiis et corpore celesti totaliter moto". The reading "taliter" (in this way) does not seem appropriate in this context, as the preceding sentences did not expand upon celestial motions. The reading "totaliter" fits much better, since it implies a distinction that Oresme will employ in a subsequent question (I.6), namely the distinction between the particular motions of the planets and the general, unique motion of the celestial sphere, which is the ultimate cause of all of the effects in the sublunar region.

Question I.8, which concerns the relationship between motion and heat, contains another passage in which a quite uncommon word of *P*—at least, in the field of meteorology—has been replaced in manuscripts of family β by a more widely used one in this context. One of the arguments against the principle that motion generates heat is that a hot soup cools down once it is moved. Manuscript *P* reads: "videmus quod poreta calida, cum movetur, frigefit; unde videtur quod motus localis sit causa frigiditatis". Manuscripts of family β replace the term "poreta" (soup) with "Boreas", the Greek name of a northern wind. This substitution allows manuscripts of family β to eliminate an unexpected term in favour of another, graphically not so distant one, which seems to fit better in the semantic field of meteorology, since the term "Boreas" occurs several times in the second book of Aristotle's *Meteorology*. The reasoning is still applicable, because one could admit that as Boreas blows, it chills hot things: in this case, the adjective "calida" would no longer be a nominative feminine singular, but an accusative neuter plural. Nevertheless, the answer to the argument, at the end of the question, is better explained by the *lectio* "poreta". Manuscript *P* states in fact that once the soup is moved (let us say, with a spoon), it cools down because it comes into contact with cold particles of air. In this sense, one can say that motion generates cold ("Ad primam, de motu porete, dicitur quod, cum poreta movetur, est in alio et alio aere continue recenti et frigido, qui plus infrigidat quam ex tali motu possit calefieri"). Manuscripts of family β replace the term "poreta" with the term "Boreas" also in the answer, which makes the explanation less comprehensible.

Another example of the superiority of *P* over the manuscripts of family β can be found in question I.9, in which Oresme inquires whether light generates heat. In the answer, Oresme states that fire is one of the luminous bodies ("unum corpus de corporibus lucidis"). Instead of this expression, most of the manuscripts of family β (*Be Kl Wr Ba* Kr^1 Kr^2 *M Sg*; *Wi* omits this passage) have an incorrect reading: "minimum corpus de corporibus magnis". The reading "minimum" instead of "unum" can be explained by the graphical resemblance between the two abbreviated words in gothic handwritings. The replacement of "lucidis" by "magnis" could be the result of the first error: we can suppose that the copyists, finding this statement unusual, tried to interpret and to correct the rest of the sentence. Some fifteenth-century manuscripts present a reading that, without being completely correct, is closer to *P* than to the majority of the manuscripts in family β. The reading of *L* is namely "minimum corpus de corporibus" (without "magnis"); manuscripts *Ka*, Kr^3 *Po* and *U*, for their part, have "unum corpus de corporibus magnis".

The same question contains another passage in which manuscript *P* proves better than manuscripts from family β. A possible objection against the thesis according to which light generates heat is that some luminous bodies, such as fish scales and rotten wood, do not produce heat ("Octavo: multa lucent, sicut squame piscium et putredines quercuum, que tamen non sunt productiva caloris"). Oresme, who maintains that light generates heat, explains that it is not to be excluded that these bodies too can produce a small amount of heat. Moreover, it is possible that they have a cooling effect (*virtus frigefactiva*) that is stronger than the heating one.[64] At the end of the question, Oresme replies to the objections against the thesis according to which light generates heat. When it comes to the argument about fish scales and rotten wood, he refers to the body of the question (*corpus questionis*) instead of repeating the explanation he had already given. Manuscript *P* has the following reading: "Ad octavam, de squamis piscium, dictum est in corpore questionis". Manuscripts of family β, for their part, transmit variants of this reading: "Ad octavam, de squamis piscium, videbitur in questione sequenti". Now the following question does not deal with light, but with the process of antiperistasis, namely the intensification of a quality due to the action of the contrary quality (I.10, *Utrum contrarium circumstans suum contrarium fortificet ipsum*). As Oresme does not come back to the heat-

64 I.9, 20: "De aliis lucidis, sicut sunt squame piscium, potest dici quod etiam lux ipsarum calefacit, sed hoc est bene remisse; et forte habent quasdam alias virtutes fortiores frigefactivas, quarum actio per huiusmodi lucem vel lumen remittitur; et ergo, quia virtus eius est fortior, aliquando effectus eius apparet et non effectus lucis, quamvis forte intensius appareret si non obstaret lumen".

ing effect of fish scales and rotten wood in the rest of the text, we can consider the variant of family β as erroneous.

In question I.8, Oresme shows that the circular motion of the sphere of fire and of the upper part of the sphere of air can be considered neither as natural nor as violent, but preternatural ("preter naturam"). While speaking about comets, in question I.15, Oresme refuses to admit that the circular motion of these bodies is a proof of their celestial nature. In doing so, he refers to the example of the sphere of fire and the upper part of the sphere of air, which both have a circular motion while being of a terrestrial nature: "Nec propter hoc oportet quod cometa sit de natura celi non plus quam aer vel ignis: talis enim motus non inest istis naturaliter, sed inest eis preter naturam, sicut prius dicebatur". Only manuscript *P* has the correct reading, since he writes that the circular motion belongs to these spheres "preter naturam". All other manuscripts state that the motion of the sphere of fire and the upper part of the sphere of air is violent, which does not correspond to Oresme's opinion in question I.8.[65]

At the end of question I.15, Oresme states that sometimes a comet takes the shape of a star, particularly when it is round: "sed tunc apparet quod sit una stella celi propter eius rotunditatem". Manuscript *P* is the only witness which reads: "rotunditatem". All other manuscripts replace this term with "profunditatem", which makes no sense. If Oresme had meant to say that a comet could be mistaken for a star because of its distance from the Earth, he would have used the term "distantia".

In the following question, I.16, which also concerns comets, Oresme exposes a theory according to which the matter that constitutes these bodies is similar to the celestial one. In this sense, one could say that this substance does not shine because it is fiery, but because it is very similar to the matter of the stars, which are luminous: "Ulterius ymaginatur quod nec cometa fit per inflammationem huius exalationis ab igne, sed a natura alicuius planete, ita quod ista opinio ymaginabatur quod cometa est unum corpus valde simile nature stellarum, nec est ibi proprie *inflammatio*, sed illud corpus fit sic lucidum quia valde declinat ad naturam stellarum". Manuscripts of family β replace the term "exalatio" with the term "inflammatio".

In the same question, Oresme excludes the possibility that the comet has a celestial nature. His main argument is that if this were the case, a comet would disappear under only two circumstances: 1) if it came too close to the

65 See for instance *L*: "talis enim motus non est ignis naturalis, sed violentus, ut prius dicebatur"; *Kl*: "talis enim motus non est naturalis in istis, sed violente, sicut prius dictum est"; *U*: "cum motus non est in istis naturalis, sed violenter, ut prius dicebatur".

Sun, or 2) if it distanced itself too much from it. In the first case, the comet would be outshone by the Sun; in the second case, the comet would become invisible because it would no longer reflect the light of the Sun. Now, continues Oresme, a comet often disappears independently from its proximity to the Sun ("absque eius approximationem ad Solem"): "cometa non est de natura celi. Patet: nam, si sic, numquam deficeret a lumine nisi per approximationem et accessum eius ad Solem, sicut et alie stelle; sed hoc falsum. Patet falsitas, quia sepe cometa visus est deficere a lumine absque eius approximationem ad Solem". Manuscript *P* is the only one which has the correct reading "absque eius approximationem ad Solem". Manuscripts of family β invert the sense of the reasoning and invalidate it, since they state that comets have been seen disappearing because they came too close to the Sun or moved away from it. Manuscripts *Ka Kl L M Sg U* read for instance: "propter hoc quod approximabat ad Solem"; manuscript *Ba* has: "quando recessit a Sole".

Another interesting example can be found in question I.19, in which Oresme deals with the Milky Way. Oresme discusses an ancient theory on the Milky Way refuted by Aristotle. The adherents of this theory believed that the parts of the sky in which there are many stars attract a great quantity of exhalation. They claim that the exhalation acts as a mirror reflecting the light of the Sun ("tamquam speculum reflectens lumen Solis") and the Milky Way would result from this phenomenon. The reading "tamquam speculum reflectens lumen Solis" can be found only in manuscript *P*. Manuscripts of family β have instead: "tamquam speculum reflectens istam exalationem". This variant is incorrect, because the exhalation is not the reflected object, but the reflective one!

In question II.1, Oresme asks whether rain is produced in the middle region of the air ("media regio aeris"). Among the arguments against this thesis, he states that fiery phenomena are produced in the middle region of the air, which seems to exclude the generation of humid phenomena in the same part of the atmosphere. Having shown how rain is produced in the middle region of the air, Oresme explains that both humid and fiery phenomena can be generated in this region of the atmosphere, and that the difference concerns the process. While humid phenomena are generated through the condensation of vapor, fiery phenomena are generated through the ejection (*extrusio*) of combustible matter downwards. The technical term "extrusio" is used in the Greek-Latin translation of the *Meteorologica* when discussing lightnings and thunders.[66] The *lectio difficilior* "extrusio" is attested only in manuscript *P*: "ad secundam:

66 Aristoteles, *Meteorologica*, II, 9, 369 a 21–22, *Translatio Guillelmi de Morbeka* (translatio noua), *AL* X 2.2, 82[971]: "Sed ad contrarium inspissationis necessarium fieri extrusionem, uelut pyrenes ex digitis exilientes".

'ibi generantur impressiones ignite', dico quod est alio modo, quia iste generantur ibi per extrusionem; aquee autem per condensationem". Manuscripts *Ba, Ba¹, Kl, Kr, Kr¹, Kr², M, Wi, Wr* read "incensionem" (manuscripts *Wi* and *Ba* add "ipsius exalationis"). Manuscripts *Ka, Kr³, Po, L, Sg, U,* have variants of this *lectio*: "intensionem vel extensionem".[67] No manuscript except *P* transmits the correct *lectio difficilior* "extrusionem".

In question II.2, Oresme asks whether rain, snow, dew and frost are of the same nature. Following Aristotle, he confirms this thesis and explains that the differences between these phenomena are only accidental. Manuscript *P* is the only witness that has a good reading of this passage: "ad quartam dico quod huiusmodi bene habent descriptiones accidentales differentes, non autem substantiales". Manuscripts *Kl* and *Wr*, along with manuscript *L*, have an incorrect *lectio*. While the first reads: "huiusmodi bene habent descriptiones, non autem essentiales"; the latter has: "huiusmodi bene habent descriptiones, non autem essentiales diffinitiones". Manuscripts *Ba, Ba¹, Kr, Kr¹, Kr², Wi* have a different, but equally incorrect *lectio*, since they transmit variants of this sentence: "huiusmodi sunt bene descriptiones, sed non sunt diffinitiones essentiales".[68] Manuscript *Kr* has a different, better text: "ad quartam dico quod huiusmodi habent bene descriptiones differentes, non autem dispositiones essentiales". Nevertheless, like the majority of manuscripts of family β, it omits the key word "accidentales". This term can be found only in manuscripts *Ka, Kr³, M¹ Po* and *U*, which read: "huiusmodi bene habent descriptiones accidentales, non autem essentiales".[69] In any case, no manuscripts of family β transmits the correct reading.

In other cases, manuscripts of family β contain readings that, while not being completely wrong, can be considered as *lectiones faciliores*. This occurs for instance in the following passage of question I.3: "Sol per accessum ad nos et recessum in obliquo circulo est causa generationis et corruptionis multorum in hoc mundo". Manuscripts of family β write "a nobis" instead of "in obliquo circulo", which implies a repetition and a loss of information.

67 *Sg*: "intensionem vel extensionem"; *Be, Ka, Kr³Po, Sg, U*: "vel per intensionem vel per extensionem"; *L*: "vel per intensionem vel extensionem"; *M¹*: "extensionem".

68 *Ba*: "huiusmodi sunt bene descriptiones, sed non sunt diffinitiones essentiales"; *Wi*: "huiusmodi sunt bene descriptiones, non autem diffinitiones essentiales"; *Kr¹*: "huiusmodi bene habent descriptiones, non autem diffinitiones essentiales"; *Kr²*: "huiusmodi bene habent descriptiones, non autem diffinitiones essentiales". *Sg* and *Be*; "huiusmodi termini habent descriptiones, non autem essentiales diffinitiones".

69 The variant transmitted by *M¹* can be considered as a corruption of the variant typical of this group: "ad quartam dico quod huiusmodi bene probat quod habent descriptiones accidentales, non autem essentiales".

The Central European Group: The Opposition between Family γ and δ

The opposition between δ and γ is proved by a certain number of omissions and errors shared by γ, but not by δ. For example, in the following cases, manuscripts of group γ omit the words in italics, while manuscripts of group δ transmit them:

I.8, 6 deinde specialiter probatur quod motus celi non calefacit *speram ignis*
I.16, 3 tertio ad idem: si esset aliqua talis exalatio ignita, aliquando deberet eclipsari, quia aliquando esset in umbra Terre *causata ex obiectione Terre ad Solem.*
II.1, 8 aliquando infra mediam regionem vapor elevatus, qui de natura sua est frigidus, quamvis accidentaliter sit calidus calore Solis, potest per antiparistasim reduci ad frigiditatem sibi naturalem et condensari in nubem *et converti in aquam* et sequitur pluvia.

The first question of the second book provides us with a very interesting example of the opposition between family γ and family δ. Oresme explains that rain can be generated in three ways. The text of *P* reads as follows:

> Quantum ad secundum, patet ex iam dictis quid sit modus generationis pluvie. Sciendum tamen quod tripliciter possunt generari nubes et pluvie. Uno modo per condensationem vaporis elevati in nubem; *secundo modo per condensationem aeris in nubem* a frigiditate medie regionis; tertio modo per repulsionem vaporis a media regione.

Manuscripts *Kl* and *Wr* omit the second modality by homoioteleuton:

> Sciendum tamen quod tripliciter possunt generari nubes et pluvie. Uno modo per condensationem vaporis elevati in nubem a frigiditate medie regionis [ingrossatus *add. Wr*] secundo modo per repulsionem vaporis a media regione.

A copyist must have noticed this anomaly and tried to remedy it by substituting "dupliciter" for "tripliciter". This variant is attested by manuscripts *Ba, Ba¹, Kr, Kr¹, Kr², M, Wi*. Only manuscripts of sub-family δ transmit the correct version.[70] We can thus distinguish three groups: 1) manuscripts which have the correct

70 Interestingly, manuscript *L* exposes the three modalities of generation of rain, but announces this distinction with the adverb "dupliciter", instead of "tripliciter".

version (*P* and sub-family δ); 2) manuscripts which contain a homoioteleuton, namely *Kl* and *Wr*—the text of these manuscripts speaks of three modalities of the generation of rain, but mentions only two; 3) manuscripts which speak of two, and not three, modalities of the generation of rain. This is the case of all manuscripts of family β with the exception of manuscripts *Kl* and *Wr* and of sub-family δ. This lead us to infer that β transmitted the correct version (with the three modalities) to δ and to γ. The latter omitted the second modality by homoioteleuton, but kept the adverb "tripliciter". This version is transmitted in manuscripts *Kl* and *Wr*. The father of manuscripts *Ba, Ba*^{1}*, Kr, Kr*^{1}*, Kr*^{2}*, M, Wi* noticed this inconsistency and corrected the adverb, writing "dupliciter" instead of "tripliciter".

The Subfamily *Ba Ba*^{1} *Kl Kr*^{1} *Kr*^{2} *Wi* (η)

Within family γ, we can distinguish an important group of manuscripts, namely *Ba Ba*^{1} *Kr Kr*^{1} *Kr*^{2} *Wi*, which form a subfamily that I will call η. In what follows, I will present the list of shared accidents of this family based on the sample collated text:

Omissions of family η

1) I.1, 1 de se *om. Ba Ba*^{1} *Kr Kr*^{1} *Kr*^{2} *Wi*
2) I.1, 1 sicut patet de se *om. Ba Kr Kr*^{1} *Wi* (sicut patet *Kr*^{2})
3) I.1, 3 de se *om. Ba Ba*^{1} *Kr Kr*^{1} *Kr*^{2} *Wi*

Variants of family η

1) I.1, 1 de impressionibus metheorologicis] de eis *Ba Ba*^{1} *Kr Kr*^{1} *Kr*^{2} *Wi*
2) I.1, 1 nota est] patet *Ba Ba*^{1} *Kr Kr*^{1} *Kr*^{2}
3) I.1, 2 de impressionibus metheorologicis non potest simul haberi scientia et opinio] questio falsa *Ba Ba*^{1} *Kr Kr*^{1} *Kr*^{2} *Wi*
4) I.1, 2 aristotelem] philosophum *Ba Ba*^{1} *Kr Kr*^{1} *Kr*^{2} *Wi*
5) I.1, 3 impressionibus metheorologicis] eis *Ba Ba*^{1} *Kr Kr*^{1} *Kr*^{2} *Wi*
6) I.1, 3 non est possibile de eis simul esse scientiam et opinionem] questio falsa *Ba Ba*^{1} *Kr Kr*^{1} *Kr*^{2} *Wi*
7) I.1, 4 impressionibus metheorologicis] ipsis *Ba Ba*^{1} *Kr Kr*^{1} *Kr*^{2} *Wi*
8) I.1, 4 tenet] patet *Ba Ba*^{1} *Kr*^{1} *Kr*^{2}
9) I.1, 4 impressiones metheorologice] ipse *Ba Ba*^{1} *Kr Kr*^{1} *Kr*^{2} *Wi*
10) I.1, 5 inordinate] inordinata *Ba Ba*^{1} *Kr Kr*^{1} *Kr*^{2} *Wi*
11) I.1, 5 est nota] patet *Ba Ba*^{1} *Kr Kr*^{1} *Kr*^{2} *Wi*
12) I.1, 5 aristotelem] philosophum *Ba Ba*^{1} *Kr Kr*^{1} *Kr*^{2} *Wi*
13) I.1, 6 assensus autem] et talis assensus *Ba Ba*^{1} *Kr Kr*^{1} *Kr*^{2} *Wi*
14) I.1, 6 allegabatur] allegatum est *Ba Ba*^{1} *Kr Kr*^{1} *Kr*^{2} *Wi*

15) I.1, 6 aristotelem] philosophum *Ba Ba¹ Kr Kr¹ Kr² Wi*
16) I.1, 6 syllogismus dyalecticus et syllogismus demonstrativus dyalecticus quidem qui generat opinionem demonstrativus vero qui generat scientiam qui generat scientiam] syllogismus dyalecticus qui generat opinionem et syllogismus demonstrativus qui generat scientiam igitur etc. *Ba Ba¹ Kr Kr¹ Wi* (*om. Kr²*)
17) II.1, 2 ex eo quod] eo quod *Ba Ba¹ Kr Kr1 Wi*] ex se quod *M*] ex eo quia *Be L Ka Kr³ Po Sg*
18) II.1, 3 exalationis] exalationum *Ba Kr Kr²*
19) II.1, 4 aliquando] quandoque *Ba Ba¹ Kr Kr¹ Kr² Wi*
20) II.1, 4 nam] quia *Ba K Kr¹ Kr² Wi*
21) II.1, 5 in primo tractatu et primo capitulo huius secundi] in primo capitulo tractatus primi secundi huius *Ba Kr Kr¹ Kr² Wi*] in primo capitulo tractatus primi huius secundi *Kl*] in primo capitulo tractatu primo huius secundi libri *Wr*] in primo capitulo tractatus primi secundi *L*] in primo capitulo primo tractatu secundi primi libri secundi *M*] libro secundo capitulo primo tractatus primi *Be Sg*] in primo capitulo tractatus primi huius secundi libri *Po U*] primo in primo capitulo tractatus primi secundi huius Ba¹] in primo capitulo tractatus primi huius secundi *Ka*
22) II.1, 6 primo exponende sunt alique conclusiones pro quesito] primo videndum est de loco generationis pluviarum *Ba Ba¹ Kr Kr¹ Kr² Wi*
23) II.1, 6 quoad figuram] de figura *Ba Ba¹ Kr Kr¹ Kr² Wi*
24) II.1, 6 de diversitate pluviarum quod quantitatem] de quantitate pluviarum *Ba Ba¹ Kr Kr¹ Kr² Wi*
25) II.1, 6 de diversitate pluviarum quoad modum cadendi guttarum] de modo cadendi pluviarum *Ba Ba¹ Kr Kr¹ Kr² Wi*
26) II.1, 6 de pluviis quoad loca et regiones ubi fiunt] de regionibus in quibus solent cadere pluvie *Ba Ba¹ Kr Kr¹ Kr² Wi*
27) II.1, 6 in media regione aeris] ibi *Ba Ba¹ Kr¹ Kr² Wi*

Additions of family η

II.1, 3 et] deinde *add. Ba Ba¹ Kr Kr¹ Kr² Wi*

Looking at the most significant variants, we notice that the members of family η often present certain abbreviated forms in order to avoid repetitions. For example, they employ the pronominal forms *eis* and *ille* instead of repeating *impressionibus metheorologicis* and *impressiones metheorologice* (examples 5, 7). Similarly, this subfamily writes "questio falsa", instead of repeating the final formula for the conclusion (example 6). Another feature of the manuscripts of this group is that they always refer to "Aristoteles" as "Philosophus" (examples 4, 12 and 15).

But what is more interesting is that the manuscripts of group η frequently organize the text differently from the other witnesses. This is already evident in the first question, as these manuscripts omit the *divisio textus* at the beginning of the answer (*P*: "In ista questione primo videndum quid est scientia et quid opinio; secundo, quot modis aliquid dicitur 'scibile' et 'opinabile'; tertio, an idem sit scibile et opinabile; quarto respondendum est ad quesitum"). Consistently with this choice, manuscripts of group η do not introduce the different articles of the answer with expressions like "quantum ad primum" and "quantum ad secundum", as the other witnesses do, but only with generic formulas such as "notandum est". Sometimes, the difference in the articulation of the text between subfamily η and the other witnesses goes deeper, as shown by the following example taken from question I.2:

Text of the other witnesses[71]	**Text of family η**
Quantum ad tertium, sit prima conclusio: universum et omnia que sunt in ipso sunt bene ordinata loco, tempore et proportione, comparando quodlibet eorum ad omnia alia.	*Tunc etiam ponuntur alie conclusiones.* Prima est: universum et omnia que sunt in ipso sunt bene ordinata loco, tempore et proportione. Patet, quia dicitur duodecimo *Metaphysice*: entia nolunt male disponi.
Probatur, quia omnia sunt bene disposita que sunt in universo; igitur omnia que sunt in universo sunt bene ordinata. Consequentia clara est de se. Antecedens patet, quia duodecimo *Metaphysice* dicitur quod entia nolunt male disponi.	*Secundo probatur* eadem conclusio, nam elementa sunt bene disposita que sunt in mundo et in universo; igitur conclusio est vera. Consequentia tenet et antecedens patet per auctoritatem allegatam ex duodecimo *Metaphysice*.
Secundo: omnia que sunt in universo sunt disposita secundum quandam consonantiam et armonicam proportionem; igitur omnia que sunt in universo sunt bene ordinata.	*Tertio*: omnia que sunt in universo sunt disposita secundum consonantiam et armonicam proportionem; igitur conclusio vera.

Just as in question I.1, the *divisio textus* at the beginning of the answer is missing (*P*: "In ista questione primo videndum est quid sit intelligendum per 'impres-

71 I have followed the text of *P*, but I collated it with that of the other witnesses.

siones metheorologice'; secundo videndum est qualiter fiant secundum naturam; tertio videndum est quali ordine fiant"). As we have seen above, this fact determines the omission of introductory formulas such as "quantum ad primum" and "quantum ad secundum", in manuscripts of family η. The conclusion referred to as the first of the third article by the other manuscripts is therefore presented in manuscripts of family η as the first among "alie conclusiones". But there is more. In manuscripts of family η, the demonstration of the first conclusion is subdivided into three arguments rather than two. However, if we compare the third argument in family η with the second argument in the other manuscripts, we notice that this difference does not affect the content of the text, but only its structure.

The members of this family also omit the *divisio textus* of questions I.3, I.5, I.6, I.18. In question I.10, the *divisio textus* is reorganized and slightly reworded. In this case the *divisio textus* consists of two paragraphs (the first beginning with the words: "in ista questione primo videndum est"; the second beginning with the words: "quantum ad primum"). The order of the two paragraphs in family η is inverted with respect to the rest of the tradition. Moreover, the text has been slightly modified:

Text of the other witnesses	**Text of family η**
⟨1⟩ *In ista questione primo videndum est* an unum contrarium possit movere aliud localiter; secundo videndum est de modo per quem unum contrarium movet suum contrarium localiter; tertio videndum est an ex condensatione alicuius subiecti caliditas vel frigiditas illius subiecti intendantur.	⟨2⟩ *Notandum primo quod* questio et ea que dicentur in questione debent intelligi de qualitatibus primis, sicut sunt caliditas, frigiditas, humiditas et siccitas, et non de qualitatibus secundis, sicut sunt albedo, nigredo et consimilia. Unde albedo posita iuxta nigredinem non facit ipsam intensiorem, sed facit eam bene magis apparere: contraria enim iuxta se posita magis elucescunt.
⟨2⟩ *Quantum ad primum, primo sciendum est* quod questio et illud quod dicetur in ista questione debet intelligi de qualitatibus primis, sicut sunt caliditas, frigiditas etc., et non de qualitatibus secundis, sicut sunt albedo, nigredo et consimilia. Unde albedo posita iuxta nigredinem non facit ipsam intensiorem, sed bene facit quod	⟨1⟩ *In ista questione taliter intellecta* primo videndum est utrum contrarium possit movere aliud localiter et fugare ipsum; secundo videndum est quomodo unum contrarium movet aliud localiter; tertio, utrum ex condensatione alicuius subiecti fiat fortificatio et intensio qualitatis et virtutis existentis in ipso.

(*cont.*)

Text of the other witnesses	Text of family η
apparet intensior: contraria enim iuxta se posita magis elucescunt.	

While in manuscript *P* and in the other witnesses of family β the *divisio textus* is followed by an explanation of the qualities concerned by the mechanism of antiperistasis, manuscripts of subfamily η invert these two points. Consistently with this inversion, the explanation concerning the qualities is not introduced by the expression "quantum ad primum", but, more generally, by the verb "notandum". Moreover, having introduced this explanation before the *divisio textus*, the latter opens with the words "in ista questio taliter intellecta". We thus notice that the manuscripts of subfamily η not only structure the text differently, but also reformulate it accordingly.

In question I.12, the *divisio textus* transmitted by the manuscripts of subfamily η differs slightly from the version that can be found in the other witnesses:

Text of the other witnesses	Text of family η
Breviter in ista questione videndum est quid sit exalatio et quid vapor; secundo videndum est de comparatione istorum ad invicem; tertio respondendum est ad quesitum.	In ista questione videndum est de natura exalationum et vaporum; secundo de comparatione ipsorum ad invicem; tertio qualiter et quarum impressionum ipse sunt materie.

The same applies to the following question, I.13, in which the gap between the manuscripts of family η and the other witnesses further increases:

Text of the other witnesses	Text of family η
In ista questione videndum est quomodo diversificantur huiusmodi impressiones; secundo videndum est de causis harum	In ista questione numerande sunt experientie; secundo dicetur de causis illarum impressionum; tertio de modo genera-

(*cont.*)

Text of the other witnesses	Text of family η
impressionum et in generali et in speciali, quibus visis patet faciliter responsio ad quesitum.	tionis et variationis; quarto movebuntur dubia.

In conclusion, the differences between family η and the other witnesses do not concern doctrinal elements, but only the argumentative structure. These variants are not isolated and did not occur by accident during the copying process, but derive from a systematic reworking of the text. In some cases, the structure of the text is made more explicit and more formulaic, so that every paragraph begins with a piece of metadiscourse allowing the student and teacher to know exactly where in the text they are. It is therefore possible that the original text of this family represents a slightly modified version of Oresme's *Questions*. As indicated by the *colophona* of the manuscripts, this version circulated at the University of Prague.

Within family η, manuscripts *Ba* and *K* contain some variants which stand out from the other witnesses:

1) I.1, 2 probatur] etiam *add. Ba Kr*
2) I.1, 3 cum] quia *Ba Kr*
3) I.1, 3 sunt complexa que qualitercumque sit sic significent] nec qualitercumque est fuit vel erit significent *Ba Kr*
4) I.1, 5 non videtur posse haberi] non potest haberi *Ba Kr*
5) I.1, 5 autem *om. Ba Kr*
6) I.1, 6 circa finem primi posteriorum] primo posteriorum circa finem *Ba Kr*
7) I.1, 6 per subiectum *om. Ba Kr*
8) II.1, 2 generationis impressionum ignitarum] impressionum ignitarum *Ba Kr*
9) II.1, 2 generationis *om. Ba Kr*

Manuscripts *Ba*[1] and *Kr*[1] are, for their part, very close to each other, since they share some unique variants:

1) I.1, 2 eo] istis *Ba*[1] *Kr*[1]
2) I.1, 3 non sint complexa que qualitercumque sit sic significant] non sint complexe nec qualitercumque fuit est vel erit sic significant *Ba*[1] *Kr*[1]

Manuscripts *Kl* and *Wr*

Manuscripts *Kl* and *Wr* display a certain number of shared accidents.

Variants shared by manuscripts *Kl* and *Wr*

1) I.1, 2 sed non est idem quod possibile est aliter se habere et quod impossibile est aliter se habere] sed non est idem impossibile aliter se habere et possibile aliter se habere *Kl Wr*
2) I.1, 6 quod autem de ipsis potest esse opinio patet] et patet *Kl*] ut patet *Wr*
3) I.1, 6 syllogismus dyalecticus et demonstrativus dyalecticus quidem qui generat opinionem demonstrativus vero qui generat scientiam] syllogismus dyalecticus qui generat opinionem et syllogismus demonstrativus qui generat scientiam *Kl Wr*
4) II.1, 6 pluviarum] pluviali *Kl Wr*
5) II.1, 6 diversitate] varietate *Kl Wr*
6) II.1, 7 frigiditas] frigus *Kl Wr*

Additions shared by manuscripts *Kl* and *Wr*

1) I.1, 6 alii] scilicet quod de impressionibus metroloycis potest esse scientia et opinio *add. Kl Wr*[72]
2) I.1, 6 opinio] igitur etc. *add. Kl Wr*
3) II.1, 1 debet] primo *add. Kl Wr* (this variant is shared also by manuscript *M*)
4) II.1, 1 sibi] non *add. Kl Wr* (manuscript *M* has "non" instead of "sibi")

Omissions shared by manuscripts *Kl* and *Wr*

I.1, 6 posito enim] posito *Kl Wr*

Inversions shared by manuscripts *Kl* and *Wr*

I, 6 a te scite] scite a te *Kl Wr*

In some cases, *Kl* and *Wr* are the only witnesses sharing the variants of *P*:

I.1, 5 inordinatiorem] inordinatam *P Wr Kl*

In addition to the variants of *Kl*, *Wr* presents some individual accidents:

I.1, 4 corruptibiles] corporales *Wr*

I.1, 6 possibile sit de eis] de ipsis possibile sit *Kl*] de impressionibus possibile sit *Wr*

It is therefore possible that *Wr* was copied from *Kl*.

72 It is worth noticing that in this passage *L* has a similar addition: "scilicet quod de impressionibus metheorologicis posset esse opinio et scientia".

Manuscript *M*

Within group γ, manuscript *M* has a special place, since it contains a lot of unique accidents, the majority of which can be considered as errors:

1) I.1, 2 impressionibus] scientia *M*
2) I.1, 3 sunt complexa que qualitercumque sit sic significant] non sunt complexa nec qualitercumque fiunt sicut fiunt *M*
3) I.1, 4 esse] haberi *M*
4) I.1, 5 est nota] nota *M*
5) I.1, 5 non debent fieri inordinate extra semper et frequenter] non debent fieri ordinate extra semper et frequenter *M*
6) I.1, 5 patet] arguitur *M*
7) I.1, 6 opinionem *om. M*
8) I.1, 6 propositionis] talis *add. M*
9) I.1, 6 ipse enim est] quia pater est M (all other manuscripts of family β read: quia pater tuus est)
10) I.1, 8 videndum est *om. M*
11) II.1, 1 primo quod non *om. M*
12) II.1, 1 sibi] non *M* (sibi non *Kl Wr*)
13) II.1, 2 idem non est] non videtur esse idem *M* (idem non videtur esse *Ba* Ba^1 *Be Kl Kr* Kr^2 Kr^3 *U Sg Wi Wr*)
14) II.1, 2 hoc esset per istum modum quod *om. M*
15) II.1, 4 aeris *om. M*
16) II.1, 4 generationis] generatio *M*
17) II.1, 4 videt] videns *M*]
18) II.1, 4 et hoc] secundo *M*
19) II.1, 4 generari] generare *M*
20) II.1, 7 sed *om. M*
21) II.1, 7 condensare] congelare *M*

Curiously, manuscript *M* often adds or omits certain prefixes that invert the sense of the text. Sometimes these inversions are the result of the substitution of one word by another. The italicized words in the following examples indicate the major differences between the other witnesses and *M*:

1) I.1, 5 non debent fieri *inordinate* extra semper et frequenter] non debent fieri *ordinate* extra semper et frequenter *M*
2) I.1, 14 quarta conclusio *impossibile* est ab aliquo idem complexum simul sciri et opinari] quarta conclusio *possibile* est de aliquo idem simul sciri et opinari *M*
3) I.2, 1 et dicit ibidem commentator *impossibile* est ut famosum sit falsum secundum totum] et ibidem dicit commentator quod *possibile* est ut famosum sit falsum secundum totum *M*

4) I.2, 10 secundo sciendum est quod huiusmodi impressiones fiunt sub orbe lune in regione elementari et ut in pluribus fiunt *super* nos] secundo est sciendum quod impressiones fiunt sub orbe lune in regione elementari et ut in pluribus fiunt *sub* nos *M*
5) I.3, 15 negetur *antecedens*] negetur *consequentia M*
6) I.3, 19 quia lumen est qualitas *sensibilis*] quia lumen est qualitas *insensibilis M*
7) I.3, 22 ad primam concedo quod mundus iste inferior *non est* continuus lationibus superioribus] ad primam concedo quod iste mundus inferior *est* continuus celo *M*
8) I.5, 5 quinto alique veritates sunt ita evidentes quod intellectus non potest eis assentire cum formidine ergo circa illas non contingit *opinio*] quinto alique veritates sunt ita evidentes quod intellectus non potest eis assentire cum formidine ergo circa illas non contingit *scientia M*
9) I.8, 7 sed quia igni naturaliter inest moveri sursum sequitur quod sibi non inest naturaliter moveri circulariter et per consequens *inest* sibi violenter] sed quia igni naturaliter inest moveri sursum sequitur quod sibi non inest naturaliter moveri circulariter et per consequens *non inest* sibi violenter *M*

Because these transformations occur systematically, one is tempted to believe that Iohannes Krickpaum, the scribe, did this on purpose. In fact, his goliardic temperament is witnessed by the colophon, in which he claims to have copied the text with his feet, instead of with his hands (see above, the description of the manuscript).

The Group *Be Ka Kr*^{3} *L M*^{1} *Po Sg U* (δ)

As we have seen above, a certain number of accidents within family β counterpose manuscripts *Ba Ba*^{1} *Kl Kr Kr*^{1} *Kr*^{2} *M Wi* (γ) to manuscripts *Be Ka Kr*^{3} *L M*^{1} *Po Sg U* (δ), which were all copied in the fifteenth century. The collated passage from the first question confirms the existence of family δ:

1) I.1, 3 esse scientia] haberi scientiam *Be Ka Kr*^{3} *L Po U*
2) I.1, 5 non videtur posse haberi] non videtur posse esse *Be Ka Kr*^{3} *Po Sg U*
3) I.1, 5 in littera] in primo illius libri *Be Ka Kr*^{3} *L Po Sg U*
4) I.1, 6 est] esset *Ba Be Ka Po Sg U* (esse *Kr*^{3})
5) I.1, 6 esset] sit *Be Ka Kr*^{3} *Po Sg U*
6) I.1, 6 simul esse] esse *Ka Kr*^{3} *L Po U*

Manuscripts *L Sg* (ε)

Within this group, *L* displays a quite individual behaviour. This manuscript is particularly close to *Sg*, as shown by these unique shared accidents:

I.2, 20 coniunctio] effectus *L Sg*
I.2, 21 fiunt] difformiter se habent *L Sg*
I.11, 19 basse] spisse vel spissiores sive bassiores *L Sg* (spisse et spissiores *Ba Kl*)
I.11, 19 figure] figure sive dispositionis *L Sg* (dispositionis *Ba Kl*)
I.16, 21 coniunctione] vel ex commixtione *add. L Sg*
II.4, 1 tardior] minor *L Sg*
II.7, 4 nec hoc facit alias calor elevans illam aquam sursum] nec hoc fit per aliquem calorem trahentem vel elevantem illam aquam sursum *L Sg*

Manuscripts *Ka Kr³ L Po U* (ζ)

Within family δ, manuscripts *Ka Kr³* (*M¹*) *Po U* (ζ) are very close to each other, as is evident from the list of their common shared accidents:

Additions of family ζ

1) I.1, 1 de se] sed *add. Ka U*
2) I.1, 2 secundo] arguitur sic *add. Ka Po U*
3) I.1, 4 quarto] arguitur sic *add. Ka U* (sic *add. Po*)
4) I.1, 4 de] sed de *Ka K³* (et de *Be Sg U*)
5) I.1, 5 quinto] arguitur sic *add. Ka Kr³ U*] sic *add. Po*

Variants of family ζ

1) I.1, 2 sed non est idem quod impossibile est aliter se habere et quod possibile est aliter se habere] sed non est idem quod possibile est aliter se habere et quod impossibile est aliter se habere *Ka Kr³ Po U*
2) I.1, 3 esse scientia] haberi scientiam nec esse Ka Kr³ Po U
3) I.1, 3 non sunt complexa que qualitercumque sit sic significant] non sint complexa quia qualitercumque se habent sic significant *Ka Po U*
4) I.1, 6 de ipsis] de impressionibus metheorologicis *Ka Kr³ Po U*
5) I.1, 6 haberes opinionem de patre tuo] haberes de patre tuo scientiam *Ka Kr³ Po*] haberes scientiam de patre tuo scientia *U*
6) I.1, 6 tamen hoc non obstante] illo tamen non obstante *Ka Kr³ U*
7) II.1, 1 debet ibi] ibi debent *Ka M¹ Po U*
8) II.1, 6 pluviis] pluvia *Ka Kr³ Po U*
9) II.1, 7 ecce causas sufficientes in media regione aeris] ergo sunt in media regione aeris cause sufficientes *Ba Ba¹ Kl Kr M Wi*] igitur in media regione

aeris sunt cause sufficientes *Be Kr Kr¹ Kr² Sg*] ergo in media regione aeris cause sufficientes *Wr*] ideo sunt cause sufficientes in media regione aeris *Ka Kr³ M M¹ Po U*

10) II.1, 7 etiam] et *Ka Kr³ M¹ Po*

Omissions of family ζ

1) I.1, 3 ex primo] primo *Ka Po U* (*om. Kr³*)
2) I.1, 4 posse *om. Ka Kr³ Po*
3) I.1, 6 in oppositum est aristoteles et albertus et alii] in oppositum est aristoteles et alii *Ka Kr³*] in oppositum autem est aristoteles et alii *U*
4) I.1, 6 hoc maxime] maxime *Ka Kr³ Po*
5) I.1, 6 sed *om. Ka Kr³ Po*
6) I.1, 6 simul esse] esse *Ka Kr³ Po U* (manuscript *L* also shares this variant)
7) I.1, 6 et demonstrativus dyalecticus quidem qui generat opinionem demonstrativus vero qui generat scientiam] qui generat scientiam *Be Kr³ Po Sg U* (manuscript *M* also shares this variant)

Inversions and transpositions of family ζ

1) I.1, 2 consequentia tenet *inv. Ka Po U*
2) I.1, 2 sed non est idem quod impossibile est aliter se habere et quod possibile est aliter se habere] sed non est idem quod possibile est aliter se habere et quod impossibile est aliter se habere *Ka Kr³ U*
3) I.1, 3 non est possibile de eis simul esse] non est possibile de eis esse simul *Ka Kr³ U*
4) II.1, 1 debet ibi] ibi debent *Ka M¹ Po U*

Within this group, *Kr³* and *Po* are particularly close, as they share some unique variants:

1) I.1, 1 igitur questio est falsa] igitur etc. *Kr³ Po*
2) I.1, 6 iret] ambulat *Kr³ Po*] ambularet *Be L M Sg*] ambulans *Ba Ba¹ Kl Kr Kr¹ Kr² Wi Wr*] ambulet *Ka U*
3) I.1, 6 conclusio] propositio *Kr³ Po*
4) I.1, 6 istam conclusionem] de ista propositione *Kr³ Po* (de ista conclusione *Ka Kl Kr² U Wr*)
5) I.1, 6 posset esse] sit *K³ Po* (esset *Ba Ba¹ Be Ka Kl Kr Kr¹ Kr² L M Sg U Wi Wr*)
6) II.1, 3 est *om. Kr³ M¹ Po*

Manuscript *Kr³* seems to be the worst witness of this group.

Manuscripts *Be* and *Sg*

Manuscripts *Be* and *Sg* are particularly close to each other:

Additions shared by manuscripts Be Sg

1) I.1, 2 secundo] arguitur *add. Be Sg*
2) I.1, 2 ratione] sic *add. Be Sg*
3) I.1, 4 quarto] sic *add. Be Sg*
4) I.1, 4 autem ut videtur non possunt formari propositiones necessarie] non est scientia nec possunt formari propositiones necessarie *Be Ka Kr*[3] *Po Sg U*
5) I.1, 5 quinto] sic *add. Be Sg*
6) II.1, 1 non] nam *add. Be Sg*
7) II.1, 3 est] verum *add. Be Sg*

Omissions shared by manuscripts Be Sg

1) I.1, 1 igitur etc. *om. Be Sg*
2) I.1, 2 simul haberi] haberi *Be Sg*
3) I.1, 2 autem *om. Be Sg*
4) I.1, 3 generationis *om. Be Sg*
5) II.1, 6 guttarum *om. Be Sg*
6) II.1, 7 maior est nota *om. Be Sg*

Variants shared by manuscripts Be Sg

1) I.1, 0 circa librum metheororum qui est quartum naturalium queritur primo] circa primum metheororum queritur primo *Be Sg*
2) I.1, 1 sicut patet de se] sicut clarus est *Be Sg*
3) I.1, 3 non est possibile de eis simul esse] non est possibile de eis simul habere *Sg*] non est possibile simul de eis habere *Be*
4) I.1, 3 cum] nam *Be Sg*
5) I.1, 5 inordinatam] inordinatissimam *Be Sg*
6) I.1, 6 sed hoc non quia de eadem re possibile est] sed hoc non est impossibile ymo possibile est de eadem re *Sg*] sed hoc non est impossibile ymo possibile est de aliqua eadem re *Be*
7) I.1, 6 haberes opinionem de patre tuo] haberes de patre tuo opinionem *Be Sg*
8) II.1, 1 locum sibi naturalem] suum locum naturalem *Be Sg*
9) II.1, 4 existens] steterat *Be Sg*] stans *Ba*[1] *Ka Kl Kr*[1] *Kr*[2] *Kr*[3] *M U Wi Wr* (steteat *L* stetit *Ba*)
10) II.1, 4 videt] et vidit *Be Sg* (et videt *L*)
11) II.1, 6 ipsius pluvie] ipsarum *Be Sg* (and *L*)

12) II.1, 6 pluviarum quoad quantitatem] guttarum ipsius pluvie *Ka Kl Kr*^{3} *M U Wr* guttarum *Be Sg*
13) II.1, 6 quoad loca et regiones] quoad regiones et loca *Be Sg*] quoad regiones et ad loca *Kl L M Wr*
14) II.1, 6 prima] hec *Be Sg*
15) II.1, 7 vaporem elevatum] vapores *Be Sg* (and *Wr*)
16) II.1, 7 elevatum in nubem et convertere in aquam] et convertere in nubem et sic sequitur pluvia *Ba Ba*^{1} *Kl Kr*^{1} *Kr*^{3} *L Wi*] et convertere in nubem et ut sic sequitur pluvia *Kr*] et convertitur in nubem et sic sequitur pluvia *Kr*^{2}] et convertere in nubem et sequitur pluvia *Ka M U*] et vertere in pluvias et sequitur inde pluvia *Be Sg*
17) II.1, 7 pluvie] pluviarum *Be Kl Sg Wr*
18) II.1, 7 et fieri] quod fieret *Ba Ba*^{1} *Ka Kl Kr Kr*^{1} *Kr*^{2} *Kr*^{3} *L M U Wi Wr*] quod fiat *Be Sg*
19) II.1, 7 potest reliquum generari] potest alterum generari *Ba Ba*^{1} *Ka Kl Kr Kr*^{1} *Kr*^{3} *L M U Wi Wr* (aliud *Kr*^{2})] potest generari alterum *Be Sg*

Besides sharing the accidents of *Sg*, *Be* has some unique accidents, the majority of which can be regarded as *lectiones deteriores* (examples 1, 3, 5):

1) I.1, 3 non sint complexa que qualitercumque sit sic significant] non sunt complexa que qualitercumque sint sic significant *Be*
2) I.1, 6 possibile sit de eis] de ipsis possibile sit *Sg*] de ipsis sit possibile *Be*
3) I.1, 6 esset *om. Be*
4) I.1, 6 sed hoc non quia de eadem re possibile est] sed hoc non est impossibile ymo possibile est de eadem re *Sg*] sed hoc non est impossibile ymo possibile est de aliqua eadem re *Be*
5) I.1, 6 super manus et pedes] per pedes et manus *Be*
6) II.1, 6 primo exponende sunt] primo ponende sunt *Sg*] ponende sunt primo *Be*
7) II.1, 7 in illo loco possunt generari pluvie in quo *Sg*] in nullo locu possunt generari pluvie nisi in quo *Be*

As *Be* shares all accidents of *Sg* and adds its own ones, it is likely that it derives from *Sg*.

Stemma Codicum and Manuscripts Chosen for the Edition

In light of all of these considerations, we can propose the following *stemma codicum*:

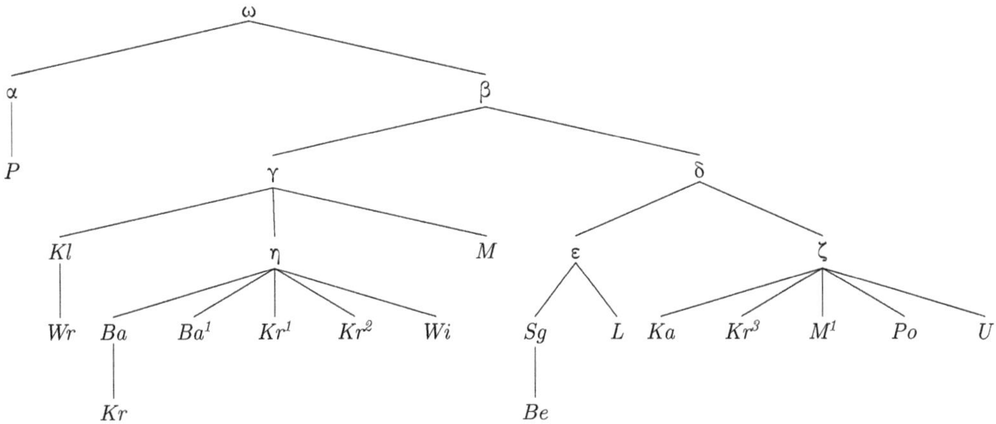

As a result of the study of the manuscript tradition of the second redaction of Oresme's *Questions on Meteorology*, I chose *P* as the base text and *Ba*, *Kl*, *L*, and *U* as representatives of the different groups of family β. The collation has shown that within their groups, these manuscripts have relatively few accidents not shared by the whole family.

Location of the Questions in the Manuscripts

The following list indicates the place of each question in the manuscripts transmitting the second redaction of Oresme's *Questions on Meteorology*. Its aim is to help scholars working directly with one of these codices to quickly locate specific passages in Oresme's text.

I.1 Utrum possibile sit de impressionibus metheorologicis habere simul scientiam et opinionem
Ba 4ra Ba_1 2ra *Be* 39^{r} *Ka* 2ra *Kl* 1ra *Kr* 59va Kr_1 3ra Kr_2 245ra Kr_3 195ra *L* 181ra *M* 1ra *P* 226^{r} *Po* 1ra *Sg* 1^{r} *U* 2ra *Wi* 49ra *Wr* 1^{r}

I.2 Utrum impressiones metheorologice fiant secundum naturam inordinatiorem quam sit natura celi
Ba 5ra Ba_1 2vb *Be* 40^{r} *Ka* 2vb *Kl* 2va *Kr* 60rb Kr_1 3va Kr_2 245vb Kr_3 196rb *L* 182rb *M* 1vb *P* 227^{v} *Po* 2rb *Sg* 3^{r} *U* 3ra *Wi* 49va *Wr* 2^{v}

I.3 Utrum iste mundus inferior sit continuus lationibus superioribus ut virtus eius inde gubernetur
Ba 5vb Ba_1 3ra *Be* 41^{r} *Ka* 4ra *Kl* 4ra *Kr* 61va Kr_1 4ra Kr_2 246vb Kr_3 197rb *L* 183va *M* 2va *P* 228^{v} *Po* 3vb *Sg* 3^{v} *U* 4rb *Wi* 50ra *Wr* 4^{v}

I.4 Utrum, cessante motu celi, cessarent motus in isto mundo inferiori
Ba 7rb Ba_1 4va *Be* 42^{r} *Ka* 5rb *Kl* 6ra *Kr* 61ra Kr_1 4vb Kr_2 248ra Kr_3 199ra *L* 185rb *M* 3va *P* 230^{v} *Po* 5va *Sg* 7^{v} *U* 5vb *Wi* 51ra *Wr* 7^{v}

I.5 Utrum eedem opiniones infinities reiterentur
Ba 8rb Ba_1 5ra *Be* 43^{r} *Ka* 6ra *Kl* 7ra *Kr* 61rb Kr_1 5rb Kr_2 248vb Kr_3 200ra *L* 186ra *M* 4ra *P* 232^{r} *Po* 6va *Sg* 9^{r} *U* 6va *Wi* 51va *Wr* 8^{v}

I.6 Utrum elementa sint continue proportionalia ad invicem
Ba 9rb Ba_1 5vb *Be* 43^{v} *Ka* 7ra *Kl* 8vb *Kr* 61vb Kr_1 5vb Kr_2 249vb Kr_3 201rb *L* 187rb *M* 4vb *P* 233^{v} *Po* 7rb *Sg* 11^{r} *U* 7va *Wi* 52ra *Wr* 10^{v}

I.7 Utrum quatuor elementa semper et immutabiliter habeant eandem proportionem ad invicem
Ba 11ra Ba_1 7ra *Be* 45^{r} *Ka* 8vb *Kl* 11rb *Kr* 62vb Kr_1 6vb Kr_2 251ra Kr_3 203rb *L* 189ra *M* 5vb *P* 235^{v} *Po* 10rb *Sg* 14^{v} *U* 9rb *Wi* 53ra *Wr* 14^{r}

I.8 Utrum motus celi sit causa calefactionis ignis in sua spera et etiam aeris superioris
Ba 12ra Ba_1 7vb *Be* 46^{r} *Ka* 9va *Kl* 12va *Kr* 63ra Kr_1 7rb Kr_2 251vb Kr_3 204va / 207^{v}[73] *L* 190ra *M* 6va *P* 237^{r} *Po* 11va *Sg* 16^{v} *U* 10rb *Wi* 53vb *Wr* 15^{v}

I.9 Utrum lumen sit productivum caloris
Ba 13va Ba_1 8vb *Be* 47^{r} *Ka* 11ra *Kl* 13va *Kr* 63vb Kr_1 8ra Kr_2 253ra Kr_3 206ra / 208^{v} *L* 191va *M* 7ra *P* 238^{v} *Po* 13ra *Sg* 18^{v} *U* 11va *Wi* 54vb *Wr* 18^{v}

I.10 Utrum contrarium circumstans suum contrarium fortificet ipsum
Ba 14vb Ba_1 9va *Be* 48^{r} *Ka* 12rb *Kl* 16va *Kr* 64rb Kr_1 8vb Kr_2 254ra Kr_3 209^{v} *L* 191vb *M* 8ra *P* 240^{r} *Po* 14va *Sg* 21^{v} *U* 13ra *Wi* 55va *Wr* 20^{v}

---

73 This question and the following one can be found twice in this codex. See above the description of this manuscript.

I.11 Utrum semper media regio aeris sit frigida
Ba 16rb *Ba*$_1$ 10va *Be* 49^{r} *Ka* 13vb *Kl* 18va *Kr* 65ra *Kr*$_1$ 9va *Kr*$_2$ 256ra *Kr*$_3$ 211^{r} *L* 194ra *M* 9ra *P* 241^{v} *Po* 16rb *Sg* 24^{r} *U* 14rb *Wi* 56va *Wr* 23^{r}

I.12 Utrum omnium impressionum metheorologicarum vapor et exalatio sint principium materiale
Ba 18va *Ba*$_1$ 11va *Be* 50^{v} *Ka* 16ra *Kl* 21va *Kr* 65vb *Kr*$_1$ 10va *Kr*$_2$ 258ra *Kr*$_3$ 213^{r} *L* 196ra *M* 10va *P* 244^{r} *Po* 19rb *Sg* 28^{r} *U* 16rb *Wi* 57vb *Wr* 27^{r}

I.13 Utrum impressiones ignite, seu ille que fiunt per inflammationem, fiant naturaliter in aere
Ba 19va *Ba*$_1$ 12rb *Be* 52^{r} *Ka* 17va *Kl* 23rb *Kr* 66va *Kr*$_1$ 11ra *Kr*$_2$ 259ra *Kr*$_3$ 214^{v} *L* 197ra *M* 11rb *P* 245^{r} *Po* 20vb *Sg* 30^{r} *U* 17va *Wi* 58vb *Wr* 29^{v}

I.14 Utrum de nocte, serenitate existente, debeant apparere hyatus et voragines et sanguinei colores in celo
Ba 20vb *Ba*$_1$ 12vb *Be* 52^{v} *Ka* 18va *Kl* 25vb *Kr* 66vb *Kr*$_1$ 11vb *Kr*$_2$ 259vb *Kr*$_3$ 215^{r} *L* 197vb *M* 11vb *P* 246^{r} *Po* 22ra *Sg* 31^{v} *U* 18rb *Wi* 59rb *Wr* 31^{r}

I.15 Utrum cometa sit de natura celi vel elementari
Ba 21va *Ba*$_1$ 13va *Be* 53^{v} *Ka* 19va *Kl* 26ra *Kr* 67rb *Kr*$_1$ 12ra *Kr*$_2$ 260va *Kr*$_3$ 216^{r} *L* 198va *M* 12va *P* 247^{r} *Po* 23rb *Sg* 33^{r} *U* 19rb *Wi* 60ra *Wr* 32^{v}

I.16 Utrum cometa sit exalatio calida et inflammata
Ba 22vb *Ba*$_1$ 14rb *Be* 54^{v} *Ka* 20va *Kl* 27vb *Kr* 67vb *Kr*$_1$ 12vb *Kr*$_2$ 261va *Kr*$_3$ 217^{r} *L* 199va *M* 13ra *P* 248^{v} *Po* 24vb *Sg* 35^{r} *U* 20rb *Wi* 60vb *Wr* 34^{v}

I.17 Utrum motus comete sit naturalis vel violentus
Ba 23va *Ba*$_1$ 14vb *Be* 55^{r} *Ka* 21rb *Kl* 28vb *Kr* 68ra *Kr*$_1$ 13ra *Kr*$_2$ 262rb *Kr*$_3$ 218^{r} *L* 200rb *M* 13va *P* 249^{r} *Po* 25vb *Sg* 36^{v} *U* 20vb *Wi* 61ra *Wr* 36^{r}

I.18 Utrum comete significent mortem principum, siccitatem et ventos et motus terre
Ba 24ra *Ba*$_1$ 15rb *Be* 56^{r} *Ka* 22rb *Kl* 30ra *Kr* 68va *Kr*$_1$ 13va *Kr*$_2$ 263ra *Kr*$_3$ 255^{r} *L* 201rb *M* 14rb *P* 250^{r} *Po* 26vb *Sg* 38^{r} *U* 21va *Wi* 61vb *Wr* 37^{r}

I.19 Utrum galaxia sit de natura celi vel de natura elementari
Ba 25ra *Ba*$_1$ 15vb *Be* 56^{v} *Ka* 23ra *Kl* 31rb *Kr* 69ra *Kr*$_1$ 14ra *Kr*$_2$ 263vb *Kr*$_3$ 255^{v} *L* 202rb *M* 14vb *P* 251^{r} *Po* 27vb *Sg* 39^{v} *U* 22va *Wi* 62rb *Wr* 38^{v}

II.1 Utrum locus generationis pluvie sit media regio aeris
Ba 26vb *Ba*$_1$ 17ra *Be* 58^{v} *Ka* 25ra *Kl* 33rb *Kr* 69vb *Kr*$_1$ 14rb *Kr*$_2$ 265ra *Kr*$_3$ 257^{v} *L* 203vb *M* 15vb *P* 253^{v} *Po* 29vb *Sg* 42^{v} *U* 24ra *Wi* 63rb *Wr* 41^{v}

II.2 Utrum ros et pruina, nix et pluvia, sint eiusdem speciei
Ba 28ra *Ba*$_1$ 17vb *Be* 59^{v} *Ka* 26rb *Kl* 35rb *Kr* 70rb *Kr*$_1$ 15va *Kr*$_2$ 266rb *Kr*$_3$ 259ra *L* 205ra *M* 16vb *P* 254^{r} *Po* 31rb *Sg* 44^{v} *U* 25ra *Wi* 64ra *Wr* 43^{v}

II.3 Utrum grandines magis debeant generari in hieme quam in autumno
Ba 29ra *Ba*$_1$ 18va *Be* 60^{v} *Ka* 27va *Kl* 36vb *Kr* 71ra *Kr*$_1$ 16rb *Kr*$_2$ 267rb *Kr*$_3$ 260^{v} *L* 206ra *M* 17va *P* 256^{v} *Po* 32vb *Sg* 47^{r} *U* 26rb *Wi* 64vb *Wr* 45^{v}

II.4 Utrum aqua calida applicata frigori congelanti citius congeletur quam aqua frigida
Ba 30rb *Ba*$_1$ 19va *Be* 61^{v} *Ka* 28vb *Kl* 38rb *Kr* 71va *Kr*$_1$ 16vb *Kr*$_2$ 268va *Kr*$_3$ 262^{r} *L* 207ra *M* 18rb *P* 257^{v} *Po* 34ra *Sg* 49^{r} *U* 27rb *Wi* 65va *Wr* 47^{v}

II.5 Utrum rubedo matutina sit signum pluvie
Ba 31ra *Ba*$_1$ 20ra *Be* 62^{r} *Ka* 29vb *Kl* 39va *Kr* 72ra *Kr*$_1$ 17rb *Kr*$_2$ 269rb *Kr*$_3$ 263^{r} *L* 207vb *M* 18vb *P* 258^{v} *Po* 35rb *Sg* 50^{v} *U* 28ra *Wi* 66ra *Wr* 49^{r}

II.6 Utrum caligo sit signum pluvie future
Ba 32rb *Ba*$_1$ 20vb *Be* 62bis^{r} *Ka* 31rb *Kl* 41ra *Kr* 72va *Kr*$_1$ 18ra *Kr*$_2$ 270rb *Kr*$_3$ 264^{v} *L* 209ra *M* 19vb *P* 260^{r} *Po* 36va *Sg* 53^{r} *U* 29rb *Wi* 66vb *Wr* 51^{r}

II.7 Utrum aqua naturaliter ascendat ad orificia fontium
Ba 33ra *Ba*$_1$ 21rb *Be* 62bis^{v} *Ka* 32ra *Kl* 41vb *Kr* 73ra *Kr*$_1$ 18va *Kr*$_2$ 271ra *Kr*$_3$ 265^{v} *L* 209vb *M* 20rb *P* 261^{r} *Po* 38rb *Sg* 54^{r} *U* 29vb *Wi* 67rb *Wr* 52^{v}

II.8 Utrum aque fontium generentur in terra
Ba 33vb *Ba*$_1$ 21vb *Be* 63^{v} *Ka* 33ra *Kl* 43ra *Kr* 73rb *Kr*$_1$ 19ra *Kr*$_2$ 271vb *Kr*$_3$ 266^{v} *L* 210va *M* 20vb *P* 262^{r} *Po* 39vb *Sg* 55^{v} *U* 30va *Wi* 67vb *Wr* 54^{r}

II.9 Utrum mare sit perpetuum vel aliquando fuit factum
Ba 35ra *Ba*$_1$ 22va *Be* 64^{v} *Ka* 34rb *Kl* 44va *Kr* 74ra *Kr*$_1$ 19vb *Kr*$_2$ 272vb *Kr*$_3$ 244^{r} *L* 211vb *M* 21va *P* 263^{v} *Po* 41ra *Sg* 58^{r} *U* 31vb *Wi* 68va *Wr* 56^{r}

II.10 Utrum mare debeat fluere et refluere
Ba 36rb *Ba*$_1$ 23va *Be* 65^{v} *Ka* 35vb *Kl* 46ra *Kr* 74vb *Kr*$_1$ 20va *Kr*$_2$ 274ra *Kr*$_3$ 245^{v} *L* 213ra *M* 22rb *P* 265^{r} *Po* 41ra *Sg* 60^{r} *U* 33ra *Wi* 69rb *Wr* 58^{v}

II.11 Utrum aqua maris debeat esse salsa
Ba 37vb *Ba*$_1$ 24va *Be* 67^{r} *Ka* 37rb *Kl* 48ra *Kr* 75va *Kr*$_1$ 21ra *Kr*$_2$ 275rb *Kr*$_3$ 247^{v} *L* 214vb *M* 23rb *Po* 42vb *Sg* 63^{r} *U* 34rb *Wi* 70rb *Wr* 61^{r}

II.12 Utrum aque puteales debeant esse salse
Ba 39ra *Ba*$_1$ 25rb *Be* 68^{r} *Ka* 38vb *Kl* 48vb *Kr* 76ra *Kr*$_1$ 22ra *Kr*$_2$ 276ra *Kr*$_3$ 249^{r} *L* 216ra *M* 24rb *Po* 44rb *Sg* 65^{v} *U* 45va *Wi* 71rb *Wr* 63^{v}

III.1 Utrum ventus sit exalatio calida et sicca
Ba 40ra *Ba*$_1$ 25vb *Be* 69^{r} *Ka* 39vb *Kl* 50va *Kr* 76va *Kr*$_1$ 22va *Kr*$_2$ 276vb *Kr*$_3$ 250^{r} *L* 216vb *M* 24vb *Po* 45ra *Sg* 67^{r} *U* 36ra *Wi* 71vb *Wr* 64^{v}

III.2 Utrum Sol faciat cessare ventos et moveat eos
Ba 41ra *Ba*$_1$ 26vb *Be* 69^{v} *Ka* 41ra *Kl* 52ra *Kr* 77ra *Kr*$_1$ 23ra *Kr*$_2$ 277va *Kr*$_3$ 251^{v} *L* 218ra *M* 25va *Po* 46rb *Sg* 69^{r} *U* 37ra *Wi* 72rb *Wr* 66^{v}

III.3 Utrum Boreas veniat a polo artico et Auster a polo antartico
Ba 41vb *Ba*$_1$ 27rb *Be* 70^{v} *Ka* 42ra *Kl* 53ra *Kr* 77va *Kr*$_1$ 23va *Kr*$_2$ 278rb *Kr*$_3$ 252^{v} *L* 218vb *M* 26ra *Po* 47rb *Sg* 70^{v} *U* 37vb *Wi* 72vb *Wr* 68^{v}

III.4 Utrum motus terre sit possibilis
Ba 42vb *Ba*$_1$ 28ra *Be* 71^{v} *Ka* 43rb *Kl* 54va *Kr* 78ra *Kr*$_1$ 24ra *Kr*$_2$ 279rb *Kr*$_3$ 254^{r} *L* 219vb *M* 26vb *Po* 48va *Sg* 72^{r} *U* 38vb *Wi* 73va *Wr* 70^{r}

III.5 Utrum tranquillitas sit signum terre motus
Ba 44ra *Ba*$_1$ 29ra *Be* 72^{v} *Ka* 44va *Kl* 55ra *Kr* 78vb *Kr*$_1$ 24vb *Kr*$_2$ 280rb *Kr*$_3$ 200^{r} *L* 220vb *M* 27va *Po* 50ra *Sg* 74^{v} *U* 40ra *Wi* 74va *Wr* 72^{v}

III.6 Utrum tonitruum sit extinctio ignis in nube
Ba 44vb *Ba*$_1$ 29vb *Be* 73^{r} *Ka* 45vb *Kl* 57rb *Kr* 79rb *Kr*$_1$ 25rb *Kr*$_2$ 281ra *Kr*$_3$ 221^{v} *L* 221vb *M* 28rb *Po* 51ra *Sg* 76^{r} *U* 40vb *Wi* 75rb *Wr* 74^{v}

III.7 Utrum exalatio ex qua fiunt tonitrua et corruscationes debeat descendere versus terram
Ba 46rb *Ba*$_1$ 30vb *Be* 74^{r} *Ka* 47ra *Kl* 58vb *Kr* 80ra *Kr*$_1$ 26ra *Kr*$_2$ 282ra *Kr*$_3$ 222^{v} *L* 223ra *M* 29ra *Po* 52va *Sg* 79^{v} *U* 41vb *Wi* 76ra *Wr* 76^{r}

III.8 Utrum corruscatio sit possibilis
Ba 47ra *Ba*$_1$ 31vb *Be* 75^{r} *Ka* 48rb *Kl* 60ra *Kr* 80va *Kr*$_1$ 26va *Kr*$_2$ 282vb *Kr*$_3$ 225^{r} *L* 224ra *M* 29va *Po* 53vb *Sg* 81^{r} *U* 42vb *Wi* 76vb *Wr* 78^{r}

III.9 Utrum fulmen aliquod sit penetrans et aliud adurens
Ba 48ra *Ba*$_1$ 32va *Be* 76^{r} *Ka* 49rb *Kl* 61ra *Kr* 81ra *Kr*$_1$ 27ra *Kr*$_2$ 283vb *Kr*$_3$ 226^{v} *L* 224vb *M* 30rb *Po* 54vb *Sg* 83^{r} *U* 43va *Wi* 77rb *Wr* 79^{v}

III.10 Utrum indifferenter quodlibet posset percuti fulmine
Ba 49ra *Ba*$_1$ 33vb *Be* 77^{r} *Ka* 50va *Kl* 62vb *Kr* 81va *Kr*$_1$ 27vb *Kr*$_2$ 284vb *Kr*$_3$ 228^{r} *L* 225vb *M* 31ra *Po* 56rb *Sg* 85^{r} *U* 44vb *Wi* 78rb *Wr* 80^{v}

III.11 Utrum tiphon et ecnephia et incensio sint exalationes calide et sicce
Ba 49va *Ba*$_1$ 34ra *Be* 77^{r} *Ka* 51ra *Kl* 63rb *Kr* 81vb *Kr*$_1$ 28ra *Kr*$_2$ 285ra *Kr*$_3$ 229^{r} *L* 226rb *M* 31rb *Po* 57ra *Sg* 86^{r} *U* 45ra *Wi* 78vb *Wr* 82^{r}

III.12 Utrum visus frangatur in occursu medii densioris vel rarioris, puta aque vel aeris
Ba 50rb *Ba*$_1$ 34vb *Be* 78^{r} *Ka* 52ra *Kl* 64rb *Kr* 82rb *Kr*$_1$ 28va *Kr*$_2$ 285vb *Kr*$_3$ 230^{r} *L* 227ra *M* 32ra *Po* 58vb *Sg* 87^{v} *U* 45vb *Wi* 79rb *Wr* 83^{v}

III.13 Utrum cuiuslibet corporis superficies reflectat virtutes super ipsam incidentes, sicut sunt lumen, species coloris et huiusmodi
Ba 52ra *Ba*$_1$ 36vb *Be* 79^{v} *Ka* 54ra *Kl* 66vb *Kr* 83va *Kr*$_1$ 29va *Kr*$_2$ 287vb *Kr*$_3$ 233^{r} *L* 229ra *M* 33rb *Po* 60rb *Sg* 91^{v} *U* 47va *Wi* 80vb *Wr* 87^{r}

III.14 Utrum virtutes reflexe sint debiliores virtutibus non reflexis
Ba 54ra *Ba*$_1$ 38ra *Be* 81^{r} *Ka* 56ra *Kl* 68vb *Kr* 84vb *Kr*$_1$ 30rb *Kr*$_2$ 289ra *Kr*$_3$ 235^{v} *L* 230rb *M* 34rb *Po* 61vb *Sg* 94^{v} *U* 48vb *Wi* 81vb *Wr* 89^{v}

III.15 Utrum aliqua specula representent solum colorem et non figuram, accipiendo 'colorem' prout extendit se ad lucem et ad colorem proprie dictum
Ba 55ra *Ba*$_1$ 39ra *Be* 81^{v} *Ka* 57rb *Kl* 70rb *Kr* 85ra *Kr*$_1$ 30vb *Kr*$_2$ 289vb *Kr*$_3$ 237^{r} *L* 231rb *M* 35ra *Po* 63rb *Sg* 96^{v} *U* 49vb *Wi* 82va *Wr* 91^{r}

III.16 Utrum in apparitione ipsius halonis interponatur vapor inter visum et astrum
Ba 56va *Ba*$_1$ 40va *Be* 83^{r} *Ka* 59rb *Kl* 72rb *Kr* 86ra *Kr*$_1$ 31vb *Kr*$_2$ 291va *Kr*$_3$ 239^{v} *L* 232vb *M* 36rb *Po* 65rb *Sg* 100^{r} *U* 51rb *Wi* 83vb *Wr* 94^{r}

III.17 Utrum halo fiat per refractionem vel per reflexionem radiorum visualium super vaporem medium inter astrum sub quo halo apparet et visum nostrum
Ba 57ra *Ba*$_1$ 41va *Be* 84^{r} *Ka* 60rb *Kl* 73vb *Kr* 86va *Kr*$_1$ 32va *Kr*$_2$ 293ra *Kr*$_3$ 241^{r} *L* 233vb *M* 37ra *Po* 66rb *Sg* 102^{r} *U* 52rb *Wi* 84va *Wr* 96^{r}

III.18 Utrum halo debeat apparere secundum circuli periferiam vel circumferentiam
Ba 58va Ba_1 42vb *Be* 85^{r} *Ka* 62ra *Kl* 75va *Kr* 87va Kr_1 33ra Kr_2 294ra Kr_3 267^{r} *L* 235ra *M* 38ra *Po* 68ra *Sg* 104^{v} *U* 53va *Wi* 85vb *Wr* 98^{v}

III.19 Utrum radius visualis debeat reflecti ab aere uniformi non condensato nec mixto vaporibus ad oculum videntis existentis in eodem aere sine adiutorio speculorum
Ba 60ra Ba_1 44rb *Be* 86^{v} *Ka* 64ra *Kl* 78ra *Kr* 88va Kr_1 34va Kr_2 295rb Kr_3 269^{v} *L* 236vb *M* 39va *Po* 70ra *Sg* 108^{r} *U* 55rb *Wi* 87ra *Wr* 102^{r}

III.20 Utrum colores yridis sint colores secundum rei veritatem
Ba 61vb Ba_1 45va *Be* 88^{r} *Ka* 66va *Kl* 80vb *Kr* 89va Kr_1 35vb Kr_3 272^{v} *L* 239ra *M* 41ra *Po* 72va *Sg* 112^{r} *U* 57rb *Wi* 88va *Wr* 105^{v}

III.21 Utrum yris sit forma realis nubi impressa vel sit solum forma ymaginaria
Ba 62vb Ba_1 46rb *Be* 89^{v} *Ka* 68ra *Kl* 82va *Kr* 90ra Kr_1 36va Kr_3 274^{v} *L* 240rb *M* 41vb *Po* 74ra *Sg* 114^{v} *U* 58va *Wi* 89rb *Wr* 107^{v}

III.22 Utrum yris sit forma dyaphanalis vel specularis, vel utrum yris fiat per reflexionem vel refractionem
Ba 63vb Ba_1 47ra *Be* 90^{v} *Ka* 69rb *Kl* 84ra *Kr* 90vb Kr_1 37rb Kr_3 276^{r} *L* 241va *M* 42vb *Po* 75va *Sg* 117^{r} *U* 59va *Wi* 90rb *Wr* 110^{r}

III.23 Utrum, supposito quod yris fiat per reflexionem radiorum, utrum tunc fiat in nube vel super guttulas roridas vel super stillicidia
Ba 64vb Ba_1 47vb *Be* 91^{v} *Ka* 70vb *Kl* 86ra *Kr* 91rb Kr_1 38ra Kr_3 278^{r} *L* 242vb *M* 43vb *Po* 77va *Sg* 119^{v} *U* 61ra *Wi* 91rb *Wr* 112^{v}

III.24 Utrum omnis yris debeat esse tricolor
Ba 65va Ba_1 48ra *Be* 92^{r} *Ka* 71va *Kl* 87ra *Kr* 91va Kr_1 38rb Kr_3 278^{v} *L* 243va *M* 44rb *Sg* 121^{r} *U* 61vb *Wi* 92vb *Wr* 114^{r}

III.25 Utrum yris solum dupliciter et non multipliciter potest apparere
Ba 66rb Ba_1 48vb *Be* 93^{r} *Ka* 72va *Kl* 88rb *Kr* 92ra Kr_1 39ra Kr_3 280^{r} *L* 244va *M* 45ra *Sg* 122^{v} *U* 62vb *Wi* 93va *Wr* 115^{v}

III.26 Utrum semper, apparentibus duabus yridibus, superior yris debeat habere colores conversim positos
Ba 67rb *Ba*$_1$ 49va *Be* 93^{v} *Ka* 74rb *Kl* 90ra *Kr* 92vb *Kr*$_1$ 39va *Kr*$_3$ 281^{v} *L* 245vb *M* 46ra *Sg* 125^{r} *U* 64rb *Wi* 93rb *Wr* 118^{r}

III.27 Utrum yris superior vel secundaria necessarie debeat apparere et esse remissior in coloribus quam principalis yris
Ba 68rb *Ba*$_1$ 50rb *Be* 94^{v} *Ka* 75va *Kl* 91va *Kr* 93rb *Kr*$_1$ 40rb *Kr*$_3$ 283^{r} *L* 246vb *M* 46vb *Sg* 127^{r} *U* 65va *Wi* 94ra *Wr* 120^{r}

III.28 Utrum yris debeat apparere secundum circuli periferiam si non sit impedimentum
Ba 69ra *Ba*$_1$ 50vb *Be* 95^{r} *Ka* 76va *Kl* 92vb *Kr* 93vb *Kr*$_1$ 40vb *Kr*$_3$ 284^{v} *L* 247vb *M* 47rb *Sg* 128^{v} *U* 66va *Wi* 94vb *Wr* 121^{v}

III.29 Utrum semper, tempore apparitionis yridis, necesse sit centrum Solis et centrum horizontis et centrum yridis et polos eorumdem esse in eadem linea recta
Ba 70va *Ba*$_1$ 51va *Be* 96^{v} *Ka* 78va *Kl* 94vb *Kr* 94vb *Kr*$_1$ 41vb *Kr*$_3$ 287^{r} *L* 249vb *M* 48va *Sg* 131^{v} *U* 68va *Wi* 96ra *Wr* 125^{r}

III.30 Utrum semper, tempore apparitionis yridis et halonis, dyameter yridis sit precise dupla ad dyametrum halonis
Ba 71vb *Ba*$_1$ 52va *Be* 98^{r} *Ka* 80vb *Kl* 97rb *Kr* 95vb *Kr*$_1$ 42vb *Kr*$_3$ 289^{r} *L* 251va *M* 49vb *Sg* 135^{r} *U* 70va *Wi* 97rb *Wr* 128^{r}

III.31 Utrum, omni hora diei artificialis oculo existente in superficie horizontis, posset aliqua portio yridis apparere ubicumque fuerit homo vel oculus videntis
Ba 73ra *Ba*$_1$ 53rb *Be* 99^{r} *Ka* 82vb *Kl* 99rb *Kr* 96vb *Kr*$_1$ 43va *Kr*$_3$ 291^{r} *L* 253ra *M* 50vb *Sg* 138^{r} *U* 72rb *Wi* 98rb *Wr* 131^{r}

III.32 Utrum yris Lune potest pluries apparere quam bis in quinquaginta annis in consimilibus coloribus et proprietatibus sicut yris Solis
Ba 74ra *Ba*$_1$ 54ra *Be* 100^{r} *Ka* 84ra *Kl* 101ra *Kr* 97va *Kr*$_1$ 44rb *Kr*$_3$ 293^{r} *L* 254va *M* 51va *Sg* 140^{r} *U* 73va *Wi* 99ra *Wr* 133^{r}

III.33 Utrum, visa yride et Sole vel astro elevato notabiliter super horizontem, apparet minor portio maioris circuli quam appareat Sole existente in fine horizontis
Ba 75ra *Ba*$_1$ 54va *Be* 101^{r} *Ka* 85va *Kl* 103ra *Kr* 98rb *Kr*$_1$ 45ra *Kr*$_3$ 295^{v} *L* 255vb *M* 52va *Po* 78vb *Sg* 142^{v} *U* 75rb *Wi* 99vb *Wr* 135^{v}

III.34 Utrum virge et parelii fiant ex reflexione vel refractione radiorum sicut yris vel halo
Ba 76ra *Ba*$_1$ 55ra *Be* 102^{r} *Ka* 87rb *Kl* 105rb *Kr* 99rb *Kr*$_1$ 45vb *Kr*$_3$ 298^{r} *L* 257rb *M* 53va *Po* 80va *Sg* 145^{v} *U* 77ra *Wi* 100vb *Wr* 138^{r}

III.35 Utrum per iuvamen artis possunt fieri metalla
Ba 76vb *Ba*$_1$ 55vb *Be* 103^{r} *Ka* 88va *Kl* 107ra *Kr* 99vb *Kr*$_1$ 46ra *Kr*$_3$ 299^{v} *L* 258va *M* 54rb *Po* 82ra *Sg* 147^{v} *U* 78rb *Wi* 101va *Wr* 140^{r}

IV.1 Utrum tantum quatuor sunt qualitates prime, scilicet caliditas, humiditas, frigiditas, siccitas
Ba 78rb *Ba*$_1$ 57rb *Be* 105^{v} *E* 158vb *Ka* 92ra *Kl* 109va *Kr* 101rb *Kr*$_1$ 47va *Kr*$_2$ 298ra *Kr*$_3$ 304^{v} *L* 261va *M* 56rb *Po* 85ra *Sg* 152^{v} *U* 81vb *Wi* 103ra *Wr* 145^{r}

IV.2 Utrum quatuor qualitates, videlicet caliditas, frigiditas, etc., sint principia activa in generatione cuiuslibet mixti, accipiendo tamen 'generationem' pro alteratione precedente introductionem forme substantialis mixti, eo modo quo Aristoteles accipit in secundo capitulo huius quarti
Ba 79ra *Ba*$_1$ 57vb *Be* 106^{r} *E* 159rb *Ka* 93rb *Kl* 110vb *Kr* 102ra *Kr*$_1$ 48rb *Kr*$_2$ 298vb *Kr*$_3$ 306^{r} *L* 262vb *M* 57ra *Po* 86rb *Sg* 155^{r} *U* 83rb *Wi* 103vb *Wr* 146^{v}

IV.3 Utrum putrefactio causetur a calido extrinseco fugiente intrinseco
Ba 79vb *Ba*$_1$ 58ra *Be* 107^{r} *E* 160rb *Ka* 94rb *Kl* 112ra *Kr* 102va *Kr*$_1$ 48vb *Kr*$_2$ 299va *Kr*$_3$ 309^{v} *L* 264ra *M* 57vb *Po* 87ra *Sg* 157^{r} *U* 84va *Wi* 105ra *Wr* 148^{r}

IV.4 Utrum frigus preservat a putrefactione
Ba 80ra *Ba*$_1$ 58vb *Be* 108^{r} *E* 161ra *Ka* 95va *Kl* 113rb *Kr* 103ra *Kr*$_1$ 49rb *Kr*$_2$ 300va *Kr*$_3$ 310^{v} *L* 265ra *M* 58va *Po* 88ra *Sg* 159^{r} *U* 85vb *Wi* 105ra *Wr* 150^{r}

IV.5 Utrum aliqua animalia possunt generari per putrefactionem
Ba 80vb *Ba*$_1$ 59va *Be* 108^{v} *E* 161vb *Ka* 97ra *Kl* 114va *Kr* 103vb *Kr*$_1$ 49vb *Kr*$_2$ 301rb *Kr*$_3$ 312^{r} *L* 266rb *M* 59rb *Po* 89rb *Sg* 161^{r} *U* 87rb *Wi* 105vb *Wr* 151^{v}

IV.6 Utrum diffinitio digestionis sit bona in qua dicitur: 'digestio est perfectio a calido naturali et proprio ex oppositis passivis'
Ba 81rb *Ba*$_1$ 59vb *Be* 109^{v} *E* 162va *Ka* 98ra *Kl* 115vb *Kr* 104va *Kr*$_1$ 50rb *Kr*$_2$ 302ra *Kr*$_3$ 313^{v} *L* 267rb *M* 59vb *Po* 89vb *Sg* 162^{v} *U* 88rb *Wi* 106rb *Wr* 153^{r}

IV.7 Utrum balnea conferant ad digestionem
Ba 82ra *Ba*$_1$ 60ra *Be* 110^{r} *E* 163rb *Ka* 99va *Kl* 116vb *Kr* 105rb *Kr*$_1$ 50vb *Kr*$_2$ 302vb *Kr*$_3$ 315^{r} *L* 268rb *M* 60va *Po* 90va *Sg* 164^{v} *U* 89va *Wi* 106vb *Wr* 154^{v}

IV.8 Utrum sint tantum tres digestiones
Ba 82vb Ba_1 61ra *Be* 111^{r} *E* 164ra *Ka* 101ra *Kl* 118va *Kr* 106rb Kr_1 51rb Kr_2 303vb Kr_3 317^{r} *L* 269rb *M* 61rb *Po* 91va *Sg* 167^{r} *U* 91ra *Wi* 107vb *Wr* 156^{v}

IV.9 Utrum frixata debeant magis dici assata quam elixata
Ba 83vb Ba_1 61vb *Be* 112^{r} *E* 165rb *Ka* 103va *Kl* 121va *Kr* 107va Kr_1 52ra Kr_2 305ra Kr_3 319^{r} *L* 271rb *M* 62va *Po* 92rb *Sg* 170^{r} *U* 93rb *Wi* 108va *Wr* 159^{r}

IV.10 Utrum vermes qui generantur in ventribus animalium generentur in stomaco vel in intestinis
Ba 84va Ba_1 62rb *Be* 113^{r} *E* 166ra *Ka* 105ra *Kl* 122va *Kr* 108va Kr_1 52va Kr_2 305vb Kr_3 320^{v} *L* 272rb *M* 63ra *Po* 93ra *Sg* 172^{r} *U* 94va *Wi* 108vb *Wr* 160^{v}

IV.11 Utrum in quolibet mixto dominetur terra vel aqua vel ambo
Ba 84vb Ba_1 62va *Be* 113^{v} *E* 166va *Ka* 105vb *Kl* 123rb *Kr* 109ra Kr_1 52vb Kr_2 306ra Kr_3 321^{r} *L* 273ra *M* 63va *Po* 94vb *Sg* 173^{r} *U* 95rb *Wi* 109rb *Wr* 161^{v}

Editorial Principles

Text. As we have seen, the manuscript tradition of the second redaction of Oresme's *Questions on Meteorology* is clearly divided into two families: on the one side manuscript *P*, which is the only Parisian witness of Oresme's text; on the other side, all other copies, which have a Central European origin. We have also seen that, in case of divergence between the two traditions, *P* generally provides the better text, while the Central European copies contain important errors and omissions which are not shared by *P* and which I do not think could be ascribed to Oresme. I therefore chose *P* as base for the edition and adopted the text of the other family only in cases in which *P* has obvious errors.

As I have already pointed out, *P* abruptly stops at question II.10, in the middle of a sentence. After a blank leaf, the text resumes with question II.11 from Albert of Saxony's *Questions on Meteorology*, which bears no relevance to the reconstruction of Oresme's commentary. Since the text of the other manuscripts derives from a different and, in many cases, corrupt, tradition, I preferred not to mix the two versions. As a consequence, I decided to publish only the portion of the text transmitted in *P*. The rest of the commentary—which, at least in the transmitted form, cannot be ascribed to Oresme—will be the object of another edition with different goals and different editorial principles.

Orthography. As a rule, I have tried to stick to the manuscript tradition as closely as possible. For this reason, I have adopted a medieval orthography, spelling *e* in cases in which Classical Latin would have spelled *œ* or *æ*. For instance, I wrote "celum", and not "cœlum" or "cælum", as these forms are not attested in any of the manuscripts. In cases where the manuscript tradition, or even the same manuscript, attests different forms for the same word (*eclipsis*/*eclypsis*; *hyemps*/*hyems*/*hiems*; *galasia*/*galaxia*; *autumno*/*automno*/*autumpno*/*autompno*), I have adopted a standardized orthography.[1] In order not to clutter up the apparatus, I have not mentioned these orthographical peculiarities. I have also ignored the variance in the forms *meteorologice*, *metheorologice*, *metroloyce*, *metroholoyce*, since they reflect the writing habits of the scribes and alternative forms sometimes appear within the same manuscript. Moreover, because of the gothic system of abbreviations, it is often difficult to determine exactly which form was adopted by a given scribe, since it is

1 Following this principle, I have written *mundus*, *descensus*, *ascensus*, *distractio*, *superior*, *polita*, instead of *mondus*, *desscensus*, *asscensus*, *disstractio*, *supperior*, *pollita* (forms that are attested in manuscript *P*).

possible to expand the abbreviation "me^{ce}" as any of the preceding forms. I therefore always adopted the form "metheor*", which is the most frequently attested one, in the text of the edition, keeping however the spelling "meteor*" for the title of the volume. The latter spelling is also attested in medieval manuscripts, as proven for instance by the colophon of manuscript *Ba*^{1} (see above), though it appears less frequently, at least in the fourteenth and fifteenth century manuscripts transmitting Oresme's text. However, in the absence of an autograph of Oresme's commentary, it seemed to me a more neutral choice to adopt for the title of the volume. This spelling, which is the correct transliteration of the Greek word, is currently employed to refer to Aristotle's *Meteorology* in other critical editions of the medieval *corpus aristotelicum*, as for instance Vuillemin-Diem's edition of Moerbeke's translation.

Even though I followed the text of *P*, I did not adopt its orthographical peculiarities, such as the use of *o* instead of *u* in some nasals ("mondus", for "mundus") or the doubling of certain labial sounds ("supperior" for "superior"). I have also distinguished between *u*—for vowels—and *v*—for consonants.

All abbreviations have been solved, except for "etc." I have always spelled out the numbers, although in the manuscripts they are sometimes also written in Roman, sometimes in Arabic numerals. For ease of reading, I have, however, employed Arabic numerals in the questions about the proportionality of the elements (I.6–7).

Internal divisions. The division into books and questions is attested in the manuscripts. For easier consultation, I have numbered the paragraphs in angular brackets and adopted a line numbering starting from 1 at the beginning of every new question. New columns in the collated manuscripts are indicated by a vertical line which is referenced by the folio in the margin.

Figures. At question I.19, manuscript *P* announces a figure, which however cannot be found neither in this codex nor in the other witnesses of Oresme's text. I have reconstructed the figure based on Oresme's explanation and on the manuscripts of the first redaction of Oresme's commentary, which transmit the same figure in the same context.

Critical apparatus. The apparatus is positive. I mention the adopted variant before the square brackets, and the variants of the other witnesses after them. The apparatus contains all readings of *P* that I do not adopt in the text, including repetitions and obvious mistakes. As for the variants from the other manuscripts, I have adopted a principle of economy and have concentrated exclusively on variants that affect the sense of the text. I have therefore ignored purely synonymical variants such as "ergo/igitur/ideo"; "vel/aut"; "sic/ita"; "sive/seu"; "nec/neque"; "scilicet/videlicet"; "aliquando/quandoque"; "plus/magis"; "super/supra", the variability of pronouns ("iste/ille", "illud/istud" for example) and variants in word order and verb tenses that do not affect the meaning. For the same reason, the apparatus ignores the differences in order and formulation in the structural parts of the text and in the transition between articles and questions. For instance, at the end of the first article of question I.2, manuscript *P* reads: "de quo diffusius apparebit in processu", while manuscript *Ba* has: "de quibus postea patebit", manuscript *L* has: "de quibus plenius postea patebit", and manuscript *Sg* reads: "de quibus postea". As we have seen above, while discussing the differences between manuscript *P* and manuscripts of family β, analogous or even more substantial variants occur throughout the whole text. This variance is due to the circulation of Oresme's text, which was used to teach Aristotle's *Meteorology* at the Arts Faculty: masters and students constantly readapted and reformulated it according to their needs. We have seen that this is particularly evident in manuscripts of sub-family γ, which result from a *pronunciatio* given at the University of Prague. In light of these considerations, I found it useless to mention this kind of variants in the apparatus.

Source apparatus. In the source apparatus, I have identified explicit references and, whenever possibile, *loci paralleli* in Classical and Medieval authors, as well as in other works by Oresme. I have identified references to Aristotle's texts by giving the title of the work, the book, the chapter, and the pagination in Bekker's edition.[2] I have referred to the volumes of the *Aristoteles latinus* series in case of literal quotations. In the same way, I have provided only general references to the works and chapters indicated by Oresme and I have referred to a particular edition only in the case of literal quotations. I have also referred to the *Auctoritates Aristotelis*[3] when the passages could be found in this compilation. Internal references within the second redaction of Oresme's *Questions on Meteorology* are indicated by book, question and paragraph. References to Averroes's commentaries on *De celo*, *De anima*, and *De somno et vigilia* follow the modern editions of these texts.[4] For all other commentaries by Averroes, I have resorted to the Renaissance Venice edition.[5]

---

2 Aristoteles, *Opera*, ed. I. Bekker, Berlin 1831–1870 (5 vols.).

3 J. Hamesse (ed.), *Les Auctoritates Aristotelis: un florilège médiéval. Étude historique et édition critique*, Louvain-la-Neuve/Paris 1974 (Philosophes médiévaux, 17).

4 J.F. Carmody and R. Arnzen (eds.), *Averrois Commentaria Magna in Aristotelem. De celo et mundo*, Leuven 2003 (Recherches de Théologie et Philosophie médiévales, Bibliotheca, 4) (2 vols.); F.S. Crawford (ed.), *Commentarium magnum in Aristotelis De anima libros*, Cambridge (MA) 1953 (Corpus commentariorum Averrois in Aristotelem. Versiones latinae, 6/1); A.L. Shields (ed.), *In De somno et vigilia, Averrois Cordubensis compendia librorum Aristotelis qui 'Parva Naturalia' vocantur*, Cambridge (MA) 1949 (Corpus commentariorum Averrois in Aristotelem Versiones latinae, 7).

5 Averroes, *Aristotelis opera cum Averrois commentariis*, Venice 1562–1574, reprint Frankfurt am Main 1962 (9 vols.).

Abbreviations and Signs Adopted

Signs[6]

In the text:

| beginning of a new column in the manuscript indicated in the margin

In the critical apparatus:

] announces the variants

... substitutes non repeated words in a lemma

Abbreviations

In the critical apparatus:

ac	ante correctionem
add.	addidit
a.m.	alia manu
cod.	codex
codd.	codices
col.	columna
corr.	correxit
del.	delevit
des.	desinit [definitive interruption of a manuscript]
eras.	erasit
f.	folium
fort.	fortasse
inv.	invertit
inf.	inferior
iter.	iteravit
lec. dub.	lectio dubia
lib.	liber
l.	linea
mg.	in margine
ms.	codex manu scriptus
mss.	codices manu scripti
n.	numerus
om.	omisit

6 According to the conventions recommended by the S.I.E.P.M. See A. Dondaine, "Abbréviations latines et signes recommandés pour l'apparat critique des éditions de textes médiévaux", *Bulletin de la Société Internationale pour l'Etude de la Philosophie Médiévale* (S.I.E.P.M.) 2 (1960), 142–149.

om. hom.	omisit per homoeoteleuton
pc	post correctionem
spat. vac.	spatium vacuum
sqq.	sequentes
sub. lin.	sub linea
transp.	transposuit

In the source apparatus:

ac	ante correctionem
AL	Aristoteles Latinus
cap.	caput (capitulum)
cf.	confer (conferas, conferatur, etc.)
comm.	commentarium
ed.	editio
ed. Colon.	Editio Coloniensis
f.	folium
ff.	folia
ibid.	ibidem
inc.	incipit
lib.	liber
lin.	linea
n.	numerus (lineae, paginae)
q.	quaestio
sqq.	sequentes
tract.	tractatus

Nicole Oresme,
Questiones in Meteorologica de ultima lectura, recensio parisiensis

∵

Liber I

I.1 Utrum possibile sit de impressionibus metheorologicis habere simul scientiam et opinionem

P 226^{r} *Kl* 1ra *Ba* 4ra *L* 181ra *U* 2ra

Circa librum *Metheororum*, qui est quartum naturalium, queritur primo utrum possibile sit de impressionibus metheorologicis habere simul scientiam et opinionem.

⟨1⟩ Et arguitur primo quod non: de impressionibus metheorologicis non contingit habere scientiam nec etiam opinionem; igitur questio est falsa. Consequentia nota est de se. Antecedens probatur, quia nichil potest sciri nec opinari nisi propositio; impressiones autem metheorologice non sunt propositiones, sicut patet de se; igitur etc.

⟨2⟩ Secundo: idem non est scibile et opinabile; igitur de impressionibus metheorologice non potest simul haberi scientia et opinio. Consequentia tenet de se. Antecedens patet per Aristotelem in primo *Posteriorum* circa finem[1]. Et probatur ratione, nam scientia est de illo quod impossibile est aliter se habere, opinio autem de eo quod possibile est aliter se habere; sed non est idem quod impossibile est aliter se habere et quod possibile est aliter se habere; et per consequens non est idem scibile et opinabile.

⟨3⟩ Tertio: de impressionibus metheorologicis non potest esse scientia, et ergo non est possibile de eis simul esse scientiam et opinionem. Consequentia nota est de se. Antecedens probatur, quia quod non est verum non potest sciri, | *Kl* 1rb ex primo *Posteriorum*[2]; sed impressiones metheorologice non sunt vere, cum non sint complexa que, qualitercumque sit, sic significant.

⟨4⟩ Quarto: de impressionibus metheorologicis non possunt formari propositiones necessarie; igitur etc. Consequentia tenet, quia de illo de quo debet esse scientia, debent posse formari propositiones necessarie. Et ante-

---

18–19 quod non est verum non potest sciri] nichil contingit sciri nisi verum *Ba Kl L U*   19 sed] *om. P*   20 sint complexa que qualitercumque sit sic significant] sint complexa nec qualitercumque est fuit vel erit significant *Kl* sunt propositiones nec sunt complexa nec qualitercumque est fuit vel erit significant *Ba* sunt complexa que qualitercumque stant sic significent *L* sunt complexa que qualitercumque se habent sic significent *U*

---

1 Aristoteles, *Analytica Posteriora* I, 33, 89 a 7–10.

2 Ibid., I, 33, 88 b 34–37; *Auct. Arist.*, 319, n. 95.

[*L* 181rb] cedens patet, quia | impressiones metheorologice sunt corruptibiles; de corruptibili autem, ut videtur, non possunt formari propositiones necessarie.

⟨5⟩ Quinto: que fiunt inordinate, de illis non videtur posse haberi scientiam; [*U* 2rb] sed impressiones metheorologice sunt huiusmodi; | igitur etc. Maior est nota, quia de quibus debet esse scientia, non debent fieri inordinate extra semper et frequenter. Minor patet per Aristotelem in littera; dicit enim: 'hec autem sunt que accidunt secundum naturam quidem inordinatiorem'[3].

⟨6⟩ In oppositum est Aristoteles et Albertus et alii qui tradiderunt nobis scientiam de impressionibus metheorologicis. Et sic patet auctoritatibus quod de impressionibus metheorologicis potest esse scientia. Quod autem de ipsis potest esse opinio patet, quia de eis possunt formari propositiones quibus aliquis assentit cum formidine; assensus autem cum formidine est opinio. Et confirmatur quod possibile sit de eis simul esse scientiam et opinionem, quia si non, hoc maxime esset ex eo quod non est idem scibile et opinabile, sicut allegabatur per Aristotelem circa finem primi *Posteriorum*[4]. Sed hoc non, quia de eadem re possibile est simul esse scientiam et opinionem. Posito enim quod pater tuus esset indutus pelle asinina et iret super manus et pedes, ista conclusio esset a te opinata: 'iste est asinus' demonstrando patrem tuum. Et opinando istam conclusionem, haberes opinionem [*Ba* 4rb] de patre tuo, nam ipse est res significata | per subiectum propositionis a te opinate; tamen hoc non obstante haberes scientiam de eodem in sciendo istam propositionem: 'omnis homo est risibilis'; ipse enim est res significata per subiectum istius propositionis a te scite. Similiter de eodem, puta de eadem conclusione, potest simul esse syllogismus dyalecticus et demonstra-

---

2 ut videtur non possunt formari] non videntur posse formari *Ba Kl L* non est scientia nec possunt formari *U* 6 in littera] in primo huius *Ba Kl* in primo illius libri *L* in primo huius libri *U* ‖ dicit enim] ubi dicit *Ba Kl L U* 7 accidunt] contingunt *Ba Kl L U* ‖ quidem inordinatiorem] inordinatam *Kl* 8–10 qui tradiderunt nobis scientiam de impressionibus metheorologicis et sic patet auctoritatibus quod de impressionibus metheorologicis potest esse scientia] *om. Ba Kl U* scilicet quod de impressionibus metheorologicis posset esse opinio et scientia *L* 12 formidine] ad oppositum *add. Ba Kl L U* 17 et iret] ambulans *Ba Kl* et ambularet *L* et ambulet *U* 18–19 demonstrando patrem tuum] *om. Ba Kl L U* 22 propositionem] conclusionem *Ba U om. P* ‖ ipse enim est] quia pater tuus est *Ba Kl L U* 23 propositionis] conclusionis *Ba Kl U*

---

3 Aristoteles, *Meteorologica*, I, 1, 338 b 20–22, translatio Guillelmi de Morbeka (translatio noua), *AL* X 2.2, 9^{8-10}: "Hec autem sunt quecumque accidunt secundum naturam quidem, inordinatiorem tamen ea que primi elementi corporum"; cf. II, 3, 358 a 26.

4 Ibid., I, 33, 89 a 23–39.

tivus—dyalecticus quidem, qui | generat opinionem; demonstrativus vero, *Kl* 1va qui generat scientiam. Similiter eandem conclusionem quam Sortes opinatur, Plato potest scire, | quod non esset nisi idem potest esse scibile et opina- *P* 226^{v} bile.

⟨7⟩ In ista questione primo videndum est quid sit scientia et quid opinio; secundo, quot modis aliquid dicitur 'scibile' et 'opinabile'; tertio, an idem sit scibile et opinabile; quarto respondendum est ad quesitum.

⟨8⟩ Quantum ad primum, sciendum est quod scientia est habitus adhesivus firmus quo aliquis alicui | complexo firmiter vero seu necessario assen- *L* 181va tit sine formidine ad oppositum propter aliquam evidentiam sumptam per principia vel per aliqua ex principiis deducta. Per istam descriptionem scientia distinguitur ab errore quorundam hereticorum, qui quidem error, quamvis sit assensus firmus—immo aliquando ita firmus quod tales heretici mallent mori quam assentire opposito—, tamen non est scientia, quia non est habitus adhesivus complexo vero. Similiter per istam descriptionem scientia distinguitur a fide, quia, quamvis fides sit assensus firmus alicui complexo firmiter, tamen non est scientia, quia non est firmus | propter aliquam *U* 2va evidentiam sumptam per principia vel per ex principiis deducta. Similiter per istam descriptionem scientia distinguitur ab opinione, quia opinio est assensus infirmus seu cum formidine ad oppositum; scientia autem sine formidine. Per istam etiam descriptionem scientia distinguitur ab intellectu seu ab habitu principiorum quia, quamvis habitus primi principii sit habitus adhesivus quo aliquis assentit alicui complexo firmiter vero sine formidine ad oppositum, et quamvis hoc sit propter aliquam evidentiam, tamen non est propter evidentiam sumptam per principia vel per ex principiis deducta. Sic igitur patet descriptio scientie.

⟨9⟩ Opinio autem est habitus quo aliquis assentit alicui complexo cum formidine ad oppositum. Circa quod est sciendum quod | talis formido vel *L* 181vb potest esse ex parte opinantis | seu opinionem habentis de aliquo, vel potest *Kl* 1vb esse ex parte ipsius medii, vel potest esse ex parte ipsiusmet complexi. Verbi gratia, de primo, si esset aliqua conclusio necessaria et scibilis, et Sortes pro

11 ex principiis deducta] experientiis inducta *L U* 13 heretici] *om. Ba Kl L U* 13–14 mallent mori] prius vellent mori *Ba Kl* potius vellent mori *L U* 16–17 alicui complexo firmiter] *om. Ba Kl L U* 17–18 propter aliquam evidentiam sumptam per principia vel per ex principiis deducta] per evidentiam deductam *Ba Kl L* ex experientia adducta *U* 20 infirmus seu] *om. Ba Kl L U* 23 adhesivus] firmus *add. Ba Kl L U* 24 hoc sit] assentit *Ba L U* 25 sumptam per principia vel per] *om. Ba Kl L U* ‖ deducta] deductam *Ba Kl L U* 26 sic igitur patet descriptio scientie] ideo scientia differt ab eis *Ba Kl U om. L* 27 habitus] assensus *P L om. U* 29 esse] vel *add. P*

ista conclusione haberet premissas necessarias, et tunc Sortes cum formidine assentiret et conclusioni et premissis, ista formido diceretur attendi ex parte ipsius opinantis. Exemplum secundi sic: si esset aliqua conclusio necessaria et Sortes pro ista conclusione haberet premissas non necessarias, sed contingentes, tunc Sortes illi conclusioni assentiret cum formidine, que quidem formido esset ex parte medii. Exemplum tertii sic: si esset aliqua conclusio que non esset necessaria, sed contingens, et Sortes assentiret illi *Ba* 4va cum formidine, illa formido diceretur attendi | ex parte ipsiusmet complexi.

⟨10⟩ Quantum ad secundum, sciendum est quod quoddam dicitur 'scibile propinquum', quoddam vero dicitur 'scibile remotum'. 'Scibile propinquum' dicitur illud complexum firmiter verum cui intellectus assentit sine formidine ad oppositum, propter aliquam evidentiam sumptam ex principiis, et similiter significatum illius complexi quod solet vocari 'significabile complexe' et antiquitus vocabatur 'modus rei'. 'Scibile' vero 'remotum' dicitur subiectum scibilis propinqui, accipiendo 'scibile propinquum' pro ipsomet complexo firmiter vero, et non pro significato illius complexi, seu significabili complexe, seu modo rei. Et similiter 'scibile remotum' dicitur res significata per subiectum talis scibili propinqui. Et sicut est distinctum de scibili, ita proportionaliter potest distingui de opinabili.

⟨11⟩ Quantum ad tertium, sit prima conclusio quod omnis res mundi est scibilis, accipiendo 'scibile' pro re significata per subiectum propositionis scibilis. Probatur, quia omnis res mundi est significabilis per istum termi- *U* 2vb *Kl* 2ra num 'ens', | qui quidem | terminus est subiectum alicuius complexi sciti, sicut istius: 'omne ens est substantia vel accidens'.

P 227^{r} ⟨12⟩ Secunda conclusio: omnis res mundi est opinabilis, accipiendo 'opinabile' pro re significata per subiectum conclusionis opinate. Patet, quia omnis res mundi est significabilis per istum terminum 'ens', qui quidem terminus potest esse subiectum alicuius propositionis opinate.

⟨13⟩ Tertia conclusio: accipiendo 'scibile' et 'opinabile' pro complexo scito vel opinato, adhuc idem est scibile et opinabile, prout opinio est habitus infirmus ex parte habentis eam vel ex parte medii. Patet, quia idem complexum potest sciri a Sorte et opinari a Platone. Similiter Sortes potest primo

---

1 haberet] habeat *P* 2 attendi] accidere *Ba Kl* 7 sed contingens] *om. Ba Kl* 11 firmiter] *om. Ba Kl* ‖ verum] et necessarium *add. Ba Kl* ‖ intellectus] firmiter *add. Ba Kl* 12 sumptam ex principiis] etc. *P* 13–14 complexe] complexum *Ba Kl L* 16 vero] etc. *add. P* et necessarium *add. Ba Kl* 20 sit prima conclusio quod] *om. P* 23 est] potest esse *Ba Kl L U* 26 conclusionis] propositionis *Kl* 27 res mundi] res *Ba L U* terminus *Kl* 31 infirmus] *om. Ba Kl L U* ‖ eam] *om. P*

opinari aliquod complexum et postea scire illud idem; potest enim primo habere pro aliquo complexo syllogismum dyalecticum et postea demonstrativum.

⟨14⟩ Quarta conclusio: impossibile est ab aliquo idem complexum simul sciri et opinari. Patet, quia impossibile est quod aliquis assentiat alicui sine formidine et cum formidine. | Hoc enim implicat contradictionem. *L* 182ra

⟨15⟩ Quinta conclusio: accipiendo 'scibile' et 'opinabile' pro complexo scito vel opinato et prout infirmitas assensus est ex parte ipsiusmet complexi, non idem est scibile et opinabile. Patet, quia non idem est contingens et necessarium. Et ita intellexit Aristoteles in primo *Posteriorum*[5]. Ex hiis patet qualiter idem est scibile et opinabile et qualiter non.

⟨16⟩ Quantum ad quartum sit prima conclusio: de impressionibus metheorologicis potest esse scientia. Probatur, nam, per primam conclusionem tertii articuli, omnis res mundi est scibilis; sed impressiones metheorologicis sunt aliqua res mundi; igitur etc.

⟨17⟩ Secunda conclusio: de impressionibus metheorologicis potest esse opinio. Patet per secundam conclusionem tertii articuli, quia omnis res mundi est | opinabilis. *Ba* 4vb

⟨18⟩ Tertia conclusio, que est responsalis | ad quesitum: possibile est *Kl* 2rb de impressionibus metheorologicis simul habere scientiam et opinionem. Patet, quia impressiones simul possunt significari per subiectum conclusionis scite et per subiectum conclusionis opinate. Similiter de impressionibus metheorologicis Sortes potest habere scientiam quando Plato habet de eis opinionem.

⟨19⟩ Quarta conclusio: possibile est quod aliquis de impressionibus metheorologicis habeat scientiam et opinionem. Patet, quia aliquis potest formare in mente sua duas propositiones, unam scibilem et unam opinabilem, que sic se habent quod per utriusque propositionis subiectum impressiones metheorologice significentur. Sed diceres: iam dixisti in una conclu-

---

1 opinari] ymaginari *L* 4 ab aliquo idem complexum] de aliquo idem *Ba* de aliquo *Kl L* idem *U* 6 et] pro eodem tempore assentiat eidem *add. Ba* pro eodem tempore assentiat *add. L U* ‖ formidine] ad oppositum *add. Kl* ‖ contradictionem] *om. P* 10 in primo posteriorum] cum dixit quod impossibile esset pro eodem et in eodem tempore simul habere scientiam et opinionem *Ba Kl U lect. dub. L* 13–14 tertii articuli] *om. Ba L U* 17 secundam] primam *P* ‖ tertii articuli] *om. Ba Kl L U* 25 possibile] *om. P* 26 aliquis] idem *Ba Kl L U* 27–28 unam scibilem et unam opinabilem] unam scitam et unam opinatam *Ba Kl L U*

---

5 Ibid., I, 33, 89 a 23–39.

sione tertii articuli quod impossibile est idem ab eodem scire et opinari. Concedo capiendo ‘scibile’ et ‘opinabile’ pro complexo noto, scito vel opinato; nichilominus possibile est ab eodem sciri et opinari | *U* 3ra accipiendo ‘scibile’ vel ‘opinabile’ pro ‘scibili’ vel ‘opinabili remoto’, videlicet pro re significata per subiectum conclusionis scibilis vel opinabilis, sicut arguebatur in ratione de patre tuo pelle asinina induto.

L 182rb ⟨20⟩ Ad rationes. Ad primam dico quod ista bene probat quod de impressionibus metheorologicis nec est scientia nec opinio, prout ‘scibile’ et ‘opinabile’ accipiuntur pro ‘scibili’ et ‘opinabili propinquo’; nichilominus de eis potest esse scientia et opinio prout ‘scibile’ et ‘opinabile’ accipiuntur pro ‘scibili’ et ‘opinabili remoto’.

⟨21⟩ Ad secundam dico quod ista bene probat quod non est idem scibile et opinabile accipiendo ‘scibile’ et ‘opinabile’ pro ‘scibili’ et ‘opinabili propinquo’, prout infirmitas assensus attenditur ex parte ipsiusmet complexi.

⟨22⟩ Ad tertiam: ‘quod non est verum non contingit scire’; verum est pro scibili propinquo; nichilominus potest sciri tamquam scibile remotum.

⟨23⟩ Ad quartam solet dici quod, quamvis de impressionibus metheorologicis, sicut de aliis rebus corruptibilibus, non possunt formari propositiones necessarie categorice, tamen de eis possunt formari propositiones ypotetice necessarie, | *P* 227^{v} sicut sunt propositiones conditionales: | *Kl* 2va ‘si tonitruum est, tonitruum est sonus factus in nubibus’. Aliter potest dici quod immo de impressionibus metheorologicis possunt formari propositiones necessarie, saltem ex terminis sicut isti: ‘ens’, ‘aliquid’, etc. Verum est tamen quod hoc non intentum est ad propositum.

⟨24⟩ Ad quintam bene conceditur, sicut et Aristoteles dicit in littera, quod impressiones metheorologice secundum naturam inordinatiorem fiunt et minus regularem quam sit naturam corporum celestium[6]. Hoc tamen non

1 tertii articuli] prius posita *add. Ba Kl om. L* ‖ ab eodem] de eodem *Ba Kl U* 2 noto] *om. Ba Kl L U* 3–4 scibile] *om. P* 4–5 videlicet pro re significata per subiectum conclusionis scibilis vel opinabilis] *om. Ba Kl L U* 6 induto] scilicet possibile est idem scire et opinare capiendo scire vel opinare pro scito remoto vel opinato remoto *add. Kl* 7–8 impressionibus] *om. P* 9 et opinabili] *om. L P* 14 attenditur] est *Ba Kl L U* ‖ complexi] opinati *add. Ba Kl L U* 20 sicut sunt propositiones] *om. Ba Kl* 20–21 si tonitruum est tonitruum est sonus factus in nubibus] *om. Ba Kl L U* 23 etc] unum de terminis communibus *Kl* verum *Ba L U* 25 in littera] *om. Ba Kl L U* 26 naturam] et *add. P*

6 Aristoteles, *Meteorologica*, I, 1, 338 b 20–22, translatio Guillelmi de Morbeka (translatio noua), AL X 2.2, 9^{8-10}: “Hec autem sunt quecumque accidunt secundum naturam quidem, inordinatiorem tamen ea que primi elementi corporum”.

obstante, bene fiunt ita ordinate quod sufficit ad hoc quod de eis habeatur scientiam, et de hoc diffusius in alia questione dicetur[7].

I.2 Utrum impressiones metheorologice fiant secundum naturam inordinatiorem quam sit natura celi

Queritur consequenter utrum impressiones metheorologice fiant secundum *Ba* 5ra naturam inordinatiorem quam sit natura celi.

⟨1⟩ Et arguitur primo quod non: impressiones metheorologice non fiunt secundum naturam; igitur non fiunt secundum naturam inordinatiorem. Consequentia tenet. Antecedens probatur, quia impressiones metheorologice fiunt supernaturaliter, nam communiter dicitur quod Deus tonat et quod deus pluit. Sed Aristoteles dicit in *De somno et vigilia*: 'quod omnes vel plures existimant, prestat fidem tamquam ab experientia dictum'; et Commentator ibidem dicit: 'impossibile est ut famosum sit falsum secundum totum'[8].

⟨2⟩ Secundo ad idem: que fiunt secundum naturam, fiunt ab agente pro- *U* 3rb prio et sibi simili preter agens commune, ut homo ab homine, equus ab equo. Sic autem non | est de impressionibus metheorologicis: nix non gene- *L* 182va rat nivem, nec grando grandinem.

⟨3⟩ Tertio ad idem, nam impressiones fiunt per motum violentum; non igitur naturaliter. Consequentia tenet. Antecedens probatur, nam alique impressiones fiunt per ascensum vaporis, qui est de natura aque; ascensus autem vaporis est motus violentus. | Similiter fulmen fit per descensum ignis; *Kl* 2vb descensus autem ignis est motus violentus.

---

2 et de hoc in alia questione dicetur] *om. Ba Kl L U* ‖ diffusius] confusius *P* 4 celi] supercelestium *Ba* corporum supercelestium *Kl* 6 inordinatiorem] quam sit naturam corporum supercelestium *add. Kl L U om. Ba* 8–9 deus tonat et quod deus pluit] deus pluit fulminat et tonat et sic de aliis *Ba Kl U* deus pluit tonat et sic de aliis *L* deus pluit fulminat tonitruat *U* 9 omnes vel] *om. Ba Kl* 14 preter agens commune] quod non agit ut agens commune *Ba Kl* et non ut agens commune *L U*

---

7 I.2.

8 Aristoteles, *De insomniis et de divinatione per somnum*, 3, 1462 b 14–16, ed. H.J. Drossaart Lulofs in *De insomniis et de divinatione per somnum. A New Edition of the Greek Text with the Latin Translation*, Leiden 1947 (Aristoteles Latinus, XV 1–2), 33^{5-7}: "nam quod omnes quidam vel plures existimant habere aliquam significationem sompnia prestat fidem tamquam ab experientia dictum"; Averroes, *In De somno et vigilia*, ed. Shields, 95^{32-33}: "impossibile enim est ut famosum sit falsum secundum totum"; cf. *Auct. Arist.*, 204, nr. 101.

⟨4⟩ Quarto, nam, dato quod impressiones metheorologice fiunt secundum naturam, probatur quod non secundum naturam inordinatiorem; nam sequitur: fiunt secundum naturam inordinatiorem, igitur fiunt secundum naturam inordinatam. Consequens est falsum. Consequentia tenet ex eo quod omnis comparativus presupponit positivum. Falsitas consequentis patet, nam natura non est inordinata. Tertio enim *Celi* et octavo *Physicorum* dicitur: 'nichil est inordinatum eorum que secundum naturam fiunt'[9].

⟨5⟩ Quinto ad idem: fluxus et refluxus maris est quedam impressio metheorologica et fit ita ordinate sicut motus celi; igitur non omnes impressiones fiunt secundum naturam inordinatiorem quam sit natura celi.

⟨6⟩ Sexto ad idem: in universo nichil est inordinatum, ex eo quod universum est pulcherrimum; unde Tullius: 'nichil universo pulchrius vel amabilius', etc.[10].

⟨7⟩ In oppositum est Aristoteles primo huius, ubi dicit quod impressiones fiunt secundum naturam inordinatiorem ea que est primi elementi corporum[11], et per 'primum elementum corporum' intendit celum; unde Commentator tertio *Celi*: 'elementum primum naturaliter dicitur de materia et de forma, sed ex consequenti dicitur de quolibet corpore simplici'[12].

---

4 falsum] ergo et antecedens *add. Kl U* 6–7 tertio enim celi et octavo physicorum dicitur] ut dicitur in tertio huius *Ba Kl L U* 7 nichil est inordinatum eorum que secundum naturam fiunt] nichil eorum est inordinatum *Ba Kl* 16 per] ly *add. Ba* ‖ corporum] corpora *L U* ‖ celum] corpus *Ba* corpus celeste *Kl U* ea scilicet corpus celeste celum et alia *L* 17 primum] proprie *Ba Kl L U* ‖ naturaliter] solummodo *Ba L om. Kl U*

---

9 Aristoteles, *De celo*, III, 2, 300 b 5–6; *Physica*, VIII, 1, 252 a 12–13, translatio Iacobi Veneti (translatio vetus), *AL* VII 1, eds. Bossier and Brams, 1: 282^{11}: "At vero nichil inordinatum est eorum que natura et secundum naturam sunt"; translatio Guillelmi de Morbeka (translatio nova), AL VII 3, 373^{10}: "At vero nihil inordinatum eorum quae natura et secundum naturam sunt".

10 Marcus Tullius Cicero, *De natura deorum*, II, 40, ed. M. van den Bruwaene, Brussels 1978 (Collectio Latomus, 154), 127^{30-31}: "quo spectaculo nihil potest admirabilius esse, nihil pulchrius".

11 Aristoteles, *Meteorologica*, I, 1, 338 b 20–22, translatio Guillelmi de Morbeka (translatio noua), *AL* X 2.2, 9^{8-10}: "Hec autem sunt quecumque accidunt secundum naturam quidem, inordinatiorem tamen ea que primi elementi corporum".

12 Averroes, *Commentarium magnum in De celo*, IV, comm. 37, ed. J.F. Carmody and R. Arnzen, Leuven 2003 (Recherches de Théologie et Philosophie médiévales, Bibliotheca 4) (2 vols.), 2: 735^{19-21}: "*Non dico primum elementum*—quoniam hoc nomen elementum aliquando dicitur de hiis corporibus quatuor, aliquando de prima materia".

Modo celum est primum corpus simplex, sicut patet per Aristotelem in primo *Celi*[13].

⟨8⟩ In ista questione primo videndum est quid sit intelligendum per 'impressiones metheorologicas'; secundo videndum est qualiter fiant secundum naturam; tertio videndum est quali ordine fiant.

⟨9⟩ Quantum ad primum, sciendum quod 'impressio' accipitur uno modo communiter pro quacumque alteratione per quam imprimitur vel inducitur aliqua qualitas, sive sit albedo, sive caliditas, sive sit figura, etc. Et ita communiter non accipitur hic. | Secundo modo accipitur magis stricte pro alteratione | ad illas qualitates que immediate consequuntur qualitates primas vel actionem earum aut elementorum, sicut sunt inflammatio, | congelatio, condensatio, digestio et coagulatio, et sic de similibus. Tertio modo accipitur pro mixto immediate consequente | talem alterationem, sicut | est nubes, grando, fulmen, etc., et etiam accidentia eorum, sicut est color nubis aut motus exalationis, sicut sunt venti aut soni, sicut est tonitruum. Et istis duobus ultimis modis sumitur in proposito.

U 3va
Kl 43vb
P 228r
L 182vb
Ba 5rb

⟨10⟩ Secundo sciendum est quod huiusmodi impressiones fiunt sub orbe Lune in regione elementari, et ut in pluribus super nos in aere, sicut est generatio ventorum, nubium et huiusmodi; et etiam aliqua earum fiunt sub terra, sicut fontium generationes.

⟨11⟩ Tertio sciendum est quod alique earum sunt maioris durationis, alique vero minoris, de quo diffusius apparebit in processu.

⟨12⟩ Quantum ad secundum, sciendum est primo quod hoc nomen 'natura' non solum dicitur de principio activo motus, verum etiam de passivo, et non solum de principio activo motus particulari, verum etiam de principio activo motus universali, sicut est celum et intelligentia. Nec etiam solum dicitur de principio activo motus principali, verum etiam de principio activo motus instrumentali, sicut est gravitas vel levitas. Similiter non solum dicitur de principio passivo motus remoto, sicut est materia prima, verum etiam de principio passivo motus propinquo, sicut est ipsum mobile quod movetur, sicut ignis vel aqua, et hoc large accipiendo hoc nomen 'natura'.

---

1 primum corpus simplex] primum corporum simplicium *Ba* corporum simplicium primum *Kl* primum corpus et simplicissimum *L* ‖ per aristotelem in] *om. Ba Kl L U* 9 magis stricte] *om. Ba Kl L U* 10 alteratione] prima *add. Ba Kl L U* 13 immediate] *om. Ba Kl L U* 14 color] calor *L* 18–19 est generatio] grandines *Kl* tumores *L* 24 de] *om. P*

---

13 Aristoteles, *De celo*, I, 2, 269 a 18–269 b 17.

⟨13⟩ Tunc sit prima conclusio quod impressiones metheorologice fiunt secundum naturam quantum est ex parte agentis. Patet, quia fiunt a celo et elementis, sicut dicit in littera; vult enim quod celum est causa activa impressionum[14].

Kl 3rb ⟨14⟩ Secunda conclusio: alique impressiones non solum fiunt secundum naturam quantum est ex parte agentis, verum etiam quantum est ex parte passi, sicut est descensus pluvie; iste enim est cum inclinatione naturali ipsius passi.

⟨15⟩ Tertia conclusio: alique impressiones fiunt violente quantum est ex *U* 3vb parte | passi. Patet, nam descensus fulminis et elevatio vaporis, quamvis quantum est ex parte agentis fiant naturaliter, tamen fiunt violente quantum est ex parte passi. Patet ergo qualiter impressiones metheorologice fiunt secundum naturam et qualiter non.

⟨16⟩ Quantum ad tertium, sit prima conclusio: universum et omnia que *L* 183ra sunt in ipso sunt bene ordinata loco, tempore et proportione, comparando | quodlibet eorum ad omnia alia. Probatur, quia omnia sunt bene disposita que sunt in universo; igitur omnia que sunt in universo sunt bene ordinata. Consequentia clara est de se. Antecedens patet, quia duodecimo *Metaphysice* dicitur quod entia nolunt male disponi[15].

⟨17⟩ Secundo: omnia que sunt in universo sunt disposita secundum quandam consonantiam et armonicam proportionem; igitur omnia que sunt in universo sunt bene ordinata. Consequentia tenet; antecedens patet per Boe- *Ba* 5va thium in sua *Musica*, qui dispositionem et ordinem universi vocat | 'musicam mundanam', que non est auribus sensibilis, sed bene intelligibilis, et est debita ordinatio et proportio numero, pondere et mensura[16].

⟨18⟩ Sed diceret aliquis contra: multa sunt in universo que non sunt bene ordinata, sicut sunt monstrua et peccata et alia mala. Respondetur quod *Kl* 3va quamvis ista secundum se videntur esse inordinata, tamen in ordine ad |

---

3 in littera] *om. Ba Kl L U* 4 impressionum] et qualitates elementorum active concurrunt ad productionem impressionum *add. Ba Kl* et similiter qualitates elementorum active concurrunt ad productionem impressionum *add. L U* 21 et armonicam proportionem] et armoniam et proportionem *L* 24 sensibilis] perceptibilis *L U* 25 proportio] dispositio *Ba Kl U* ‖ mensura] et figura *add. Kl* 27 et peccata] *om. Ba Kl U* 28 ordine] comparatione *U*

---

14 Aristoteles, *Meteorologica*, I, 2, 339 a 22–32.

15 Aristoteles, *Metaphysica*, XII, 10, 1076 a 3–5, translatio Guillelmi de Morbeka (translatio noua), *AL* XXV 3.2, $269^{535-536}$: "entia uero non uolunt disponi male"; cf. *Auct. Arist.*, 138, n. 279.

16 Boethius, *De institutione musica*, I, 2.

totum sunt ita bene ordinata quod non possent melius ordinari. Et sciendum est quod ex hoc universum est pulchrius quam si talia non essent; unde, si omnia equaliter essent bona, universum non esset ita pulchrum sicut modo. Bene hoc potest patere exemplo: | si aliquis paries esset totaliter albus, non *P* 228^{v} tamen ex hoc delitiaretur visus sicut si in una parte esset albus et in alia niger et in alia rubeus, et sic diversis coloribus coloratus. Ita consimili modo ymaginandum est universum. Similiter si essent aliqui qui eque alte sonarent, auditus in audiendo illum sonum non tantum delectaretur sicut delectaretur quando unus cantat alte et alius basse, dummodo tamen hoc fit cum ordine; consimili modo de universo.

⟨19⟩ Secunda conclusio: | impressiones metheorologice fiunt secundum *U* 4ra naturam ordinatam. Patet, nam per precedentem conclusionem omnia entia mundi sunt bene ordinata. Secundo ad idem: motus celi, qui sunt causa istarum impressionum, fiunt ordinate; igitur et huiusmodi impressiones fiunt ordinate. Consequentia tenet, quia ordinatio in causa arguit ordinem in effectu. Tertio: quod huiusmodi impressiones fiant | ordinate patet; *L* 183rb quedam enim impressiones presciuntur ex aliis et consequuntur alias, sicut patebit in processu[17]. Istud autem videtur arguere ordinem debitum huiusmodi impressionum.

⟨20⟩ Tertia conclusio: in isto ordine huiusmodi impressionum est maior difformitas quam sit in motibus celi. Patet per experientiam, quia non ita faciliter nec ad tantum tempus potest presciri pluvia sicut potest | presciri *Kl* 3vb coniunctio aliquarum planetarum.

⟨21⟩ Quarta conclusio: huiusmodi impressiones non fiunt ita ordinate sicut motus celi. Istam conclusionem intendebat Aristoteles cum dixit: 'hec autem fiunt secundum naturam inordinatiorem'[18], ita quod li 'in' non debet

1 totum] universum *add. Ba Kl* 2 essent] in isto mundo *add. Kl* 4 albus] depictus sicut dealbatus totaliter *Ba* dealbatus vel totaliter denigratus *Kl* dealbatus *L U* 4–5 non tamen ex hoc delitiaretur visus] non esset ita pulcher *Ba Kl L U* 5 albus] dealbatus *L* 6 et sic diversis coloribus coloratus] et sic de aliis mediis coloribus *Ba U* et sic de aliis *L* similiter si hic esset depictus passer hic aquila et hic mustela et hic milnus *add. Kl* 7 universum] in proposito *Ba Kl L U* ‖ sonarent] cantarent *Ba Kl L U* 8 auditus] intellectus *Ba Kl U* intelligentes *L* ‖ delectaretur] delectarentur *L* 8–9 delectaretur] delectarentur *L* 13 mundi] *om. Ba Kl L U* ‖ ordinata] disposita *Ba* et bene disposita *add. Kl* 15 tenet] et antecedens patet *add. Kl L U* 22 faciliter] finaliter *Ba* ‖ nec] ita cito nec *add. Kl* 23 coniunctio] motus seu *add. Kl* effectus *L*

17 II.6.

18 Aristoteles, *Meteorologica*, I, 1, 338 b 20–22, translatio Guillelmi de Morbeka (translatio noua), *AL* X 2.2, 9^{8-10}: "Hec autem sunt quecumque accidunt secundum naturam quidem, inordinatiorem tamen ea que primi elementi corporum".

exponi pro ‘non’ in proposito, sed ad istum intellectum quod impressiones metheorologice fiunt secundum naturam inordinatiorem, id est minus ordinate quam motus celi, modo quo dicimus unum equum esse peiorem alio, *Ba* 5vb hoc est minus bonum, | quia ubi est maior difformitas et minus uniformiter fiunt, dicuntur minus ordinate fieri. Sed ita est de impressionibus metheorologicis in respectu ad motum celi; igitur etc.

⟨22⟩ Ad rationes. Ad primam negatur quod huiusmodi impressiones fiant supernaturaliter. Et cum dicebatur: ‘tamen communiter dicitur’, etc., hoc verum est, sed non a sapientibus. Et igitur ad auctoritatem Aristotelis et Commentatoris est dicendum quod illud quod plures—supple: sapientes—dicunt, prestat fidem; si tamen hoc plures ignorantes dicerent, non propter hoc fidem prestaret.

⟨23⟩ Ad secundam: ‘que fiunt secundum naturam fiunt a sibi simili’, breviter: hoc non est universaliter verum, immo hoc oportet intelligi de generatione et factione substantiarum perfectarum.

⟨24⟩ Ad tertiam dico quod ista bene probat quod alique impressiones non fiunt naturaliter, sed violente, quantum est ex parte passi; nichilominus bene fiunt naturaliter quantum est ex parte agentis, sicut dicebatur in secundo articulo questionis[19].

⟨25⟩ Ad quartam dico quod predicta auctoritas debet exponi: ‘secundum *Kl* 4ra naturam inordinatiorem’, id est: ‘minus ordinate’, et hoc in | comparatione ad motus celi; nec tunc sequitur: ‘igitur fiunt secundum naturam inordinatam’.

⟨26⟩ Ad quintam respondetur quod non omnis fluxus et refluxus maris fit *L* 183va ita ordinate sicut motus celi; immo aliquando | fit maior, aliquando minor.

⟨27⟩ Ad sextam: illa probat quod impressiones metheorologice non fiunt inordinate; tamen, hoc non obstante, fiunt minus ordinate quam motus celi.

Hoc de questione.

1–2 ad istum intellectum quod impressiones meteorologice] *om. P* 2 secundum naturam inordinatiorem] inordinate *P* 4 maior difformitas] minor uniformitas *Ba Kl L U* ‖ minus uniformiter] magis difformiter *Ba Kl* maior difformitas *L U* 5 fiunt] difformiter se habent *L U* 20 predicta auctoritas] *om. P* 22 sequitur] ista fiunt secundum naturam inordinatiorem *add. Kl*

---

19 I.2, 14.

I.3 Utrum iste mundus inferior sit continuus lationibus superioribus ut virtus eius inde gubernetur

Queritur consequenter utrum mundus iste inferior sit continuus lationibus *U* 4rb superioribus ut virtus eius inde gubernetur.

⟨1⟩ Ista questio querit primo utrum iste mundus inferior sit continuus lationibus superioribus; secundo querit utrum virtus istius mundi inferioris gubernetur a lationibus superioribus.

⟨2⟩ De primo probatur primo quod iste mundus non est continuus lationibus superioribus; nam continua sunt quorum motus est unus secundum se; sed sic non est de isto mundo inferiori et celo, seu lationibus superioribus; igitur etc. Maior | patet per diffinitionem continuorum[20]. Minor declaratur, quia *P* 229^{r} celum movetur circulariter; ista autem inferiora, que sunt gravia vel levia, moventur motu recto.

⟨3⟩ Secundo, continuorum extrema debent esse unum[21]; sed sic non est de isto mundo inferiori et ipso celo; nam extremum istius mundi inferioris et extremum celi sunt diversarum specierum, et per consequens non potest ex eis fieri per se unum, quia non se habent ad invicem sicut actus et | potentia. *Ba* 6ra

⟨4⟩ Tertio probatur quod nec iste mundus celo est contiguus; nam contiguorum extrema debent esse simul[22]. Sed extrema istius mundi inferioris et ipsius celi non sunt simul. Probatur, quia non sunt in eodem | loco, quia *Kl* 4rb tunc requireretur ad hoc quod essent simul simultate locali. Quod non sunt in eodem loco patet, quia extremum istius mundi inferioris est in loco ignis; extremum autem ipsius celi est super locum ignis.

⟨5⟩ Deinde probatur quod non totus iste mundus inferior gubernatur a lationibus superioribus seu a motibus celi, quia voluntas humana est pars istius mundi inferioris et tamen non gubernatur a motibus celi, quia si sic, sequeretur quod non esset virtus libera potens libere in opposita[23]. Etiam sapiens vir dominabitur astris, quod non esset si totus iste mundus inferior gubernaretur et regeretur | a motibus celi. *L* 183vb

---

7 secundum se] in se *L om. Ba Kl U* 18 sunt] possunt esse *Ba Kl L U* 19 sunt] possunt esse *Ba Kl L U* 25 opposita] quod est contra aristotelem secundo libro ethicorum *add. Kl*^{pc} quod est contra aristotelem tertio ethicorum *add. Ba* quod est contra aristotelem in libro ethicorum *add. Kl*^{ac} *L U* 26 dominabitur] predominabitur *L U*

---

20 Cf. Aristotelem, *Physica*, V, 4, 228 a 20–22.

21 Cf. Ibid., V, 4, 228 a 22–24.

22 Cf. Ibid., IV, 3, 227 a 6–7.

23 Cf. Aristotelem, *Ethica*, III, 1–3.

⟨6⟩ Secundo ad idem, nam, si totus iste mundus inferior gubernaretur a motibus celi, sequeretur quod ista inferiora fierent ita ordinate sicut sunt lationes superiores; sed hoc est falsum et contra illa que dicebantur in questione precedenti[24]. Consequentia declaratur a simili, nam, si sunt duo | [*U* 4va] horologia et motus unius regit motum alterius, tunc tanta esset ordinatio motus unius sicut motus alterius.

⟨7⟩ Tertio ad idem: gubernare et regere denotat cognitionem, ut patet primo *Politice*[25]. Motus autem celestes, cum sint accidentia, non cognoscunt, et per consequens non regunt nec gubernant ista inferiora; et similiter celum est unum corpus simplex non cognoscens.

⟨8⟩ Quarto ad idem: gubernare ista inferiora est agere in ea; sed modo celum non agit in ista inferiora, quia que sic se habent quod unum agit in aliud, debent communicare in materia. Celum autem et ista inferiora non communicant in materia, ex eo quod materia celi non est | [*Kl* 4va] subiectum generationis et corruptionis; materia autem istorum inferiorum est subiectum talium; quare una earum est alterius rationis ab alia.

⟨9⟩ Quinto ad idem: aliqua fiunt infra terram ad quorum factionem non concurrit ipsum celum; igitur non omnia que fiunt in isto mundo inferiori fiunt virtute celi. Antecedens patet, nam multa fiunt infra terram ad que non potest attingere lumen celi; sed celum illa que agit in istis inferioribus, producit mediante suo lumine; si ergo ad aliqua lumen celi non attingit, videtur quod ad productionem talium non concurrat.

⟨10⟩ Sexto: posito quod celum non moveretur, adhuc lapis molaris existens in concavo orbis Lune, remoto prohibente, descenderet. Similiter, si celum non moveretur, non est verisimile quin, si ignis approximaretur stuppe, eam accenderet et combureret[26]. Igitur | [*Ba* 6rb] videtur quod non ad cuiuslibet effectus productionem in isto mundo inferiori concurrat motus celi, et per consequens non quelibet actio in isto mundo inferiori gubernatur et regitur a motibus celi.

---

5 regit] tangeret *L U* 10 celum] corpus celeste *Ba Kl L U* 11 gubernare] regere *Ba Kl L U* 12 inferiora] ergo nec gubernat seu regit ista inferiora minor patet *add. Ba Kl* ergo nec gubernat inferiora probatur *add. L* ergo nec gubernat inferiora *add. U* 16 una earum est alterius rationis ab alia] ista est alterius rationis a materia celesti *Ba Kl* ipsa est alterius rationis quam materia celestium *L U* 21 attingit] potest pertingere *Ba* potest attingere *Kl U* 23 lapis molaris] homo *Kl* lapis *Ba L U* 24–25 si celum non moveretur] *om. Ba Kl* 25 si ignis approximaretur stuppe] ignis applicatus stuppe *Ba Kl L U*

---

24 I.2, 21–22.

25 Aristoteles, *Politica*, I, 19, 1254 b 5–9; cf. *Auct. Arist.*, 253, n. 28.

26 Cf. Condamnationem parisiensem (anno 1277), art. 156, ed. D. Piché, Paris 1999, 127: "Quod si celum staret, ignis in stupam non ageret, quia natura deesset".

⟨11⟩ In oppositum est Aristoteles in primo huius, tractatu primo, capitulo secundo, ubi dicit ista verba: 'est autem ex necessitate continuus iste mundus lationi|bus superioribus, ut omnis ipsius virtus gubernetur inde'[27]. *L* 184ra

⟨12⟩ In ista questione primo videndum est utrum iste mundus inferior sit continuus vel contiguus ipsi celo; secundo videndum est utrum omnis virtus istius mundi inferioris gubernetur a celo.

⟨13⟩ Quantum ad primum, sciendum est primo quod per 'mundum inferiorem' intenditur ista tota massa que est infra | orbem Lune, seu in concavitate celi, que quidem massa | aggregata est ex mixtis et quatuor elementis. *P* 229^{v} *Kl* 4vb

⟨14⟩ Secundo sciendum est quod potest ymaginari primo de isto mundo inferiori quod tota materia istorum inferiorum esset una massa continua, extensa circumquaque usque ad orbem Lune, ita quod materia aeris circumstantis Sortem esset continua materie ipsius Sortis; nam, si Sortes corrumpitur et ad ymaginationem maneat materia eius, sic tamen quod de eius potentia non educeretur alia forma, et similiter si forma aeris circumstantis Sortem corrumperetur absque hoc quod alia forma de potentia materie ipsius aeris educeretur, ille due materie essent ad invicem continue et non essent propinquiores tunc quam prius erant; ergo videtur quod etiam prius erant continue, et ita argueretur de aliis. Et secundum istam ymaginationem potest teneri quod iste mundus inferior esset una massa continua recte, sicut unus paries est continuus, quamvis eius una pars sit informata albedine, alia modo nigredine et alia rubedine, etc. Et ita proportionaliter de tota massa materie istius mundi inferioris et formis substantialibus potest dici. Et ulterius secundum istam ymaginationem potest dici quod materia totius mundi—accipiendo 'mundum' pro aggregato ex celo et isto mundo inferiori—esset una massa continua, quamvis una pars materie, puta celum, esset satiata forma quam habet et materia istorum inferiorum non, et quod propter hoc materia celi non est subiectum generationis et corruptionis, qualiter tamen est materia istorum inferiorum. Et secundum istam ymagina- *U* 4vb

---

2 ubi] unde *P* ‖ ex necessitate] *om. Ba Kl L U* 13 sortes] forma sortis *Ba Kl L U* 14 maneat materia eius sic tamen quod] *om. Ba Kl L U* 14–15 de eius potentia] de potentia materie eius *Ba Kl L U* 18 ergo] etiam *add. P* 20 quod] totus *add. Ba Kl L U* ‖ una massa continua recte] continuus ipsi celo recte *Ba Kl L U* 26 pars materie puta celum] eius pars puta materia celi *Ba Kl L U*

---

27 Aristoteles, *Meteorologica*, I, 2, 339 a 22–23, translatio Guillelmi de Morbeka (translatio noua), *AL* X 2.2, 10^{34-35}: "Est autem ex necessitate continuus iste superioribus lationibus, ut omnis ipsius virtus gubernetur inde"; cf. *Auct. Arist.*, 171, n. 2.

tionem ulterius potest dici quod iste mundus inferior continuus est ipsi celo, nec oportet quod omnes partes eiusdem continui moveantur uno motu con- *L* 184rb similí; | nam Secana est una massa continua, et tamen una eius pars potest moveri versus unam partem et alia versus aliam.

Kl 5ra ⟨15⟩ Oportet veruntamen istud esse | contra communiter loquentes; communiter enim dicitur quod ad hoc quod aliqua sint continua, requiritur quod ultima eorum sint unum per se; sed modo, quia ultimum istius mundi inferioris et ultimum ipsius celi non sunt unum per se nec constituunt unum *Ba* 6va per se, cum sint diversarum specierum et unum non est in potentia | ad reliquum, sequitur istum mundum inferiorem non esse continuum ipsi celo, et consimili ratione nec aer circumstans Sortem est continuus Sorti. Et cum arguebatur: 'materia Sortis et materia aeris circumstantis Sortem sunt continue; ergo Sortes et aer sunt continui', negetur antecedens. Et cum argueba- *U* 5ra tur: 'si starent sine formis, | essent continue', admittatur. Et cum dicebatur: 'nunc sunt ita propinque sicut tunc', admittatur. Et cum ulterius dicebatur: 'igitur nunc sunt continue', negatur consequentia, quia continuitas non attenditur solum penes propinquitatem partium, sed etiam penes hoc quod ultima eorum sint unum illorum (supple: que sunt continua). Sed modo, cum partes materie sunt informate formis distinctis specie, non constituunt unum per se, cum ex distinctis specie quorum unum non est in potentia ad aliud non potest fieri unum per se. Finaliter igitur dicendum est secundum communiter loquentes quod iste mundus inferior non est continuus ipsi celo, sed contiguus. Patet, nam ultima eorum non sunt unum sed simul, ex eo quod inter ea nullum est medium.

⟨16⟩ Quantum ad secundum, sciendum est primo quod per 'virtutem istius mundi inferioris' debemus intelligere non solum virtutem activam, verum etiam passivam, et per 'gubernare' debemus intelligere idem quod agere, sicut principale agens intelligendo bonum finem. Secundo sciendum est quod illud dicitur fieri a motibus supercelestibus quod fit ab intelligentiis et corpore celesti totaliter moto.

Kl 5rb *P* 230r ⟨17⟩ Tunc | sit prima conclusio: celum | per suum motum agit in ista inferiora. Patet per experientiam, nam videmus quod Sol per accessum ad nos

2 omnes] *om. Ba Kl L U* ‖ continui] simul *add. Ba Kl L U* 3 secana] lancea mota ad latus *Ba Kl* lapis *L* lacus *U* 3–4 potest moveri] movetur *Ba Kl L U* 8–9 per se nec constituunt unum per se] *om. Ba Kl L U* 13 antecedens] consequentia *Ba Kl L U* 14 starent sine formis] materie essent separate a formis *Ba Kl* si essent separate a formis *L U* 20 per se] continuum *Ba Kl* compositum seu continuum *L U* 27 agere] regere *Ba Kl L U* 28 sicut principale agens intelligendo] *om. Ba* ‖ bonum finem] finem intentum *Kl L U* 29 supercelestibus] celi *Ba Kl L U* 30 totaliter] taliter *Ba Kl L U*

et recessum in obliquo circulo est causa generationis et corruptionis multorum in hoc mundo. Cum enim Sol accedit ad nos, sicut in vere, incipiunt | *L* 184va florere arbores et generantur de novo multe albe; et iterum per recessum a nobis, sicut in autumno, huiusmodi e contrario tendunt ad corruptionem; unde Aristoteles , in secundo *De generatione*: 'motus Solis in obliquo circulo est causa generationis et corruptionis in istis inferioribus'[28]. Similiter in secundo *Physicorum*: 'homo generat hominem et Sol'[29]. Item experimur quod mare fluit et refluit secundum motum Lune, quod non esset nisi ageret in ista inferiora. Item aliqui planete per eorum accessum causant frigiditatem in istis inferioribus, aliqui vero caliditatem. Item in certis temporibus est generatio aliquorum animalium et similiter aliquarum plantarum, | *Ba* 6vb sic quod non in aliis temporibus, sicut experimur. Ex omnibus istis sequitur quod celum agit in ista inferiora, immo adhuc astrologi dicunt quod celum | *U* 5rb agit in ista inferiora fortunia et infortunia, propter quod ipsi sciunt, scita nativitate alicuius hominis, an talis in futuro prosperabitur vel non.

⟨18⟩ Secunda conclusio: celum agit in ista inferiora triplici instrumento, videlicet motu, lumine et influentia. Motu quidem, quia motu celo movetur spera ignis et superior pars aeris una cum ipso, sicut postea videbitur[30]. Ex quo apparet quod celum motu suo agit in ista inferiora. Similiter celum agit in ista inferiora suo | *Kl* 5va lumine. Patet, quia celum per lumen causat calorem in istis inferioribus, de quo etiam postea dicetur[31]. Similiter celum agit in ista inferiora influentia. Patet, nam aliqua fiunt in terra, sicut metalla et consimilia, ad que non attingit lumen celi; igitur oportet dicere quod celum sua influentia concurrat ad talium productionem. Similiter etiam mare fluit et refluit quando Luna est in puncto medie noctis, nec tunc eius lumen attingit mare; ideo oportet dicere quod Luna sua influentia sit causa talis fluxus maris.

---

1 in obliquo circulo] a nobis *Ba Kl L U* 3 multe albe] multi fructus *Ba Kl L U* 8 secundum motum lune] *om. Ba Kl L U* ‖ ageret] propter influentiam lune *Ba Kl L U* 14–15 propter quod ipsi sciunt scita nativitate alicuius hominis] *om. P* 20 celum per lumen] lumen *Kl* lumen suum *Ba L U* 20–21 in istis inferioribus] in mundo isto inferiori *Ba Kl L U* 22 aliqua] multa *Ba Kl L U* ‖ metalla] gemme *add. Ba Kl L U* 24–25 et refluit] *scrips. et del. P* 25 quando] quia *P* ‖ medie] medio *P* 26 luna sua influentia] influentia lune *Ba* lune influentia *Kl* ‖ sua influentia] *om. P*

---

28 Aristoteles, *De generatione et corruptione*, II, 10, 336 a 15–b 15; cf. *Auct. Arist.*, 170, n. 41: "motus solis et aliorum planetarum in obliquo circulo est causa generationis et corruptionis rerum inferiorum".

29 Aristoteles, *Physica*, II, 2, 194 b 13, *Translatio Iacobi Veneti* (translatio vetus), *AL* VII 1, 55^{8}.

30 I.8, 23.

31 I.9, 18.

⟨19⟩ Est sciendum quod talis influentia est una qualitas distincta a lumine et est qualitas insensibilis, diffusa per totum medium, recte sicut est species coloris. Differt a lumine, quia lumen est qualitas sensibilis; huiusmodi autem influentia est qualitas insensibilis. Similiter influentia non impeditur per aliquid opacum; lumen autem impeditur per aliquid opacum, sicut per murum vel per aliquid | (*L* 184vb) huiusmodi. Similiter lumen causatur solum ab astro; influentia autem causatur a qualibet parte celi.

⟨20⟩ Tertia conclusio: omnis virtus naturalis huius mundi inferioris gubernatur a celo. Patet, quia aliqua virtus naturalis huius mundi inferioris gubernatur a celo, sicut patet ex dictis, et non plus una quam alia; igitur etc. Secundo, quia motus celi est primus motuum; igitur omnes alii motus ab ipso dependent, et celum mediante suo motu est causa cuiuslibet motus in istis inferioribus; ergo etc[32]. Tertio, quia solus motus celi est eternus; igitur videtur quod sit regula istorum motuum inferiorum, et per consequens motus istorum inferiorum et actiones gubernantur et reguntur a motu celi[33]. Ex ista conclusione sequitur quod oportet istum mundum inferiorem esse contiguum ipsi | (*Ba* 7ra) celo. Patet; aliter enim celum non posset agere in ipsum mundum inferiorem. Si enim inter istum mundum inferiorem et ipsum celum | (*Kl* 5vb) esset spatium vacuum, celum non posset agere in ista inferiora, ex eo quod in vacuo | (*U* 5va) non esset aliquod subiectum receptivum luminis et influentie celi quod | (*P* 230^{v}) esset delativum istorum in mundum inferiorem. Nichil enim agit in distans nisi per medium: si igitur deficeret medium receptivum actionis, non posset fieri actio in distans.

⟨21⟩ Quarta conclusio: celum non agit in voluntatem cogendo ipsam, sed bene inclinando ipsam. Patet, aliter enim actus voluntatis non essent liberi, quod est falsum, quia experimur quod libere fiunt in nobis actus voluntatis.

⟨22⟩ Ad rationes. Ad primam concedo quod mundus iste inferior non est continuus lationibus superioribus, sicut due prime rationes probant. Et cum Aristoteles dicit quod iste mundus inferior est continuus celo, etc., oportet exponi pro li 'continuus' idem quod 'contiguus', et ideo utitur isto nomine

---

5–6 per murum] puta a muro vel ferro *L U* 11 motus] istorum inferiorum *add. Kl L U* 13 eternus] necessarius *Ba Kl L U* 15 et actiones] *om. Ba Kl L U* 21 delativum] et portativum *add. U* ‖ mundum inferiorem] ista inferiora *Ba Kl L U* 22–23 deficeret medium receptivum actionis] illud medium non esset receptivum actionis *Ba Kl L U* 24 voluntatem] humanam *add. Ba Kl L U* 25 liberi] actus puri liberi arbitrii *Ba* actus liberi arbitrii *Kl* actus pure liberi *L* pure liberi *U* 28 lationibus superioribus] celo *Ba Kl L U* ‖ rationes] ante oppositum *add. Ba Kl L U* 30 li] ly *Ba Kl L U*

---

32 Cf. Aristotelem, *Metaphysica*, II, 2, 994 a 18–20.

33 Cf. Aristotelem, *Meteorologica*, I, 2, 339 a 24–27.

'continuus' improprie. Vel forte littera habet 'contiguus', et potest esse ex vitio translatoris vel scriptoris, unde communiter expositores pro 'continuus' exponunt 'contiguus'[34].

⟨23⟩ Ad tertiam: 'quecumque sunt contigua, eorum ultima debent esse simul', concedo. Et quando dicitur quod sic non est de ultimo celi et istius mundi, negatur: bene sunt simul propter hoc quod inter ea nullum est medium.

⟨24⟩ Ad alias rationes quibus probatur quod iste mundus non gubernatur a celo—nam voluntas, etc.—, de hoc dictum est in ultima conclusione[35].

⟨25⟩ Ad secundam: 'sequitur quod ista inferiora fierent ita ordinate sicut *L* 185ra motus celi', negatur, quia non omnia reguntur a lationibus superioribus, puta voluntas humana et actus eius. Etiam posito quod omnia indifferenter et equaliter | gubernarentur a celo, tamen, quia est maior difformitas in *Kl* 6ra istis inferioribus, non apparet ibi tanta ordinatio sicut in motibus celi.

⟨26⟩ Ad tertiam conceditur quod nec motus celi nec celum, circumscripto motore, cognoscit. Nichilominus intelligentia movens celum et producens motum cognoscit, et hoc sufficit ad hoc quod large loquendo dicamus celum regere et gu|bernare ista inferiora. *Ba* 7rb

⟨27⟩ Ad quartam: 'celum non agit', etc., negatur. Et cum dicebatur: 'quia celum et ista inferiora non communicant in materia', admittatur. Et propter hoc bene conceditur quod celum non agit in ista inferiora sic quod ista inferiora reagant in ipsum celum, seu non agit in ista inferiora mediante qualitate habente contrarium in materia. Et propter hoc non oportet quod celum et ista inferiora communicent in materia. Unde, si aliqua agerent et paterentur ad invicem, tunc bene oporteret quod | communicarent in mate- *U* 5vb ria; sed sic non est de celo et de istis inferioribus: ista enim inferiora non agunt in celum, quamvis tamen bene e contrario.

⟨28⟩ Ad quintam respondetur quod quamvis celum non concurrat ad productionem eorum que fiunt in terra suo lumine, tamen concurrit ad productionem eorum sua influentia.

---

6 mundi] inferioris *add. Ba Kl U* 8 gubernatur] regularetur *Ba Kl* regeretur *L U* 9 voluntas] humana *add. Ba Kl L U* 10–11 ita ordinate sicut motus celi] inordinate *P* 11 non] *om. P* ‖ omnia] ista inferiora *add. Ba Kl L U* 13 gubernarentur] regerentur *Ba Kl U* 27 e contrario] celum agat in ipsa *add. Ba Kl L U*

---

34 Cf. Thomam de Aquino, *Super Meteora*, lib. 1, cap. 2, lect. 3, in *Opera omnia* iussu impensaque Leonis XIII P. M. edita, Rome 1886, 3: 329: "*continuum* autem hic accipit pro *contiguo*".

35 I.3, 19.

⟨29⟩ Ad ultimam respondetur quod si celum quiesceret, forte, naturaliter loquendo, lapis in concavo orbis Lune non descenderet, nec ignis applicatus stuppe in eam ageret. Sed finaliter, secundum veritatem loquendo, quamvis Commentator non dixisset, sic dicendum est: quod quamvis celum quiesceret, et tamen prima causa influentiam suam non retraheret, adhuc lapis | [Kl 6rb] descenderet et ignis combureret stuppam. Sed si prima causa omnem suam influentiam retraheret, omnia subito adnichilarentur. Unde ymaginandum est quod omnia bene et semper ab ipso primo dependent in esse et in conservari sicut lumen dependet a corpore luminoso; | [L 185rb] unde quam cito corpus luminosum retraheretur, tam cito lumen in medio desinit esse. Ita, tam cito omnia entia a primo in nichil caderent, quam cito prima causa omnem suam influentiam retraheret.

I.4 Utrum, cessante motu celi, cessarent motus in isto mundo inferiori

Queritur consequenter utrum, cessante motu celi, cessarent motus in isto inferiori.

⟨1⟩ Et arguitur primo quod non, nam, si sic, sequeretur quod si cessaret motus celi, ista inferiora essent magis perfecta quam sunt modo et melius ordinata; sed hoc est falsum et contra prius | [P 231r] dicta[36]. Et consequentia probatur, nam, si cessante motu celi cessarent motus istorum inferiorum, sequeretur quod cessante motu celi ista inferiora manerent perpetua et si sic, essent | [Ba 7va] perfectiora quam modo et melius ordinata, ut videtur. Cum enim non essent

1–2 naturaliter loquendo] *om. Ba Kl* 2 lapis] molaris existens *add. Ba Kl L* existens *add. U* 3 ageret] ut dicunt philosophi *add. Ba Kl L* 4 commentator non dixisset] aristoteles et commentator sic non dixissent *Ba Kl L U* 5 influentiam suam non retraheret] influeret istis inferioribus *Ba Kl L* influeret in ista inferiora *U* ‖ lapis] molaris de concavo orbis lune *add. Kl* molaris existens in concavo orbis lune *add. Ba L* de concavo orbis lune *add. U* 6 combureret] applicatus stuppe in eam ageret *Kl U* applicatus stuppe eam combureret *Ba* appositus stuppe in eam ageret *L* ‖ stuppam] quia omnia dependent ab ipso primo in esse et conservari *add. Ba Kl U* quia omnia dependent ab ipso primo in esse et consiliari et conservari *add. L* 7 subito] penitus *Ba Kl L U* 8–9 quod omnia bene … a corpore luminoso] de prima causa et sua influentia sicut de corpore luminoso et lumine *Ba Kl L U* 10 in medio desinit esse] deficeret *Ba Kl L U* 10–12 ita tam cito omnia entia a primo in nichil caderent quam cito prima causa omnem suam influentiam retraheret] et sic est de prima causa *Kl U om. Ba* etiam de prima causa *L* 16 et melius ordinata] *om. Ba Kl L U*

36 I.2, 21–22.

aliqui motus in istis inferioribus, non esset alteratio, et per consequens nec generatio nec corruptio.

⟨2⟩ Secundo, nam, cessante motu celi, non ex hoc cessaret illuminatio et influentia celestis; sed illuminatione et influentia celesti manente, adhuc possunt fieri motus in istis inferioribus, sicut alie alterationes et huiusmodi.

⟨3⟩ Tertio sic: cessante motu celi, non est verisimile quin, si adhuc ignis approximaretur stuppe, ageret in ipsam et unum contrarium applicatum alteri | [*Kl* 6va] ageret in ipsum[37].

⟨4⟩ Quarto ad idem: intelligentie et celum mediante solo motu regunt ista inferiora et motus et actiones istorum. Igitur, quamvis | [*U* 6ra] cessante motu celi non ita ordinate fierent motus et actiones istorum inferiorum, non tamen propter hoc desinerent esse totaliter, scilicet si celum non moveretur. Et potest exemplificari de hoc, nam in absentia rectoris alicuius politie non totaliter desinit regimen illius politie, quamvis illa politia non ita bene regatur in absentia illius rectoris politie sicut in presentia.

⟨5⟩ Oppositum videntur dicere dicta posita in precedenti questione: dicebatur enim quod celum per suum motum regit et gubernat ista inferiora, et similiter motus et actiones ipsorum[38].

⟨6⟩ Secundo ad idem: cessante causa, cessat effectus; sed motus celi est causa istorum motuum inferiorum; igitur etc. Et precipue in essentialiter ordinatis, cessante primo, cessant postrema[39]. Modo, quia cause effective secundarie, sicut sunt ista agentia | [*L* 185va] inferiora, sunt subordinate causis universalibus, sicut intelligentiis, igitur, cessante actione intelligentiarum qua movent suos orbes, videntur etiam cessare actiones istorum agentium inferiorum.

⟨7⟩ Breviter, ista propositio: 'cessante motu celi, cessarent motus istorum inferiorum', secundum quod habemus ex logica, equivalet uni conditionali, vel causali vel temporali; et ergo intellectus questionis est: utrum, si cessarent motus celi, etiam cessarent motus istorum inferiorum, vel quia cessarent, vel dum cessarent motus celi, cessarent motus istorum inferiorum.

---

1 alteratio et per consequens nec] *om. Ba Kl L U* 6 sic] sicut tangebatur in precedenti questione *add. Ba Kl L U* 6–7 si adhuc ignis approximaretur stuppe] ignis applicatus stuppe *Ba Kl L U* 9 et celum] universaliter *Ba Kl* et celum ultimum *U* ‖ mediante solo motu] per suum motum *Ba Kl L U* 21 ordinatis] subordinatis *Kl L*

---

37 I.3, 28.

38 I.3, 16–17.

39 Cf. Aristotelem, *Metaphysica*, II, 2, 994 b 19–20.

Ba 7vb Modo aliqua conditionalis potest esse vera cuius tamen antecedens | est impossibile, ut ista: 'si asinus volat, asinus habet pennas'[40]; et ergo, stante quod impossibile est naturaliter cessare motum celi, tamen potest queri, si cessaret, quid fieret de motu istorum inferiorum.

Kl 6vb ⟨8⟩ Et ad videndum hoc suppono primo | quod celum agit in ista inferiora triplici instrumento, scilicet motu, lumine et influentia—hoc prius declaratum est[41].

⟨9⟩ Et ulterius sciendum est quod celum per motum agit in ista inferiora dupliciter. Uno modo immediate, scilicet movendo ignem secum sibi propinquum. Alio modo celum agit in ista inferiora per suum motum mediate, isto modo quod quia celum movetur localiter, aliquando una pars celi est presens alicui parti quiescenti in isto mundo, et deinde alia, et sic aliter et aliter influit illi parti secundum quod est alia et alia influentia et aliud et aliud lumen secundum alterius et alterius partis celi presentiam.

U 6rb ⟨10⟩ Tunc sit prima conclusio: | si celum cessaret a quolibet istorum trium, videlicet tam a motu quam a lumine quam ab influentia, sic quod non moveretur nec ageret aliquod lumen in ista inferiora nec aliquam influentiam, *P* 231v nulla fieret actio neque motus | in istis inferioribus. Patet, nam in essentialiter ordinatis, cessante primo, cessant postrema. Similiter, si essent decem rote et prima moveret secundam, nec secunda moveretur aliunde, et secunda tertiam, etc., stante prima, starent alie. Sic in proposito ymaginandum est de celo et isto mundo inferiori.

⟨11⟩ Secunda conclusio: si celum cessaret a motu solum, et non a lumine *L* 185vb nec ab influentia, non propter hoc cessarent | motus et actiones istorum inferiorum. Patet, quia adhuc Sol suo lumine calefaceret ista inferiora et similiter sua influentia alteraret ea, quibus alteratis, alterarent alia. Verum est tamen quod, si celum cessaret a motu solum, aliter fierent actiones in istis inferioribus quam modo fiant. Patet, quia Sol non sic diversimode applicaretur istis inferioribus sicut si celum moveretur.

⟨12⟩ Sed contra conclusionem diceret aliquis: motus celi localis est primus *Kl* 7ra omnium motuum | istorum inferiorum[42]; igitur videtur quod ipso cessante,

---

5 primo] istud quod dictum est in precedenti questione *add. Ba Kl L* 9–10 propinquum] illud enim quod movetur localiter movet corpus sibi propinquum *add. Ba Kl L U* 25 suo lumine] *om. Ba Kl L U* 25–26 similiter sua influentia] *om. Ba Kl L U* 28 applicaretur] applicatur *P* 31 istorum inferiorum] aliorum *Ba Kl L U*

---

40 Cf. Averroem, *In Physicam*, VII, comm. 2, ed. apud Iunctas, 4: f. 308 BC; *Auct. Arist.*, 156, n. 195.

41 I.3, 17.

42 Cf. Aristotelem, *Meteorologica*, I, 2, 339 a 23–24; *Auct. Arist.*, 171, n. 4.; *Physica*, VIII, 9, 265 b 8–12; *Auct. Arist.*, 158, n. 224.

omnes alii motus | *Ba* 8ra inferiores cessent. Respondetur quod bene verum est quod motus celi est causa et primus omnium motuum aliorum qui fiunt isto modo qualiter iam fiunt; tamen, hoc non obstante, si ipse cessaret, adhuc fierent motus et actiones istorum inferiorum, sed aliter tunc quam modo. Et igitur, si celum quiesceret, stante lumine et influentia, nullum mixtum perfectum generaretur, cum ad generationem mixti perfecti requiritur quod Sol et planete aliter et aliter se habeant secundum accessum et recessum ad ista inferiora. Et ergo fetus stat in ventre tempore determinato per motum Solis; et etiam, sicut videmus, arbores per accessum Solis pullulant, per recessum vero Solis arescunt. Similiter, stante celo in quiete, mixta perfecta bene durarent per aliquod tempus, deinde corrumperentur in alia mixta imperfecta, deinde tandem resolverentur in elementa; et tunc fieret continua generatio et corruptio elementorum, ita quod tandem elementa calida essent sub Sole et elementa frigida fluerent ad alia loca ad oppositum.

⟨13⟩ Tertia conclusio: si celum moveretur unico motu, scilicet motu diurno, adhuc fierent actiones et motus, generationes et corruptiones in istis inferioribus, licet tunc aliter quam modo. Patet, quia Sol aliquando accederet et aliquando recederet. Verum est tamen | *U* 6va quod tunc non esset hiems nec estas, nec esset distinctio temporum nisi per diem et noctem.

⟨14⟩ Et ergo sequitur quarta conclusio, quod ad hoc quod fiant actiones et motus, generationes et corruptiones in istis inferioribus, sicut modo fiunt requiruntur duo motus, scilicet motus diur|*L* 186^{ra}nus et obliquus—scilicet planetarum. Et secundum hoc Sol accedit et recedit ad nos vel a nobis.

⟨15⟩ Quinta conclusio: si celum esset uniforme in suis partibus ita quod Sol esset consimilis influentie sicut pars orbis sibi propinqua, et celum moveretur, nihil de novo generaretur. Patet, quia a voluntate antiqua non provenit actio nova, secundum naturaliter loquentes, nisi mediante | *Kl* 7rb celo aliter et aliter se habente[43]. Sed modo, si celum esset uniforme in suis partibus, non se haberet aliter nec in toto nec in parte quoad lumen et influentiam, sed semper consimili modo; et igitur bene factum est | *Ba* 8rb quod celum est dissimile in

5–6 nullum mixtum perfectum generaretur] nichil generaretur saltem in mixtis perfectis *Ba* nichil generaretur saltem in mixtis *Kl L U* 8 fetus] animalis *add. Ba Kl L U* 9 sicut] ut sicut *P* 10 in quiete] *om. Ba Kl L U* 12 elementa] alia *P* 17–18 patet quia sol … et aliquando recederet] *om. Ba Kl L U* 21 sicut modo fiunt] *om. Ba Kl L U* 22–23 scilicet planetarum] *om. Ba Kl L U* 23 sol accedit et recedit] sol et planete accedunt et recedunt *Ba Kl L U* ‖ vel a nobis] *om. Ba Kl L U* 25–26 moveretur] localiter adhuc *add. Ba Kl L U* 28 se habente] influente *Ba Kl L*

43 Cf. Averroem, *In Physicam*, VIII, comm, 4, 341 I; comm. 8, f. 344 I–K; comm. 15, f. 349 L–350 M; comm. 43, f. 383 G; cf. *Auct. Arist.*, 159, n. 237.

suis partibus, sic scilicet quod Sol habeat aliam influentiam quam alia pars orbis et unus planeta aliam quam alius.

⟨16⟩ Sexta conclusio: si celum cessaret a lumine et influentia, non tamen a motu locali, nihil ageret in ista inferiora nisi forte quod moveret ignem secum. Patet: quod agit in ista inferiora, hoc est mediante lumine et influentia, et movetur localiter ut aliter et aliter secundum alias et alias partes pos- *P* 232^{r} sit | influere.

⟨17⟩ Ad rationes. Ad primam dico quod si omnia ista inferiora essent perpetua, non essent ita bene ordinata sicut modo, nec universum esset ita pulchrum sicut modo. Decor enim universi est quod modo generentur aliqua et postea corrumpuntur et postea generantur altera.

⟨18⟩ Ad secundam bene concedo quod si celum cessaret a motu, non tamen a lumine et influentia, adhuc fierent actiones et motus in istis inferioribus, quamvis non eodem modo sicut nunc, sicut dictum est in questione.

⟨19⟩ Ad tertiam ex eadem radice.

⟨20⟩ Ad quartam consimili modo.

Et sic patet ad questionem.

1.5 Utrum eedem opiniones infinities reiterentur

Queritur consequenter utrum eedem opiniones infinities reiterentur.

⟨1⟩ Et arguitur primo quod non, nam, si sic, sequeretur quod mundus esset *Kl* 7va eternus. Consequens est falsum: est enim contra veritatem; veritas | enim *L* 186rb est quod mundus | incepit per creationem. Consequentia tenet, nam impossibile esset opiniones easdem infinities reiterari in hominibus nisi mundus esset eternus.

U 6vb ⟨2⟩ Secundo: intellectus non variat suum iudicium de eodem; ergo opiniones non variantur nec reiterantur. Consequentia tenet, nam ad hoc quod intellectus varietur in opinione de eadem re, oportet eum variare suum iudicium. Sed antecedens patet ex eo quod sic videtur esse de intellectu sicut de sensu; sed sensus non variat iudicium suum de suo sensibili: non enim quod modo iudicat albedinem, postea iudicat nigredinem.

---

2 orbis] sibi propinqui *add. Ba Kl L U* 4 locali] *om. Ba Kl L U* 4–5 nisi forte quod moveret ignem secum] *om. Ba Kl L U* 12 motu] solum *add. Ba Kl L U* 13 motus] passiones *Ba Kl L U* 22 in hominibus] *om. Ba Kl L U* 25 nec reiterantur] *om. Ba Kl L U*

⟨3⟩ Tertio sic: nulle opiniones reiterantur; ergo falsum est quod eedem opiniones infinities reiterentur in hominibus. Consequentia tenet. Antecedens probatur, nam, si aliqua opinio reiteraretur, sequeretur quod idem numero post eius corruptionem posset redire, quod est contra philosophice loquentes.

⟨4⟩ Quarto ad idem: scientie sunt eterne; ergo, postquam habita est scientiam de uno, numquam corrumperetur; et per consequens non amplius de illo erit opinio, sed semper scientia.

⟨5⟩ Quinto: alique veritates sunt ita evidentes quod intellectus non potest | *Ba* 8[va] eis assentire cum formidine; ergo circa illas non contingit esse opinio; ergo non circa quamcumque veritatem contingit infinities opiniones reiterari; primum enim principium est ita evidens quod intellectus non potest ei dissentire nec assentire cum formidine.

⟨6⟩ Sexto ad idem: si questio esset vera, sequeretur quod aliquando esset vel fuisset vel foret aliquod tempus in quo quasi omnes homines errarent. Hoc falsum est. Falsitas patet per Aristotelem in primo *Rhetorice*, ubi dicit sic: 'autem homines, ut plurimum, attingunt ad veritatem'[44]. Consequentia tenet, quia illo tempore quo ista opinio communiter variaretur quod mundus esset eternus, communiter homines errarent.

⟨7⟩ In oppositum est Aristoteles in littera, dicens quod contingit easdem opiniones infinities reiterari[45].

⟨8⟩ In ista questione videndum est circa quas materias contingit fieri variationem opinionum et circa quas non; secundo videbitur quomodo contingit fieri variationem huiusmodi opinionum; tertio videndum est de causa efficiente universali huiusmodi variationis; quarto videndum est de causa | *Kl* 7[vb] efficiente particulari huiusmodi variationis; quinto, de causa finali huiusmodi variationis.

10 eis] dissentire nec *add. L U* ‖ esse] *om. P* 12 principium] in aliqua scientia *add. Ba* ‖ evidens] in sua veritate *add. Ba Kl L U* 15 quasi omnes homines] homines communiter *Ba Kl L U* 16 dicit] *om. P* 18 variaretur] *om. P* 20 aristoteles] volens reddere causam cuiusdam dicti *add. Ba Kl L U* 21 infinities] bis vel ter ymmo infinities reverti *Ba* ‖ reiterari] in hominibus *add. Kl L U*

44 Aristoteles, *Rhetorica*, I, 1, 1355 b 14–16, translatio Guillelmi de Morbeka, *AL* XXXI 2, 161[26]: "autem et homines ad veritatem apti nati sunt sufficienter et in pluribus adipiscuntur veritatem".

45 Aristoteles, *Meteorologica*, I, 3, 339 b 27–30.

⟨9⟩ Quantum ad primum, sciendum est primo quod quedam sunt veritates ita evidentes quod circa eas non contingit esse | [*L* 186va] opiniones, sed scientiam vel intellectum, sicut est veritas primi principii. Ipsum enim est ita evidens in sua veritate quod quam cito intellectus circa ipsum considerat, ei assentit nec potest ei dissentire, | [*U* 7ra] cum intellectus naturaliter inclinatus sit ad assentiendum vero sibi evidenti, | [*P* 232v] sicut lapis naturaliter inclinatus est ad descendendum. Et quia circa huiusmodi veritates non contingit esse opiniones, etiam circa huiusmodi veritates non contingit opiniones reiterari.

⟨10⟩ Secundo est sciendum quod alique partes alicuius contradictionis sic se habent quod non est maior ratio pro una parte quam pro alia, quamvis in veritate una earum sit vera; que tamen sit illa, latet homines. Et circa tales partes etiam non contingit esse opiniones, nec circa tales contingit opiniones reiterari. Verbi gratia, nam sic se habent partes istius contradictionis, 'omnes stelle sunt pares; non omnes stelle sunt pares' sic intelligendo: 'tota multitudo stellarum est numerus par; non tota multitudo stellarum est numerus par'; quelibet enim istarum est equaliter dubia.

⟨11⟩ Tertio est sciendum quod alique partes eiusdem contradictionis sic se habent quod una pars est magis probabilis quam alia; immo aliquando falsa est magis probabilis quam vera, cum nihil prohibeat quedam falsa esse probabiliora quibusdam veris[46]. Et circa tales partes contradictionis contingit esse opiniones, et circa eas contingit opiniones reiterari. | [*Ba* 8vb] Et alique | [*Kl* 8ra] de illis sic se habent quod circa unam istarum solum contingit esse duas opiniones, et sic se habet ista: 'anima intellectiva est indivisibilis'. Aliqua autem sic se habet quod circa eam contingit esse plures opiniones duabus, et sic se habet ista: 'mundus est perpetuus'. Unde aliqui opinabantur quod mundus esset perpetuus et a parte ante et a parte post[47]; aliqui quod esset perpetuus a parte post et non a parte ante; aliqui autem quod nec esset perpetuus a parte ante nec a parte post[48].

⟨12⟩ Quarto est sciendum quod aliqua sunt circa que reiterantur opiniones diversimode absque mutatione illarum rerum circa quas variantur huius-

---

3 principii] et sicut patet de isto primo omne totum sua parte est maius *add. Ba Kl L U* 4 in sua veritate] *om. Ba Kl L U* 7 descendendum] existens extra suum locum *add. Ba Kl L U* 8 opiniones] infinities *add. Ba Kl L U* 13 opiniones] infinities *add. Ba Kl L U* 19 est] *om. P* ‖ nihil] non *codd.* 21 reiterari] et sic se habet ista mundus est eternus similiter sic se habet ista anima intellectiva est indivisibilis *add. Ba Kl L U* 22–23 opiniones] partes opinionis *Kl* 25 perpetuus] eternus *Ba Kl L U* 30 diversimode] *om. Ba Kl* ‖ mutatione] variatione *Ba Kl L U*

---

46 Cf. Aristotelem, *Topica*, VII, 11, 161 a 30–31; *Auct. Arist.*, 331. n. 121.

47 Cf. Aristotelem, *De celo*, I, 12.

48 Cf. Ibid., I, 10, 279 b 12–17.

modi opiniones et reiterantur, sicut est de eternitate mundi; similiter de indivisibilitate anime intellective: absque enim mutatione veritatum circa ista, reiterantur et variantur | *L* 186vb opiniones circa ea. Alia autem sunt ex quorum mutatione variantur opiniones et assensus circa ea; unde ex variatione rei aliquando opinamur quod talis cibus sit sanus in tali tempore et insanus in alio. Similiter quod in tali loco expedit esse talia edificia et in tali talia, et sic de consimilibus.

⟨13⟩ Quinto est sciendum quod non solum fit opinionum variatio in speculativis, verum etiam in practicis, ut in legibus, in artibus mecanicis, in ydeomatibus, in modo tradendi scientias | *U* 7rb et in consimilibus; unde nos videmus quod in artibus mecanicis magis fit variatio: tunice enim que sic forte aliquando placebant, nunc displacent, et e converso. Similiter ydeomata in regionibus per processum temporis valde variantur. Similiter sepe fit variatio in modo tradendi scientiam; hoc videmus, nam Antiqui alio modo tradiderunt scientias et in alio modo loquendi quam Moderni, sicut patet in libris Antiquorum. Sic ergo patet circa quam materiam contingit huiusmodi variatio et opinionum reiteratio.

⟨14⟩ Quantum ad secundum, sciendum est quod opinio aliquando mutatur in opinionem contrariam, aliquando autem simpliciter in ignorantiam. Hoc patet, nam quandoque in aliquo studio magistri habent aliquam opinionem et postea per processum temporis opinionem contrariam; et aliquando simpliciter illam opinionem dimittunt, non habentes de ea opinionem, nec circa eam amplius considerant. Et sic patet qualiter diversimode contingit opinionem mutari. Etiam aliquando aliqua opinio mutatur in una regione et non in aliis; | *P* 233^{r} aliquando in pluribus.

⟨15⟩ *Kl* 8rb Quantum ad tertium, sit ista conclusio: huiusmodi variatio fit effective a corpore celesti tamquam ab agente universali et generali. Patet ex precedentibus duabus questionibus, nam ista inferiora reguntur ab illis superioribus, sicut prius dicebatur[49]; ergo causa variationis in istis inferioribus est influentia | *Ba* 9ra celestis. Istud etiam patet per astrologos, ponentes quod aliquando fiunt

6 loco] tempore *L*   10 scientias] in mutatione vestium *add. Ba Kl L U*   10–11 unde nos videmus quod in artibus mecanicis magis fit variatio] *om. Kl* videmus enim quod multiplex fit variatio vestium *Ba L U*   11 tunice] vestes *Ba Kl L U*   15 et in alio modo loquendi] *om. Ba Kl L U*   15–17 sicut patet in … et opinionum reiteratio] *om. Ba Kl L U*   24 mutatur] servatur *Ba Kl L U*   27 ab agente universali generali] a causa efficiente universali et generali *Ba Kl L U* ‖ et] in *P*

49 I.3, 19; I.4, 10.

alique magne coniunctiones que sunt cause inceptionis quarundam sectarum et opinionum que diu durant, si non sint mediocres; si autem sint parve, tunc sunt cause inceptionis sectarum et opinionum que per modicum tempus durant.

L 187ra ⟨16⟩ Quantum ad quartum, dico quod causa agens | particularis variationis opinionum potest esse depopulatio que potest fieri per diluvium vel per mortalitatem, ita quod postquam homines perierunt per diluvium vel per mortalitatem, postea novi novas inveniunt opiniones. Similiter variatio opinionum in aliquo regno potest fieri per guerras in illo regno; unde, cum sunt guerre in aliquo regno, homines non possunt insistere studio, sed oportet eos insistere aliis, et sic potest accidere variatio opinionum in regno. Etiam potest fieri propter sterilitatem; unde, acquisitis necessariis, Antiqui inceperunt philosophari[50]. Similiter potest accidere variatio opinionum propter variationem complexionum et inclinationum hominum; unde (*Kl* 8va) aliquando homines magis inclinantur ad praxim | quam ad speculandum, et hoc aliquando accidit propter malitiam hominum. Cum enim speculativi (*U* 7va) non honorentur, | qui bene insisterent speculationibus ponunt se ad practicam, sicut tempore hodierno experimur. Plures enim se dant legibus quam artibus, et isto modo opiniones variantur in ignorantiam et non in opinionem contrariam.

⟨17⟩ Quantum ad quintum, potest dici quod causa finalis huiusmodi variationis et reiterationis opinionum est decor universi, unde propter talem variationem universum est pulchrius. Unde propter aliquam rem in se turpem

6 depopulatio] destructio seu mortalitas hominum *Ba* depopulatio hominum *Kl* destructio hominum *U*   7–8 postquam homines perierunt per diluvium vel per mortalitatem] post talem mortalitatem *Ba* post mortem hominum per diluvium vel per mortalitatem *Kl* post obitum hominum per diluvium vel per mortalitatem *L lect. dub. U*   9 in aliquo regno] in aliqua regione *Ba Kl L U* ‖ per guerras] propter errores *L* ‖ in illo regno] *om. Ba Kl L U*   9–10 unde cum sunt guerre in aliquo regno] nam in regionibus hominum propter errores *L U*   11–12 sed oportet eos insistere aliis et sic potest accidere variatio opinionum in regno] *om. Ba Kl L U*   13 philosophari] sicut recitat aristoteles in prohemio metaphysice ubi dicit quod antiqui sacerdotes in egypto iam habentes omnia vite necessaria invenierunt scientias speculativas *add. Ba Kl L U*   14 variationem complexionum et inclinationum hominum] naturalem inclinationem hominum ad alias opiniones *Ba* naturalem inclinationem hominum ad opiniones *Kl* naturalem inclinationem hominum ad opiniones aliquas *L U*   15 praxim] scientias practicas *Ba* id est ad scientias practicas *add. Kl L U*   17–19 qui bene insisterent … legibus quam artibus] ponunt se ad leges *Kl L U om. Ba*   19–20 opinionem contrariam] scientiam ut si artiste student leges *Ba Kl L U*

50 Aristoteles, *Metaphysica*, I, 1, 918 b 20–25; cf. *Auct. Arist.*, 116, n. 20.

una tota multitudo potest fieri pulchrior, sicut potest apparere picturis. Alia causa similiter huiusmodi variationis potest esse exercitium et intellectus hominum delectatio, quia delectatur in novitatibus et variationibus.

⟨18⟩ Ad rationes. Ad primam dico quod secundum Philosophum conceditur quod mundus est eternus[51], et isto sumpto infinities eedem opiniones reiterantur; cuius causa dicta est in questione, quia infinities consimiles fiunt coniunctiones planetarum que aliquando erant, que sunt causa huiusmodi variationis et reiterationis opinionum[52].

⟨19⟩ Ad secundam dico quod non est simile de sensu et in|tellectu; nam *L* 187rb bene contingit intellectum errare circa intelligibile, non autem contingit | *Ba* 9rb sensum errare circa suum proprium sensibile, ut dicit Aristoteles secundo *De anima*, appositis tribus condicionibus quas Temixtius ibi apponit[53].

⟨20⟩ Ad tertiam dicendum est quod dictum Aristotelis debet intelligi quod 'eedem' opiniones, hoc est 'consimiles', ita quod dictum Aristotelis debet intelligi de eisdem opinionibus in specie, non autem in numero.

⟨21⟩ Ad quartam concedo quod | scientie sunt eterne secundum speciem, *Kl* 8vb non autem secundum numerum, immo non adhuc simpliciter et continue scientie eterne sunt secundum speciem, licet intercise, ita quod ante quodlibet tempus fuit scientiam et post quodlibet tempus erit scientiam.

⟨22⟩ Ad quintam patet ex dictis in questione.

⟨23⟩ Ad sextam respondetur quod dictum Aristotelis, cum dicit 'homines ut plurimum attingunt ad veritatem', hoc debet intelligi quod hoc est verum

1 picturis] ut si aliquis paries est depictus colore albo viridi et nigro cum hoc uno alio colore hoc est in se turpis *add. Ba* ut si aliquis paries est depictus albo colore et viridi et nigro qui color tamen in se turpis est et sic de aliis coloribus tamen ex hoc ille color redderetur pulchrior *add. Kl* ut si aliquis paries est depictus albo colore et viridi et nigro qui color tamen in se turpis est et sic de aliis coloribus tamen ex hoc ille color redderetur pulchrior *add. L* ut si aliquis paries est depictus colore albo et viridi et nigro qui color in se turpis est et sic de aliis coloribus tamen ex hoc ille paries redderetur pulchrior *add. U* 6–8 quia infinities consimiles … et reiterationis opinionum] *om. Ba Kl L U* 13–15 dictum aristotelis debet … autem in numero] eedem opiniones infinities reiterentur sed non eedem in numero sed eedem in specie *Ba Kl L U* 18 intercise] discontinue *Ba Kl L U*

51 Cf. Aristotelem, *De celo*, I, 12, 281 a 28–283 b 22; II, 1, 283 b 26–284 b 5.

52 I.5, 15.

53 Aristoteles, *De anima*, II, 2, 418 a 14–15; Themistius, *In De anima*, ed. G. Verbeke, *Commentaire sur le traité de l'âme d'Aristote. Traduction de Guillaume de Moerbeke. Édition critique et étude sur l'utilisation du Commentaire dans l'œuvre de saint Thomas*, Louvain-la-Neuve/Paris 1957 (Corpus Latinum Commentariorum in Aristotelem Graecorum, 1), 132–133; cf. *Auct. Arist.*, 179, n. 66; Nicolaum Oresme, *Questiones in De anima*, II.10, ed. B. Patar, Louvain-la-Neuve 1995, 194[15–17].

in agibilibus et in factis particularibus, de quorum numero non est eternitas mundi, et ex hoc patet solutio argumenti.

Hoc de questione.

1.6 Utrum elementa sint continue proportionalia ad invicem

P 233[v] Queritur consequenter utrum elementa sint continue proportionalia ad invicem.

⟨1⟩ Et arguitur quod sic. Et primo quod sint continue proportionalia in quantitate; nam elementa sunt equalia in quantitate; ergo sunt continue proportionalia in quantitate. Consequentia tenet, nam, si sint quatuor termini qui sic se habent quod qualis est proportio primi ad secundum, talis etiam est secundi ad tertium et similiter tertii ad quartum, illi sunt continue proportionales ad invicem. Sed si sint equalia in quantitate, tunc qualis est proportio ignis ad aerem, talis est proportio aeris ad aquam, et similiter aque ad terram. Antecedens probatur, nam si elementa essent inequalia in quantitate et magnitudine, tunc etiam essent inequalia | *U* 7[vb] in virtute, cum in maiori magnitudine sit maior virtus. Sed hoc est falsum, nam si essent inequalia in virtute, tunc illud quod esset maioris virtutis cetera corrumperet, cum elementa sint ad invicem contraria.

⟨2⟩ Secundo videtur quod sint continue proportionalia in raritate, nam, in quanto ignis est rarior aere, in tanto aer est rarior aqua, et aqua est rarior terra; et sic consimiliter in densitate, cum ex uno pugillo terre fiunt decem aque, et ex uno aque decem aeris, et sic ultra. Ex hoc etiam argueretur quod continue essent proportionalia in magnitudine; nam videtur quod taliter se habeant ad invicem in magnitudine qualiter se habent ad invicem | *Kl* 9[ra] in raritate et densitate.

⟨3⟩ Tertio videtur quod sint proportionalia in gravitate et levitate: in quantum enim ignis est levior | *L* 187[va] aere, in tantum aer est levior aqua et in tantum aqua est levior terra; et in quantum terra est gravior aqua, in tantum aqua | *Ba* 9[va] est gravior aere, et sic ultra.

⟨4⟩ In oppositum arguitur probando quod elementa non sint continue proportionalia in quantitate, nam, si sint, sequitur quod in quanto ignis esset maior aere et aer maior aque, quod in tantum aqua esset maior terra; sed

1 in agibilibus et] *om. Ba Kl L U* 14 et magnitudine] *om. Ba Kl L U* 24 et densitate] *om. Ba Kl L U* 28 et sic ultra] et aer igne *Kl*

hoc est falsum. Consequentia tenet ex diffinitione continue proportionalium. Falsitas consequentis patet, nam aqua non est maior terra, et ad minus non in tantum in quantum aer est maior aqua; nam, si sic, sequeretur quod, sicut aer circuit, ambit et cooperit totam aquam, ita etiam aqua ambiret et cooperiret totam terram. Hoc autem est falsum, sicut patet per experientiam.

⟨5⟩ In ista questione primo videndum est de intellectu quorundam terminorum; secundo videndum est an elementa sint equalia; tertio videndum est de quesito principali, an sint continue proportionalia.

⟨6⟩ Quantum ad primum, sciendum est quod per 'elementa', prout 'elementum' accipitur in proposito, intelligo elementa pura vel corpora eis similia. Secundo sciendum est quod continua proportionalitas et geometrica, de qua intendimus in proposito, dicitur similitudo proportionum terminorum sic se habentium ad invicem quod qualis est proportio primi ad secundum, talis est secundi ad tertium, et sic deinceps. Verbi gratia: 8, 4, 2, 1. Isti termini dicuntur ad invicem continue proportionales, nam sicut 8 ad 4 est proportio dupla, ita similiter 4 ad 2, et similiter 2 ad 1. Proportionalitas autem geometrica discontinua est quando non est eadem proportio primi ad secundum | *Kl* 9rb sicut secundi ad tertium, licet bene sit eadem proportio primi ad secundum | *U* 8ra sicut tertii ad quartum. Verbi gratia: 6, 3, 2, 1. Isti termini sunt proportionales discontinue: sunt enim proportionales, quia sicut proportio 6 ad 3 est dupla, ita similiter proportio 2 ad 1 est dupla; sed non sunt proportionales continue, quia non se habent 3 ad 2 sicut 6 ad 3: 3 enim ad 2 est proportio sesquialtera; 6 autem ad 3 est | dupla. *P* 234^{r}

⟨7⟩ Et differt proportionalitas continua a discontinua, quia minimus numerus terminorum in quibus potest reperiri proportionalitas continua est numerus ternarius, sed minimus numerus terminorum in quibus reperiri potest proportionalitas discontinua est numerus quaternarius. Conveniunt autem in hoc quod non est dare maximum numerum terminorum in quibus aliqua earum potest reperiri, sed quocumque dato, dabilis est maior in quo quelibet predictarum proportionum potest inveniri. Disproportionalitas autem | dicitur proportionum dissimilitudo, verbi gratia sicut hic: 4, 3, *L* 187vb 2, 1. Sicut enim se habent 4 ad 3, ita non se habent 2 ad 1. Similiter, sicut se habent 4 ad 3, ita non se habent 3 ad 2.

---

7 intellectu] expositione *Kl L U* 19 sicut secundi ad tertium licet bene sit eadem proportio primi ad secundum] *om. hom. P* 31–32 disproportionalitas] terminorum *add. Ba Kl U*

⟨8⟩ Quantum ad secundum, sciendum est quod elementa possunt ymaginari equalia quadrupliciter: primo modo in extensione; | (*Ba* 9vb) secundo modo in virtute; tertio modo in materia; quarto modo in spissitudine.

⟨9⟩ De primo ergo sit prima conclusio: elementa non sunt equalia in quantitate seu extensione. Patet, quia ignis est maior aere et aer maior aqua et aqua terra; unde, nisi aer et ignis essent maiores terra et aqua, non sufficerent ad implendum speram inter nos et celum, quod tamen est necessarium ne sit vacuum.

⟨10⟩ Secunda conclusio: nec sunt equalia in virtute. Patet, nam videmus quod modicus ignis consumit multum aerem; ergo, si modicus ignis est maioris | (*Kl* 9va) virtutis quam sit multus aer, et cum orbis ignis sit maior quam sit orbis aeris, sequitur etiam quod sit maior in virtute. Secundo: si elementa essent equalia in virtute, sequeretur quod terra esset tante virtutis sicut aliquod aliorum elementorum; quod est falsum, quia, accepta aliqua terra equalis quantitatis cum alio elemento, ipsa habet minorem virtutem quam illud elementum, et propter hoc terra in mixtis habundat quantum ad quantitatem. | (*U* 8rb) Per istam conclusionem excluditur una opinio que dicit omnia elementa esse equalia in virtute, sed non in magnitudine.

⟨11⟩ Tertia conclusio est quod elementa non sunt equalia secundum materiam, id est non quodlibet elementum habet tantum de materia sicut aliud. Patet, nam si sic, cum aqua sit rarior quam terra, si cum hoc haberet tantum de materia sicut terra, oporteret quod esset multo maior quam sit terra. Modo hoc est falsum, quia tunc terra totaliter esset aquis cooperta.

⟨12⟩ Sed diceret aliquis: immo aqua est maior terra, sed tamen celum elevat aquam maris plus quam terra sit elevata et preservat terram ex una parte sicca et non aquis cooperta propter generationem et conservationem animalium[54]. Breviter hoc non valet, nam si hoc esset verum, demonstratio Aristotelis in secundo *Celi* non valeret, qua demonstrat spericitatem aque[55], nam in ista demonstratione presupponitur quod aqua | (*L* 188ra) semper tendit ad locum decliviorem et bassiorem; modo hoc esset falsum si aqua maris sic per celum esset elevata.

⟨13⟩ Per istam conclusionem excluditur una opinio que ymaginatur elementa esse equalia secundum materiam. Ymaginatur quod terra tantum

6 aqua] *om. P*   26 conservationem] salvationem *Ba Kl L U*   30 et bassiorem] *om. Ba Kl L U* ‖ esset] *om. P*

54 Cf. Iohannem Buridanum, *Questiones in Meteorologica*, I.3, ed. S. Bages, *Les Questiones super tres libros Metheororum Aristotelis de Jean Buridan: étude suivie de l'édition du livre I*, Thèse de doctorat de l'École des Chartes, 1986 (2 vols.), 2: 34–35.

55 Aristoteles, *De celo*, II, 4, 287 a 30–287 b 22.

habeat de materia sicut aqua, et aqua tantum sicut aer et sic ultra. Sed quia terra est magis densa quam sit aer, | ipsa est minus extensa quam sit aer, non *Ba* 10ra obstante quod tantum habeat de materia sicut aer. Unde ista opinio ymaginatur quod tota materia prima equaliter sit divisa in materias elementorum, excepta illa parte materie que est materia mixtorum[56].

⟨14⟩ Quarta conclusio: nec spissitudines | orbium elementorum sunt *Kl* 9vb equales; nam, si sic, tunc tantum esset de concavo aeris usque ad concavum ignis quantum est de concavo ignis usque ad concavum orbis Lune; et similiter tantum esset de centro Terre usque ad concavum aeris, quantum esset a concavo aeris usque ad concavum orbis Lune. Sed hoc est falsum. Consequentia primi patet ex eo quod spissitudo orbis aeris est spatium inter concavum aeris et concavum ignis, et similiter spissitudo orbis ignis est spatium inter concavum orbis ignis et concavum | orbis Lune. Con- *P* 234^{v} sequentia autem secundi patet eo quod distantia a centro Terre usque ad concavum aeris est spissitudo duorum elementorum, scilicet terre et aque, sicut distantia a concavo aeris usque ad concavum orbis Lune, scilicet aeris et ignis.

⟨15⟩ Per istam conclusionem | excluditur opinio illorum qui ymaginaban- *U* 8va tur quod tota semidyameter istius mundi inferioris esset divisa in quatuor partes equales quarum una esset distantia a centro Terre usque ad convexum Terre, secunda autem esset distantia a concavo aque usque ad eius convexum, tertia vero esset distantia a concavo aeris usque ad eius convexum, quarta vero esset distantia a concavo ignis usque ad eius convexum. Ista opinio exclusa est per predictam conclusionem.

⟨16⟩ Et est sciendum quod si ista opinio esset vera, possent alique conclusiones curiales deduci ex principiis et conclusionibus geometrie. | Unde, *L* 188rb sumpta veritate opinionis, posset inveniri et geometrice demonstrari in qua proportione se habeat aggregatum ex terra et aqua ad terram solum; similiter, in qua proportione se habeat aggregatum ex terra, aqua et aere ad terram solum; similiter, in qua proportione se habeat aggregatum ex terra, aqua, aere et igne ad terram solum. Et ad hoc demonstrandum oportet supponere penultimam | duodecim *Elementorum* Euclidis, videlicet quod propor- *Kl* 10ra tio sperarum est tripla ad proportionem dyametrorum istarum sperarum

1 et sic ultra] et aer sicut ignis *Ba* 2 aer] aliquod aliorum elementorum *Ba* aqua *L* vel aliquod aliorum elementorum *add. Kl* || aer] vel aliquod aliorum elementorum *add. Ba Kl* 3 aer] aqua *L* aliquod ipsorum *Ba* vel aliquod aliorum elementorum *add. Kl* 22 a concavo] *om. P* 26 curiales] *om. Ba Kl L U*

56 Cf. Iohannem Buridanum, *Questiones in Meteorologica*, I.3, ed. Bages, 2: 20.

Ba 10rb vel semidyametrorum[57], quia qualis est proportio dyame|trorum ad invicem, talis est proportio semidyametrorum ad invicem, cum qualis est proportio totorum ad invicem, talis etiam est proportio medietatum istorum totorum. Sicut igitur proportio dyametri ad dyametrum est proportio spere ad speram proportio triplicata, ita etiam proportio semidyametri ad semidyametrum est proportio sperarum illarum semidyametrorum proportio triplicata. Ista ergo proportione presupposita, statim, supposita dicta opinione, inveniretur proportio aggregati ex terra et aqua ad terram. Nam, cum semidyameter terre esset equalis distantie a concavo aque usque ad eius convexum, sequitur statim quod proportio semidyametri aggregati ex terra et aqua est dupla ad semidyametrum terre; et si sic, sequitur per propositionem allegatam quod proportio aggregati ex terra et aqua—puta illius spere aggregate—ad speram terre est proportio que est tripla duple; et cum proportio octupla sit tripla duple, sequitur proportionem spere aggregate ex terra et aqua ad speram terre esse octupla, qualis est proportio 8 ad 1. Quod autem proportio 8 *U* 8vb ad 1 sit tripla duple patet faciliter signatis quatuor terminis continue | proportionalibus proportione dupla, scilicet 1, 2, 4, 8. Hic sunt quatuor termini continue proportionales proportione dupla; igitur, per unam propositionem quinti *Elementorum* Euclidis[58], proportio primi ad ultimum est tripla ad proportionem primi ad secundum, secundi ad tertium, tertii ad quartum. Et hoc *Kl* 10rb etiam | patet ex alio, nam, cum aliquod totum componitur ex tribus partibus equalibus, tunc istud totum est triplum ad quamlibet istarum partium. *L* 188va Cum ergo proportio 8 ad 1 componatur ex tribus duplis, videlicet | ex proportione 2 ad 1 et ex proportione 4 ad 2 et ex proportione 8 ad 4, sequitur proportionem 8 ad 1, que est octupla, esse tripla ad quamlibet istarum quarum quelibet est dupla. Sic ergo patet quod octupla est tripla duple et per consequens, si semidyameter aggregati ex terra et aqua est dupla ad semidyametrum terre, sequitur proportionem totius spere aggregate ex terra et aqua ad terram esse proportionem 8 ad 1.

5–6 ita etiam proportio … semidyametrorum proportio triplicata] *om. hom. Ba Kl* 7 proportione] suppositione *Ba Kl L U* 9 terre] *om. Ba Kl L U* ‖ convexum] sicut a concavo terre usque ad eius convexum *add. Ba Kl L U* 11 propositionem] euclidis prius *add. Ba Kl* euclidis *add. L U* 16 faciliter] *om. Ba Kl L U* 18–19 propositionem quinti elementorum euclidis] quinti *P*

---

57 Ibid., 365^{16}–366^{12}: "Spere ad se invicem in triplici proportione sunt propriorum diametrorum".

58 Campanus de Novara, *Liber Elementorum Euclidis*, XII.18, ed. H.L.L. Busard, Wiesbaden 2005, 365^{16}–366^{12}: "Spere ad se invicem in triplici proportione sunt propriorum dyametrorum".

⟨17⟩ Similiter ex eadem radice potest inveniri | [*P* 235^{r}] in qua proportione se habeat aggregatum ex terra et aqua et aere ad terram solum. Nam ex quo semidyameter totius aggregati ad | [*Ba* 10va] semidyametrum terre habet se in proportione tripla, qualis est proportio 3 ad 1, sequitur ex propositione Euclidis allegata quod proportio aggregati ex terra, aqua et aere ad terram solum est tripla triple, qualis est proportio 27 ad 1, quod faciliter potest videri si ponantur quatuor termini continui proportionales proportione tripla, verbi gratia: 1, 3, 9, 27. Hic sunt quatuor termini continue proportionales proportione tripla, ergo proportio primi—hoc est maximi—ad ultimum est tripla ad quamlibet istarum proportionum quarum quelibet est tripla. Ex isto ergo patet quod proportio 27 ad 1 est tripla triple.

⟨18⟩ Ex eadem radice potest inveniri in qua proportione se habeat aggregatum ex omnibus quatuor elementis ad terram solum. Nam, stante predicta opinione, | [*Kl* 10va] proportio semidyametri aggregati ex quatuor elementis est quadrupla ad proportionem semidyametri terre, et per consequens proportio totius aggregati ex quatuor elementis ad terram est proportio que est tripla quadruple, qualis est proportio 64 ad 1. Hoc enim patet quod illa sit tripla quadruple si signentur quatuor termini continue proportionales proportione quadrupla isto modo: 1, 6, 16, 64. Hic enim sunt quatuor termini continue proportionales; ergo proportio maximi ad minimum est tripla ad proportionem primi ad secundum, secundi ad tertium, tertii ad quartum. Et cum quelibet istarum | [*U* 9ra] proportionum sit quadrupla, sequitur proportionem 64 ad 1 esse tripla quadruple. Ecce ergo primos tres numeros cubitos: 8, 27, 64. Et ita aliqui putabant proportionem elementorum iam dictam se habere, et hoc esset verum supposita veritate illius opinionis.

⟨19⟩ Ex hoc tunc ulterius potest inveniri quanta esset spera aeris sola et similiter spera ignis sola et similiter spera aque sola. Supposito quod spera terre esset sicut 1, tunc, secundum dicta, aqua esset sicut | [*L* 188vb] 7, aer vero sicut 19, ignis vero sicut 37. Et secundum hoc tunc elementa non essent continue proportionalia, sed disproportionalia.

⟨20⟩ Ulterius, supposita veritate predicte opinionis, posset inveniri in qua proportione se haberet aggregatum ex aere, aqua et terra ad aggregatum ex aqua et terra, | [*Ba* 10vb] et similiter in qua proportione se haberet aggregatum ex omnibus elementis ad aggregatum ex terra, aqua et aere. Primum patet, nam proportio semidyametri aggregati ex terra, aqua et aere habet se in proportione sesquialtera ad semidyametrum aggregati ex terra et aqua. Ergo

---

1 ex] *iter. P* 3 habet se] *om. P* 6 faciliter] *om. Ba Kl L U* 26–27 et similiter spera ignis sola] *om. Ba Kl* 29 37] 27 *codd.*

proportio aggregati ex illis tribus ad aggregatum ex illis duobus est tripla-[*Kl* 10vb] sesquialtere. Et cum proportio semi|dyametri aggregati ex omnibus quatuor elementis habeat se in proportione sesquitertia ad semidyametrum aggregati ex terra, aere et aqua, sequitur proportionem totius aggregati ex omnibus elementis ad aggregatum ex tribus esse proportionem tripla sesquitertie proportionis, cum proportio sperarum sit tripla ad proportionem semidyametrorum.

⟨21⟩ Quantum ad tertium, sciendum est quod aliqui posuerunt elementa esse continue proportionalia secundum eorum magnitudinem, sic quod sicut se habet ignis ad aerem in extensione, ita aer ad aquam et aqua ad terram; vel sicut ignis se habet ad aggregatum ex tribus aliis elementis, ita aer se habet ad aggregatum ex aliis duobus elementis, et sic se habet aggregatum ex aliis duobus ad terram solum. Et crediderunt hoc esse ideo, quia sicut [*P* 235v] se habent elementa | in eorum transmutatione ad invicem, ita et in eorum quiete. Sed in transmutatione eorum ad invicem sunt continue proportionalia, nam, sicut ex uno pugillo terre fiunt dęcem pugilla aque, ita ex uno aque pugillo fiunt decem pugilli aeris, et ex uno aeris decem ignis. Et isti ulterius [*U* 9rb] dix|erunt quod elementa in eorum magnitudine essent continue ad invicem proportionalia proportione decupla, ex quo sic … in eorum transmutatione.

⟨22⟩ Alii autem dixerunt quod elementa essent continue proportionalia in proportione qua se habent 33 ad 1, dimissis fractionibus, et hoc quoad eorum dyametros vel semidyametros. Unde ymaginabantur quod, si semidyameter [*L* 189ra] terre esset sicut unum, tunc semidyameter aque esset sicut | 33, et ulterius semidyameter aeris esset signanda per unum numerum qui haberet se in [*Ba* 11ra] [*Kl* 11ra] eadem proportione ad 33 in qua se habent 33 ad 1. Et ulterius | semi|dyameter ignis esset signanda per unum numerum qui sic se haberet ad numerum quo signanda est semidyameter aeris sicut 33 ad 1, ita quod voluerunt quod semidyameter spere ignis esset trigintuplatripla ad semidyametrum spere aeris, et quod semidyameter spere aeris esset trigintuplatripla ad semidyameter aque, et semidyameter aque trigintuplatripla ad semidyametrum terre. Et

10 in extensione] *om. Ba Kl L U* 11–12 ita aer se … aliis duobus elementis] *om. P* 13–14 sicut se habent elementa] sicut elementa sunt continue proportionalia *Ba Kl L U* 14 eorum] magnitudine et in *add. Ba* 18 magnitudine] transmutatione ad invicem *Ba Kl L U* 19 ita etiam proportio … semidyametrorum proportio triplicata] quia sicut se habent ad invicem in eorum magnitudine sic etiam se habent ad invicem in eorum transmutatione *Ba Kl* quia sicut se habent ad invicem in eorum magnitudine sic etiam in eorum transmutatione sicut ipsi dixerunt *L U* 22 semidyameter] diameter *P* 28–29 spere aeris et … trigintuplatripla ad semidyameter] *om. hom. P*

ex proportione istarum semidyametrorum ad invicem ulterius inveniebant proportionem sperarum elementorum ad invicem mediante ista propositione: ‘proportio sperarum est tripla ad proportionem semidyametrorum’[59].

⟨23⟩ Contra istas duas opiniones pono istam conclusionem quod elementa non sunt continue proportionalia secundum eorum magnitudines. Probatur, nam, si sic, tunc oporteret quod in eadem proportione aqua esset maior terra in qua ignis est maior aere et aer maior aqua. Modo hoc est falsum, nam, si sic, sequeretur quod aqua cooperiret totam terram. Modo hoc est falsum, igitur etc.

⟨24⟩ Ad rationes. Ad primam: ‘elementa sunt equalia in quantitate’, nego. Et quando dicebatur: ‘si non, sequeretur quod unum esset maioris virtutis quam reliquum’, concedo. Et quando dicebatur: ‘qualiter ergo istud quod est maxime virtutis inter alia non corrumpit alia?’, dico quod ratio huius est quia quanto aliquod elementorum est maioris activitatis, tanto est minoris resistentie et maioris passivitatis; et quanto aliquod elementum est minoris activitatis, tanto est maioris resistentie et minoris passivitatis; et ergo, quamvis inter elementa unum eorum sit maxime activitatis, non tamen propter hoc corrumpit alia propter magnam resistentiam aliorum.

⟨25⟩ De secunda autem et tertia, an sint continue proportionalia in raritate vel in gravitate vel levitate, videbitur in questione sequenti[60]. *Kl* 11rb

I.7 Utrum quatuor elementa semper et immutabiliter habeant eandem proportionem ad invicem

Queritur consequenter utrum quatuor elementa semper et immutabiliter habeant eandem proportionem ad invicem, sic quod una vice ignis non sit maior | quam alia, nec similiter aer; etc. *L* 189rb

⟨1⟩ Et arguitur quod sic: | continue tantum est de raritate sicut modo est, *U* 9va et continue tantum est de densitate sicut modo, ita quod semper raritas et

8 terram] sicut aer cooperit totam aquam *add. Ba Kl L U* 9 igitur etc] quod patet ad experientiam *Ba Kl* ut patet de se *L U* 13 inter alia] et activitatis *Kl U om. Ba L* 14 activitatis] virtutis in agendo *Ba* 23 maior] *scrips.* maior *corr. sup. lin.* rarior *Ba* rarior *L*

59 Campanus de Novara, *Libri Elementorum Euclidis*, XII.18, ed. Busard, 365^{16}–366^{12}: “Spere ad se invicem in triplici proportione sunt propriorum dyametrorum”.

60 I.7, 7.

densitas habent eandem proportionem ad invicem; igitur continue equaliter est de elementis gravibus et levibus; et per consequens, cum omnia elementa sint gravia vel levia, immutabiliter omnia elementa obtinebunt eandem proportionem ad invicem. Antecedens probatur, nam tota massa *Ba* 11rb istorum corruptibilium | replet orbem Lune; si igitur aliquando raritas et *P* 236r densitas haberent se inequaliter ad invicem quam modo se | habent, oporteret vel quod latera celi magis appropinquarentur ad invicem, vel quod fieret vacuum, quorum utrumque est impossibile naturaliter.

⟨2⟩ Secundo ad idem sic: si aliquid condensatur, necesse est quod aliquid rarefiat; igitur continue equaliter est de raritate et densitate, et per consequens de caliditate et frigiditate, cum huiusmodi qualitates mutuo se consequuntur ad invicem; et sic continue elementa calida et frigida obtinebunt eandem proportionem ad invicem.

⟨3⟩ Tertio: continue caliditas et frigiditas habent se equaliter; ergo elementa calida et frigida continue habebunt se equaliter, et hoc quoad eorum proportionem. Consequentia tenet; sed antecedens probatur, nam, cum Sol causat calorem super terram, generatur frigiditas infra; et e contrario, cum recedit et est frigiditas super terram, remittitur frigiditas infra terram[61], quod videtur esse signum quod caliditas et frigiditas equaliter se habent ad invicem quoad earum proportionem.

Kl 11va ⟨4⟩ Quarto ad idem: Sol per eius accessum | ad nos causat caliditatem circa nos et per eius recessum a nobis causatur circa nos frigiditas. Ita etiam videtur consimili modo circa illos qui morantur nobis opposito modo; et si sic, sequeretur quod quando Sol accedit versus nos et habemus circa nos calorem, ipsi habent frigiditatem, et e contrario. Quod videtur esse signum quod continue se habeant equaliter ad invicem caliditas et frigiditas.

⟨5⟩ In oppositum arguitur, nam aliquis potest multam caliditatem generare et frigiditatem corrumpere per combustionem aliquorum terrestrium et frigidorum; et per consequens non equaliter se habebunt ad invicem caliditas et frigiditas; et per consequens nec elementa calida et frigida; et per consequens non immutabiliter quatuor elementa obtinebunt eandem proportionem ad invicem.

17 generatur] intenditur *Ba Kl L U* 17–18 cum recedit et est … frigiditas infra terram] *om. Ba Kl*

61 Cf. I.10.

⟨6⟩ In ista | questione, quia visum erat in alia questione de proportione ele- *U* 9vb mentorum ad invicem quoad magnitudinem[62], nunc primo videndum est de proportione elementorum ad invicem quoad eorum raritatem et densitatem, | gravitatem et levitatem. Secundo videndum est de quesito principali. *L* 189va Tertio videndum est qualiter elementa possunt ad invicem perpetuari, cum tamen sint ad invicem contraria activa et passiva mutuo se corrumpentia.

⟨7⟩ Quantum ad primum, sit prima conclusio: non semper elementa habent se in continua proportione ad invicem in raritate et densitate. Patet, nam aliquando nec terra nec aqua condensata vel rarefacta, ignis est densior et aer rarior, vel e contrario, quod non potest esse si semper quatuor elementa | *Ba* 11va essent ad invicem proportionalia in raritate et densitate.

⟨8⟩ Secunda conclusio: nec semper | sunt continue proportionalia ad invi- *Kl* 11vb cem in gravitate et levitate. Patet, nam ignis nullo modo est gravis, vocando 'grave' quod tendit versus medium infra orbem Lune; igitur nulla est proportio gravitatis aliorum elementorum ad gravitatem ignis, cum nullam gravitatem habeat.

⟨9⟩ Quantum ad secundum, sit prima conclusio: quatuor elementa non semper habent se in eadem proportione ad invicem quoad eorum magnitudines. Patet per rationem qua arguebatur ad oppositum: aliquis enim potest comburere multa terrestria et destruere frigiditatem et generare caliditatem, cum non necesse sit quod semper caliditas et frigiditas equaliter habeant se ad invicem; nec est simile de raritate et densitate propter repletionem loci infra orbem Lune. Secundo, nam aliqui planete causant in istis inferioribus caliditatem, aliqui vero frigiditatem; sed aliquando sunt coniuncti plures planete frigidi quam aliquando, et aliquando plures planete calidi; igitur aliquando causatur in istis inferioribus plus de frigiditate et aliquando | minus; *P* 236^{v} et similiter est de caliditate.

⟨10⟩ Quantum ad tertium, quomodo elementa possunt perpetuari, cum tamen sint contraria et activa et passiva ad invicem, oportet primo supponere quod si verbi gratia ignis rarefiat, quod oportet aerem condensari; aliter enim oporteret quod latera celi ab invicem plus recederent, quod est impossibile.

9 rarefacta] rarificata *P* 14 orbem] speram *Kl* 22 repletionem] capacitatem *Kl* 26 frigiditate] caliditate *P* 29 contraria] inequalia et contraria *Kl* inequalia contraria *Ba U*

62 I.6, 9.

⟨11⟩ Oportet secundo supponere quod si ignis augetur, quod oportet ipsum condensari vel aerem sibi propinquum propter eandem rationem.

U 10ra *Kl* 12ra ⟨12⟩ Oportet tertio supponere quod si ignis diminuatur, | oportet quod ignis rarefiat vel aer sibi contiguus vel aliquid aliorum elementorum, ne fiat *L* 189vb vacuum infra orbem Lune vel ne oporteat latera celi plus ad invicem | appropinquari.

⟨13⟩ Oportet supponere quarto quod quodlibet elementorum determinat sibi certam latitudinem raritatis et densitatis quam non transgreditur nisi violenter[63].

⟨14⟩ Oportet supponere quinto quod est dare unam certam latitudinem raritatis et densitatis quam non transgreditur elementum nisi cum eius corruptione: est enim aliqua latitudo raritatis sub qua non potest esse aqua de sua natura, hoc est illam formam aqua non sibi determinat; nichilominus sub illa potest esse aqua per actionem ignis in eam. Similiter est una raritas maior sub qua aqua non potest esse neque naturaliter neque violenter, sic *Ba* 11vb quod quam cito illam attingit, | ipsa corrumpitur.

⟨15⟩ Tunc sit prima conclusio: impossibile est quod ignis omnia infra concavum orbis Lune corrumperet. Patet, quia cum celum vel latera celi non possunt extendi, oporteret ignem attingere tantam densitatem sub qua esse non potest. Et per consequens, cum ipse ignis continue entia mundi corrumperet alia ab ipso, continue plus et plus condensaretur et magis violentaretur, cum stare sub densitate sit sibi violentum. Et per consequens continue in actione eius debilitaretur et ante combustionem omnium aliorum a se tantum debilitaretur in eius actione quod non posset plus agere et per consequens nec alia corrumpere.

⟨16⟩ Ex ista conclusione sequitur quod ignis in spera sua tantum posset augeri quod non posset plus agere et quod, si stuppa sibi approximaretur, ipsa se non combureret. Patet, quia tantum posset augeri quod tantum condensaretur quod eius actio totaliter remitteretur; unde ex condensatione ignis, ignis remittitur in agendo, sicut patet cum aliquis ignis quod non

---

2 propinquum] contiguum *Kl* 3 diminuatur] condensetur vel diminuatur *L* 11 elementum] nec naturaliter nec violente *add. Ba Kl L U* 15 maior] *om. Ba Kl L U* 17–18 impossibile est quod … orbis lune corrumperet] quod ignis posset tantum corrumpere quod impossibile esset quod naturaliter plura infra orbem lune corrumpeat *Ba* quod ignis non posset tantum corrumpere quod impossibile esset naturaliter quod plura infra orbem lune corrumperet *Kl* 17 est] *om. P* 18 celum vel] *om. Ba Kl L U*

---

63 Cf. Nicolaum Oresme, *Questiones super De generatione et corruptione*, I.4, ed. Caroti, $29^{154-157}$.

potest libere se dilatare non comburit ita intense sicut cum libere se dilatare | [*Kl* 12rb] potest.

⟨17⟩ Sed diceret aliquis: hoc videtur esse contra Aristotelem in secundo *De anima*, ubi dicit quod si infinita combustibilia igni apponerentur, ignis cresceret ad infinitum[64]. | [*U* 10rb] Breviter respondetur quod bene hoc verum est cum esset in loco in quo libere ad quamcumque distantiam posset se dilatare; sed sic non est infra orbem Lune; igitur non quantumcumque infra orbem Lune potest augeri.

⟨18⟩ Ex isto etiam potest reddi causa quare diluvia non diu durant, nam [*L* 190ra] per talem augmentationem et multiplicationem unius elementi, elementa non sunt sub raritate et densitate sibi naturalibus, sed violentis; et quia violentum non est diu durativum, sequitur huiusmodi diluvia non posse diu permanere.

⟨19⟩ Ex istis patet ad tertium articulum, nam, quamvis elementa sint contraria et activa et passiva ad invicem, tamen quodlibet eorum tantum potest augeri quod eius actio debilitatur et non ulterius agit, sicut dicebatur de augmentatione ignis et remissione sue actionis.

⟨20⟩ Ad rationes. Ad primam dico quod quamvis equaliter sit continue de raritate et densitate, | [*Ba* 12ra] tamen ex hoc non oporteret quod continue equaliter sit de gravitate et levitate. Et quando dicebatur: | [*P* 237r] 'cum mutuo se consequuntur', respondetur quod quamvis ita sit, hoc tamen non est proportionaliter, quia, cum ferrum ignitur, ipsum non tantum acquirit de levitate sicut de caliditate.

⟨21⟩ Ad secundam patet eodem modo quod quamvis caliditas et frigiditas, raritas et densitas, mutuo se consequantur, tamen hoc non est proportionaliter, sicut patet de ferro ignito.

⟨22⟩ Ad tertiam respondetur quod non tantum proportionaliter generatur de frigiditate infra terram quantum de caliditatem super terram, nec e contrario. Et ex hoc patet ad articulum.

⟨23⟩ Ad quartam respondetur quod non semper proportionaliter | [*Kl* 12va] tantum generatur circa nos caliditas, quantum frigiditas generatur nobis opposito modo, solum per accessum Solis et recessum. Et ratio est quia non solum Sol productivus est caliditatis in istis inferioribus, verum etiam alii planete; et ideo possibile est quod quando Sol generat circa nos calorem per eius acces-

5 est] quod *add. P* 7 quantumcumque] in infinitum *Ba* qualitercumque *Kl U* 12 est] *om. P* 20 de gravitate et levitate] de caliditate et frigiditate *Ba Kl L U*

64 Aristoteles, *De anima*, II, 4, 416 a 15–16.

sum ad nos, quod alii planete in loco opposito etiam generant calorem; et sic non proportionaliter, quantum est hic calidius, est ibi frigidius.

Hoc de questione.

I.8 Utrum motus celi sit causa calefactionis ignis in sua spera et etiam aeris superioris

L 190rb Queritur consequenter utrum motus celi sit causa calefactionis ignis | in sua spera et etiam aeris superioris.

⟨1⟩ Et arguitur primo in generali quod motus localis non est causa caloris sic: id quod est causa frigiditatis non est causa caloris; sed motus localis est huiusmodi; ergo etc. Maior est nota, quia idem non videtur esse oppositorum causa[65]. Minor probatur, quia videmus quod poreta calida, cum movetur, frigefit; unde videtur quod motus *U* 10va | localis sit causa frigiditatis.

⟨2⟩ Secundo, nam experimur quod aqua currens frigidior est quam aqua stans, que tamen deberet esse calidior si motus esset causa caloris.

⟨3⟩ Tertio ad idem: si motus esset causa caloris, sequeretur quod, cum aliquod mobile descenderet—puta aliquod grave, ut lapis—propter eius descensum calefieret; sed hoc non fit, nam, si sic, sequeretur etiam quod grave in descendendo leve fieret, cum levitas consequatur caliditatem; et per consequens grave in fine tardius descenderet quam in principio, quia in fine esset minus grave. Et ex hoc tunc sequeretur quod motus naturalis non esset in fine velocior quam in principio, quod est contra philosophice loquentes[66].

⟨4⟩ Quarto ad idem, nam aliqui sunt habentes manus tremulas, quibus tamen non propter hoc manus sunt *Ba* 12rb | calidi *Kl* 12vb |ores. Similiter aliqui febricitantes tremunt et fortiter per totum eorum corpus moventur; non tamen propter hoc calefiunt.

⟨5⟩ Quinto ad idem: si motus localis calefaceret, vel hoc esset accipiendo 'motum' pro 'fluxu', vel accipiendo 'motum' pro 'forma fluente', vel acci-

1 in loco opposito] circa nos *Ba Kl L U* 2 non] semper *add. Ba Kl L U* 9 poreta] boreas *add. Ba Kl L U* 17 grave] gravius *P* 19 philosophice loquentes] communiter loquentes *Ba* plures *Kl L U*

65 Cf. Aristotelem, *De generatione et corruptione*, II, 10, 336 a 30–31; *Auct. Arist.*, 170, n. 42.

66 Cf. Aristotelem, *De celo*, II, 6, 288 a 27–33; Averroem, *Commentarium magnum in De celo*, II, comm. 35, ed. Carmody, Arnzen, 333$^{88-96}$: f. 119 C; Nicolaum Oresme, *Questiones in De celo*, II.7 *Utrum motus naturalis sit velocior in fine quam in principio*, ed. C. Kren, unpub. Ph.D. Dissertation, University of Wisconsin, 1965, 525.

piendo 'motum' pro 'mobili': non pro 'fluxu', quia sic esset de genere passionis et non esset agens productivum alicuius. Nec etiam potest dici quod pro 'forma fluente', nam forma fluens in motu est ipsum ubi; illud autem non est causa caloris, quia potest esse frigidum. Similiter nec motus accipiendo 'motum' pro 'mobili', quia potest esse mobile quod nec formaliter nec virtualiter est calidum.

⟨6⟩ Deinde specialiter probatur quod motus celi non calefacit speram ignis. Patet, nam hoc esset quia celum movet speram ignis secum; sed hoc non, quia ipsum celum est corpus politissimum, et ergo in nulla sui parte ignis potest sibi adherere ut circumducatur vel rapiatur una cum celo.

⟨7⟩ Secundo, certum est quod ille motus circularis ignis non esset igni naturalis, cum ex primo *Celi* uni corpori simplici non insunt plures motus naturales[67]. Sed quia igni naturaliter inest moveri sursum, sequitur quod sibi non inest naturaliter moveri circulariter, et per consequens inest sibi violenter; et cum violenta non diu durant, cum nullum violentum | sit perpetuum, *L* 190va non est verisimile quod tam diu durasset motus circularis ipsius ignis una cum celo.

⟨8⟩ Tertio: ignis est calidus in summo; ergo non potest calefieri, et per consequens celum per suum motum non calefacit ignem in spera sua. *P* 237^{v}

⟨9⟩ Deinde | probatur quarto quod celum per suum motum non ca|lefa- *U* 10vb cit aerem superiorem, quia inter ipsum et celum mediat | spera ignis, igitur *Kl* 13ra ille aer et celum distant; sed quia agens et passum debent esse immediata[68], videtur quod celum non calefaciat aerem superiorem.

⟨10⟩ In oppositum est Aristoteles in primo huius, cum dicit: 'lato enim circulariter primo elemento', etc.[69]

---

1 mobili] nullum istorum potest dici ergo etc. maior patet per sufficientem divisionem minor probatur *add. Ba Kl U* nullum istorum potest dici ergo etc. maior patet per sufficientem divisionem minor probatur *add. L* ‖ esset] est *P* 2 esset] est *P* 3 motu] locali *add. Ba Kl L U* 5 quod] nec naturaliter *add. Ba* 7 speram ignis] *om. Ba Kl* 10 vel rapiatur una cum celo] *om. Ba Kl* 11 circularis] localis *Kl* 18 summo] gradu *add. Kl L U* 20 motum] *om. P* 24 huius] tractatu primo capitulo secundo *add. Kl* tractatu primo *Ba L U*

---

67 Aristoteles, *De celo*, I, 268 b 30–269 a 2.

68 Cf. Aristotelem, *Physica*, III, 2, 202 a 5–8; *De celo*, II, 7, 289 a 29–30; *De generatione et corruptione*, I, 6, 322 b 19–25; *Auct. Arist.*, 168, n. 14.

69 Aristoteles, *Meteorologica*, I, 3, 340 b 11, translatio Guillelmi de Morbeka (translatio noua), *AL* X 2.2, 13^{125}: "Lato autem primum elementum circulariter".

Ba 12va ⟨11⟩ Breviter in ista questione videndum est qualiter motus localis | sit causa caloris; secundo videndum est qualiter celum movet speram ignis et aeris—utrum scilicet rapiendo vel alio modo—; tertio videndum est qualiter per hoc ista calefiunt a celo.

⟨12⟩ Quantum ad primum, sciendum est quod fuit opinio Sancti Thome quod motus localis est causa calefactionis quia in genere motuum, motus localis est primus. Sed in decimo *Metaphysice* dicitur quod in quolibet genere, primum debet esse causa aliorum in illo genere, et specialiter priorum post ipsum[70]. Cum igitur post motum localem sit calefactio—que ceteris paribus est prior et nobilior, quia est productiva nobilioris qualitatis, ut puta caliditatis—sequitur motum localem esse causam calefactionis.

⟨13⟩ Alia erat opinio quod motus localis est causa calefactionis quia proprium est calidi moveri et movere, et ergo motus est perfectio ipsius calidi. Cum igitur unam perfectionem consequatur alia, et cum caliditas etiam sit perfectio ipsius calidi, sequitur quod, habito motu locali, consequenter habebitur caliditas; et sic motus dicetur causa caloris quia, ipso habito, habetur caliditas[71].

⟨14⟩ Tertia erat opinio quod motus localis, sicut in animali, est causa caloris propter duplicem inclinationem, que quidem inclinationes sunt ad invicem contrarie; et ex tali contrarietate inclinationum causatur calor.

⟨15⟩ Breviter prima opinio non valet, quia tunc quilibet motus localis deberet calefacere; modo hoc est falsum. Falsitas patet quia, cum aliquid
Kl 13rb movetur, sicut terebellum, partes | extremales calefiunt citius quam partes centrales, et tamen ita bene partes centrales moventur localiter sicut partes extremales; centrales enim non calefiunt nisi a partibus extremalibus primo
L 190vb calefactis. Sed consequentia tenet, cum non plus ipse intendat de uno motu | locali quam de alio.

---

1 est] in generali *add. Ba Kl* 3 utrum scilicet rapiendo vel alio modo] *om. Ba Kl* 5 quod] fuerunt plures opiniones prima *add. Ba Kl* 7 quod] *om. P* 9 igitur] immediate *add. Ba Kl L U* 20 calor] in animali *add. Ba Kl L U* 26 tenet] per opinionem *add. Ba Kl*

---

70 Thomas de Aquino, *In De celo*, lib. 2, tract. 7, lect. 10, in *Opera omnia* iussu impensaque Leonis XIII P.M. edita, Rome 1886, 3: 158, n. 10. Cf. Aristotelem, *Metaphysica*, X, 1, 1052 b 18–19; 31–32; Averroem, *In Metaph.*, X, comm. 2, ed. apud Iunctas, 8: f. 251 I; *Auct. Arist.*, 135, n. 239.

71 Cf. Averroem, *Commentarium magnum in De celo*, II, comm. 42, eds. Carmody, Arnzen, 2: 355[197–205]; Thomam de Aquino, *In De celo*, lib. 2, cap. 7, lect. 10, ed. Leonina, 3: 158, n. 10.

⟨16⟩ Nec secunda opinio valet, per eandem rationem. Nec tertia valet, quia grave in descendendo solum habet unam inclinationem, quo tamen non obstante ipsum calefit in descendendo. Secundo patet quod tertia opinio non valet nec sufficit, | quia quod in motu animalis generatur caloris, hoc *Ba* 12vb est propter emissionem spirituum.

⟨17⟩ Alia est opinio Alberti. Sed premittatur primo una distinctio, quod motus localis | potest fieri dupliciter: uno modo potest fieri per quem fit *U* 11ra distractio et confricatio partium ipsius mobilis; alio modo potest fieri per quem non fit aliquid tale, sicut si aliquid movetur in vacuo[72].

⟨18⟩ Tunc secundum opinionem Alberti sit prima conclusio: motus localis qui fit primo modo est causa caloris et calefacit, nisi sit aliquid impediens. Patet de ferro in molendino circa | quod volvitur lapis molaris, et cum ibi *Kl* 13va apponitur aliquid viscosum vel pingue, videtur quod consumitur, quod non videtur nisi per calorem generatum. Similiter videmus quod terebellum calefit postquam diu est motum in foramine. Similiter, cum sagitta plumbata movetur, expertum est quod plumbum fuit liquefactum, quod fuit ex calore liquefaciente causato ex motu talis sagitte[73]. Item cum animal movetur localiter, tunc calefit, et hoc quia in motu eius fit confricatio membrorum et musculorum, quamvis etiam alia causa concurrat, videlicet quod cor mittit calores et spiritus ad exteriora membra ipsius animalis. Sic ergo motus qui fit cum confricatione est causa caloris. Unde ymaginandum est quod ex tali confricatione fit quedam rarefactio et quedam partium distractio quam consequitur caliditas.

⟨19⟩ Sed diceret aliquis contra: si hoc esset verum, sequeretur quod caliditas non esset | qualitas prima, propter hoc quod ipsa caliditas causatur ex *P* 238^{r} alia et sequitur alteram, puta raritatem. Respondetur quod caliditas bene dicitur qualitas prima activa; hoc tamen non obstante ipsa potest consequi unam aliam qualitatem, puta raritatem[74]. Vel potest dici quod ut plurium et

---

3–5 secundo patet quod … propter emissionem spirituum] *om. P* 6 opinio] modernorum et est opinio *add. Kl* 8 fieri] localis *add. P* 9 vacuo] tunc dicit albertus quod ratio est quod ex tali confricatione et collisione fit rarefactio violente et istam rarefactionem violentam consequitur calefactio et hec opinio concordat cum omnibus experientiis unde videmus quod ex tali confricatione et calefactione corrumpuntur quedam partes circa locum confricationis et etiam videmus quod quanto sunt aliqua corpora magis polita tanto est minor confricatio et calefactio *add. Ba Kl* 13 viscosum vel pingue] unctuosum *Ba Kl L U* 20 membra] *om. P L* 25–26 causatur ex alia et] *om. Ba Kl L U*

---

72 Albertus Magnus, *Meteora*, lib. 1, tract. 1, cap. 11, ed. Colon. VI/1, 15^{30-41}.

73 Cf. Aristotelem, *De celo*, II, 7, 289 a 25–30.

74 Cf. Nicolaum Oresme, *Questiones in Physicam*, IV.14, ed. Caroti [et al.], $517^{133-135}$.

pluries raritas sequitur caliditatem quam e converso; ideo caliditas dicitur qualitas prima, et non raritas.

⟨20⟩ Secunda conclusio: si aliquis sit motus localis sine confricatione et partium distractione, ille non est causa caloris. Patet, quia partes centrales *L* 191ra terebelli circa quas non fit confricatio | non calefiunt nisi a partibus extrema- *Ba* 13ra libus | calefactis propter confricationem. Similiter, si illud ferrum in molendino circa quod volvitur lapis molaris moveretur una cum illo lapide versus eandem differentiam positionis sic quod non fieret confricatio, non calefieret.

Kl 13vb ⟨21⟩ Ex isto sequitur quod si aliquod grave quantumcumque | velociter moveretur in vacuo, vel etiam sagitta plumbata, non calefieret, quia ibi non fieret confricatio ipsius ad aerem. Similiter, si esset aliqua sagitta plumbata et plumbum esset involutum in aliquo ligno vel aliquo huiusmodi, et si moveretur in aere, non propter hoc illud plumbum liquefieret, cum tunc non *U* 11rb fieret confricatio eius ad aerem. Similiter, | cum aliquis lapis percutitur ferro, non generantur scintille nisi ubi fit talis confricatio. Sic igitur patet qualiter motus localis est causa caloris et qualiter calefacere non inest omni motu locali, sed illi per quem fit confricatio et partium distractio.

⟨22⟩ Quantum ad secundum, sit ista conclusio quod ignis non ex hoc movetur circulariter cum celo quod celum trahat ignem secum. Patet, quia celum est corpus ita politum quod in nulla parte ignis potest sibi adherere sic quod ignis una cum ipso circumducatur et ab ipso celo rapiatur. Secundo a simili: unus orbis celestis motu suo non vehit secum alium; ergo nec in perpetuo celum rapit et trahit secum ignem. Consequentia tenet de se; sed antecedens patet per Commentatorem in secundo *Celi*, ubi dicit quod ibi non est raptus nec violentia, nisi fictitie[75].

⟨23⟩ Secunda conclusio: ignis sic movetur, videlicet circulariter una cum celo, per virtutem sibi impressam a celo, eo modo quo ferrum movetur insequendo magnetem. Patet, quia ex quo non movetur motu raptus, sicut dicebat prima conclusio, non videtur esse alius modus dicendi nisi dicatur quod

3 secunda] quarta *P* 4 centrales] *om. Ba Kl L U* 8 eandem] eam *P* 10 velociter] *om. Ba Kl L U* 11 plumbata] *om. Ba Kl L U* 13 esset involutum] poneretur *Ba Kl L U* 14 in aere] *om. Ba Kl L U* ‖ plumbum] calefieret vel *add. Ba Kl L U* 20 trahat] rapiat *Ba Kl L U* 21 corpus] *om. P* 23 vehit] trahit *Ba Kl L* rapit vel trahit *U* ‖ in perpetuo] *om. Ba Kl L U* 30 modus dicendi] motus *Ba Kl* modus *L*

75 Averroes, *Commentarium magnum in De celo*, II, comm. 3, eds. Carmody, Arnzen, 2: 274–275.

sic movetur per virtutem sibi impressam a celo. Nec ille motus est igni naturalis, quia non sequitur suam naturam, scilicet formam eius, nec etiam est ei violentus, quia motui circulari nullus motus est contrarius[76]; sed cuilibet motui violento alius motus est contrarius, puta motus naturalis illius | *Kl* 14ra cui talis motus est violentus.

⟨24⟩ Sed diceres: quomodo ergo ille motus inest igni? Respondetur quod inest sibi preter naturam, sicut etiam dicit Aristoteles in primo *Celi*[77]. Et potest dici ei impertinens; impertinens enim in arte obligatoria dicitur | *L* 191rb quod nec sequitur ad positum nec repugnat posito[78]: ita motus circularis ignis, quia nec inest ei naturaliter nec inest ei violenter, ergo potest dici impertinens sibi.

⟨25⟩ Quantum ad tertium, dico quod ignis non calefit a celo per confricationem eius ad celum. Patet, quia ibi nulla | *Ba* 13rb est confricatio, propter hoc quod ipsum celum est corpus politissimum. Sed tamen, quia virtus celi supra est fortior quam infra, partes superiores ipsius ignis moventur velocius quam partes inferiores, et similiter partes superiores aeris moventur velocius quam partes inferiores. Et ergo, quia partes inferiores non ita velociter insequuntur partes superiores, ibi fit quedam distractio partium et rarefactio quam consequitur caliditas, et sic ibi causatur calor.

⟨26⟩ Ad rationes. Ad primam de motu porete dicitur quod cum poreta movetur, est in alio et alio aere continue recenti et frigido, qui plus infrigidat quam ex tali motu possit calefieri.

⟨27⟩ Ad secundam respondetur quod aqua stans est calidior quam aqua currens, | *P* 238^{v} non propter hoc quod est stans, sed propter hoc quod radii Solis magis unite reflectuntur super eam quam super aquam currentem. | *U* 11va Et quando dicebatur quod si motus localis est causa caloris, aqua currens deberet calefieri, conceditur quod bene aliqualiter calefieret nisi esset aliquid

2 suam naturam scilicet formam eius] formam ignis *Ba Kl L U*   6 igni] celo *P*   18 rarefactio] sicut in confricatione *add. Kl*   19 et sic ibi causatur calor] *om. Kl* consimili modo ymaginandum est de aere *Ba*   20 porete] boree *Ba Kl L U* ‖ poreta] boreas *Ba Kl L U*   24 currens] fluens *Ba Kl L U*   25 reflectuntur] refranguntur *P* ‖ eam] ea *P*

76 Cf. Aristotelem, *De celo*, I, 3, 270 a 19–20; 4, 270 b 32; *Auct. Arist.*, 160, n. 8.

77 Aristoteles, *De celo*, I, 2, 269 a 9–269 b 16.

78 Cf. Martinum Anglicum, *De obligationibus—Über die Verpflichtungen*, Lateinisch-Deutsch, ed. F. Schupp, Hamburg 2014 (Philosophische Bibliothek, 462) 18; Paulum Venetum, *Logica magna*, ed. E.J. Ashworth, Oxford 1988 (Classical and Medieval Logic Texts, 8) 180.

Kl 14rb impediens, videlicet quod continue est circa alium et alium aerem recentem frigidum qui potest re|conpensari caliditati que posset causari ex tali motu.

⟨28⟩ Ad tertiam concedo quod lapis in descendendo calefit. Et quando dicebatur: 'igitur leve fieret', concedo—hoc est, fit minus gravis. Et cum dicebatur: 'igitur in fine moveretur tardius quam in principio', concedo nisi aliud obesset; sed modo in eius descensu acquirit quemdam impetum de quo dicebatur super primum *Celi*[79], qui magis permovet ad velocitatem descensus quam sit illa levitas impeditiva que acquiritur ex calefactione illius mobilis in eius descensu.

⟨29⟩ Ad quartam concedo quod non quilibet motus localis calefacit, nisi qui fit cum confricatione; et aliquando aliqua possunt adesse que tantum vel plus de frigiditate generant quam ille motus possit generare de caliditate. Et ex hoc dicatur ad argumentum.

L 191va ⟨30⟩ Ad quintam dico quod motus calefacit accipiendo | 'motum' pro 'mobili taliter se habente'; sed utrum mobile sic se habere sit ipsum mobile vel aliud, vel que res sit, dictum est in libro *Physicorum*[80].

⟨31⟩ Ad alias rationes quibus probatur quod ignis non movetur circulariter cum celo, concedo quod non sic quod celum trahat ipsum secum—nec aliud probat articulum—, sed bene sic quod movetur sic per virtutem celi sibi impressam—nec contra hoc fuit articulum.

⟨32⟩ Ad secundam concedo quod ille motus non inest igni naturaliter, nec etiam violenter, sed inest sibi preter naturam.

---

6 aliud] aliunde *Ba Kl L* 7–8 de quo dicebatur super primum celi] *om. Ba* de quo dicebatur in primo celi *U* 8 permovet] pro motu illius *P* 9 levitas] caliditas *Ba Kl L U* 15 calefacit] *om. P* 16 mobile sic se habere sit ipsum mobile] motus sit mobile *Ba* 17 vel aliud vel que res sit] *om. Ba Kl L* motus taliter se habeat sicut in ipso mobili *U* ‖ dictum est in libro physicorum] non modo *P* 22 motus] circularis *add. Kl L U*

---

79 Nicolaus Oresme, *Questiones super De celo*, II.7, *Utrum motus naturalis sit velocior in fine quam in principio*, ed. Kren, 559$^{343–346}$: "Quarto modo quod est ad propositum, ex velocitatione motus per quam acquiritur quedam habilitas vel impetus et quedam fortificatio accidentalis ad velocius movendum". Cf. Anonymum (Nicolaum Oresme?), *Questiones super De celo*, ms. München, Bayerische Staatsbibliothek, Clm 4375, ff. 47ra–76ra, I.22, *Utrum motus naturalis sit velocior in fine quam in principio*, f. 61va: "omne motum naturaliter ab intrinseco in velocitando motum acquirit fortitudinem et habilitatem eo adiuvantem ipsum motum, quod potest dici impetus vel inclinatio accidentalis".

80 Nicole Oresme, *Questiones in Physicam*, III.2–III.8, ed. Caroti [et al.], 304–348.

⟨33⟩ Ad tertiam | concedo quod celum per suum motum non calefacit ignem, cum est calidus in summo; nichilominus, si esset calidus remisse, calefieret ex tali motu. *Ba* 13va

⟨34⟩ Ad quartam dico quod non oportet quod quodlibet passum sit immediatum agenti principali, sed bene oportet quod cuilibet agenti aliquid sit immediatum, si debet agere in illud et mediante illo, seu per illud, in aliquod non sibi immediatum potest agere. Et ex hoc potest dici ad formam argumenti.

Hoc de questione.

I.9 Utrum lumen sit productivum caloris

Queritur consequenter utrum lumen sit productivum caloris. *Kl* 14va

⟨1⟩ Et arguitur quod non. Omne quod fit et generatur fit a sibi simili in specie; sed calor est alterius speciei a lumine; ergo calor non producitur a lumine, et per consequens conclusio falsa.

⟨2⟩ Secundo, nam, si sic, sequeretur quod lumen esset qualitas prima, immo prior quam caliditas. Consequentia tenet ex eo quod | lumen esset productivum caloris. Falsitas consequentis patet, quia solum sunt quatuor qualitates prime, videlicet caliditas, frigiditas, etc. *U* 11vb

⟨3⟩ Tertio ad idem: si sic, sequeretur quod, posito lumine, ponetur caliditas; sed hoc est falsum. Consequentia tenet ex eo quod, posita causa, ponitur effectus. Falsitas consequentis patet, quia lumen fit subito[81], caliditas autem fit successive; ergo non ita cito ponitur caliditas sicut lumen.

⟨4⟩ Quarto ad idem: probatur quod lumen Solis non calefacit ista inferiora. Nam, cum inter speram Solis et ista inferiora sint aliqui orbes celestes, oporteret quod Sol suo lumine prius calefaceret illos quam ista inferiora. Consequentia tenet ex eo quod agens non agit in remotum nisi prius agat in propinquum. Falsitas consequentis patet, quia orbes celestes non sunt susceptivi caloris.

⟨5⟩ Quinto: sequeretur etiam quod Sol suo lumine calefaceret mediam regionem aeris; sed hoc non, cum semper media regio aeris sit frigida. Con-

1 celum per suum motum] *om. P* 3 ex tali motu] a celo *Ba Kl U* a celo quia propter unumquodque tale etc. *L* 6 agere] et *add. P* 19 est] *om. P* 23 aliqui] alique *P*

81 Cf. Aristotelem, *De anima*, II, 7, 418 b 23–26; *Auct. Arist.*, 180, n. 70.

L 191vb sequentia tenet, quia non potest agere calorem hic inferius iuxta terram nisi | prius ageret calorem in media aeris regione.

⟨6⟩ Sexto ad idem: si questio esset vera, sequeretur quod ubi esset plus de *Kl* 14vb lumine, ibi esset plus de caliditate; sed hoc est falsum. | Consequentia tenet ex eo quod lumen esset productivum caloris: ubi autem esset plus de productivo caloris, ibi esset plus de calore. Falsitas consequentis patet, nam tunc in *Ba* 13vb media aeris regione esset plus de calore | quam sit inferius prope terram; sed hoc est falsum, cum media regio aeris sit semper frigida. Consequentia tenet ex eo quod ibi est plus de lumine.

⟨7⟩ Septimo ad idem: alique stelle non producunt in istis inferioribus caliditatem, non obstante quod producunt lumen in istis inferioribus; igitur non omne lumen est productivum caloris.

P 239r ⟨8⟩ Octavo: multa | lucent, sicut squame piscium et putredines quercuum, que tamen non sunt productiva caloris[82].

⟨9⟩ In oppositum est Aristoteles in primo huius: vult enim quod celum per motum et lumen calefacit ista inferiora et elevat exalationes que sunt materia impressionum[83].

⟨10⟩ Breviter in ista questione primo videndum est de generatione luminis; secundo videndum est si omne lumen sit productivum caloris; tertio videndum est de causa et modo quo lumen producit calorem.

⟨11⟩ Quantum ad primum, potest notari primo differentia inter 'lucem' et 'lumen' et 'splendorem'. Unde 'lux' est qualitas incorporata corpori lucido, *U* 12ra et movetur | ad motum corporis lucidi. 'Lumen' autem dicitur species illius qualitatis causata in medio, sicut species coloris. Unde species lucis voca- *Kl* 15ra tur 'lumen' et non movetur ad motum corporis | lucidi; alias enim accidens migraret de subiecto in subiectum, cum ipsum lumen non sit in corpore

---

7 sit inferius prope terram] hic circa nos *Ba* hic prope nos *L* 8 est] *om. P* 8–9 consequentia tenet ex eo quod ibi est plus de lumine] *om. Ba Kl L U* 10–11 non producunt in istis inferioribus caliditatem] producunt in istis inferioribus frigiditatem *Ba Kl L U* 13 quercuum] et aliqui vermiculi *add. Ba* et oculi catorum et aliqua alia de nocte *add. Kl* et vermiculi aliqui *add. L U* 15 celum] sol *Ba Kl L U* 23–24 species illius qualitatis] illa qualitas *Ba Kl* 24 medio] perspicuo *add. Ba Kl* 25 lucidi] luminosi *Ba Kl L U* 26 migraret] iret *Ba Kl U*

---

82 Cf. Aristotelem, *De anima*, II, 7, 419 a 1–7; Averroem, *Commentarium magnum in Aristotelis De anima libros*, II, comm. 72, ed. Crawford, 239–240.

83 Aristoteles, *Meteorologica*, I, 3, 341 a 13–b 36; I, 9, 346 b 22–25.

lucido, sed in medio. Sed ‘splendor’ dicitur lumen reflexum supra aliquo corpore terso et polito, sicut supra auro vel speculo. Unde tale dicitur ‘splendere’; corpus autem luminosum ‘lucere’ dicitur; medium autem ‘illuminare’. Verum est tamen quod, large utendo istis vocabulis, solemus accipere unum pro alio. Secundo, quia lumen multiplicatur directe, nos dicimus corpus lucidum ‘radiare’.

⟨12⟩ Potest notari quod quidam dicitur ‘radius rectus’, quidam vero ‘reflexus’, quidam autem ‘refractus’. ‘Rectus’ dicitur qui transit libere ipsum medium sine fractione et reflexione notabili. Radius autem ‘reflexus’ dicitur qui non transit per medium sine reflexione, sed reflectitur versus eandem partem unde venit, sicut pila cum proicitur super terram. Et si iste radius | *L* 192ra incidens incidit perpendiculariter super ipso reflectente, tunc idem fit radius incidens et reflexus; si autem radius incidens incidit oblique, tunc alius est radius incidens et alius est radius reflexus. Radius autem refractus | *Ba* 14ra dicitur qui non totaliter reflectitur nec etiam totaliter libere transit non fractus per medium, sed aliqualiter frangitur—aliquando a perpendiculari, aliquando ad perpendicularem—, sicut contingit quando aliquis radius incidit supra aliqua superficie media diversarum rationum coniungente, sicut patet cum denarius positus in aqua videtur, oculo existente in aere, maior[84].

⟨13⟩ Tunc sit prima conclusio: lumen generatur per reflexionem vel refractionem. Patet, nam omne medium est aliqualiter | *Kl* 15rb densum[85]; ergo omne medium aliqualiter reflectit; ergo ubicumque multiplicatur lumen, ibi reflectitur: nichil enim est aliud multiplicatio luminis nisi illa reflexio continua.

1 lucido] luminoso *Ba Kl L om. U* 3 corpus autem luminosum] lumen autem *Ba Kl L U* 8 dicitur] *om. P* 10 sine reflexione] *om. P* 11 super terram] ad parietem *Ba L* in parietem *Kl* in parieta *U* 14 incidens] incidentie *P* 16–17 frangitur aliquando a … aliquando ad perpendicularem] propinquat ad perpendicularem *Ba L* appropinquat ad perpendicularitatem *Kl* appropinquat ad perpendicularem *U* 18–19 patet cum denarius … in aere maior] cum res est posita in aqua *Ba Kl*

84 Cf. Alhazen, *Perspectiva*, lib. 7, cap. 5, prop. 14, ed. P. Pietquin, *Le septième livre du traité De aspectibus d'Alhazen, traduction latine médiévale de l'Optique d'Ibn al-Haytham*, Brussels 2010 (Mémoires de la Classe des lettres, Collection in-8o; 3e série, 5), 151–165; Witelonem, *Perspectiva*, lib. 10, prop. 4, ed. F. Risner, *Opticae Thesaurus: Alhazeni arabis libri septem, nunc primum editi. Eiusdem Liber de crepusculis et nubium ascensionibus. Item Vitellonis Thuringopoloni libri X*, Basel 1572, reprint New York 1972 (The Source of Science, 94), 407–408; Robertum Grossatestam, *De lineis, angulis et figuris*, ed. L. Baur, *Die philosophischen Werke des Robertus Grosseteste, Bischofs von Lincoln*, Münster i.W. 1912 (Beiträge zur Geschichte der Philosophie des Mittelalters, 9), 1912, 63^{11-19}; *De iride seu de iride et speculo*, ed. Baur, 74^{9-10}; Iohannem Pecham, *Perspectiva communis*, pars 1, prop. 15–16, ed. D. Lindberg, *John Pecham and the Science of Optics*, Madison 1970, 89–91, pars 2, prop. 2, ed. Lindberg, 212.

85 Cf. Alhazen, *Perspectiva*, lib. 7, cap. 2, prop. 8, ed Pietquin, 101^{11-13}.

⟨14⟩ Sequitur correlarie quod ceteris paribus, ubi medium est magis rarum, ibi est minus lumen, et ubi est magis densus—non tamen nimis—, ibi est plus de lumine recte, ut de caliditate, que non est ita intensa in raro sicut in denso.

⟨15⟩ Sed diceres contra, nam videtur quod quanto aer est rarior et serenior, tanto est maius lumen circa terram; sed hoc non esset verum si in medio *U* 12rb raro esset | minus de lumine quam in medio denso. Secundo sequeretur quod prope Solem vel prope celum esset valde modicum de lumine, eo quod medium est ibi valde rarum.

⟨16⟩ Ad primum respondetur: concedo quod quanto aer est magis rarus, tanto est maius lumen circa terram, sed non maius in aere, propter hoc quod est rarior; sed tunc, quia aer est rarus, omnes radii Solis transeunt sine magna reflexione et tota activitas terminatur ad terram. Ad secundam concedo quod prope celum est modicum lumen in actu, sed tamen prope celum dicitur magnum lumen in potentia, nam, si prope celum esset corpus opacum bene potens reflectere radios solares, maius causaretur ibi lumen quam si idem corpus opacum reflecteret hic inferius radios solares.

⟨17⟩ Quantum ad secundum, sciendum est quod duo sunt corpora lucida in *P* 239^{v} mundo, videlicet Sol et ignis. Omnes enim alie stelle et planete alii a | Sole *Kl* 15va recipiunt lumen | a Sole; unde lumen Lune non est aliud quam lumen Solis *L* 192rb reflexum supra | corpore lunari; et sic similiter de lumine aliorum planetarum et stellarum.

⟨18⟩ Et breviter sit ista prima conclusio: lumen Lune et etiam lumen alia- *Ba* 14rb rum stellarum calefacit. Patet: lumen | Solis calefacit; ergo et lumen aliarum stellarum. Antecedens patet per experientiam. Notandum est tamen quod quamvis Sol suo lumine calefaciat, non tamen Sol est formaliter calidus. Si quis enim Solem digito tangeret, non sentiret calorem, cum corpora celestia a qualitatibus habentibus contrarium sint exempta, cum sint perpetua, secundum quod probatum est in primo *Celi*[86]. Consequentia principalis patet, nam lumen aliarum stellarum non est nisi lumen Solis reflexum; si ergo lumen Solis est calefactivum, et lumen aliarum stellarum erit calefactivum. Et confirmatur, nam calefacere est de ratione ipsius luminis; ergo, si est aliquod lumen quod est calefactivum, sicut lumen Solis, sequitur quod omne

---

2 et ubi] *iter. P* ‖ non tamen nimis] *om. Ba Kl L* 3 ita] *om. P* 5 aer] medium *Kl* 7–9 secundo sequeretur quod … ibi valde rarum] *om. Ba Kl* 10 aer est] est aer est *P* 14 celum] solem *L* ‖ modicum] modicus *P* ‖ celum] solem *L U* 15 celum] solem *L U* 16 radios solares] *om. P* 17 inferius] prope terram *add. Ba Kl U* circa terram *add. L*

---

86 Aristoteles, *De celo*, I, 3, 270 a 14–26; *Auct. Arist.*, 160, n. 16.

lumen erit calefactivum. Item specialiter de Luna patet, quia lumen Lune est calefactivum auctoritate Aristotelis in *De animalibus*, ubi dicit quod noctes in plenilunio sunt calidiores[87], quod non videtur esse ex alio nisi quia tunc est maius lumen Lune. Item patet de stellis fixis auctoritate Commentatoris secundo *Celi*, qui dicit quod omnes stelle calefaciunt, | licet secundum (*U* 12va) magis et minus[88]. Item hoc potest experiri per magna specula comburentia, congregando radios Lune vel radios aliarum stellarum.

⟨19⟩ Secunda conclusio: ignis, | qui est unum corpus de corporibus lucidis (*Kl* 15vb) qui sunt in mundo, calefacit et per lumen et per calorem; tamen differenter, quia magis per calorem quam per lumen. Item quia per lumen calefacit ita cito remote sicut propinque; sed per calorem prius calefacit partem sibi propinquam quam partem remotam.

⟨20⟩ De aliis lucidis, sicut sunt squame piscium, potest dici quod etiam lux ipsarum calefacit, sed hoc est bene remisse; et forte habent quasdam alias virtutes fortiores frigefactivas, quarum actio per huiusmodi lucem vel lumen remittitur; et ergo, quia virtus eius est fortior, aliquando effectus eius apparet et non effectus lucis, quamvis forte intensius appareret si non obstaret lumen.

⟨21⟩ Quantum ad tertium, dico quod lumen est productivum caloris. Patet, | (*Ba* 14va) quia lumen est una qualitas valde nobilis; et ergo, cum multiplicatur per medium, de potentia illius medii aptum natum est educere qualitatem nobiliorem de qualitatibus activis et passivis ad invicem; et cum inter istas | (*L* 192va) caliditas sit nobilissima, sequitur quod lumen, cum multiplicatur per medium, educit de ipso caliditatem. Similiter, cum ipsum lumen sit una qualitas ita nobilis, ipsum aptum natum est disponere ad generationem elementi nobilissimi, quod quidem elementum est ignis. Et cum generando caliditatem disponit ad generationem ignis, sequitur quod lumen producit caliditatem. Quod autem ignis inter cetera elementa sit nobilissimus | patet signo (*Kl* 16ra)

---

4 patet de stellis fixis] specialiter de aliis patet *P* 6 comburentia] concava *Ba Kl L U* 7 lune] vel solis *add. Kl* 8 unum corpus de corporibus lucidis] minimum corpus de corporis magnis *Ba Kl* minimum corpus de corporibus *L* unum corpus de corporibus magnis *U* 11 propinque] remoto impedimento *add. Kl* 14 et] quia *Ba Kl L U* 18 lumen] lux vel lumen *Ba Kl* 22 istas] *om. P* 26–27 generando caliditatem disponit ad generationem ignis] generando caliditatem sequitur formam substantialem ipsius ignis *Ba* generando caliditatem sequatur formam ignis *Kl L U*

---

87 Aristoteles, *De partibus animalium*, IV, 5, 680 a 34–35.

88 Averroes, *Commentarium magnum in De celo*, II, comm. 42, eds. Carmody, Arnzen: 2, 350^{61}–351^{86}; $352^{102-117}$.

et ratione. Signo quidem, quia quidam, sicut Caldei, adoraverunt ignem tamquam Deum propter eius nobilitatem. Patet hoc etiam ratione, quia inter cetera elementa ignis est magis activus. Finaliter igitur concludo quod lumen est productivum caloris, non tamen tamquam agens principale, sed instrumentale. Corpus enim lucidum est productivum principale ipsius caloris; lumen autem est instrumentum mediante quo corpus lucidum huiusmodi calorem producit.

⟨22⟩ Ad rationes. Ad primam: 'simile', etc., dico quod sufficit similitudo virtualis, ita quod agens sit potens facere quod passum aptum natum est recipere.

[*U* 12vb] ⟨23⟩ Ad secundam concedo quod lumen est qualitas prima, immo prior qualitatibus activis et passivis ad invicem. Et cum dicitur: 'qualitates prime sunt quatuor qualitates elementorum', concedo, qualitates prime habentes contrarium, sed sic non est de lumine.

⟨24⟩ Ad tertiam potest negari quod lumen fiat subito. Si tamen concederetur quod fieret subito, non tamen propter hoc oporteret concedere quod caliditas fieret subito; unde non oportet quod quam cito incipit esse causa sufficiens alicuius effectus, quod ita cito incipiat ille effectus, sed sufficit quod ita cito incipiat posse produci; et sic esset in proposito de lumine et calore si lumen produceretur subito.

[*Ba* 14vb] ⟨25⟩ Ad quartam dico quod lumen Solis bene recipitur in orbibus celesti- [*P* 240^{r}] bus, sed tamen, quia orbes celestes non sunt apti nati recipere calorem, | inde [*Kl* 16rb] est quod mediante lumine non generatur in eis calor. Cum | tamen ipsum lumen sit multiplicatum per orbes celestes usque ad ista inferiora, et cum ista inferiora sint receptiva caloris, inde est quod mediante lumine causatur in istis inferioribus calor et non in orbibus celestibus. Et ulterius con- [*L* 192vb] cedo quod Sol non posset agere lumen in ista inferiora | nisi primo ageret in orbes celestes, quia tunc Sol non immediate producit calorem, sed mediante lumine, secundum dispositionem recipientis, lumen potest aliquando agere calorem et aliquando non, et aliquando intensiorem et aliquando remissiorem.

⟨26⟩ Ad quintam dico ex eadem radice quod potest agere calorem hic inferius et non in media regione aeris, quia, quantum ibi adduceretur de calore,

---

9–10 recipere] pati *Ba Kl L U* 19 posse produci] produci *P* 29 agere] producere *Kl L U* 32–33 hic inferius] in istis inferioribus prope terram *Kl* hic inferius prope terram *Ba L U* 33 adduceretur] educeretur *Ba Kl L U*

plus remitteretur de frigiditate medie regionis; et qua sit causa frigiditatis medie regionis postea videbitur in una questione[89].

⟨27⟩ Ad sextam ex eadem radice.

⟨28⟩ Ad septimam potest dici quod alique stelle habent unam aliam virtutem, seu multiplicant de se unam virtutem insensibilem aliam a lumine, que aliquando est productiva frigiditatis et aliquando caliditatis una cum lumine, et aliquando plus frigiditatis quam lumen aliarum stellarum sit productivum caliditatis. Et inde est quod alique stelle dicuntur frigefacere, non tamen per lumen, sed per unam aliam qualitatem que potest vocari 'influentia'. Alii dicunt quod omnes | stelle calefaciunt; et cum dicitur eis: 'alique [*Kl* 16va] frigefaciunt', exponunt: 'id est minus calefaciunt'[90].

⟨29⟩ Ad octavam de squamis piscium dictum est in corpore questionis[91]. Et sic est finis.

I.10 Utrum contrarium circumstans suum contrarium fortificet ipsum

Queritur consequenter utrum contrarium circumstans suum contrarium [*U* 13ra] fortificet ipsum.

⟨1⟩ Et arguitur primo quod non: contrarium circumstans suum contrarium agit in ipsum destruendo ipsum; ergo non fortificat ipsum. Consequentia tenet; antecedens patet si aqua approximetur igni.

⟨2⟩ Secundo: si sic, sequeretur quod aqua ex aere posset generare ignem; sed hoc est impossibile. Impossibilitas patet ex eo quod unum contrarium non generat reliquum. Sed consequentia patet, nam, si aqua frigida sit approximata aeri calido, si eius caliditatem intenderet, tamen non posset eam intendere quod ex illo aere fieret ignis.

---

5 insensibilem] in istis inferioribus *Kl om. Ba* 6 frigiditatis et aliquando] *om. Ba Kl U* 7–8 lumen aliarum stellarum sit productivum caliditatis] caliditatis sit productivum lumen earum *P* 9 qualitatem] insensibilem *add. Ba L U* seu virtutem insensibilem *add. Kl* 12 dictum est in corpore questionis] patebit in questione sequenti *Ba Kl L* patebit quid sit dicendum in una alia questione *U* 21 frigida] frigido *P* 22 non] *om. P Ba L U*

---

89 I.11, 11–16.

90 Cf. Averroem, *Commentarium magnum in De celo*, II, comm. 42, eds. Carmody, Arnzen, 2: $356^{216-224}$.

91 I.9, 20.

⟨3⟩ Tertio: si sic, sequeretur quod si una gutta aque poneretur ad medium ignis, quod plus frigefieret; et similiter, si panis poneretur in furno, deberet plus frigefieri. Ista sunt falsa, | [*Ba* 15ra] sicut patet per experientiam. Consequentia vel consequentie sunt bone, si questio est vera.

⟨4⟩ Quarto ad idem: sequitur quod ignis, quia circumdat alia elementa, deberet ipsa in qualitatibus eorum intendere; et sic ignis alia elementa fortificaret, et per consequens non corrumperet. Et sic ex hac augmentatione ipsius ignis in spera sua non posset argui aliorum elementorum corruptio, | [*L* 193ra] sed magis fortificatio, quod est contra Aristotelem in primo huius[92].

⟨5⟩ Quinto ad idem: si questio esset vera, hoc esset isto modo quod qualitas | [*Kl* 16vb] que esset in partibus extremalibus per contrarium circumstans pelleretur versus partes interiores et quod ex tali unione qualitatis in partibus interioribus subiecti illa qualitas intenderetur. Sed probo quod non fiat isto modo, nam, si sic, sequeretur quod accidens migraret de subiecto in subiectum; sed hoc est impossibile. Consequentia videtur tenere ex eo quod ista eadem qualitas que fuit in partibus extremalibus esset pulsa ad partes interiores. Sequeretur etiam secundo quod unum contrarium fugaret alterum et moveret ipsum localiter, quod iterum non videtur verisimile.

⟨6⟩ In oppositum est Aristoteles, quasi per totum processum istius libri: in reddendo enim multorum effectuum causas presupponit quod subiectum contrarium circumstans aliud fortificat ipsum et intendit eius qualitatem sue qualitati contrariam[93]. Et istud etiam bene patet ad experientiam, nam in estate quidam fontes sunt magis frigidi quam in hieme, cuius ratio non videtur esse alia nisi quod in estate tales fontes circumstat | [*U* 13rb] calor, qui est contrarius frigiditati, ex cuius circumstantia illa frigiditas intenditur. Similiter nam videmus quod si aliqua gutta aque proiciatur super mappam siccam, partes eius tenent se simul et aliqualiter illa gutta rotundatur, ac si humiditas illius gutte timeat et fugiat siccitatem mappe illius. Et ex tali unione partium videtur quod illa humiditas intendatur, cum virtus unita sit fortior se ipsa dispersa[94]. Cum autem gutta | [*P* 240v] aque proicitur super mappam humidam vel mensam, fluit et dilatat se, quia simile applaudit suo simili.

---

2 si panis poneretur] *om. P* ‖ in furno] ad furnum calefactum *Ba Kl L U* 11–12 pelleretur] moveretur *Ba Kl* tenderetur *L* 19 quasi] *om. Ba Kl L U* 26 nam] non *P* 28 timeat et] *om. Ba Kl L U* 30 proicitur] producitur *P* 31 mensam] super aliquid huiusmodi *Ba Kl L U*

---

92 Aristoteles, *Meteorologica*, I, 3, 340 a 1–3.

93 Ibid., I, 4, 342 a 1–16; I, 12.

94 Cf. Anonymum, *Liber de causis*, prop. XVI, ed. A. Pattin, *Le Liber de causis. Édition établie à l'aide de 90 manuscrits avec introduction et notes*, Leuven 1966 (Extrait du *Tijdschrift voor filosofie*, 28), 138^{15-16}; *Auct. Arist.*, 232, n. 13.

⟨7⟩ In ista questione primo videndum est an unum contrarium possit movere *Kl* 17ra aliud localiter; secundo videndum est de modo per quem unum contrarium movet suum contrarium localiter; tertio videndum est an ex condensatione alicuius subiecti caliditas vel frigiditas illius subiecti intendantur.

⟨8⟩ Quantum ad primum, primo sciendum est quod questio et illud quod dicetur in ista questione debet intelligi de qualitatibus primis, sicut sunt caliditas, frigiditas, etc., | et non de qualitatibus secundis, sicut sunt albedo, *Ba* 15rb nigredo et consimilia. Unde albedo posita iuxta nigredinem non facit ipsam intensiorem, sed bene facit quod apparet intensior: contraria enim iuxta se posita magis elucescunt[95].

⟨9⟩ Et sit breviter una conclusio quod unum contrarium potest movere aliud localiter, verbi gratia ignis potest movere aquam localiter. Patet, nam, si aliquod lignum viride ponatur ad ignem, videmus quod aqua vel humor aqueus qui est in illo ligno pellitur | et fugatur, et movetur usque ad extre- *L* 193rb mitatem illius ligni, que quidem extremitas ligni sita est extra ignem; igitur etc.

⟨10⟩ Secunda conclusio: cum unum contrarium circumdat aliud, puta ignis aquam, condensat ipsum. Patet, nam ex quo unum contrarium movet aliud localiter, per primam conclusionem partes circumferentiales aque, cum circumdantur igne, moventur omnes indifferenter versus partes centrales illius aque, et ex hoc tunc fit condensatio illius aque.

⟨11⟩ Tertia conclusio: contrarium circumstans suum contrarium intendit ipsum. Probatur, quia per secundam conclusionem condensat ipsum et facit quod partes eius fiunt propinquiores quam prius erant; sed talem condensationem et approximationem partium sequitur intensio virtutis et qualitatis illius subiecti. Confirmatur | ista conclusio signo, nam propter | fri- *U* 13va *Kl* 17rb gus circumstans in hieme aliqui fontes sunt minus frigidi quam in estate; et in estate, propter calorem circumstantem sunt magis frigidi. Secundo, nam propter hoc etiam ventres hominum in hieme sunt calidiores quam in estate[96], et similiter homines et cetera animalia in regionibus frigidis sunt

---

6 qualitatibus] contrariis *P* 7 qualitatibus] contrariis *P* 13 viride] *om. Ba Kl L U* 14 et movetur] *om. Ba Kl L U* 24 eius] fugiunt intra et *add. Kl* 29 hominum] animalium *Ba om. Kl L U* 30 estate] ut dicit ypocras *add. Ba Kl L U* ‖ in regionibus frigidis] in hieme *Kl L om. Ba U*

---

95 Cf. Aristotelem, *De sophisticis elenchis*, 15, 174 b 5–6, *Auct. Arist.*, 333, n. 17; *Rhetorica*, III, 3, 1405 a 12–13, *Auct. Arist.*, 267, n. 57.

96 Cf. Aristotelem, *Meteorologica*, I, 12, 348 b 2–5; *Auct. Arist.*, 172, n. 9.

magis condensata et compacta, et in regionibus calidis magis rara; et calor in animalibus habitantibus in regionibus frigidis est magis intensus quam *Ba* 15va sit in animalibus habitantibus in regionibus calidis, | et propter hoc dicitur quod homines de regionibus frigidis sunt magis audaces quam homines de regionibus calidis.

⟨12⟩ Quantum ad secundum, videlicet quomodo fit quod unum contrarium movet sic aliud, breviter potest dici uno modo quod hoc fit isto modo, nam cuiuslibet subiecti qualitas apta nata est se diffundere et dilatare; cum ergo sic se dilatando invenit aliquid sibi contrarium resistens, non potens ulterius se dilatari, reflectitur versus partem unde venit, recte sicut lumen inveniens obstaculum reflectitur et similiter pila proiecta ad terram movetur e contrario versus partem unde venit, et ita consimili modo in proposito.

⟨13⟩ Breviter potest dici aliter, supponendo quod omne ens diligit se permanere: quia ergo unum contrarium est corruptivum et destructivum alte- *Kl* 17va rius, quodlibet contrariorum ex naturali in|clinatione fugit suum contrarium et movetur ab eo; unde in tali motu forma illius quod movetur est movens principale, reliquum autem contrarium est movens occasionale. Unde, si aqua circumdaretur igne, partes circumferentiales aque moverentur versus *L* 193va partes centrales et movens principale esset forma ipsius aque; | ignis autem esset movens occasionale. Si enim aqua non circumdaretur igne, forma aque non moveret partes circumferentiales aque versus centrales.

⟨14⟩ Dubitatur hic utrum talis motus localis quo partes circumferentiales aque moventur versus partes centrales eiusdem propter ignem circumstantem sit aque naturalis vel violentus. Et videtur quod sit aque naturalis, quia *U* 13vb principaliter est a principio intrinseco, | scilicet a forma ipsius aque. Probatur autem quod sit aque violentus et non naturalis, quia aliter eiusdem corporis simplicis essent plures motus naturales, quod est contra Aristote- *P* 241r lem in primo *Celi*[97]. Respondetur quod talis motus est naturalis, recte sicut | motus aque sursum ad replendum vacuum est naturalis, quamvis bene pre- *Ba* 15vb supponat violentiam alicuius corporis. | Et quando dicebatur: 'solum unius corporis est unus motus naturalis', conceditur absolute et non facta violen-

1 condensata] dempsata *P* ‖ compacta] propter frigiditatem *add. Ba Kl L U* ‖ rara] raro *P* propter calorem *add. Ba Kl L U*   1–3 et calor in … in regionibus calidis] similiter animalia habitantia in regionibus frigidis sunt magis condensata quam animalia habitantia in regionibus calidis *Ba Kl L U*   8 diffundere] discendere *L*   9 sic se dilatando] *om. Ba Kl L U*   12 proposito] sed non est verisimile quod ex tali reflexione potest fieri intensio *add. Ba Kl L U*

97 Aristoteles, *De celo*, I, 2, 268 b 30–269 a 2.

tia alicui corpori; nichilominus, ex presuppositione, unius corporis simplicis possunt esse plures motus naturales. Unde, cum aer in fistula posita in aqua secundum unum eius extremum extrahitur et movetur violenter sursum, aqua naturaliter insequitur istum aerem ne fiat vacuum, et ista insecutio et motus aque sursum potest dici naturalis ipsi aque, quia inest sibi a principio intrinseco, puta forma substantiali ipsius aque, quamvis illa forma non sic moveret illam aquam | nisi esset facta violentia alicui corpori, scilicet aeri. *Kl* 17[vb]

⟨15⟩ Item dubitatur, posito quod duo contraria, puta ignis et aqua, ad invicem approximentur et ignis sit multo fortior quam aqua, utrum ista aqua fugat ignem sicut ignis fugat aquam. Respondetur quod sic, quia quamvis ignis sit fortior in agendo quam aqua, potest tamen esse debilior in resistendo quam ipsa aqua sit in agendo; et ergo, quia quodlibet eorum est fortius in agendo quam sit reliquum in resistendo, per actionem et reactionem eorum ad invicem quodlibet fugat aliud et movet ipsum localiter. Si tamen ignis et aqua tantum distarent quod ignis bene posset attingere sua actione aquam, aqua autem non posset e contrario attingere ignem, tunc ignis moveret aquam localiter et aqua non moveret ignem localiter.

⟨16⟩ Item dubitatur, postquam per fugam unius contrarii ab alio contraria ab invicem elongantur ab alio, sic quod non amplius se tangant, utrum unum adhuc fugat aliud. Potest dici quod sic, quia quamvis non se tangant, tamen adhuc quodlibet eorum potest attingere aliud sua virtute, mediante aere intermedio.

⟨17⟩ Item dubitatur an semper sic fiat, videlicet quod unum contra|rium *L* 193[vb]
approximatum alteri fugat ipsum. Respondetur quod non semper sic fit: aliquando enim propter nimiam gravitatem unius contrarii reliquum non potest ipsum fugare, verbi gratia sicut cum ferrum frigidum ponitur ad ignem, non propter hoc, saltem notabiliter, fugantur partes extremales | ver- *U* 14[ra]
sus | centrales. Notandum est hic quod intensio contrarii que fit ex tali pulsu *Kl* 18[ra]
partium circumferentialium vel lateralium versus partes centrales per contrarium circumstans vocatur 'antiparistasis'.

⟨18⟩ Secundo est sciendum quod aliquando contrario approximato suo contrario, sicut ignis aque tantum ipsum excedit quod quamvis in principio sic moveat partes circumferentiales versus partes centrales et intendat illud contrarium, tamen postea sua virtute penetrat | istud contrarium, puta illam *Ba* 16[ra]
aquam, et e contrario ipsam rarefacit et virtutem eius prius intensam remit-

1 simplicis] *om. P* 4 istum aerem] *om. P* 19 non amplius se tangant] unum non potest attingere aliud *Ba Kl L U* 23 sic fiat videlicet quod] *om. Ba Kl L U* 26 fugare] nec movere *add. Ba Kl L U* 27 extremales] circumferentiales *Ba Kl L U* 29–30 per contrarium circumstans] *om. Ba Kl* 32 aque] aliquando *add. P*

tit. Et ita, quamvis in principio actionis unum contrarium intendat aliud, tamen postea ipsum destruit et corrumpit, postquam fugavit ipsum tantum et similiter condensavit quod ulterius fugere non potest, nec sub maiori densitate forma eius esse potest.

⟨19⟩ Quantum ad tertium, videlicet utrum ex condensatione alicuius subiecti caliditas vel frigiditas illius subiecti intendatur, breviter respondeo quod sic: istud enim patet ad experientiam, quia aqua congelata est frigidior quam aqua non congelata, quod videtur esse signum quod ex istius aque condensatione frigiditas eius intenditur. Similiter hoc patet ex alio, nam ignis in materia densa, sicut in ferro, fortius agit quam in materia rara, sicut in stuppa. Et potest confirmari hoc idem ratione, quia in condensatione partes propinquius iacent et magis sunt unite; virtus autem unita fortior est ipsa dispersa[98]. Sed diceres: tunc sequitur quod idem esset causa contrariorum, quia condensatio esset causa frigiditatis et caliditatis; hec autem sunt contraria. Breviter respondeo quod non est impossibile quod idem in alia *Kl* 18rb et alia materia sit causa alterius et alterius, quamvis non in eadem; unde | condensatio, quamvis in materia calida non sit causa intensionis caliditatis et frigiditatis, tamen in materia calida est causa intensionis caliditatis, in materia autem frigida est causa intensionis frigiditatis. Sed contra: posito quod in eodem subiecto sit simul caliditas et frigiditas et illud subiectum condensetur, tunc videtur quod condensatio in eodem subiecto sit causa intensionis caliditatis et frigiditatis. Respondetur quod casus est impossibilis. Ipso tamen concesso diceretur quod si caliditas et frigiditas essent *P* 241^{v} equalium | graduum, tunc ex tali condensatione nec intenderetur caliditas nec frigiditas, quia continue per eorum actionem ad invicem manerent sub *L* 194ra equalibus gradibus, | quamvis illud subiectum condensaretur. Vel posset dici *U* 14rb quod utrumque intenderetur, sed non posset intendi usque ad summum. | Si autem unum illorum esset intensius et aliud remissius, diceretur quod intensius remitteretur et non remissius.

⟨20⟩ Dubitabit aliquis a quo generabitur huiusmodi gradus qualitatis, cum ex condensatione qualitas subiecti condensati intenditur. Respondetur quod occasionaliter generatur a contrario circumstante, principaliter autem a forma intrinseca illius subiecti una cum virtute celesti.

10 fortius] potius *P* 15 idem] condensatio *P* 28 esset] *om. P* 28–29 intensius remitteretur et non remissius] intensius remitteret remissius *P* 30 qualitatis] caliditatis *P* 32 contrario] frigido *Kl L* 33 celesti] illius elementi *P*

98 Cf. Anonymum, *Liber de causis*, ed. Pattin, 138^{15-16}; *Auct. Arist.*, 232, n. 13.

⟨21⟩ Ad rationes. Ad primam: 'unum contrarium agit in aliud', etc., admittatur; cum hoc tamen stat quod in principio illius actionis istud contrarium intenditur, sicut dicebatur in questione[99].

⟨22⟩ Ad secundam dico quod non est impossibile quod unum contrarium generet aliud | occasionaliter et per accidens, quamvis hoc sit impossibile | pure naturaliter et per se—de quo etiam tangebatur in questione[100]. *Kl* 18va *Ba* 16rb

⟨23⟩ Ad tertiam dico quod illa gutta aque in principio bene intenditur in eius frigiditate; postquam tamen penetrata est calore ignis, e contrario dilatatur et eius frigiditas remittitur et corrumpitur. Et sic similiter de pane in furno.

⟨24⟩ Ad quartam dico quod ignis posset tantum augeri quod quamvis in principio alia elementa fortificaret, tamen postea, postquam calore ignis ipsa elementa penetrarentur, maior eorum pars ab igne corrumperetur.

⟨25⟩ Ad quintam: 'sequitur quod accidentia migrarent de subiecto', etc., dico negando consequentiam, quia non solum caliditas que est in partibus extremalibus movetur versus partes centrales, verum etiam partes extreme in quibus est ipsa caliditas.

Et sic est finis questionis.

I.11 Utrum semper media regio aeris sit frigida

Queritur consequenter utrum semper media regio aeris sit frigida.

⟨1⟩ Et arguitur primo quod non: semper media regio aeris est calida; igitur non frigida. Consequentia tenet; antecedens patet ex eo quod aer per se et naturaliter est calidus et humidus.

⟨2⟩ Secundo: si semper media regio aeris esset frigida, vel ergo illa frigiditas inesset sibi naturaliter, vel violente. Non potest dici quod naturaliter, cum contrarium frigiditatis insit sibi naturaliter, puta caliditas; nec potest dici quod violente, | quia tunc non semper media regio esset frigida, cum nullum violentum semper duret[101]. *Ba* 16va

⟨3⟩ Tertio: si sic, sequeretur quod media regio aeris sit semper densa; sed hoc est falsum. Consequentia tenet ex quo frigiditas est causa condensatio-

13 maior eorum pars ab igne corrumperetur] maior eorum virtus corrumperetur et etiam pars *L*

99 I.10, 11.

100 I.10, 13.

101 Cf. Aristotelem, *De celo*, I, 2, 269 b 9.

nis. Falsitas consequentis patet, nam tunc impediret ne videremus celum et astra, nam propter eius densitatem species visibiles celi et etiam astrorum non possent multiplicari usque ad visum nostrum.

⟨4⟩ Quarto sequeretur quod aer qui est in media regione aeris esset gravior quam aer qui est hic inferius; hoc est falsum. Consequentia tenet ex eo quod | *L* 194rb magis frigida | *Kl* 18vb minus frigidis sunt graviora; aer autem hic inferius esset minus frigidus aere existente in media regione, si semper media regio aeris esset frigida. Falsitas consequentis patet, quia tunc aer qui est in media regione descenderet et aer qui est hic inferius ascenderet, cum ordo gravium et levium sit graviora esse sub levioribus et leviora esse super graviora.

⟨5⟩ Quinto: si sic, sequeretur quod in media regione non generarentur impressiones ignee; sed hoc est falsum. Consequentia tenet propter nimiam frigiditatem medie regionis ipsius aeris. | *U* 14va Falsitas consequentis patet, nam ibi generantur corruscationes et fulgura; ista autem sunt impressiones ignee.

⟨6⟩ Sexto ad idem: ille aer qui est in media regione est magis remotus ab elementis frigidis—scilicet a terra et aqua—quam sit iste aer hic inferius, et similiter est magis propinquus igni et motui celi, quorum est calefacere, quam sit aer inferior; igitur est magis calidus quam sit aer inferius, et per consequens non semper media regio aeris est frigida.

⟨7⟩ In oppositum est Aristoteles in primo huius, ubi solvit illam questionem quare nubes non fiant in suprema regione aeris, sed in media: dicit huius causam esse quia | *Kl* 19ra suprema regio aeris semper est calida; et ergo, quamvis sufficiat elevare, tamen non potest convertere vaporem elevatum in nubem et in aquam, ipsam convexando, sed ipsam dissolvit et consumit | *P* 242^{r} ut non generentur nubes ex eo. Sed quia in media regione aeris est frigiditas, cum vapores calore Solis elevantur usque ibi per frigiditatem aeris existentis in media regione, illi vapores condensantur—frigidi enim est condensare—et ex tali condensatione vaporis generantur nubes[102]. Ecce qualiter Aristoteles in solvendo predictam dubitationem supponit mediam regionem aeris esse frigidam. Et per 'mediam regionem aeris' debet intelligi non ille aer qui est immediate prope terram, nec ille qui est immediate prope ignem, sed ille qui est in medio per aliquam distantiam a terra et similiter per aliquam distantiam ab igne. Et vocatur a philosophis 'medium interstitium aeris' ab

---

6 magis] *om. P* 12 ignee] in aere *P* 20 huius] tractatu primo capitulo secundo *add. Kl* tractatu primo capitulo sexto *add. Ba L* capitulo sexto tractatus primi *add. U* ‖ ubi solvit illam questionem] ubi dicit tres causas *Ba Kl* ubi reddit tres causas *L U* 24 ipsam convexando] *om. Ba Kl L U* 25–26 cum vapores] *iter. P* 30 non] *iter. P*

---

102 Aristoteles, *Meteorologica*, I, 3, 340 b 29–32.

'inter' et 'sto, stas', quasi stans inter partes extremas aeris. Et ille aer qui est infra medium interstitium potest vocari 'inferior regio aeris'; ille autem qui est super vocatur 'suprema regio aeris', ita quod orbis ipsius aeris in eius spissitudine ymaginetur dividi in tres partes, scilicet in inferiorem, | mediam et *L* 194va supremam.

⟨8⟩ In ista questione videndum est primo de qualitate medie regionis aeris; secundo de causa qualitatis illius medie regionis; tertio incidentaliter videndum est de quantitate medie regionis aeris; quarto de figura eius; quinto de variatione eius secundum diversa tempora.

⟨9⟩ Quantum ad primum, sit prima conclusio quod semper media regio aeris *Kl* 19rb *U* 14vb est frigida. Patet, quia vapores calore Solis elevati usque ibi convertuntur in aquam, que est frigida, quod est signum quod media regio aeris est frigida. Similiter ibi generantur impressiones frigide, sicut nix, grando et similia, quod non esset nisi media regio aeris esset frigida.

⟨10⟩ Secunda conclusio est quod media regio aeris non est frigida in summo. Patet, quia | tunc esset aqua. Ista consequentia tenet per Aristotelem *Ba* 16vb in primo et in secundo *De generatione*, ubi vult quod frigiditas inest primo aque[103]; nichil ergo potest esse frigidum in summo nisi sit aqua. Sed contra: immo probatur quod media regio aeris sit ita frigida sicut est aqua, quia condensando et infrigidando vaporem elevatum generat aquam et generat tantam frigiditatem quanta est frigiditas aque; sed nichil agit nec generat intensiorem gradum alicuius qualitatis quam ipsummet habeat; ergo, si media regio aeris generat tantam frigiditatem quanta est frigiditas aque, videtur etiam quod tantam habeat. Quod autem nichil agat intensiorem gradum alicuius qualitatis quam ipsummet habeat probatur, nam alias simile ageret in simile sibi, quod est contra Aristotelem in primo *De generatione*[104], et contra naturalem appetitum ipsius agentis, quia solum agens appetit passum sibi assimilare. Respondetur negando | consequentiam, quod aliquid scili- *Kl* 19va cet ageret intensiorem gradum alicuius qualitatis quam ipsummet haberet. Et cum dicebatur: 'media regio aeris non est frigida in summo', sicut dicit secunda conclusio, concedo. Et cum ulterius dicebatur: 'tamen generat frigiditatem in summo in ipso vapore', nego; sed bene generat in ipso tantum gradum frigiditatis quantum ipsamet habet, et isto habito et generato, ille

17 primo] celi *add. Kl L*   27 appetit] intendit *Ba Kl L U*

103 Aristoteles, *De generatione et corruptione*, I, 329 a 25–35; II, 3, 331 a 4–5.
104 Ibid., I, 7, 323 b 19–25.

vapor, cum sit de natura aque, ulterius per formam eius intrinsecam ducitur ad frigiditatem aque, recte sicut aqua calefacta se ipsam reducit ad frigiditatem pristinam.

⟨11⟩ Quantum ad secundum, videlicet que sit causa quod media regio aeris est frigida, breviter una potest esse causa distantia eius a Terra una cum quibusdam aliis concurrentibus. Unde ymaginandum est quod aer hic inferius prope Terram calefit ex reflexione luminis Solis a Terra. Et cum radius incidens et reflexus prope Terram sint magis propinqui, aer prope Terram est magis calidus. Et cum illi radii recedendo a Terra magis et magis elongen- *L* 194vb tur ab invicem, continue ex tali elongatione | istorum radiorum ab invicem *Ba* 17ra aer qui est magis et magis distans a Terra | minus et minus calefit, ex eo *U* 15ra quod | propinquitas radii incidentis et reflexi est causa calefactionis. Et cum isti radii in media regione aeris sint valde distantes ab invicem, modicam vel nullam generant ibi caliditatem. Et cum alia sint que generant ibi fri- *P* 242^{v} giditatem, de quibus statim | dicetur, que etiam generarent frigiditatem in inferiori regione ipsius aeris, si non esset propinquitas radii incidentis et reflexi, sequitur ipsam mediam regionem aeris esse frigidam. Sed suprema regio aeris, quamvis non calefaciat ex reflexione luminis Solis a Terra, tamen *Kl* 19vb calefit aliunde, videlicet ab igne existente in spera sua, | quia illi igni suprema regio aeris est immediata. Et tunc, ubi ignis calefaciendo supremam regionem aeris terminat suam actionem, ibi incipit media regio aeris et terminatur ibi ubi radius incidens et reflexus tantum distant quod non sufficiunt calefacere.

⟨12⟩ Alia causa frigiditatis potest esse remissio luminis, quia, cum lumen sit productivum caloris, sicut prius dicebatur[105], sequitur quod ubi est remissius lumen, ibi minus de calore producitur; sed in media regione aeris est minus de lumine, quia lumen causatur per reflexionem speciei lucis; igitur, ubi est minor reflexio, ibi est minus de lumine. Sed cum in media regione aeris sit minor reflexio quam hic prope terram, cum aer sit minus densus et minus reflectens quam terra, sequitur quod est minus de lumine et per consequens minus de calore. Et cum sint ibi alia valde istum aerem infrigidantia, sequitur quod ille aer est frigidus. Sed diceres: videtur quod istud non valeat, quia in nocte, cum non est lumen Solis apud nos, aer inferior adhuc est magis calidus quam aer qui est in media regione. Respondetur quod ille

---

7 a terra] ad terram *P* 21–23 et terminatur ibi … non sufficiunt calefacere] *om. Ba Kl L U*
25 est] *iter. P*

---

105 I.9, 18.

calor aeris inferioris, quamvis non generatur in nocte ex reflexione luminis Solis a Terra, tamen generabatur in die. Et huius signum est quod, quando noctes sunt longe, propter absentiam luminis Solis quasi totus calor aeris inferioris remittitur. Unde semper de mane aer hic inferius est magis frigidus quam de sero, ceteris paribus.

⟨13⟩ Tertia causa frigiditatis medie regionis potest | *Ba* 17rb esse quia vapores frigidi elevantur calore Solis, in tantum quod veniunt iuxta supremam regionem aeris, qui est valde calida ab igne; et cum unum contrarium fugat aliud, sicut predictum est[106], illi vapores | *Kl* 20ra fugantur a suprema aeris regione et condensantur et infrigidantur, tendendo ad eorum | *L* 195ra dispositionem naturalem. Et huiusmodi vapores | *U* 15rb fugantur versus mediam regionem aeris et aerem ibi existentem infrigidant, et ex isto aer ibi existens est frigidus.

⟨14⟩ Quarta causa frigiditatis medie regionis ipsius aeris potest assignari talis, quod aer qui est in suprema regione aeris movetur circulariter cum igne, et ex tali motu calefit et rarefit. Et cum aer in suprema regione rarefit, oportet aerem sub eo condensari ne dimensiones se penetrent; et ex tali condensatione una cum aliis causis causatur frigiditas in media regione aeris.

⟨15⟩ Quinta causa potest assignari quod aer prope terram calefactus quosdam vapores fugat a terra, cum unum contrarium fugat aliud, et quosdam vapores fugat sursum versus mediam regionem. Et cum vapor de natura sit frigidus, quia tendit ad naturam aque, isti vapores, fugati ad mediam regionem, aerem ibi existentem infrigidant; et tunc, cum illum aerem circumstet aer supreme regionis, qui est calidus, et etiam aer inferioris regionis, qui etiam est calidus, frigiditas medie regionis intenditur; nam ex predictis contrarium circumstans suum contrarium fortificat ipsum[107].

⟨16⟩ Sexta causa potest assignari quod influentia celestis causat frigiditatem in media regione aeris ad istum finem, ut fiat ibi generatio et corruptio mixtorum imperfectorum; impressiones enim generate in media regione, sicut est pluvia et consimiles, multum faciunt ad generationem et augmentationem | *Kl* 20rb mixtorum perfectorum. Dicit enim Aristoteles secundo *Physicorum*: pluvia cadit ut crescant frumenta[108].

---

3 absentiam] et umbrationem *add. Kl L* 9 sicut predictum est] *om. Ba Kl L U* 9–10 condensantur] inspissantur *Ba* 19 a terra] ad terram *P U* 21–22 fugati ad mediam ... ibi existentem infrigidant] expellunt et fugant istum aerem *Ba* fugant istum aerem *Kl L U* 24 calidus] propter quam causam *add. P* 30 secundo] tertio *P* 31 frumenta] gramina *L*

---

106 I.10, 9.

107 I.10, 11; cf. *Auct. Arist.*, 172, n. 10.

108 Aristoteles, *Physica*, II, 8, 198 b 16–23; cf. Nicolaum Oresme, *Questiones in Physicam*, II, 16, ed. Caroti [et al.], $291^{204-205}$.

⟨17⟩ Quantum ad tertium restat videre de quantitate et situ medie regionis aeris. Unde sciendum est quod media regio aeris est bene spissa et profunda; | *Ba* 17va unde Alhazen dicit in tractatu suo *De crepusculis* quod vapores ex quibus fiunt nubes bene possunt elevari a terra per quinquaginta duo miliaria[109]. Et certum est quod illi vapores, si ex illis fiunt nubes, non elevantur ultra mediam regionem aeris; | *P* 243r igitur tante spissitudinis, scilicet quinquaginta duorum miliarum, est media regio aeris, excepto illo spatio quod est inter terram et illum locum aeris ubi primo possunt generari nubes, quod spatium est bene parvum. Ex hoc patet quod media regio aeris est valde spissa et profunda. Quod autem illud spatium quod est inter terram et locum aeris in quo | *U* 15va possunt generari nubes sit parvum, patet, nam visum est quod aliquis stans super montes vidit nubes inferius, et hoc est signum quod tale spatium non est multum magnum. Ex hoc ulterius patet quod media regio aeris non est aer purus, sed multis grossis vaporibus permixtus; | *L* 195rb nec est ita rarus sicut est aer supreme regionis. Et ergo, quia videmus Solem per mediam regionem, radii | *Kl* 20va refranguntur supra illo aere ibi existente et non veniunt non fracti ad visum nostrum. Et propter hoc dicit Alhazen in tractatu allegato quod videmus lumen Solis antequam Sol sit super nostrum horizontem[110]. Unde, cum videmus aliquod visibile per media diversarum rationum in raritate et densitate, tunc videmus istud visibile per radios refractos supra superficie illa duo media coniungente. Et per talem refractionem apparet visibile maius quam si videretur per radios non fractos, aliquando etiam minus, et similiter apparet in alio loco quam sit. Hoc patet, nam denarius in fundo aque apparet maior quam extra aquam. Et aliquando, si poneretur in aliquod vas, non videretur in aliqua distantia visus ab eodem, et si superfunderetur aqua, videretur in illa distantia[111]. Sic etiam propter interpositionem medie regionis inter Solem et visum nostrum—que quidem media regio multis vaporibus grossis est permixta—aliquando apparet nobis Sol citius super horizontem quam in veritate | *Ba* 17vb sit in eius ortu, et in eius occasus diutius apparet nobis esse super horizontem quam sit. Et propter hoc etiam de mane, cum oritur, propter pluralitatem vaporum inter Solem et visum nostrum, super quibus vaporibus refranguntur radii Solis, apparet Sol

---

3 crepusculis] crepusculo *P* 14 multis grossis vaporibus permixtus] grossus quasi vapore permixtus *Ba Kl L* grossus quasi vaporibus permixtus *U* 28 aliquando] enim *add. P*

---

109 Ibn Mu'had, *De crepusculis et nubium ascensionibus*, ed. M. Smith, in "The Latin Version of Ibn Mu'ādh's Treatise on Twilight and the Rising of Clouds", *Arabic Sciences and Philosophy* 2 (1992), 83–132, at 100^{91}–116^{418}.

110 Ibid., 97^{20}–99^{67}.

111 Cf. I.9, 12.

maior quam in meridie. Item patet spissitudo medie regionis ex eo quod aliquando videmus nubem bene alte super nubem, et tam nubes inferior quam superior sunt in media regione, ex quo potest argui spissitudo et profunditas medie regionis.

⟨18⟩ De situ autem medie regionis sciendum est quod media regio non dicitur 'media' propter equalem distantiam a terra | *Kl* 20vb et a concavo orbis ignis; unde punctus medius qui est inter terram | *U* 15vb et inter concavum orbis ignis non est signandus in medio medie regionis aeris—immo forte nec in aliquo puncto medie regionis—, sed in puncto supreme regionis, ita quod ymaginandum est quod suprema regio aeris est spissior quam sit totus aer inter ipsam et terram. Et ideo, sicut dixi, non est media quia equaliter distet ab igne et terra, sed dicitur 'media' per interpositionem eius inter supremam regionem aeris et illam regionem aeris in qua non possunt generari nubes propter nimiam eius propinquitatem ad terram. Quod autem aer qui est ultra mediam regionem sit spissior quam totus ille qui est inter supremam regionem aeris et terram patet ex eo quod dicitur quod quidam montes sunt ita alti quod plus et maior portio | *L* 195va illorum montium est in suprema regione aeris quam sit totus aer qui est inter terram et supremam regionem aeris, et illi montes adhuc non attingunt speram ignis; modo hoc non esset verum nisi suprema regio aeris esset multo spissior quam totus aer infra eam. Et si sic, tunc media regio aeris non est in medio per equalem distantiam a terra et a suprema regione aeris. Hoc etiam patet ex alio, quia comete qui generantur in suprema regione aeris plus ascendunt super convexum medie regionis quam sit distantia a terra usque ad convexum medie regionis; hoc iterum non esset verum nisi suprema regio aeris esset multo spissior quam totus aer inferior.

⟨19⟩ Quantum ad quartum | *Ba* 18ra restat videre de figura medie regionis aeris. Et dico primo quod versus polos est multo spissior quam versus equinotialem. Patet, quia versus polos est minor caliditas quam versus equinotialem. | *Kl* 21ra Primo, quia ibi radii Solis magis oblique reflectuntur, et ergo statim tantum elongantur ab invicem | *P* 243^{v} radius incidens et reflexus quod non possunt generare caliditatem in aere, et sic ibi bassius incipit media regio aeris quam si radius incidens et reflexus non tantum statim ab invicem elongarentur. Similiter circa polos partes celi moventur tardius quam circa equinotialem, et cum ex motu causetur calor, sequitur quod si motus est minus velox, quod

---

20 infra eam] a terra usque ad illam regionem *Ba* inter terram et ipsam *Kl L U* 23 ascendunt] elevantur *Ba Kl L U* 24 convexum medie regionis] supremam regionem aeris *Ba Kl L U*

U 16ra est causativus minoris caloris; et ergo suprema | regio aeris non est ibi ita spissa sicut circa equinotialem, sed est ibi magis tenuis, et per consequens media regio magis spissa. Hoc etiam patet signo, quia in illis partibus nubes fiunt magis basse quam in partibus versus equinotialem. Ex hoc potest ymaginari qualis figure sit media regio aeris et in convexo et in concavo.

⟨20⟩ Quantum ad quintum, restat videre de variatione medie regionis. Pro quo dico quod in hieme media regio aeris super nos est magis spissa. Patet, quia radii solares incidunt magis oblique super terram quam in estate; et ergo radius incidens et reflexus statim tantum elongantur quod non sufficiunt calefacere aerem; et ergo bassius incipit media regio aeris, et propter hoc in hieme nubes sunt bassiores quam in estate.

⟨21⟩ Secundo dico quod in estate media regio aeris est magis tenuis, et una illius causa est quia in estate radii solares minus oblique incidunt super terram, et propter hoc radius incidens et reflexus recedendo a terra minus *Kl* 21rb *L* 195vb elongantur | ab invicem quam si radii | solares inciderent oblique, et igitur altius calefaciunt aerem, et per consequens concavum medie regionis plus distabit a terra; per consequens media regio aeris non erit ita spissa. Alia causa est quia in superiori parte aeris in estate propter accessum Solis causatur maior calor, et per consequens suprema regio aeris augetur. Sed ad augmentationem supreme regionis aeris sequitur diminutio medie regio- *Ba* 18rb nis; igitur in estate media regio aeris diminuitur. Tertia causa potest esse | quia in estate suprema regio aeris est magis calida et similiter inferior; ergo media regio aeris, que est frigida, circumdatur contrariis fortioribus; et per consequens, cum unum fugat aliud, partes extreme medie regionis moventur versus partes interiores medie regionis, et per consequens fit condensatio medie regionis, et per consequens fit minus profunda, sed tamen magis frigida quam in hieme, per antiparistasim, quia circumdatur fortioribus contrariis. Et huius signum est quod in estate aliquando generantur grandines tante quantitatis quod tante non possunt generari in hieme.

⟨22⟩ Ad rationes. Ad primam respondetur quod quamvis aer de natura sua sit calidus, tamen potest infrigidari violente per aliquid extrinsecum; et sic fit de aere in media regione, sicut prius dicebatur.

---

4 basse] spisse et spissiores *Ba Kl* spisse vel spissiores sive bassiores *L* 5 figure] dispositionis *Ba Kl* figure sive dispositionis *L* 10 bassius incipit media regio aeris] ibi spissius est tunc interstitium medium *Ba* ibi spissius incipit media regio aeris *Kl* 11 bassiores] spissiores *Ba Kl L* 26 fit] *om. P* 28 est] *om. P*

⟨23⟩ Ad secundam dico quod illa frigiditas est sibi violenta. Et cum dice- *U* 16rb batur: 'violentum non diu durat', concedo; et ergo bene concedo quod ibi continue non manet idem aer, sed bene alius et alius; et quamvis nullum | *L* 196ra violentum sit perpetuum, tamen semper est aliquod violentum. Et ideo, quamvis nullus aer in media regione sit semper frigidus, tamen semper aliquis aer in media regione est frigidus.

⟨24⟩ Ad tertiam: 'sequeretur quod media regio aeris esset valde densa', conceditur. Et cum dicitur: 'ergo impediret ne videmus celum et stellas', dico quod densitas non impedit visionem, sed bene opacitas. Unde per unum vitrum quod est densius quam sit unum lignum, | bene videmus, et tamen *Kl* 21va per lignum non possumus videre. Sed verum est quod ista densitas medie regionis bene facit quod non quamlibet partem celi videmus per radium rectum, sed plures partes celi videmus per radium refractum.

⟨25⟩ Ad quartam: 'sequitur quod ille aer fiet magis gravis quam aer inferior', concedo. Et ergo, postquam est magis gravis, descendet, et inferior ascendet et infrigidatur, et sic continue unus post alium.

⟨26⟩ Ad quintam dico quod ibi non fiunt impressiones ignite propter caliditatem medie regionis, sed fiunt ibi per antiparistasim propter frigiditatem medie regionis, de quo postea videbitur[112].

⟨27⟩ Ad sextam: 'ille aer est propinquior causis calefacientibus', etc., nego consequentiam quod ergo debeat esse calidior, quia cause infrigidantes illum | aerem sunt fortiores, de quibus dicebatur in secundo articulo huius *Ba* 18va questionis.

Et sic finitur questio.

I.12 Utrum omnium impressionum metheorologicarum vapor et exalatio fuerit principium materiale

Queritur consequenter utrum omnium impressionum metheorologicarum *P* 244^{r} vapor et exalatio sint principium materiale.

⟨1⟩ Arguitur primo quod non: impressionum | ignitarum nullum istorum est *Kl* 21vb principium materiale; ergo questio falsa. Consequentia tenet. Antecedens probatur, nam non potest dici quod talium impressionum vapor sit materia, nam Aristoteles in primo huius vult oppositum: vult enim quod talium

---

21 infrigidantes] agentes frigiditatem *Ba Kl L U*

---

112 I.12, 15; I.13, 12–14.

impressionum que generantur per inflammationem non vapor, sed exalatio sit principium[113]. Nec etiam potest dici quod talium impressionum exalatio sit materia; cum sit de natura terre per Aristotelem[114], videtur esse tante gravitatis quod non potest attingere locum illum sursum ubi generantur huiusmodi impressiones. Et confirmatur hoc: materia impressionum ignitarum debet esse bene inflammabilis; sed exalatio non est huiusmodi, quia exalatio est sicca, per Aristotelem[115]. Siccum autem non est bene inflammabile, sicut patet de cimiteribus; inflammabile enim faciliter debet esse pingue. Et confirmatur adhuc quod exalatio non sit materia talium impressionum; *U* 16va nam materia impressionum aquearum segregatur ab aqua, | sicut est vapor, et etiam materia impressionum ignearum debet segregari ab igne; sed exalatio non segregatur nec extrahitur ab igne, immo magis a terra; igitur.

⟨2⟩ Secundo ad principale: vapor non est materia impressionum generatarum sursum, sicut pluvie, rores, grandines et pruine, etc.; igitur questio falsa. Consequentia tenet, quia non potest dici quod huiusmodi impressionum exalatio sit materia, cum sit calida et sicca, huiusmodi autem impressiones sunt frigide et humide. Antecedens patet, nam vapor, cum sit de natura aque, est tante gravitatis quod non potest moveri sursum usque ad illum locum ubi generantur huiusmodi impressiones.

⟨3⟩ Tertio probatur quod impressionum aquearum vapor non sit materia nec impressionum que generantur per inflammationem exalatio sit materia, *Ba* 18vb quia, cum vapor segre|gatur ab aqua et exalatio a terra, sequitur quod supra in aere non fierent nisi impressiones humide et aquee, et huiusmodi *Kl* 22ra *L* 196rb impressiones | non fierent super terram, cuius oppositum ex|perimur. Consequentia tenet eo quod vapor solum segregatur ab aqua, exalatio solum a terra.

⟨4⟩ Oppositum vult Aristoteles in primo huius, tractatu secundo, capitulo primo[116].

⟨5⟩ Breviter in ista questione videndum est quid sit exalatio et quid vapor; secundo videndum est de comparatione istorum ad invicem; tertio respondendum est ad quesitum.

---

3 per aristotelem] *om. Ba Kl U* 15 tenet] quia non potest dici ergo questio falsa consequentia tenet *add. P* 24 fierent] fiunt *P*

---

113 Aristoteles, *Meteorologica*, I, 4, 341 b 10–25.

114 Ibid.

115 Ibid., I, 4, 341 b 10.

116 Ibid., I, 4, 341 b 10–25.

⟨6⟩ Quantum ad primum, sciendum est quod 'exalatio' dicitur quoddam corpus subtile, seu fumus calore Solis a terra vel ab aqua terra permixta elevatus. 'Vapor' autem dicitur quoddam corpus subtile, seu fumus calore Solis ab aqua vel a terra aqua permixta elevatus. Ex isto patet quod exalatio est sicca siccitate terrestri et calida accidentaliter a Sole, et vapor est humidus humiditate aquea et accidentaliter calidus a Sole.

⟨7⟩ Hic dubitatur utrum vapor sit eiusdem speciei cum aqua et exalatio eiusdem speciei cum terra. Ad istum dubium diversi diversimode respondent; unde dicit unus expositor quod exalatio que generatur ex corpore simplici, sicut a terra, est eiusdem speciei cum eo corpore, differens ab eo solum secundum dispositiones accidentales[117]. Et assignat rationem illius quod omne corpus naturale quod derelictum nature sue proprie ex se reddit ad dispositionem naturalem alicuius corporis et sequitur motum naturalem eius, est eiusdem speciei et eiusdem nature specifice cum illo corpore; sed sic est quod exalatio dimissa nature sue proprie reddit ad naturam terre ex qua extrahebatur; igitur etc. Maior est nota. Minor patet, | *U* 16vb nam, cum exalatio a terra segregata derelinquitur nature sue proprie, per absentiam caloris | *Kl* 22rb Solis descendit ad terram unde venit. Similiter sic est de vapore extracto ab aqua, qui, cum est elevatus et derelinquitur nature sue, descendit; ergo etc.

⟨8⟩ Alii dicunt quod in tali segregatione exalationis a terra est vera generatio, et similiter in segregatione vaporis ab aqua, ita quod isti volunt quod exalatio est alterius speciei ab illo a quo segregatur, scilicet a terra; et similiter vapor est alterius speciei ab aqua. Unde volunt quod quodlibet istorum sit unum mixtum imperfectum habens formam substantialem distinctam a forma elementi, puta aque vel terre[118]. Et quando dicitur eis quod Aristoteles dicit exalationem esse de natura terre et vaporem | *P* 244^{v} esse de natura aque[119], dicunt quod Aristoteles hoc dicit propter hoc quod materia exalationis immediate ante eius generationem stetit sub forma terre et habet siccitatem terream; similiter sic dicunt de vapore in ordine | *Ba* 19ra ad ipsam aquam.

---

9 expositor] *om. Ba Kl* 11 dispositiones] proprietates *Ba Kl L U* 12 quod] *om. P*

117 Cf. Iohannem Buridanum, *Questiones in Meteorologica*, I.11, ed. Bages, 2: 9–10; 21.

118 Cf. Iohannem Buridanum, *Questiones in Meteorologica*, I.11, ed. Bages, 2: 8; Albertum de Saxonia, *Questiones in Meteorologica*, I.12, in A. Panzica, "Albert of Saxony's *Questions on Meteorology*: Introduction, Study of the Manuscript Tradition and Edition of Book I–II.2", *AHDLMA* 86 (2019), 12; Themonem Iudeum, *Questiones in Meteorologica*, I.12, ed. Octavianus Scotus, Venice 1522, f. 95ra.

119 Aristoteles, *Meteorologica*, I, 4, 341 b 9–10.

⟨9⟩ Breviter concedo quod prima opinio sit verior propter ipsam rationem quam adducit, nec illi de secunda opinione solvunt illam, nec video aliquam rationem pro eis quin faciliter possit solvi tenendo primam opinionem.

⟨10⟩ Quantum ad secundum, videlicet de comparatione istorum ad invicem, dico | [*L* 196va] quod utrumque istorum, tam vapor quam exalatio, est calidus accidentaliter, videlicet a calore ipsum elevante; tamen exalatio est calidior vapore, quia exalatio est de natura terre, et ergo intensius et fortius incorporatur sibi calor | [*Kl* 22va] quam vapori, qui est de natura aque, cum aqua sit minus densa quam terra.

⟨11⟩ Secundo dico quod utrumque istorum est leve, non a se, sed a calore elevante, sed tamen magis exalatio quam vapor. Patet, quia exalatio est magis calida et habet calorem magis potentem elevare.

⟨12⟩ Tertio dico quod utrumque eorum elevatur sursum per calorem in eo generatum, sed magis sursum elevatur exalatio. Patet etiam quia est magis calida.

⟨13⟩ Sed diceres: sicut se habet terra ad aquam, ita se habet exalatio ad vaporem; sed terra est gravior aqua, et ergo exalatio est gravior vapore. Respondetur: concedo quod quantum ad naturam specificam et propriam, exalatio est gravior vapore sicut terra est gravior aqua; sed tamen accidentaliter exalatio per calorem elevantem fortius sibi incorporatum quam vapori, potest dici levior, et propter hoc magis sursum potest moveri, recte sicut ferrum intensius recipit calorem | [*U* 17ra] ab igne quam stuppa, propter tenacitatem materie.

⟨14⟩ Verum est tamen quod duplex potest distingui exalatio: una que est elevata a terra pingui et unctuosa; et talis elevatur ut plurimum in uno globo, propter tenacitatem partium ad invicem, et illi calor accidentalis tantum incorporatur quod potest valde alte ascendere, et habet tantum de calore quod sine repulsu potest ambulare per mediam regionem aeris usque ad supremam; et ex tali per inflammationem eius generantur impressiones ignee. | [*Kl* 22vb] Alia autem est exalatio que elevatur a terra non pingui nec unctuosa; et illa non sic elevatur in uno globo, nec partes eius propter siccitatem tenent se ad invicem, nec sibi fortiter incorporatur calor, nec ita etiam alte ascendit, | [*Ba* 19rb] sed ipsa cito dissolvitur et ex tali generantur venti.

2–3 aliquam rationem pro eis] aliquam positionem illius opinionis *Ba* aliquam rationem probabiliorem *Kl* aliquam rationem probatam *L* 15 calida] et habet calorem magis potentem elevare *add. Ba Kl L* 20 incorporatum] incorporatur *P* 21 hoc] *om. P* ‖ potest] posset *P* 22 tenacitatem] densitatem *Ba* paucitatem *P* 32–33 ascendit] elevatur *Ba* movetur *Kl L U*

⟨15⟩ Quantum ad tertium, sit prima conclusio: omnium impressionum siccarum vel ignitarum exalatio est materia. Patet, nam exalatio que elevatur, vel elevatur ultra mediam regionem aeris, vel manet in media regione, vel manet infra mediam regionem, vel manet in concavitatibus ipsius terre. Si elevatur usque ad supremam regionem aeris, vel ergo quasi usque ad speram ignis, vel infra. Si primum et ibi inflammatur, fiunt comete vel fit cometa. Si autem citra et ibi inflammatur, fiunt capre saltantes vel trabes ignite et consimilia. Si autem exalatio elevatur usque ad mediam regionem aeris et includitur in nube et inflammatur, fiunt corruscationes et fulmina. Si autem non includitur | [*L* 196vb] in nube nec inflammatur et est multitudo talis exalationis et movetur a frigido medie regionis, fiunt ex ea venti. Si autem talis exalatio est conglobata et non inclusa in nube et repellitur versus terram a frigiditate medie regionis, fiunt ex ea sydera cadentia, cum ex tali motu et pulsu incenditur. Si autem talis exalatio non movetur notabiliter directe versus terram vel ad latus nec inflammatur, sed coloratur lumine celesti, fiunt hyatus et voragines et colores sanguinei secundum aliam et aliam dispositionem illius exalationis. Si autem illa exalatio manet infra mediam regionem et dissolvitur et cadit ad mare, fit salsedo maris. Si autem talis exalatio non pertingit | [*Kl* 23ra] ad aerem, sed manet in mari vel in | [*U* 17rb] concavitatibus terre, tunc, cum manet in mari, fit ex hoc salsedo maris, et cum manet in concavitatibus terre, fit motus terre.

⟨16⟩ Secunda conclusio: omnium impressionum humidarum vel frigidarum vapor est materia. Patet, nam vapor, | [*P* 245r] si elevatur usque ad mediam regionem et ibi condensatur in nubem et in aquam, fit pluvia. Si autem congelatur et conglutinatur multum, fit grando. Si autem congelatur et conglutinatur, sed non tantum, fit nix. Si autem vapor elevatus manet infra mediam | [*Ba* 19va] regionem et de nocte infrigidatur et non congelatur, fit ros. Si autem congelatur, fit pruina. Si autem vapor manet in concavitatibus terre et propter frigus condensatur, fit aqua et fiunt rivuli. Si autem vapor elevetur supra in mediam regionem et fiunt nubes quiescentes vel etiam ante generationem nubium vapores ibi elevati quiescunt, tunc, per reflexionem vel refractionem luminis Solis vel Lune supra illis fit halo. Et quandoque in elevatione talium vaporum per reflexum luminis fit iris. De hiis plenius apparebit postea.

⟨17⟩ Tertia conclusio: quod questio est vera. Patet inducendo in singulis impressionibus: omnes enim vel generantur ex exalatione vel ex vapore.

---

6 comete] *om. P* 13 cadentia] volantia *Ba Kl L U* ‖ incenditur] calor eius intendatur *Ba* intendatur *Kl L U* 15 celesti] solis *Ba Kl L* 24 conglutinatur] conguletur *P* 28 fiunt rivuli] fiunt fontes et rivi *Ba Kl L U*

⟨18⟩ Ad rationes. Ad primam 'exalatio non est materia', etc., negatur. Et quando probabatur: 'exalationes non possunt ita alte ascendere, cum sint *Kl* 23rb de natura terre', concedo ex se; | nichilominus hoc potest fieri per calorem elevantem ipsas exalationes. Et ad confirmationem: 'materia impressionum ignitarum debet esse inflammabilis', concedo; et ulterius dico quod alique *L* 197ra exalationes sunt inflammabiles, sicut exalationes segregate | a terra pingui et unctuosa, nichilominus iste exalationes adhuc tenent terrestrem siccitatem. Ad aliam confirmationem nego quod si materia impressionum aquearum segregatur ab aqua, quod ergo materia impressionum ignitarum debet segregari ab igne, quia ignis non est ad hoc dispositus quod ab eo segregatur materia impressionum ignitarum, sed aqua bene est ad hoc disposita ut ab ea segregatur materia impressionum aquearum.

⟨19⟩ Ad secundam dico quod immo vapor est materia impressionum aquearum. Et cum dicitur: 'non potest ita alte elevari sicut generantur huiusmodi impressiones', concedo quantum est de se, sed tamen potest per calorem ipsum elevantem.

⟨20⟩ Ad tertiam dico quod immo a mari potest segregari et elevari exalatio, propter terrestrem mari admixtum. Similiter ex terra potest segregari et extrahitur vapor propter aqueum sibi permixtum. Et sic patet ad argumentum.

Et sic finitur questio.

I.13 Utrum impressiones ignite, seu ille que fiunt per inflammationem, fiant naturaliter in aere

U 17va Queritur consequenter utrum impressiones ignite, seu ille que fiunt per inflammationem, fiant naturaliter in aere.

Kl 23va ⟨1⟩ Et arguitur quod non, nam impressiones ignite videntur esse de natura ignis; igitur non fiunt naturaliter in aere. Consequentia tenet ex eo quod *Ba* 19vb aer vel locus aeris non est locus naturalis | illius vel illorum que sunt de natura ignis; nam impressiones ignite habent pro materia exalationem, que est calida et sicca, quod est proprium igni.

⟨2⟩ Secundo: si huiusmodi impressiones naturaliter fierent in aere, tunc generarentur ab aere; sed hoc non, quia aer non est ita calidus sicut sunt

---

1 ad primam] *om. P* 14 elevari] levari *P L* 26–27 illius vel illorum que sunt de natura] *om. Ba Kl L U*

huiusmodi impressiones; et quia nichil agit gradum intensiorem quam ipsummet habeat, sequitur quod huiusmodi impressiones non generantur ab aere.

⟨3⟩ Tertio ad idem: illa que fiunt violente in aere non fiunt naturaliter; sed impressiones ignite sunt huiusmodi; ergo etc. Maior patet. Minor probatur, quia huiusmodi impressiones fiunt ex exalatione elevata ad aerem; sed elevatio exalationis est violenta, cum exalatio sit de natura terre, per Aristotelem in primo huius[120]; ergo etc.

⟨4⟩ In oppositum arguitur: primo, quod huiusmodi impressiones fiunt in aere, patet sensibiliter; sed quod fiant ibi naturaliter probatur, quia fiunt ibi a causis naturalibus et a causa naturaliter agente, quod quidem agens est ipse calor.

⟨5⟩ In ista questione videndum est quomodo diversificantur huiusmodi impressiones; secundo videndum est de causis harum impressionum et in generali et in speciali, quibus visis patet faciliter responsio ad quesitum.

⟨6⟩ Quantum ad primum, | sciendum est quod huiusmodi impressiones *Kl* 23vb diversificantur in magnitudine et in parvitate: alique enim sunt magne, alique parve, secundum magnitudinem et parvitatem exalationis.

⟨7⟩ Secundo etiam diversificantur | in figura: alique enim apparent figure *L* 197rb oblonge; alique etiam apparent figure late; alique etiam apparent sicut globi. Et ista diversitas provenit ex diversitate figure ipsius exalationis elevate.

⟨8⟩ Tertio etiam diversificantur qualitate, sicut raritate et | densitate et *P* 245^{v} colore. Alique enim fiunt ex exalatione magis densa, alique vero ex exalatione magis rara. Et secundum hoc etiam alique apparent coloris rubei, alique coloris albi. Alique etiam earum sunt venenose et infective, alique vero non, immo aliquando alique inficiunt homines vel alia animalia.

⟨9⟩ Quarto diversificantur etiam in motu. Alique enim apparent velociter moveri, alique vero tardo. Similiter alique apparent moveri sursum, alique ad latus. Similiter alique aliquando motu suo faciunt sonum | in aere, recte *Ba* 20ra sicut esset sonus aliquorum armatorum.

⟨10⟩ Quinto etiam diversificantur tempore, quia alique diu durant, alique cito evanescunt; | et alique aliquando durant per unum mensem, alique per *U* 17vb

6 ad aerem] ab aqua *Ba Kl L U* 11 a causa] ab agente *P* 24 hoc] *om. P* 30 armatorum] almatorum *P* armorum *Ba*

120 Aristoteles, *Meteorologica*, I, 4, 341 b 10.

annum, alique per noctem, et alique ut plurimum fiunt tempore venenoso et tempore epidimie.

⟨11⟩ Sexto diversificantur loco, nam alique fiunt prope terram, ita quod apparent stare super terram vel ire; alique etiam fiunt modicum altius; alique etiam fiunt prope mediam (*Kl* 24ra) | regionem aeris; alique etiam fiunt in media regione aeris—et de istis postea dicetur, et non modo[121]. Alique etiam fiunt super mediam regionem aeris. Similiter alique ut plurimum fiunt in locis ubi sunt cadavera mortuorum, sicut iuxta patibula et in cimiteriis.

⟨12⟩ Quantum ad secundum, sciendum est quod exalatio calida et viscosa et inflammabilis est materia harum impressionum. Causa efficiens earum est ipsum calidum, et aliquando frigidum circumstans comprimens materiam, scilicet exalationem unde debet fieri impressio. Causa autem formalis est quia habent formam mixti imperfecti, que potest dici quedam flamma, ex quo patet quod omnes sunt eiusdem speciei, licet differant accidentaliter propter causas extrinsecas. Causa finalis est propter generationem mixtorum perfectorum, etiam ad pronosticandum quedam futura. Et ideo dicitur quod quando est descensus stellarum in aliqua parte celi, tunc ex ista parte veniunt postea magni venti et fiunt procelle in mare.

⟨13⟩ Eundo autem magis ad causas particulares, sciendum est quod quando exalatio est in continua segregatione a terra et illa exalatio est pinguis et unctuosa et sic quod adhuc unum eius extremum est prope terram et aliud extremum in aere, aliquando de nocte ista exalatio frigiditate noctis comprimitur et condensatur, et calor interius tunc intenditur et incendit et inflammat illam exalationem. Et quando sic fit, tunc apparet sicut candela (*Ba* 20rb) | stans ardens, et huiusmodi impressiones accidunt et contingunt super sepulcra mortuorum. Aliquando autem exalatio (*L* 197va) | pinguis et inflammabilis elevatur modicum altius, sic quod eius extremum inferius non immediate est prope terram. Et tunc, quando de nocte talis exalatio inflammatur et aere moto movetur, apparet sicut lucerna vel candela ambulans vel currens super terram, et homines talibus impressionibus non possunt bene appropinquare;

1–2 et alique ut plurimum fiunt tempore venenoso et tempore epidimie] et hoc secundum naturam ipsius exalationis *Ba* secundum preiacentiam ipsius exalationis *Kl U* et hoc secundum potentiam ipsius exalationis *L* 7–8 ubi sunt cadavera mortuorum] in locis pinguibus et unctuosis *Ba Kl L U* 10 earum] istarum impressionum que fiunt in suprema regione aeris prope ignem *Ba L U* 12–18 causa autem formalis … procelle in mare] *om. P Kl* 20–21 pinguis et unctuosa et sic quod] calida et inflammabilis bene et quando sic est disposita *Ba* calida et inflammatur sic quod *Kl* calida et inflammabilis sic quod *L U* 26 exalatio] inflammatio *P*

121 I.13, 14.

nam, cum homines volunt eis appropinquare, | [*Kl* 24rb] movent aerem, et aere moto iste impressiones moventur et deportantur in aere continue antecedendo homines volentes eis appropinquare. Aliquando etiam accidit quod huiusmodi impressiones fiunt supra auribus equorum ambulantium, nam, | [*U* 18ra] cum de auribus ipsius equi continue extrahitur huiusmodi exalatio unctuosa, ipsa potest incendi et inflammari et apparebunt huiusmodi impressiones. Unde bene visum est huiusmodi impressiones de nocte supra auribus equorum apparere.

⟨14⟩ Quando autem exalatio elevatur usque ad mediam regionem, aliquando frigiditate medie regionis repellitur et tunc, secundum quod illa exalatio est aliter et aliter disposita et aliter et aliter in repulsu movetur, secundum hoc fiunt alie et alie impressiones. Si enim exalatio est longa et non multum lata et, cum repellitur a frigiditate medie regionis, uritur, fit una impressio ad modum trabis ardentis. Si autem illa exalatio est ad modum globi et, cum pertingit ad mediam regionem, | [*Kl* 24va] in repulsu eius descendit totaliter versus terram, quamvis oblique, et ignitur in eius descensu, apparet tamquam stella cadens. Si autem huiusmodi exalatio est ad modum globi et, cum pertingit ad mediam regionem, repellitur et, cum modicum distat a media regione iterum ascendit et tunc iterum repellitur, per inflammationem talis exalationis | [*Ba* 20va] apparet ignis ludens. Si autem ista exalatio in medio est grossa et in extremis constricta et, cum pertingit ad mediam aeris regionem, repellitur et incenditur, apparet draco volans.

⟨15⟩ Si autem exalatio non impedita forte per nubes medie regionis et aliquid huiusmodi transit per mediam regionem usque ad supremam et ibi | [*P* 246r] incenditur, et si ista res est magna et lata, apparet tamquam magna res accensa. Si autem ista exalatio est dispersa per globos, qui tamen globi adherent per exalationes subtiles et raras intermedias, tunc, cum uno globo incenso et consumpto reliquus incenditur, apparet ignis saltare sicut capra. Cum autem ista exalatio habet longitudinem sine magna latitudine et in extremo superiori est magis subtilis et magis inflammabilis quam in extremo inferiori, tunc ista exalatio in extremo superiori primo incenditur, et parte | [*L* 197vb] extrema incensa et consumpta, pars prope incenditur, et tunc iterum apparet sicut stella cadens. Si autem illa exalatio tota simul incenditur, apparet sicut lancea ardens. Si autem illa exalatio habet longitudinem et etiam aliqualem latitudinem et tota simul incenditur, apparet sicut columna

---

13 uritur] *om. Ba Kl U* 14 trabis] turbis *P* 15–18 in repulsu eius … ad mediam regionem] *om. hom. Ba L U* 18 ad] *om. P* 20 ludens] saltans *Ba* saltare *Kl* 22 incenditur] inflammatur *Ba Kl L U* 25 lata] et *add. P* 28 sicut capra] *om. Ba Kl L U* 29 magna] *om. Ba Kl L*

ardens, et inferior pars illius exalationis, que est grossior, apparet tamquam basis illius columne.

U 18rb ⟨16⟩ Dubitaret aliquis utrum huiusmodi impressiones | ignite, si tangerentur manu, comburerent manum. Respondetur quod ille que sunt in materia multum rara non comburerent, sed possent esse ex materia ita densa quod utique comburerent manum. Istis visis, patet quod questio est vera.

⟨17⟩ Ad rationes. Ad primam, cum dicitur: 'exalationes sunt de natura ignis', nego, immo exalatio que est materia harum impressionum est de natura
Kl 24vb terre, quamvis sit | calida calore accidentali ipsam elevante.

⟨18⟩ Ad secundam dico quod huiusmodi materia, que in se iam habet calorem bene fortem, bene potest incendi cum movetur in aere, quia est materia valde inflammabilis.

⟨19⟩ Ad tertiam dico quod illa exalatio elevatur sursum a calore sibi incorporato. Et quamvis iste motus sit violentus quantum est ex parte passi,
Ba 20vb tamen est naturalis | quantum est ex parte agentis; unde ista elevatio est motus naturalis, quamvis eam precessit motus violentus.

Et sic patet ad questionem.

I.14 Utrum de nocte, serenitate existente, debeant apparere hyatus et voragines et sanguinei colores in celo

Queritur consequenter utrum de nocte, serenitate existente, debeant apparere hyatus et voragines et sanguinei colores in celo.

⟨1⟩ Et arguitur quod huiusmodi impressiones non debeant apparere in celo, quia celum non est receptivum huiusmodi impressionum; ergo huiusmodi impressiones non sunt in celo, sed in aere; igitur non deberent apparere in celo, sed magis in aere, cum res magis debet apparere ubi est quam ubi non est.

⟨2⟩ Secundo: huiusmodi impressiones non apparent in celo nec etiam in aere de die; ergo minus debent apparere de nocte. Antecedens patet ad sensum; sed consequentia probatur, quia illa que per Aristotelem possunt movere visum, magis videntur in maiori lumine[122]; sed cum huiusmodi impressiones sunt huiusmodi, ergo, cum maius sit lumen de die quam de

---

9 calida] sicca et calida *Ba Kl L U*

---

122 Aristoteles, *De anima*, II, 7, 419 a 7–9; cf. *Auct. Arist.*, 180, n. 72.

nocte, sequitur huiusmodi impressiones magis debere apparere in die quam in nocte.

⟨3⟩ Tertio videtur quod non debeant apparere serenitate existente, quia, cum huiusmodi impressiones fiant per lumen diruptum ex obscuro | *Kl* 25ra et nigro, sicut dicit Aristoteles[123], sed cum in tempore nebuloso maior fiat diruptio luminis ex obscuro, videtur quod magis debeant apparere in tempore | *L* 198ra nebuloso quam sereno.

⟨4⟩ Quarto ad idem: si sic, hoc esset isto modo, sicut dicit Aristoteles, quod nigrum dirumpens lumen apparet remotius[124]. Sed propter hoc non debet esse, quia, cum istud nigrum sit ita propinquum visui sicut residuum, debet apparere ita propinquum visui sicut residuum; aliter enim sensus visus sic iudicando deciperetur. Et cum sic iudicando iudicet de colore—puta de nigredine, de albedine vel rubedine—, et cum color sit proprium obiectum visus, | *U* 18va tunc sensus circa suum proprium obiectum deciperetur, quod est contra Aristotelem in secundo *De anima*[125].

⟨5⟩ Quinto videtur quod non debeant apparere colores sanguinei, quia non videtur quod magis debeant apparere illi quam alii; sed ita est quod alii colores non apparent, sicut est color viridis vel color croceus vel aliquis talis.

⟨6⟩ In oppositum est Aristoteles[126], et etiam patet ad sensum, quia aliquando de nocte, serenitate existente, apparent huiusmodi impressiones.

⟨7⟩ In ista questione primo exponendi sunt quidam termini in questione positi; secundo ponende sunt quedam | *P* 246^{v} suppositiones; tertio ponende sunt quedam conclusiones; quarto videndum est de causa unius effectus qui etiam aliquando contingit de nocte, serenitate existente.

⟨8⟩ Quantum ad primum, sciendum est quod 'hyatus' et 'voragines' dicuntur *Ba* 21ra quedam aperture quibus celum apparet apertum, et differunt secundum magis et minus. Unde quidam dicunt quod cum illa apertura apparet magna et profunda, tunc dicitur 'vorago'. Cum autem illa apertura apparet minus | *Kl* 25rb

10 visui] *om. Ba Kl L U* 11 visui] *om. Ba Kl L U* 14 sensus] visus *Ba Kl L lect. dub. U* 19 aristoteles] primo huius tractatu secundo capitulo secundo conclusione prima *add. Ba Kl L U* 20 huiusmodi impressiones] hyatus voragines et sanguinei colores *Ba Kl L U* 26 quibus celum apparet apertum] celi *Ba Kl L U*

123 Aristoteles, *Meteorologica*, I, 5, 342 a 34–b 21.

124 Ibid., I, 5, 342 a 14–16.

125 Aristoteles, *De anima*, II, 6, 418 a 14–15; cf. *Auct. Arist.*, 179, n. 66.

126 Aristoteles, *Meteorologica*, I, 5.

profunda, tunc dicitur 'hyatus'. Alii autem dicunt quod cum illa apertura apparet cum rubedine, ac si esset ibi ardor, dicitur 'vorago'; cum autem apparet sub quadam obscuritate, tunc dicitur 'hyatus'[127].

⟨9⟩ Quantum ad secundum, sit prima suppositio: quando aliqua duo equaliter distant a visu, quorum unum est coloratum vel saltem apparens sub colore magis claro, reliquum vero sub colore minus claro, tunc istud quod apparet sub colore clariori apparet esse propinquius quam istud quod est coloratum colore minus claro. Et istud sciunt pictores: partem enim que debet apparere supereminere colorant colore magis claro et illam que debet apparere densa colore minus claro, sicut colore nigro vel alio tendente ad nigredinem. Et propter istam causam potest contingere quod aliqua superficies plana potest apparere concava vel convexa. Si enim continue versus medium coloretur colore claro et versus extremum colore obscuro, iudicaretur convexa, quia partes medie apparerent propinquiores, quia essent colorate colore clariori. Si autem partes versus medium alicuius superficiei plane essent colorate colore obscuro et versus extremum colore claro, tunc illa superficies apparet concava, quamvis esset plana, quia partes extreme apparent propinquiores quam partes medie; nam essent colorate colore clariori. Sic similiter superficies convexa posset apparere plana, sicut si partes magis versus nos essent colorate colore obuscuro et remotiores a nobis colore claro, tunc posset fieri recompensatio claritatis et distantie.

⟨10⟩ Secunda suppositio: cum lux vel album videtur per nigrum, apparet rubeum. Ista suppositio patet cum videmus Solem per fumum[128], et propter hoc ignis etiam in carbone apparet coloris rubei; | *L* 198rb carbo enim est nigri coloris, ignis autem est lucidus.

⟨11⟩ Quantum ad tertium, sit prima conclusio quod possibile est quod aliquando appareat vorago in celo. Probatur, nam possibile est quod aliquando elevetur exalatio in medio densior et profundior, in extremo autem eius | *U* 18vb

1 apertura] et profunditas *add. Ba L* est profunda et *add. U* 8–10 partem enim que … colore minus claro] quando enim volunt aliquam partem magis facere apparere colorant ipsam colore magis claro quam illam quam volunt minus apparere *Ba Kl L U* 10–11 sicut colore nigro vel alio tendente ad nigredinem] *om. Ba Kl L U* 12 vel convexa] et profunda *Ba Kl L U* 12–21 si enim continue … claritatis et distantie] *om. Ba Kl L U* 23 cum videmus solem per fumum] quia cum solem videmus per fumum tunc apparet nobis rubeus *Ba Kl L U* 26–27 aliquando] de nocte serenitate existente *add. Ba Kl L U* 28 extremo] extrema *P*

127 Ibid., I, 5, 342 b 15–18, translatio Guillelmi de Morbeka (translatio noua), *AL* X 2.2, $18^{258-261}$.

128 Cf. ibid., III, 374 a 3–8; 347 b 8–11.

minus densa et minus profunda. Et cum | [*Kl* 25va] exalatio sit de natura terre, etiam participat colorem terre, videlicet nigredinem. Cum autem huiusmodi exalatio est inter nos et celum, lux celestis minus clare videtur per medium illius exalationis et magis clare videtur per extremum illius exalationis. Igitur, per secundam suppositionem, celum visum per medium illius exalationis apparet coloris minus clari, videlicet rubedinis minus clare, et per extremum apparet rubedinis magis clare. Ergo, per primam suppositionem, partes extreme illius exalationis apparent magis propinque, partes autem intermedie apparent magis remote, ergo iudicamus ibi profunditatem in medio, | [*Ba* 21rb] et concavitatem, et tamen propter nimiam distantiam celi et exalationis et propter luciditatem celi in comparatione ad illam exalationem iudicamus illam exalationem non magis propinqua quam celum. Cum ergo talis fantasia et apparitio fiat in exalatione, nos iudicamus hoc esse in celo, et sic celum iudicamus apertum et concavatum.

⟨12⟩ Secunda conclusio est: possibile est quod appareat nobis hyatus in celo. Patet, quia exalatio potest esse in medio ita densa quod per eius medium non videmus lumen et per eius extremum videmus, quia partes in extremo sunt rariores et minus dense; et tunc in medio iudicatur obscuritas et in extremo claritas, et illa apparitio est hyatus.

⟨13⟩ Tertia conclusio: quod possunt apparere sanguinei colores. Patet, nam possibile est quod exalatio | [*P* 247^{r} *Kl* 25vb] sit uniformis in raritate et densitate in partibus eius extremis et mediis. Et quando sic est, equaliter lumen vel lux videtur per huiusmodi exalationem, et tunc partes extreme iudicantur equaliter prope sicut medie. Et cum lux visa per nigrum faciat apparere rubeum, sequitur illam exalationem apparere rubeam, et illam rubedinem iudicamus in celo, propter causam prius dictam. Ex isto est quod Sol, cum de mane videtur et fiunt vapores et exalationes inter Solem et visum, quod Sol apparet rubei coloris, quia lux Solis visa per nigrum, sicut per vapores et exalationes, facit apparere rubeum.

⟨14⟩ Quarta conclusio: huiusmodi impressiones, videlicet hyatus, vorago, etc., non apparent | [*L* 198va] de die. Patet, quia ex nimio | [*U* 19ra] lumine Solis opacitas et obscuritas ipsius exalationis totaliter supervincitur, et sic apparet totum clarum.

⟨15⟩ Quinta conclusio: huiusmodi impressiones fiunt in tempore sereno et non in tempore nebuloso. Patet, quia in tempore sereno possunt elevari exalationes usque ad supremam regionem aeris non impedite a nubibus;

---

11 propter] magnam *add. Ba Kl L U*   15 conclusio] est *add. P* ‖ quod] de nocte serenitate existente *add. Ba Kl L U*   16 in medio] *om. Ba Kl L U*   31 etc.] et sanguinei colores *Ba Kl L U*   36 exalationes] vapores *Ba Kl L*

in tempore autem nebuloso non possunt, quia tunc impediuntur a nubibus, que non possunt pertingere usque ad supremam regionem.

Ba 21va ⟨16⟩ Quantum ad quartum, videlicet qualiter aliquando in nocte, | tempore existente sereno, audiuntur aliqui soni, ac si essent voces hominum vel brutorum: causa illius solet dari quod, cum in aliquo vapore elevato aliqua exalatio includitur, isto vapore compresso, illa exalatio petit exitum, et tunc causatur talis sonus, sicut ferrum calidum quando ponitur in aqua.

Kl 26ra ⟨17⟩ Sed diceres: quomodo est quod aliquando apparet sicut vox hominis vel vox alterius animalis? Solet dici quod hoc est ex virtute constellationis et dispositionis materie, recte sicut ex constellatione grandines descendentes aliquando sunt figurate figura hominis, aliquando alterius animalis. Ita etiam in proposito dicatur de diversificatione et varietate illius soni sic causati.

⟨18⟩ Ad rationes. Ad primam concedo quod huiusmodi impressiones non sunt in celo, nec celum est receptivum huiusmodi impressionum; bene tamen apparent in celo propter causam prius dictam[129].

⟨19⟩ Ad secundam dicatur quod non fiunt de die, sed de nocte, propter causam prius dictam[130].

⟨20⟩ Ad tertiam dico quod non fiunt in tempore nebuloso propter nimiam spissitudinem vaporis elevati et etiam quia nubes est bene cumstricta, et propter hoc non facit talem fantasiam.

⟨21⟩ Ad quartam dico quod visus bene decipitur in iudicando de distantia vel situ; situs autem non est proprium obiectum visus, sed est sensibile commune. Circa autem sensibilia communia sensus plures decipitur, sicut habet videri circa secundum *De anima*[131].

⟨22⟩ Ad quintam dico quod causa quare ibi non apparent colores crocei vel virides est quia lux visa per nigrum non est apta nata facere apparere huiusmodi colores.

Et sic finitur questio.

---

2 que] quia *P*   4–5 brutorum] animalium *add. Ba Kl L U*   15 receptivum huiusmodi impressionum] aptum natum huiusmodi impressiones recipere *Ba Kl L U*   24–25 sicut habet videri circa secundum de anima] *om. Ba Kl L U*

---

129 I.14, 11.

130 I.14, 14.

131 Cf. Aristotelem, *De anima*, II, 6, 418 a 14–15.

I.15 Utrum cometa sit de natura celi vel elementari

Queritur consequenter utrum cometa sit de natura celi vel elementari. *U* 19rb

⟨1⟩ Et arguitur quod sit de natura celi: illud est de natura celi | cui convenit *Kl* 26rb motus naturalis ipsius celi; sed sic est quod motus circularis ipsius celi convenit comete. Maior est nota, | ex eo quod motus arguit formam. Sed minor *L* 198vb patet, quia motus circularis est motus naturalis ipsius celi; ille autem convenit comete: cometa enim est aliquando visus moveri motu diurno circa Terram.

⟨2⟩ Secundo: si cometa non esset de natura celi, sed fieret ex exalatione vel vapore elevato, sicut fiunt alie impressiones, vel igitur fieret in aere, vel in igne. Nullum istorum | potest dici. Et primum non potest dici quod in aere, *Ba* 21vb et supponitur quod quia cometa diu durat, oporteret quod fieret de materia grossa et compacta; isto sumpto, probo quod non potest fieri in aere: non enim in media regione, quia, que fiunt in media regione vel infra per inflammationem, oportet quod sint de materia minus grossa et de materia magis rara, ut cito et faciliter possit inflammari; sed cum materia comete sit valde grossa et compacta, caliditas infra mediam regionem et etiam antiparistasis non sufficerent ad inflammationem talis materie ex qua dicitur fieri cometa. Nec etiam potest dici quod fiat | in summa regione aeris vel in igne, nam, *P* 247v quamvis materia ita grossa, si ipsa esset ibi, bene posset inflammari, tamen videtur quod propter eius nimiam grossitiem non potest ita alte elevari.

⟨3⟩ Tertio ad idem: si esset de natura elementari et non celi, tunc aliquando posset fieri sub alia figura quam solet fieri, videlicet sine cauda vel sine coma vel sine barba. Consequentia tenet, quia exalatio de qua tunc generaretur, si non esset de natura celesti, posset taliter fieri vel disponi quod fieret sub figura rotunda.

⟨4⟩ Quarto ad idem rationibus Senece, nam, si esset de alia natura quam *Kl* 26va de natura celi, tunc fieret in aere, sicut alie impressiones; sed quecumque impressiones fiunt in aere sunt parve durationis, sicut videmus de nubibus et consimilibus[132]; sed comete non sunt parve durationis, immo magne; igitur etc.

1 elementari] elementorum *Ba Kl* elementali *L* 15–16 materia comete sit valde grossa et compacta] exalatio de qua fit cometa sit de materia grossa *Ba* exalatio sit de materia comete *Kl L* exalatio sit de materia grossa *U* 24 fieri vel] *om. Ba Kl L U*

132 Seneca, *Questiones naturales*, VII, 22, 1.

⟨5⟩ Quinto: impressiones que fiunt in aere non durant nec manent sub eodem astro nec moventur ad motum illius astri, et hoc si fiunt per inflam- *U* 19va mationem; | sed visum est quod cometa movebatur ad modum astri.

⟨6⟩ Sexto: comete sunt magne significationis et significant aliquos effectus futuros qui non statim eveniunt; sed impressiones alie, si significant aliquos effectus proprios, illi statim eveniunt, sicut patet de rubedine matutina, postquam statim videtur sequi pluvia. Ex hiis videtur quod comete sint de *L* 199ra natura celi et non elementari, | de qua sunt alie impressiones.

⟨7⟩ In oppositum est Aristoteles, qui ostendens modum generationis comete, dicit quod fit per inflammationem exalationis elevate usque ad summam *Ba* 22ra regionem | aeris bene alte[133].

⟨8⟩ In ista questione primo enunciande sunt quedam experientie quibus apparet diversitas cometarum; secundo sunt recitande opiniones dicentium ipsum esse de natura celi; tertio post ista in alia questione videndum est de qua natura sit secundum veritatem[134].

Kl 26vb ⟨9⟩ Quantum ad primum, sciendum est quod comete differunt | primo magnitudine: aliquando enim sunt visi valde magni; etiam aliquando visi sunt parvi, et aliquando ita parvi quod de difficili propter eorum parvitatem possunt videri[135].

⟨10⟩ Secundo ipsi comete differunt multitudine: quia aliquando visus est unus, aliquando vero plures simul.

⟨11⟩ Tertio differunt in figura: nam aliquando apparet caudatus, aliquando barbatus et aliquando comatus, secundum quod coma apparet retro, vel ante, vel supra[136]. Unde visus est cometa habens caudam ita magnam quod extendebatur usque ad medium celi, ut dicit Aristoteles[137].

⟨12⟩ Quarto differunt comete in colore: nam aliqui apparent coloris rubei, aliqui vero coloris candidi[138].

---

2 astro] erratico *add. Ba Kl L U*   5 si] qui *P*   6 proprios] futuros *Ba Kl L U*   17 magni] immo aliquando tanti vel maiores sicut sol *add. Ba Kl L U*

---

133 Aristoteles, *Meteorologica*, I, 7, 344 a 5–344 a 33.

134 I.16, 10–11.

135 Cf. Senecam, *Questiones naturales*, VII, 17, 3.

136 Cf. Senecam, *Questiones naturales*, VII, 11, 2.

137 Aristoteles, *Meteorologica*, I, 6, 343 b 1; 343 b 17–25.

138 Cf. Senecam, *Questiones naturales*, VII, 17, 3.

⟨13⟩ Quinto differunt motu et quiete: aliquando enim cometa visus est quiescere, aliquando visus est etiam moveri, aliquando transversaliter, aliquando etiam motu diurno, aliquando etiam ad motum unius stelle, ita quod apparuit insequi stellam, et aliquando velocius et aliquando tardius.

⟨14⟩ Sexto differunt loco: quidam enim generantur altius, quidam vero bassius. Similiter generantur sub diversis partibus ipsius celi, sed ut plurimum generantur sub parte celi septentrionalis et solent apparere circa Viam lacteam.

⟨15⟩ Septimo differunt tempore: quia aliquando sunt multi anni quod non videtur cometa, aliquando autem in parvo tempore, sicut in duobus annis, visi sunt duo vel tres. Et etiam differunt tempore sicut quod aliqui diu durant, aliqui autem durant minus. Unde quidam dicunt quod ad minus durant per septem dies et ad plus per sex menses[139].

⟨16⟩ Octavo differunt in variatione coloris: quidam enim visi sunt variare colorem, quidam autem visi sunt non variare colorem. Similiter quidam visi sunt deficere propter eorum approximationem ad Solem; quidam autem visi sunt | deficere sine huiusmodi approximatione. *Kl* 27ra

⟨17⟩ Nono differunt in significatione: quidam enim significant hoc et qui- *U* 19vb dam aliud, sicut patet postea in una questione[140].

⟨18⟩ Decimo differunt quia aliqui fiunt sub aliquo astro fixo vel erratico; aliqui autem fiunt per se, non sub aliquo astro.

⟨19⟩ Quantum ad secundum, sciendum est quod dicentes cometas esse de natura celi diversificabantur | in suis opinionibus; unde quidam dixerunt *L* 199rb quod cometa esset unus de planetis[141]. Sed breviter hoc non valet, | quia ali- *P* 248^{r} quando visus est cometa aliis | planetis apparentibus, quod est signum quod *Ba* 22rb non est aliquis planetarum[142].

⟨20⟩ Alii autem dixerunt quod cometa esset una stella que ad tempus apparet habere comam propter vapores sub illa stella elevatos[143]. Sed illud

---

1 quinto] aliquando *add. P* 2 transversaliter] motu retrogrado *Ba Kl U* retrograde *L* 4 stellam] motum unius stelle *Ba Kl L U* ‖ et aliquando velocius et aliquando tardius] *om. Ba Kl L U* 5–6 quidam vero bassius] *om. Ba Kl L U* 9 sunt] transeunt *Ba Kl L U* 10 annis] vel tribus *add. L* 11 diu] *om. P* 13 per sex menses] per octo dies et alii dicunt quod per sex menses *Ba* per octo menses *Kl* 17 deficere] *om. P* 20 fiunt] moventur *Ba Kl L U* 21 fiunt] moventur *Ba Kl L U* 25 planetis] omnibus *add. Ba L*

---

139 Plinius, *Naturalis historia*, II, 22, 90.

140 I.18.

141 Cf. Aristotelem, *Meteorologica*, I, 6, 342 b 30–35.

142 Cf. ibid., I, 6, 343 a 30–35.

143 Cf. ibid., I, 6, 342 b 35–343 a 20.

dictum non valet, nam, quamvis aliquando sic fiat, tamen non semper, quia non semper cometa apparet moveri ad motum alicuius stelle. Sed tamen, si semper predicto modo fieret et non alio, semper deberet apparere moveri ad motum alicuius stelle.

⟨21⟩ Alii dixerunt quod ex coniunctione plurium stellarum apparet cometa[144]. Istud iterum non valet, quia semper cometa deberet apparere moveri, ex quo semper stelle moventur—sive sint coniuncte, sive non; sed consequens est falsum, quia aliquando apparet quiescere. Secundo, si sic, sequeretur quod statim, postquam cometa deficeret, stelle deberent ad invicem apparere divise ex quarum coniunctione apparuit cometa; modo hoc non est verum: non enim semper, ubi cometa apparet deficere, apparent alique stelle divise[145]. Tertio, nam non est verisimile quod stelle ita diu essent coniuncte sicut apparet cometa et durat. Quarto, nam, cum stelle non sunt coniuncte, apparent quantitatis valde parve, et non est verisimile quod ex *Kl* 27rb coniunctione | duarum vel trium stellarum que per se apparent ita parve debeant apparere ita magne[146] sicut cometa.

⟨22⟩ Alia erat opinio Senece quod cometa non est aliqua de stellis fixis nec est aliquis de planetis, sed est una alia stella que habet motum valde diversum et incognitum, et aliquando est hic et aliquando est alibi[147].

⟨23⟩ Sed contra: non est verismile quod aliqua stella habeat talem figuram qualem habet cometa; ergo nec est verisimile quod cometa sit aliqua *U* 20ra stella. Antecedens patet per Commentatorem in secundo *Celi*, | dicentem quod omnes stelle sunt figure sperice[148]. Sed diceretur quod talis stella non est talis figure, sed apparet talis ex lumine et radiis eius—propter hoc enim apparet coma. Hoc non valet, quia, cum talis stella sit sperica et equaliter cir- *Ba* 22va cum se diffundit | lumen et radios, sequitur quod non magis propter radios et lumen eius ex una parte apparet coma quam ex alia. Aliter diceretur quod forte illa stella apparet habere comam propter exalationes sub ipsa eleva-

---

5 coniunctione] vel ex commixtione *add. L*   9 postquam] post postquam *P*   15 apparent] apparet *P*   16 magne] magna *P* in coniunctione *add. Ba Kl L U*   19 hic et aliquando est alibi] idem *P*   25 equaliter] aliqualiter *P*

---

144 Cf. ibid., I, 6, 342 b 27–29.

145 Cf. ibid., I, 6, 343 b 25–27.

146 Aristoteles, *Meteorologica*, I, 6, 343 b 33–344 a 2.

147 Seneca *Questiones naturales*, VII, 22, 1; VII, 24, 2; VII, 27, 6.

148 Averroes, *Commentarium magnum in De celo*, II, comm. 26, eds. Carmody, Arnzen, 2: 313–314.

tas[149]. Hoc non valet, quia sic illa stella aliquando deberet apparere sine tali coma, cum aliquando deficiant ex|alationes sub ipsa[150]. *L* 199va

⟨24⟩ Secundo; quia cometa aliquando movetur ita diversimode quod non est verisimile quod ita diversimode movetur aliqua stella celi, sive sit fixa sive erratica. Similiter cometa aliquando quiescit, sed numquam aliqua stella celi quiescit.

⟨25⟩ Tertio, nam cometa ita prope aliquando accedit ad polos quod nulla *Kl* 27va stella erratica ita prope potest accedere.

⟨26⟩ Quarto, nam aliquando cometa est alterius et alterius coloris; nulla autem pars celi est modo unius coloris et postea alterius.

⟨27⟩ Quinto, nam aliquando apparent simul plures comete, et igitur non potest dici quod cometa sit una stella celi; isti enim qui dixerunt non posuerunt plures stellas una.

⟨28⟩ Sexto, nam aliquando cometa deficit absque hoc quod intret sub radiis Solis; numquam autem aliqua stella apparet deficere nisi subintret radios ipsius Solis; ergo finaliter non est aliqua stella celi.

⟨29⟩ Ad rationes. Ad primam: 'cometa movetur circulariter', dico quod hoc est ad motum aeris continentis, qui movetur circulariter ad motum ignis, et ignis ad motum celi. Nec propter hoc oportet quod cometa sit de natura celi plus quam aer vel ignis: talis enim motus non inest istis naturaliter, sed inest eis preter naturam, sicut prius dicebatur[151].

⟨30⟩ Ad secundam: 'vel fieret in aere, vel in igne', dico quod fit in aere. Et quando dicitur: 'materia de qua generatur est ita grossa quod non potest ita alte elevari', dico quod immo propter calorem inclusum et incorporatum in ipsa materia.

⟨31⟩ Ad tertiam: 'tunc aliquando deberet', etc., respondetur quod forte pluries sic fit, sed tunc apparet quod sit una stella celi propter | eius rotunditatem. Sed quando apparet cum cauda vel barba, tunc iudicatur differentia inter ipsum cometam et alias stellas celi. *Kl* 27vb

1 apparere] *om. P* 3 quia cometa] *iter. P* 8 erratica] celi *Kl L U om. Ba* 16 celi] cometa *P* 19 celi] non *add. P* 20 inest istis naturaliter] est ignis naturalis *L* 20–21 inest eis preter naturam] violenter *Ba Kl U* violentus *L* 24 propter] fortem *add. Kl L Ba U* ‖ et incorporatum] *om. Ba Kl L U* 27 propter] nimiam *add. Ba Kl L U* 27–28 rotunditatem] profunditatem *Ba Kl L U* 29 alias stellas celi] planetas *Ba Kl U* planetas vel stellas *L*

149 Cf. Aristotelem, *Meteorologica*, I, 6, 342 b 35–343 a 20.

150 Cf. ibid., I, 6, 343 a 27–30.

151 I.8, 24.

⟨32⟩ Ad quartam: 'quecumque impressiones', etc., verum est in media regione vel infra. Si tamen fiant super mediam regionem, tunc tamen propter multitudinem materie et compacitatem et grossitiem | [*U* 20rb] possunt diu | [*P* 248v] durare.

⟨33⟩ Ad quintam | [*Ba* 22vb] concedo quod alique impressiones que fiunt in aere non diu apparent sub aliquo astro. Cum hoc stat quod alique impressiones que fiunt magis alte, ubi aer movetur circulariter, possunt continue moveri sub aliquo astro et moveri motu diurno una cum aere in quo sunt.

⟨34⟩ Ad sextam de significatione comete videbitur seorsum in una questione[152].

I.16 Utrum cometa sit exalatio calida et inflammata

Queritur consequenter utrum cometa sit exalatio calida et inflammata.

⟨1⟩ Et arguitur quod non, nam talis exalatio vel esset | [*L* 199vb] multa vel pauca. Si dicatur quod talis exalatio sit pauca, tunc cito per ignitionem et inflammationem eius consumitur, et sic cometa non ita diu duraret sicut visus est durare. Si autem dicatur quod illa exalatio sit multa, tunc sequitur quod cometa debeat apparere continue crescere et augeri; nam, primo una parte ignita, alia igniretur; vel si simul omnes eius partes ignirentur, non ita diu duraret sicut visus est durare.

⟨2⟩ Secundo, nam, si esset exalatio calida ignita, non deberet apparere in suprema regione aeris, quod est contra experientiam. Consequentia probatur, nam talis exalatio non potest transire per mediam | [*Kl* 28ra] regionem aeris: cum ipsa sit frigida et illa exalatio sit calida, sequitur quod cum pertingit ad mediam regionem aeris, pellitur et fugatur per viam a qua venit; nam prius dicebatur quod unum contrarium movet suum contrarium localiter[153].

⟨3⟩ Tertio ad idem: si esset aliqua talis exalatio ignita, aliquando deberet eclipsari, quia aliquando esset in umbra Terre, causata ex obiectione Terre ad Solem; sed consequens est falsum, quia numquam visus est cometa eclipsari.

---

7 ubi aer movetur circulariter] *om. Ba Kl L U* ‖ continue moveri] diu durare *Ba Kl L U* 8 sub aliquo astro] *om. Ba Kl L U* 13 ignitionem] ignem *P* 16 ignita] consumpta *Ba Kl L U* 23 per viam a qua] ad locum a quo *Ba U* a via a qua *Kl* ad viam a qua *L* 26–27 causata ex obiectione terre ad solem] *om. Ba Kl*

---

152 I.18.

153 I.10, 13.

⟨4⟩ In oppositum est Aristoteles, qui opinionibus Antiquorum reprobatis dicentium cometam esse de natura celi, dicit quod cometa est de natura elementari; et ponens modum generationis eius, dicit ipsum esse exalationem calidam et ignitam et inflammatam[154].

⟨5⟩ In ista questione primo ponende sunt opiniones aliquorum dicentium cometam esse de natura elementari; secundo ponende sunt conclusiones secundum opinionem Aristotelis.

⟨6⟩ Quantum ad primum, sciendum est quod illi qui posuereunt cometam esse de natura elementari erant diversificati, sicut et illi qui posuerunt cometam de natura celi. Unde quidam posuerunt quod cometa esset ignis condensatus; unde ymaginabantur quod aliqua pars ignis supra aliquando condensetur in tantum quod sic lucet ad modum ad quem apparet nobis cometa. Et secundum quod iste ignis secundum suas partes alias et alias est condensatus, apparet nobis aliquando barbatus, aliquando caudatus.

⟨7⟩ Breviter ista opinio non videtur | vera, nam ignis non | potest stare sub tanta densitate quod sic luceret, immo nec adhuc aer ad modum ad quem cometa apparet lucere. *Ba* 23ra *U* 20va

⟨8⟩ Alia erat opinio quod materia comete | bene est exalatio, sed non talis qualis est | exalatio aliarum impressionum. Unde ista opinio ymaginabatur quod exalatio que est materia comete virtute stellarum purgatur et subtiliatur ab omni terrestri. Ulterius ymaginatur quod nec cometa fit per inflammationem huius exalationis ab igne, sed a natura alicuius planete, ita quod ista opinio ymaginabatur quod cometa est unum corpus valde simile nature stellarum, nec est ibi proprie inflammatio, sed illud corpus fit sic lucidum quia valde declinat ad naturam stellarum[155]. *Kl* 28rb *L* 200ra

⟨9⟩ Ista opinio non posset multum formaliter impugnari, tamen potest contra eam argui, nam, cum illud corpus non sit lucidum per inflammationem, oportet quod sit lucidum a Sole, sicut et alie stelle; et si sic, sequitur quod cum illud corpus esset in umbra Terre, quod tunc deficiat a lumine; et si sic, sequitur cometam aliquando eclipsari. Hoc tamen numquam visus erat;

---

7 aristotelis] ponentis cometam esse de natura elementari *add. Kl L* U^{ac} 24 inflammatio] exalatio *Ba Kl L U*

---

154 Ibid., I, 6, 343 a 22–344 a 4; I, 7, 344 a 5–344 a 33.

155 Robertus Grossatesta, *De cometis*, ed. C. Panti, in *Moti, virtù e motori celesti nella cosmologia di Roberta Grossatesta. Studio ed edizione dei trattati De sphera, De cometis, De motu supercelestium*, Florence 2001 (Corpus Philosophorum Medii Aevi. Testi e Studi, 16), 324^{61}–326^{117}.

ergo videtur magis quod sit unum corpus lucidum per inflammationem, et non per lumen Solis, sicut alie stelle.

⟨10⟩ Quantum ad secundum, sit prima conclusio: cometa non est de natura celi. Patet, nam, si sic, numquam deficeret a lumine nisi per approximationem et accessum eius ad Solem, sicut et alie stelle; sed hoc est falsum. Patet falsitas, quia sepe cometa visus est deficere a lumine absque eius approximationem ad solem.

⟨11⟩ Secunda conclusio: cometa est de natura elementari, quia vel cometa est de natura celesti, vel elementari. Sed non est de natura celesti per pri- *Kl* 28va mam conclusionem; ergo de natura elementari. Maior patet ex eo | quod omne corpus est de natura celi vel elementari.

⟨12⟩ Tertia conclusio: cometa est exalatio calida elevata ex istis inferiori- *P* 249^{r} bus, inflammata in superiori regione | aeris propter caliditatem ibi habundantem ex eius proximitate ad ignem. Probatur, quia non fit per refractionem luminis celestis supra illa exalatione, sicut Aristoteles probat in littera[156]; ergo sequitur quod cometa sit unum corpus realiter lucidum. Sed illud non est aer neque ignis, quia ista propter nimiam eorum raritatem non *Ba* 23rb *L* 200rb possent sic lucere. Nec etiam illud | corpus lucidum est terra vel aqua, quia | huiusmodi propter eorum gravitatem non possent ita alte elevari. Ergo oportet quod illud corpus lucidum sit condensatius quam sit ignis et aer et sit subtilius quam terra et quam sit aqua. Tale autem corpus medio modo se *U* 20vb habens erit vapor vel exalatio. Sed non potest dici quod sit vapor, | quia ille non potest ita alte elevari sicut ubi est cometa. Ergo oportet dici quod illud corpus sic lucidum et inflammatum sit exalatio; exalatio enim altius potest elevari quam vapor, quia calor potest sibi fortius incorporari, quia talis exalatio est de natura terrestri.

⟨13⟩ Ex isto potest reddi causa quare cometa aliquando est maior, aliquando minor, nam hoc est secundum multitudinem et paucitatem illius exalationis. Et propter illud cometa aliquando durat diutius, aliquando minus diu. Per hoc etiam potest reddi causa quare cometa aliquando apparet coloris rubei, aliquando coloris magis tendentis ad albedinem, nam hoc est secundum quod illa exalatio est minus vel magis densa vel rara, vel magis *Kl* 28vb vel minus terrestris. Potest etiam reddi causa quare aliquando | apparet cau-

5 est] *om. P* 6–7 absque eius approximationem ad solem] quando recessit a sole *Ba* propter hoc quod approximabat ad solem *Kl L U* 19 gravitatem] grossitiem et gravitatem *Kl L U* 33–205.1 caudatus et non barbatus et aliquando e contrario] caudatus barbatus *P*

156 Aristoteles, *Meteorologica*, I, 6, 343 a 26–343 b 7.

datus et non barbatus, et aliquando e contrario, nam, si exalatio inferius inflammata est ante, apparet barbatus; si retro, apparet caudatus.

⟨14⟩ Ad rationes. Ad primam dico quod ista exalatio ignita, quamvis non sit in tanta quantitate quod ipsa non augmentata aliunde posset tam diu lucere sicut apparet cometa, tamen continue nove et nove exalationes elevantur ad istum locum ubi est cometa. Inde est quod cometa potest tam diu apparere sicut apparet; et quando non amplius ille exalationes nove elevantur, cometa incipit deficere.

⟨15⟩ Ad secundam dico quod illa exalatio propter magnum calorem sibi incorporatum potest transire per mediam regionem usque ad supremam, absque hoc quod repellatur a frigiditate medie regionis.

⟨16⟩ Ad tertiam dico quod numquam debet apparere eclipsari, quia ex inflammatione eius habet lumen, et non | *Ba* 23va a Sole; et ergo, non obstante quod sit in umbra Terre, adhuc apparet lucere.

I.17 Utrum motus comete sit naturalis vel violentus

Queritur consequenter utrum motus comete sit naturalis vel violentus.

⟨1⟩ Et arguitur primo quod non sit naturalis. Corpora elementaria non moventur naturaliter motu circulari; | *L* 200va sed cometa est corpus elementare—est enim de natura elementari, sicut prius dicebatur[157]; ergo etc. Maior est nota, quia corpora elementaria | *Kl* 29ra moventur naturaliter motu recto, scilicet sursum vel deorsum, et non motu circulari; nam solum illo modo movetur celum naturaliter. Et cum idem motus in specie non conveniat naturaliter corporibus distinctis specie, sequitur motum circularem aliis corporibus quam celo non esse naturalem; | *U* 21ra et per consequens non naturaliter inest comete.

⟨2⟩ Secundo: motus naturalis alicuius est solum secundum unam differentiam positionis; sed sic non est de motu comete, nam aliquando movetur circulariter ab oriente versus occidentem, aliquando dimittit illum motum et movetur ab occidente versus orientem, et aliquando descendit, aliquando ascendit.

17–18 est enim de natura elementari] quia non est de natura celi *Ba Kl L U* 20 vel] *om. P* 22 quam] cum *P* 26 circulariter] *om. Ba Kl L U*

---

157 I.16, 10–11.

⟨3⟩ Tertio ad idem: nullum corpus grave vel leve naturaliter circumfertur; sed cometa est corpus grave vel leve; igitur etc.

⟨4⟩ Quarto: omnis motus qui inest alicui ab extrinseco non est naturalis; sed sic est de motu comete: ille enim sibi inest a celo vel ab aere ipsum continente.

⟨5⟩ Quinto: nichil movetur naturaliter motibus contrariis; sed cometa movetur motibus contrariis; ergo cometa non movetur naturaliter, nam cometa aliquando movetur ascendendo, aliquando movetur descendendo.

⟨6⟩ Sexto ad idem: ille locus in quo est cometa vel est sibi naturalis vel violentus. Si est sibi naturalis, ergo debet in eo quiescere et per consequens, si movetur extra ipsum, hoc non est motu naturali. Si autem dicatur quod ille locus sit sibi violentus, et cum omne corpus habeat locum sibi naturalem, sequitur quod cometa habet alium locum sibi naturalem. Vel igitur ille *P* 249v locus est supra cometam vel infra: si dicitur quod supra, ergo debet moveri | ascendendo; si dicitur quod infra, tunc debet moveri descendendo.

⟨7⟩ In oppositum arguitur quia tali motu visi sunt omnes comete moveri; igitur videtur quod talis motus ad minus comete non sit violentus.

Kl 29rb ⟨8⟩ In ista questione primo videndum est quibus motibus | comete visi sunt moveri et que sint cause talium motuum. Secundo videndum an tales motus comete sint naturales vel violenti vel preter naturam insint sibi.

Ba 23vb ⟨9⟩ Quantum ad primum, sciendum est | quod erat una opinio que dixit cometam non moveri, quamvis appareat nobis quod movetur. Unde ista opinio ymaginabatur quod exalatio que est materia comete esset ad modum *L* 200vb corporis girativi, vertens unum extremum versus orientem, | aliud vero versus occidentem. Et quando ista exalatio in extremo versus orientem inflamaretur, tunc ista parte consumpta, reliqua incenderetur, et sic continue apparet moveri versus occidentem tam diu quod illa exalatio est consumpta. *U* 21rb Et secundum | istum modum etiam potest salvari quare aliquando apparet ascendere et aliquando descendere. Si sic diceretur, non potest convinci aerem superius moveri circulariter ad motum celi, quia non videtur ex alio posse probari quod aer superius movetur circulariter nisi per motum comete; si autem negetur motus comete talis, non habebimus modum probandi aerem superius moveri circulariter. Sed breviter ista opinio non valet,

---

7–8 movetur motibus contrariis … aliquando movetur descendendo] est huiusmodi igitur scilicet ascendendo vel descendendo quare etc. *P* 17 ad minus] *om. Ba Kl L U* 23 materia] de natura *Ba Kl U* natura *L*

quia secundum talem modum non possemus probare quod cometa aliquando occidit, aliquando oritur, aliquando movetur versus septentrionem, aliquando versus meridiem, aliquando versus occidentem, et dimittit illum motum et incipit moveri versus orientem.

⟨10⟩ Dicendum est ergo quod cometa vere movetur, et aliquando movetur motu diurno, et aliquando velocitat | *Kl* 29va motum suum, et aliquando retardat motum suum, scilicet diurnum, et aliquando movetur versus occidentem, et dimittens illum motum movetur versus orientem, quasi motu reflexionis, et aliquando ascendit et aliquando descendit. Causa primi, quod movetur motu diurno, est propter motum ipsius aeris continentis cometam. Sicut enim vinum movetur ad motum dolii continentis ipsum, ita cometa movetur ad motum aeris continentis ipsum; sed quia aer movetur motu diurno ad motum celi, sequitur, etc. Sed | *Ba* 24ra aliquando velocitat illum motum, ita quod velocius circuit quam aer totus. Causa est quia, quod aer movetur circulariter, est a virtute celi sibi impressa; et quia exalatio que est materia comete recipit fortius illam virtutem, inde est quod cometa velo|*L* 201^{ra}cius circuit quam aer. Et aliquando visum est quod cometa tantum velocitavit motum suum quod uno vespere erat retro Solem et post lapsum parvi temporis, ut tertium vel quartum dierum, erat ante Solem. Dicitur enim quod tempore quo Karolus intravit Ceciliam, uno vespere videbatur cometa post occasum Solis et postea per aliquos dies videbatur ante ortum Solis ille idem cometa. Et causa huius potest esse quia exalatio que fuit materia illius comete virtute alicuius astri precedentis Solem erat elevata et in ascensu illius exalationis forte per ventum vel aliquid huiusmodi impedita, quod non poterat elevari directe sub illo astro, et propterea prepulsa ad aliquem locum post Solem et ibi inflata tunc apparuit post Solem. Et cum continue virtute astri | *U* 21va precedentis Solem insequitur illud astrum et trahitur ab illo astro ad modum | *Kl* 29vb ad quem ferrum trahitur a magnete, tandem movetur quod venit ad locum | *P* 250^{r} aeris infra illud astrum. Et quia illud astrum est ante Solem, ille cometa aliquando precedit Solem quem primo sequebatur.

⟨11⟩ Aliquando etiam cometa movetur versus orientem vel versus unam partem et postea, dimittens illum motum, movetur versus aliam. Et huius causa potest dari quod exalatio virtute coniunctionis aliquarum stellarum elevatur, et postquam elevata est et inflammata, movetur ad motum stelle

12 motu diurno] circulariter *Ba Kl L U* 15 que est materia comete] *om. Ba Kl L U* 16 virtutem] a celo *add. Kl L U* 18–19 ut tertium vel quartum dierum] *om. Ba Kl L* 22 que fuit] *om. P* 25–26 et propterea prepulsa ... tunc apparuit post solem] *om. Ba Kl L* 26 inflata] et *add. P* 27 solem] ista exalatio *add. Kl L U* ipse cometa *add. Ba* 28 ferrum] *om. P*

que tunc habet dominium. Et postea, quando virtus illius stelle et influentia debilitatur et una alia incipit habere dominium, tunc dimmitens primam insequitur secundam.

⟨12⟩ Aliquando etiam cometa movetur ascendendo et descendendo, et huius causa est quia taliter movetur planeta in suo epiciclo, cuius planete motum cometa insequitur. Vel aliter posset dici quod in principio inflammationis, cum est ibi multum de materia subtili et inflammabili, cometa ascendit et postea, materia subtiliori consumpta et grossiori remanente, e converso descendit.

Ba 24[rb] ⟨13⟩ Quantum | ad secundum, breviter dicendum quod motus comete non est sibi violentus nec est sibi naturalis nisi ad modum ad quem motus ferri ad magnetem dicitur ferro naturalis, sed inest comete preter naturam. Patet, quia motus diurnus et motus quo insequitur stellam vel planetam est sibi impertinens; similiter motus quo movetur ascendendo vel descendendo ad *L* 201[rb] motum planete in epiciclo. Sed si dicitur quod sic ascendit | vel descendit *Kl* 30[ra] propter levitatem vel gravitatem materie, tunc | debet dici quod illi motus sunt sibi pure naturales ad modum ad quem descendere est gravi naturale et ascendere levi.

⟨14⟩ Ad rationes hinc inde facte: non probant nisi quod motus comete non est sibi naturalis nec violentus; non tamen probant quin insit sibi preter naturam.

Hoc de questione.

I.18 Utrum comete significent mortem principum, siccitatem et ventos et motus terre

Queritur consequenter utrum comete significent mortem principum, siccitatem et ventos et motus terre.

U 21[vb] ⟨1⟩ Et arguitur | primo quod non significent mortem principum, quia indifferenter respiciunt homines populares sicut principes, propter quod non plus debent significare mortem principum quam mortem popularium[158].

17 gravi] gravis *P* 21 naturam] id est propter aliquam virtutem sibi impressam *add.* *L U*

158 Cf. Albertum Magnum, *Meteora*, lib. 1, tract. 3, cap. 11, ed. Colon. VI/1, 32^{55-57}.

⟨2⟩ Secundo: galaxia non significat ista; ergo nec cometa. Consequentia tenet ex eo quod galaxia est de natura comete. Patet per Aristotelem in littera[159]. Antecedens est notum de se.

⟨3⟩ Tertio: cometa aliquando insequitur planetas benivolos sicut malivolos; ergo non semper debet significare ista mala. Confirmatur, nam aliquando visus est cometa qui ista mala non significavit; unde Seneca dicit quod sub Nerone, id est Nerone regnante, apparuit quidam cometa qui detraxit aliorum cometarum infamiam[160]. Unde post apparitionem unius comete non secutum fuit aliquid malum cuius quidem mali ille cometa fuisset signum.

⟨4⟩ Quarto: probatur quod cometa non significet siccitates. Nam, per Aristotelem in littera, cometa significat inundationes aquarum[161]; et si sic, non videtur significare siccitates: aliter enim significaret opposita.

⟨5⟩ Quinto: patet quod non significat ventos et motus terre; | nam, si significat ventos, non significabit motum terre, quia motus terre fit tranquillitate | existente, sicut postea videbitur[162]. Cum autem est ventus, aer non est tranquillus; ergo non potest simul significare motum terre et ventos. *Ba* 24va *Kl* 30rb *P* 250^{v}

⟨6⟩ Oppositum dixerunt plures philosophi[163] et oppositum est expertum. Illud autem quod omnes vel plures dixerunt, prestat fidem tamquam ab experientia dictum, sicut dicitur in *De somno et vigilia*, supra quo Commentator dicit: 'impossibile est ut aliquod dictum famosum sit falsum secundum totum'[164]. Fama enim quam omnes clamant non omnino deprehenditur[165].

---

5 ista] *om. P* 6–7 dicit quod] *om. P* 8 detraxit] destruxit *P* destraxit *Ba* traxit *Kl L* 16 sicut postea videbitur] *om. Ba Kl L U* 22 fama enim quam … non omnino deperditur] *om. Ba Kl L U*

---

159 Aristoteles, *Meteorologica*, I, 8, 345 b 33–36.

160 Seneca, *Questiones naturales*, VII, 17, 2, ed. P. Oltramare, Paris 1961, 2, 319: "Nec est quod putemus eundem uisum esse sub Claudio quem sub Augusto uidimus, nec hunc qui sub Nerone Caesare apparuit et cometis detraxit infamiam illi similem fuisse qui post excessum diui Iulii ludis Veneris Genetricis circa undecimam horam diei emersit".

161 Aristoteles, *Meteorologica*, I, 6, 343 b 1–4.

162 III.5, *Utrum tranquillitas sit signum terre motus*. Cf. Aristotelem, *Meteorologica*, II, 8, 366 a 5–6.

163 Cf. Aristotelem, *Meteorologica*, I, 7, 344 b 28–345 a 10; Senecam, *Questiones naturales*, VII, 28; Albertum Magnum, *Meteora*, ed. Colon. IV/1, lib. 1, tract. 3, cap. 11.

164 Aristoteles, *De insomniis et de divinatione per somnum*, 3, 1462 b 14–16, ed. Drossaart Lulofs, 33^{5-7}: 'nam quod omnes quidam vel plures existimant habere aliquam significationem sompnia prestat fidem tamquam ab experientia dictum'; Averroes, *In De somno et vigilia*, ed. Shields, 95^{32-33}: "impossibile enim est ut famosum sit falsum secundum totum"; cf. *Auct. Arist.*, 204, n. 101.

165 Cf. Aristotelem, *Ethica*, VII, 14, 1153 b 27–28; *Auct. Arist.*, 241, n. 126.

⟨7⟩ In ista questione primo ponenda est una distinctio significationis cometarum; secundo, prosequendo de una, ponende sunt conclusiones.

⟨8⟩ Quantum ad primum, sciendum est quod duplex est significatio comete: quedam generalis, que convenit quasi omnibus cometis indifferenter; et ista *L* 201va significatio est verior, et etiam de ista significatione considerare pertinet | ad presentem scientiam. Alia est significatio comete specialis, de qua considerare pertinet ad astrologos. Unde astrologi ponunt diversas species cometarum, quibus imposita sunt ab eis nomina distincta, et quilibet illorum cometarum habet unam significationem specialem a significatione aliorum, de quibus significationibus specialibus non in proposito.

⟨9⟩ Sed quantum ad significationes generales cometarum sit prima conclusio: omnis cometa est signativus alicuius mali. Patet, nam omnes philosophi hoc dixerunt. Et quando dicitur de cometa viso sub Nerone quod ille non significavit aliquid malum[166], respondetur quod immo, quia tunc fuit *U* 22ra ille pessimus homo Nero. Similiter | confirmatur auctoritate Plinii, dicen- *Kl* 30va tis quod cometa est sydus | terribile[167], et non capit illum sydus pro celesti sydere, sed vocat cometam 'sydus' quia apparet quod sit sydus. Similiter Ptolemeus in secundo sui *Quadripartiti*, capitulo secundo et tredecimo, declarat quod cometa est signus multorum malorum[168].

⟨10⟩ Secunda conclusio: cometa est signativus magnorum ventorum. Patet quia cometa, sicut prius dictum est, fit ex elevatione exalationis calide et sicce, et hoc siccitate terrestri[169]; sed talis exalatio est materia ventorum, sicut postea videbitur[170]; ergo, cum tunc sit sufficiens materia, fit multitudo ventorum et fiunt venti magni. Similiter, cum subtile et pingue illius exala-

---

7 astrologos] astronomos *Ba* ‖ astrologi] astronomi *Ba* 9 unam] *om. P* 13 viso sub nerone] neronis *Ba Kl* nerone regnante *L U* 15 plinii] cuiusdam dicti *Ba* 17 sit] *om. P* 18 tredecimo] tertio *Ba Kl L U* ‖ declarat] declaret *P* 23 materia] tunc *add. P*

---

166 Cf. Senecam, *Questiones naturales*, VII, 17, 2.

167 Plinius, *Naturalis historia*, II, 22, 92, ed. J. Beaujeu, Paris 1950, 40: "Terrificum magna ex parte sidus".

168 Ptolemeus, *Tetrabiblos*, translatio Guillelmi de Morbeka, eds. G. Vuillemin-Diem and C. Steel, *Ptolemy's Tetrabiblos in the Translation of William of Moerbeke. Claudii Ptolemaei 'Liber Iudicialium'*, Leuven 2015 (Ancient and Medieval Philosophy. Series 1, 19), 222–223: 3; II.14, 230: 10.

169 I.16, 12. Cf. Aristotelem, *Meteorologica*, I, 7, 344 b 18–26.

170 III.1, *Utrum ventus sit exalatio calida et sicca*. Cf. Aristotelem, *Meteorologica*, II, 4, 360 a 10–14.

tionis per inflammationem est consumptum, grossiori terrestri remanente et cometa deficiente de illo grosso, fiunt venti magni.

⟨11⟩ Tertia conclusio: cometa est signativus motus terre. Nam, cum talis exalatio a terra elevata calore Solis calefacta continue ascendit, aliquando propter frigus adveniens pori terre clauduntur, et exalatio et caliditas repellitur, que sunt causa motus terre, sicut postea videbitur[171].

⟨12⟩ Quarta conclusio: cometa est signativus inundationum aquarum. Patet, quia est signativus magnorum ventorum. Cum autem fiunt magni venti, aqua maris et fluviorum repellitur super terram et fiunt inundationes | *Ba* 24vb aquarum. Similiter cometa est signativus magni motus terre. Cum autem fit magnus motus terre, fiunt casus et eversiones terrarum et montium; et aliquando illi montes cadentes in fluvios faciunt effluere fluvios ad terram siccam, et sic fiunt inundationes aquarum. Et sic patet qualiter cometa per accidens est signum inundationis aquarum.

⟨13⟩ Quinta conclusio: cometa est signativus sterilitatis terre. Patet, quia *P* 251^{r} ad generationem comete | requiritur multa | exalatio, et in elevatione talis *Kl* 30vb *L* 201vb exalationis humiditates extrahuntur a terra unde deberent generari fructus terre; illa autem deficiente, terra redditur sterilis.

⟨14⟩ Sexta conclusio: cometa significat mortalitates et epidimias. Patet, quia alique partes illius exalationis sunt venenose et ipse extracte a terra ad aerem reddunt aerem venenosum, a quo aere inficiuntur animalia, et sic fiunt aliquando mortes subitanee.

⟨15⟩ Septima conclusio: cometa significat guerras et homicidia et huiusmodi. Patet, quia per talem exalationem siccam et calidam existentem in aere homines excitantur et fiunt quasi colerici. | Colerici autem pervi sunt *U* 22rb ad guerras et homicidia et huiusmodi. Etiam per talem infectionem hominum ab exalatione illa existente in aere homines quasi perdunt sensus et semper sunt pervi ad guerras et iracundias, quas consequuntur homicidia.

⟨16⟩ Octava conclusio: comete precipue significant mortem principum. Patet, quia significant guerras et bella. In bellis autem principes sunt primi, vel saltem deberent esse. Confirmatur, quia principes magis delicate sunt

---

1 grossiori] siccitate *Ba Kl L U* 2 cometa deficiente de illo grosso] tunc de illa siccitate terrestri grossa *Ba* tunc siccitate terrestri manente *Kl* cometa de siccitate terrestri grossiori manente *L U* 4 elevata] *om. Ba Kl L U* 11 eversiones] scissiones *L* 17 humiditates extrahuntur] pinguedo extrahitur *Ba Kl L U* 30–31 sunt primi, vel saltem deberent esse] perduntur *Ba Kl L U* 31–212.2 principes magis delicate … et confirmatur quia] *om. hom. Ba Kl* magis delicati aliis sunt interfici ideo citius interficiuntur et moriuntur propter aerem existentem venenosum *L*

---

171 III.4, *Utrum motus terre sit possibilis*. Cf. Aristotelem, *Meteorologica*, II, 8, 366 a 3–5.

nutriti, et igitur citius inficiuntur per aerem existentem venenosum. Et confirmatur, quia, sicut dicunt astrologi, ad principes est specialis influentia ab illa que est ad populares et cometa apparente illa influentia est vigorosior, et est illa que est causa mortalitatum principum.

⟨17⟩ Nona conclusio: cometa est signum mutationum regnorum. Probatur, quia est signum guerrarum et bellorum regni contra regnum. In talibus autem bellis aliquando unum regnum obtinet contra aliud; et sic gentes de uno regno incipiunt habitare aliud regnum, sicut accidit cum Saxones expulerunt Anglicos.

⟨18⟩ Notandum quod cometa non significat ista sic quod sit causa istorum *Kl* 31ra effectuum, | sed quia ista quodam ordine fiunt ex consimili causa ex qua fit cometa.

⟨19⟩ Ad rationes. Ad primam patet ex dictis in questione.

⟨20⟩ Ad secundam negatur quod galaxia sit de natura comete. Et cum dicitur quod hoc dicit Aristoteles, potest dici quod non; licet hoc ponitur in nova translatione[172], forte est ex vitio scriptoris, quia stabat in margine exemplari *Ba* 25ra et posuerunt hoc ad textum. | Unde in antiqua translatione Aristoteles non videtur hoc dicere[173]. Et adhuc, supposito quod galaxia esset de consimili *L* 202ra exalatione, tamen, quia illa ex|alatio non est tanta nec ita venenosa nec ita a profundo terre extrahitur sicut exalatio ex qua fit cometa, igitur galaxia non significat talia.

⟨21⟩ Ad tertiam: quamvis aliquando sequatur planetam benivolum, nichilominus est ex exalatione tali de qua dictum est[174], unde fiunt multa mala, sicut dictum est in questione. Et forte quod talis cometa qui insequitur planetam benivolum non est signum tantorum malorum sicut cometa qui insequitur planetam malignum, sicut Martem vel aliquid talem.

⟨22⟩ Ad quartam patet ex questione qualiter significat aquarum inundationes.

---

3 ad] *om. P* 6–9 in talibus autem … saxones expulerunt anglicos] *om. Ba Kl L U* 10 ista sic] ita *P* 14 negatur quod] *om. P* 15 nova] antiqua *Ba Kl om. P* 17 et posuerunt hoc ad textum] *om. P* 20 igitur] et *add. P Kl* 23 de qua] quali *P*

---

172 Aristoteles, *Meteorologica*, I, 8, 345 b 36–346 a 6, translatio Guillelmi de Morbeka (translatio noua), *AL* X 2.2, 26$^{465-470}$.

173 Aristoteles, *Meteorologica*, translatio Gerardi Cremonensis, I, 8, in P.L. Schoonheim (ed.), *Aristotle's Meteorology in the Arabico-Latin Tradition: A Critical Edition of the Texts, with Introduction and Indices*, Leiden/Boston/Cologne 2000 (Aristoteles Semitico-latinus, 12), 20–21.

174 I.16, 12.

⟨23⟩ Ad quintam respondetur quod significat motum terre et magnitudinem ventorum disiunctive, ita quod si exalatio tota extrahitur non manente aliqua, saltem magna et multa in concavitatibus terre, tunc fiunt multi et magni venti propter exalationem extractam a terra ad aerem, et tunc non fit motus terre. Quando autem non tota extrahitur, sed multa manet in terra, non tunc fiunt venti, sed motus terre.

I.19 Utrum galaxia sit de natura celi vel de natura elementari

Queritur consequenter utrum galaxia sit de natura celi vel de natura elementari. [*Kl* 31rb *U* 22va]

⟨1⟩ Et arguitur primo quod sit de natura elementari auctoritate Aristotelis dicentis eandem esse materiam galaxie et comete[175]; sed quia materia | [*P* 251^{v}] comete est elementaris, sicut prius dicebatur[176], sequitur etiam quod natura galaxie sit elementari, et per consequens galaxia non erit de natura celi, sed elementari. Quod autem eadem sit natura comete et galaxie Aristoteles probat a signo, nam pauci fiunt comete propter hoc quod exalatio ex qua deberent fieri comete convertitur in galaxiam[177].

⟨2⟩ Secundo: si una stella erratica potest tantam exalationem elevare quod fiat cometa, tunc etiam multe stelle possunt tantam exalationem elevare quod per eius inflammationem fiat galaxia. Sed ita est quod una stella erratica potest tantam exalationem elevare quod fit cometa; ergo etiam plures stelle propinque possunt tantam exalationem elevare quod fit galaxia per inflammationem talis exalationis. Et ideo, ut videtur, est quod semper galaxia apparet sub illa parte celi ubi sunt multe stelle propinque seminate, ac si huiusmodi stelle talem exalationem elevent per cuius inflammationem fit galaxia.

⟨3⟩ Tertio: si galaxia esset de natura celi et non elementari, hoc esset | [*Ba* 25rb *Kl* 31rb] per illum modum, sicut postea plenius declarabitur, quod una pars celi esset densior et alia pars minus densa[178]. Sed hoc est inconveniens, nam celum

2 extrahitur] a terra *add.* *Ba Kl L U* 16 tantam] latam *P* 26 postea plenius declarabitur] prius declarabatur *Ba Kl L*

175 Aristoteles, *Meteorologica*, I, 8, 345 b 36–346 b 10.
176 I.16, 11–12.
177 Aristoteles, *Meteorologica*, I, 8, 346 b 7–10.
178 I.19, 17.

L 202rb videtur esse uniforme | in raritate et densitate, cum in celo non debeat esse aliqua difformitas propter eius nobilitatem.

⟨4⟩ Quarto ad idem: si galaxia esset de natura celi, tunc de ea non esset determinandum hic in isto libro, cuius oppositum facit Aristoteles[179]. Consequentia tenet, quia in isto libro solum determinandum est de impressionibus aeris et impressionibus que sunt de natura elementari.

Kl 31va ⟨5⟩ Oppositum arguitur: nulla impressio elementaris semper durat; | sed galaxia semper durat; ergo galaxia non est impressio elementaris. Minor nota est ad sensum. Antecedens patet, vel maior, quia propter habere materiam, quod est causa corruptionis, nulla impressio elementaris semper durat.

⟨6⟩ Secundo: galaxia movetur ad motum celi; ergo galaxia videtur esse de natura celi. Consequentia tenet, quia motus arguit naturam sicut formam, ut dicitur secundo *Physicorum*[180]. Antecedens patet de se, quia circumducitur cum celo.

⟨7⟩ Tertio: si galaxia esset nature elementaris, tunc esset consimilis nature cum cometa, sicut dicit Aristoteles[181]; sed hoc est falsum, cum galaxia sit semper, cometa autem raro.

U 22vb ⟨8⟩ Et ex alio, quia cometa significat mortem principum et consimilia mala sicut predictum est[182], galaxia autem non.

⟨9⟩ In ista questione primo ponende sunt opiniones de galaxia quid sit et improbationes earum; secundo ponende sunt conclusiones; tertio videndum est secundum veritatem de galaxia quid sit et quomodo causetur.

⟨10⟩ Quantum ad primum, sciendum est quod erat una opinio Antiquorum qui dixerunt quod quondam Sol errando in motu eius exivit viam suam consuetam, et sic errando in suo motu transivit per illam partem celi et combussit eam, propter quod adhuc illa pars celi apparet alba[183]. Breviter ista opinio non valet, quia celum non est receptivum impressionum corruptivarum sicut est combustio; igitur, posito adhuc quod Sol per illam viam

---

13–14 ut dicitur] cum dicitur natura *P* 20 sicut predictum est] *om. Ba Kl L U*

---

179 Aristoteles, *Meteorologica*, I, 8.

180 Aristoteles, *Physica*, 192 b 20–23; cf. *Auct. Arist.*, 144, n. 50.

181 Aristoteles, *Meteorologica*, I, 8, 345 b 36–346 b 10.

182 I.18, 9–17.

183 Cf. Aristotelem, *Meteorologica*, I, 8, 345 a 11–18.

transisset quam dicimus esse Viam lacteam, non tamen propter hoc illam partem celi combussisset, cum celum non possit comburi nec calefieri, et etiam cum Sol non sit formaliter calidus. Et ex alio, quia tunc magis illa pars celi per quam continue transit Sol, sicut zodiacus seu circulus animalium, deberet apparere magis alba vel quasi iam tota combusta, quod est falsum[184].

⟨11⟩ Secunda opinio de galaxia erat quod aliquando de nocte, serenitate existente, Terra interponitur | Soli et illis | astris in multitudine seminatis; igitur ibi apparet Via lactea, | ita quod tunc Sol non radiat super illas stellas et tunc ille stelle sic lucent, et sic illa pars celi apparet sic albescere. Sed alie partes celi que respiciuntur a Sole | et non sunt in umbra Terre non sic lucent nec sic apparent nobis albescere propter hoc quod lumen Solis, quod est maius, lumen eorum offuscat. Sed quia lumen illius partis celi ubi apparet Via lactea non est offuscatum per lumen Solis, cum Sol illas partes non respiciat propter interpositionem Terre, ille partes celi apparent nobis albescere[185]. Sed breviter ista opinio non valet, quia ille partes celi et ille stelle ubi apparet galaxia sunt ita remote quod ad ipsas non attingit umbra Terre, et per consequens semper respiciuntur a Sole. Nec obstat interpositio Terre; ergo non magis deberet ibi apparere galaxia quam alibi, nisi aliter dicatur quam dicit ista opinio. Secundo, dato quod aliquando Sol non respiceret illam partem celi propter interpositionem Terre, tamen propter motum Solis aliquando illam partem celi respiceret; et tunc secundum predictam opinionem non debet apparere albescere[186].

Ba 25va *Kl* 31vb *L* 202va *P* 252^r^

⟨12⟩ Tertia opinio erat ista, quod galaxia causaretur ex reflexione luminis a Sole ab humiditate vel exalatione elevata. Unde ymaginabantur | quod sub illa parte celi ubi apparent tam multe stelle elevantur multe exalationes, et quod de nocte Sol existens sub nobis proicit lumen super illas exalationes, quod quidem lumen ibi reflectitur versus visum nostrum. Et quia sic illa exalatio | propter reflexionem luminis super ipsam apparet albescere, et quia talis exalatio est sub illis stellis, nos dicimus ibi esse galaxiam et Viam lacteam[187]. Breviter ista opinio non valet, nam ad eam sequitur quod non

U 23ra *Kl* 32ra

---

2 nec calefieri] *om. Ba Kl L U* 6 de nocte] *om. P* 8 ibi] ubi *P* 16 remote] alte *Ba Kl L U* 18 ibi] *om. P* 20 propter interpositionem terre] *om. Ba Kl L U* 22 albescere] sed hoc est falsum quia continue apparet ibi galaxia ubi stelle sunt seminate in multitudine *add. Ba Kl L U* 24 humiditate] humore *Ba Kl L U*

---

184 Cf. ibid., I, 8, 345 a 19–25.
185 Cf. ibid., I, 8, 345 a 26–31.
186 Cf. ibid., I, 8, 345 a 32–b 9.
187 Ibid., I, 8, 345 b 10–12.

semper galaxia apparet nobis sub eadem parte celi. Sed consequens est fal- *Ba* 25vb sum, nam continue apparet nobis | in illa parte celi ubi sunt tam multe stelle. Consequentia probatur, nam, speculo et oculo quiescente, re visa mutante locum, non semper iudicamus rem visam in eadem parte speculi; nec iudi- *L* 202vb cium manet uniforme, sed mutatur. Cum ergo | Sol continue moveatur, sequitur quod nobis quiescentibus et similiter illa exalatione quiescente que ab illis dicitur tamquam speculum reflectens lumen solis ad nos, illa Via lactea, seu illud lumen reflexum, non debeat nobis apparere continue in eodem loco nec sub eadem parte celi, cuius oppositum experimur[188].

⟨13⟩ Quarta opinio erat quorundam dicentium, sicut recitat Albertus, galaxiam esse concursum duorum luminum, scilicet luminis ignis in spera sua et luminis stellarum sic sparsarum in multitudine. Unde ymaginabantur quod cum ille stelle luceant versus speram ignis et ignis versus celum, quod ibi appareat galaxia et talis albedo ubi concurrunt illa duo lumina[189]. Contra istam opinionem argumentat Albertus quia, si ignis in spera sua concederetur lucere, tamen valde modicum luceret; aliter enim prohiberet visionem stellarum. Et ergo, quia valde modicum luceret, obfuscaretur lumen suum a lumine illarum stellarum, et per consequens una cum lumine talium stellarum non causaret talem albedinem qualis apparet ibi ubi apparet Via lactea. Secundo, nam, secundum Avicennam, secundum Algazelem, secundum Alexandrum et plures alios[190], ignis in spera sua non lucet; et quia ista opinio hoc supponit, ipsa est falsa[191]. Quod autem ignis in spera sua non luceat, *Kl* 32rb huius ratio est sua | nimia raritas; unde Avicenna: 'nichil lucet nisi sit aliqualiter densum'. Unde aer propter sui raritatem non lucet; sed ignis est multum *P* 252^{v} *U* 23rb rarior aere; igitur nec ignis | propter sui raritatem lucet. Ad idem: | ille partes celi que non sunt stelle habent plus de lumine quam ignis, et tamen

4 speculi] celi *Kl om. Ba* 7 lumen solis] istam exalationem *Kl L U om. Ba* 10 sicut recitat albertus] *om. Ba Kl L U* 13 speram ignis] ignem *Ba Kl L U* 14 galaxia] via lactea *Ba Kl L U* 16 prohiberet] prohibet *P* 20 algazelem] galienum *Ba Kl L U* 21 alios] philosophos *Ba Kl L U* 25 igitur nec ignis] *iter. P* 25–26 ille partes celi que non sunt stelle] alique partes celi sunt que *Ba Kl*

188 Cf. ibid., I, 8, 345 b 28.

189 Albertus Magnus, *Meteora*, lib. 1, tract. 2, cap. 6, ed. Colon. VI/1, 22^{54-63}.

190 Alexander Aphrodisiensis, *Expositio libri Meteorologicorum Aristotelis*, I, 4, ed. A.J. Smet, Leuven 1968 (Corpus Latinum Commentariorum in Aristotelem Græcorum, 4), 34^{9}sqq.; Algazel, *Metaphysica*, pars 2, trac. 3, ed. J.T. Muckle, Toronto 1933 (St. Michael's Mediaeval studies), 156^{69-81}. Cf. Albertum Magnum, *De causis proprietatum elementorum*, lib. 1, tract. 2, cap. 11, ed. Colon. V/2, 81^{17-22}.

191 Albertus Magnus, *Meteora*, lib. 1, tract. 2, cap. 6, ed. Colon. VI/1, 22^{63}–23^{6}.

propter earum raritatem non lucent; igitur videtur quod ignis in sua spera non luceat[192].

⟨14⟩ Quinta opinio, quam pretendit nova translatio, est ista: quod galaxia sit inflammata exalatio. Unde ymaginatur quod sub illa parte celi in qua ille stelle sunt in multitudine seminate que nobis apparent ubi apparet nobis Via lactea, continue virtute illarum stellarum elevatur exalatio, que in suprema | *L* 203ra aeris regione inflammatur; et ipsa | *Ba* 26ra inflammata, apparet nobis Via lactea in celo[193]. Et si dicatur contra istam opinionem: sequeretur quod galaxia potius deberet apparere sub circulo animalium, hoc est sub zodiaco, propter hoc quod ibi sunt plures stelle potentes elevare et extrahere exalationem que posset inflammari in suprema regione aeris et facere huiusmodi fantasiam et apparitionem, tenentes istam opinionem respondent quod quamvis sub circulo animalium fiant multe elevationes exalationum, tamen nimio calore Solis ibi vigente huiusmodi exalationes dissolvuntur, sic quod non faciunt ibi huiusmodi apparitionem, quia statim consumuntur, et propter hoc etiam ibi comete raro apparent[194].

⟨15⟩ Quantum ad secundum, sit primo ista suppositio: quod eadem impressio aeris a diversis hominibus in diversis regionibus existentibus non apparet sub eadem parte celi, sed sub diversis. Et ratio huius est, nam, sit unus eorum Sortes et alius Plato, | *Kl* 32va et videant eandem impressionem aeris; tunc radii visuales istorum duorum intersecantur se in illo visibili et proceduntur ad diversas partes celi. Et tunc Sortes iudicat illam impressionem esse sub illa parte celi ad quam terminatur suus radius visualis, et Plato iudicabit illud idem visibile esse sub illa parte celi ad quam terminatur suus radius visualis. Et quia huiusmodi radii visuales ad diversas partes celi terminantur, idem visibile ab illis duobus iudicatur esse sub diversis celi partibus, sicut faciliter patet in figura. Propter hoc etiam existentes in diversis Terre partibus iudicant Lunam esse sub diversis partibus ultimi celi et iudicant Lunam non esse sub illa parte ultimi celi sub qua in rei veritate est preterquam illi quibus Luna est in Zenith capitis.

---

6 stellarum] continue *add. P* 10–11 que posset inflammari] a terra et inflammare *Ba Kl L U* 15 hoc] *om. P* 26–27 sicut faciliter patet in figura] *om. Ba Kl L U* 30 capitis] *om. Ba Kl L U*

---

192 Cf. Albertum Magnum, *Meteora*, lib. 1, tract. 2, cap. 6, ed. Colon. VI/1, 23^{6-34}.

193 Aristoteles, *Meteorologica*, I, 8, 345 b 33–346 b 10.

194 Cf. ibid., I, 8, 345 b 11–16.

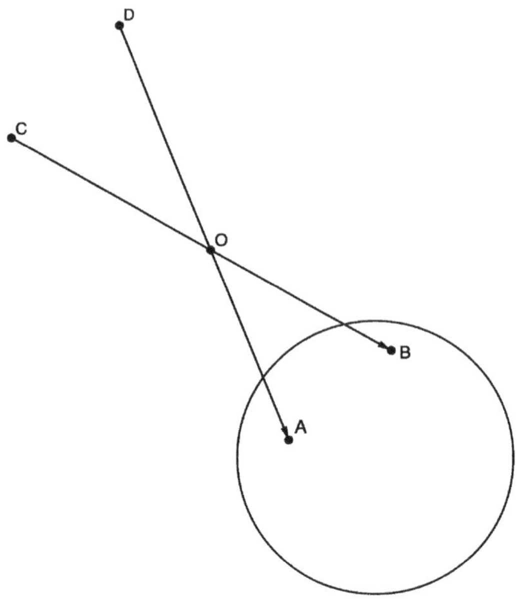

⟨16⟩ Tunc sit prima conclusio: galaxia non est in aere. Probatur, quia tunc a diversis hominibus in diversis regionibus existentibus non appareret sub eadem parte celi stellati, sed sub diversis partibus eius. Sed hoc est falsum. Consequentia tenet per suppositionem; sed falsitas consequentis patet, quia in quacumque regione homo est, semper apparet sibi galaxia sub eadem parte celi stellati. | *L* 203rb Semper enim apparet ibi ubi sunt | *U* 23va tam multe stelle propinque sine ordine seminate.

⟨17⟩ Secunda conclusio: galaxia non est impressio elementaris, nam, si sic, tunc esset in aere, quamvis appareat in celo. Sed hoc est contra primam conclusionem. Et confirmatur, nam, si esset impressio causata ex inflammatione exalationis sub | *Ba* 26rb illis stellis in tam magna multitudine seminatis, sequitur quod non semper appareret ita uniformis sicut apparet in sua claritate, sed aliquando eius claritas intenderetur et aliquando remitteretur, cuius oppositum videmus. Consequentia tenet ex eo quod non semper illa exalatio est equaliter subtilis vel equaliter grossa vel rara, et quia etiam virtus elevativa talis exalationis non semper est uniformis, immo aliquando maior, aliquando minor secundum adiutorium | *P* 253r *Kl* 32vb aliarum stellarum. Verum est quod aliqui, volentes sustinere quod galaxia fieret ex elevatione exalationis sub illis stellis, possent dicere quod illa exalatio sic elevata non apparet

15 grossa] vel equaliter spissa *add. Ba Kl L U*   18 fieret] non esset in celo sed in aere *Ba Kl L U*

esse Via lactea propter hoc quod ipsa sit inflammata, sed propter hoc quod reflectit lumen illarum stellarum e contrario versus celum, et propter hoc celi partes interposite apparent albescere. Et sic potest impediri argumentum factum pro prima conclusione, non tamen argumentum factum pro secunda.

⟨18⟩ Tertia conclusio: quod galaxia est de natura celi. Probatur, quia vel est de natura celi vel elementari: non elementari, per secundam conclusionem; igitur relinquitur quod de natura celi. Secundo: vel exalatio seu fantasia seu Via lactea est in aere, quamvis appareat in celo, vel est in celo, sicut apparet in celo. Non potest dici quod sit in aere, quia tunc a diversis hominibus in diversis regionibus existentibus appareret sub diversis partibus celi, quod est falsum, quia continue et ubicumque homo sit, apparet sibi Via lactea ibi ubi apparent sibi ille stelle in tam magna multitudine seminate. Si autem dicatur quod est in celo, vel igitur est in celo propter reflexionem luminis illarum stellarum supra exalationem elevatam, vel non propter hoc. Non potest dici primum, quia secundum quod illa exalatio aliquando plus vel minus illud lumen reflecteret secundum quod esset plus vel minus densa vel rara, secundum hoc Via lactea appareret magis vel minus clara et alba, | cuius *L* 203va oppositum apparet, quia semper apparet uniformis claritatis et albedinis. Si enim dicatur quod non propter hoc est in celo quod lumen illarum stellarum | supra illa exalatione reflectitur versus celum, sequitur quod vere est *Ba* 26va *U* 23vb de natura celi, et hec est conclusio.

⟨19⟩ Dicendum est ergo quod galaxia est una pars celi; unde ymaginandum est quod alique partes celi sunt densiores et alique rariores, alique medio modo se habentes. | Que sunt densiores, lucent et sunt stelle[195]; que *Kl* 33ra vero rariores, non lucent, et sunt ille partes que sunt prope stellas et inter illas. Alique sunt medio modo se habentes, ita quod non sunt ita dense sicut stelle nec sunt ita rare sicut alie partes celi que non lucent. Et sic se habent ille partes celi que sunt interposite stellis ibi existentibus ubi apparet Via lactea. Et propter hoc quia medio modo se habent in densitate et raritate, non omnino lucent sicut stelle nec omnino carent luce sicut alie partes celi, sed medio modo se habent, et propter hoc apparent albe. Et ergo illam partem

1 via lactea] galaxia *add. Ba Kl L U* 2 reflectit lumen] reflectitur lumen *Ba Kl* reflectitur a lumine *L U* 8 vel] *om. P* ‖ exalatio seu] *om. Ba Kl* 14 celo] sicut apparet *add. Ba* 15 elevatam] versus celum *add. Ba* 31 luce] lumine *Ba Kl L U*

195 Cf. Averroem, *Commentarium magnum in De celo*, II, comm. 42, eds. Carmody, Arnzen, 2, 350^{61}–351^{86}; *Auct. Arist.*, 165, n. 72.

celi ubi sunt ille stelle in magna multitudine dicimus esse Via lacteam seu Viam albam seu galaxiam.

⟨20⟩ Ad rationes. Ad auctoritatem Aristotelis respondetur quod illa non fuit opinio Aristotelis, sed erat interposita eius textui ex vitio scriptoris vel translatoris, quia hodierno tempore, facientes scribere aliquos textus, videntes glosam quod eis placet in margine, dicunt suis scriptoribus quod illam glosam apponant textui, et ita potuit accidere textui aristotelis.

⟨21⟩ Ad secundam: 'nam, si una stella', etc., concedo quod ille multe stelle bene possent elevare talem exalationem. Sed quod continue talis exalatio haberet se uniformiter in magnitudine et parvitate, in densitate et raritate, non concedo esse verum. Hoc tamen apparet propter hoc quod Via lactea semper apparet consimilis claritatis.

⟨22⟩ Ad tertiam concedo quod una pars celi est densior et alia rarior, quod etiam celum non est uniforme in raritate et densitate, sed difforme, et illa *Kl* 33rb difformitas eius in raritate et densitate facit ipsum magis | pulchrum.

⟨23⟩ Ad quartam: 'si esset de natura celi, de ea non esset determinandum', concedo, sed tamen Aristoteles determinavit de ea hic propter hoc quod plures Antiquorum crediderunt galaxiam esse de natura elementari[196].

Et sic patet ad questionem.

1 multitudine] seminate *add. Ba Kl L U* 3 aristotelis] dicentis eandem esse materiam galaxie et comete *add. Ba Kl L U* 5–7 quia hodierno tempore … accidere textui aristotelis] *om. Ba Kl L U* 6 quod] *om. P* 10 in magnitudine et parvitate] *om. Ba Kl L U* 12 claritatis] et per consequens non est dicendum quod ista exalatio in superiori regione aeris inflammata sit via lactea *add. Ba Kl L U* 14 in raritate et densitate] in suis partibus *Ba Kl L U* 15 in raritate et densitate] *om. Ba Kl L U* 16 si] sed *P*

196 Aristoteles, *Meteorologica*, I, 8, 345 a 14–30.

Liber II

II.1 Utrum locus generationis pluvie sit media regio aeris

Ba 26[vb] *L* 203[vb] *P* 253[v] *U* 24[ra]

Circa secundum librum *Metheororum* queritur utrum locus generationis pluvie sit media regio aeris.

⟨1⟩ Et arguitur primo quod non: nullum elementum debet generari extra locum sibi naturalem; sed media regio aeris est locus innaturalis aque et pluvie; ergo nec aqua nec pluvia debet ibi generari.

⟨2⟩ Secundo ad idem: media regio aeris est locus generationis impressionum ignitarum; ergo non est locus generationis impressionum aquearum, et per consequens non est locus generationis pluvie. Consequentia tenet ex eo quod idem non videtur esse locus generationis oppositorum. Antecedens patet, quia in media regione aeris generantur corruscationes et fulgura, que sunt impressiones ignite.

⟨3⟩ Tertio ad idem: si questio esset vera, hoc esset per istum modum quod vapor elevatus usque ad mediam regionem per frigiditatem medie regionis condensaretur in nubem et inde converteretur in aquam; sed hoc non est, quia consimili ratione | videtur quod per condensationem exalationis debe- *Kl* 33[va]
ret fieri nubes et conversio illius in pulveres, et ita deberet pluere pulveres sicut aquas, quod non videmus.

⟨4⟩ Quarto ad idem: aliquando pluvie et nubes generantur in inferiori regione aeris; ergo non semper media regio aeris est locus generationis pluvie. Consequentia tenet. Antecedens probatur; nam visum est quod aliquis existens in monte vidit infra se in vallibus nubes[1], et hoc est signum quod

1 metheororum] secundum distinctionem in principio positam ab alberto *add. Kl* 5 nec pluvia] *om. P* 8 tenet] videtur tenere *P* 9 videtur esse] est *P* 14 converteretur in aquam] sequeretur pluvia *Ba Kl L U* 16 in] *om. P* 21 in vallibus] *om. Ba Kl L U*

1 Cf. Iohannem Buridanum, *Questiones in Meteorologica*, I.18, ed. Bages, 2: 240–241, nr. 8–9: "Et ego hiis dictis credidi et credo quia bis expertus sum hec manifestissime, primo semel eundo ad Avinionem, ultra montem Pesatium in Ricordania, supra alteram villam sittam in valle profundissima que vocatur Leval: nos enim plures existentes in monte alto supra eramus in aere sereno et claro et etiam ultra illam villam videbamus montes illustratos sole, et tamen in fundo, supra illam villam, videbamus horribiles nubes et audiebamus magna tonitrua; et cum descendissemus in vallem, invenimus copiosam multitudinem grandinum nondum liquefactarum et gentes adhuc ex tempestate stupentes et aquas per rivos insolitos fluentes ita quod cum difficultate transiremus eas".

infra mediam regionem possunt generari nubes et pluvie. Montes enim illi non sunt ita alti quod mediam regionem aeris vel ultra attingerent.

⟨5⟩ In oppositum est Aristoteles in primo tractatu et primo capitulo huius secundi[2].

⟨6⟩ In ista questione primo exponende sunt alique conclusiones pro quesito; secundo videndum est de modo generationis pluviarum; tertio videndum est de diversitate pluviarum quoad quantitatem guttarum; quarto, | (*L* 204ra) quoad figuram guttarum ipsius pluvie; quinto videndum est de diversitate pluviarum quoad quantitatem; sexto videndum est de diversitate pluviarum quoad modum cadendi guttarum; septimo videndum est de pluviis quoad loca et regiones ubi fiunt; octavo videndum est de duratione pluviarum.

⟨7⟩ Quantum ad primum, sit prima conclusio: media regio aeris est locus generationis pluviarum. Probatur: in illo loco possunt generari pluvie in quo possunt esse et concurrere cause sufficientes ad generationem | (*U* 24rb) pluvie; sed hoc potest esse in media regione aeris; igitur etc. Maior est nota. | (*Ba* 27ra) Minor declaratur, quia usque ad mediam regionem aeris possunt calore Solis vel virtute aliarum stellarum elevari vapores qui sunt materia ipsius pluvie, et similiter in media regione aeris est sufficiens frigiditas potens condensare vaporem elevatum in nubem et convertere in aquam. Ecce causas sufficientes in media regione aeris | (*Kl* 33vb) ad generationem pluvie. Etiam, si non esset vapor elevatus, adhuc propter frigiditatem medie regionis aer ibi existens posset condensari et converti in nubem et fieri pluvia. Ubicumque enim, si sunt cause sufficientes, ex uno elemento potest reliquum generari.

⟨8⟩ Secunda conclusio: aliquando infra mediam regionem potest generari pluvia. Patet, quia aliquando infra mediam regionem vapor elevatus, qui de natura sua est frigidus, quamvis accidentaliter sit calidus calore Solis, potest per antiparistasim reduci ad frigiditatem sibi naturalem et condensari in nubem et converti in aquam et sequitur pluvia. Cum enim vapor, qui de natura sua est frigidus, elevatur infra mediam regionem aeris, ubi viget

---

1–2 montes enim illi … vel ultra attingerent] quia sunt tam alti quod mediam regionem aeris vel ultra attingunt *Kl* quia aliqui montes sunt tam alti quod mediam regionem aeris attingunt vel ultra *L* 4 secundi] secundum distinctionem factam in principio huius primi libri *add. Kl* 17 vapores] sufficientes *add. Kl L U* vel *add. P* 19 elevatum in nubem et convertere in aquam] et convertere in nubem et sic sequitur pluvia *Ba Kl L U* 21 aer] *om. P* 28 et converti in aquam] *om. Ba Kl*

---

2 Aristoteles, *Meteorologica*, I, 9, 346 b 15–31.

intensa caliditas, calor sibi accidentaliter inherens exalat propter caliditatem circumstantem, cum simile applaudat suo simili, et derelinquit vaporem elevatum nature proprie, et tunc statim, cum ille vapor sit de natura frigidus, propter caliditatem circumstantem per antiparistasim | frigiditas *P* 254^{r} in eo intenditur, et tunc ille vapor condensatur et in nubem convertitur et sequitur pluvia. Sic ergo non solum in media regione, verum etiam infra potest generari pluvia. Et ex hoc patet solutio ad quartum argumentum.

⟨9⟩ Quantum ad secundum, patet ex iam dictis quis sit modus generationis pluvie. Sciendum tamen quod tripliciter possunt generari nubes et pluvie: uno modo per condensationem vaporis elevati in nubem; secundo modo per condensationem aeris in nubem a frigiditate | medie regionis; tertio modo *L* 204rb per repulsionem vaporis a media regione, et ita pluvia fit modo consimili sicut fit descensus guttarum ab aliquo coopertorio | olle fumantis. Videtur *Kl* 34ra enim, cum aliqua olla est plena vel semiplena aqua calida fumante, et si super illam ollam ponitur aliquod coopertorium, tunc ex repulsu illius fumi ascendentis ad coopertorium ibi fiunt quedam guttule infra illud coopertorium, quod patet experimentaliter si ipsum elevetur de olla. Istud etiam patet experimentaliter in sublimationibus aquarum. Et ymaginandum est | *U* 24va circa istum modum quod cum vapores eleventur usque ad mediam regionem aeris, aliquando repelluntur, et | ex tali repulsu appropinquantur et *Ba* 27rb condensantur et fit nubes et sequitur pluvia.

⟨10⟩ Ulterius notandum est quod propter consimilem repulsionem vaporum in animalibus in hieme est plus de flemmate et in estate est plus de colera. In hieme enim, cum pori sunt clausi, vapores ascendentes repelluntur et generatur flemma. Sed in estate, cum pori sunt aperti, vapores exalant et dimittunt siccum terrestre et generatur colera.

⟨11⟩ Sciendum est etiam quod aliquando ex vaporibus elevatis fiunt nubes steriles quas non sequitur pluvia per dissolutionem earum a vento vel aliunde; aliquando autem sequitur pluvia, et iste nubes possunt dici fertiles.

⟨12⟩ Quantum ad tertium, sciendum quod pluvie differunt secundum diversitatem quantitatis guttarum. Aliquando enim fiunt pluvie secundum guttas magne quantitatis; aliquando vero secundum guttas minoris quantitatis. Et

---

1 accidentaliter] incidentaliter *P* 4 antiparistasim] antifrasim *P om. Ba* 6 infra] hoc est subtus *add. Ba Kl L U* 9 tripliciter] dupliciter *Ba* 10 per condensationem vaporis elevati in nubem] *om. hom. Ba Kl* 13 videtur] *om. P* 25 exalant] ascendunt *Ba Kl L U* 29 et iste nubes possunt dici fertiles] et tunc magis inspissantur *Ba Kl* et iste nubes possunt dici utiles vel fructifere *L*

quod aliquando cadunt gutte magne, huius potest multiplex esse causa. Uno modo hoc potest esse propter vehementem condensationem et coagulationem ipsius vaporis in media regione propter frigiditatem medie regionis vel infra per antiparistasim. Alia causa est | [*Kl* 34rb] quia aliquando vapor est viscosus, et tunc conglobatur in descensu per maiorem quantitatem ex eo quod propter viscositatem partes se magis tenent ad invicem. Alia causa potest esse quia aliquando nubes est multum profunda, et tunc aliquando gutte superiores in velocitando motum suum continue attingunt guttas inferiores et coniunguntur eis, et inde fit gutta magna.

⟨13⟩ Quantum ad quartum, videlicet de figura guttarum ipsius pluvie, dicitur quod semper sunt rotunde figure, sicut videmus. Quamvis enim in principio bene essent alterius figure, tamen in descendendo continue volvuntur et ex tali volutione rotundantur, sicut dicit Albertus[3]. Et potest esse alia causa, nam res melius conservatur in figura sperica et rotunda quam in alia; et ergo aqua guttaliter | [*L* 204va] descendens tendit ad figuram rotundam propter meliorem sui conservationem. Tertia causa: quia ad eandem figuram tendit pars alicuius homogenei ad quam tendit suum totum; sed sic est quod aqua tota in suo loco naturali tendit ad spericitatem, | [*Ba* 27va] sicut probat Aristoteles in secundo *Celi*[4], et ergo etiam quelibet gutta. Et ista potest esse una ratio quare gutte sunt figure rotunde. | [*U* 24vb] Quarto dico: nisi aliud obstaret, deberent esse figure concave, et quanto plus descenderent, tanto magis deberent esse gibbose, quia deberent tendere ad unam figuram qua apparet portio orbis aque circumdantis terram | [*P* 254^{v}] in tali loco in quo essent ille gutte. Unde nubes est talis figure, et quanto magis est inferius, tanto est maioris gibbositatis, et quanto magis superius, tanto minoris, quia quanto magis | [*Kl* 34va] est inferius, tanto est portio orbis minoris, et quanto magis superius, tanto est portio orbis maioris; sed eadem linea intersecat ab orbe minori portionem magis gibbosam quam ab orbe maiori. Et inde est quod idem vas in basso plus tenet de aqua quam in alto, quia in basso aqua existens in ipso habet maiorem gibbum superius, in alto autem minorem, quia in basso gibbus aque superius in vase est portio minoris orbis, in alto autem maioris.

---

1 gutte] pluvie *P*   2–3 coagulationem] congelationem *Ba Kl*   3–4 vel infra per antiparistasim] *om. Ba Kl*   5 maiorem quantitatem] guttas maiores *Ba Kl L U*   13 volutione] et rotatione *add. Ba*   14 sperica et] *om. Ba Kl L*   22 qua apparet portio orbis] portionis *Ba Kl L U*   25 minoris] orbicularitatis *add. Ba* gibbositatis *add. Kl L U*

---

3 Albertus Magnus, *Meteora*, lib. 2, tract. 1, cap. 22, ed. Colon. VI/1, 59^{13-20}.

4 Aristoteles, *De celo*, II, 4, 287 b 4–14.

⟨14⟩ Quantum ad quintum, de qualitate pluvie, sciendum quod quedam est pluvia que est dulcis et pinguis et magis aerea, et ista secundum Albertum multum proficit plantis[5]. Alia autem est pluvia que propter mixtionem alicuius terrestris et propter adustionem illius est magis pontica. Et ista quandoque exsiccat terre nascentia et multum nocet eis, et recte tantum aliquando sicut urina vel acetum. Et etiam aliquando pluvia est venenosa propter putrefactionem, in tantum quod aliquando ex ea generantur parve rane, et hoc aliquando in nube et aliquando prope terram. Etiam aliquando pluvia habet restringere fluxum ventris, et ratio huius est quia aliquando sibi multum est permixtum de sicco terrestri combusto, cuius est exsiccare et restringere. Item pluvie differunt aliquando colore. Communiter enim est talis coloris qualem communiter videmus. Aliquando etiam pluvia visa est coloris rubei, sicut | sanguinis, et causa eius erat magna adustio terrestris sibi *Kl* 34vb admixta. Et forte ex simili causa fit vinum rubeum. Aliquando etiam visa est pluvia alba ad modum lactis, et | hoc erat propter grossitiem materie. *L* 204vb Et aliquando cum pluviis cadunt quedam monstrua de quibus difficile esset reddere causam. Et dicit Albertus quod | cum pluvia cecidit semel unus vitu- *Ba* 27vb lus, et causa illius fuit virtus celi nobis ignota[6].

⟨15⟩ Quantum ad sextum, videlicet de modo cadendi, sciendum est quod quandoque cadunt recte, cum non propelluntur a vento vel ab aliquo simili. Aliquando cadunt oblique, et hoc est ideo quia sic ducuntur vel propelluntur a vento, vel quia sic sunt propulse a frigiditate medie regionis. Aliquando etiam propter spissitudinem nubis congregantur gutte in tanta multitudine et quantitate quod cadit aqua ac si funderetur de magno vase.

⟨16⟩ Quantum ad septimum, sciendum est quod quedam sunt loca in quibus fiunt pluvie valde raro, et hoc est propter defectum materie, quia illa loca sunt valde sicca, sic quod nulli sunt ibi vapores elevandi ex quibus fiat pluvia, vel ibi est tantus calor quod omnes vapores elevati consumuntur; | *U* 25ra unde dicunt auctores quod in Egipto raro fiunt pluvie. Alia autem sunt loca

---

13–14 terrestris sibi admixta] permixta cum sicco terrestri *Kl L U om. B*   15 alba ad modum lactis] coloris lactei *Ba Kl* coloris albi vel lactei *L*   17 cum] de *P*   28 elevati consumuntur] quas elevat consumit *P*

---

5 Albertus Magnus, *Meteora*, lib. 2, tract. 1, cap. 20, ed. Colon. VI/1, $58^{23–26}$.

6 Ibid., lib. 3, tract. 3, cap. 20, ed. Colon. VI/1, 171^{80}–172^{3}: "Corpora autem animalium perfectorum raro formantur in nube, licet hoc semel dicat Avicenna contingisse quod corpus vituli cecidit, et hoc ipse maxime attribuit virtuti stellarum in tempore illo formam vituli imprimentium".

in quibus sepe fiunt pluvie, sicut sunt illa loca ubi est sufficientia vaporis et sufficientia caloris potentis illos elevare et eos segregare a corporibus humidis.

Kl 35ra ⟨17⟩ Quantum ad ultimum, | videlicet de duratione pluvie, sciendum est quod cum aliqui vapores sunt elevati et conversi in nubem et non continue novus et novus vapor advenit, tunc, illa nube consumpta, deficiunt pluvie; sed si continue novus vapor eveniret, continue pluvia duraret.

⟨18⟩ Ad rationes. Ad primam dico quod quamvis media regio aeris non sit locus naturalis aque, nichilominus ibi potest bene generari aqua. Falsum enim est quod solum in loco naturali alicuius elementi istud elementum possit generari, immo in quolibet loco, si cause productive talis elementi fue- *P* 255r rint approximate, | tale elementum potest generari.

⟨19⟩ Ad secundam: 'ibi generantur impressiones ignite', dico quod est alio modo, quia iste generantur ibi per extrusionem; aquee autem per condensationem.

⟨20⟩ Ad tertiam concedo quod vapor ibi convertitur in nubem et in *L* 205ra aquam. Et quando | dicebatur: 'igitur exalatio deberet ibi converti in terram', negatur, quia exalatio apta est ad generationem aliarum impressionum, scilicet ignitarum. Et ideo non oportet quod si vapor convertitur in aquam, *Ba* 28ra quod ex|alatio convertatur in terram. Veruntamen sciendum est quod pluries cadunt cum pluvia pulveres terre elevati cum vapore a calore Solis.

⟨21⟩ Ad quartam patuit in questione, quia elementum bene generatur extra locum suum naturalem sue generationis.

Et sic patet ad questionem.

II.2 Utrum ros et pruina, nix et pluvia, sint eiusdem speciei

Kl 35rb Queritur consequenter utrum ros et pruina, nix et pluvia, sint eiusdem speciei.

⟨1⟩ Et probatur quod non: illud quod est naturaliter calidum et humidum non est eiusdem speciei cum illo quod est naturaliter frigidum et humidum;

---

14 extrusionem] incensionem ipsius exalationis *Ba* incensionem *Kl* intensionem vel extensionem *L U* 17 terram] pulveres *Ba Kl L U* 20 terram] cineres *Ba* terram vel pulveres *Kl L U* 22–23 quia elementum bene generatur extra locum suum naturalem sue generationis] *om. Ba Kl L U*

sed sic est de rore et pluvia, quia ros naturaliter est calidus et humidus, pluvia autem naturaliter frigida et humida; etc. Hoc patet per effectus roris et pluvie.

⟨2⟩ Secundo: si ista essent eiusdem speciei, sequeretur quod deberent generari eisdem temporibus; sed hoc non, quia aliqua eorum generantur tempore sereno, alia autem tempore nebuloso, sicut patet per experientiam.

⟨3⟩ Tertio ad idem: ros et materia roris est subtilis substantie; pluvia autem et eius materia sunt grosse | substantie; ergo videtur quod ros et pluvia non *U* 25rb debent esse eiusdem speciei. Antecedens patet, nam ros est substantia ita subtilis et ita elevabilis quod si una testa ovi esset repleta rore et foramen per quod infunderetur ros bene obstrueretur et poneretur ad calorem Solis iuxta aliquam hastam, illa testa ovi rore plena ascenderet prope hastam propter hoc quod ros, que est faciliter consumabilis calore Solis elevaretur. Hoc autem non fieret si testa ovi esset repleta aqua. Illud autem designat quod ros est subtilioris substantie quam sit aqua; unde videtur quod distinguuntur specie.

⟨4⟩ Quarto ad idem: differentia specifica sumitur ex diversitate rationum diffinitivarum; sed ita est quod roris et pruine, nivis et pluvie sunt diverse rationes diffinitive; ergo. Maior est nota; minor patet per Aristotelem in littera[7].

⟨5⟩ Quinto: eorum que sunt eiusdem speciei debet esse idem locus natu- *Kl* 35va ralis generationis ipsorum; sed sic non est de rore et pruina, nive et pluvia; ergo. Maior nota; minor patet, quia naturalis locus generationis nivis et pluvie est altior quam sit naturalis locus generationis roris et pruine.

⟨6⟩ In oppositum est Aristoteles, qui dictis impressionibus assignat easdem causas materiales et efficientes et solum differentias accidentales. | Dicit *Ba* 28rb enim rorem et pruinam fieri inferius propter easdem causas propter quas superius fiunt pluvia et nix, sed differre secundum magis et minus secundum multitudinem et paucitatem materie, raritatis et condensationis, caloris et frigiditatis concurrentis ad istorum productionem[8].

9 et eius materia] *om. Ba Kl L U* 14 faciliter consumabilis] substantia elevabilis *Ba Kl L U* 15–16 illud autem designat quod ros est subtilioris substantie] *om. L U* 16 substantie] materie *Ba Kl* 29 superius] *om. Ba Kl L U*

7 Aristoteles, *Meteorologica*, I, 11, 347 a 14–24.

8 Ibid., I, 11, 347 b 11–24.

⟨7⟩ In ista questione primo videndum est de comparatione nivis ad pluviam; [*L* 205rb] secundo de comparatione roris ad pluviam, | et postea in alia questione videbitur de comparatione pruine ad pluviam[9]; tertio respondendum est ad quesitum.

⟨8⟩ Quantum ad primum, sciendum est quod primo nix et pluvia in hoc conveniunt quod utrumque eorum fit ex vapore multo; qui quidem vapor, quando congelatur antequam perfecte convertatur in nubem, fit nix; quando autem condensatur et convertitur in nubem absque hoc quod congeletur, fit pluvia. Secundo conveniunt in hoc quod utriusque materia calore Solis elevatur ad locum in quo huiusmodi generantur. Tertio conveniunt in hoc quod utriusque materia, tam nivis quam pluvie, a frigiditate condensatur. Quarto conveniunt in hoc quod utrumque eorum fit in media regione ipsius [*P* 255v] aeris. Quinto conveniunt in hoc quod ambo possunt fieri ex eodem vapore; | nam possibile est quod, aliquo vapore converso in nivem in superiori regione [*Kl* 35vb] aeris, | quod ista nive cadente usque hic inferius liquefacit et fit aqua antequam veniat ad terram, quamvis in superiori parte aeris cadens super altos [*U* 25va] montes maneat nix. Et ideo aliquotiens ad montes cadit nix | et ad valles cadit pluvia. Et ista potest esse una causa quare in montibus aliquando est nix et non in vallibus, immo aliqui sunt montes super quos continue manet nix. Sic ergo visum sit in quibus conveniunt nix et pluvia.

⟨9⟩ Sed ad videndum in quibus differunt, sciendum est primo quod differunt loco et tempore, ad istum intellectum quod nix fit in locis et temporibus magis frigidis et pluvia in locis et temporibus minus frigidis; unde ad condensationem nivis requiritur frigus congelans, quod non requiritur ad generationem pluvie.

⟨10⟩ Differunt secundo in modo generationis, quia nix fit per congelationem vaporis; pluvia autem per conversionem eius in aquam.

⟨11⟩ Tertio differunt colore, quia pluvia habet colorem aque, nix autem [*Ba* 28va] habet colorem valde album. Nix enim est ex materia perspicua et rara, | aliqualiter tendens ad luciditatem, et propter hoc nix est alba. Et ex alio, quia in generatione nivis frigiditas est agens in materiam humidam et obtinens super illam, et frigiditas sic se habens est causa albedinis. Solet enim dici quod homines albi raro reperiuntur audaces, nam quod sunt albi est causa defectus caloris, qui quidem calor est causa audacie.

---

23 minus frigidis] magis calidis *Ba Kl L U*   24 condensationem] generationem *Ba Kl L U*

---

9 II.3, 16.

⟨12⟩ Quarto differunt quia pluvia diutius durat quam nix, ad istum intellectum quod longiori tempore solet pluere quam nivigere. Et huius ratio est, nam cum pluit, terra magis madefit, et iterum erit apta ad hoc ut per calorem Solis ab ea extrahatur vapor et elevetur, qui iterum est materia nove pluvie. Sed cum nivigit, propter frigiditatem pori terre clauduntur, ita quod non possunt extrahi vapores | *Kl* 36ra ita cito quod statim sint materia nove nivis. Sed diceres: si propter hoc quod tu dicis diutius plueret, sequeretur quod semper deberet pluere, quia pluvia cadente terra madefit et iterum elevaretur vapor et iterum plueret | *L* 205va et iterum elevaretur, et sic semper plueret. Respondetur quod non, quia aliquando, propter aliam et aliam influentiam Solis et alias et alias dispositiones mundi, iste humor diu stat antequam elevetur.

⟨13⟩ Quinto differunt in figura, nam pluvia cadit per guttas que sunt figure rotunde, sed nix non cadit secundum partes rotunde figuratas, immo aliquando in quibusdam particulis sunt figure quarundam stellarum figuris similes, et causa istius est influentia celi. Communiter tamen cadit secundum partes dispositas ad modum lane, ut patet ad sensum.

⟨14⟩ Quantum ad secundum, videlicet de comparatione roris ad pluviam, sciendum quod materia roris est vapor modicus et subtilis; materia autem pluvie est vapor multus et grossus, | *U* 25vb et propter hoc materia roris citius exalatur et evaporatur et consumitur quam materia pluvie. In hoc ergo primo differunt ros et pluvia.

⟨15⟩ Secundo differunt quia ros causatur inferius, pluvia autem superius. Patet ratione et signo. Ratione enim patet, quia si vapor pro roribus elevaretur alte, propter eius subtilitatem consumeretur vel converteretur et condensaretur in nubem, et fieret pluvia. Patet hoc etiam signo, quia aliquando, homine existente in campis, ros | *Ba* 28vb cadit super pedes et non super caput, ac si vapor pro roribus non elevaretur ad altitudinem unius hominis. Nec | *P* 256^{r} est ymaginandum propter hoc quod ros exiens a terra sic madefaciat herbas et gramina. Patet, nam si ponantur stramina super terram, supra inveniuntur rore madefacta et non infra, et hoc est signum quod ros non | *Kl* 36rb exit a terra; sed vapor pro roribus modicum elevatus frigiditate noctis condensatur et

---

2 longiori tempore] diutius *Ba Kl L U* 4 qui] quia *P* 5 ita] *om. P* 8 terra madefit et] *om. P* 9 vapor] *om. P* ‖ et iterum elevaretur] *om. Ba Kl L U* 11 solis] celi *Ba Kl L U* ‖ humor] vapor *Ba Kl L U* 15–16 figure quarundam stellarum figuris similes] figure sensibiles figuratis quorundam florum *Ba Kl* figure similes figure florum *L* figure similes figuris quorundam florenorum *U* 21 quam] pruina *P* 27 et] aliquando *add. P*

generatur ros, et cadit guttatim insensibiliter super herbas et gramina. Sed cum ille multe partes super herbas congregantur ad invicem, fit ex eis una gutta sensibilis. Et causa quare ille particule non manent disperse super foliis herbarum sicut cadunt, sed congregantur ad invicem, est ista, quia propter siccitatem folii herbe particule roris humide fugiendo congregant se ad invicem propter antiparistasim humidi et sicci et fit ex eis gutta sensibilis.

⟨16⟩ Differunt tertio quia ros non fit nisi tempore temperato caliditate et frigiditate; pluvia autem potest fieri alio tempore. Et causa huius est quia materia roris est vapor modicus et subtilis, et igitur in tempore multum calido consumitur; in tempore autem frigido congelatur et tunc fit pruina. Et ergo in estate, quando tempus est multum calidum, non rorat, nec hieme, sed in vere, cum tempus est calidus temperate. *L* 205vb | Pluvia autem, cum fiat ex vapore multo et grosso, qui non cito consumitur, potest fieri in aliis temporibus quam temperatis.

⟨17⟩ Differunt quarto quia ros non fit nisi tempore tranquillo et sereno; pluvia autem bene fit in aliis temporibus. Cum enim tempus non est tranquillum, materia roris dissolvitur et exsiccatur per ventum, et sic non fit generatio roris. Etiam, cum tempus est nebulosum, tunc elevantur vapores ex quibus generatur pluvia. Si tamen aliquo tempore nebuloso, rorat, *U* 26ra | hoc est propter consumptionem et dissolutionem nubis, *Kl* 36va | et ergo quando rorat tempore nebuloso est signum serenitatis.

⟨18⟩ Quinto differunt quia cum Boreas flat, non generatur ros, sed tamen tunc bene generatur pluvia. Et huius ratio est quia, flante Borea, tempus est frigidum; in tempore autem frigido non generatur ros, sicut dictum est; ergo etc.

⟨19⟩ Differunt sexto quia ros cadit per guttas *Ba* 29ra | insensibiles; pluvia autem per guttas sensibiles.

⟨20⟩ Adhuc notandum est de rore unum speciale quod dicit Albertus. Dicit enim quod aliquando tempore calido elevatur materia rorida cum sicco terrestri, qua elevata, istud humidum, seu ille vapor, caliditate aeris digeritur digestione bona et permiscetur ei fortiter illud siccum terrestre. Et postea frigido temperato pinguescit illud humidum et cadit ros fluens et viscosus ad modum mellis ad folia arborum et ad flores. Et postea, Sole agente in illud, evaporatur istud humidum et comburitur siccum terrestre quod postea manet super folia herbarum ad modum farine, quod sua

---

1 herbas et gramina] folia arborum et herbarum *Ba* foliis herbarum *Kl L U* 3–4 disperse super foliis herbarum sicut cadunt] super folia herbarum sic ab invicem distantes *Ba Kl* 5 herbe] *om. Ba Kl* 12 vere] et autumno *add. L U* 20 est] *om. P* 22 non] nos *P* 31 ei] eius *P* 33 ad flores] florum *P* ‖ sole] calefaciente et *add. Ba Kl L U*

siccitate exsiccat herbas et nocet eis. Et quando talis ros cadit super gramina et nondum illud humidum est evaporatum, sed adhuc est sicut mel, animalia, et precipue oves, vehementi appetitu comedunt illa gramina et herbas. Et quia mel non decoctum solvit ventrem, inde est quod oves comedentes huiusmodi herbas fluxu ventris moriuntur[10].

⟨21⟩ Quantum ad tertium, dico quod huiusmodi non differunt specie. Probatur, quia non differunt nisi solum differentiis accidentalibus | et secundum [P 256v] magis et minus, sicut dicit Aristoteles[11]. Et iam | apparuit qualiter differunt [Kl 36vb] differentiis accidentalibus.

⟨22⟩ Ad rationes. Ad primam dico quod verum est quod ros dicitur calidus et humidus in comparatione ad plu|viam. Ecce qualiter differunt secundum [L 206ra] magis et minus que non solent diversificari specie in caliditate et frigiditate et humiditate.

⟨23⟩ Ad secundam et tertiam dico quod bene probant differentiam accidentalem inter ista, non autem specificam et substantialem.

⟨24⟩ Ad quartam dico quod huiusmodi bene habent descriptiones accidentales differentes, non autem substantiales.

⟨25⟩ Ad quintam dico quod aliqua eiusdem speciei, differentes tantum secundum magis et minus, sic se habent quod aliqua eorum naturaliter habent generari in uno loco et alia naturaliter in alio, nec hoc est aliquid inconveniens.

Et sic sit dictum ad questionem.

II.3 Utrum grandines magis debeant generari in hieme quam in autumno

Consequenter queritur utrum grandines magis debeant generari in hieme [U 26rb] quam in autumno.

⟨1⟩ Et arguitur quod magis in | hieme, nam grando est aqua fortiter congelata; [Ba 29rb] fortis autem congelatio fit in hieme et non in vere nec in autumno.

12 diversificari] diversificare *P* 15 substantialem] essentialem *Ba Kl L U* 17 substantiales] essentiales *Ba Kl U* essentiales diffinitiones *L* 23 hieme] vel estate *add. Ba* 24 quam] in vere vel *add. Ba Kl L U*

10 Albertus Magnus, *Meteora*, lib. 2, tract. 1, cap. 14, ed. Colon. VI/1, 53–54.

11 Aristoteles, *Meteorologica*, I, 11, 347 b 11–24.

⟨2⟩ Secundo, grando vel generatur a frigiditate medie regionis vel generatur per antiparistasim. Si dicatur primum, cum maior sit frigiditas in hieme quam in autumno, sequitur quod magis debet generari in hieme. Si autem dicatur secundum, sequitur quod magis generabitur in estate, quod est falsum. Consequentia tenet quia in estate est maior antiparistasis propter intensiorem caliditatem.

⟨3⟩ Tertio, nix et grando habent convenientiam in hoc quod utrumque eorum generatur per congelationem vaporis; sed nix magis generatur in hieme quam in vere vel in autumno; igitur videtur quod sic debet esse de grandine.

⟨4⟩ Quarto probatur quod non possunt fieri grandines, nam si sic, hoc oporteret esse per congelationem vaporis. Vel igitur in generatione grandinum vapor prius convertitur in aquam | [*Kl* 37ra] quam congelatur, vel prius congelatur quam convertitur in aquam. Non potest dici primum quia, cum vapor conversus est in aquam, statim ista aqua descendet et non expectabit quod congelatur, et per consequens sic non erunt grandines. Si autem dicatur secundum, tunc fiet nix et non grando, quia nix fit quando vapor prius congelatur quam convertatur in aquam.

⟨5⟩ In oppositum est Aristoteles in littera, qui vult quod temporibus temperatis, sicut in vere et in autumno, magis fiant grandines quam in hieme et similiter quam in estate[12].

⟨6⟩ In ista questione primo videndum est de loco in quo generantur grandines et modo generationis grandinum; secundo videndum est de figura grandinum; tertio, de tempore in quo pluries fiunt; quarto, de pruina, de qua relinquebatur in precedenti questione[13].

⟨7⟩ Quantum ad primum, | [*L* 206rb] sciendum est quod erat una opinio de loco in quo generantur grandines, videlicet quod solum generantur in media regione aeris propter hoc | [*Ba* 29va] quod grandines fiunt per congelationem; et cum frigiditatis sit congelare et solum media regio aeris est frigida inter regiones aeris, videtur quod solum grandines generantur ibi.

3 quam] in vere vel *add. Ba Kl L U* 15 descendet] propter eius gravitatem *add. Ba Kl L U*
19 in littera] *om. Ba Kl L U* ‖ vult] videtur velle *Ba Kl L U*

12 Aristoteles, *Meteorologica*, I, 12, 347 b 34–348 a 4.

13 II.2, 7.

⟨8⟩ Sed breviter istam opinionem Aristoteles in nova translatione nititur destruere ex intentione[14]. Primo, nam videmus quod in summitatibus altorum montium non generantur grandines, cum tamen bene generantur inferius. Hoc non esset si solum grandines generarentur in media regione aeris et non alibi. Unde visum est aliquibus existentibus in summitati altorum montium circumquaque esse serenitatem et Solem lucere et splendere, et cum venerunt inferius, interim quod ipsi fuerint in montibus, ciciderunt grandines magne quantitatis et interim | inferius magne erant tempestates[15]. *U* 26va Istud videtur esse signum grandines non solum generari in media | regione *Kl* 37rb aeris, verum etiam in loco bassiori.

⟨9⟩ Secundo, nam aliquando nubibus existentibus auditus est sonus magnus et terribilis in illis nubibus, quod est signum | illas nubes esse vel fuisse *P* 257^{r} bene prope nos, et bassius multo quam in media regione aeris. Et postea statim cediderunt grandines magne quantitatis. Istud iterum est signum grandines non solum generari in media regione aeris, verum in loco bassiori.

⟨10⟩ Tertio, nam vise sunt grandines figure angularis. Hoc iterum est signum quod iste grandines non a loco ita alto ciciderunt sicut est media regio aeris; unde non potuissent descendere a tanta altitudine quin per earum motum et confricationem ad aerem fuissent rotundate.

⟨11⟩ Unde finaliter grandines possunt generari in aere satis propinquo terre et multum calido propter reflexionem radiorum Solis a terra, et hoc per talem modum quod frigiditas in nube patitur antiparistasim ab aere calido circumstante; et ergo fugiendo ab illo congregatur et unitur et vehementer fortificatur, ut deinde fiat non solum nubis in aquam conversio, verum aque congelatio, et citior congelatio quam aqua descendat. Et quia quanto aer est calidior, tanto est antiparistasis maior, ergo grandines magne et non figure rotunde fiunt tempore calido et in loco satis propinquo terre. Unde vide-

---

2 ex intentione] *om. Ba Kl L U* 12 esse vel] *om. Ba Kl L U* 15 solum] grandines *add. P* 21 reflexionem] refractionem *P*

---

14 Aristoteles, *Meteorologica*, I, 12, 348 a 21–348 b 2.

15 Cf. Iohannem Buridanum, *Questiones in Meteorologica*, I.18, ed. Bages, 2: 240–241, nr. 8–9: "Et ego hiis dictis credidi et credo quia bis expertus sum hec manifestissime, primo semel eundo ad Avinionem, ultra montem Pesatium in Ricordania, supra alteram villam sittam in valle profundissima que vocatur Leval: nos enim plures existentes in monte alto supra eramus in aere sereno et claro et etiam ultra illam villam videbamus montes illustratos sole, et tamen in fundo, supra illam villam, videbamus horribiles nubes et audiebamus magna tonitrua; et cum descendissemus in vallem, invenimus copiosam multitudinem grandinum nondum liquefactarum et gentes adhuc ex tempestate stupentes et aquas per rivos insolitos fluentes ita quod cum difficultate transiremus eas".

mus, licet pluries fiant grandines in vere et autumno quam in estate, tamen, *Kl* 37va quando in estate, fiunt multo maiores, | et non figure rotunde, quam quando fiunt in vere et autumno.

⟨12⟩ Sed diceret aliquis: quare ergo non immediate prope terram generantur grandines, cum prope terram sit maior antiparistasis, cum ibi sit maior calor propter minorem elongationem radii incidentis et radii reflexi ad invi- *Ba* 29vb cem? Respondetur quod huius causa est | quia nondum illa materia, sicut est ille vapor, est sufficienter alterata quod ex ea fiat nubes. Aliter potest dici ymaginando per istum modum quod, cum simile applaudit suo simili, calor incorporatus vapori, qui de natura sua est frigidus, quamvis accidentaliter calidus, cum ille vapor venit ad aerem calidum, calor elevans istum vapo- *U* 26vb rem aliquantulum incipit se dilatare propter simile circumstans. | Et cum *L* 206va in principio ille vapor primo elevatur a terra, non statim | ille calor totaliter exalat, sed adhuc elevat vaporem ulterius. Sed postea, cum ille calor exalatur in tantum quod illum vaporem ulterius elevare non potest propter aerem calidum circumstantem, aliqui gradus frigiditatis qui adhuc erant in illo vapore incipiunt intendi et fortificari, et vapor incipit uniri et condensari et in aquam converti. Et postea propter magnam antiparistasim tantum fortificatur frigiditas inferius et intenditur quod sequitur congelatio et generantur grandines. Ex hoc patet ad dubium, videlicet quare non immediate prope terram generantur grandines, sed in loco altiori, tamen bassiori quam sit media regio aeriss.

⟨13⟩ Ulterius diceret aliquis: nonne etiam grandines generantur in media regione? Dico quod sic, et hoc est per istum modum quod cum vapor elevatus et calefactus pertingit ad mediam regionem aeris, tunc frigiditate medie *Kl* 37vb regionis convertitur in aquam | et congelatur et fiunt grandines. Et ad hoc quod illa congelatio cito fiat, multum iuvat hoc quod iste vapor prius erat calefactus. Dicit enim Aristoteles aquam calefactam citius posse congelari quam aquam frigidam[16], cuius rationem quidam assignant quia aqua calefacta est rarior et passibilior et a frigore penetrabilior[17].

⟨14⟩ Diceret adhuc aliquis: quare est quod ille vapor non prius congelatur quam convertatur in aquam? Respondetur quod hoc est propter habundantiam humiditatis et grossitiem ipsius vaporis. Et ideo, quia vapor ex quo generatur nix non est ita habundanter humidus nec est ita grossus sicut

---

9 cum] *om. P*   19 inferius] ulterius *Ba Kl L U*

---

16 Aristoteles, *Meteorologica*, I, 12, 348 b 31–34.

17 Cf. II.4.

vapor ex quo generatur grando, inde est quod prius potest congelari quam converti in aquam, et tunc generatur nix.

⟨15⟩ Quantum ad secundum, dico quod grandines aliquando sunt figure rotunde, et tunc veniunt a loco alto; aliquando autem sunt figure angularis, et tunc veniunt a loco bassiori. | Et cause istarum figurarum possunt elici [*P* 257v] ex iam dictis. Et aliquando in aliquibus apparent figure animalium, sed hoc referendum est ad specialem virtutem celestem[18].

⟨16⟩ Quantum ad tertium, dicendum est quod grandines pluries fiunt in vere et in autumno quam in estate et in hieme. Patet, quia in estate est defectus vaporis qui deberet esse | materia grandinum; | in hieme autem deficit [*U* 27ra] [*Ba* 30ra] antiparistasis sufficiens in loco bassiori, et deficit calor potens elevare materiam grandinum (qui est vapor grossus) usque ad mediam regionem. In vere autem et in autumno est | sufficiens materia, scilicet vapor humidus; simi- [*L* 206vb] liter est sufficiens calor potens elevare usque ad mediam regionem aeris; et similiter | in inferiori regione est sufficiens antiparistasis, etc. [*Kl* 38ra]

⟨17⟩ Quantum ad quartum, primo dico quod pruina convenit cum rore in materia et in quantitate materie et in qualitate; nam tam ros quam pruina generantur ex vapore modico et subtili. Secundo conveniunt in loco generationis eorum, quia utrumque eorum generatur in inferiori regione aeris. Sed differunt ex parte agentis, quia pruina fit ab intensiori frigiditate congelante et ros fit a remissiori frigiditate condensante vaporem roridum. Et in hoc ros convenit cum pluvia et pruina cum nive, quia sicut pluvia fit a minori frigiditate condensante quam nix, ita ros a minori frigiditate quam pruina. Et ulterius ros et pluvia in hoc conveniunt quod utrumque est vapor condensatus et non congelatus; nix autem et pruina conveniunt in hoc quod utrumque est vapor congelatus.

⟨18⟩ Ad rationes. Ad primam, cum dicebatur: 'grando est aqua fortiter congelata', concedo. Et cum dicitur: 'sed congelatio fortiter non potest fieri nisi a frigido', concedo vel circumstante vel intrinseco. Modo quamvis in vere et in autumno non fiat congelatio vaporis propter frigus circumstans, tamen

---

12 grossus] *om. Ba Kl L U* 13 scilicet vapor humidus] grandinum *Ba Kl L U* 18 vapore] rore *P* 20 intensiori] inferiori *Ba Kl L om. U* 28 et cum dicitur] *om. P* 28–29 nisi a frigido] *om. P*

---

18 Cf. II.2, 13.

bene fit per frigus intrinsecum acquisitum et intensum per antiparistasim, scilicet circumiacentiam aeris calidi, sicut dicebatur in questione.

⟨19⟩ Ad secundam: ‘vel generaretur per antiparistasim’, etc., dico quod generatur utroque modo. Et quando dicitur: ‘tunc etiam et maxime deberet generari in estate’, verum est nisi aliud obesset, videlicet defectus materie ipsarum grandinum. Et similiter, cum dicitur: ‘si a frigiditate, ergo maxime deberet generari in hieme’, verum est nisi aliud obesset, videlicet defectus *Kl* 38rb caliditatis potentis | elevare vaporem grossum qui est materia grandinum.

⟨20⟩ Ad tertiam: ‘nix et grando habent convenientiam in hoc quod utrumque fit per congelationem’, concedo. Et cum dicitur: ‘nix fit in hieme’, con- *U* 27rb cedo; ‘ergo et grandines’, nego consequentiam. Et ratio est | quia materia nivis est vapor non ita grossus sicut est materia grandinum; ergo aliquando *Ba* 30rb *L* 207ra est bene calor sufficiens ad elevandum vaporem qui est materia | nivis, | qui non esset sufficiens ad elevandum materiam grandinum. Et igitur, licet tam nix quam grando fiant per congelationem, tamen aliquando fit nix quando non potest fieri grando, propter defectum caloris non potentis elevare materiam ita grossam et multam sicut est materia grandinum.

⟨21⟩ Ad quartam dico quod in generatione grandinum vapor prius convertitur in aquam quam congeletur. Et quando tunc dicebatur: ‘ergo aqua descendet et non expectabit congelationem’, dico quod congelatio prevenit descensum, ita quod antequam aqua possit descendere, ipsa congelatur. Et inde est quod aliquando ante casum grandinum cadunt gutte magne; et iste sunt vel fuerunt de inferiori parte nubis, ita quod iste, quia non tantum distabant a terra, citius descendebant quam potebant congelari; alie autem que magis distant citius congelantur quam descendant (sicut sunt partes in superiori parte nubis), et ipsis congelatis fiunt grandines.

II.4 Utrum aqua calida applicata frigori congelanti citius congeletur quam aqua frigida

Consequenter queritur utrum aqua calida applicata frigori congelanti citius congeletur quam aqua frigida.

P 258r *Kl* 38va ⟨1⟩ Et arguitur primo quod non: ubi est maior resistentia passi ad aliquod agens, ibi est | tardior actio; | sed in actione frigoris congelantis in aquam

2 circumiacentiam] circumstantiam *Kl L U om. Ba* 22 gutte] *om. P* 23 vel fuerunt] *om. Ba Kl L U* 30 tardior] minor *L* ‖ actio] alteratio *Kl Ba*

calidam, congelans invenit maiorem resistentiam quam si eadem aqua esset frigida propter contrarietatem caliditatis ad frigiditatem; ergo etc.

⟨2⟩ Secundo: si questio esset vera, videretur a simili quod aqua frigida applicata calefacienti citius calefieret quam si esset calida; sed hoc est falsum. Videmus enim, si aqua remisse calida et aqua frigida in eadem quantitate applicentur igni, quod aqua remisse calida citius attingit aliquem datum gradum caliditatis quam aqua frigida.

⟨3⟩ Tertio ad idem, nam signatis duabus aquis eiusdem quantitatis—*a* aqua calida, *b* aqua frigida—et applicentur alicui frigori congelanti, tunc antequam istud frigus congelet *a* aquam calidam, oportet quod prius reducat ipsam ad illum gradum frigiditatis quem habet *b* in principio congelationis; et tunc iam *b* reductum est ultra illum et habet ante primum gradum; igitur *b* citius congelabitur quam ipsum *a*; ergo questio falsa.

⟨4⟩ Quarto ad idem: caliditas prohibet congelationem; sed cum aqua calida approximatur | *L* 207rb frigori congelanti, caliditas eius intenditur. Dictum enim est prius quod contrarium applicatum contrario intendit ipsum[19]; igitur videtur quod aqua calida tardius debet congelari quam aqua frigida.

⟨5⟩ Oppositum vult Aristoteles in littera, capitulo primo, tractatu secundo, | *U* 27va ubi sic ait: 'confert autem ad celeritatem congelatonis aquam precalefactam esse'[20]. Ex quibus verbis manifeste videtur velle quod ad citiorem | *Ba* 30va congelationem iuvat aquam esse calefactam. Hoc non esset nisi aqua calida citius congelaretur quam frigida.

⟨6⟩ Breviter respondendo ad istam questionem sit prima conclusio quod questio est vera. Patet ad experientiam si aqua calida et aqua frigida frigori | *Kl* 38vb congelanti applicentur.

5–6 si aqua remisse calida et aqua frigida in eadem quantitate applicentur igni] *om. Ba Kl* 6 quod] *om. P* 7 gradum caliditatis] gradum intensiorem caliditatis *Ba* gradum certum et caliditatis intensioris *Kl* gradum certum et intensiorem caliditatis *L U* 8 eiusdem quantitatis] *om. Ba Kl L U* 10 aquam calidam] *om. Ba Kl L U* 12 primum gradum] *om. P* 14 prohibet] impedit *Ba Kl L U* 18 in littera capitulo primo tractatu secundo] secundo huius tractatu primo capitulo secundo *Ba Kl L U* ‖ secundo] secundum nostram distinctionem *add. Kl* secundum illam distinctionem *add. L U*

19 I.10, 11.

20 Aristoteles, *Meteorologica*, I, 12, 348 b 30–32, translatio Guillelmi de Morbeka (translatio noua), *AL* X 2.2, 33^{656}: "Confert autem adhuc ad celeritatem coagulationis et precalefactam esse aquam".

⟨7⟩ Sed de causa quare hoc sit est difficultas; unde Albertus videtur reddere causam illius dicens quod causa huius est quia calor rarefacit ipsam aquam et facit ipsam paratam ad recipiendum multum frigus et cito; aqua autem frigida densa est et clausa et non potest ita cito recipere multum frigus[21]. Contra istud alii obiciunt, quia ymaginatio Alberti videtur esse quod aqua talis congelatur per frigiditatem ab extrinseco intrante ipsam, et hoc non videtur esse verum, quia tunc accidens migraret de subiecto in subiectum; et cum corpora mutuo se non penetrent, non potest dici quod frigiditas existens in quibusdam corporibus, manens in illis, intret illam aquam calidam. Potest responderi ad hoc quod partes alique calide intercluse inter partes aque expelluntur a frigiditate aeris circumstantis, et loco earum subintrant partes frigide ipsius aeris circumstantis. Sed contra, nam queritur quo fugiant ille partes calide expulse; circumquaque enim inveniunt frigidum obvians. Ad hoc autem Albertus videtur respondere quod ille partes calide simul fugiunt ad centrum illius aque congelande; et hoc declarat per signum in hiis verbis: 'quando aqua calida congelanti exposita congelatur, in medio sui accipit glacies eius ebulitionem et eminentiam, quod non accipit aqua frigida congelata'. Causa autem huius, ut dicit, est ista, quia cum frigus intraverit ibi aliquantulum, fugit calor ad medium aque et ibi comprehensus exalavit modicum aque, et sic residua aqua frigiditate congelata est[22].

⟨8⟩ Alii dicunt quod calor aereus et calor igneus non sunt eiusdem rationis. *Kl* 39ra Modo ymaginantur quod aque frigide permixtus | sit calor aereus et *L* 207va aque calide permixtus sit calor igneus. Modo, cum aqua calefit, | ille calor aereus exalatur et inducitur calor igneus. Ulterius ymaginatur quod cum *Ba* 30vb aqua calida et frigida | frigori congelanti sunt applicate, tunc calor aereus *U* 27vb permixtus aque frigide prohibet ne illa aqua | subito congeletur; sed quia in aqua calida non est calor aereus, sed igneus, qui multum contrariatur frigori circumdanti, ibi fit fortis actio frigoris in illum calorem. Et quia caliditas est modice resistentie, cito illa caliditas ignea expellitur vel extinguitur, qua

10 calide] terrestres *add. Ba Kl L U* 12 aeris] *om. P* 13 expulse] *om. Ba Kl L U* 23 aque calide permixtus sit calor igneus modo] *om. P* 25 frigori] *om. P* 28 fortis] magna *Ba Kl L* maxima *U*

21 Albertus Magnus, *Meteora*, lib. 2, tract. 1, cap. 31, ed. Colon. VI/1, 64^{19-21}: "causa autem huius est, quia calor rarificat aquam et facit eam patulam ad frigus multum et subito recipiendum".

22 Ibid., lib. 2, tract. 1, cap. 31, ed. Colon. VI/1, 64^{21-28}: "quando aqua calida, frigori congelanti exposita congelatur, in medio sui accipit glacies eius ebullitionem et eminentiam, quam non accipit aqua frigida congelata. Causa autem est, quia cum frigus intravit in aquam calidam, fugit calor ad medium aquae et elevavit secum aquam; et sic frigore congelata".

expulsa | et extincta, ulterius aqua congelabitur. Sed quia aereus calor non *P* 258v tantum contrariatur frigori cicumstanti sicut calor igneus, nec etiam modice resistit sicut calor igneus, tardius per frigus expellitur circumstans vel extinguitur. Et quia ille est permixtus aque frigide et prohibet eius congelationem, aqua frigida tardius congelatur quam si esset calida.

⟨9⟩ Istis visis apparet michi quod posset adhuc dari alia causa illius, supponendo primo quod aqua calida est rarior aqua frigida. Ulterius supponendo quod frigidum multiplicat prope se unam qualitatem que non est frigiditas—et est una qualitas non habens contrarium, recte sicut a corpore luminoso multiplicatur una qualitas, puta lumen, que non habet contrarium, et ita consimili modo ymaginor de caliditate—, ulterius ymaginor illam qualitatem non habentem contrarium esse productivam frigiditatis in subiecto apto nato, et non in quocumque, recte sicut lumen, quamvis non sit caliditas, tamen est eductivum et productivum caliditatis in subiecto apto nato[23].

⟨10⟩ Istis suppositis dico quod quia aqua calida est magis rara quam aqua frigida, illa qualitas multiplicata a frigore circumstante melius potest se in aqua calida dilatare et de ea frigiditatem educere et congelationem | quam *Kl* 39rb in aqua frigida, propter hoc quod aqua frigida est magis densa et magis compacta; et inde est quod tardius congelatur aqua frigida quam aqua calida. Secundum istum modum dicendi non oportet dicere quod accidens migret de subiecto in subiectum, nec oportet concedere penetrationem corporum quorum unum non informetur alterum.

⟨11⟩ Ad rationes. Ad primam: 'aqua calida plus', etc., hoc negarent illi de secunda causa posita in questione, | immo dicerent quod aqua frigida plus *Ba* 31ra resisteret; et hoc est propter calorem aereum admixtum | aque frigide, qui *U* 28ra est maioris resistentie quam sit calor igneus admixtus aque calide. Aliter | *L* 207vb potest dici quod quamvis aqua calida plus resistat congelanti quando esset equaliter densa, non tamen quando est inequaliter densa; sed quia semper aqua calida est magis rara et minus densa quam aqua frigida, ipsa minus resistit quam frigida propter aliqua que concurrunt.

⟨12⟩ Ad secundam: 'sequitur', etc., negatur. Et ratio negationis est densitas aque frigide. Propter densitatem enim aque frigide qualitas multiplicata a calido non potest ita libere se dilatare et multiplicare per aquam densam

---

17 frigida] rara *P* ‖ melius] *om. Ba Kl L U* 18–19 quam in aqua frigida] quod non ita libere nec ita cito potest facere in aqua frigida *Ba Kl L U* 28 calida] frigida *P*

---

23 Cf. I.9, 21.

sicut qualitas multiplicata a frigido potest se dilatare et multiplicare per aquam raram, que est calida. Et ergo, quamvis aqua calida citius congeletur quam aqua frigida propter eius raritatem, non tamen aqua frigida citius calefit quam aqua calida propter eius densitatem.

⟨13⟩ Ad tertiam dico quod quia mobile velox potest attingere mobile tardum et anticipare, non est impossibile quod aqua calida citius congeletur quam aqua frigida, quamvis aqua frigida per tempus | [*Kl* 39va] habeat ante multos gradus frigiditatis. Ex hoc intuenti patet ad argumentum.

⟨14⟩ Ad quartam dico quod quamvis in principio per satis modicum tempus caliditas aqua calefaciente intenditur, tamen, cum qualitas eductiva frigiditatis de subiecto apto nato multiplicata a frigore circumstante per aquam calefactam se dilataverit, amplius caliditas aqua calefaciente remittitur et frigiditas educitur et tandem congelabitur, et citius quam si fuisset frigida propter causas predictas.

Et sic patet ad questionem.

II.5 Utrum rubedo matutina sit signum pluvie

Consequenter queritur utrum rubedo matutina sit signum pluvie.

⟨1⟩ Et arguitur primo quod non: rubedo vespertina non est signum pluvie, sed magis signum serenitatis; ergo etiam videtur quod sic sit de rubedine matutina, cum non videtur esse maior ratio de rubedine | [*P* 259r] vespertina quod sit signum serenitatis quam de rubedine matutina.

⟨2⟩ Secundo ad idem: rubedo matutina causatur ex interpositione exalationis terrestris | [*U* 28rb] inter visum nostrum et ipsum celum vel Solem; sed cum exalatio sicca est elevata, non debet sequi pluvia; ergo etc. Maior est nota | [*L* 208ra] ex eo quod, cum lucidum videtur per nigrum, apparet rubeum, sicut Sol per fumum[24]; ergo, quandocumque | [*Ba* 31rb] apparet nobis rubedo in Sole, est signum quod inter Solem et nos est exalatio terrestris et nigra. Minor patet de se.

⟨3⟩ Tertio ad idem: color rubeus circa Lunam non significat pluviam; ergo nec color rubeus circa Solem; ergo, cum Sol de mane videtur rubeus, non videtur esse signum pluvie. Antecedens patet, nam dicitur quod Luna pal-

7–8 multos gradus frigiditatis] *om. P* 11 de subiecto apto nato] *om. Ba Kl L U* 12 aqua calefaciente] aque a calefactione *Ba Kl L U* 13 educitur] introducitur *L U* 21 secundo] *om. P*

24 Cf. Aristotelem, *Meteorologica*, III, 374 a 3–8; 347 b 8–11.

lida pluit, rubicunda flat, sed alba serenat[25]; igitur rubedo circa Lunam non significat pluviam, sed magis | ventos. *Kl* 39vb

⟨4⟩ Quarto ad idem: si rubedo matutina pluviam significaret, hoc esset propter hoc quod essent aliqui vapores elevati, qui essent causa talis apparitionis, ex quibus generatur pluvia; sed propter vapores tales elevatos non potest argui quod fiat pluvia, quia possunt calore Solis consumi et non erit pluvia. Similiter possunt per ventos ad aliam regionem deduci, et sic iterum in illa regione non erit pluvia in qua Sol de mane apparet rubeus.

⟨5⟩ Oppositum vult opinio vulgarium dicentium quod rubedo vespertina circa Solem significat serenitatem diei sequentis; rubedo autem matutina circa Solem significat pluviam in illa die.

⟨6⟩ Breviter in ista questione primo ponende sunt alique distinctiones; secundo videndum est de quesito et de quibusdam aliis signis pluvie.

⟨7⟩ Quantum ad primum, sciendum est quod quedam sunt signa aliquorum que cum hoc quod sunt signa eorum, etiam sunt cause illorum, sicut alique coniunctiones et constellationes speciales sunt signum pluvie vel alicuius alterius effectus, et cum hoc sunt cause. Alia autem sunt aliquorum signa que non sunt cause istorum, sed solum effectus, sicut fumus est signum ignis: ubi enim est fumus, signum est quod ibi sit ignis; sed fumus non est causa ignis, sed effectus. Alia autem sunt signa aliquorum que non sunt cause illorum nec effectus illorum nec illa dependent ab invicem, sed solent evenire consequenter, videlicet unum post aliud. Adhuc quedam sunt presentis, quedam preteriti, quedam futuri, et adhuc quedam sunt signa preteriti vel futuri longe vel prope. Et adhuc quedam sunt signa aliquorum de quorum significatione potest reddi causa; quedam autem de quorum significatione non potest a nobis reddi causa, sed est nobis occulta.

⟨8⟩ Iterum signorum significantium pluvias futuras quedam sunt in celo et apparent in celo, sicut coniunctiones planetarum; alia autem sunt | que *L* 208rb

---

4 elevati] interpositi *Ba Kl L U* 8 de mane] actu *P* 10 circa solem] *om. Ba Kl L U* 11 circa Solem] *om. Ba Kl L U* 15 illorum] et non effectus *add. Ba Kl* et effectus *add. L U* 16 speciales] *om. Ba Kl L U* 17 et cum hoc sunt cause] *om. Ba Kl L U* 17–20 alia autem sunt … ignis sed effectus] *om. P* 26 occulta] ignota *Ba Kl L U*

---

25 Cf. H. Walther, *Initia carminum ac versuum Medii Aevi Posterioris Latinorum. Alphabetisches Verzeichnis der Versanfänge mittelateinischer Dichtungen*, Göttingen, 1969 (Carmina medii aevi posterioris Latina, 1), nr. 20577: "pallida Luna pluit, rubicunda flat, alba serenat".

sunt in aere, licet appareant in celo—sicut est rubedo matutina (quamvis appareat in celo, tamen est in aere); alia sunt que sunt in aere et etiam appa- *Ba* 31va rent in aere, sicut sunt aliqui venti | qui sunt signa pluviarum.

Kl 40ra ⟨9⟩ Alia autem sunt signa que apparent in animalibus, sicut est motus aurium asini: cum enim movet et fricat aures suas, est signum future plu- *U* 28va vie. Similiter cantus | corvorum et plura talia que apparent in animalibus sunt signa pluviarum futurarum.

⟨10⟩ Sciendum est quod signa pluvie future que sunt et apparent in celo sunt signa astrologica et non pertinent ad istam scientiam; et sciendum est quod sunt minus certa quam sunt signa pluvie existentia in regione elementari et que pertinent ad istam scientiam. Et de illis signis astrologicis pluviarum sunt multi speciales libri, de quorum numero unus satis utilis vocatur *Liber de impressionibus aeris*[26].

⟨11⟩ Sed diceres: cum dicis quod illa signa sunt minus certa illis que sunt in regione elementari, sed contra: illa signa sunt magis certa que cum hoc quod sunt signa alicuius effectus sunt etiam cause, quam sunt illa que sunt signa et non cause; sed sic est de signis pluvie que apparent et sunt in celo in comparatione ad illa que sunt in regione elementari; ergo etc. Respondetur quod bene verum est quod illa signa essent magis certa si quis sciret ea; sed modo non scimus naturam corporum supercelestium nisi per propositiones universales, et igitur difficile est mediantibus illis iudicare de particularibus. Et etiam ex alio, quia significant suos effectus nimis remote; signa autem que sunt in regione elementari significant suos effectus magis prope. *P* 259^{v} Et ex alio, quia non | solum respiciendum est ad ipsas causas, sed etiam ad dispositionem illius passi; et ergo astrologi aspicientes ad causas agentes alicuius pluvie vel alterius effectus, non respicientes ad materiam, sepe decipiuntur. Si enim super aliquam regionem siccam et deficientem vaporibus esset coniunctio aliqua signans pluviam futuram in illa regione et astrologus respiceret ad coniunctionem et non ad dispositionem regionis, iudicaret de *Kl* 40rb pluvia et deciperetur, quia quamvis ex tali coniunctione, | supposita sufficiente humiditate illius regionis, sequitur pluvia, tamen, deficiente humiditate et materia pluvie future, non oportet sequi pluviam. Et etiam quia aliquando constellatio est super unam regionem et non super aliam, et tunc

---

3 pluviarum] futurarum *add. Ba Kl L U* 4 signa] pluviarum *add. Ba Kl L U* 9–10 sunt signa astrologica et non pertinent ad istam scientiam et sciendum est quod] *om. hom. Ba Kl L U* 12 satis utilis] *om. Ba Kl L U* 25 passi] et ipsius materie *add. Ba Kl L U* 26 non respicientes ad materiam] *om. Ba Kl L U*

---

26 Robertus Grossatesta, *De impressionibus aeris*, ed. Baur, 41–51.

non significat pluviam in quacumque regione. Et adhuc, supposito | [*Ba* 31vb] quod astrologus iudicaret pluviam futuram in regione supra qua | [*L* 208va] est illa constellatio, adhuc posset decipi, quia quamvis in illa regione vapores sufficienter elevarentur, tamen, ipsis elevatis, per ventum possent deduci ad aliam regionem; et sic potest contingere quod regio iudicata ab astrologis esse sicca esset pluviosa et regio iudicata pluviosa esset sicca. Et ergo oportet astrologum tot et tanta respicere quod non | [*U* 28vb] est bene possibile quod certe iudicaret. Per signa autem que fiunt in regione elementari pertinentia ad istam scientiam, quia magis prope significant suos effectus, possumus certius iudicare.

⟨12⟩ Quantum ad secundum, dico primo quod rubedo vespertina prope Solem significat serenitatem diei sequentis; cuius ratio est quia, sicut prius dicebatur, cum lucidum videtur per nigrum, apparet rubeum[27]. Cum ergo de sero, cum Sol est in occasu, Sol appareat nobis rubeus, signum est quod tunc vapores sunt interpositi inter visum nostrum et Solem. Et cum Sol in occasu est prope terminum horizontis, signum est etiam quod illi vapores sunt prope terram et descenderunt. Et cum Sole occidente de nocte deficiat calor elevans, illi vapores ex quibus potuisset fuisse generata pluvia descendunt ad terram, et per consequens die sequenti erit serenitas.

⟨13⟩ Secundo dico quod | [*Kl* 40va] rubedo matutina prope Solem significat pluviam futuram. Et ratio huius est, nam cum Sol de mane apparet rubeus, signum est quod inter Solem et nos vapores qui sunt materia pluvie sunt interpositi. Et quia Sol continue ascendit super nostrum horizontem, elevat illos vapores secum et cum hoc plures et plures ex quibus erit pluvia.

⟨14⟩ Tertio dico quod cum Sol de mane oriens apparet maior quam communiter solet apparere, est signum pluvie future. Patet, nam, sicut prius dicebatur, denarius in aqua apparet maiorem quam extra aquam[28]. Cum ergo Sol oriens apparet nobis maior quam communiter solet apparere, signum est quod inter Solem et visum nostrum sunt vapores intermedii per quos Sol videtur, qui sunt causa quod Sol apparet nobis maior quam solet apparere. Et quia Sol continue ascendit super nostrum horizontem, elevat illos vapores secum ex quibus erit pluvia.

6 pluviosa] humida *Ba Kl L U* ‖ pluviosa] humida *Ba Kl L U* 7 tot et] *om. Ba Kl L U* 16 quod] *om. P* 30 videtur] *om. P* 32 pluvia] in illa die *add. Ba Kl L U*

27 II.5, 2; cf. I.14, 10.
28 I.11, 17.

⟨15⟩ Quarto dico quod quando Sol magis mane oritur quam solet, est L 208vb signum pluvie future. Et | ratio huius est quia, quando vapores sunt inter- Ba 32ra positi inter visum nostrum et Solem, tunc propter refractionem ra|diorum ipsius Solis, Sol citius apparet nobis super nostrum horizontem quam appareret si illi vapores non essent interpositi. Quando igitur videmus Solem magis mane surgere quam soleat, signum est quod tunc sunt tales vapores interpositi inter Solem et nos ex quibus erit pluvia in illa die.

U 29ra ⟨16⟩ Quinto dico quod, quando apparent plures Soles unus prope alium, est signum pluviarum. Et ratio huius est, nam inter nos et Solem multi vapo- Kl 40vb res, qui sunt alte elevati, sunt interpositi, ex quibus erunt magne | pluvie.

⟨17⟩ Sexto dico quod, quando Sol apparet pallidus, est signum pluvie, nam hoc est signum quod inter Solem et nos sunt vapores aquei interpositi qui, si P 260r essent terrestres et nigri, facerent | Solem nobis apparere rubeum, qui nunc faciunt apparere album pallidum.

⟨18⟩ Septimo dico quod, quando halo apparet circa Lunam vel Solem, est signum pluvie future, et magis quando apparet circa Solem, et hoc etiam fit propter vapores interpositos. Et quando apparet circa Lunam et disgregatur ex una parte, tunc ex ista parte significat ventum futurum; quando non disgregatur, significat pluviam futuram.

⟨19⟩ Octavo dico quod, quando in novitate Lune cornua apparent nigra et obtusa, est signum quod erit pluvia per istum mensem. Huius causa est quia, quando fit talis apparitio, est interpositio multorum vaporum elevatorum ex quibus erit pluvia.

⟨20⟩ Nono dico quod quando stelle apparent rare et plus distantes quam soleant, est signum pluvie, sed non statim, sed per tres vel quatuor dies postea. Et ratio est quia quando stelle apparent rare seminate in celo, tunc signum est quod in media regione aeris sunt multi vapores per quos appa- L 209ra rent tales stelle, qui sunt causa talis apparitionis, qui | continue disponunt ad pluviam futuram. De aliis signis pluviarum futurarum videbitur in questione sequenti[29].

---

3 refractionem] reflexionem *Ba Kl L U* 4 nobis] ascendere *Ba Kl L U* 7 in illa die] *om. Ba Kl L U* 9 pluviarum] magnarum futurarum *add. Ba Kl L U* 13 facerent] *iter. P* ‖ solem] solis *P* 17 interpositos] etc. *P* 21 obtusa] tortuosa *P* ‖ quod erit pluvia per istum mensem] pluvie *Ba U* future pluvie *Kl L* 22 interpositio] signum *P* 24 dico] *om. P* ‖ distantes] ab invicem *add. Ba Kl L U* 27 multi vapores] vapores interpositi *Ba Kl L U*

---

29 II.6.

⟨21⟩ Ad rationes. Ad primam dico quod non est idem de rubedine matutina et serotina. Et causa dicta est in questione[30].

⟨22⟩ Ad secundam dico quod cum Sol de mane apparet rubeus, bene visum est quod non solum vapores aquei sunt inter nos et Solem—nam tunc Sol appareret nobis pallidus—, sed cum vaporibus aqueis sunt ibi quedam terrestres exalationes que faciunt nobis apparere Solem rubeum. Et quia non solum sunt illi, sed cum eis sunt vapores aquei, ex siccitate | talium | exala- *Kl* 41ra *Ba* 32rb tionum terrestrium non potest argui serenitas futura, quia plures sunt ibi vapores aquei ex quibus erit pluvia.

⟨23⟩ Ad tertiam dico quod quia rubedo apparens circa Lunam est quasi similis rubedini serotine apparenti in alii partibus celi, ergo sicut rubedo serotina non est signum future pluvie, ita nec rubedo que apparet circa Lunam.

⟨24⟩ Ad quartam dico: concedo quod ista signa non sunt ita certa quin *U* 29rb possunt aliquando impediri, sed tamen hoc fit valde raro.

Hoc de questione.

II.6 Utrum caligo sit signum pluvie future

Consequenter queritur utrum caligo sit signum pluvie future.

⟨1⟩ Et arguitur quod sic, quia caligo causatur ex elevatione vaporum; vapores autem elevati sunt causa pluvie.

⟨2⟩ Oppositum videtur velle Aristoteles, cum dicit: ‘caligo est velut nebula sterilis’[31].

⟨3⟩ In ista questione primo videndum est adhuc de quibusdam signis pluvie future que sunt in regione elementari; deinde videndum est de quibusdam signis existentibus in aliis rebus, sicut in animalibus vel in pulveribus, in caligo et in consimilibus.

2 serotina] vespertina *L U* 6 solem] solum *P* 10–13 quia rubedo apparens … apparet circa lunam] quia luna apparet de nocte non est signum pluvie future ita nec rubedo apparens de nocte circa lunam *P* 14 signa non sunt ita certa] non sunt talia signa *P* 20 velut] ut *Ba Kl L* 24 animalibus] lapidibus *add. Kl L* ‖ pulveribus] plantis *U*

30 II.5, 12–13.

31 Aristoteles, *Meteorologica*, I, 9, 346 b 35, translatio Guillelmi de Morbeka (translatio noua), *AL* X 2.2, 28^{524}: “est enim caligo uelut nebula sterilis”.

⟨4⟩ Quantum ad primum, sciendum est quod nebula est nubes rara. Quedam est preveniens et non subsequens pluvias; quedam vero est subsequens. De nebula non subsequente dicendum est quod, cum talis apparet, est signum pluvie, et maxime si ascendit et non descendit per guttas insensibiles. Et ratio huius est, nam tunc est signum quod materia pluvie incipit congregari et elevari. Sed nebula subsequens pluvias est serenitatis | *L* 209rb signum; nam post casum pluvie illud quod est magis rarum et magis leve manet sursum et ex isto fit nebula. Et quando illud | *Kl* 41rb apparet post pluvias, signum est quod materia pluvie cecidit et non amplius elevatur nec congregantur vapores qui sunt causa pluvie, nec amplius fit conversio elevati in nubem et in aquam. Similiter est de caligine: cum apparet ante pluviam et non descendit, est signum pluvie future; cum autem descendit per guttulas insensibiles, est signum serenitatis; et similiter, cum subsequitur pluviam, est signum serenitatis, et ratio eadem est sicut de nebula. Ex hoc apparet ad quesitum et ad rationem et ad auctoritatem.

Ba 32va ⟨5⟩ Aliud signum pluvie future est cum Sol est in ortu et super ipsum apparet nubes magna et nigra. Tunc, si Sol statim absconditur sub illa nube, est signum pluvie et statim. Quando autem Sol non absconditur sub illa nube, tunc non est signum | *P* 260^{v} pluvie. Et causa primi est quod cum Sol absconditur sub illa nube, signum est quod non obtinet super illam nubem ad consumendum et ad dissolvendum ipsam, et inde sequitur | *U* 29va pluvia. Causa autem secundi est quia, cum Sol non absconditur sub illa nube, signum est quod Sol obtinet super illam nubem ad dissolvendum et ad consumendum ipsam; nube autem consumpta et dissoluta non sequitur pluvia.

⟨6⟩ Aliud signum pluvie future est frequens mutatio ventorum, et maxime quando ventus girando insequitur Solem; nam tunc ventus congregat vapores in unum, ex quibus generatur pluvia.

⟨7⟩ Aliud signum pluvie future est defectus roris tempore debito; cum enim non rorat tempore debito, signum est quod ille vapor qui est materia roris | *Kl* 41va elevatur cum aliis sursum et fiunt nubes et sequuntur pluvie. Etiam, cum non rorat tempore debito, signum est quod calor est fortis et intensus consumens vaporem roridum, qui est subtilis, et per eius formationem elevantur vapores grossi quos non potest consumere, ex quibus fiunt nubes et sequitur pluvia.

---

4 per guttas insensibiles] *om. Ba Kl L U* 6 elevari] sursum et ex isto generatur pluvia *add. Ba* sursum *add. Kl L U* 10 pluvie] statim future *add. Ba Kl L U* ‖ nec amplius] nam amplior *P* 11 descendit] per guttulas insensibiles sed magis ascendit *add. Ba Kl L U* 12 per guttulas insensibiles] *iter. P* 15 et] *om. P* 27 in unum] movendum *Kl* 30 sequuntur] frequenter *Ba Kl* frequenter *add. L U* 33 grossi] multi et grossi *Ba Kl L U*

⟨8⟩ Aliud signum pluvie future est asperitas radiorum Solis; nam talis significat calorem fortem potentem elevare vaporem multum et grossum, ex quo fiunt pluvie.

⟨9⟩ Aliud signum est quod in occasu Solis apparent nubes grosse et nigre. Ille enim de nocte propter absentiam Solis non possunt consumi et convertuntur in aquam et sequitur pluvia.

⟨10⟩ Aliud signum, sicut quidam dicunt, est apparitio iridis[32]. Illa enim significat pluviam, sed non significat pluvias magnas. Unde dicitur quod iris excludit cataclismum, id est magnum diluvium. Propter hoc enim vulgares dicunt | quod semel, scilicet tempore Noe, omnia in isto mundo per dilu- *L* 209va vium fuerunt submersa, et quod, iride apparente, adhuc mundus durabit per triginta annos. Apparitio igitur iridis non significat pluviam | magnam. *Ba* 32vb Etiam dicitur quod, si ventus veniat contra Solem, iride apparente, designat pluviam statim futuram; si autem e converso, designat pluviam statim preteritam, et causa istius videbitur postea[33].

⟨11⟩ Aliud signum est quando lapides apparent humidi tamquam sudantes serenitate | preexistente. Et ratio huius est, nam hoc est signum quod iam *Kl* 41vb cadunt guttule, licet insensibiles, propter quas lapides sic apparent humectati, et quod iam materia pluvie est sursum | et preparatur et disponitur *U* 29vb statim ad pluviam futuram.

⟨12⟩ Quantum ad secundum, unum est signum pluvie future quando pulveres elevantur plus quam est consuetum. Hoc enim significat quod fortis fit elevatio vaporum multorum et grossorum, qui sunt materia pluvie future. Aliud signum est cum folia arborum in nemoribus moventur plus quam est consuetum propter eandem rationem. Aliud signum est magna diminutio fontium et fluviorum et velox; nam hoc est signum quod fit elevatio multorum vaporum, ratione cuius fontes et fluvii diminuuntur, qui quidem vapores elevati sunt causa pluvie. Aliud signum pluvie future est agitatio et tumor ipsius maris; tunc enim sequuntur pluvie magne et tempestates, et causa illius potest esse iam dicta. Aliud signum pluvie future est scintilla-

---

4 grosse et nigre] longe nigre et spisse *Ba Kl L* longe magne et spisse *U* 5 solis] et defectus caloris *add. Ba L U* 5–6 et convertuntur in aquam] nec disgregari *Ba Kl L U* 11 apparente] diluvium cessavit et ergo adhuc solent dicere quod iride apparente *add. Ba* 15 et causa istius videbitur postea] *om. L U* 18 cadunt] *om. P* 24 folia arborum] arbores *Ba Kl L U* 25 est consuetum] in aliis diebus *Ba Kl L U* ‖ est] *om. P*

---

32 Cf. ibid., III, 2, 372b19–20.

33 III.16.

tio olei in lampade, et istud signum ponit Virgilius[34]. Aliud signum pluvie future est quando calibs falcis secantis gramina apparet nigra et rubescens. Et plura talia consimilia sunt cause pluvie future de quibus difficile esset reddere rationem.

⟨13⟩ Aliud signum pluvie future est quando cornices volando clamant et *L* 209vb *Ba* 33ra in volando alis suis sonant[35], | et similiter multarum aliarum avium | cantus est signum pluvie future. Aliud signum pluvie future est quando yrundines volant basse et volant girando[36]. Aliud signum est quando musce intensius mordent quam consueverant. Aliud signum pluvie future est quando animalia sicut sunt boves et vacce avidius comedunt quam consueverant. Aliud *P* 261^r^ signum est confricatio animalium et similiter motus et fricatio aurum asini. | De hiis enim et consimilibus, nisi fingeremus, satis difficiliter possemus reddere rationem.

II.7 Utrum aqua naturaliter ascendat ad orificia fontium

Consequenter queritur utrum aqua naturaliter ascendat ad orificia fontium.

⟨1⟩ Et arguitur quod sic: illud quod movetur a loco sibi innaturali ad locum *Kl* 42ra sibi naturalem, movetur naturaliter; sed aqua ascendens | ad orificia fontium movetur a loco sibi innaturali versus locum sibi naturalem; ergo. Maior est nota. Minor probatur, quia locus naturalis aque est circa terram et super terram, et non sub terra, ex eo quod terra est gravior aqua; sed locus naturalis gravioris est sub loco naturali minus gravis.

⟨2⟩ Secundo: aqua non quiescit naturaliter in terra; ergo exiens et ascendens ad orificia fontium movetur naturaliter. Consequentia tenet; antece- *U* 30ra dens patet, quia nullus minus grave quiescit naturaliter sub magis | gravi.

⟨3⟩ Tertio ad idem: quia videmus in quibusdam fontibus aqua ascendere scaturiendo et quasi bulliat; sed non potest dici quod hoc sit violente, quia nullum violentum tam diu durat quam diu durat scaturies talis ipsius aque; igitur hoc est naturaliter.

---

8 musce] et pulices *add. L U* 8–9 intensius mordent] frequentius incidunt et intensius *Ba* 10 sicut sunt boves et vacce] puta boves *Kl* puta oves *L* puta boves vel vacce *U om. Ba* 10–11 aliud signum est confricatio animalium et similiter motus et fricatio aurum asini] *om. Ba Kl L U*

---

34 Virgilius, *Georgica*, I, vv. 391–392.

35 Cf. ibid., I, vv. 388–389.

36 Cf. ibid., I, v. 377.

⟨4⟩ Quarto ad idem: aliqui fontes sunt in montibus altis, in quorum montium cacumine vel prope scaturit aqua et ascendit de terra; modo hoc non esset nisi aliqua aqua ascenderet. Nullum enim videtur ibi esse violentans quod violente faciat illam aquam ascendere: nec enim hoc facit ventus, qui propellit illam aquam sursum, nec hoc facit aliquis calor elevans illam aquam sursum, nec hoc fit per aliquid artificium, sicut per rotas vel aliquid huiusmodi, mediantibus quibus illa aqua trahatur sursum.

⟨5⟩ In oppositum arguitur: nullum grave naturaliter ascendit; sed ita est quod aqua est gravis; ergo etc.; et per consequens | nec naturaliter ascendit ad ori- [*L* 210ra] ficia fontium.

⟨6⟩ In ista questione primo ponenda est una distinctio; secundo ponende sunt conclusiones.

⟨7⟩ Quantum ad primum, aqua potest ascendere sursum simpliciter violente, ut si aqua proiceretur sursum vel si aqua traheretur sursum et altius quam esset locus sue originis. Secundo modo aqua potest ascendere naturaliter ad compressionem alicuius gravioris. Verbi gratia, si esset aliquod vas plenum aqua et interponeretur terra pulverizata, aqua ascenderet et staret super terram, et hoc naturaliter, | quia talis est ordo naturalis gravium et levium et [*Kl* 42rb] inclinatio naturalis | ipsorum quod graviora appetunt esse sub levibus et [*Ba* 33rb] leviora appetunt esse supra gravibus. Unde, si in medio aeris esset aliquod vas plenum aqua et interponeretur illi aque lignum, ita nataret ibi super aquam sicut hic. Et si esset in fundo illius aque, naturaliter ascenderet et tamen, si non haberet sub se immediate illam aquam, descenderet. Tertio modo aqua ascendit quodammodo naturaliter quando plus est ab una parte quam ab alia, et istud non est proprie moveri sursum, sed proprie est adequari, ita quod quando ex una parte est | alta et ex alia depressa, tunc ubi [*U* 30rb] est altior descendit usque ad illam partem ubi est depressior, et tunc illa que est in parte magis depressa ascendit, sic quod ibi fit quedam adequatio. Quarto aqua ascendit naturaliter per alterationem, sicut contingit ex elevatione vaporum.

---

3 ascenderet] de terra *add. Ba Kl L U* 4 ascendere] de terra *add. Ba Kl L U* 5–6 nec hoc facit aliquis calor elevans illam aquam sursum] nec hoc fit per aliquem calorem trahentem vel elevantem illam aquam sursum *L om. Ba Kl* 14–15 et altius quam esset locus] vel aliquo alio tali modo et ita violenter ascenderet ad locum *Ba Kl L U* 18 naturalis] *om. Ba Kl L U* 19 naturalis] *om. Ba Kl L U* 21 aqua] *om. P* 26 depressa] bassa *Ba Kl L U* 27 ubi est depressior] bassiorem *Ba* ubi est bassior *Kl* que est bassior *L U* 28 depressa] bassa *Ba Kl L U* 29 contingit] communiter fit *Ba Kl L U*

⟨8⟩ Quantum ad secundum, sit prima conclusio: aqua non ascendit ad orificia fontium primo modo, videlicet violente, quia talis ascensus ipsius aque ita diu durat quod si esset violentus nullo modo ita diu duraret. Secundo, quia sicut arguebatur, non videtur ibi esse aliquid violentans quod violente faciat aquam sic ascendere ad orificia fontium.

⟨9⟩ Secunda conclusio: aqua non ascendit naturaliter ad orificia fontium secundo modo, videlicet per compressionem alicuius gravioris, puta terre. Nam, ex quo terra est solida et non fluxibilis nec pulverisata, per quam transit aqua que ascendit in orificiis | [*L* 210rb] fontium, ipsa non comprimit aquam, sed aqua | [*P* 261v] libere transit per meatus terre quas invenit, | [*Kl* 42va] et non compressa usque ad orificia fontium. Si autem diceretur quod terra comprimeret aquam quia non esset ita solida sicut dicis, sed purgata et fluxibilis, sequeretur quod terra deberet plus et plus esse aquis cooperta, sed hoc est falsum. Consequentia tenet ex eo quod si terra esset taliter fluxibilis et purgata, descenderet et comprimeret aquam, et appareret esse sub aqua et aqua super ipsam. Et quia non potest esse impedimentum, cum terra esset sic purgata, satis cito terra congregaret se in unum in medio aque et circumquaque poneret se aqua et cooperiret terram, cuius oppositum videmus. Et ideo, quia terra est solida et partes terre se tenent ad invicem, pori et meatus | [*Ba* 33va] terre manent per quos aqua transit per terram non compressa. Nec ipsa terra facit sic ascendere aquam per compressionem, sicut tamen aliqui ymaginabantur; et propter hoc concesserunt quod tandem erit diluvium universale et terra totaliter aquis cooperta et in medio aquarum congregata.

⟨10⟩ Tertia conclusio: aqua ascendit naturaliter ad orificia fontium, sed numquam superius nec altius quam sit locus sue originis, sed ad quemlibet infra. Istut patet experimentaliter in conductibus aquarum per canales de uno loco ad alium. Unde, cum aqua sic ducitur et in aliquo loco altiori in aliquo canali ascendit, non potest altius ascendere quam est locus sue originis, sed in quolibet loco citra potest ascendere. Et per istud possent inveniri altitudines aliquorum locorum. Et ita etiam est in montibus: si in aliquo loco scaturit aliquis fons, signum est quod talis aqua venit a loco altiori et principium sue originis est in loco altiori. Et ita | [*U* 30va] aqua potest | [*Kl* 42vb] duci per canales sic quod ascenderet in cacumine alicuius montis dum tamen principium sue

---

5 ad orificia fontium] ad locum sue originis *Kl L U om. Ba* 10 per] orificia et *add. Kl* 12 purgata et fluxibilis] magis pulverizata *Ba L* pulverizata *Kl U* 14 purgata] pulverizata *Ba Kl L U* 16 purgata] pulverizata *Ba Kl L U* 20 per terram non compressa] *om. Ba Kl L U* 21 aquam] terram aqua *P* 23 et in medio aquarum congregata] *om. Ba Kl L U* 26 infra] *om. P* 27–28 altiori in aliquo] *om. P* 33 dum tamen] tamen dum *P*

originis esset in loco altiori, sed impossibile est aliquem fontem esse in aliquo loco nisi principium sue originis sit in loco altiori. Et propter hoc in terra plana non solent esse fontes scaturientes: quia enim una pars terre non est altior alia, si aliqua aqua generatur in una parte illius terre, non fluit ad aliam, propter hoc quod illa pars terre non est declivior; sed aqua semper fluit ad locum decliviorem. Et est sciendum quod talis motus et scaturies aque in orificiis | fontium non est proprie ascensus, sed est magis descensus, ex quo *L* 210va oritur in loco altiori. Et igitur, cum volumus habere fontes in aliqua terra plana, fodimus in ea foveas profundas, et tunc aqua que generatur in poris in superiori parte descendit versus fundum illius fovee tamquam ad locum decliviorem et ibi scaturit.

⟨11⟩ Ex hiis sequitur quod fontes non veniunt a mari. Patet, quia fontes non veniunt a loco decliviori ad locum altiorem quam sit locus originis ipsorum; sed locus maris est locus declivior quam sint loca ipsorum fontium; igitur sequitur etiam quod fontes et fluvii magis descendunt ad marem. Patet, quia semper petunt | locum decliviorem; locus autem maris est declivior. *Ba* 33vb

⟨12⟩ Sed diceret aliquis: communiter tamen dicitur fluvios omnes venire a mari. Breviter hoc negatur; sed bene conceditur aliquos fieri ex mari propter exalationem et evaporationem que fit a mari, quam elevationem sequitur pluvia ex qua aliquando fiunt quidam fluvii, vel saltem faciunt ad augmentationem fluviorum.

⟨13⟩ Quarta conclusio: aqua etiam potest ascendere ultra quarto modo. Sed tamen sciendum quod illa non est aqua pura, neque substantialiter neque qualitative, | sed in via alterationis ad aliud elementum. Sic igitur *P* 262^{r} patet finaliter quod semper oportet principium originis fontium et aquarum esse altiorem quam sunt scaturies ipsorum.

⟨14⟩ Ad rationes. Ad primam: 'omne elementum movetur naturaliter', etc., dico quod quantum est de natura aque, locus naturalis aque esset in terra et in medio mundi, ex quo cuiuslibet gravis locus naturalis est medium mundi; sed tamen, quia terra est gravior aqua, terra existente in medio mundi, non est naturale aque esse in medio mundi, sed est sibi naturale quod sit super terram, cum minus gravia naturaliter sint super magis gravia. Et igitur dico

1–2 sed impossibile est aliquem fontem esse in aliquo loco nisi principium sue originis sit in loco altiori] *om. hom. Ba Kl L U* 10 fundum illius fovee] illam foveam *Ba Kl L U* 20 fluvii] fontes *Ba Kl L U* 24 alterationis] generationis *Ba Kl L U* 25 principium] *om. P* 28 quod] ista ratione quod *add. P* ‖ locus naturalis aque] ipsa *Ba U* aqua *Kl U* 30 terra existente in medio mundi] *om. Ba Kl L U* 31 in medio mundi] in terra *Ba L U om. Kl*

Kl 43ra quod rebus stantibus sicut stant, aqua naturaliter | exit terram et ascendit naturaliter in orificiis fontium, dum tamen locus in quo sit sit altior.

⟨15⟩ Ad secundam consimili modo. Ad tertiam et quartam similiter.

Hoc de questione.

II.8 Utrum aque fontium generentur in terra

U 30vb Consequenter queritur utrum aque fontium generentur in terra.

⟨1⟩ Et arguitur primo quod non: aque fontium fiunt ex aqua pluviali; ergo non de novo generantur in terra. Consequentia tenet. Antecedens patet ex eo quod quando sunt multe et magne pluvie, sicut in hieme, aque fontium sunt maiores; quando autem sunt pauciores, sicut in estate, aque fontium sunt minores.

⟨2⟩ Secundo: terra de natura sua est sicca; aqua autem de natura sua est *L* 210vb humida; | ergo non videtur quod aqua fontium generetur in terra, cum sic terra et aqua contrarientur ad invicem.

⟨3⟩ Tertio ad idem: terra non generatur in loco aque; ergo nec aqua generatur in loco terre, et per consequens aque fontium non generantur in terra.

⟨4⟩ Quarto ad idem: aqua fontium provenit ex mari; igitur non generatur in terra. Consequentia tenet. Antecedens probatur, quia cum tot et tanti fluvii fluant ad mare, mare tantum augeretur quod esset ultra modum magnum nisi e contrario a mari procederent et pervenirent fontes qui sunt principia fluviorum.

⟨5⟩ Quinto ad idem: si sic, hoc esset ex vapore elevato in terra; sed hoc non, quia deficit ibi elevans.

⟨6⟩ Oppositum dicit Aristoteles[37].

⟨7⟩ In ista questione primo ponende sunt alique opiniones et improbande; secundo videndum est de vero modo generationis fontium; tertio videndum est quomodo aque generate in terra exeunt de terra.

1–2 exit terram et ascendit naturaliter in orificiis fontium] ascendit super terram et naturaliter est super eam *Ba Kl L U* 2 in quo sit] originis *Ba Kl L U* 18 tantum augeretur quod esset ultra modum magnum] super modum augeretur *Ba Kl L U* 26 exeunt] fluunt *Kl* effluunt *L U*

---

37 Aristoteles, *Meteorologica*, I, 13, 349 b 23–35.

⟨8⟩ Quantum ad primum, | sciendum est quod erat una opinio que ponit *Ba* 34ra in terra | aliqua susceptacula aquarum pluvialium[38]. Et ista opinio ponit in *Kl* 43rb terra non esse novam generationem aque, sed cum plueret, illa aqua absorberetur a terra et congregaretur in illis susceptaculis et cavernis, a quibus susceptaculis ymaginabantur effluere fluvios. Et cum istud susceptaculum esset tantum quod in eo posset tantum recipi de aqua quantum posset ab eo effluere donec iterum de novo in eo reciperetur aqua, tunc esset principium fluvii qui non exsiccaretur per annum, sed de uno anno flueret in alium. Cum autem istud susceptaculum non esset tantum, non posset esse principium fluvii qui non exsiccaretur in anno. Et ulterius ista opinio ymaginabatur quod ista aqua que exit ab illo susceptaculo in estate, restauratur in hieme. Et talia susceptacula aquarum non posuerunt unum vel duo, sed multa, secundum quod in terra multi fluunt fluvii.

⟨9⟩ Sed breviter ista opinio non valet, quia aliquando ex pede | unius mon- *U* 31ra tis per annum tantum exit de aqua quod, si tota esset simul, esset maior quam totus ille mons; ergo non potest poni quod ista tota aqua fuisset simul in illo monte. Sed diceres: illa aqua non solum fuit in illo monte vel in illa terra que est super locum effluxus illius fluvii, sed etiam | infra. Hoc non *L* 211ra valet, quia tunc oporteret dicere quod aqua altius ascenderet quam esset locus sue originis, quod est contra dicta in alia questione[39]. Secundo, nam dicit Seneca esse visum per experientias aquam pluvialem non intrare terram ultra decem pedes[40], et tamen, si inferius foderetur, aqua inveniretur; ergo non potest dici quod | omnis aqua que est in terra veniat a pluvia. Ter- *Kl* 43va tio, quia cum pluit, modicum de pluvia imbibitur et absorbetur a terra, et residuum fluit ad fluvios et ex hoc fluvii augentur; ergo non potest dici quod solum ex aqua pluviali proveniat illa aqua que exit | a terra. Quarto: hoc *P* 262^{v} maxime apparet in montibus ex quibus proveniunt fontes et multi fluvii et magni. Cum pluit, quasi tota aqua que cadit super illos montes descendit de superficie illorum montium, propter hoc quod semper aqua petit locum decliviorem, et quasi nichil absorbetur a montibus de illa aqua; ergo non potest dici quod ex aqua pluviali fiat quod fontes et fluvii fluant de pedibus

---

11 in estate] *om. Ba Kl L U* ‖ restauratur] per pluvias *add. Ba Kl L U* 18 que est super locum effluxus] super quam est fluxus *Ba* que est super principium effluxus *Kl L* que est super principium fluxus *U* 20 locus] principium *Ba Kl L U* 21 non] *P* 24 pluvia] aqua pluviali *Ba Kl L U* 29 de] in *P*

---

38 Ibid., I, 13, 349 b 2–15; cf. Senecam, *Questiones naturales*, III, 8; Albertum Magnum, *Meteora*, lib. 2, tract. 2, cap. 2, ed. Colon. VI/1, $66^{46–54}$.

39 II.7, 10.

40 Seneca, *Questiones naturales*, III, 7, 1.

montium. Verum est tamen quod bene aliqualiter aqua pluvialis facit ad hoc *Ba* 34rb quod fontes et fluvii augentur, sed numquam aqua | pluvialis sufficeret ad continuationem fontium et fluviorum nisi etiam de novo aqua generaretur in terra.

⟨10⟩ Alia erat opinio quod fontes venirent ex mari. Et motivum illorum erat istud, quia cum tot et tanti fluvii fluant ad mare, mirabiliter mare fieret magnum, nisi e contrario ex mari provenirent fontes et fluvii[41]. Modus autem per quem fierent ex mari fontes et fluvii secundum eos erat iste, quod terra esset cincta ipso mari tamquam una corrigia, et terra de mari imbibit aliquam partem maris, et tunc per calorem Solis elevantur vapores in terra ab illa aqua imbibita. Et cum illi vapores inveniunt poros terre frigidos, ex illa frigiditate condensantur et convertuntur in aquam et fiunt gutte aquarum. *U* 31rb Et cum ille in multitudine congregantur, | faciunt sibi exitum a terra et fiunt fontes et fluvii. Et etiam aliquando, cum iam aliqui vapores sunt conversi in aqua, adhuc per calorem Solis continue vapores elevantur et aliquando *L* 211rb *Kl* 43vb vapor posterior aquam | ante genitam expellit, | et sic aliquando aqua ex terra ascendit per modum ebullitionis.

⟨11⟩ Contra istam opinionem arguitur primo auctoritate Senece, qui dicit quod ultra profunditatem trecentarum altitudinum hominis invenitur terra solida sine aqua[42]. Modo hoc non esset verum si a mari aqua imbiberetur a terra, que elevaretur usque huc ubi generantur fontes et fluvii (istud quod dicit Seneca dicit esse visum per fossores). Secundo, si opinio esset vera, tunc maxime in parvis montibus prope maria deberent esse plures fontes quam in magnis montibus remotis a mari. Oppositum autem videtur ad experientiam. Consequentia probatur, quia ex quo illi montes essent magis prope mare, plus deberent habere de aqua imbibita a mari et elevata in eis per calorem Solis pro generatione fontium et fluviorum. Tertio, si vapores haberent potestatem expellendi aquam ante se, que ascenderet a terra per modum ebullitionis, sequitur quod etiam haberent potestatem impediendi ingressum aque pluvie; sed hoc est falsum.

⟨12⟩ De isto plures possent recitari opiniones, quas dimitto propter brevitatem; sed opiniones iam recitate inter alias non tenendas sunt principaliores, propter quod eas recitavi.

2 aqua] *om. P* 6 tot] *om. P* 6–7 mirabiliter mare fieret] iam mare esset *Ba Kl L U* 13 fiunt] fluunt *Ba Kl L U* 14 cum] tunc *P* 16 ex terra] *om. Ba Kl L U* 21–22 istud quod dicit seneca dicit esse visum per fossores] *om. Ba Kl L U* 23 fontes] et plures *add. P* 27 haberent] *om. P Kl* 28 ante se que ascenderet a terra] ascendentem ad terram *Ba Kl L U*

41 Cf. Aristotelem, *Meteorologica*, II, 2, 355 b 21–25; Senecam, *Questiones naturales*, III, 5.

42 Seneca, *Questiones naturales*, III, 7, 3.

⟨13⟩ Quantum ad secundum, sit prima conclusio: aqua generatur in terra. Patet, quia in terra recipitur aqua et de terra effluit aqua; sed ista non provenit ex aqua pluviali nec ex mari, sicut iam probatum est. Nec potest dici vere alio modo preter per nove | aque generationem, quod si dicatur faciliter *Ba* 34va improbabitur; ergo dicatur quod aqua generatur in terra.

⟨14⟩ Secunda conclusio: terra usque bene inferius est mixta, et non pura. Patet, quia cum fiunt fossiones in terra, inveniuntur minere et aliqua mixta in terra, quod non esset si terra totaliter inferius esset pura et non mixta.

⟨15⟩ Tertia conclusio: terra versus superficiem est porosa. Patet, quia cum pluit, aqua pluta imbibitur et absorbetur pro parte | a terra, quod non esset *P* 263r nisi terra esset porosa; unde alicubi etiam in terra inveniuntur ita magne concavitates et voragines quod aliquando, | sicut dicit Seneca, absorbuntur *Kl* 44ra civitates et ville[43]. Secundo, nam visum est quod fossores fodientes in terra inveniunt concavitates et voragines in terra. Omnia ista sunt signa quod terra est porosa.

⟨16⟩ Quarta conclusio: in huiusmodi poris terre sunt aliqua | corpora que *L* 211va non sunt terra, sicut vapor vel aer vel aliquid huiusmodi. Patet, quia aliter esset vacuum in natura.

⟨17⟩ Istis conclusionibus positis, videtur modus generationis aquarum in *U* 31va terra. Et est talis quod vapor vel aer existens in huiusmodi poris terre per frigus terre condensatur et in aquam convertitur, et postea, hiis corporibus conversis in aquam, aqua petente descensum, in locum decliviorem descendit, et illi pori novo corpore replentur, quod iterum per frigiditatem terre condensatur et in aquam convertitur. Et quando huiusmodi aque sic generate in terra aggregantur in aliquo loco terre, faciunt ibi unam bonam molem et quantitatem aque. Et est recte ymaginandum de isto sicut contingit in minori mundo, scilicet in homine; cum enim vapores ascendunt ad caput, convertuntur ibi, sicut dicitur, propter frigiditatem cerebri in reumata et aliquid simile aque, quod descendens exit per nares.

2 et] quia *P* 3 dici] *om. P* 4 quod si dicatur] quia si aliter dicatur *Ba* 6 et non pura] *om. Ba Kl L U* 7 minere et] multa alia que non sunt de conditione terre sicut *in marg. Ba* 9 superficiem] superiorem partem *Ba Kl L* 12 seneca] quod aliquando *add. P* 19 videtur] verus *add. Ba Kl L U* 23–24 pori novo corpore replentur quod iterum per frigiditatem terre condensatur et in aquam convertitur] vapores iterum condensatur et in aquam convertuntur et sic fiunt fluvii *Ba Kl L U* 25 unam bonam] magnam *Ba Kl L U* 28 sicut dicitur] sicut dicunt medici *Ba Kl L U* 29 aque] simile *P* ‖ quod] propter gravitatem *add. Ba Kl L U*

43 Ibid., III, 16, 4.

⟨18⟩ Quantum ad tertium, sciendum est quod aliquando multe aque generantur in terra et congregantur in terra et manent in terra per longum tempus propter hoc quod non inveniunt locum decliviorem per quem possint exire et fluere a terra. Et aliquando tam diu manent in terra quod in eis generantur pisces; unde, sicut recitat Seneca, per fossionem inventa sunt stagna in terra *Ba* 34vb in quibus erant pisces, et | homines comedentes ex illis moriebantur[44]. Cuius ratio est quia ipsi propter carentiam ipsius Solis erant venenosi, et propter hoc quedam anguille que generantur et manent in locis obscuris sunt venenose.

Kl 44rb ⟨19⟩ Aliquando autem, cum multe gutte aque generate | in terra congregantur et possunt habere exitum ad locum decliviorem, exeunt a terra, et sic fiunt fontes et fluvii. Et propter hoc ut in pluribus in pedibus et lateribus montium exeunt et oriuntur fontes et fluvii et continuantur per continuationem generationis aque in terra[45].

⟨20⟩ Ad rationes. Ad primam: 'aqua fontium generatur ex aqua pluviali'; negatur. Et quando dicitur: 'in hieme, cum multe cadunt pluvie, fontes', etc., respondetur quod hoc est ideo quia per casum pluvie terra plus disponitur et habilitatur ad hoc quod in terra possit generari aqua. Et propter hoc in *L* 211vb hieme fontes sunt maiores, | quia tunc vapores ex quibus generatur aqua in terra non exsiccantur calore Solis sicut in estate, et propter hoc in estate fontes fiunt minores.

⟨21⟩ Ad secundam dico quod quamvis terra naturaliter sit sicca, tamen, hoc non obstante, aqua potest ibi generari, cum ibi est materia ex qua potest generari aqua; unde nichil prohibet unum contrarium in loco alterius posse generari.

U 31vb ⟨22⟩ Ad tertiam dico quod immo terra | potest generari in loco aque; sed verum est quod plus generatur in terra de aqua quam generatur de terra in aqua, sed hoc est propter salutem mixtorum.

⟨23⟩ Ad quartam: 'videtur quod a mari veniant'; nego. Et quando probatur: 'tot et tanti fluvii ingrediuntur mare', etc., dico quod quia illi fluvii ingredientes mare disperguntur multum late calore Solis, tantum elevatur et consumitur de mari seu de fluviis ingredientibus mare quantum posset

2 et congregantur in terra] *om. Ba Kl* 5 seneca] quod *add. P* || stagna in terra] magna foramina in terra *Ba* magna stagna et flumina *L U* 6 homines] omnes *P* 7 ipsius solis] lucis *Ba Kl L U* 10 multe gutte aque generate] aque generate in multitudine *Ba Kl L U* 13 exeunt et oriuntur fontes et fluvii] effluunt fontes *Ba Kl L U* 22 naturaliter] *om. Ba Kl L U*

44 Ibid., III, 19, 1–3.

45 Cf. Aristotelem, *Meteorologica*, I, 13, 350 b 27–30.

augeri de fluviis mare ingredientibus. Et propter hoc mare quasi in eodem statu permanet absque hoc quod fontes et fluvii | proveniant a mari. Et de hoc Aristoteles dicit in littera, et exemplificatum erat de una quantitate aque dispersa per mensam et de tanta retenta simul[46]. *P* 263v

⟨24⟩ Ad ultimam: 'ibi deficit calor elevans', nego; immo calor Solis bene est tantus quod ab aliquibus partibus terre inferius potest elevare. | Et adhuc, posito quod deficeret elevans, posset dici, sicut dicebatur circa litteram, quod cum corpora que sunt in poris terre sunt conversa in aquam, de supra intrat aer ad replendum huiusmodi poros, quibus repletis, iterum generatur aqua, et sic continuando[47]. *Kl* 44va *Ba* 35ra

Hec de questione.

II.9 Utrum mare sit perpetuum vel aliquando fuerit factum

Consequenter queritur utrum mare sit perpetuum vel aliquando fuerit factum.

⟨1⟩ Et arguitur quod non sit perpetuum. Illud quod habet materiam non est perpetuum, sed mare habet materiam. Non enim est simplex forma nec simplex materia, sed compositum ex materia et forma.

⟨2⟩ Secundo: mare vel est mixtum vel est simplex elementum; sed quodcumque dicatur, mare non est perpetuum.

⟨3⟩ Tertio: partes maris sunt corruptibiles; ergo mare non est perpetuum. Consequentia tenet, nam quam cito aliqua pars maris corrumpitur, non est illud mare quod ante erat, cum totum non sit aliud | quam sue partes simul sumpte. Maior patet, quia calore Solis alique partes maris elevantur et tantum calefiunt quod convertuntur in aliam materiam, puta in aerem vel in ignem. *L* 212ra

⟨4⟩ Quarto: mare non est in loco sibi naturali; igitur non est perpetuum. Consequentia tenet, nam si est in loco sibi non naturali, cum nullum violentum sit perpetuum, sequitur quod non diu durabit in illo loco. Antecedens probatur, nam locus naturalis ipsius maris vel ipsius aque, cum sit grave, non est circa terram, sed in medio mundi, cum omnium gravium medium mundi sit locus naturalis.

3 una quantitate aque] aqua *Ba Kl L U* 8 de] *iter. P* 9 poros] terre *add. Ba Kl L U* ‖ repletis] *om. P* 10 et sic continuando] *om. Ba Kl* 22–23 tantum] tamen non *P* 23 quod] quia *P* 28 vel ipsius aque] *om. Ba Kl L U*

46 Ibid., II, 2, 355 b 20–32.

47 Ibid., I, 13, 349 b 23–35.

⟨5⟩ In oppositum arguitur secundum Aristotelem: mundus est perpetuus et corpora celestia sunt perpetua[48], et ista inferiora a celo reguntur[49], | *Kl* 44vb et ipsum mare pertinet ad integritatem ipsius mundi; ergo videtur quod mare sit perpetuum. Et ideo, si fuisset factum, | *U* 32ra non fuisset ratio quare plus in uno tempore quam in alio, sicut arguit Commentator in octavo *Physicorum* et Aristoteles probans eternitatem mundi[50].

⟨6⟩ In ista questione primo videndum est de quesito; secundo de terra mari cooperta et de loco naturali ipsius maris; tertio videndum est de refluxu aquarum.

⟨7⟩ Quantum ad primum, sit primo ista distinctio: mare manere vel fuisse perpetuum dupliciter intelligitur: uno modo secundum identitatem specificam; alio modo secundum identitatem naturalem—et hoc dupliciter, nam vel simpliciter vel secundum quid. 'Materiale simpliciter' voco si istud idem non mutatur secundum eius partes et | *Ba* 35rb semper fuit et semper erit, quomodo Aristoteles concederet celum esse sempiternum ab eterno et esse perpetuum. 'Identitatem' autem 'materialem secundum quid' voco prout identitas attenditur ex continuatione partium sibi invicem succedentium, quomodo diceremus quod idem Secana iam fluit Parisius qui fluxit ante duos vel tres annos.

⟨8⟩ Tunc sit prima conclusio: secundum identitatem materialem simpliciter nec terra nec mare etc. est perpetuum. Patet, nam a mari alique eius partes aliquando segregantur et corrumpuntur, et aliquando sibi alique de novo eveniunt, et per consequens non est idem in numero simpliciter nec quod erat iam sunt duo anni. Et hoc probant bene rationes ante oppositum.

⟨9⟩ Secunda conclusio: mare est perpetuum secundum identitatem materialem secundum quid. Patet, quia continuatur per continuationem partium sibi in perpetuo succedentium.

⟨10⟩ Tertia conclusio: mare est eternum secundum speciem ad istum intellectum: nunc est mare, semper fuit, semper erit. Hoc probabatur cum arguebatur ad oppositum. Nec obstat istis fides nostra, | *L* 212rb quia sic est dicendum loquendo naturaliter. Nichilominus tamen, loquendo supernaturaliter, con-

---

14 non mutatur] *iter. P* 18 duos vel tres] tres *Ba Kl* centum *L U* 22 segregantur et corrumpuntur] elevantur *Ba Kl L U* 22–23 de novo eveniunt] fluunt *Ba Kl L U* 24 anni] et similiter de terra *add. L* 28 eternum] perpetuum *Ba Kl L U*

---

48 Aristoteles, *De celo*, II, 1, 283 b 25–30.

49 Ibid., I, 2, 339 a 22–23.

50 Aristoteles, *Physica*, VIII, 1, 251 a 10–253 a 22; Averroes, *In Phys.*, ad locum.

cedendum est quod mare est creatum sicut alia, sed de hoc non est curandum | in proposito; unde dicit Albertus: 'quid michi de miraculis Dei, | cum de naturalibus dissero'?[51]. *P* 264^{r} *Kl* 45ra

⟨11⟩ Secundum istas conclusiones patet illam opinionem quorundam philosophorum non valere qui dixerunt quod in principio terra fuit cooperta aquis et alique partes istarum aquarum essent consumpte calore Solis, et ubi erant ille, terra esset modo discooperta, et residuum modo diceretur 'mare', et tandem totum exsiccaretur, et tunc amplius non esset mare, et tunc terra esset totaliter aquis discooperta[52]. Et movebantur quia videntes aliqua stagna vel fluvios viderunt ea aliquantulum tunc diminuta a longis temporibus; ita crediderunt | quod sic etiam fieret de mari. *U* 32rb Hoc motivum non valet, quia si aliqua stagna hic diminuuntur et tandem exsiccantur, alibi incipit terra humectari, et tandem fiunt ibi stagna ubi non erant prius. Nec etiam valet opinio illorum qui dixerunt | mare non esse aliud quam sudorem terre[53], nec *Ba* 35va etiam qui dixerunt quod generaretur ex fontibus. Et hec satis improbata sunt in littera[54].

⟨12⟩ Quantum ad secundum, scilicet de terra mari cooperta et de loco naturali maris, non curando nisi de magno mari, Albertus opinatur quod terra existens sperica cingitur a magno mari quasi una corrigia lata, procedente a polo artico per orientem ad polum antarticum et egrediendo per occidentem. Et ymaginatur ex utraque parte terram esse mari discoopertam, quamvis bene ibi sint stagna et maria parva fluentia ad mare magnum, et ita ipse ponit antipodes nobis, scilicet homines habitantes ex opposito nobis, habentes pedes suos contra pedes nostros, et zenith oppositum zenith nostro[55], et quando est nobis meridies, est eis media nox. Et istam opinionem Albertus sibi assumpsit propter aliquos antiquos philosophos et poetas

---

1 est creatum sicut alia] fuit factum de novo et creatum ex nichilo *Ba Kl L U* 2 albertus] in primo de generatione *add. Ba Kl L U* ‖ dei] divinis *Kl L* 3 dissero] loquor *Ba Kl L* 5 cooperta] totaliter *add. Ba Kl L U* 10 a longis] *iter. P* 12 stagna] *om. P*

51 Albertus Magnus, *De generatione et corruptione*, lib. 1, tract. 1, cap. 22, ed. Colon. V/2, 129^{15-16}: "nihil ad me de Dei miraculis, cum ego de naturalibus disseram". Cf. Sigerum de Brabantia, *Questiones in tertium De anima*, ed. B. Bazan in *Siger de Brabant, Quaestiones in tertium De anima. De anima intellectiva. De aeternitate mundi*, Louvain-la-Neuve/Paris 1972 (Philosophes médiévaux, 13), 83: "Sed nihil ad nos nunc de Dei miraculis, cum de naturalibus disseramus"; Iohannem de Ianduno, *Questiones in Metaphysicam*, I.16, Venice 1553, reprint Frankfurt am Main 1966, f. 13ra; *Auct. Arist.*, 168, n. 21.

52 Cf. Aristotelem, *Meteorologica*, II, 1, 353 b 6–11.

53 Cf. ibid., II, 1, 353 b 11–13; 3, 357 a 24–26.

54 Ibid., II, 1, 353 b 17–354 a 5.

55 Albertus Magnus, *Meteora*, lib. 2, tract. 2, cap. 11, ed. Colon. VI/1, 75.

qui locuti fuerunt de antipodibus[56]. Et probat hanc opinionem quia non *Kl* 45rb esset ratio quare ista pars terre | esset discooperta aquis et alia non; igitur videtur quod sic debeat dici. Etiam ista opinio videtur patere per describentes mappam mundi. Similiter videtur quod terra debeat sic esse cincta aquis propter naturalem ordinem elementorum. Naturalis enim ordo elementorum est quod aqua circumdet terram. Sed quia propter salutem animalium *L* 212va et plantarum non circumdat eam totam, | ad minus debet eam circuire ad modum talis corrigie ymaginate.

⟨13⟩ Sed videtur, si verum sit quod dicitur, quod ista opinio non sit vera, nam si mare magnum esset ymaginandum isto modo, clarum est quod non esset multum magne latitudinis, et per consequens posset transnavigari. Sed hoc est falsum, nam dicitur quod magni principes inceperunt transnavigare et navigaverunt per tres annos, et non potuerunt transnavigare ipsum, in cuius signum adhuc sunt quedam columne fixe in occidente que dicuntur 'columne eraclee', quas Aristoteles nominat in littera[57]. Secundo: cum continue cum fluviis fluat terra ad fundum maris magni, iam fundus maris esset tantum elevatus quod quasi iam nulla esset profunditas maris magni ibi *Ba* 35vb ubi ymaginatur Albertus | cincturam[58], et per consequens per talem fluxum terre una cum fluviis ad mare, mare continue plus et plus elevaretur et continue plus et plus de terra cooperiret.

⟨14⟩ Videtur ergo quod opinio Alberti, salva eius reverentia, non sit bona. Et ad rationes suas, cum dicit de antipodibus, negatur. Sed tamen bene *U* 32va conceduntur antipodes | de principio orientis ad istos de fine occidentis. Ad aliam de describentibus mappam mundi respondetur quod ipsi fuerunt *Kl* 45va eiusdem opinionis. Ad aliam: | 'non est ratio quare magis ex una parte', etc., respondetur quod ista ratio vadit contra istam cincturam, sicut contra alium modum ymaginandi de mari, nam similiter potest argui: non est ratio quare magis terra ex una parte est cincta mari quam ex alia. Dicatur igitur ad rationem quod sic fuit ab eterno quod una pars esset cooperta aquis et alia discooperta propter ordinem elementorum et salute animalium.

---

3 patere] facere *P* 3–4 describentes] describentem *P* 15 eraclee] eracliti *P* 21 salva eius reverentia] *om. Ba Kl L U* 28 quam] *iter. P* 30 animalium] et planctarum *add. Ba Kl L U*

---

56 Cf. Platonem, *Timeus*, 62 D–63 A; Aristotelem, *De celo*, IV, 1, 308 a 17–21; Plinium, *Naturalis historia*, VII, 9–11; Maurum Servium Honoratum, *Ad Georgica*, I, v. 235, eds. G. Thilo and H. Hagen, Leipzig 1881, $186^{9–20}$; *Ad Aeneidem*, VI, 127, ed. E. Jeunet-Mancy, Paris, 2012, 48.

57 Aristoteles, *Meteorologica*, I, 14, 350 b 1–3; II, 1, 354 a 12; 22.

58 Albertus Magnus, *Meteora*, lib. 2, tract. 2, cap. 11, ed. Colon. VI/1, $75^{1–46}$.

⟨15⟩ Alia | [*P* 264^{v}] erat opinio que ymaginabatur quod mare et terra sunt eccentrica, ita quod non est idem centrum maris et centrum terre. Et propter hoc est quod mare in loco est et quiescit naturaliter et alii fluvii fluunt ad mare tamquam ad locum sibi naturalem. Et sic erat ab eterno eccentrica ut una pars terre esset discooperta aquis propter salutem animalium et planctarum. Ista opinio non valet, nam si sic, sequeretur quod si una pars terre et una pars aque dimitterentur eadem de uno loco, deberent naturaliter tendere ad diversa loca, cuius oppositum videmus, quia ambo tenderent ad medium mundi. Consequentia probatur, quia quodlibet tenderet ad centrum sui loci naturalis; si igitur aliud est centrum terre et aliud centrum aque circa que quiescunt naturaliter, | [*L* 212vb] sequitur istud quod dictum est. Secundo, nam tunc non esset idem centrum totius gravitatis, quod est falsum. Tertio, nam tunc terra discooperta aquis esset circularis figure, sed hoc est falsum, quia est longior quam sit lata. Consequentia tenet ex natura sperarum secantium se.

⟨16⟩ Aliter igitur dicendum est quod aliqua portio terre bene | [*Ba* 36ra] magna est discooperta magno mari, licet secundum eius multas partes sit cooperta stagnis. | [*Kl* 45vb] Alia autem pars terre est cooperta magno mari. Pro quo ymaginandum quod si terra esset uniformiter gravis, ita quod idem esset centrum sue magnitudinis et sue gravitatis, ipsa iaceret in medio mundi et esset circumquaque aqua circumdata. Sed quia ipsa non est uniformiter gravis, quia portio terre discooperta aquis est magis levis, propter hoc centrum magnitudinis terre non est idem cum centro gravitatis terre, et centrum gravitatis terre multo plus distat a superficie convexa aquis discooperta quam distet a superficie convexa terre aquis cooperte, propter hoc quia ista superficies propinquior medio mundi in aqua naturaliter locatur tamquam in loco decliviori. Et ad istum locum fluit omnis aqua naturaliter tamquam ad locum decliviorem. Et ex isto statim | [*U* 32vb] patet quod ille locus quem occupat mare est locus naturalis aque.

⟨17⟩ Et ymaginandum est quod centrum gravitatis ex terra et aqua est medium mundi. Ulterius ymaginandum est quod aliquando centrum magnitudinis terre et similiter aliquando centrum gravitatis terre et similiter aliquando superficies convexa terre mari cooperta erit in superficie convexa terre mari discooperta et e contrario, per istum modum quod cum fluviis fluentibus ad mare aliquando partes terre fluunt ad fundum maris, et sic ex illa parte fit gravior terra, et illa terra que prius erat centrum gravitatis pel-

---

3–4 est quod mare … eterno eccentrica ut] *om. Ba Kl L U* 13 terra] pars terre *Ba Kl L U* ‖ figure] figurative *P* 17 stagnis] et maribus parvis ab ymbribus *add. Kl L* et parvis maribus *add. Ba U* 20 uniformiter gravis] uniformis *Ba Kl L U*

litur versus nos et tandem per continuum fluxum aliquarum partium terre cum fluviis ad mare erit in superficie convexa terre mari discooperta, et e contrario.

⟨18⟩ Secundum etiam istam ymaginationem ostendi potest qualiter fiunt *L* 213ra montes, quia alique partes terre | sunt magis solide et lapidose, que non ita *Kl* 46ra fluunt cum fluviis ad mare sicut partes prope minus solide, sed manent, | et sic fiunt montes. Et propter hoc bene credendum est quod de die in diem augentur montes, licet insensibiliter.

Ba 36rb ⟨19⟩ Quantum ad tertium, scilicet de fluxu fluviorum et aquarum, quare fluvii plus fluunt a septentrione versus austrum quam e contrario, ratio est quia versus septentrionem est magnum frigus et est locus aptus per generationem aquarum, et versus austrum, quia ibi est magnus calor; ideo non est locus aptus per generationem aquarum[59]. Et ideo quia versus septentrionem est habundantia aquarum, et non tanta versus austrum, inde est quod fluvii fluunt a septentrione versus austrum, et non e contrario. Et iste fluxus modicum apparet in fluviis magne latitudinis et plus apparet in fluviis magis *P* 265r strictis et ubi fiunt minus late propter strictitudinem loci. Verum est | tamen quod in quibusdam locis aliquando fit e contrario, scilicet quod fluvii fluunt ab austro versus septentrionem, et hoc est ibi ubi est tantus calor, sicut sub via Solis, quod aque quasi fervent et bulliunt, et ibi aqua tantum rarefit et extenditur quod movetur versus septentrionem, unde auctor unus in quo- *U* 33ra dam tractatu de fluxu maris | dicit quod in aliquibus locis mare bulliendo movetur in altum.

⟨20⟩ Et sic rationes vadunt viis suis.

Hec de questione.

II.10 Utrum mare debeat fluere et refluere

Consequenter queritur utrum mare debeat fluere et refluere.

⟨1⟩ Et videtur quod non, nam si sic, cum hoc fiat ordinate et continue, non videtur quod hoc fiat violente; ergo videtur quod hoc fiat secundum naturam. Sed probo quod non, quia omne agens naturale determinatum est ad alterum oppositorum, non potens in ambo opposita—nono *Metaphysice*[60].

---

6 partes] parte *P* 12–13 et versus austrum … per generationem aquarum] *om. hom Ba Kl P*

---

59 Cf. Aristotelem, *Meteorologica*, II, 1, 354 a 23–26.

60 Aristoteles, *Metaphysica*, IX, 2, 1046 b 4–6; cf. *Auct. Arist.*, 134, n. 222.

Sed mare fluendo et refluendo movetur motibus oppositis; ergo non videtur quod hoc sit naturaliter. Si ergo mare non fluit nec refluit violente nec etiam naturaliter, videtur quod non debeat | fluere et refluere. *Kl* 46rb

⟨2⟩ Secundo: contra naturam maris est ascendere; sed mare non posset sic fluere et refluere nisi ascenderet. Cum enim efflueret, ascenderet, et cum reflueret, descenderet.

⟨3⟩ Tertio: unumquodque ens | in loco sibi nautrali debet quiescere; sed *L* 213rb ita est quod locus quem occupat mare est sibi naturalis, sicut predictum est[61]. Ergo in isto loco debet quiescere et non debet sic moveri fluendo et refluendo.

⟨4⟩ Quarto si mari esset naturale sic fluere et refluere, tunc talis fluxus et refluxus magis deberet competere | magnis maribus quam parvis, que *Ba* 36va 'brachia parva maris' nominantur. Sed consequens patet falsum per experientiam.

⟨5⟩ Oppositum patet per experientiam. Videmus enim mare fluere et refluere.

⟨6⟩ Breviter per 'fluxum' et 'refluxum' maris intenditur accessus et recessus maris versus nos, de quo primo videndum est; secundo sunt quedam dubitationes; tertio videndum est de quibusdam confortantibus fluxus et refluxus ex quibus apparebunt solutiones dubitationum.

⟨7⟩ Quantum ad primum, sciendum quod quidam ymaginabantur quod motus diurnus efficiat istum motum maris. Ymaginabantur enim quod sicut motu diurno movetur ignis circulariter et aer qui est ultra periferias altissimorum montium movetur similiter isto modo[62], sic etiam ymaginabantur de motu ipsius maris, cum omnia ista inferiora fluxibilia nata sunt assequi istum motum, scilicet motum diurnum, quantum patitur eorum natura. Et ergo dixerunt quod aqua non potest perficere suum circuitum circa terram propter hoc quod non potest tantum ascendere quantum est elevata terra mari discooperta. Et quando usque ibi est mota per eius effluxus | quod non *Kl* 46va potest ulterius | ascendere, tunc e converso refluit ad suum locum. Et ita isti *U* 33rb

2 non] *om. P* 8–9 sicut predictum est] ex precedenti questione *Ba* sicut dicebatur in precedenti questione *Kl L U* 13 brachia parva maris] parva stagna maris *Ba Kl* quedam parva stagna maria *L U* 21 quidam] antiqui *Kl L U* 29 discooperta] cooperta *P*

61 II.9, 16.

62 Cf. I.8, 23.

causam fluxus et refluxus maris assignant motum diurnum[63]. Sed hoc non valet, quia fluxus et refluxus maris non regulantur in succedendo sibi secundum motum diurnum ita quod fiant bis vel semel precise in die naturali, sed regulantur secundum motum Lune, ut post dicetur[64]. Ergo causa propria fluxus et refluxus maris non debet assignari motus diurnus, sed magis motus Lune.

⟨8⟩ Aliter ergo dicendum est. Pro quo primo videndum est quid intelligatur per 'circulum meridianum', et tunc ponende sunt tres suppositiones, deinde una conclusio. Breviter per 'circulum meridianum' intelligitur ille circulus qui dividit horizontem et intesecat ipsum ad angulos rectos sperales, *Ba* 36vb *L* 213va transiens de uno polo ad alium, qui | sic se habet quod cum Sol fuerit in aliquo eius puncto, est meridies illis super quorum emisperio est Sol.

⟨9⟩ Tunc sit prima suppositio quod Luna habet dominium super humiditates aquosas in moveando eas et faciendo eas tumescere. Propter hoc enim videmus quedam animalia quorum complexio consistit in humiditate quasi fluente, sicut sunt conche et ostriche, ad augmentationem virtutis *P* 265v Lune augeri | et in diminutione virtutis Lune diminui. Istam suppositionem pluribus aliis signis declarat Albertus, capitulo septimo, tractatus secundi *De proprietatibus elementorum*[65].

⟨10⟩ Secunda suppositio est quod quilibet planeta fortius agit in aliquam regionem cum est in circulo meridiano et in illo puncto in quo est Sol cum est in meridie, quam quando non est in illo circulo. Et huius ratio est quia tunc magis de directo respicit ista que sunt sub illo circulo, et per consequens *Kl* 46vb Luna tunc fortius agit in ista inferiora cum | sit in isto circulo.

⟨11⟩ Tertia suppositio, quod quilibet planeta etiam magnam virtutem habet cum est in puncto opposito sub Terra. Patet, quia tunc radii eius reflectuntur ab octava spera et a stella sibi opposita versus illas partes que sunt sub meridiano talis regionis.

⟨12⟩ Istis suppositis sit ista conclusio quod motus maris, scilicet fluxus et refluxus, est effective a motu Lune. Patet, quia ad adventum radiorum Lune

3 semel] vel ter *add. Ba Kl L U* 11 transiens de uno polo ad alium qui] procedendo ab oriente versus occidentem ita quod meridianus *Ba Kl L U* 16 ad augmentationem] *om. P* 17 istam suppositionem] *iter. P* 24 sit] tunc *add. P* 25 quilibet planeta] *om. P* 25–26 etiam magnam virtutem habet] fortius agit *Ba Kl L U*

63 Cf. Albertum Magnum, *De causis proprietatum elementorum*, lib. 1, tract. 2, cap. 7, ed. Colon. v/2, 72^{79-88}.

64 I.10, 12.

65 Albertus Magnus, *De causis proprietatum elementorum*, lib. 1, tract. 2, cap. 7, ed. Colon. v/2, 73^{66}–74^{67}.

super nostrum emisperium mare augetur tumescendo, et cum sic tumescit movetur et diffunditur ad latera eius, et sic fit effluxus maris. Sed per recessum radiorum Lune, que specialiter habet dominium super humiditates, per primam suppositionem, aqua | maris relinquit istum tumorem et minoratur, *U* 33va et ita partes maris que effluxerunt revertuntur ad locum maris refluendo. Cum enim Luna descendit sub nostro emisperio et directo aspectu aspicit stellas sibi oppositas super nostrum emisperium, eis emittit vim suam motivam aquarum, propter quam reflexam ab illis stellis super mare, secundum ymaginationem tertie suppositionis, mare incipit iterum tumescere, sicut fecit ad presentiam Lune existentis in meridiano talis regionis in qua tumescit sic, et iterum mare effluit, licet motu debiliori. Sic ergo patet qualiter Luna existens super nostrum emisperium est causa fluxus et refluxus | maris *Ba* 37ra et etiam sub nostro emisperio: super nostrum enim | emisperium existens *L* 213vb est causa fluxus et refluxus maris mediantibus radiis rectis; sub emisperio autem existens est causa fluxus et refluxus maris mediantibus radiis reflexis.

⟨13⟩ Ex hiis patet quare duorum fluxuum existentium in die naturalis unus est maior altero. Luna enim existente super nostrum emisperium plus facit mare tumescere propter radios rectos; sub emisperio autem existens minus propter radios reflexos. Potest etiam apparere ex hiis quare effluxus sunt maiores in plenilunio et similiter in coniunctione Lune | cum Sole quam in *Kl* 47ra quadris: in plenilunio enim Luna maioris est virtutis propter hoc quia tunc multum habet lumen; in coniunctione autem eius cum Sole augetur eius virtus a Sole.

---

7 vim suam] *iter. P* 10–11 existentis in meridiano talis regionis in qua tumescit sic] respicientis illam regionem *Ba Kl L U* 15 reflexis] obliquis *Ba Kl L U* 17 nostrum] *om. P* 22 lumen] a sole *add. Ba Kl L U* ‖ cum] a *P* 23 sole] *des. P*

Appendix

The following table shows the correspondences between the questions of the second redaction of Oresme's *Questions on Meteorology* and the passages in Aristotle's *Meteorology*. References to Albert the Great's commentary are given when Oresme's texts is closer to Albert than to Aristotle.

Oresme		Aristotle
I.1	Utrum possibile sit de impressionibus metheorologicis habere simul scientiam et opinionem	I, 1, 339 a 1–2
I.2	Utrum impressiones metheorologice fiant secundum naturam inordinatiorem quam sit natura celi	I, 1, 338 b 20–24
I.3	Utrum iste mundus inferior sit continuus lationibus superioribus ut virtus eius inde gubernetur	I, 1, 339 a 22
I.4	Utrum, cessante motu celi, cessarent motus in isto mundo inferiori	I, 3, 340 b 15–19
I.5	Utrum eedem opiniones infinities reiterentur	I, 3, 339 b 28–30
I.6	Utrum elementa sint continue proportionalia ad invicem	I, 3, 340 a 11–17
I.7	Utrum quatuor elementa semper et immutabiliter habeant eandem proportionem ad invicem	I, 3, 340 a 11–17
I.8	Utrum motus celi sit causa calefactionis ignis in sua spera et etiam aeris superioris	I, 3, 340 a 24–29; 340 b 10–13; 340 b 29–341 a 5
I.9	Utrum lumen sit productivum caloris	I, 3, 340 a 24–32
I.10	Utrum contrarium circumstans suum contrarium fortificet ipsum	I.4, 341 b 36–342 a 10; I.12, 348 b 4–7; I.12, 349 a 4–8; I.12, 348 b 31–349 a 3; I.10, 347 a 36–b 10
I.11	Utrum semper media regio aeris sit frigida	I, 3, 340 a 26–27
I.12	Utrum omnium impressionum metheorologicarum vapor et exalatio sint principium materiale	I, 4, 341 b 1–12
I.13	Utrum impressiones ignite, seu ille que fiunt per inflammationem, fiant naturaliter in aere	cf. I, 4, 341 b 13–24
I.14	Utrum de nocte, serenitate existente, debeant apparere hyatus et voragines et sanguinei colores in celo	I, 5, 342 a 34–35

(*cont.*)

Oresme		Aristotle
I.15	Utrum cometa sit de natura celi vel elementari	I, 6–7
I.16	Utrum cometa sit exalatio calida et inflammata	I, 7
I.17	Utrum motus comete sit naturalis vel violentus	I, 7, 344 a 25–33
I.18	Utrum comete significent mortem principum, siccitatem et ventos et motus terre	I, 7, 344 b 18–20; 344 b 26–30; Albertus Magnus, *Meteora*, lib. 1, tract. 3, cap. 11
I.19	Utrum galaxia sit de natura celi vel de natura elementari	I.8
II.1	Utrum locus generationis pluvie sit media regio aeris	I, 11, 347 b 11–14
II.2	Utrum ros et pruina, nix et pluvia sint eiusdem speciei	I, 11, 347 b 16–18
II.3	Utrum grandines magis debeant generari in hieme quam in autumno	I, 12, 348 b 12–26
II.4	Utrum aqua calida applicata frigori congelanti citius congeletur quam aqua frigida	I, 12, 348 b 30–349 a 2
II.5	Utrum rubedo matutina sit signum pluvie	
II.6	Utrum caligo sit signum pluvie future	I, 9, 246 b 35–36
II.7	Utrum aqua naturaliter ascendat ad orificia fontium	cf. Albertum Magnum, *Meteora*, lib. 2, tract. 2, cap. 11
II.8	Utrum aque fontium generentur in terra	I, 13, 349 b 24–35
II.9	Utrum mare sit perpetuum vel aliquando fuerit factum	II, 3, 357 b 26–358 a 2
II.10	Utrum mare debeat fluere et refluere	II, 1, 354a5–35

Bibliography

Primary Sources

Agrimi, J. (ed.), *Le Quaestiones de sensu attribuite a Oresme e Alberto di Sassonia*, Florence 1983 (Pubblicazioni della Facoltà di lettere e filosofia dell'Università di Pavia, 29).

Albertus de Saxonia, *Questiones super Meteorologica*, in A. Panzica, "Albert of Saxony's *Questions on Meteorology*: Introduction, Study of the Manuscript Tradition, and Edition of Book I–II.2", *Archives d'histoire doctrinale et littéraire du Moyen Âge* 86 (2019), 231–356.

Albertus Magnus, *De generatione et corruptione*, ed. P. Hossfeld, Münster i.W. 1980 (Alberti Magni Opera omnia. Editio Coloniensis V/2).

Albertus Magnus, *Meteora*, ed. P. Hossfeld, Münster i.W. 2003 (Alberti Magni Opera omnia. Editio Coloniensis VI/1).

Alexander Aphrodisiensis, *Expositio libri Meteorologicorum Aristotelis*, ed. A.J. Smet, Leuven 1968 (Corpus Latinum Commentariorum in Aristotelem Graecorum, 4).

Algazel, *Metaphysica*, ed. J.T. Muckle, Toronto 1933 (St. Michael's Mediaeval studies).

Alhazen, *Perspectiva*, lib. I–III, in A. Mark Smith (ed.), *Alhacen's Theory of Visual Perception: a Critical Edition, with English Translation and Commentary, of the First Three Books of Alhacen's "De aspectibus", the Medieval Latin Version of Ibn al-Haytham's "Kitāb al-manāzir"*, Philadelphia 2001 (Transactions of the American Philosophical Society, 1).

Alhazen, *Perspectiva*, lib. VII, ed. P. Pietquin, *Le septième livre du traité De aspectibus d'Alhazen, traduction latine médiévale de l'Optique d'Ibn al-Haytham*, Brussels 2010 (Mémoires de la Classe des lettres, Collection in-80; 3e série, 5).

Anonymus, *Liber de causis*, ed. A. Pattin, *Le Liber de causis. Édition établie à l'aide de 90 manuscrits avec introduction et notes*, Leuven 1966 (Extrait du *Tijdschrift voor filosofie*, 28).

Anonymus, *Les Auctoritates Aristotelis: un florilège médiéval. Étude historique et édition critique*, ed. J. Hamesse, Louvain-la-Neuve/Paris 1974 (Philosophes médiévaux, 17).

Aristoteles, *Opera*, ed. I. Bekker, Berlin 1831–1870 (5 vols.).

Aristoteles, *De insomniis et de divinatione per somnum*, ed. H.J. Drossaart Lulofs in *De insomniis et de divinatione per somnum. A New Edition of the Greek Text with the Latin Translation*, Leiden 1947 (Aristoteles Latinus, XV 1–2).

Aristoteles, *Physica*, translatio Iacobi Veneti (translatio vetus), eds. F. Bossier and J. Brams, Leiden/New York 1990 (Aristoteles Latinus, VII 1–2) (2 vols.).

Aristoteles, *Meteorologica, translatio Gerardi Cremonensis*, ed. P.L. Schoonheim, *Aristotle's Meteorology in the Arabico-Latin Tradition: A Critical Edition of the Texts, with

Introduction and Indices, Leiden/Boston/Cologne 2000 (Aristoteles Semitico-latinus, 12).

Aristoteles, *Meteorologica. Recensio et Translatio Guillelmi de Moerbeka*, ed. G. Vuillemin-Diem, Brussels 2008 (Aristoteles Latinus, x 2.1–2) (2 vols.).

Averroes Cordubensis, *Aristotelis opera cum Averrois commentariis*, Venice 1562–1574, reprint Frankfurt am Main 1962 (9 vols.).

Averroes Cordubensis, *Commentarium magnum in Aristotelis De anima libros*, ed. F.S. Crawford, Cambridge (MA) 1953 (Corpus commentariorum Averrois in Aristotelem. Versiones latinae, 6/1).

Averroes Cordubensis, *In De somno et vigilia*, ed. A.L. Shields, *Averrois Cordubensis compendia librorum Aristotelis qui 'Parva Naturalia' vocantur*, Cambridge (MA) 1949 (Corpus commentariorum Averrois in Aristotelem. Versiones latinae, 7).

Averroes Cordubensis, *Commentarium magnum in De celo et mundo*, eds. J.F. Carmody and R. Arnzen, Leuven 2003 (Recherches de Théologie et Philosophie médiévales, Bibliotheca 4) (2 vols.).

Calcidius, *Timeus*, ed. J.H. Waszink, London/Leiden 1962 (Corpus Platonicum Medii Aevi. Plato latinus, 4).

Campanus de Novara, *Libri Elementorum Euclidis*, ed. H.L.L. Busard, Wiesbaden 2005 (Texte und Abhandlungen zur Geschichte der exakten Wissenschaften, 1).

Iohannes Buridanus, *Questiones super Meteorologica*, ed. S. Bages, *Les Questiones super tres libros Metheororum Aristotelis de Jean Buridan: étude suivie de l'édition du livre I*, Thèse de doctorat de l'École des Chartes, 1986 (2 vols.).

Iohannes Buridanus, *Questiones super De generatione et corruptione*, in M. Streijger, P.J.J.M. Bakker, and J.M.M.H. Thijssen (eds.), *John Buridan Quaestiones super libros De generatione et corruptione Aristotelis*, Leiden/Boston 2010 (Medieval and Early Modern Philosophy and Science, 14).

Ibn Mu'had, *De crepusculis et nubium ascensionibus*, ed. M. Smith, in "The Latin Version of Ibn Mu'ādh's Treatise on Twilight and the Rising of Clouds", *Arabic Sciences and Philosophy* 2 (1992), 83–132.

Iohannes de Ianduno, *Questiones in Metaphysicam*, Venice 1553, reprint Frankfurt am Main 1966.

Iohannes de Sacrobosco, *De sphaera*, in L. Thorndike (ed.), *The "Sphere of Sacrobosco" and Its Commentators*, Chicago 1949, 76–142.

Iohannes Pecham, *Perspectiva communis*, in D. Lindberg, *John Pecham and the Science of Optics*, Madison 1970 (The University of Wisconsin Publications in Medieval Science, 14).

Marcus Tullius Cicero, *De natura deorum*, ed. M. van den Bruwaene, Brussels 1978 (Collectio Latomus, 154).

Martinus Anglicus, *De obligationibus—Über die Verpflichtungen*, Lateinisch-Deutsch, ed. F. Schupp, Hamburg 2014 (Philosophische Bibliothek, 462).

Maurus Servius Honoratus, *Ad Georgica*, eds. G. Thilo and H. Hagen, Leipzig 1881.

Nicolaus Oresme, *Le livre de Éthiques d'Aristote*, in A.D. Menut (ed.), *Maistre Nicole Oresme: Le livre de Éthiques d'Aristote, Published from the Text of* MS. *2902, Bibliothèque Royale de Belgique, with a Critical Introduction and Notes*, New York 1940.

Nicolaus Oresme, *Questiones super De anima*, ed. C. Kren, unpublished Ph.D. dissertation, University of Wisconsin 1965.

Nicolaus Oresme, *De proportionibus proportionum and Ad pauca respicientes*, ed. E. Grant, Madison 1966 (The University of Wisconsin Publications in Medieval Science).

Nicolaus Oresme, *Questiones super De anima*, ed. B. Patar, Louvain-la-Neuve 1995 (Philosophes médiévaux, 32).

Nicolaus Oresme, *Questiones super De generatione et corruptione*, ed. S. Caroti, Munich 1996 (Bayerische Akademie der Wissenschaften. Veröffentlichungen der Kommission für die Herausgabe ungedruckter Texte aus der mittelalterlichen Geisteswelt, 20).

Nicolaus Oresme, *Questiones super Physicam*, eds. S. Caroti, J. Celeyrette, S. Kirschner, and E. Mazet, Leiden/Boston 2013 (Studien und Texte zur Geistesgeschichte des Mittelalters, 112).

Paulus Venetus, *Logica magna*, ed. E.J. Ashworth, Oxford 1988 (Classical and Medieval Logic Texts, 8).

Plato, *Timeus* (translatio Calcidii), ed. J.H. Waszink, London/Leiden 1962 (Corpus Platonicum Medii Aevi. Plato latinus, 4).

Plinius, *Naturalis historia*, ed. J. Beaujeu, Paris 1950.

Pseudo-Scotus, *Questiones super Meteorologica*, ed. L. Wadding, Lyon 1639, reprint Hildesheim 1968 (Iohannes Duns Scotus, Opera omnia) (12 vols.), 1: 1–125.

Ptolemeus, *Tetrabiblos*, translatio Guillelmi de Morbeka, in G. Vuillemin-Diem and C. Steel (eds.), *Ptolemy's Tetrabiblos in the Translation of William of Moerbeke. Claudii Ptolemaei 'Liber Iudicialium'*, Leuven 2015 (Ancient and Medieval Philosophy. Series 1, 19).

Robertus Grossatesta, *De lineis, angulis et figuris*, ed. L. Baur, *Die philosophischen Werke des Robertus Grosseteste, Bischofs von Lincoln*, Münster i.W. 1912 (Beiträge zur Geschichte der Philosophie des Mittelalters, 9), 59–65.

Robertus Grossatesta, *De cometis*, ed. C. Panti, in *Moti, virtù e motori celesti nella cosmologia di Roberta Grossatesta. Studio ed edizione dei trattati De sphera, De cometis, De motu supercelestium*, Florence 2001 (Corpus Philosophorum Medii Aevi. Testi e Studi, 16).

Seneca, *Questiones naturales*, ed. P. Oltramare, Paris 1961 (2 vols.).

Sigerus de Brabantia, *Questiones in tertium De anima*, ed. B. Bazan in *Siger de Brabant, Quaestiones in tertium De anima. De anima intellectiva. De aeternitate mundi*, Louvain-la-Neuve/Paris 1972 (Philosophes médiévaux, 13).

Statuta et Acta rectorum Universitatis Carolinae Pragensis: 1360–1614, eds. F. Šmahel and G. Silagi, Prague 2018 (Documenta Historica Universitatis Carolinae Pragensis, 1).

Themistius, *In De anima*, ed. G. Verbeke in *Commentaire sur le traité de l'âme d'Aristote. Traduction de Guillaume de Moerbeke. Édition critique et étude sur l'utilisation du Commentaire dans l'œuvre de saint Thomas*, Louvain-la-Neuve/Paris 1957 (Corpus Latinum Commentariorum in Aristotelem Graecorum, 1).

Themo Iudeus, *Questiones in Meteorologica*, ed. O. Scotus, Venice 1522.

Thomas de Aquino, *In De celo*, in *Opera omnia* iussu impensaque Leonis XIII P.M. edita, Rome 1886, 3: 1–257.

Thomas de Aquino, *Super Meteora*, in *Opera omnia* iussu impensaque Leonis XIII P.M. edita, Rome 1886, 3: 323–421.

Virgilius, *Aeneis*, ed. J. Perret, Paris 2007 (2 vols.).

Virgilius, *Georgica*, ed. E. De Saint-Denis, Paris 1982.

Witelo, *Perspectiva*, lib. II–III, in S. Unguru (ed.), *Witelonis perspectivae, liber secundus et liber tertius. Books II and III of Witelo's Perspectiva: a Critical Latin Edition and English Translation with Introduction, Notes and Commentaries*, Wrocław/Warsaw/Krakow 1991 (Studia Copernicana, 28).

Witelo, *Perspectiva*, in F. Risner (ed.), *Opticae Thesaurus: Alhazeni arabis libri septem, nunc primum editi. Eiusdem Liber de crepusculis et nubium ascensionibus. Item Vitellonis Thuringopoloni libri X*, Basel 1572, reprint New York 1972 (The Source of Science, 94).

Secondary Sources

Agrimi, J., "Les Questions de sensu attribuées à Albert de Saxe. Quelques remarques sur les rapports entre philosophie naturelle et médecine chez Buridan, Oresme et Albert", in J. Biard (ed.), *Itinéraires d'Albert de Saxe. Paris-Vienne au XIVᵉ siècle. Actes du colloque organisé le 19–22 juin 1990 dans le cadre des activités de l'URA 1085 du CNRS à l'occasion du 600ᵉ anniversaire de la mort d'Albert de Saxe*, Paris 1991 (Études de philosophie médiévale, 69), 191–204.

Andersson-Schmitt, M., H. Hallberg, and M. Hedlund, *Mittelalterliche Handschriften der Universitätsbibliothek Uppsala, Katalog über die C-Sammlung*, 6: Handschriften C, Stockholm 1995 (Acta Bibliothecae R. Universitatis Upsaliensis, 26).

Babbitt, S.M., *Oresme's Livre de Politiques and the France of Charles V*, Philadelphia 1985 (Transactions of the American Philosophical Society, 75).

Bages, S., *Les Questiones super tres libros Metheororum Aristotelis de Jean Buridan: étude suivie de l'édition du livre I*, Thèse de doctorat de l'École des Chartes, 1986 (2 vols.).

Bammesberger, A., "Anmerkungen zum Baseler Epigramm", in Id., *Baltistik: Aufgaben und Methoden. Akten der Tagung vom 10. bis 11. November 1995 in Eichstätt*, Heidelberg 1968, 121–126.

Bénédictins du Bouveret, *Colophons de manuscrits occidentaux des origines au XVIe siècle*, Fribourg (Switzerland) 1979 (Spicilegii Friburgensis Subsidia) (6 vols.).

Birkenmajer, A., *Études d'histoire des sciences en Pologne*, choix d'articles par J.B. Korolec, A.M. Birkenmajer, textes polonais trad. par C. Brendel et al., revus par J. Wolf, Wrocław/Warsaw/Krakow 1972 (Studia Copernicana, 4).

Burckardt, M., "Aus dem Umkreis der ersten Basler Universitätsbibliothek", *Basler Zeitschrift für Geschichte und Alterumskunde*, 58/59 (1959), 155–191.

Caroti, S., "*Modi rerum* and Materialism: a Note on a Quotation of a Condemned Articulus in Some Fourteenth-Century Parisian *De anima* commentaries", *Traditio* 55 (2000), 211–234.

Celeyrette, J., "Les *Questions sur la Physique* dans l'œuvre de Nicole Oresme", in J. Celeyrette and C. Grellard (eds.), *Nicole Oresme philosophe. Philosophie de la nature et philosophie de la connaissance à Paris au XIVe siècle*, Turnhout 2014 (Studia Artistarum, 39), 63–82.

Combes, A., "Note sur les *Sententiae Magistri Joannis Gerson* du manuscrit B.N.LAT. 15156", *Archives d'histoire doctrinale et littéraire du Moyen Âge* 14 (1939), 365–385.

Courtenay, W.J., "The Early Career of Nicole Oresme", *Isis* 91 (2000), 542–548.

Courtenay, W.J., "The University of Paris at the Time of Jean Buridan and Nicole Oresme", *Vivarium* 42 (2004), 3–17.

Delisle, L., *Inventaire des manuscrits de l'Abbaye de Sainct-Victor conservés à la Bibliothèque Impériale, sous les numéros 14232–15175 du fonds latin*, Paris 1869.

Denifle, H. and E. Chatelain, *Chartularium Universitatis Parisiensis*, 1200–1452, Paris 1889–1897 (4 vols.).

Dondaine, A., "Abbréviations latines et signes recommandés pour l'apparat critique des éditions de textes médiévaux", *Bulletin de la Société Internationale pour l'Etude de la Philosophie Médiévale* (*S.I.E.P.M.*) 2 (1960), 142–149.

Döring, D., *Die Bestandsentwicklung der Bibliothek der Philosophischen Fakultät der Universität zu Leipzig von ihren Anfängen bis zur Mitte des 16. Jahrhunderts: ein Beitrag zur Wissenschaftsgeschichte der Leipziger Universität in ihrer vorreformatorischen Zeit*, Leipzig 1990 (*Beiheft 99 zum Zentralblatt für Bibliothekswesen*).

Duhem, P., *Le système du monde: histoire des doctrines cosmologiques de Platon à Copernic*, Paris 1913–1959 (10 vols.).

Ebbesen, S., C. Thomsen-Thörnqvist, and V. Decaix, "Questiones on De sensu et sensato, De memoria, and De somno et vigilia. A Catalogue", *Bulletin de Philosophie Médiévale* 57 (2015), 59–116.

Faral, E., "Jean Buridan: notes sur les manuscrits, les éditions et le contenu de ses ouvrages", *Archives d'histoire doctrinale et littéraire du Moyen Âge* 35 (1960), 1–53.

Feller, L.J., *Catalogus codicum manuscriptorum Bibliothecae Paulinae in Academia Lipsiensi*, Leipzig 1686.

Glorieux, P., "Le *Commentaire sur les Sentences* attribué à Jean Gerson", *Recherches de théologie ancienne et médiévale* 18 (1951), 128–139.

Gorochov, N., *Le Collège de Navarre: de sa fondation (1305) au début du XVème siècle (1418): histoire de l'institution, de sa vie intellectuelle et de son recrutement*, Paris 1997 (Études d'histoire médiévale, 1).

Gossner, J.W., *Le Quadripartit Ptholomee*, unpublished Ph.D. dissertation, Syracuse University 1951.

Göber, W. and J. Klapper, *Katalog der Handschriften der ehemaligen Universitätsbibliothek Breslau*, ca. 1920–1944 [manuscript catalogue].

Halm, K., G. von Laubmann, and W. Meyer, *Catalogus codicum latinorum Bibliothecae Regiae Monacensis. IV.3, Codices num. 15121–21313 complectens*, Munich 1878 (Catalogus codicum manu scriptorum Bibliothecae Regiae Monacensis IV.3).

Halm, K., G. von Laubmann, and W. Meyer, *Catalogus codicum latinorum Bibliothecae Regiae Monacensis. I.2, Codices num. 2501–5250 complectens*, Munich 1894 (Catalogus codicum manu scriptorum Bibliothecae Regiae Monacensis III.2).

Henschel, A.G.E.Th., *Catalogus codicum Medii Aevi medicorum ac physicorum qui manuscripti in bibliothecis Vratislaviensibus asservantur*, Wrocław 1847 (2 vols.).

Jacquart, D., *Supplément à E. Wickersheimer, Dictionnaire biographique des médecins en France au Moyen Âge*, Geneva 1979 (Hautes études médiévales et modernes, 35).

Kejř, J., "Pronuntiatio", *Studie o rukopisech* 36 (2005–2006), 57–65.

Kirschner, S., "Eine weitere Fassung eines lateinischen 'De caelo-Kommentars' von Nicolaus Oresme?", in B. Fritscher and G. Brey (eds.), *Cosmographica et Geographica. Festschrift für Heribert M. Nobis zum 70. Geburtstag*, 1, Munich 1994 (Algorismus, 13), 209–222.

Kowalczyk, M., A. Kozłowska, M. Markowski, S. Włodek, and M. Zwiercan, *Catalogus codicum manuscriptorum medii aevi latinorum qui in Bibliotheca Jagellonica Cracoviae asservantur, 5, Numeros continens inde a 668 usque ad 771*, Wrocław/Warsaw 1992.

Kristeller, P.O., *Iter Italicum, 5: Alia itinera III and Italy III, Sweden to Yugoslavia, Utopia, Supplement to Italy (A–F)*, London 1963.

Launoy (de), J., *Regii Navarre Gymnasii Parisiensis Historia*, Paris 1677 (2 vols.).

Lejbowicz, M., "Guillaume Oresme, traducteur de la *Tétrabible* de Claude Ptolémée", *Pallas* 30 (1983), 107–133.

Lejbowicz, M., "Nicole Oresme dans la lumière de l'urbanité", in P.J.J.M. Bakker, E. Faye, and C. Grellard (eds.), *Chemins de la pensée médiévale: Études offertes à Zénon Kaluza*, Turnhout 2002 (Textes et Études du Moyen Âge, 20), 675–708.

Lejbowicz, M., "Nicole Oresme, spectateur engagé", in J. Celeyrette and C. Grellard (eds.), *Nicole Oresme philosophe. Philosophie de la nature et philosophie de la connaissance à Paris au XIVᵉ siècle*, Turnhout 2014 (Studia Artistarum, 39), 21–61.

Le Clerc, J.V., B. Hauréau, and P. Meyer (eds.), *Histoire littéraire de France, XIIIᵉ siècle*, 18, Paris 1835.

Lisiecki, A., *Katalog rękopisów bibliotecki seminarynej w Poznaniu*, Poznań 1905.

Lohr, Ch., "Medieval Latin Aristotle Commentaries. Authors: Jacobus-Johannes Juff", *Traditio* 26 (1970), 135–216.

Lohr, Ch., *Aristotelica Helvetica*, Fribourg (Switzerland) 1994 (Scrinium Friburgense, 6).

Lohr, Ch., "Aristotelica Berolinensia", *Traditio* 54 (1999), 353–423.

Lohr, Ch., *Latin Aristotle Commentaries*, I.2. *Medieval Authors M–Z*, Florence 2010 (Corpus Philosophorum Medii Aevi. Subsidia, 18).

Markowski, M., *Burydanizm w Polsce w okresie przedkopernikańskim: studium z historii filozofii nauk ścisłych na Uniwersytecie Krakowskim w XV wieku*, Wrocław 1971.

Markowski, M., *Buridanica quae in codicibus manu scriptis bibliothecarum Monacensium asservantur*, Wrocław/Warsaw 1981.

Markowski, M., *Repertorium commentariorum medii aevi in Aristotelem Latinorum quae in Bibliothecis Wiennae asservantur*, Wrocław 1985.

Markowski, M., *Repertorium commentariorum medii aevi in Aristotelem Latinorum quae in Bibliotheca Amploniana Erffordiae asservantur*, Wrocław 1987.

Markowski, M., "Aristotelica Poznaniensia", *Acta mediaevalia* 17 (2004), 281–293.

Markowski, M., *Repertorium commentariorum medii aevi in Aristotelem Latinorum qui in bibliothecis austriacis: Admont, Furt bei Göttweig, Graz, Heiligenkreuz, Klagenfurt, Klosterneuburg, Kremsmünster, Linz, Melk, Salzburg, Sankt Florian, Vorau, Zwettl asservantur*, Krakow 2008.

Markowski, M., *Repertorium commentariorum medii aevi in Aristotelem Latinorum, qui in Bibliotheca Universitatis Lipsiensis asservantur*, Krakow 2012.

Markowski, M. and Z. Włodek, *Repertorium commentariorum medii aevi in Aristotelem Latinorum quae in Bibliotheca Jagellonica Cracoviae asservantur*, Wrocław 1974.

Makowski, T., *Rękopisy w zbiorach kościelnych*, Warsaw 2014 (Zbiory rękopisów w Polsce, 2).

McCluskey, S.C., *Nicole Oresme on Light, Color, and the Rainbow: an Edition and Translation, with Introduction and Critical Notes, of Part of Book III of his "Questiones super IV libros Meteororum"*, unpublished Ph.D. dissertation, University of Wisconsin 1974.

Menhardt, H., "Die Handschriftensammlung der bischöflichen Bibliothek in Klagenfurt", *Zentralblatt für Bibliothekswesen* 39 (1922), 363–381.

Menhardt, H., *Handschriftenverzeichnis der Kärntner Bibliotheken 1: Klagenfurt*, Vienna 1927 (Handschriftenverzeichnisse Österreichischer Bibliotheken 1: Kärnten).

Meunier, F., *Essai sur la vie et les ouvrages de Nicole Oresme*, Paris 1857.

Michael, B., *Johannes Buridan: Studien zu seinem Leben, seinem Werken und zur Rezeption seiner Theorien im Europa des Späten Mittelalters*, Berlin 1985 (2 vols.).

Moonan, L., "The Scientific Writings of Lawrence of Lindores (d. 1437)", *Classica et Mediaevalia* 39 (1988), 273–317.

Neveux, F., "Nicole Oresme et le clergé normand du XIVe siècle", in J. Quillet (ed.), *Autour de Nicole Oresme. Actes du colloque Oresme organisé à l'Université de Paris XII*, Paris 1990 (Bibliothèque d'histoire de la philosophie), 9–36.

Oergel, G., "Das Collegium zur Himmelspforte während des Mittelalters", *Mitteilungen des Vereins für Geschichte und Altertumskunde von Erfurt* 19 (1898), 19–114.

Panzica, A., "Une nouvelle rédaction des *Questions sur les Météorologiques* de Nicole Oresme", *Bulletin de philosophie médiévale* 57 (2015), 257–264.

Panzica, A., "Nicole Oresme à la Faculté des Arts de Paris: les *Questions sur les Météorologiques*", *Archives d'histoire doctrinale et littéraire du Moyen Âge* 84 (2017), 7–89.

Panzica, A., "L'hypothèse de la cessation des mouvements célestes au XIVe siècle: Nicole Oresme, Jean Buridan et Albert de Saxe", *Vivarium* 56 (2018), 83–125.

Panzica, A., "Albert of Saxony's *Questions on Meteorology*: Introduction, Study of the Manuscript Tradition, and Edition of Book I–II.2", *Archives d'histoire doctrinale et littéraire du Moyen Âge* 86 (2019), 231–356.

Panzica, A., "Les *Questions sur les Météorologiques* du manuscrit Vat. Lat. 4082: Blaise de Parme, Nicole Oresme et l'*Inter omnes impressiones*", *Bulletin de philosophie médiévale* 61 (2019), 153–182.

Panzica, A., "Commenter les *Météorologiques* à l'Université de Cracovie: de l'assimilation des modèles parisiens à la naissance d'une tradition polonaise", *Recherches de Théologie et Philosophie Médiévales* 87/1 (2020), 77–166.

Panzica, A., "Henricus Totting de Oyta's and Nicole Oresme's Commentaries on *Meteorology*: Some New Identifications in Eastern Europe", *Bulletin de philosophie médiévale* 62 (2020), 195–211.

Panzica, A., *De la Lune à la Terre. Les débats sur le premier livre des Météorologiques d'Aristote au Moyen Âge latin (XIIe–XVe siècles)*, Turnhout (Studia Artistarum), forthcoming.

Ouy, G., *Les manuscrits de l'Abbaye de Saint-Victor: catalogue établi sur la base du répertoire de Claude de Grandrue (1514)*, Turnhout 1999 (Bibliotheca Victorina, 10) (2 vols.).

Petit, D., *Untersuchungen zu den baltischen Sprachen*, Leiden/Boston 2010 (Brill's Studies in Indo-European Languages and Linguistics, 4).

Petrescu, L., "The Threefold Object of the Scientific Knowledge. Pseudo-Scotus and the Literature on the *Meteorologica* in Fourteenth-century Paris", *Franciscan Studies* 72/1 (2014), 465–502.

Piccard, G., *Die Wasserzeichenkartei im Hauptstaatsarchiv Stuttgart*, Stuttgart 1961–1997 (17 vols.).

Scarpatetti, B.M. von, R. Gamper, and M. Stähli, *Katalog der datierten Handschriften in der Schweiz in lateinischer Schrift vom Anfang des Mittelalters bis 1550*, 1: *Die Handschriften der Bibliotheken von Aarau, Appenzell und Basel*, Dietikon/Zurich 1977.

Scarpatetti, B.M. von, R. Gamper, and M. Stähli, *Katalog der datierten Handschriften in der Schweiz in lateinischer Schrift vom Anfang des Mittelalters bis 1550, 3: Die Handschriften der Bibliotheken St. Gallen-Zürich*, Dietikon/Zurich 1991.

Scherrer, G., *Verzeichniss der Handschriften der Stiftsbibliothek von St. Gallen*, Halle 1875.

Schum, W., *Beschreibendes Verzeichnis der Amplonianischen Handschriften-Sammlung zu Erfurt*, Berlin 1887.

Seńko, W., *Repertorium commentariorum Medii Aevi in Aristotelem Latinorum quae in bibliothecis publicis Parisiis asservantur (Bibliothèque Nationale, Arsenal, Mazarine, Sorbonne, Ste Geneviève)*, Warsaw 1982.

Streijger, M., P.J.J.M. Bakker, and J.M.M.H. Thijssen (eds.), *John Buridan, Quaestiones super libros De generatione et corruptione Aristotelis*, Leiden/Boston 2010 (Medieval and Early Modern Philosophy and Science, 14).

Suter, H., "Eine bis jetzt unbekannte Schrift des Nic. Oresme", *Zeitschrift für Mathematik und Physik*, Hist.-Lit. Abteilung 27 (1882), 121–125.

Thorndike, L., "Oresme and Fourteenth Century Commentaries on the *Meteorologica*", *Isis* 45 (1954), 145–152.

Thorndike, L., "More Questions on the *Meteorologica*", *Isis* 46 (1955), 357–360.

Thorndike, L. and P. Kibre, *A Catalogue of Incipits of Scientific Writings in Latin*, Cambridge (MA), 1963 (The Medieval Academy of America, 29).

Unterkircher, F., *Die datierten Handschriften der Österreichischen Nationalbibliothek von 1501 bis 1600*, Vienna 1976 (Katalog der datierten Handschriften in lateinischer Schrift in Österreich, 4).

Verboon, A.R., *Lines of Thought. Diagrammatic Representation and the Scientific Texts of the Arts Faculty, 1200–1500*, unpublished Ph.D. dissertation, Leiden University 2010.

Verboon, A.R. and E.P. Bos, *Master Albert: Termini naturales*, in C. Angotti, M. Calma, and M. Teeuwen (eds.), *Portraits de Maîtres offerts à Olga Weijers*, Turnhout 2012 (Textes et Études du Moyen Âge, 65), 335–360.

Walz, D., *Die historischen und philosophischen Handschriften der Codices Palatini Latini in der Vatikanischen Bibliothek* (Cod. Pal. Lat. 921–1078), Wiesbaden 1999 (Kataloge der Universitätsbibliothek Heidelberg, 3).

Weijers, O., and M. Calma, *Le travail intellectuel à la Faculté des arts de Paris: textes et maîtres (ca. 1200–1500). Répertoire des noms commençant par L-M-N-O*, Turnhout 2005 (Studia Artistarum, 13).

Wickersheimer, E., *Dictionnaire biographique des médecins en France au Moyen Âge*, Geneva 1979 (Hautes études médiévales et modernes, 34/1–2) (2 vols.).

Wisłocki, W., *Catalogus codicum manuscriptorum Bibliothecae Universitatis Jagellonicae Cracoviensis*, Krakow 1877 (2 vols.).

Wöhler, H.-U., "Die erste philosophische Fakultät in Sachsen bis zum Beginn der

Reformation in localen, regionalen und überregionalen Kontext", *Bochumer Philosophisches Jahrbuch für Antike und Mittelalter* 13 (2008), 217–240.
Zathey, J., A. Lewicka-Kamińska, and A.L. Hajdukiewicz, *Historia Biblioteki Jagiellonskiej*, 1, 1364–1775, Krakow 1966.

Index codicum

Manuscripts mentioned in the introduction.

Index fontium

Sources referenced in this edition. Page numbers in italics indicate the sources referenced by the editor.

Index rerum

Notable things referenced in the edition.

Index nominum antiquorum

Ancient and Medieval names referenced in the introduction.

Index nominum modernorum

Modern names referenced in the introduction.

www.ingramcontent.com/pod-product-compliance
Ingram Content Group UK Ltd.
Pitfield, Milton Keynes, MK11 3LW, UK
UKHW022224051026
15624UKWH00002B/340

9 789004 461406

New Frontiers in Regional Science: Asian Perspectives

Volume 75

This series is a constellation of works by scholars in the field of regional science and in related disciplines specifically focusing on dynamism in Asia.

Asia is the most dynamic part of the world. Japan, Korea, Taiwan, and Singapore experienced rapid and miracle economic growth in the 1970s. Malaysia, Indonesia, and Thailand followed in the 1980s. China, India, and Vietnam are now rising countries in Asia and are even leading the world economy. Due to their rapid economic development and growth, Asian countries continue to face a variety of urgent issues including regional and institutional unbalanced growth, environmental problems, poverty amidst prosperity, an ageing society, the collapse of the bubble economy, and deflation, among others.

Asian countries are diversified as they have their own cultural, historical, and geographical as well as political conditions. Due to this fact, scholars specializing in regional science as an inter- and multi-discipline have taken leading roles in providing mitigating policy proposals based on robust interdisciplinary analysis of multifaceted regional issues and subjects in Asia. This series not only will present unique research results from Asia that are unfamiliar in other parts of the world because of language barriers, but also will publish advanced research results from those regions that have focused on regional and urban issues in Asia from different perspectives.

The series aims to expand the frontiers of regional science through diffusion of intrinsically developed and advanced modern regional science methodologies in Asia and other areas of the world. Readers will be inspired to realize that regional and urban issues in the world are so vast that their established methodologies still have space for development and refinement, and to understand the importance of the interdisciplinary and multidisciplinary approach that is inherent in regional science for analyzing and resolving urgent regional and urban issues in Asia.

Topics under consideration in this series include the theory of social cost and benefit analysis and criteria of public investments, socio-economic vulnerability against disasters, food security and policy, agro-food systems in China, industrial clustering in Asia, comprehensive management of water environment and resources in a river basin, the international trade bloc and food security, migration and labor market in Asia, land policy and local property tax, Information and Communication Technology planning, consumer "shop-around" movements, and regeneration of downtowns, among others.

Researchers who are interested in publishing their books in this Series should obtain a proposal form from Yoshiro Higano (Editor in Chief, higano@jsrsai.jp) and return the completed form to him.

Yasushi Asami • Yukio Sadahiro • Ikuho Yamada •
Kimihiro Hino
Editors

Studies in Housing and Urban Analysis in Japan

Springer

Editors
Yasushi Asami
Department of Urban Engineering
The University of Tokyo
Bunkyo-ku, Tokyo, Japan

Yukio Sadahiro
Interfaculty Initiative in Inform. Stud.
The University of Tokyo
Bunkyo-ku, Tokyo, Japan

Ikuho Yamada
Center for Spatial Information Science
The University of Tokyo
Kashiwa-shi, Chiba, Japan

Kimihiro Hino
Department of Urban Engineering
The University of Tokyo
Bunkyo-ku, Tokyo, Japan

ISSN 2199-5974 ISSN 2199-5982 (electronic)
New Frontiers in Regional Science: Asian Perspectives
ISBN 978-981-99-8026-0 ISBN 978-981-99-8027-7 (eBook)
https://doi.org/10.1007/978-981-99-8027-7

This Springer imprint is published by the registered company Springer Nature Singapore Pte Ltd.
The registered company address is: 152 Beach Road, #21-01/04 Gateway East, Singapore 189721, Singapore

Paper in this product is recyclable.

Preface

This book contains a collection of papers published in Japanese by researchers closely associated with our laboratory. They are translated into English with the permission of the academic societies where the original papers were published and compiled into a single volume on the occasion of my retirement from the University of Tokyo. Hopefully, this book provides an overview of the research activities of our laboratory.

The laboratory is called the Laboratory of Housing and Urban Analysis and has long operated with urban housing theory and urban analysis as the mainstay of its research. The first professor of the laboratory was Dr. Masahiko Honjo, who studied the spatial configuration of residential areas. The second professor was Dr. Kaoru Shimofusa, who studied housing issues from the perspective of housing economics, including the appropriate setting of rents. Dr. Shimofusa was also interested in analytical research methods and offered a course entitled "Urban Analysis." Dr. Kozo Okudaira, an associate professor under Dr. Shimofusa, was also involved in the production of architectural models and compiled a book on urban analysis methods entitled *Urban Engineering Reader: Analyzing the City* (Okudaira 1976), which is still a bible for researchers who aim at urban analysis. However, Dr. Okudaira passed away suddenly in 1979. The third professor of the laboratory was Professor Atsuyuki Okabe, who explored the methodology of spatial analysis, especially summarized in *Spatial Tessellations. Concepts and Applications of Voronoi Diagrams* (Okabe et al. 1992) and *Spatial Analysis along Networks: Statistical and Computational Methods* (Okabe and Sugihara 2012) and others to establish spatial analysis methods. Then, Yasushi Asami became the fourth professor.

It was in 1981, when I was a senior in the Department of Urban Engineering at the University of Tokyo, that I decided to join the Laboratory of Housing and Urban Analysis organized by Professor Shimofusa and Associate Professor Atsuyuki Okabe. At the time, I was strongly attracted by the approach to research that considered urban events analytically and captured the city on a quantitative basis. Dr. Shimofusa gave me advice on the direction of my research from a broad perspective, and Dr. Okabe specifically guided me on methods and logical thinking, and I completed my graduation thesis, "Factor Analysis of Urbanization: Using Takasaki City Mesh Data" (Asami 1982). In my graduation thesis, I considered population growth to be a major indicator of urbanization, and used multiple regression analysis to explore what factors promote urbanization. I then went on to a master's program in the Department of Urban Engineering, Graduate School of Engineering, the University of Tokyo, where I conducted my master's research in the same laboratory. I received guidance under the same system, and in particular, Dr. Okabe taught me in detail the basic concepts of research, how to construct analytical methods, how to handle actual data, and how to summarize my research. My current approach to research is largely due to the teachings of Dr. Okabe. I summarized my master's research and presented it as my master's thesis, "Distribution and Location Analysis of Population and Facilities in Urban Areas" (Asami 1984). During that time, he encouraged me to study in the USA and I was able to decide to study in the Department of Regional Science at the University of Pennsylvania.

At the University of Pennsylvania, I studied under the supervision of Dr. Tony Smith and Dr. Masahisa Fujita, who led a regional science workshop where researchers from inside and outside the university presented and discussed their research every week. I also presented my research every semester, and the comments I received were very helpful in improving my research. In my doctoral research, I was interested in land use theory, which considers the spatial organization of cities through the real estate market, and I reexamined the concept of land rent as a key concept in the theory. In doing so, I applied game theory, which was relatively new at the time, in an attempt to establish an academic foundation for land use theory. My doctoral dissertation, "Game-Theoretic Approaches to Bid Rent" (Asami 1987), was reviewed by Dr. Tony Smith, Dr. Masahisa Fujita, and Dr. Xavier Vives. I owe a great deal to Dr. Tony Smith and Dr. Masahisa Fujita for my axiomatic methods and urban economics approach.

After receiving Ph.D. degree, I returned to my old "home," the Laboratory of Housing and Urban Analysis, as an assistant professor in the Department of Urban Engineering, Faculty of Engineering, the University of Tokyo in 1987. At that time, Dr. Okabe and I ran the laboratory together and taught students. I asked myself how to establish my area of expertise and decided to utilize my strengths. My background in urban planning and urban economics led me to try to open up a new field by fusing these two fields.

When I returned to Japan in 1987, Japan was in the midst of a real estate price bubble that soared, and shortly thereafter the bubble burst, fundamentally damaging the urban structure. The power of the real estate market was clearly demonstrated. Under these circumstances, urban planning scholars argued that strong planning regulations were necessary to break the market's malady. On the other hand, urban economists argued that too much regulation distorts the market and leads to inefficient urban structures. The two disciplines on which I was relying had reached completely opposite conclusions. Since I was relying on them, I had to resolve this contradiction. Therefore, I considered what the premises of both theories were and what differences led to their different conclusions. At the time, urban planning studies was weak in its conception of planning that incorporates the market, and urban economics was weak in its model that fully incorporates external diseconomies among land uses. If we can compensate for these weaknesses, we can obtain a consistent form of real estate for both fields. This became the focus of my subsequent research activities.

Some of the papers in this book are studies in which I was directly involved, while others were conducted independently by the respective authors. However, I expect that some of the authors may have been influenced in some small way by the way they approached their research in the process of research guidance and discussion of their research.

This book consists of two parts: Part I, Housing analysis, which contains research related to housing and residential life, and Part II, Urban analysis, which contains research related to urban analysis and spatial analysis. Our laboratory is called the Laboratory of Housing and Urban Analysis, and we have put up two signs in the front of the book. Part I, Housing analysis, contains papers on housing markets, housing environments, the lives of the older adults, urban crime, and photovoltaic panels installed on houses. The second part, Urban analysis, contains papers on methodology development, analysis, urban activity analysis, and traffic analysis.

An important area of Housing Analysis is the analysis of the housing market. Housing is a special good that has heterogeneity (no two houses are alike), durability (it is used for a long period of time), and high price (it is expensive and therefore not something that can be purchased many times). For this reason, they form a field of study in economics, as in the field of housing economics. Suzuki-Asami paper analyzes the price and market residence time of existing detached houses. In the case of existing houses, if demand is high, the house can be sold normally, but when demand falls, sellers are faced with the choice of selling at a lower price or giving up on the sale. This creates a somewhat unique housing market. Baba-Shimizu-Asami paper also analyzes the probability of rebuilding older condominiums. In Japan, each owner of a condominium unit has a special ownership right called strata title. In order to rebuild a condominium, it is necessary to obtain the agreement of the unit owners by a special majority vote of 80% or more. Therefore, it is difficult to rebuild a condominium unless the circumstances are very favorable for rebuilding, which is directly related to the high demand in the housing market.

In Housing Analysis, the living environment around a house is also an important subject of analysis. The term "living environment" covers a wide range of factors, including safety, health, convenience, amenity, and sustainability. This field analyzes how these factors ultimately affect people's housing behavior. For this purpose, it is necessary to comprehensively measure the living environment. Ai paper analyzes the relationship between housing environment scores and population growth.

As Japan's population ages, it is important to accurately understand the lives of the older adults and consider necessary housing policies. Kim et al. paper analyzes the relationship between social participation of the older adults and the density of neighborhood facilities. Sekiguchi paper analyzes the problem of accessibility to stores for the older adults by examining not only distance but also road conditions in detail. This kind of steady analysis of the behavior of the older adults produces hints for building a society that takes the older adults into consideration.

Crime is one of the most important safety-related issues in the living environment. One of the crimes involving housing is residential burglary. There are certain characteristics of houses and areas where residential burglary is most likely to occur. Conversely, if we can analyze trends, we can determine what countermeasures should be taken to reduce residential burglary, and this information can be used to improve safety. This book includes Hino-Kojima paper, which constructs a model to predict the risk of residential burglary, and Uesugi-Hino paper, which analyzes the relationship between the socioeconomic status of a neighborhood and residential burglary.

The use of renewable energy has been increasing in recent years due to heightened environmental awareness. For homes, a relatively easy renewable energy source to use is solar panels. The installation of solar panels has advanced considerably due to the feed-in tariff system, but at the same time, it has caused landscape problems. In Okazawa-Hino-Asami paper, the relationship between the installation of residential solar panels and the home and surrounding environment is analyzed.

Urban Analysis deals with the problem of analyzing spatial arrangements and spatial representations. A time map is a map that shows the time required between points intuitively in terms of the distances between each other by changing the location of the points themselves so that the time required is as proportional as possible to the distances between the points. In general, the method differs depending on what and how much importance is placed on actual position of points and precise representation of time since it is not possible to make the time required completely proportional to the distance between points. Therefore, it can be said that the methods of creating time maps are diverse and still in the process of development. Nishi-Asami paper develops a method for constructing time maps and shows examples of their application.

Predicting how a building will be constructed on a given site is important for analyzing the formation process of urban areas. The owner of a site will develop the site in a way that is optimal for himself based on the site conditions. In Taima et al. paper, a method is constructed to predict what kind of building will be built in a given site in a highly commercial area, and examples of the prediction results are presented.

One method often used to estimate housing prices is hedonic analysis. In hedonic analysis, a regression analysis is conducted with housing land prices and housing prices as the objective variables and variables representing the status of factors expected to affect prices as explanatory variables. Hiruta-Asami paper analyzes the severity of multicollinearity and shows how to obtain appropriate regression results.

In urban planning, the setback of a building (how far a building is set back from the street), the length of frontage of site to the adjacent road, and the depth of the site have been considered and analyzed as important indicators to express the urban environment. Usui-Asami paper presents a framework that allows for a comprehensive analysis of these relationships and analyzes their relationship with building density.

The vitality of a city center depends on the location and activity of commercial establishments. Therefore, how shopping areas develop or decline is an important concern in urban planning. Inasaka paper develops a method to express which direction commercial facilities are spreading and presents the results of this analysis.

Urban transportation is an important factor in determining the convenience of an urban area. Urban transportation has been undergoing major changes in recent years due to the development of ICT technology, including the spread of sharing services and the development of demand-responsive transit. Hasegawa-Suzuki paper theoretically analyzes what kind of public transportation system is desirable.

Looking at the topics of the papers gathered in this way, I am deeply moved by a renewed appreciation of the breadth of research that our laboratory has conducted. I also feel the joy of having nurtured so many researchers. I hope that the younger members of our laboratory will continue to develop and deepen their research fields in the future.

When my former teacher, Dr. Atsuyuki Okabe, retired, we translated Japanese papers by researchers who had been under his care into English and compiled them into a book (Asami et al. 2009). This stimulated us to plan this book. I am grateful to Yukio Sadahiro, Ikuho Yamada, Kimihiro Hino, and Hiroyuki Usui, for their support in the planning of this project. I would like to express my deep gratitude to them.

We hope that this book will help readers learn more about the current status of housing and urban analysis research in Japan and assist in its future development.

Tokyo, Japan Yasushi Asami

References

Asami Y (1982) Factor analysis of urbanization: using mesh data of Takasaki City. Graduation Thesis, Department of Urban Engineering, Faculty of Engineering, The University of Tokyo

Asami Y (1984) Distribution and location analysis of population and facilities in Tokyo. Master's thesis, Department of Urban Engineering, Faculty of Engineering, The University of Tokyo

Asami Y (1987) Game-theoretic approaches to bid rent. Ph.D. dissertation, Department of Regional Science, University of Pennsylvania University Microfilms Inc., Ann Arbor, Michigan

Asami Y, Sadahiro Y, Ishikawa T (eds) (2009) New frontiers in urban analysis. In Honor of Atsuyuki Okabe. CRC Press, Boca Raton

Okabe A, Boots B, Sugihara K (1992) Spatial tessellations. Concepts and applications of Voronoi diagrams. J. Wiley and Sons

Okabe A, Sugihara K (2012) Spatial analysis along networks: statistical and computational methods. John Wiley & Sons

Okudaira K (1976) Toshi Kogaku Dokuhon: Toshi wo Kaisekisuru [Urban Engineering Reader: Analyzing the City]. Shokoku-sha Publishing Co.

Contents

Part I
Housing Analysis

Chapter 1
Demand–Supply Relationship in the Resale Housing Market in the Suburbs of Tokyo

Masatomo Suzuki and Yasushi Asami

Abstract Vacant houses are currently being promoted for use in the resale housing market, and the purchase of resale houses has become an important option in society. This chapter investigates the demand–supply relationship in the resale housing market on the outskirts of the Tokyo metropolitan area. Focusing on the two established market-clearing factors: "price" and "time required to sell a property," we investigated the manner in which potentially disadvantageous properties–in terms of building and locational characteristics–are sold in the market. In suburbs less than 30 km from central Tokyo, we show that such properties tend to be sold quickly or fetch high prices, suggesting a strong demand for resale houses for young households. In addition, in the outer suburbs (40 km or farther away from central Tokyo), potentially disadvantageous properties required longer time to sell or fetch low prices. Moreover, certain owners were found attempting to dispose their properties quickly at a low price, suggesting that they consider it costly to maintain their unnecessary properties.

Keywords Housing market analysis · Resale housing market · Price · Time on the market · Population decline

---

The contents of this chapter are based on the following paper originally published in a Japanese journal: Suzuki M, Asami Y (2017) The demand-supply relationship in the resale housing market: Evidence from the outskirts of the Tokyo metropolitan area. Journal of the City Planning Institute of Japan 52(3): 514–520 (in Japanese).

M. Suzuki (✉)
Yokohama City University, Yokohama, Kanagawa, Japan
e-mail: suzuki.mas.ke@yokohama-cu.ac.jp

Y. Asami
Department of Urban Engineering, The University of Tokyo, Tokyo, Japan
e-mail: asami@csis.u-tokyo.ac.jp

Y. Asami et al. (eds.), *Studies in Housing and Urban Analysis in Japan*,
New Frontiers in Regional Science: Asian Perspectives 75,
https://doi.org/10.1007/978-981-99-8027-7_1

1.1 Introduction

The outskirts of the Tokyo metropolitan area is experiencing a decrease in the population (or number of households). Thus, the number of vacant houses is increasing. For spatial planning to enhance the long-term sustainability of the area, the relationship between neighborhood characteristics and population dynamics, including decreasing population (or number of households) (Ai 2014; Miyake et al. 2014), generation balance (Ai 2016; Fujii 2008; Fujii and Oe 2005, 2006), and increasing number of vacant houses (Sakamoto and Yokohari 2016; Ujihara et al. 2006, 2016a, 2016b; Yamanshita and Morimoto 2015), has been investigated.

Recently, vacant houses have been promoted for use in the resale housing market, and purchasing resale houses has become an important option in society (Hirayama 2011). Several city-planning studies have investigated owners' willingness to utilize vacant houses (e.g., Ito and Kaido 2013). However, real estate markets have rarely been investigated. Among the few who have investigated these markets, Todoroki (2011) and Umeda and Hirayama (2002) only investigated special cases such as foreclosure sales and properties whose prices had dropped significantly.

The resale housing market is crucial for clarifying the building and locational characteristics that new residents consider when purchasing resale houses, thus clarifying the boundaries of the urban area that are expected to be sustained in the long run. Consequently, the resale housing market becomes inactive before the population (or number of households) starts to decrease, and vacant houses begin to increase in number. Focusing on utilizing resale houses, the similarities and differences between the building/locational characteristics that determine the demand–supply relationship of resale housing markets and the neighborhood dynamics documented in the literature must be clarified.

The demand–supply relationship in real estate markets is characterized by two indicators: "price" and "time on the market" (TOM) (Han and Strange 2015). As in the hedonic analysis, price represents the value of the building or locational characteristics. However, the TOM has rarely be investigated in the city planning literature. It is a liquidity measure that measures the time from the initial listing in the market to the completion of sales. A short TOM indicates a "hot" market with more buyers than sellers. For buyers searching for desirable houses across several properties/regions, there is a trade-off between price and TOM. Their determination in the market is dependent on the micro-level segments of the building or locational characteristics (Piazzesi et al. 2020). Thus, both must be investigated to understand the housing resale market. Ong and Koh (2000) investigated the effect of building or locational characteristics on price and TOM, focusing on heterogeneities in the floor levels of high-rise apartments.

This chapter investigates the relationship between building or locational characteristics in the context of population shrinkage in Japan and price and TOM, which are the basic components of real estate markets. It clarifies the demand–supply relationship of the resale housing markets in the suburbs of the Tokyo metropolitan area. We focus on the area along the Tobu Tojo/Ogose lines in Saitama Prefecture because certain of these areas are now facing population shrinkage (or a decreasing number of households) (Miyake et al. 2014). We categorized the building characteristics by age, floor area, etc. and categorized their location based on distance from central Tokyo, time from the nearest station, etc. From the estimated price/TOM equations, we obtained the following demand–supply relationship in the resale housing market. Overall, relative to supply, the demand for resale houses was low (high) in areas with typical characteristics of population shrinkage (growth). However, the demand–supply relationship for potentially disadvantageous properties changed with the distance from central Tokyo. In other words, demand strongly exists even for potentially disadvantageous properties in the suburbs relatively close to central Tokyo; on the outskirts farther away from central Tokyo, such properties exhibit long TOM, and certain owners may reduce prices to quickly dispose of their properties.

The remainder of this chapter is organized as follows. Section 1.2 describes the target area and data. Section 1.3 describes the empirical methods considered in this study. Section 1.4 presents and discusses the results. Finally, Sect. 1.5 concludes the paper with certain policy implications.

1.2 Target Area and Data

In the Tokyo metropolitan area, characteristics of areas vary significantly across train lines even within the same distance from central Tokyo. To compensate for the factors that affect price and TOM, we focused on the area along the Tobu Tojo/Ogose lines in Saitama Prefecture.[1] Certain of these areas, particularly those 40 km or further from central Tokyo, are now facing population shrinkage (and a decreasing number of households) (Miyake et al. 2014). The area covers the town of Moroyama, where vacant houses are rapidly increasing in number (Hasegawa 2005). Therefore, this area is suitable for our analysis in the context of population shrinkage.

We employed the list price data of resale houses from a real estate market database collected by At Home Co., Ltd. The data included the list price, date of listing, building characteristics, and (geocoded) locational characteristics for each property listed in the database between January 1, 2011 and August 31, 2014. We employed a time interval that includes sufficient samples to capture recent trends in the resale housing market, whose importance is growing.[2] As shown in Fig. 1.1,

[1] The cities include Wako, Niiza, Asaka, Shiki, Fujimi, Fujimino, Kawagoe, Tsurugashima, Hidaka, Sakado, Higashi-matsuyama, and Kumagaya. The towns include Ogose, Miyoshi, Moroyama, Kawashima, Yoshimi, Hatoyama, Tokigawa, Namerigawa, Ranzan, Ogawa, and Yorii.

[2] After the bubble burst, the resale housing market became important in Japan. Indeed, the proportion of resale houses to all the transacted houses (resale houses and newly built ones) is exhibiting an increasing trend (MLIT 2017). The transaction volume of newly built houses fluctuates over time, while that of resale houses are stable over time; this suggests that the macroeconomic effect on the resale housing market in the analysis periods is relatively small. Notably, we control the macroeconomic trend by season/year dummies.

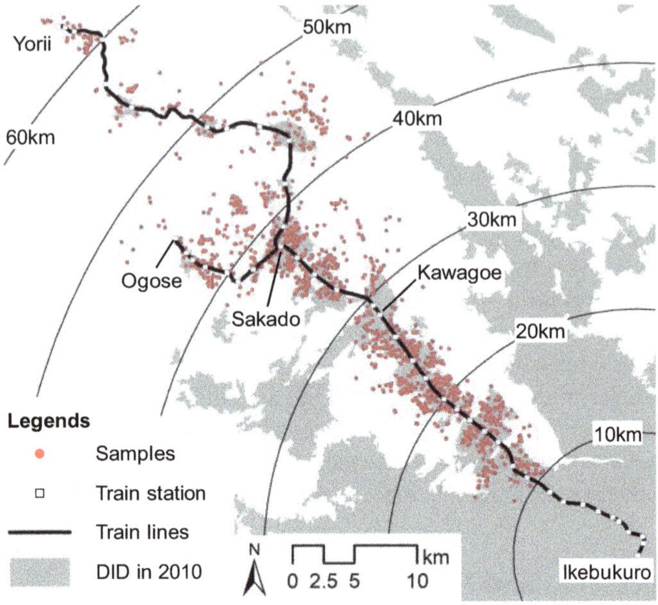

Fig. 1.1 Spatial distribution of resale houses on the market

we restricted the sample to houses whose nearest train station was on the Tobu Tojo/Ogose lines. To investigate suburban areas, the samples were restricted to those located 10 km or farther from central Tokyo (in this case, Ikebukuro station). In particular, within a distance of 30–50 km from central Tokyo, many houses in the market are far away from the nearest station. We measured the TOM by employing houses listed on or after April 1, 2011 with the listing ending on or before May 31, 2014. We assumed that (i) transactions were made at the list price one day after the final listing[3] and (ii) if the listing interval became 3 months or greater, the property was sold once and again listed for sale (note that the properties are usually listed once a month if they are not yet sold). We omitted samples with TOM longer than 2 years (730 days).

Table 1.1 summarizes the variables used in the analysis. The explanatory variables include the markup ratio (M), which is the ratio of list price to expected sale price; variables of macroeconomic environments (**E**); variables of building characteristics (**B**); and variables of locational characteristics (**R**), where bold indicates the vectors.

The markup ratio (M) is the ratio of the list price to the expected sale price (see Sect. 1.3 for details), showing individual circumstances. Engelhardt (2003) and Genesove and Mayer (1997, 2001) provide an example. In the US, households

[3] Most of the previous literature assumes that properties once listed on the real estate market database are finally transacted, at which time the listing ends (Shimizu et al. 2004).

Table 1.1 Summary statistics

Variables	Mean	S.D.	Min.	Max.
Price/TOM				
List price (1000 JPY)	17,366	11,261	1300	120,000
Sale price (1000 JPY)	16,718	10,848	1300	120,000
Time on the market (day)	94.8	126.2	1.0	727.0
Markup ratio (*M*)				
Markup ratio	1.098	0.374	0.187	4.331
Variables of macroeconomic environment (**E**)				
Fall	0.272	0.445	0	1
Year 2013	0.316	0.465	0	1
Year 2014	0.083	0.276	0	1
Variables of building characteristics (**B**)				
Lot size: less than 50 m^2	0.050	0.219	0	1
Lot size: 250 m^2 or more	0.082	0.275	0	1
Lot size (m^2)	143.5	132.2	25.3	1939.8
Floor space (m^2)	94.7	34.6	23.8	405.1
Building age: less than 5 years	0.076	0.265	0	1
Building age: 30 years or more	0.279	0.449	0	1
Building age (year)	21.6	12.1	0.0	66.0
Variables of locational characteristics (**R**)				
Time to nearest station: less than 10 min	0.262	0.440	0	1
Bus to nearest station	0.167	0.373	0	1
Max. slope: 10 degrees or more	0.057	0.232	0	1
Proportion of elderlies: 25% or more	0.052	0.222	0	1
Detached owner-occupied housing district	0.376	0.484	0	1
Densely inhabited districts in 1960	0.021	0.145	0	1
Densely inhabited districts in 2010	0.662	0.473	0	1
Low-rise residential zones	0.279	0.449	0	1
Medium-to-high-rise residential zones	0.170	0.376	0	1
Distance from central Tokyo: 20–30 km	0.250	0.433	0	1
Distance from central Tokyo: 30–40 km	0.271	0.445	0	1
Distance from central Tokyo: 40–50 km	0.225	0.418	0	1
Distance from central Tokyo: 50–70 km	0.071	0.256	0	1

that have remaining mortgage debt are likely to experience a longer TOM to sell their houses at high prices. However, in the suburbs of the Tokyo metropolitan area, typical owners who wish to sell their resale houses are typically elderly and have already finished mortgage payments (Hirayama 2011); thus, such a situation in the US is not likely to apply to our analysis.

The macroeconomic environment variables (**E**) include a seasonality dummy (fall: July–September) and year dummies (2013 and 2014) for the listing date.

The variables of building characteristics (**B**) included land area, floor area, and building age. We also prepared the following special dummy variables to capture the differences from typical houses. As the average land area is 143.5 m^2, dummy variables for "less than 50 m^2" and for "250 m^2 or greater" are constructed. For building age, dummy variables are constructed for "less than five years" and for "30 years or greater" (i.e., built under the former earthquake-resistance standards). We omitted samples with a building age of 100 years or older as an exception.

For the variables of locational characteristics (**R**), we prepared the following special dummy variables that captured differences from typical houses. Previous literature has investigated neighborhood characteristics of population shrinkage using the following variables: distance from the nearest station, slope of the residential district, balance of the age of residents, variety of building types, periods of development, zoning, and distance from the metropolitan center (Fujii 2008; Miyake et al. 2014; Ujihara et al. 2016a, 2016b). Therefore, for the distance from the nearest station, we constructed dummy variables for "within a 10 min walk to the nearest station" and "more than a 30 min walk to the nearest station (i.e., need to take a bus)."

For the slope of the residential district, a dummy variable was constructed if the maximum slope was 10 degrees or greater. For the balance the age of residents, a dummy variable was constructed if the proportion of elderly residents (i.e., the ratio of households whose members are all 65 years old or older to all households) was 25% or greater. For varieties of building types, we constructed a dummy variable for "detached owner-occupied housing districts," wherein the detached housing ratio was 80% or greater and the owner-occupied housing ratio was 80% or greater, following the method of Fujii (2008). For the development periods, dummy variables were constructed for densely inhabited districts (DIDs) in 1960 and 2010. For zonings that protect the nice residential environments, we constructed dummy variables for "category I exclusive low building residential zone" (hereafter, "low-rise residential zones") and "category I exclusive high and medium building residential zone" (hereafter, "medium-to-high-rise residential zones").

For the distance from the metropolitan center, we categorized the area by 10 km intervals from central Tokyo; here, the 50–70 km interval was used to maintain sufficient samples for analyses. The previous literature on population dynamics in the Tokyo metropolitan area (Fujii 2008; Fujii and Oe 2005, 2006; Miyake et al. 2014) employed 10 km intervals from central Tokyo to discuss the characteristics of the area. Thus, we employed the same distance intervals to discuss our results in Sect. 1.4 along with the previous literature. To prepare the variables, we used property geocodes to overlap with the spatial data.[4]

[4] We calculated the time distance from the nearest station (we employed the digital national land information, MLIT) by measuring the Euclidean distance and dividing it by the walking speed (80 m/min). For maximum slopes (250 m mesh) and DID areas, we employed the digital national land information (MLIT). For the proportion of elderly persons, detached housing ratio, and owner-occupied housing ratio, we employed the district-level 2010 Population Census.

Table 1.2 Price and TOM in each distance interval from central Tokyo

	10–20 km	20–30 km	30–40 km	40–50 km	50–70 km
Sale price (1000 JPY)	24,154	17,749	16,182	12,413	9492
	(14,587)	(10,987)	(7962)	(7029)	(5352)
Time on the market (day)	88.9	82.1	104.1	94.5	105.5
	(118.5)	(116.9)	(137.1)	(130.4)	(117.8)
Number of observations	284	386	419	348	109

Notes: For the sales price and TOM, we show the average (standard deviations in parentheses)

We used 1546 houses after omitting samples with missing information. As shown in Table 1.1, the average TOM was 94.8 days, and the list price was, on average, 9.8% higher than the sale price. The average age of the buildings was 21.6 years. Table 1.2 summarizes the price and TOM for each distance interval from central Tokyo. At the metropolitan level, sales prices declined and TOM increased farther away from central Tokyo.

1.3 Empirical Methods

Following Ong and Koh (2000), we investigated the effect of the building/locational characteristics of housing units on their prices and TOM.[5]

Step 1: A hedonic analysis (OLS) was conducted using building characteristics (**B**) and locational characteristics (**R**) as explanatory variables. The dependent variable is the log of sale price, *Ps*. We then estimated the expected sale price.

$$\ln \hat{P}_S = \hat{\alpha}_0 + \mathbf{B}'\hat{\boldsymbol{\alpha}}_B + \mathbf{R}'\hat{\boldsymbol{\alpha}}_R \tag{1.1}$$

The "dash" for the vectors shows the transposed matrix, α_0 is the constant, $\boldsymbol{\alpha}_i$ ($i \in \{B, R\}$) is the corresponding vector of coefficients, and the "hat" represents the estimated value.

Step 2: From Eq. (1.1), we define the markup ratio (M) as the ratio of the list price P_L to the expected sale price:

$$M \equiv \frac{P_L}{\hat{P}_S} \tag{1.2}$$

[5] Housing markets are determined with price and TOM, and the following two types of assumptions exist: (i) the TOM is independently determined as a function of M, **E**, **B**, and **R**; or (ii) the TOM is determined simultaneously with price as a function of **E**. In this study, we employed the former assumption to investigate the effect of building characteristics (**B**) and locational characteristics (**R**) on both price and TOM, as in the analysis by Ong and Koh (2000).

Step 3: TOM is explained as a function of the markup ratio (*M*), macroeconomic environment variables (**E**), building characteristic variables (**B**), and locational characteristic variables (**R**).

$$\text{TOM} = \Phi\,(M, \mathbf{E}, \mathbf{B}, \mathbf{R}) \tag{1.3}$$

While prices are determined from the "universal" components such as building and locational characteristics, the TOM becomes short if the list price is lower than its original value (i.e., its expected sale price) and if the market is in the active period (Eq. (1.3)). Specifically, we conducted a survival analysis that analyzed the time since the event (i.e., final sales) occurred, assuming a Weibull distribution.[6] Let $f(t)$ be the probability density function of the TOM in period t, and its cumulative distribution (i.e., the probability that the TOM is shorter than t) becomes

$$F(t) = \int_0^t f(s)ds = \Pr\,(\text{TOM} \leq t) \tag{1.4}$$

However, the probability that the properties are still in the market in period t (i.e., the probability that the TOM is t or longer) is

$$S(t) = 1 - F(t) = \Pr\,(\text{TOM} \geq t) \tag{1.5}$$

From Eq. (1.5), the probability (or hazard) that properties that are still in the market in period t will be sold in the next period (or instance) becomes

$$h(t) = \frac{f(t)}{S(t)} \equiv \lambda p t^{p-1} \tag{1.6}$$

which is assumed to follow a Weibull distribution. The hazard function $h(t)$ is an increasing (decreasing) function of time t if parameter $p > 1$ ($p < 1$); that is, properties are more likely to be sold (properties are more likely to remain listed) with increase in the TOM. Therefore, the estimation of Eq. (1.3) becomes the estimation of parameters p and λ in Eq. (1.6). Note that the coefficients of explanatory variables for the TOM and hazard of sales, λ, are reversed:

$$\lambda = \exp\left(\gamma_0 + \gamma_M M + \mathbf{E}'\boldsymbol{\gamma}_E + \mathbf{B}'\boldsymbol{\gamma}_B + \mathbf{R}'\boldsymbol{\gamma}_R\right) \tag{1.7}$$

where γ_0 is the constant, γ_M is the coefficients of the markup ratio, and $\boldsymbol{\gamma}_i$ ($i \in \{E, B, R\}$) are the corresponding coefficient vectors.

[6] Compared to the direct estimate of the TOM by OLS, we confirm that the survival analysis better fits the model with larger log-likelihood. In addition, we employed the Weibull distribution in the survival analysis after comparing the log-likelihoods in exponential and log-normal distributions.

1.4 Empirical Results

Table 1.3 presents the estimation results of the sale price for the entire sample and the subsamples at each distance interval from central Tokyo.[7] If the coefficients of explanatory variables are positive (negative) and statistically significant, the variables exert an effects on pushing up (down) the price. The basic determinants included lot size, floor area, building age, and distance from central Tokyo.

Table 1.4 shows the estimation results of TOM (specifically, the hazard of sale, λ, in Eq. (1.7)) for the entire sample and for the subsamples of each distance interval from central Tokyo.[8] In each estimation, the parameter p satisfies $p < 1$, showing the reasonable trend that properties are more likely to remain listed with increase in the TOM.[9] If the coefficients of the explanatory variables are positive (negative) and statistically significant, the variables have an effect on pushing down (up) the TOM. The basic determinants include the seasons and years of the initial listing and the markup ratio. For building or locational characteristics, variables such as lot size, building age, bus access to the nearest station, detached owner-occupied housing districts, and distance from central Tokyo had a statistically significant effect on the TOM.[10] In general, the TOM is determined within the search and matching framework. Thus, the TOM typically contains considerable larger random components, and the accuracy of its estimation is relatively low compared with the price (Yavas and Yang 1995). Even in the study by Ong and Koh (2000), which employs methods similar to our analysis, only certain building or locational characteristics affected the TOM compared to price. Thus, Table 1.4 presents reasonable estimation results. In particular, in the 50–70 km interval from central Tokyo, a clearer effect on TOM than on price was observed; this implies that the market is adjusted both by prices and TOM. Therefore, we confirmed the importance of both price and TOM in capturing the demand–supply relationship in the resale housing market.

At the metropolitan level, sales prices declined and TOM increased in areas farther away from central Tokyo (see Table 1.2). Considering this metropolitan-level trend, from Tables 1.3 and 1.4, we summarize the effect of each property characteristic on the demand–supply relationship within each distance interval

[7] Since the VIF is below 5, a multi-collinearity problem does not exist.

[8] In contrast to the estimation of price, in the survival analysis on TOM, we do not include the controls such as lot size, floor area, and building age, as these variables do not exhibit statistical significance.

[9] This confirms that the Weibull distribution is more appropriate than exponential distribution ($p = 1$).

[10] We compared the log-likelihood to choose the survival analysis with Weibull distribution to estimate TOM. In addition, to estimate TOM with the entire sample, we constructed or selected the appropriate explanatory variables after comparing the log-likelihoods between each estimated model. For the entire sample, the likelihood ratio test (the null hypothesis is without any explanatory variables) yielded $\chi^2 = 154.08$ with statistical significance (at 1% level); thus, we confirmed the validity of the entire estimates.

Table 1.3 Determinants of sale price

Distance from central Tokyo:	Entire sample			10–20 km			20–30 km		
Explanatory variables	Coef.	*t*-value		Coef.	*t*-value		Coef.	*t*-value	
Lot size: less than 50 m^2	−0.326	−8.01	***	−0.036	−0.44		−0.248	−5.24	***
Lot size: 250 m^2 or more	0.023	0.54		−0.556	−2.61	**	−0.827	−6.58	***
Building age: less than 5 years	0.041	1.08		−0.027	−0.35		−0.032	−0.54	
Building age: 30 years or more	0.046	1.39		0.184	1.94	*	0.107	1.81	*
Time to nearest station: less than 10 min	0.069	3.36	***	0.200	4.34	***	0.050	1.33	
Bus to nearest station	−0.073	−2.82	***	−0.102	−1.32		−0.228	−3.66	***
Max. Slope: 10 degrees or more	−0.119	−3.03	***	−0.008	−0.03				
Proportion of elderlies: 25% or more	0.054	1.42		0.095	1.02		−0.052	−0.43	
Detached owner-occupied housing district	−0.112	−5.20	***	−0.095	−1.21		−0.153	−4.65	***
Densely inhabited districts in 1960	0.169	2.91	***	0.043	0.50		0.064	0.60	
Densely inhabited districts in 2010	0.150	6.05	***	0.363	3.81	***	0.181	3.40	***
Low-rise residential zones	0.094	4.60	***	0.065	1.06		−0.035	−1.02	
Medium-to-high-rise residential zones	0.095	3.79	***	0.080	1.74	*	−0.039	−0.86	
Lot size (m^2)	0.001	12.06	***	0.006	7.67	***	0.006	11.93	***
Floor space (m^2)	0.007	22.23	***	0.007	7.40	***	0.004	6.78	***
Building age (year)	−0.021	−15.88	***	−0.022	−6.96	***	−0.024	−10.78	***
Distance from central Tokyo: 20–30 km	−0.195	−7.26	***						
Distance from central Tokyo: 30–40 km	−0.359	−13.53	***						
Distance from central Tokyo: 40–50 km	−0.631	−20.31	***						
Distance from central Tokyo: 50–70 km	−0.758	−16.98	***						
Constant	9.430	195.75	***	8.774	61.25	***	9.139	113.48	***
Adjusted-R^2	0.74			0.69			0.83		
Number of observations	1546			284			386		

Distance from central Tokyo:	30–40 km			40–50 km			50–70 km		
Explanatory variables	Coef.	*t*-value		Coef.	*t*-value		Coef.	*t*-value	
Lot size: less than 50 m^2	−0.300	−1.61		−0.382	−1.84	*			
Lot size: 250 m^2 or more	−0.058	−0.75		0.230	3.93	***	0.195	1.54	
Building age: less than 5 years	0.028	0.38		0.096	1.12		0.220	1.04	
Building age: 30 years or more	0.005	0.09		0.042	0.70		−0.192	−1.84	*
Time to nearest station: less than 10 min	0.066	1.87	*	−0.075	−1.65		0.194	2.27	**
Bus to nearest station	−0.121	−2.12	**	−0.044	−1.18		−0.086	−0.92	
Max. Slope: 10 degrees or more				−0.193	−4.20	***	−0.075	−1.16	
Proportion of elderlies: 25% or more	−0.034	−0.41		0.024	0.38		0.123	1.26	
Detached owner-occupied housing district	−0.077	−2.00	**	−0.118	−2.76	***	−0.138	−1.45	
Densely inhabited districts in 1960	−0.036	−0.11		0.247	2.17	**			
Densely inhabited districts in 2010	0.199	4.67	***	0.141	3.23	***	−0.050	−0.30	
Low-rise residential zones	0.043	1.14		0.155	3.64	***	0.433	4.23	***
Medium-to-high-rise residential zones	0.115	2.31	**	0.185	3.07	***	0.139	0.85	
Lot size (m^2)	0.002	8.31	***	0.001	4.98	***	0.001	4.75	***
Floor space (m^2)	0.004	7.29	***	0.006	11.04	***	0.004	2.63	**
Building age (year)	−0.020	−7.91	***	−0.023	−8.76	***	−0.019	−2.70	***
Constant	9.168	109.23	***	8.982	103.10	***	8.875	34.99	***
Adjusted-R^2	0.60			0.73			0.74		
Number of observations	419			348			109		

Notes: The dependent variable is the log of the sale price (1000 JPY). Significance level: ***1%, **5%, *10%

Table 1.4 Determinants of TOM

Distance from central Tokyo:	Entire sample			10–20 km			20–30 km		
Explanatory variables	Coef.	*z*-value		Coef.	*z*-value		Coef.	*z*-value	
Lot size: less than 50 m^2	0.235	1.90	*	0.039	0.16		0.356	2.13	**
Lot size: 250 m^2 or more	0.180	1.84	*	0.493	1.06		0.428	1.07	
Building age: less than 5 years	0.208	2.09	**	0.447	2.21	**	0.139	0.72	
Building age: 30 years or more	−0.047	−0.79		−0.008	−0.06		−0.141	−1.16	
Time to nearest station: less than 10 min	−0.074	−1.15		−0.009	−0.06		−0.195	−1.31	
Bus to nearest station	−0.148	−1.81	*	−0.131	−0.53		−0.375	−1.57	
Max. Slope: 10 degrees or more	−0.058	−0.47		0.504	0.70				
Proportion of elderlies: 25% or more	0.042	0.35		−0.268	−0.96		0.863	1.86	*
Detached owner-occupied housing district	−0.116	−1.69	*	0.115	0.46		−0.215	−1.68	*
Densely inhabited districts in 1960	0.115	0.63		−0.035	−0.13		1.074	2.66	***
Densely inhabited districts in 2010	−0.093	−1.17		0.033	0.11		−0.269	−1.33	
Low-rise residential zones	−0.066	−1.04		0.012	0.06		−0.138	−1.07	
Medium-to-high-rise residential zones	0.026	0.34		0.117	0.79		−0.005	−0.03	
Distance from central Tokyo: 20–30 km	0.141	1.68	*						
Distance from central Tokyo: 30–40 km	−0.017	−0.20							
Distance from central Tokyo: 40–50 km	0.099	1.03							
Distance from central Tokyo: 50–70 km	−0.007	−0.05							
Fall	−0.106	−1.80	*	0.031	0.21		0.053	0.43	
Year 2013	0.267	4.67	***	0.298	2.11	**	0.388	3.35	***
Year 2014	0.925	9.32	***	0.887	3.87	***	1.141	6.03	***
Markup ratio	−0.384	−5.33	***	−0.166	−1.13		−0.366	−2.28	**
Constant	−2.224	−15.60	***	−2.841	−7.00	***	−1.804	−5.92	***
Parameter *p*	0.63 ***			0.65 ***			0.60 ***		
Log-likelihood	−3207.68			−576.77			−814.93		
Number of observations	1546			284			386		

Distance from central Tokyo:	30–40 km			40–50 km			50–70 km		
Explanatory variables	Coef.	*z*-value		Coef.	*z*-value		Coef.	*z*-value	
Lot size: less than 50 m^2	0.564	0.93		0.155	0.21				
Lot size: 250 m^2 or more	−0.067	−0.36		0.116	0.78		0.940	3.10	***
Building age: less than 5 years	0.196	0.93		0.041	0.15		0.640	0.98	
Building age: 30 years or more	−0.086	−0.78		−0.007	−0.05		0.568	2.13	**
Time to nearest station: less than 10 min	−0.028	−0.25		−0.210	−1.29		−0.436	−1.43	
Bus to nearest station	−0.197	−1.05		−0.231	−1.76	*	0.221	0.68	
Max. Slope: 10 degrees or more				0.029	0.18		−0.214	−0.89	
Proportion of elderlies: 25% or more	−0.052	−0.20		0.269	1.19		0.240	0.63	
Detached owner-occupied housing district	0.015	0.12		−0.122	−0.77		−1.475	−4.06	***
Densely inhabited districts in 1960	0.577	0.57		−0.018	−0.05				
Densely inhabited districts in 2010	−0.129	−0.91		0.058	0.37		−2.205	−2.82	***
Low-rise residential zones	−0.105	−0.88		0.087	0.59		0.029	0.08	
Medium-to-high-rise residential zones	−0.012	−0.07		−0.036	−0.17		1.791	2.20	**
Fall	−0.197	−1.72	*	−0.233	−1.85	*	0.090	0.38	
Year 2013	0.184	1.62		0.061	0.47		0.401	1.78	*
Year 2014	0.903	4.60	***	0.758	3.30	***	1.003	1.83	*
Markup ratio	−0.496	−3.55	***	−0.524	−2.95	***	−1.303	−3.61	***
Constant	−2.163	−8.94	***	−1.822	−6.70	***	−1.311	−2.31	**
Parameter *p*	0.65 ***			0.61 ***			0.85 **		
Log-likelihood	−852.78			−728.83			−196.46		
Number of observations	419			348			109		

Notes: The dependent variable is hazard λ (probability of sale). Significance level: ***1%, **5%, *10%. The significance level for parameter p is based on the Wald test, where the null hypothesis is $\ln(p) = 0$; that is, $p = 1$

Property characteristics	10−20km	20−30km	30−40km	40−50km	50−70km
Lot size: less than 50m^2		*		-	
Lot size: 250m^2 or more	-	-		+	+
Building age: less than 5 years	+				
Building age: 30 years or more	+	+			*
Time to nearest station: less than 10 min	+		+		+
Bus to nearest station		-	-	-	
Max. slope: 10 degrees or more				-	
Proportion of elderies: 25% or more		+			
Detached owner-occupied housing district		- -	-	-	-
Densely Inhabited Districts in 1960		+		+	
Densely Inhabited Districts in 2010	+	+	+	+	-
Low-rise residential zones				+	+
Medium-to-high-rise residential zones	+		+	+	+
Market demand (relative to supply)	- -	-		+	++
	Low				High

Fig. 1.2 Demand–supply relationship for building/locational characteristics in each distance interval from central Tokyo

in Fig. 1.2. Demand relative to supply is high (low) if the price has a positive (negative) premium and/or the TOM reduces (increases). Thus, we showed the degree of demand as "++" ("− −") if both the price and TOM indicated high (low) demands with statistical significance; and "+" ("−") if either price or TOM showed high (low) demands with statistical significance, while considering the other cases as a reference. In the case wherein the price and TOM indicated demands in opposite directions, the price and TOM reduced; we indicated the state as "*." The significance level was set at 10%, and the diagonal lines indicate the cases without such samples.

From Tables 1.3 and 1.4 and Fig. 1.2, we first discuss the overall trends and then proceed to the discussion for suburbs close to central Tokyo and for outskirts farther from central Tokyo.

1.4.1 Overall Trends

Overall, relative to the supply, the demand for resale houses is low (high) in areas with typical characteristics of population shrinkage (growth).[11] First, strong demand exists in districts with convenient access to the nearest station and/or zones that protect the residential environment. For houses located within 10 min of walking distance to the nearest station, demand was relatively high in almost all distance intervals (10–20, 30–40, and 50–70 km), suggesting that convenient locations

[11] For small properties whose lot size is less than 50 m^2, demand is low within a 40–50 km interval, and price is reduced with a short TOM in an interval 20–30 km away from central Tokyo. For new properties that are less than 5 years old, strong demand exists within 10–20 km from central Tokyo.

sustain demand even on the outskirts farther away from central Tokyo. In the entire Tokyo metropolitan area, Ai (2014, 2016) shows that the population increases in areas located within 500 m of the nearest station; thus, our results suggest that purchasing a resale house may be suitable as an option to live in these convenient locations. In addition, properties in "medium-to-high-rise residential zones" tend to exhibit high demand in each distance interval (10–20 and 30–70 km).

However, the demand tends to be low in each distance interval for properties without convenient access to the nearest station and/or in detached owner-occupied housing districts. For houses that require taking a bus to the nearest station, the demand decreases at almost all distance intervals (20–50 km). Specifically, prices decline within the 20–40 km interval, and TOM increases within the 40–50 km interval.[12] In addition, in detached owner-occupied housing districts, the demand is low in all distance intervals (20–70 km), particularly in those characterized by a long TOM in the 50–70 km interval. This corresponds to a small population influx in areas that lack a variety of building types (Fujii 2008; Miyake et al. 2014).

The tendency of resale houses influencing the renewal of districts varies according to their distance from Tokyo. For properties located in the 1960 DID, demand is relatively high within 20–30 and 40–50 km intervals. For properties located in the DID of 2010, demand was relatively high in 10–50 km intervals, whereas it decreased with increased TOM in the 50–70 km interval. This may be attributed to the fact that the population declines in the newly developed residential districts within 50 km of central Tokyo. Moreover, the population shrinks in the traditional districts on the outskirts, 50 km or farther from central Tokyo (Miyake et al. 2014).

1.4.2 Suburbs Close to Central Tokyo

Demand is strong in the suburbs, relatively close to (within 30 km) central Tokyo. For houses that are 30 years old or older, demand is relatively high within 10–30 km intervals; for properties in aging districts (where the proportion of elderly residents is 25% or more), demand is relatively high, and properties are sold quickly. A potential reason for this strong demand is that, within commutable distance to central Tokyo, young households with children can purchase spacious houses at reasonable prices. This may explain why new residents who have settled in existing residential districts tend to be large households (Yoshida et al. 2007). Another potential reason for purchasing resale houses is to live close to parents within commutable distance from central Tokyo (Nakashima et al. 2011); these areas are likely to be occupied by elderly parents.

[12] There is no such effect beyond 50 km from central Tokyo, possibly because daily life does not rely on public transports.

1.4.3 Outskirts Farther Away from Central Tokyo

However, in the outskirts farther away (40 km or greater) from central Tokyo, the demand for potentially disadvantageous properties is low as the population (and number of households) declines. Specifically, houses in detached owner-occupied housing districts and/or those in the 2010 DIDs tended to exhibit an increased TOM in the 50–70 km distance interval. Furthermore, for properties located in inclined districts with a maximum slope of 10 degrees or greater, negative price premiums existed in the 40–50 km interval.

In certain cases, for old properties, the price declined while the TOM became short. Indeed, properties that were 30 years old or older exhibited a shorter TOM with a lower price. Recently, certain owners of vacant houses have found it costly to maintain their property (e.g., due to the holding tax). As reported by Hirayama (2011), the value of homeownership becomes negative unless the owner finds a buyer. Thus, our results suggest that certain owners are motivated to dispose of their property quickly by reducing their prices.

On the outskirts farther away from central Tokyo, where the demand is shrinking, the demanded properties exhibit characteristics that are suitable for the location. In other words, for large houses with lot sizes of 250 m^2 or greater, demand is low within 10–30 km of central Tokyo, whereas there is a strong demand within 40–70 km of central Tokyo. In particular, in suburbs 40 km or farther from central Tokyo, the reason may be that spacious resale houses are more reasonable to purchase than newly built houses. Furthermore, properties in "low-rise residential zones" attract demands within 40–70 km from central Tokyo, suggesting the zones with nice residential environments result in a population influx. However, these characteristics are typically observed in the low-demand "detached owner-occupied housing districts"; thus, the absolute demand may not be that high.

1.5 Conclusion

This chapter investigated the relationship between building and locational characteristics in the context of population shrinkage in Japan and price and TOM, which are the basic components of real estate markets. It clarified the demand–supply relationship of the resale housing markets in the suburbs of the Tokyo metropolitan area. Overall, relative to the supply, the demand for resale houses was low (high) in areas with typical characteristics of population shrinkage (growth). However, the demand–supply relationship for potentially disadvantageous properties varied according to distance from central Tokyo. Thus, there is a strong demand in the suburbs relatively close to (within 30 km) central Tokyo, even for old houses in areas with a large proportion of elderly residents. However, in the outskirts farther away (more than 40 km) from central Tokyo, houses in the detached owner-occupied housing area are losing demand, exhibiting an increased TOM, and certain owners may reduce their prices to quickly dispose of their properties.

The results suggest the potential of the suburbs relatively close to central Tokyo, where the resale housing market plays a role in circulating built-up residential areas where old houses and elderly residents are concentrated. In addition to the utilization of vacant properties, resale houses are a new option beyond purchasing costly newly built houses or renting tiny flats. However, the results suggest that in the outskirts farther away from central Tokyo, market demands for resale houses in the detached owner-occupied housing area are limited; thus, these property owners must continue maintenance, utilize them within their neighborhood communities, or utilize them as land after demolishing the building. The results also suggest that certain owners feel that it is costly to maintain vacant houses. An effective strategy may be getting the neighbors to agree to use the land and/or maintain houses before the properties are inherited and owners become unknown to the public.

This chapter conducted a basic analysis to capture the unexplored demand–supply relationship in the resale housing market in the suburbs of Tokyo. It is necessary to understand the market dynamics, including vacant houses that do not enter the resale housing market. Moreover, the reasons behind purchasing resale houses and the characteristics of such households in the suburbs relatively close to central Tokyo must be clarified using survey data of the population influx.

References

Ai H (2014) How living environment indices effect on population change patterns of local districts. J City Plan Inst Jpn 49(3):567–572

Ai H (2016) Indexing of living environment attracts young and productive age generations. J City Plan Inst Jpn 51(3):860–866

Engelhardt GV (2003) Nominal loss aversion, housing equity constraints, and household mobility: evidence from the United States. J Urban Econ 53(1):171–195

Fujii T (2008) A micro-level analysis on generational changes and district characteristics in the Tokyo metropolitan area. J Archit Plan 633:2399–2407

Fujii T, Oe M (2005) An analysis of the population structure in the Tokyo metropolitan area in terms of generation balance. J Archit Plan 593:123–130

Fujii T, Oe M (2006) A study on generational changes in the suburbs of the Tokyo metropolitan area: a comparative analysis from 1980 to 2020 among cohorts by GBI. J Archit Plan 605:101–108

Genesove D, Mayer C (2001) Loss aversion and seller behavior: evidence from the housing market. Q J Econ 116(4):1233–1260

Genesove D, Mayer CJ (1997) Equity and time to sale in the real estate market. Am Econ Rev 87(3):255–269

Han L, Strange WC (2015) The microstructure of housing markets: search, bargaining, and brokerage. In: Duranton G, Henderson JV, Strange WC (eds) Handbook of regional and urban economics, vol 5B. Elsevier, Amsterdam, pp 813–886

Hasegawa H (2005) Study on changes of small lots and houses in suburban detached houses estates built in mass-housing period: part 2, case study in Moroyama town, Saitama, Tokyo metropolitan area. Proceedings of the Architectural Institute of Japan (F-1), pp 1545–1546

Hirayama Y (2011) Urban conditions: housing, life, and sustainability. NTT Publishing, Tokyo

Ito S, Kaido K (2013) The transformation of the spatial uses of vacant houses and vacant lots and the demographic structures of the residences in suburban housing estates. J City Plan Inst Jpn 48(3):999–1004

Ministry of Land, Infrastructure, Transport and Tourism (2017) Housing economics data: change in the share of used houses in the housing transactions. http://www.mlit.go.jp/statistics/details/t-jutaku-2_tk_000002.html. Accessed 3 Aug 2017

Miyake R, Koizumi H, Okata J (2014) Research on the distribution of the households-decreasing districts and district characteristics in the Tokyo metropolitan area. J City Plan Inst Jpn 49(3):1029–1034

Nakashima T, Ito N, Lee YG, Tsukuda H, Otsuki T (2011) Analysis of real-estate transaction in "new town" in local city over 30 years for depopulation and aging society. Proceedings of the Architectural Institute of Japan (F-1), pp 69–72

Ong SE, Koh YC (2000) Time on-market and price trade-offs in high-rise housing sub-markets. Urban Stud 37(11):2057–2071

Piazzesi M, Schneider M, Stroebel J (2020) Segmented housing search. Am Econ Rev 110(3):720–759

Sakamoto K, Yokohari M (2016) Dynamic characteristics of emerging vacant houses and abandoned lands in residential neighborhoods. J City Plan Inst Jpn 51(3):854–859

Shimizu C, Nishimura KG, Asami Y (2004) Search and vacancy costs in the Tokyo housing market: attempt to measure social costs of imperfect information. Rev Urban Reg Dev Stud 16(3):210–230

Todoroki O (2011) A basic study on the trends of the mortgaged real estate market at the city planning area. J City Plan Inst Jpn 46(3):565–570

Ujihara T, Abe H, Murata N, Washio N (2016b) Reality of "spongy urban area" in local city: based on situation of development and loss, vacant house. J Jpn Soc Civ Eng D3 72(1):62–72

Ujihara T, Abe H, Nonaka S (2016a) Reality of spongy urban area based on state analysis of residential districts. J City Plan Inst Jpn 51(3):466–473

Ujihara T, Taniguchi M, Matsunaka R (2006) The reality of urban retreat: reverse sprawl and its relation with residential zone types. J City Plan Inst Jpn 41(3):977–982

Umeda N, Hirayama Y (2002) Crisis in the home ownership system? The spatial distribution of the financially troubled property in Kobe. Proc Archit Inst Jpn (Kinki) 42:841–844

Yamanshita S, Morimoto A (2015) Study on occurrence pattern of the vacant houses in the local hub city. J City Plan Inst Jpn 50(3):932–937

Yavas A, Yang S (1995) The strategic role of listing price in marketing real estate: theory and evidence. Real Estate Econ 23(3):347–368

Yoshida T, Koyama Y, Hasegawa H (2007) Characteristics of householders newly moving in suburban neighborhood districts of detached houses. J City Plan Inst Jpn 42(3):703–708

Chapter 2
Old Condominiums and Their Tendency to Be Rebuilt: A Case Study of the Tokyo Metropolitan Area

Hiroki Baba, Chihiro Shimizu, and Yasushi Asami

Abstract Condominiums are important housing stock in Japan, but they often need to be rebuilt. This research investigates the factors and spatial distribution of condominium rebuilding. To this end, data on rebuilt condominiums was integrated with data on old condominiums defined as the ones with old earthquake resistance standards. Using a probit model, the tendency of certain factors to contribute to rebuilding was assessed. Findings reveal that building age, land area, distance to the city center, number of housing units, and floor area possible for expansion are statistically significant, which is consistent with the findings of previous studies on the qualitative analysis of condominium rebuilding in Japan. When the marginal effects of the factors were calculated, a four-year increase in the average building age, for instance, increases the probability of rebuilding by 1.49 percentage point. We also confirmed that condominiums with high rebuild probability are concentrated in the western part of central Tokyo and are also likely to exist in suburban areas.

Keywords Condominium · Rebuilding · Real estate · Probit model · Spatial distribution · Floor area ratio

---

The contents of this paper are based on the following paper originally published in a Japanese journal: Baba, H., Shimizu, C., Asami, Y. (2020) Factors and spatial distribution of rebuilding of old condominium: Case study of Tokyo metropolitan area. Journal of City Planning Institute of Japan 55(3): 1143–1150 (in Japanese).

H. Baba (✉) · C. Shimizu
Graduate School of Social Data Science, Hitotsubashi University, Tokyo, Japan
e-mail: h.baba@r.hit-u.ac.jp

Y. Asami
Department of Urban Engineering, The University of Tokyo, Tokyo, Japan

Y. Asami et al. (eds.), *Studies in Housing and Urban Analysis in Japan*,
New Frontiers in Regional Science: Asian Perspectives 75,
https://doi.org/10.1007/978-981-99-8027-7_2

2.1 Introduction

2.1.1 Background and Research Aim

Condominiums are important housing stock in Japan, with an estimated 6.5 million units as of the end of 2018 (MLIT n.d.-a). Demand for this type of housing increases as urban agglomeration intensifies (Rosen and Walks 2013). Once a building is supplied, it deteriorates and should be ultimately rebuilt, but demolition and rebuilding are more difficult than that for detached housing (Saito and Hasegawa 2001). Condominiums have multiple owners, which raises various issues such as rebuilding consensus and handling of mortgage (Hasegawa 1999). Therefore, it is not simple to renew condominiums, and they might be left as they are, even if they deteriorate or become mismanaged in the future (Asami and Saito 2019). To deal with this problem, Baba (2020) constructed a database of condominiums from private company data and visualized the geographical distribution of aging condominiums, but it is not clear to what extent rebuilding is realistic.

Condominiums are generally durable goods, and it is difficult to make elastic quantity adjustments for future population increase and decrease (Glaeser and Gyourko 2005). Nevertheless, it is natural to rebuild condominiums to some degree when real estate property values rise significantly in areas of excess demand. Nevertheless, in Japan, as rebuilding requires the agreement of 80 percent of unit owners and eviction of tenants, the number of condominiums that have been completed or are undergoing rebuilding as of April 2019 is merely 267 (MLIT n.d.-b). Therefore, although previous studies have chosen case studies of rebuilt condominiums (Hasegawa 1999), establishing general trends is a challenging task.

This study focuses on old condominiums, defined as the ones with old earthquake resistance standards,[1] and examines the trends in rebuilding factors and the spatial distribution of rebuilding probability. Specifically, We examine the difference between existing condominiums with old earthquake resistance standards and those that have been rebuilt, and analyze the difference in trends using a probit model. In particular, the rebuilding issue is associated with the economic load of the unit owners (Tanaka and Kumata 1999). Therefore, We considered variables such as the size of land area, floor area, and expandable floor area by rebuilding. In addition, after estimating the probability of rebuilding, We demonstrated the spatial distribution of grid cells that are (or are not) expected to be rebuilt.

The target area in this study is the Tokyo metropolitan area (Tokyo, Saitama, Chiba, and Kanagawa Prefectures), where old condominiums have been supplied since early stages (Fig. 2.1). Given that approximately 60.6% of the rebuilt condominiums across Japan exist here, this area is suitable for analyzing the

[1] Condominiums with old earthquake resistance standards refer to the ones whose building permit was approved by May 1981. This is because a new earthquake resistance standard had been implemented in June 1981. In this study, considering the period of construction, condominiums completed before 1982 were used as the data for analysis.

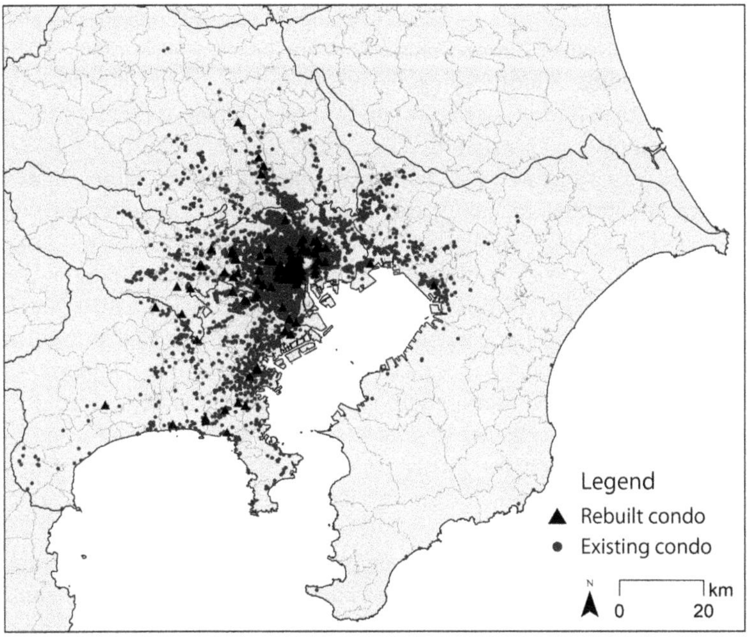

Fig. 2.1 Geographical location of rebuilt and existing condominiums with old earthquake resistance standards in the Tokyo metropolitan region

trend of rebuilding through market mechanisms. Condominiums here that have old earthquake resistance standards are being actively considered for rebuilding due to aging and seismic performance issues.

2.1.2 Current Situation of Rebuilding Condominiums

With the advent of condominiums, the housing system enables us to build dense urban areas (Rosen and Walks 2015). Particularly in areas where housing demand is high, such as Hong Kong and Singapore, most people live in either condominium or apartment housing. Despite their necessity and amenity, condominiums face many challenges. In Hong Kong, during the redevelopment of residential high-rise neighborhoods, private developers struggled with absent housing owners (Yau and Ling Chan 2008). In Singapore, the owners could not reach a consensus on the sale of sites in areas with high development potential, and the government amended the law to allow sale with 80% consensus (Sim et al. 2002). In the case of the Gold Coast, where more than 50% of the condominiums are more than 30 years old, the cityscape of the area is degraded due to aging condominiums and the difficulty of updating the area (Warnken et al. 2003). Even in Toronto, Canada, condominium neighborhoods sometimes face an anticommons problem (Webb and

Webber 2017). Lastly in Japan, while an oversupply of new condominiums is being built, the accumulation of old condominiums has become a problem (Baba 2020; Ronald and Hirayama 2006). Aging is one of the keywords considering future condominium issues, and it is appropriate to target Japanese cities for analysis as many condominiums are already old.

In Japan, previous research has focused on the aging of condominiums and the difficulties of rebuilding, including the development of a legal framework, land constraints, and consensus building in management associations (e.g., Asami et al. 2012). The pioneering work on the issue of condominium rebuilding is conducted by Takamizawa and Go (1989), where the authors anticipate that the issue of condominium rebuilding will arise in the future, and proposed that the possibility of floor space increase, location, and the floor area per unit are important factors. Hasegawa (1999) analyzed cases where reconstruction was completed at an early stage, and he identified rebuilding factors such as floor area ratio, location, the amount of self-pay burden for owners, and the number of owners. Thus, some of the rebuilding factors include the physical characteristics and location of the condominium.

Consensus building of the management association is another important factor for rebuilding. Previous studies have been conducted in this regard, including one that discusses optimal consensus building requirements from the perspective of the rebuilding benefits (Asami 2009), one that enumerates the consensus building process and describes its characteristics (Saito and Hasegawa 2001), and one that ranks the consensus building capacity of management associations (Saito et al. 1999). Although these discussions offer important findings for promoting condominium rebuilding, they do not fully consider factors that indicate the potential for physical reconstruction, such as the possibility of floor area increase, location, and floor area per unit, as advocated by Takamizawa and Go (1989) and Hasegawa (1999). This may be attributed to the fact that there are only a few examples of condominium rebuilding in Japan.

Recent increases in condominium rebuilding enable researchers to conduct quantitative analysis of their physical characteristics. For instance, Hanazato (2017) used discriminant analysis to show the difference in trend between two groups: rebuilds and massive renovations. Although the author revealed the limited nature of successful rebuilds, they cannot be representative of older condominiums as a whole, as the control group (massive renovation group) consisted of 47 cases. Therefore, although quantitative analysis of the potential for condominium rebuilding provides useful insights into what types of old condominiums are suitable for future rebuilding, previous studies have not collected sufficient number of the samples.

Considering the previous studies above, this study has the following originalities. First, this study covers most of the condominiums with old earthquake resistance standards, enabling a more general discussion of condominium rebuilding. Second, this study explicitly introduces physical characteristics such as the possibility of floor area increase and location into a model, which have been difficult to deal with in previous studies. Third, using the regression results, this study identifies areas where condominium rebuilding is likely to occur in the future and areas where condominiums are likely to remain unmaintained.

2.2 Data and Method

2.2.1 Data

We used the data on condominiums completed before 1982 by the Real Estate Economic Institute, Co., Ltd.[2] regarding the characteristics of condominiums. This is because a new earthquake resistance standard had been implemented in June 1981. Property data had been collected between 1967 and 1982, containing a variety of information associated with building and location characteristics. Pieces of information for this analysis were extracted and used as variables: property name, location, site area, total floor area, number of floors, total number of units, year of completion, and average private floor area per unit. The average private floor area per unit was categorized into single occupancy (−40 m^2), two-person household (40–55 m^2), and family (55–m^2), referring to the guideline floor area levels described in the Basic Plan for Housing and Living Standards (MLIT n.d.-c). This is because data representing a range of floor area exist for each property and cannot be expressed as an interval scale. Note that as the condominium data in this study is a compilation of information on sales, there may be multiple rows of data for the same property depending on the time of release. Therefore, all data were sorted by property name to avoid duplicate use of data.

We used 462,635 housing unit samples on condominiums with old earthquake resistance standards. Before analysis, it is necessary to consider whether these samples are representative of all the condominiums. According to the statistics issued by MLIT (n.d.-a), the national stock of condominiums with old earthquake resistance standards is approximately 1.03 million units, and Baba (2020) illustrated that approximately 56.8% of these units are located in Tokyo and three prefectures. However, the government statistics refers to the number of condominiums based on confirmation of the building code, including properties that were never built, so it might lead to an overcount of the number of condominium units. Although the data used in this study are approximately 20% less than the value presented by the government statistics, it is considered to be within a sufficient range to measure statistical properties.[3]

The reconstruction data were extracted from the cases published by the Urban Renewal Association of Japan for Tokyo, Chiba, Saitama, and Kanagawa prefectures.[4] Of the 157 reconstruction cases in Tokyo and three prefectures, 114 were selected as the sample for analysis. We considered the case of overlap with data for condominiums with old earthquake resistance standards and reconstructed condominiums, and the 14 duplicate property names were then removed.

[2] https://www.fudousankeizai.co.jp/publicationCatelist?cateId=8&id=52.

[3] As listwise deletion is performed for missing variables, the sample size is smaller than above. However, given the large sample size, this should not be a problem when analyzing statistical trends.

[4] http://m-saisei.info/index.html.

In addition to the condominium data, data related to the surrounding environment and land use regulations were also added. Specifically, distance to the city center,[5] distance to the nearest station, and land use zoning were measured from the National Land Numerical Information.[6] The land use zoning was broadly classified into exclusively residential, residential, commercial, and industrial areas. It is true that introducing land price into a model is effective for increasing the explanatory power, but the land price would explain most location factors as an important proxy for the economic value of condominiums. Therefore, land price value and location may face the problem of multicollinearity. As this study focuses on the predictive spatial distribution of condominium rebuilding, the location factors supersede the land price. This study thus proceeded with the analysis assuming that the land price is controlled by location variables such as distance to the nearest station, distance to the city center, and land use zoning dummies.

We generated data of floor area available for expansion considering both regulations designated by land use zoning and by the width of the road frontage. The land use zoning was judged by overlaying zoning polygons from the National Land Numerical Information. When considering the regulation by the width of the road frontage, the entire road width was estimated from the road centerline line data provided by the Conservation GIS-consortium Japan.[7] Road widths were classified as −3.0 m, 3.0–5.5 m, 5.5–13.0 m, 13.0–19.5 m, and 19.5–m, the maximum floor area ratio was calculated using road widths of 3.00 m, 4.25 m, 9.25 m, 16.25 m, and 20.00 m, respectively, for convenience. Although the range of road width thresholds appears to be wide, most cases where the standard floor area ratio is restricted are narrow roads with road widths of 4 m or less in Japan. As the road frontage could not be identified from the building point data, the width of the road frontage was calculated as an approximation. Specifically, assuming a square site area of $a\text{m}^2$, for the road edge overlapped with a circle whose radius is the distance $\sqrt{2a}/2\text{m}$ from the building center point to the farthest building edge, We adopted the widest road as the width of the road frontage. Finally, We compared the maximum floor area ratios by both land use zoning and the regulation by the width of the road frontage, and the smaller one was defined as the floor area ratio available for expansion.

To remove outliers from the dataset, preprocessing was then performed in the following steps. First, We omitted properties whose latitude and longitude were not identified to the street level using the geocoder provided by the University of

[5] If the city center is defined by a single point such as Tokyo Station, municipalities in the western part of Tokyo may be overmeasured. Therefore, the definition of the city center is based on the following assumptions: the distance to the city center is measured as 0 m if the location is in the six central wards of Tokyo (Chiyoda, Chuo-ku, Minato-ku, Shinjuku, Bunkyo-ku, and Shibuya-ku); otherwise, the minimum distance to five stations with a large number of passengers on the Yamanote Line (Shinjuku, Ikebukuro, Tokyo, Shinagawa, and Shibuya stations) is measured.

[6] https://nlftp.mlit.go.jp/ksj/.

[7] http://cgisj.jp. Note that this data was generated based on the Numerical Map 2500 provided by the Geospatial Information Authority of Japan (GSI) and is not a basic survey result.

Table 2.1 Descriptive statistics

Variable	Unit	Mean	S.D.	Min.	Median	Max.
Land area	1000 m^2	2.178	5.956	0.100	0.881	149.983
Total floor area	1000 m^2	4.710	7.762	0.128	2.716	170.502
Number of stories	Story	7.148	2.745	2	7	25
Number of housing units	Unit	66.182	100.733	4	40	2233
Building age as of 2020	Year	43.551	4.258	38	42	93
Dummy on single layout	–	0.180	0.384	0	0	1
Dummy on two-person layout	–	0.327	0.469	0	0	1
Dummy on family layout	–	0.493	0.500	0	0	1
Distance to the nearest station	1000 m	0.447	0.409	0.013	0.331	7.971
Distance to the city center	1000 m	10.081	11.347	0.000	6.252	79.275
Dummy on exclusively residential land use	–	0.250	0.433	0	0	1
Dummy on residential land use	–	0.228	0.419	0	0	1
Dummy on commercial land use	–	0.392	0.488	0	0	1
Dummy on industrial land use	–	0.131	0.337	0	0	1
Floor area ratio	%	277.336	142.804	60	200	800
Observation		5821				

Tokyo,[8] properties whose year of completion was after 1983, and properties whose site area or total floor space was less than 100m^2. In addition, We performed listwise deletion, by which properties with at least one of the variables missing were omitted. As a result, 5821 samples were obtained. The descriptive statistics of the data is shown in Table 2.1.

2.2.2 *Method*

First, the floor area available for expansion, add_floor$_i$, is calculated in a simplified manner as follows.

$$\text{addfloor}_i = \left(\text{FAR}_i^{\max} - \text{FAR}_i^{\text{use}}\right) \text{lotsize}_i \tag{2.1}$$

where $\text{FAR}_i^{\max}$ represents the maximum floor area ratio considering the land use and the building code regulations, $\text{FAR}_i^{\text{use}}$ represents the current floor area ratio, and lot_size$_i$ represents the land area. Equation (2.1) indicates the extent of floor area that can be increased by rebuilding, which can be an indicator related to the financial burden for owners at the time of rebuilding. During the analysis, to identify spatial

[8] https://geocode.csis.u-tokyo.ac.jp/geocode-cgi/geocode.cgi?action=start.

trends in floor area available for expansion, We visualized the floor area divided by the total number of units per municipality.

A probit model was then used to analyze the characteristics of rebuilt condominiums. The model is often used where the variables are on a nominal or ordinal scale and has been applied to various analysis targets, such as occupant satisfaction evaluation (Choi and Asami 2004) and analysis of trends in rebuilding when the building code regulations are tightened (Hanazato 2017). The model in this study used the status of a condominium with old earthquake resistance standard i as a binary explained variable, that is, whether the condominium i is rebuilt or not. The rebuilding choice of the stakeholders (e.g., condominium management association) of the condominium i is denoted as y_i, and $y_i = 1$ is defined as rebuilding, and $y_i = 0$ is otherwise. In the model, the dependent variable for the potential preference y_i^* behind the choice can be expressed linearly with J variables:

$$y_i^* = \alpha + \boldsymbol{\beta x}_i' + u_i \tag{2.2}$$

where α is the parameter, $\boldsymbol{\beta}$ is the $1 \times J$ parameter vector, $\boldsymbol{x}_i$ is the i-th $1 \times J$ variable vector, and u_i is the error term. Now assume that the choice to rebuild is taken when the potential preference is greater than 0 and not to rebuild when it is less than 0. Given the probability when $y_i = 1$:

$$\Pr(y_i = 1) = \Pr\left(y_i^* > 0\right) = \Pr\left(-\left(\alpha + \boldsymbol{\beta x}_i'\right) < u_i\right) \tag{2.3}$$

Assuming that u_i is standard normal distribution, Pr(•) can be expressed in terms of the cumulative distribution function, so let $F(\bullet)$ be the cumulative density function[9]:

$$\Pr\left(y_i^* > 0\right) = 1 - F\left(-\left(\alpha + \boldsymbol{\beta x}_i'\right)\right) = F\left(\alpha + \boldsymbol{\beta x}_i'\right) \tag{2.4}$$

In the case of no rebuilding ($y_i = 0$), it can be expressed as:

$$\Pr\left(y_i^* \leq 0\right) = 1 - F\left(\alpha + \boldsymbol{\beta x}_i'\right) \tag{2.5}$$

For more details on this model, please see Wooldridge (2016).[10]

The explanatory variables include the ones related to possibility of floor area increase, building characteristics, and location, according to previous studies. Therefore, We added land area, number of housing units, number of stories, floor

[9] The cumulative density function of the standard normal distribution can be identified as $F\left(\alpha + \boldsymbol{\beta x}_i'\right) = \int_{-\infty}^{\alpha+\boldsymbol{\beta x}_i'} \frac{1}{\sqrt{2\pi}} \exp\left(-\frac{y^2}{2}\right) dy$.

[10] Parameter estimation is performed by the maximum likelihood method. The likelihood function $L(\alpha, \boldsymbol{\beta})$ is $L(\alpha, \boldsymbol{\beta}) = \prod_{m=1}^{N_1} F\left(\alpha + \boldsymbol{\beta x}_i'\right) \prod_{n=1}^{N_2} \left(1 - F\left(\alpha + \boldsymbol{\beta x}_i'\right)\right)$ where N_1 is the sum of condominium owners that choose to rebuild and N_2 that do not, and the sum of both is the number of observations. The parameters, α, $\boldsymbol{\beta}$, are estimated by maximizing the likelihood function.

layout dummies, total floor area, and floor area available for expansion. Moreover, distances to the nearest station and to the city center are important variables describing the location characteristics of condominiums, and land use zoning is also considered an important indicator of the location, so these variables were also added. Moreover, the building age was also added as it correlates with the physical aging of the condominium, which is directly related to the likelihood of rebuilding.

2.3 Results

2.3.1 Geographic Distribution of Floor Area Available for Expansion

Figure 2.2 shows the total sum of the floor area available for expansion by municipality, and Fig. 2.3 shows the average values of the floor area available for expansion divided by the number of housing units. The former allows us to consider the potentiality of future development, whereas the latter represents the extent of rebuilding incentive for condominium owners. By observing both figures, the following findings were obtained.

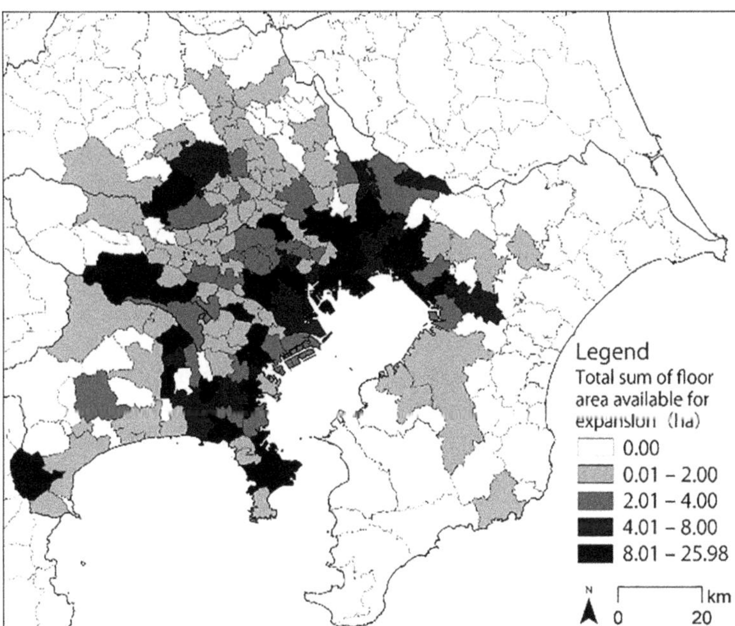

Fig. 2.2 Geographical distribution of the total sum of floor area available for expansion by municipality

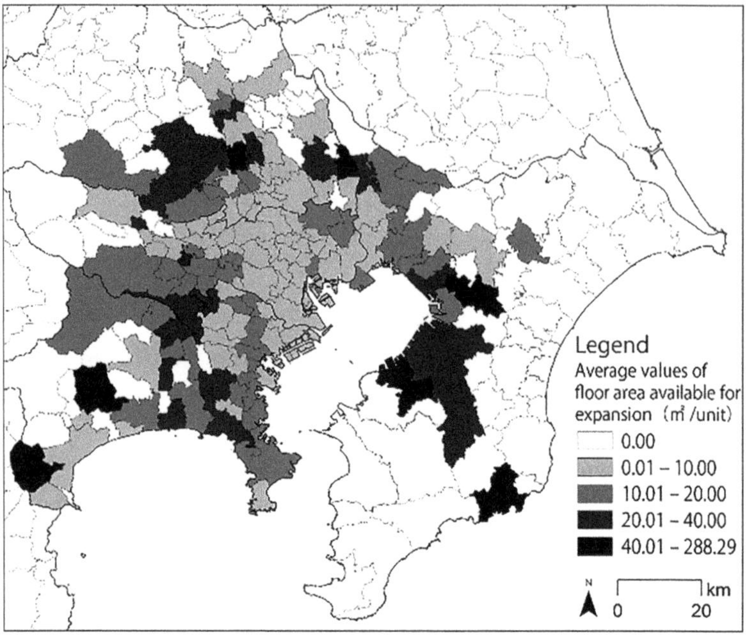

Fig. 2.3 Geographical distribution of average values of floor area available for expansion by municipality

First, the total floor area available for expansion is high in both eastern parts (e.g., Sumida and Edogawa wards) and western parts (e.g., Shibuya ward) of central Tokyo, and it is low in the southern part of Saitama prefecture. However, the average floor area available for expansion tended to be higher in the suburban areas of the Tokyo metropolitan area such as Sodegaura, Yoshikawa, and Hadano Cities, and lower in the central Tokyo area. This suggests that although there is room for development based on the total floor area available for expansion in central Tokyo, it is difficult for the owners of condominiums to benefit from floor space expansion. In suburban areas, the total sum of the floor area available for expansion is estimated to be low, but the floor area available for expansion per housing unit is large because there is a surplus land area for condominiums. Nevertheless, there is not a one-size-fits-all decision as to whether rebuilding is better than new development in the surrounding area, considering the costs of demolition and redevelopment.

2.3.2 *Difference in Averaged Characteristics Values of Condominiums with Old Earthquake Resistance Standards and Rebuilt Ones*

Rebuilt condominiums were compared to existing condominiums with old earthquake resistance standards using averaged characteristics as an indicator (Table 2.2). The results showed that the rebuilt condominiums had a relatively larger land area, lower number of stories, larger total number of housing units, and older building age. Furthermore, the distance to the nearest station was significantly closer for the rebuilt condominiums, whereas the distance to the city center was not significant. In addition, the floor area available for expansion per unit was found to be significantly larger for the rebuilt condominiums. Thus, although the number of rebuilt condominiums is relatively small, the results support the observations of previous studies. In particular, land area and floor area available for expansion per unit were approximately 2.4 times and 7.3 times larger, respectively, which is likely to be an important factor for condominium rebuilding. Nonetheless, we cannot deny the correlation and confounding among variables. For example, the assumption is that apartments located closer to the city center tend to be closer to the nearest train station. Therefore, We conducted a regression analysis using a probit model that considers the correlation of the explanatory variables in the next subsection.

2.3.3 *Statistical Difference Between Condominiums with Old Earthquake Resistance Standards and the Rebuilt Ones*

The characteristics of the rebuilt condominiums were analyzed using a probit model (Table 2.3). The results revealed the following points.

First, land area was positively correlated with the probability of rebuilding, while the number of stories was negatively correlated with the probability of rebuilding. These results imply that rebuilding occurs in condominiums with large sites and originally low-rise buildings. The condominiums planned and completed by the Japan Housing Corporation (now the Urban Renaissance Agency) are representative of such cases of rebuilding. Second, a significant negative correlation was found between the probability of rebuilding and the number of housing units. Assuming that there is one owner per unit, this could mean that rebuilding occurs in areas where the number of owners is small, which supports the results of previous studies (Asami et al. 2012). Furthermore, there was a weak positive correlation between the probability of rebuilding and the floor layout dummy. This was only for two-person households. Although the trend was not strong, the two-person household layout may be more likely to be rebuilt to accommodate for the demand to rent single-person households and buy family households. It is also conceivable that housing styles have changed over time. Although housing units of about 50 m^2 were sold for families in the 1970s, demand for small family-oriented units is thought

Table 2.2 Difference between the average values of condominiums with old earthquake resistance standards and those that were rebuilt

Variable	Unit	Mean		Standard error		*t*-value	Difference with statistical significance	
		Rebuilding		Rebuilding				
		Yes	No	Yes	No			
Land area	1000 m^2	5.14	2.12	9.54	5.85	−3.29	Diff >0	***
Number of stories	Story	5.92	7.17	2.38	2.75	5.44	Diff <0	***
Number of housing units	Unit	86.65	65.79	103.55	100.65	−2.08	Diff >0	*
Building age as of 2020	Year	54.69	43.34	11.66	3.67	−10.15	Diff >0	***
Dummy on single layout	–	0.23	0.18	0.42	0.38	−1.23		
Dummy on two-person layout	–	0.37	0.33	0.05	0.01	−0.88		
Dummy on family layout	–	0.40	0.50	0.05	0.01	1.92	Diff <0	*
Distance to the nearest station	1000 m	0.40	0.45	0.03	0.01	1.53	Diff <0	†
Distance to the city center	1000 m	8.94	10.10	1.17	0.15	0.98		
Dummy on exclusively residential land use	–	0.30	0.25	0.04	0.01	−1.21		
Dummy on residential land use	–	0.26	0.23	0.04	0.01	−0.70		
Dummy on commercial land use	–	0.38	0.39	0.05	0.01	0.34		
Dummy on industrial land use	–	0.06	0.13	0.02	0.00	2.83	Diff <0	**
Floor area available for expansion	m^2/unit	54.69	7.52	8.78	0.34	−5.37	Diff >0	***

Notes: Statistical significance ***0.1%, **1%, *5%, †10%; observation: rebuilt condo = 109, existing condo = 5712; diff = mean(rebuilt)-mean(existing)

Table 2.3 Statistical trend of rebuilt condominiums

	Estimation result				
Variable	Coeff.	Robust S.E.	Standardized Coeff.	z-value	
Ln(land area)	0.520	0.132	0.549	3.94	***
Ln(number of stories)	−0.458	0.198	−0.181	−2.32	*
Ln(number of housing units)	−0.345	0.139	−0.291	−2.48	*
Ln(building age as of 2020)	6.446	0.528	0.579	12.20	***
Dummy on two-person layout	0.312	0.170	0.146	1.84	†
Dummy on family layout	−0.018	0.178	−0.009	−0.10	
Ln(distance to the nearest station)	−0.037	0.073	−0.032	−0.51	
Ln(distance to the city center)	−0.061	0.014	−0.236	−4.35	***
Dummy on exclusively residential land use	−0.352	0.233	−0.152	−1.51	
Dummy on residential land use	0.233	0.220	0.098	1.06	
Dummy on commercial land use	0.578	0.217	0.282	2.66	**
Ln(floor area available for expansion)	0.211	0.044	0.290	4.85	***
Observation	5821				
Wald χ^2	248.54***				
Pseudo R^2 (McFadden)	0.470				

Notes: Statistical significance ***0.1%, **1%, *5%, †10%; Coeff. = Coefficient, S.E. = Standard Error

to have declined. Finally, the probability of rebuilding was significantly positively correlated with the floor area available for expansion per unit. This result is in line with previous studies (Hasegawa 1999; Takamizawa and Go 1989) and is confirmed here quantitatively.

Unlike the discussion in the previous subsection, the probability of rebuilding was not correlated with distance to the nearest station, whereas distance to the city center was significantly negatively correlated. This result indicates that rebuilding occurs when proximity to the city center takes precedence over neighborhood convenience. Furthermore, the probability of rebuilding was positively correlated with the commercial land use dummy. This implies that reaching a consensus with neighbors in commercial areas is less difficult than in residential areas. In other words, this trend could be caused by the cost of seeking a building consensus with neighbors as oppositional movements occur when a massive condominium is built in residential areas.

Comparing each factor with standardized coefficient values, the absolute values of the coefficients are larger in the order of building age, land area, number of housing units, floor area available for expansion, commercial land use dummy, distance to the city center, and number of stories. This shows that the aging of the condominiums is not only linked to rebuilding, but the possibility of floor area increase has a relatively large effect on condominium rebuilding.

In general, the above results quantitatively confirm the effects of physical rebuilding factors, including the floor area available for expansion, location, and building size, as argued by Takamizawa and Go (1989) and Hasegawa (1999). The only exception is that the influence of floor layout is not necessarily strong, and rebuilding is explained primarily by building age, size, room for floor area expansion, and location to the city center.

Having interpreted general trends, We then analyzed changes in the probability of rebuilding when the average characteristics of condominiums were varied. Specifically, We looked at what change is assumed when the standardized explanatory variables are increased in the 1σ range, and the average marginal probability effect (AMPE) was used to quantify the change.[11]

Observing AMPEs (Table 2.4), the increase in the probability of rebuilding was positively correlated with building age, land area, commercial land use dummy, and floor area available for expansion, while it was significantly negatively correlated with the number of stories, number of housing units, and distance to the city center. This is similar to the trend of the coefficients in the model of estimation results. We also examined the specific change in the probability of rebuilding, that is, the extent of the AMPE. To illustrate the interpretation in terms of building age, if We compare a condominium that is approximately 44 years old with one that is 48 years old, the probability of rebuilding is 1.49 percentage points higher for the 48-year-old condominium. The most influential variables are the building's age, land area, floor area available for expansion, number of housing units, and distance to the city center. The first three aforementioned factors increase the probability of rebuilding by 1.49, 1.42, and 0.75 percentage points, respectively, and the latter two decrease the probability of rebuilding by 0.75 and 0.61 percentage points, respectively. However, note that this is a marginal probability effect for the average of the observed values, and the marginal effect for each observation should be calculated one by one.

2.3.4 *Geographical Distribution of the Predictive Probability of Rebuilding*

Taking advantage of the probit model that was used in the previous subsection, We calculated the predictive probability of rebuilding and estimated the total floor area in areas with a high or low probability of condominium rebuilding. The geographical unit is an approximately 1 km^2 grid cell, which can indicate location trends in more detail than municipal units.

---

[11] Average marginal probability effects can be expressed by $\text{AMPE}_j = \frac{1}{n}\sum_{i=1}^{N} \frac{\partial \Pr(y_i|\boldsymbol{x})}{\partial x_{ij}} = \frac{1}{n}\sum_{i=1}^{N} F'\left(\boldsymbol{\beta x}_i'\right)\beta_j$ when we want to see the effect of the *j*th explanatory variable x_{ij} on the *i*-th observed value. Here, N is the number of observations and $F'(\bullet)$ is the derivative of the cumulative density function.

Table 2.4 AMPE of each characteristic

	Average marginal probability effect					
Variable	AMPE	S.E. (Delta method)		Mean	S.D.	Unit
Ln (land area)	0.0142	0.004	***	2178	5956	m^2
Ln (number of stories)	−0.0047	0.002	*	7.15	2.75	Story
Ln (number of housing units)	−0.0075	0.003	*	66.18	100.73	Unit
Ln (building age as of 2020)	0.0149	0.001	***	43.55	4.26	Year
Dummy on two-person layout	0.0038	0.002	†	0.33	0.47	–
Dummy on family layout	−0.0002	0.002		0.49	0.50	–
Ln (distance to the nearest station)	−0.0008	0.002		446.51	409.35	m
Ln (distance to the city center)	−0.0061	0.001	***	10,081	11,347	m
Dummy on exclusively residential land use	−0.0039	0.003		0.25	0.43	–
Dummy on residential land use	0.0025	0.002		0.23	0.42	–
Dummy on commercial land use	0.0073	0.003	*	0.39	0.49	–
Ln (floor area available for expansion)	0.0075	0.002	***	8.41	29.21	m^2
Observation						
Wald χ^2						
Pseudo R^2 (McFadden)						

Notes: Statistical significance ***0.1%, **1%, *5%, †10%; S.E. = Standard Error, S.D. = Standard Deviation; AMPE = Average Marginal Probability Effect; All the variables are standardized for measuring AMPE.; Means and S.Ds are modified from logarithmic values

Figure 2.4 represents the number of rebuilt condominiums when the entire sample is divided into 10 equal parts, with the range of predictive probability of rebuilding on the horizontal axis. The average predictive probability for condominiums with old earthquake resistance standards is about 1.88%, and the predictive probability for the bottom 80% is less than 1%. It is likely that condominium rebuilding occurs when the predictive probability of rebuilding is low due to factors other than physical characteristics, such as the ability of the management association to form a consensus and the presence of troublesome tenants. Furthermore, cases of reconstruction due to exogenous shocks, such as the designation of urban redevelopment promotion zones and seismic disguises, are also difficult to predict with this model. Nevertheless, given that approximately 83% of all reconstruction cases are included in the top 10% sample, the accuracy of the model is still guaranteed at a certain level.

We observed the geographical distribution of the total floor area in the top 10% of the predictive probability of rebuilding (Fig. 2.5). On the one hand, the condomini-

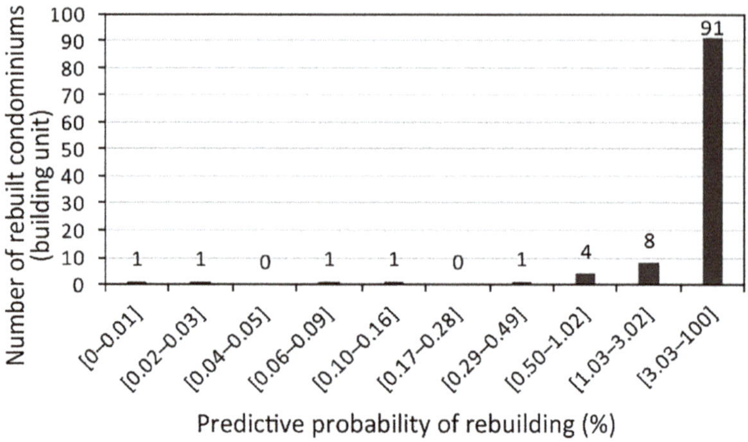

Fig. 2.4 Relationship between the number of rebuilt condominiums and the predictive probability of rebuilding (All samples are divided into 10 equal parts)

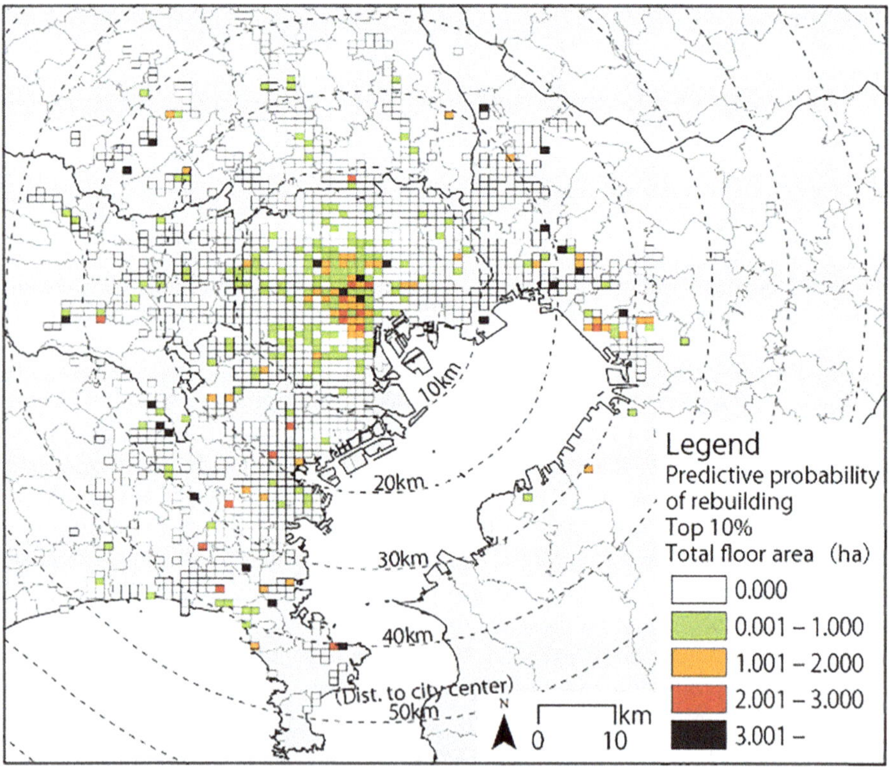

Fig. 2.5 Geographical distribution of total floor area in the top 10% of predictive probability of rebuilding

ums with a high probability of rebuilding are concentrated in Minato, Shibuya, and Shinjuku wards, indicating that the possibility of rebuilding is high in the central and western parts of the central Tokyo. On the other hand, even in the suburban areas 20–30 km away from the city center, there are grid cells of condominiums with a large total floor area. It is conceivable that such condominiums are planned and sold during a period of high economic growth; these condominiums have a relatively high floor area ratio and a low number of floors, reflecting a high probability of rebuilding. However, the model does not include information on the demand side conditions; thus, the condominiums located in suburban areas might not be rebuilt in anticipation of a future decline in demand.

In terms of the total floor area in the bottom 10% of the predictive probability of rebuilding (Fig. 2.6), the values are particularly high in the eastern and southern parts of central Tokyo. In some areas of Koto and Sumida wards, condominiums have already exploited most of the floor area ratio, and the number of housing units is likely to be high, resulting in a pushdown for the probability of rebuilding in this model. In the suburban area, grid cells with a total floor area of 1 ha or less

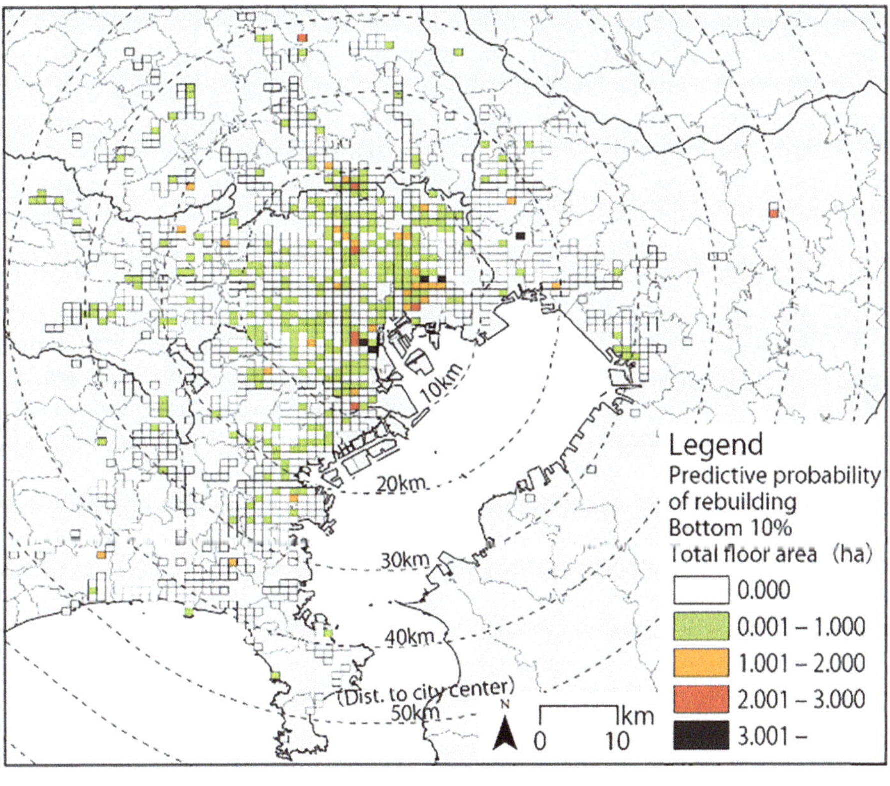

Fig. 2.6 Geographical distribution of total floor area in the bottom 10% of predictive probability of rebuilding

are widely distributed. Therefore, areas where condominiums will be rebuilt and the ones will accumulate are expected to occur simultaneously in central Tokyo. Moreover, as many grid cells with a low probability of rebuilding are also distributed in the suburban areas, the accumulation of old condominiums is expected to progress widely and gradually.

2.4 Conclusion

This study compared rebuilt condominiums to existing condominiums with old earthquake resistance standards. This revealed the trends, predicted the geographical distribution, and the probability of condominium rebuilds in the future. The following are the main findings.

We found that the trend of condominiums rebuilt with old earthquake resistance standards follows the results of previous studies. This shows that the potential of rebuilding is associated with building and location characteristics, such as the floor area available for expansion, location, and floor area per unit. This is as per the arguments of Takamizawa and Go (1989), Hasegawa (1999), and Hanazato (2017). The correlation of floor layout to the probability of rebuilding in this study was weak, and condominium rebuilding was largely influenced by building age, land and building size, room for expansion, and its distance to the city center. In addition, We showed the marginal effects of condominium characteristics on rebuilding. Specifically, We found that building age, land area, and floor area available for expansion increased the probability of rebuilding by 1.49, 1.42, and 0.75 percentage points, respectively, whereas the number of housing units and distance to the city center decreased the probability of rebuilding by 0.75 and 0.61 percentage points, respectively.

Furthermore, We visualized the areas with a high or low probability of rebuilding using a 1 km grid cell and showed that the central Tokyo and suburban areas tend to differ. Condominiums with a high probability of rebuilding were concentrated in the central and western parts of central Tokyo, whereas in the eastern and southern parts of central Tokyo, the concentration of condominiums with a low probability of being rebuilt was small. In the suburban areas, however, areas with a high probability of rebuilding are scattered throughout the city, whereas areas with a low probability of rebuilding are not concentrated but are widely distributed. There is concern that hotspots of old condominiums may emerge in the eastern and southern parts of central Tokyo in the future.

Although this study discussed the probability of rebuilding condominiums, We only used building and location characteristics as explanatory variables. Therefore, factors such as the ability of management associations, the existence of troublesome tenants, district-specific planning regulations, and exogenous shocks are not considered in the forecast. In addition, the price of housing is inherently determined by the supply–demand relationship, and the decision to rebuild is thought to change in relation to the equilibrium. Therefore, We need to build a structural model assuming more realistic settings. The construction of such a model would deepen our discussion on the likelihood of condominium rebuilding.

Acknowledgments This work was supported by JSPS KAKENHI Grant Numbers 20H00082, 20K14898.

References

Asami Y (2009) Optimization of majority rule for consensus building: resolution for rehabilitation of condominiums. Urban Hous Sci 64:137–143. https://doi.org/10.11531/uhs1993.2009.64_137. (in Japanese)

Asami Y, Fukui H, Yamaguchi M (2012) Condominium rebuilding: how to prepare for aging of building. Nippon-Hyoron-sya. (in Japanese)

Asami Y, Saito H (Eds.) (2019) Considering the end of life in condominium. Progress. (in Japanese)

Baba H (2020) Potentialities and issues of privately-owned data on Condominium Studies. Jpn J Real Estate Sci 33(4):79–83. https://doi.org/10.5736/jares.33.4_79. (in Japanese)

Choi J, Asami Y (2004) A study of the relation between sense of values and residential evaluation. J Archit Plan 69(576):133–139. https://doi.org/10.3130/aija.69.133_2. (in Japanese)

Glaeser E, Gyourko J (2005) Urban decline and durable housing. J Pol Econ 113(2):345–375. https://doi.org/10.1086/427465

Hanazato T (2017) Feasibility evaluation for condominium reconstruction by means of the application of discriminant analysis on real estate data. Jpn J Real Estate Sci 31(3):119–128. https://doi.org/10.5736/jares.31.3_119. (in Japanese)

Hasegawa H (1999) Countermeasures for problems in condominium reconstruction projects. J Archit Plan 64(523):235–242. https://doi.org/10.3130/aija.64.235_4. (in Japanese)

MLIT (n.d.-a) Number of condominium units. https://www.mlit.go.jp/jutakukentiku/house/content/001488548.pdf (in Japanese)

MLIT (n.d.-b) Current situation of condominium rebuilding. https://www.mlit.go.jp/jutakukentiku/house/content/001488550.pdf (in Japanese)

MLIT (n.d.-c) Basic plan for housing (national plan). https://www.mlit.go.jp/common/001392030.pdf (in Japanese)

Ronald R, Hirayama Y (2006) Housing commodities, context and meaning: transformations in Japan's urban condominium sector. Urb Stud 43(13):2467–2483. https://doi.org/10.1080/00420980600970680

Rosen G, Walks A (2013) Rising cities: condominium development and the private transformation of the metropolis. Geoforum 49:160–172. https://doi.org/10.1016/j.geoforum.2013.06.010

Rosen G, Walks A (2015) Castles in Toronto's sky: condo-ism as urban transformation. J Urban Aff 37(3):289–310. https://doi.org/10.1111/juaf.12140

Saito H, Hasegawa H (2001) A study on process of consensus-building in initial stage of condominiums reconstruction project and its problems: case of large-scale condominiums in the suburb. J Archit Plan 66(543):239–245. https://doi.org/10.3130/aija.66.239_1. (in Japanese)

Saito H, Hasegawa H, Souda O, Yagisawa S (1999) Capability for consensus-building of condominium reconstruction project by homeowners' association. J City Plan Inst Jpn 34:307–312. https://doi.org/10.11361/journalcpij.34.307. (in Japanese)

Sim L, Lum S, Malone-Lee LC (2002) Property rights, collective sales and government intervention: averting a tragedy of the anticommons. Habitat Int 26(4):457–470. https://doi.org/10.1016/S0197-3975(02)00021-8

Takamizawa K, Go Y (1989) Survey of the actual condition of decrepit condominium apartment and analysis for their reconstruction. J Archit Plan Environ Eng 404:89–97. https://doi.org/10.3130/aijax.404.0_89. (in Japanese)

Tanaka M, Kumata Y (1999) Problems of decrepit condominiums furthering aggravation of urban environment is rebuilding possible by means of reverse mortgage system? Stud Reg Sci 29(2):41–61. https://doi.org/10.2457/srs.29.2_41

Warnken J, Russell R, Faulkner B (2003) Condominium developments in maturing destinations: potentials and problems of long-term sustainability. Tour Manag 24(2):155–168. https://doi.org/10.1016/S0261-5177(02)00063-8

Webb B, Webber S (2017) The implications of condominium neighbourhoods for long-term urban revitalisation. Cities 61:48–57. https://doi.org/10.1016/j.cities.2016.11.006

Wooldridge J (2016) Introductory econometrics: a modern approach. Southwestern Publishing House, Nashville

Yau Y, Ling Chan H (2008) To rehabilitate or redevelop? A study of the decision criteria for urban regeneration projects. J Place Manag Dev 1(3):272–291. https://doi.org/10.1108/17538330810911262

Chapter 3
Significance of "Living Environment Score" in Quantifying Attractiveness of Regions During Residence Selection

Hisatoshi Ai

Abstract Japan's conventional city planning is based on the assumption that population growth and urbanization will continue in the future. However, it is extremely likely that the population growth assumption will no longer hold in the future, so the city planning system is at a critical juncture. Notably, the development of compact cities maintains the quality of life in cities and administrative service efficiency even in the era of population decline. In order to make a city more compact through effective planning, municipalities should formulate policies encouraging relocation, such as developing urban areas with attractive living environments and the introduction of subsidies for relocation-associated expenses. Hence, a quantitative and objective index to assess the living environment of each area from the perspective of whether it corresponds to residents' preferences is needed.

The living environment score (LES) focuses on various living environment indices and takes into account the magnitude of the influence of these indices on residence selection in order to assess the overall attractiveness of an area as a residence. We analyzed population increases from 2000 to 2010 in the Tokyo

The contents of this paper are based on the following papers originally published in a Japanese journal: Ai, H. (2014) How living environment indices effect on population change patterns of local districts: An analysis on urban area in the Greater Tokyo Area. *Journal of City Planning Institute of Japan*, 49 (3), pp. 567–572 (in Japanese). Ai, H. (2016) Indexing of living environment attracts young and productive age generations: An analysis based on local districts within the Greater Tokyo Metropolitan Area, *Journal of City Planning Institute of Japan*, 51 (3), pp. 860–866 (in Japanese). Ai, H. (2020) Calculating living environment index forecasting population increase in a grid based spatial units: Comparison of the result between a census grid basis and a basic unit block basis, *Journal of City Planning Institute of Japan*, 55 (1), pp. 41–48 (in Japanese). Ai, H. (2021) A relationship between residential reallocation based on living environment score and disaster risk reduction: Focusing on land slide risk, flood risk, and heavily build-up areas, *Journal of City Planning Institute of Japan*, 56 (3), pp. 587–594 (in Japanese).

H. Ai (✉)
Department of Contemporary Liberal Arts, Showa Women's University, Tokyo, Japan
e-mail: hisaai@swu.ac.jp

Y. Asami et al. (eds.), *Studies in Housing and Urban Analysis in Japan*,
New Frontiers in Regional Science: Asian Perspectives 75,
https://doi.org/10.1007/978-981-99-8027-7_3

metropolitan area and attempted to calculate the LES at both the census grid and local district levels. The results showed that there was a significant expected population increase tendency in regions with a high LES. However, the accuracy rate was around 60%, and further improvements are needed in order to use this as a basis for decision-making.

Indices related to disaster safety have not been considered in the LES. Thus, when we verified the relationship between population-increase tendencies or the LES with disaster risk, it was suggested that population increases might continue in regions with flood risk or risks associated with dense urban areas. Disaster risk is something that people seek to minimize, and this should be considered as a property that is different from other living environment indices that should be pursued to an appropriate level. For the future work, it is imperative to improve the LES calculation method so that disaster risk can be reduced while improving accuracy.

Keywords City planning · Living environment · Comprehensive evaluation · Attractiveness of regions

3.1 Era of Declining Population and Compact Cities

For the first time since World War II, Japan's population registered a natural decline in 2007, and it is estimated that this trend will continue in the future (Statistics Bureau, Ministry of Internal Affairs and Communications, 2014). Most prefectures have already shown a population decline trend, and from 2015 to 2020, only 8 of 47 prefectures reported no population decline (Ministry of Internal Affairs and Communications 2021). Although the population decline trend is evident at both national and prefectural levels, it is not uniform within each area. For example, Ogawa (2011) reported that areas with population growth and decline coexist in a mosaic pattern.

Population changes can be broadly attributed to natural causes, such as births and deaths, as well as social changes driven by relocations associated with employment, personnel transfer, and advancement to higher education. Koike (2006) analyzed populations by municipality and cohort in the Hokkaido Prefecture and indicated that social changes have a major impact on population changes. Specifically, this can be attributed to not only the direct impact of immigration and emigration but also the indirect influence on birth rates due to immigration and emigration of generations that have children. Although the place of residence is restricted at the prefectural level, it can be said that there is a greater degree of freedom in residence selection at the more detailed municipality or local district level. Specifically, relocating individuals are considered to select an area that matches their own residence preferences among such options with a certain spatial extent. Additionally, the 1986 and 1991 migration surveys reported that 40% of individuals relocated citing housing conditions (Arai and Ohki 1999), and that these individuals' assessment of the living environment had an impact on residence selection. Accordingly, it is considered that population changes are largely impacted by social changes and that it is highly likely that the living environment of areas with increasing social changes is seen to match the preferences of more residents.

Next, Japan's conventional city planning is a growth management system design that is based on the assumption that population growth and urbanization will continue in the future. However, as mentioned earlier, it is extremely likely that the population growth assumption will no longer hold in the future, so the city planning system is at a critical juncture. Notably, the development of compact cities maintains the quality of life in cities and administrative service efficiency even in the era of population decline. It is envisioned that compact cities in Japan, also characterized as "compact and network," consolidate urban areas along railway lines and strengthen commerce, administration, and medical welfare around railway stations, which are transportation nodes between railway and bus networks. The localization normalization plan is introduced to make cities more compact in addition to land use zoning system which is the earlier method of land use control. If there are multiple railway stations in the same municipality or an urban area, then that station may be hierarchical depending on its importance as a transportation hub in the localization normalization plan. Objective decision-making is important, such as deciding which railway station surroundings should be set as higher-order hubs and how far the scope of urbanization extends from the railway station within the hub surroundings. The former plays an important role when formulating the localization normalization plan, and the latter while setting specific residential promotion areas. Therefore, indices are imperative for assessing regions that have already been developed and urbanized.

In order to make a city more compact through effective planning, it is imperative to follow population movement so that it corresponds with the direction of the compact city plan. In Japan's present-day social system, plan-based forced migration and mass relocation are not realistic options, and municipalities, which are the main implementers of city planning, should formulate policies encouraging relocation, such as developing urban areas with attractive living environments and the introduction of subsidies for relocation-associated expenses. Discussing cost-effectiveness is essential for projects and policies that involve expenditures such as urban development and subsidies for relocation, and a quantitative and objective assessment index is needed. Essentially, as mentioned earlier, it is imperative to assess the living environment of each area from the perspective of whether it corresponds to residents' preferences.

3.2 Concept and Features of Living Environment Score

3.2.1 Conceptualization and Calculation of Living Environment Score

The living environment score is calculated as a comprehensive assessment index that combines multiple living environment indices. This can be attributed to the fact that during the selection process, individuals do not choose a residence based on a single factor and could have multiple preferences. Furthermore, the spatial unit is assumed to be a local district or a census grid, or so-called a mesh, rather than a municipality. The details of each will be described later, although the calculation method explained here is identical regardless of the spatial unit used.

The procedure for comprehensive assessment can be broadly divided into two stages. The first is to derive the standard that leads to population increase for each living environment index. For example, metropolitan areas in Japan have developed railway networks that connect the city center and its suburbs, and urban development has advanced mainly along these railway lines; therefore, areas surrounding railway stations are often popular among people searching for homes. When areas within 1 km from the railway station have exhibited, marked population increases and areas farther away have not shown much population increase, then the 1 km-radius from the railway station is considered a demarcation point leading to population increases. These standards can be derived using statistical methods by creating a cross-tabulation table with population increase tendencies for each living environment index and performing a chi-square test. The living environment indices used in the assessment will be described later.

For the second stage, when combining multiple living environment indices and calculating the living environment score, the magnitude of impact, or so-called weight, of each index on residence selection is considered. This corresponds to the consideration of multiple living environment indices that are reported to impact the resident selection, such as distance to the railway station, presence, or absence of nearby parks, and whether the land is on a slope. Although equal importance is not given to each, some are deemed important and others are used only as a reference.

The concept of discriminant analysis is used for this stage. Discriminant analysis with population increase tendency as the independent variable and standardized version of each living environment index as the explanatory variable yields a discriminant coefficient. As the explanatory variables were standardized, the magnitude of the absolute value of this discriminant coefficient is considered to reflect the magnitude of the impact of each living environment index on population increase tendencies.

When calculating the final living environment score, each of the living environment indices in individual local districts or census grids is first judged based on whether it satisfies the standard calculated in the first stage. For indices that meet the standard, the absolute value of the discriminant coefficient obtained in the second stage is added up as the raw living environment score. Specifically, the raw score enhances when living environment indices increasingly satisfy the standard leading to population growth or when living environment indices that satisfy the standard have a greater impact on residence selection. Finally, the living environment score is obtained by converting this raw score into a deviation value for the entire target region. Since it is a deviation value, a local district or census grid with average living environment regarding leading to population increase is given a score of 50, and those with higher scores could be interpreted as having a living environment that is advantageous in terms of residence selection.

If the living environment score appropriately reflects residents' preferences in residence selection, it is believed that residential promotion can be achieved by prioritizing areas with high scores. The derivation process of the living environment score can also identify which living environment index in each local district or census grid has failed to satisfy the standard at which population increase is expected. Furthermore, the magnitude of impact of each living environment index on population increase has been calculated. Therefore, discussions can be held as to which index should be improved in each local district or census grid to effectively increase the living environment score.

3.2.2 Living Environment Indices Used for Calculating Living Environment Score

When selecting living environment indices to be used in the analysis, it is important to examine what type of living environment indices impact population increase while considering residential promotion that aims for compact cities. Table 3.1 summarizes the living environment assessment items that Maene et al. (2012) surveyed among residents. Likely important items include those marked *, which had low satisfaction levels in the surveyed area, and §, considered to be important for future residential areas. We considered that our items should correspond with

Table 3.1 Living environment assessment items surveyed among residents. *Based on Maene et al. (2012)

Public transportation *§	Retail stores of daily goods*	Access to downtown*
Medical care services§	Ambulance and firefighting§	Natural disaster prevention
Migration of young generation§	Vitality of the community	Activities of local communities
Relationship with neighborhood	Sports and health care facilities	Social welfare services
Administrative services	Education and child-rearing environment	Richness of nature
Crime prevention activities	Comprehensive comfort	

* Low satisfaction levels in the surveyed area
§ considered to be important for future residential areas

these items and also could be compared in relatively small spatial units and that they should be indices that express the socioeconomic circumstances of cities. Accordingly, we analyzed 13 living environment indices in this study. It should also be noted that the data were collected and published in spatial units at the local district level or following the census grid system as a time-series data.

Maene et al. (2012) reported low satisfaction levels for three items—public transportation, shopping, and ease of going outside the area. Essentially, the level of demand by residents for these elements was high, and they are considered to have a strong impact on living environment assessments. When calculating the living environment score, we reflect these by the *distance to the railway station* and the *number of stores*.

Maene et al. also reported that items that were considered important for future residential areas were access to medical care, emergency services, and firefighting services; disaster prevention; and housing for young people. This signifies that even residents who are not necessarily dissatisfied with the current situation focus on these elements and that they ultimately impact living environment assessments. Access to *hospitals* and *firefighting services*, *slopes* as terrain, and the *aging rate* as resident attributes are used to consider these elements when calculating the living environment score.

Asami (2001) emphasized the importance of considering sustainability in addition to convenience, safety, comfort, and health when assessing living environment indices. The above-mentioned living environment indices were mainly related to convenience, safety, and health. Regarding safety, not only disaster prevention but also crime prevention is important; thus, *access to police* is also considered in the living environment score. The *presence or absence of parks* is believed to contribute to comfort and health. Regarding sustainability, given that residents remain in a region for a certain number of years, the following factors are considered to correspond to sustainability: *aging rate*, which reflects the age-group of residents; *average number of household members* and *detached house rate*, which reflect housing styles; *population density* and *vacant lot ratio*, which reflect the land use situation and the status of urban development; and the *number of businesses*, which is related to employment opportunities.

Table 3.2 summarizes the above-mentioned living environment indices together with data sources.

3.3 Setting of Target Regions and Definition of Population Increase in This Study

3.3.1 Setting of Target Regions

The target regions in this study are "city planning areas" of the Tokyo, Saitama, Chiba, and Kanagawa Prefectures. These four prefectures form the wide and

Table 3.2 Living environment indices used for calculating living environment scores and their data sources *Edited and reproduced based on Ai (2014)

Index	Abbreviations	Unit	Data source
Distance to railway station	Station	Meters	DNLI, railway time-series data
Distance to hospital	Hospital	Meters	DNLI, public facility data (2006); hospitals, medical centers
Distance to police	Police	Meters	DNLI, public facility data (2006); police stations, police boxes
Distance to firefighting services	Firefighting	Meters	DNLI, public facility data (2006); fire departments, fire houses
Presence or absence of parks	Park		DNLI, urban park data (2011)
Number of stores	Stores		Establishment and Enterprise Census (2006); stores, restaurants
Number of businesses	Businesses		Establishment and Enterprise Census (2006); establishments, offices
Slopes	Slopes	Degrees	DNLI, elevation and slope data/fifth level census grid (2011)
Population density	Density	Person/km2	National Census (2000)
Aging rate	Aging	%	National Census (2000)
Average number of household members	Household	Person	National Census (2000)
Vacant lot rate	Vacant	%	DNLI, detailed land use data/fifth level census grid (2011)
Detached house rate	Detached	%	National Census (2000)

* DNLI: digital national land information

integrated Tokyo metropolitan area, where there is a daily flow of people commuting to work and schools across the borders of these prefectures. Considering its size, the entire region does not necessarily simultaneously fall into the set of options available during residence selection, but the urban areas are continuous. Subdividing these regions into mutually independent areas is difficult regarding population flow, so the four prefectures were treated as an integrated region. Additionally, we limited our study to "city planning areas" because there are restrictions to the scope of the provision of land use data, which is one of the living environment indices used in this analysis (Fig. 3.1).

When attempting analyses with a spatial unit that is more detailed than municipalities, local districts or census grids can be considered as candidates. A local district is an administrative place name that is more detailed than a municipality in Japan. Although it does not have legal status, it is used as an address for postal items and is a spatial unit that has been widely recognized by residents. The census grid is divided at regular intervals, such as 500 m or 1 km, and in the case of Japan, the division method system (standard region census grid) is defined based on latitude

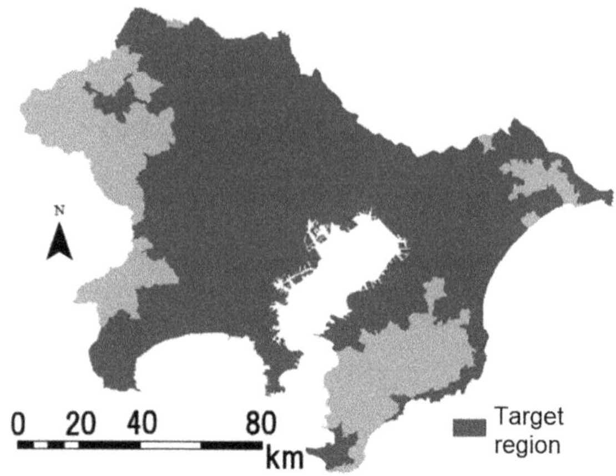

Fig. 3.1 Four prefectures and the target region of this study * Cited from Ai (2014) and translated into English

and longitude. In addition to the population census, various statistical data use the same census grid system, facilitating the superimposition of data.

One of the advantages of using the local district as the spatial unit for analysis is that natural topography, such as rivers and transportation networks, is often used as boundary lines, and movement within the same local district is relatively easier than moving in and out of the local district, and although there are exceptions, the living environment within a local district is generally homogenous. Meanwhile, local districts are often re-organized in areas undergoing development, such as land re-adjustment, and estimation processing, such as proportional division, may be required when comparing time-series data. Areas with large population fluctuations, which are of particular interest in this study, often fall under this category, and it can be indicated as a point worth noting when conducting analyses.

In the case of the census grid, one of the advantages is that it has a uniform spatial unit shape and size. Additionally, the use of standard region grids enables the seamless superimposition and analysis of population statistics, economic statistics, topographic data, and so on. The position and shape of the spatial units do not change between each time point, so time-series comparisons are also easy. Since the space is divided mechanically, it should be noted that areas that straddle rivers, major roads, administrative boundaries, etc. are often included in the same census grid and that the homogeneity of the living environment is not necessarily guaranteed within the same census grid. Residents are not aware of the division of census grids on a daily basis, so there are not many cases where the residents are aware of the census grid number of their own residences.

In this chapter, we consider the results for both when the spatial unit is set to census grids of 500 m intervals and when it is set to local districts.

3.3.2 *Definition of Population Increase and Method of Proportional Division by Population*

Population changes were calculated using population data from 2000, 2005, and 2010 population census sub-regional tabulations. The local district areas were based on the 2005 data, and the population as of 2000 of the local districts that were merged from 2000 to 2005 ("A" in Fig. 3.2) were totaled and then compared with the post-merger population in 2005 in chronological order. In contrast, the population in 2000 of local districts that were divided from 2000 to 2005 ("B" in Fig. 3.2) were estimated by proportional division by population of the 2005 data and then compared in chronological order. In the example of B in Fig. 3.2, since it is divided into two, the population of B in 2000 was halved, and this was compared with the populations of B1 and B2 after the division. The reason why proportional distribution by area was not implemented for the division was because, when dividing a local district with a concentrated population in one part, it is the densely populated zone that is divided, so it is assumed that the population of each local district after division will be roughly the same. The population as of 2010 in the local districts that were merged from 2005 to 2010 ("C" in Fig. 3.2) were estimated using the proportional division method by population and then compared in chronological order with the 2005 data. The population as of 2010 in the local districts that were divided from 2005 to 2010 ("D" in Fig. 3.2) were totaled and then compared with the 2005 data. However, cases where complex restructuring of local districts, such as the combination of divisions and mergers, were excluded from the analysis since there are limits to the estimation of population using proportional division. As mentioned above, there are no changes in the census grid in the spatial unit. Therefore, this type of proportional division by population is not implemented for this case.

In this study, "population change" was defined as follows: when comparing time-series population data of the same local district, an "increase" was defined as "when a population increased at one of two time points, and when, even for the other time point, the population of the census grid/local district exceeded the expected population based on the population change tendency of the entire target region" or "when the population of the census grid/local district was at least 2%

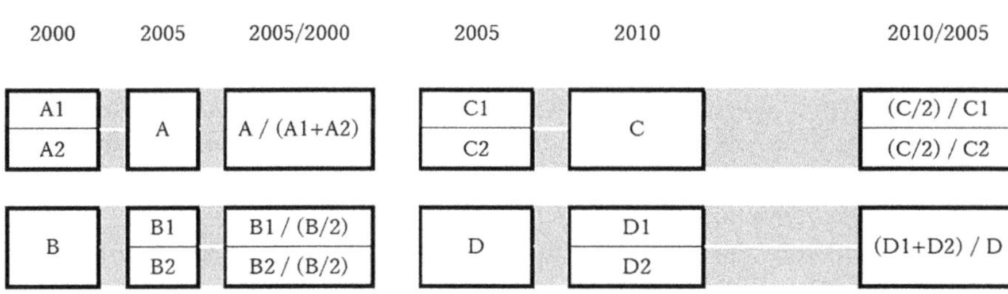

Fig. 3.2 Population estimation method when local district area changes

higher than the estimated population based on the population change tendency of the entire target region at both time points." An "increase" was set not as a simple comparison of population data between two time points, but as a comparison with the overall population increase rate as described above, in order to take into account the population change due to the population flow within as well as in and out of the entire target region (i.e., the four prefectures defined earlier), while extracting the changes that exceed those tendencies and make them the subject of discussion in this study. Additionally, comparisons of population data cannot be made if the population was zero at any single time point from 2000 to 2010 or if the population data was confidential due to few population; accordingly, these cases were excluded from the analysis.

The youth population (under the age of 18) decreased by 8.3% from 2000 to 2005, and by 0.5% from 2005 to 2010. On the other hand, the productive age population (18 to under 60) decreased by 7.1% from 2000 to 2005, and by 1.2% from 2005 to 2010. Therefore, it should be noted that, although the population decreased in terms of absolute numbers, there are cases in this study that would be classified as "increases."

3.4 Calculation of Living Environment Score

3.4.1 Cross-Tabulation of Population Growth Tendencies and Living Environment Levels

When calculating the living environment score, the population data for each local district or census grid are compared chronologically and classified according to whether the population increased or decreased. Although this can be done with any generation, here, we calculate the living environment score for both the youth and productive age population based on the findings of Ai (2016). It is expected that calculating the living environment score for each age-group based on the age-group-specific population change tendencies will enable the observation of whether residence preferences differ according to age-group or lifestyle. First, the productive age population corresponds to the generation that works and pays taxes, and it is an important generation for maintaining the vitality of a city and securing tax revenue. Meanwhile, the youth population rarely changes their place of residence alone and generally reflect the dynamics of the child-rearing generation. The youth population and productive age population are essentially mutually exclusive in terms of generational divisions, but when considering the household unit, the dynamics of the productive age population should be interpreted as partially including the dynamics of the youth population.

When calculating the living environment score, we sought to determine the necessary level for the living environment index at which population increase can be expected by conducting a chi-square test and residual analysis, verifying whether each living environment index has a significant relationship with population increase. Table 3.3, which demonstrates the results, is a series of cross-tabulation tables for each index, and the spatial units of the census grid and local district are shown simultaneously so that they can be compared. In the "increase" column for each spatial unit, the number of census grids and local districts that correspond to each spatial unit is shown in the left column and adjusted residuals are in the center column. The adjusted residuals for "non-increase," which is the case that does not correspond to an increase, have an absolute value that is equal to the "increase" values, differing only in sign; hence, they are omitted here and the number of census grids and local districts that are "non-increase" are shown in the right column.

In Table 3.3, the squares where the absolute value of the adjusted residual exceeds 1.96 and can be said to be significant at the 5% level are colored. For the classes where A.R. is over 1.96, the actual number of census grids/local districts was significantly higher than the assumed number of census grids/local districts if the living environment index and population increase were unrelated; and conversely, for the classes where A.R. is 1.96 or lower, the actual number was significantly lower. For example, for the distance to the nearest railway station in Table 3.3, it could be read that the population increase was significantly higher for the youth generation in grids within 1500 m and for the productive age generation in grids within 1000 m. When the local district was set as the spatial unit, it could be read that the population increase was significantly higher for both generations in local districts within 500 m from the railway station. A summary of whether each level of the living environment index resulted in a significantly higher population increase is listed in the standard column of Table 3.4. In Table 3.4, the left side shows the case where the census grid is the spatial unit, and the right side shows the case where the local district is the spatial unit. The interpretation here is that, if the living environment index is included in this standard, then population increase is expected in the census grid/local district from the perspective of that index.

Several indices, such as the distance to the nearest railway station or hospital, as well as the number of stores, can be seen in Table 3.4, where there are differences in the value range that significantly impacts population increase at the fourth-order census grid unit and local district unit. Differences in the area and shape between the two spatial units can be raised as one factor for this. Although there are some differences depending on the latitude, the fourth-order grid has a side length of approximately 500 m and an area of approximately 0.25 sq.-km. In contrast, local districts have considerable differences in area and shape. In general, local districts in city centers tend to have a smaller area, whereas local districts in suburbs tend to have a larger area. The number of stores and businesses are not density indices. Therefore, even if these numbers are distributed at the same density, in case the area of the tabulated spatial unit is larger, it would result in a larger number of stores and businesses that are used as indices. Additionally, distance indices such as the distance to the nearest railway station are calculated from the outer circumference of the grid or local district, excluding cases where the facility exists within the area and has a distance of zero. On this basis, if the area of the tabulated unit is larger, then the distance from the center of that spatial unit is more likely to be underestimated. For

Table 3.3 Chi-square test and residual analysis results for each living environment index *Edited and reprinted from Ai (2020) Table 1

		Census grids						Local districts					
		Youth population			Productive age population			Youth population			Productive age population		
		Increase	*A.R.*	Non-increase	Increase	*A.R.*	Non-increase	Increase	*A.R.*	Non-increase	Increase	*A.R.*	Non-increase
		$\chi^2 = 786.7$		$p = 0.00$	$\chi^2 = 1084.5$		$p = 0.00$	$\chi^2 = 191.5$		$p = 0.00$	$\chi^2 = 521.6$		$p = 0.00$
(A) Station (meters)	0	404	7.62	1008	487	18.20	930	413	4.30	921	576	9.95	770
	0–500	1633	13.45	4467	1556	20.29	4593	1981	6.51	4925	2590	14.8	4381
	500-1000	1387	10.37	4000	965	20.9	4481	1161	0.652	3242	1269	−3.38	3150
	1000–1500	855	2.84	2971	601	−2.31	3280	568	−1.13	1702	557	−7.10	1730
	1500–2000	479	−3.85	2216	364	−5.19	2382	255	−2.39	856	247	−6.55	875
	2000–3000	504	−8.43	2835	351	−10.80	3058	168	−5.77	768	185	−7.69	762
	3000–	396	−22.55	4279	345	−19.49	4447	94	−10.8	796	108	−12.4	788
		$\chi^2 = 1055.1$		$p = 0.00$	$\chi^2 = 790.0$		$p = 0.00$	$\chi^2 = 237.7$		$p = 0.00$	$\chi^2 = 257.5$		$p = 0.00$
(B) Hospital (meters)	0	423	8.35	1022	380	9.71	1081	410	3.36	965	469	2.63	915
	0–500	1818	16.25	4744	1572	17.37	5052	2041	7.69	4964	2477	10.1	4581
	500–1000	1535	10.83	4454	1147	5.26	4887	1325	3.20	3453	1509	1.05	3304
	1000–1500	893	3.13	3079	675	−0.02	3352	488	−4.37	1713	602	−3.98	1619
	1500–2000	399	−5.46	2041	318	−5.53	2164	174	−5.95	800	209	−6.58	771
	2000–3000	315	−13.83	2596	299	−10.43	2683	137	−7.15	738	162	−8.25	724
	3000–5000	191	−17.89	2449	312	−13.04	2486	47	−8.75	462	91	−6.46	421
	5000–	84	−14.57	1391	66	−13.43	1466	18	−3.29	115	13	−5.30	121
		$\chi^2 = 692.9$		$p = 0.00$	$\chi^2 = 667.4$		$p = 0.00$	$\chi^2 = 249.4$		$p = 0.00$	$\chi^2 = 514.2$		$p = 0.00$
(C) Police (meters)	0	713	8.16	1958	737	15.62	1945	745	4.41	1775	903	5.85	1624
	0-500	2685	16.97	7661	2155	13.44	8276	2830	7.24	7248	3403	9.07	6757
	500–1000	1451	0.15	5563	1065	−4.72	6051	834	−5.88	2914	930	−9.22	2850
	1000–2000	725	−18.28	5240	626	−15.46	5485	219	−8.85	1155	283	−8.73	1106
	2000–	84	−14.23	1354	86	−11.76	1414	12	−4.37	118	13	−5.22	119

		$\chi^2 = 9$[illegible]3.1		$p = 0.00$	$\chi^2 = 683.6$		$p = 0.00$	$\chi^2 = 832.3$		$p = 0.00$	$\chi^2 = 2726.7$		$p = 0.00$
(D) Firefighting (meters)	0	260	5.97	653	237	7.50	679	252	3.40	558	297	3.68	515
	0–500	1585	15.78	4081	1352	16.15	4295	1817	9.16	4197	2168	10.5	3886
	500–1000	1755	12.60	4991	1373	8.64	5434	1579	2.85	4195	1856	2.18	3974
	1000–2000	1539	−3.95	6507	1198	−6.18	6992	863	−8.30	3244	1013	−9.94	3123
	2000–	519	−26.60	5594	509	−20.89	5771	129	−11.7	1016	198	−10.4	958
		$\chi^2 = 747.7$		$p = 0.00$	$\chi^2 = 891.7$		$p = 0.00$	$\chi^2 = 73.2$		$p = 0.00$	$\chi^2 = 249.7$		$p = 0.00$
(E) Stores	0	731	−22.05	5877	638	−19.19	6244	205	−4.69	831	240	−5.96	823
	1–10	2700	−0.56	10,483	2018	−6.87	11, 290	1568	−5.29	5040	1747	−10.2	4929
	11–20	980	14.07	2290	717	8.37	2556	907	0.735	2517	1046	−0.465	2392
	21–30	411	6.36	1109	343	6.22	1177	527	1.19	1417	624	1.28	1325
	31–40	218	5.17	560	186	5.40	592	345	2.67	833	414	3.25	770
	41–	618	10.73	1457	767	25.53	1312	1088	5.77	2572	1461	13.2	2217
		$\chi^2 = 1019.2$		$p = 0.00$	$\chi^2 = 1613.3$		$p = 0.00$	$\chi^2 = 296.9$		$p = 0.00$	$\chi^2 = 698.9$		$p = 0.00$
(F) Businesses	0	102	−13.02	1335	130	−9.13	1424	327	−5.69	1300	342	−9.32	1312
	1–10	1211	−23.82	8349	1067	−19.34	8729	2033	−10.8	7009	2257	−17.4	6837
	11–20	779	−1.33	3150	566	−4.51	3394	838	2.18	2201	1030	3.91	2024
	21–30	612	5.92	1810	433	1.45	1997	415	3.22	986	570	8.30	835
	31–40	477	7.15	1269	338	2.97	1410	265	5.00	523	349	8.29	444
	41–	2477	24.56	5863	2135	25.71	6217	762	13.9	1191	984	19.2	1004
		$\chi^2 = 8$[illegible]0.0		$p = 0.00$	$\chi^2 = 774.3$		$p = 0.00$	$\chi^2 = 57.6$		$p = 0.00$	$\chi^2 = 72.3$		$p = 0.00$

(continued)

Table 3.3 (continued)

		Census grids						Local districts					
		Youth population			Productive age population			Youth population			Productive age population		
		Increase	*A.R.*	Non-increase	Increase	*A.R.*	Non-increase	Increase	*A.R.*	Non-increase	Increase	*A.R.*	Non-increase
(G) Parks	YES	3089	29.83	7196	2568	27.83	7749	2711	7.59	6865	3214	8.50	6383
	NO	2569	−29.83	14,580	2101	−27.83	15, 422	1929	−7.59	6345	2318	−8.50	6073
		$\chi^2 = 937.7$		$p = 0.00$	$\chi^2 = 708.8$		$p = 0.00$	$\chi^2 = 300.7$		$p = 0.00$	$\chi^2 = 258.4$		$p = 0.00$
(H) Density (People/km^2)	−2000	1804	−29.12	11,674	1586	−23.80	12, 294	639	−12.2	2921	748	−15.3	2924
	2000–4000	763	3.76	2539	554	−0.02	2752	491	8.11	907	487	3.23	922
	4000–10,000	1903	22.52	4268	1371	12.98	4800	1764	10.4	3928	1882	4.45	3821
	10,000–15,000	758	8.69	2060	663	10.13	2155	994	0.805	2756	1260	4.22	2492
	15,000–20,000	314	5.76	834	349	12.62	799	457	−1.66	1416	692	6.14	1181
	20,000–	116	1.03	401	146	7.05	371	295	−6.91	1282	463	−1.29	1116
		$\chi^2 = 1322.6$		$p = 0.00$	$\chi^2 = 1254.7$		$p = 0.00$	$\chi^2 = 414.3$		$p = 0.00$	$\chi^2 = 629.6$		$p = 0.00$
(I) Household (persons)	1–1.5	16	1.39	41	26	1.98	83	33	3.03	48	39	0.542	79
	1.5–2	391	11.43	760	439	19.33	735	621	10.5	1073	789	15.0	913
	2–2.5	1492	22.06	3076	1347	24.72	3268	1593	9.17	3597	2030	15.3	3176
	2.5–3	2133	13.13	6245	1392	−0.85	7063	1600	0.226	4531	1631	−8.83	4515
	3–3.5	1119	−12.59	6109	937	−10.51	6369	587	−9.22	2452	719	−9.48	2335
	3.5–4	321	−20.73	3590	334	−15.17	3627	136	−11.8	1052	221	−9.54	976
	4–	186	−14.2	1955	194	−10.64	2036	70	−6.75	4572	93	−7.17	458
		$\chi^2 = 441.6$		$p = 0.00$	$\chi^2 = 836.1$		$p = 0.00$	$\chi^2 = 91.1$		$p = 0.00$	$\chi^2 = 313.1$		$p = 0.00$
(J) Aging (%)	−10	259	3.07	804	415	18.42	711	916	−3.83	2964	1540	12.9	2394
	10–15	783	5.26	2462	796	12.42	2466	1565	6.00	3834	1693	1.02	3718
	15–20	1551	11.24	4459	1190	6.97	4838	1356	3.87	3473	1497	0.299	3344
	20–25	1586	6.59	5181	1180	1.52	5613	523	−4.36	1821	550	−8.31	1802
	25–30	803	−8.27	4122	566	−11.13	4392	179	−4.91	756	171	−8.81	782
	30–	676	−16.58	4748	522	−17.10	5151	101	−2.08	362	81	−7.08	416

		$\chi^2 = 287.6$		$p = 0.00$	$\chi^2 = 298.2$		$p = 0.00$	$\chi^2 = 87.9$		$p = 0.00$	$\chi^2 = 226.1$		$p = 0.00$
(K) Slopes (degrees)	0	162	1.59	542	149	2.97	565	26	1.33	54	43	4.47	37
	0–1	3225	9.88	10,807	2712	10.74	11, 463	2296	1.04	6419	2940	7.69	5846
	1–3	1587	1.57	5881	1291	0.80	6275	1536	4.66	3890	1729	1.71	3735
	3–5	438	−5.77	2242	341	−6.37	2397	510	−3.25	1693	584	−4.82	1634
	5–10	204	−8.97	1486	127	−10.86	1605	223	−2.06	740	190	−7.78	782
	10–	45	−11.59	818	49	−9.40	866	49	−7.66	414	46	−9.94	422
		$\chi^2 = 102.0$		$p = 0.00$	$\chi^2 = 157.4$		$p = 0.00$	$\chi^2 = 15.7$		$p = 0.01$	$\chi^2 = 44.6$		$p = 0.00$
(L) Vacant (%)	0	4518	−9.23	18,491	3685	−9.95	19, 650	3531	1.06	9950	4298	4.77	9264
	0–5	462	4.37	1419	348	1.83	1555	683	−3.00	2193	741	−6.62	2158
	5–10	245	2.82	770	215	3.66	811	221	1.61	555	246	0.412	534
	10–20	239	5.17	626	216	6.27	664	133	1.18	336	161	1.18	324
	20–30	119	3.05	332	124	5.81	339	27	−0.511	86	40	0.488	82
	30–	75	5.28	138	81	7.38	152	45	1.95	90	46	0.719	91
		$\chi^2 = 1158.7$		$p = 0.00$	$\chi^2 = 1939.4$		$p = 0.00$	$\chi^2 = 249.4$		$p = 0.00$	$\chi^2 = 1266.5$		$p = 0.00$
(M) Detached (%)	−20	714	13.73	1526	852	27.06	1455	929	10.3	1808	1332	20.9	1476
	20–40	1025	13.34	2495	1096	24.26	2442	1053	3.57	2671	1621	18.8	2117
	40–60	1086	12.89	2730	825	8.31	3029	902	0.722	2504	1083	1.36	2331
	60–80	1184	11.38	3202	750	0.38	3671	869	2.15	2289	753	−9.35	2412
	80–	1649	−33.71	11,823	1146	−37.05	12, 574	887	−14.1	3938	743	−27.4	4120

Table 3.4 Living environment indices used for calculating living environment score, and value range and weighting indices in which population increased significantly *Edited and reprinted from Ai (2020) Table 2

	Census grids				Local districts			
	Youth		Productive		Youth		Productive	
Indices	Standard	Coefficient	Standard	Coefficient	Standard	Coefficient	Standard	Coefficient
Station	<1500 m	0.162	<1000 m	0.154	<500 m	0.156	<500 m	0.321
Hospital	<1500 m	0.254	<1000 m	0.076	<1000 m	0.254	<500 m	0.046
Police	<500 m	0.148	<500 m	0.058	<500 m	0.235	<500 m	0.181
Firefighting	<1000 m	0.210	<1000 m	0.030	<1000 m	0.301	<1000 m	0.090
Parks	Present	0.077	Present	−0.007	Present	−0.214	Present	−0.057
Stores	11+	−0.042	11+	−0.008	31+	0.263	31+	0.028
Businesses	11+	−0.019	11+	−0.080	11+	−0.474	11+	−0.470
Slopes	0–1 degree	0.091	<1 degree	0.096	1–3 degrees	0.213	<1 degree	0.270
Density	2000 persons/km^2 +	0.362	4000 persons/km^2 +	−0.135	2000–10,000 persons/km^2	0.226	2000–20,000 persons/km^2	0.116
Aging	<25%	−0.166	<20%	0.395	10%–20%	−0.051	<10%	0.451
Household	1.5–3 persons	0.138	2.5人未満	0.044	1–2.5 persons	0.166	1.5–2.5 persons	0.019
Vacant	0% +	−0.091	5% +	−0.132	Not	Applicable	0%	−0.254
Detached	<80%	0.527	<60%	0.881	<40% OR 60%–80%	−0.040	<40%	0.302

both cases, the aim is not to establish absolute standards for residential promotion but rather to quantify the assessment of a relative living environment within an urban area, so even if the differences in the tabulated spatial unit impact the results, it is unlikely to be a major problem.

Next, we look at some of the items in Table 3.4 that have different results by age-group. For distances to the nearest railway station or hospital, there is a general tendency that local districts and census grids that are closer to these facilities are more likely to lead to population increase, but the result shows that population increases are expected at slightly longer distances for the youth population. These results can be interpreted as follows: the child-rearing generation, which lives with the youth population, tolerates regions slightly farther away from railway stations or hospitals as residences than the productive age populations of married couples without children or single people.

Finally, when looking at the association with the aging rate, both generations did not seem to prefer living in regions with a high aging rate, but a population increase was expected in regions with a marginally higher aging rate for the youth population. This indicates that the child-rearing generation preferred areas where relatively diverse generations, including elderly people, live, whereas married couples and single people preferred regions with a large productive age population similar to themselves. In terms of the detached house rate, neither age-group preferred regions with a high level of 80% or more, but the youth generation preferred the detached house rate to a level close to 80%. This can be interpreted as follows: the child-rearing generation preferred detached houses, whereas married couples and single people preferred apartment houses.

3.5 Relationship Between Living Environment Score and Actual Population Increase

Are there actual population increase tendencies in regions with a high living environment score derived by the above-mentioned method? Table 3.5 demonstrates the results of applying the chi-square test and residual analysis in order to create a cross-tabulation table of living environment score levels and actual population increases. As mentioned above, the results for the census grid unit and local district unit are shown together.

A significant relationship between the living environment score level and actual population increase was seen for both spatial units and for both age-groups. Specifically, there were significant population increases in census grids and local districts with a living environment score of at least 50 for the youth generation and significant population increases in census grids and local districts with a living environment score of at least 55 for the productive age generation. Furthermore, fewer population increases were registered in census grids and local districts with a living environment score of less than 50 for both age-groups. These results

Table 3.5 Living environment scores and actual population increases *Re-organized and reprinted from Ai (2016) Table 10 and Ai (2020) Table 3

		Youth population			Productive age population		
		Increase		Non-increase	Increase		Non-increase
Census grids	LES	$\chi^2 = 1373.4$		$p = 0.00$	$\chi^2 = 1682.8$		$p = 0.00$
	−50	1523	−35.95	11, 698	1624	−38.56	15, 801
	50-55	647	5.00	2010	314	1.91	1388
	55-60	1044	10.55	2825	309	4.35	1171
	60-	2444	28.53	5243	2422	38.64	5531
Local districts	LES	$\chi^2 = 448.8$		$p = 0.00$	$\chi^2 = 1261.3$		$p = 0.00$
	−50	1716	−20.53	7866	1656	−30.63	7853
	50-55	747	4.32	1974	797	0.17	2118
	55-60	845	8.99	1882	863	9.19	1612
	60-	1209	14.04	2444	1768	30.20	2016

demonstrate that a population increase can be expected when the living environment score is high, which is the fundamental property that should be satisfied by the living environment score.

The number of census grids and local districts with a living environment score of 50 or more, exhibiting a population increase, as well as those with a living environment score of less than 50, exhibiting a population non-increase, as a percentage of the total number of census grids and local districts, is set as the accuracy rate. Specifically, this is the percentage of areas where high living environment scores and population increases are consistent with each other, in line with the aim of this study. The accuracy rate was 57.7% for the youth population and 65.1% for the productive age population when the spatial unit was the census grid, and 57.1% for the youth population and 60.4% for the productive age population when the spatial unit was the local district. Although the census grid unit was marginally higher, the overall accuracy rate was generally around 60%. This result suggests that referring to the living environment score in order to determine areas where it is expected that people would continue to live in the future is better than when there is no reference data. However, the results cannot be said to have reached an accuracy rate that can be held accountable during plan formulation and decision-making. Therefore, improving this accuracy rate is a task for future study.

3.6 Relationship Between Residential Promotion by Living Environment Score and Disaster Risk

In Japan, considering that natural disasters occur frequently, disaster risk mitigation in urban areas is an important issue. Typhoons and heavy rains, in particular, have become a major concern in recent years, frequently causing river and inland water flooding and landslides. Disaster risks cannot be sufficiently considered with the living environment scores analyzed so far, but could disaster risks also be reduced if residence selection was conducted based on the living environment, such as convenience and comfort, or if residential promotion for the consolidation of urban areas based on this residence selection was conducted. This is an important issue from the perspective of living environment safety and sustainability.

A basic principle of city planning operation guidelines is that residential promotion should not include areas with major concerns with regard to disaster prevention and safety, such as landslide disasters and tsunami disaster special warning areas. However, landslide and tsunami disaster warning areas, which have lower levels of concern, are subject to comprehensive assessment not only in terms of risks but also to warning and evacuation systems, and there is nevertheless room for their inclusion in residential promotion areas under certain conditions. In practice, as reported by Nishii et al. (2019), 70–80% of all municipalities exclude areas with disaster risk from the residential promotion area setting conditions. Warabi et al. (2019) conducted interviews with municipalities and surveys of city planning zone setting records, where they indicated that disaster risk was not clearly understood on a spatial scale when urbanization promotion areas were initially set, due to location restrictions, despite disaster risks. In this backdrop, it is fully assumed that urbanization has already expanded to areas with disaster risk. Consequently, it is possible that residential promotion areas have been set even in regions with potential disaster risk. However, when considering the purpose of city planning operation guidelines, residential promotion should be planned in areas with lower disaster risk in the medium to long term.

When comparing Kiuchi et al. (2003), who investigated the availability of the living environment during residence selection, and Kita (2017), who investigated the prioritized items in a living environment, the former had about 10–20% of respondents citing natural disaster safety and earthquake/fire safety as factors, whereas the latter showed that regions with limited concerns of disasters were the third highest priority, surpassing items like transportation convenience. It cannot be assumed that the subjects of the two surveys were homogenous, but it is possible that disaster risk has assumed priority in the approximately 10-year period straddling the Great East Japan Earthquake that occurred in 2011. However, Sakamoto (2020) conducted surveys of tsunami inundation assumptions and residence reasons, where it was discussed that there was a trade-off between consideration of disaster prevention and convenience to the city center. Based on the above results, whether disaster risk can also be reduced is an important question during residential promotion based on the living environment score.

Open data of hazard maps are used to assess disaster-based damage risk. Specifically, we used the Digital National Land Information "Landslide warning areas," "Flood inundation areas," and "Dense urban areas that are extremely dangerous during earthquakes" in order to compare with the living environment score that was calculated at the fourth-order census grid spatial unit, which is approximately 500 m square.

We created a cross-tabulation table between the presence or absence of disaster risk based on the three disaster classifications (henceforth, "landslide risk," "flood risk," and "dense urban area") and the change of population by age-group and conducted a chi-square test on these values. The results showed that, with the exception of the population change for the youth generation in regions with flood risk, significant relationships were found at the 5% significance level (Table 3.6). Residual analysis also showed significant population increases in areas with no landslide risk or areas corresponding to dense urban areas for both the youth generation and productive age generation. Basically, this could be interpreted as follows: given present-day population movement tendencies, landslide risk will decrease, but dense urban area risk due to building collapses and fires during earthquakes will not abate. Regarding flood risk, it cannot be said that the risk would increase or decrease for the youth generation, whereas population movements tended to increase in higher risk areas for the productive age population. If summarizing without considering disaster risk, then the results could be interpreted as follows: steep slopes are being avoided, whereas populations are returning to urban centers on flat land with dense populations. However, as mentioned above, this is not necessarily a change in direction that will lead to a reduction in disaster damage risk.

Similar results are also obtained when limiting the analysis scope to areas with a living environment score of 50 or higher and conducting the same chi-square test and residual analysis mentioned above (Table 3.7). Considering present-day population movement tendencies, residential promotion based on the living environment score does not produce a change in the direction of reducing disaster damage risk.

Summarizing the analysis in this section based on the above results, there currently was a tendency for significant population increases in census grids with flood risk and dense urban areas, with similar tendencies noted even when limited to census grids with a living environment score of at least 50, which are believed to have an attractive environment in terms of residential promotion. Essentially, even if residential promotion is conducted with the aim of developing compact cities, it is unlikely to contribute to disaster risk in particular.

When setting residential promotion areas in location normalization plans, although regions with disaster risk are generally excluded, in practice, existing urban areas are centered around such risky locations, and it may be unavoidable to conduct residential promotion even in areas with disaster risk, such as by planning soft measures like disaster prevention activities. Assuming that the population will decline in the medium to long term, it is considered that flood risk could be effectively reduced by prioritizing disaster prevention activities in areas around railway stations and hospitals, which have a large magnitude when calculating the living environment score (i.e., has a major impact on population changes). This means that we set a high priority to improve the disaster resilience of areas where it is highly probable that residential promotion will progress. Railway stations are locations with a larger non-permanent than permanent resident population, and hospitals are bases for life-saving and rescue activities. Thus, improving the disaster resilience of these surroundings is expected to effectively reduce the scale of damage beyond just a simple comparison of estimated affected populations based on population distributions.

Table 3.6 Relationship between population change and disaster risk by age-group. *Edited and reproduced from Ai (2021) Table 4

		Youth population			Productive age population		
		Increase		Non-increase	Increase		Non-increase
Landslide		$\chi^2 = 59.2$		$p = 0.00$	$\chi^2 = 130.6$		$p = 0.00$
	YES	2087	−7.71	3657	1701	−11.44	4198
	NO	9002	7.71	12,452	8013	11.44	13,729
Flood		$\chi^2 = 3.6$		$p = 0.16$	$\chi^2 = 22.1$		$p = 0.00$
	Above floor	3859	1.63	5452	3470	4.67	5904
	Under floor	461	−1.22	719	415	−0.26	778
	None	6769	−1.08	9938	5829	−4.44	11,245
Dense urban areas		$\chi^2 = 53.0$		$p = 0.00$	$\chi^2 = 52.8$		$p = 0.00$
	YES	191	7.33	122	172	7.33	142
	NO	10,898	−7.33	15,987	9542	−7.33	17,785

Table 3.7 Relationship between population change and disaster risk in regions with high living environment score. *Edited and reproduced from Ai (2021) Table 5

		Youth population			Productive age population		
		Increase		Non-increase	Increase		Non-increase
Landslide		$\chi^2 = 49.8$		$p = 0.00$	$\chi^2 = 40.0$		$p = 0.00$
	YES	1024	−7.08	1694	754	−6.35	1154
	NO	5192	7.08	6304	4362	6.35	4825
Flood		$\chi^2 = 2.7$		$p = 0.26$	$\chi^2 = 16.1$		$p = 0.00$
	Above floor	2185	−0.04	2814	1950	4.00	2060
	Under floor	245	−1.60	359	198	−0.30	238
	None	3786	0.70	4825	2968	−3.81	3681
Dense urban areas		$\chi^2 = 37.5$		$p = 0.00$	$\chi^2 = 11.8$		$p = 0.00$
	YES	190	6.18	122	166	3.49	130
	NO	6026	−6.18	7876	4950	−3.49	5849

3.7 Conclusion

The living environment score focuses on various living environment indices and takes into account the magnitude of the influence of these indices on residence selection in order to assess the overall attractiveness of an area as a residence. In this section, we analyzed population increases from 2000 to 2010 in the Tokyo metropolitan area and attempted to calculate the living environment score at both the census grid and local district levels. The results showed that there was a significant expected population increase tendency in regions with a high living environment score. However, the accuracy rate was around 60%, and further improvements are needed in order to use this as a basis for decision-making.

Indices related to disaster safety have not been considered in the living environment score. Thus, when we verified the relationship between population-increase tendencies or the living environment score with disaster risk, it was suggested that population increases might continue in regions with flood risk or risks associated with dense urban areas. Disaster risk is something that people seek to minimize, and this should be considered as a property that is different from other living environment indices that should be pursued to an appropriate level. However, as mentioned above, it is imperative to improve the living environment score calculation method so that disaster risk can be reduced while improving accuracy.

References

Arai Y, Ohki S (1999) Recent tendency of personnel transfers in Japan: new tabulation of the data of the third migration survey. Komaba Stud Hum Geogr 13:111–136. (in Japanese)

Asami Y (ed) (2001) Residential environment: methods and theory for the evaluation. University of Tokyo Press, pp 17–22. (in Japanese)

Kita Y (2017) Analysis on residential location choice and its evaluation indexes based on modal theory. J Archit Plan (Trans AIJ) 82(732):485–495. (in Japanese)

Kiuchi N, Arita T, Tanaka N, Kawasaki N (2003) On home area definition and residential environment information acquisition of the government housing loan corporation contractors. Urban Hous Sci 43:162–167. (in Japanese)

Koike S (2006) Mechanism of population decline as seen from regions. Commun Oper Res Soc Jpn: Manag Sci 51(1):30–36. (in Japanese)

Maene M, Shimizu Y, Nakayama T (2012) Research on the change of living environment and land use in the northern Osaka suburb I: the analysis of residents' living environment. Proc AIJ Kinki Chap Archit Res Meet 2012:445–448. (in Japanese)

Ministry of Internal Affairs and Communications (2021) *2020 population census: basic statistics on population, summary of results*, p. 2 (in Japanese)

Nishii S, Manabe R, Murayama A (2019) Policies and backgrounds for setting residential promotion areas in location normalization plans: through comparison of cities with contrasting area ratio of residential promotion area to urbanization promotion area. J City Plann Inst Jpn 54(3):532–538. (in Japanese)

Ogawa H (2011) Characteristics of population dynamics of three major metropolitan area in population decline era: case study of the 2010 population census. Proc AIJ Tohoku Chap Archit Res Meet 74:223–228. (in Japanese)

Sakamoto J (2020) An empirical analysis on the relationship between awareness of tsunami hazard risk and housing choice, travel mode. J City Plann Inst Jpn 55(3):836–842. (in Japanese)
Warabi Y, Matsukawa T, Nakade B, Higuchi S (2019) Study on relation between urbanization promotion area and disaster risk area: focusing on comparison between original area and expanding area. J City Plann Inst Jpn 54(3):931–937. (in Japanese)

Chapter 4
Relationship Between Participation of Older Adults in Hobby Clubs and Sports Groups and Density of Neighborhood Facilities: A Case of Nagoya City Using JAGES Panel Data

Hongjik Kim, Kimihiro Hino, Hiroyuki Usui, Masamichi Hanazato, Daisuke Takagi, Naoki Kondo, and Katsunori Kondo

Abstract Previous studies have suggested that the closer older adults live to urban facilities, the more likely they are to participate in recreational group activities. However, close proximity can also have a negative effect on participation. This study aimed to investigate the relationship between changes in the density of neighborhood facilities and changes in the participation of older Japanese adults. The data used in this study consisted of two waves of the Japan Gerontological Evaluation Study (JAGES) survey. The survey participants were not eligible for long-term public nursing care. The facilities were categorized as (1) leisure facilities; (2) cafés, pubs, and restaurants; (3) medical and welfare facilities; and (4) food stores. A multilevel ordered logistic model was used for the analysis. The results of this study indicate that several types of neighborhood facilities are associated with increased participation by older adults. However, light-hearted going out, especially during daily use of food stores, may have some limitations in promoting participation.

Keywords Social participation · Neighborhood facility · Promotion policy · Panel data · The Japan Gerontological Evaluation Study (JAGES)

H. Kim (✉) · K. Hino · H. Usui
Department of Urban Engineering, The University of Tokyo, Tokyo, Japan
e-mail: hj-kim@e.u-tokyo.ac.jp

M. Hanazato · K. Kondo
Center for Preventive Medical Sciences, Chiba University, Chiba, Japan

D. Takagi
Graduate School of Medicine, The University of Tokyo, Tokyo, Japan

N. Kondo
Department of Social Epidemiology, Kyoto University, Kyoto, Japan

Y. Asami et al. (eds.), *Studies in Housing and Urban Analysis in Japan*,
New Frontiers in Regional Science: Asian Perspectives 75,
https://doi.org/10.1007/978-981-99-8027-7_4

4.1 Introduction

The importance of older adults' participation in group activities is increasing in countries with aging populations. Various administrative bodies, including the World Health Organization (World Health Organization 2007), are developing policies to promote the participation of older adults in group activities. Various types of group activities for older adults include hobbies, sports, community events, volunteering, and employment (Cabinet Office 2017). In particular, the effectiveness of hobbies and sports group activities in preventing care needs has been demonstrated (Kanamori et al. 2014), attracting attention from various fields, such as public health and administration. Furthermore, in Japan, although there is a high willingness among older adults to participate in hobbies and sports, the actual participation rate is low (Cabinet Office 2017).

Homes, community centers, and third places are locations where older people conduct and engage in group activities (Van den Berg et al. 2015). Third places are informal spaces where people interact socially (Oldenburg and Brissett 1982); these include public and commercial facilities, such as shopping malls (Jeffres et al. 2009). These facilities are expected to promote social interaction (Mouratidis 2018). Therefore, the establishment of these facilities near older adults' homes could aid in promoting activity participation.

Encounters outside third places also create opportunities for activity participation. According to a survey by the Ministry of Land, Infrastructure, Transport, and Tourism (Ministry of Land, Infrastructure, Transport, and Tourism 2015), older adults often go out for daily shopping, dining, socializing, and recreation. Additionally, the proportion of older adults attending medical appointments increases with age. Therefore, in addition to socialization and recreation, it is important to focus on shopping, dining, and medical appointments, which are among the main reasons for older adults going out.

Older adults tend to spend more time in their neighborhoods. The health of older adults is associated with social capital in the neighborhood, particularly social interactions. Furthermore, the distance that older adults can walk tends to decrease with age (Ministry of Land, Infrastructure, Transport, and Tourism 2015), and their walking ability is related to the frequency with which they go outdoors (Fujita et al. 2006). Owing to a decline in physical abilities, especially walking ability, the range of outings and social interactions among older adults is relatively small. Therefore, in an aging society, amenable neighborhoods are particularly important for older adults to promote walking and participating in group activities.

When considering a neighborhood environment conducive to activity participation for older adults from an urban structure perspective, the density of urban facilities (i.e., neighborhood facilities) in the vicinity is crucial. Facility density is not only related to the proximity of facilities in the neighborhood but also to the variety of choices available.

Previous studies have shown that neighborhood environment is related to the participation of older adults (Vaughan et al. 2016). In the case of neighborhood facilities, accessibility to facilities (such as grocery stores, recreational facilities, shopping malls, restaurants, cafés, and medical facilities) has been found to be positively related to older adults' activity participation (Levasseur et al. 2011; Richard et al. 2009, 2013). Thus, the more accessible the urban facilities are, the more easily older people can participate in group activities. However, if the distance to the facility is too short (e.g., if there are amenities on the ground floor of residential buildings), going to the destination may result in fewer opportunities for chance encounters or obtaining information about group activities. However, this was not considered in previous studies.

Most previous studies have used a cross-sectional design rather than a longitudinal design (Hand and Howrey 2019; Levasseur et al. 2011; Richard et al. 2009, 2013). Although cross-sectional surveys have shown that the density of neighborhood facilities is positively related to activity participation, it is not clear whether building additional facilities at the neighborhood scale results in greater participation by older adults. Therefore, a longitudinal survey is necessary to investigate how changes in the density of neighborhood facilities are related to changes in older adults' activity participation. If the increase in facility density is related to an increase in the frequency of activity participation, this provides empirical evidence that the establishment of urban facilities near older adults' homes is a policy option for facilitating participation in recreational group activities.

This study investigates the relationship between the density of neighborhood facilities and older adults' participation in hobby clubs and sports groups using panel data. The rest of the manuscript is organized as follows. Section 4.2 describes the target area and data. Section 4.3 describes the analytical methods used in the study. Section 4.4 presents and discusses the results. Finally, Sect. 4.5 presents the conclusions in the study, with additional policy implications.

4.2 Materials and Methods

4.2.1 Dataset

Panel data collected by the Japan Gerontological Evaluation Study (JAGES) in 2010 and 2016 were employed for the longitudinal analysis. The JAGES conducted a nationwide survey targeting older adults in 31 municipalities every 3 years from the 2010 survey, and the group surveyed approximately 180,000 people from 41 municipalities in 2016. The extraction was performed by municipality, and randomly sampled older adults were surveyed using questionnaires (see Kondo 2016 for more details). Given that the responses surveyed in different years were linked to each person, older adults' activity participation can be observed and traced using panel data. Among the 20 municipalities surveyed by the JAGES in both 2010

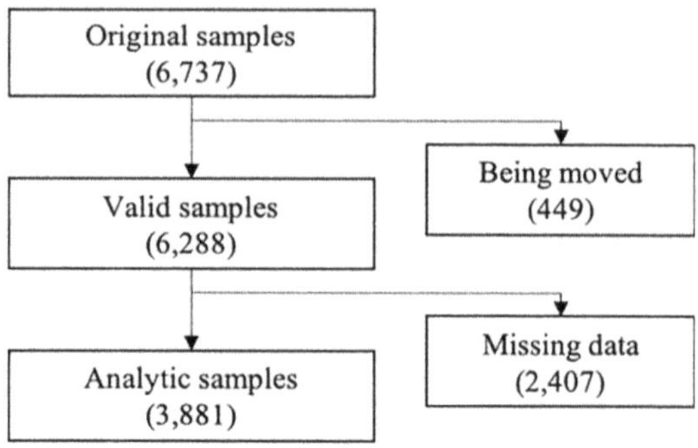

Fig. 4.1 Flowchart summarizing the selection of analytic samples

and 2016, Nagoya is one of three major metropolitan areas with a high population density (70 people/ha). This study aimed to investigate whether it is difficult to increase older adults' activity participation in areas where facility density is too high. This indicates that the distance to the nearest facility is too short (Koshizuka 1985). Thus, an area with a high level of facility density is required for the study area. Therefore, Nagoya was selected as the study area.

As shown in Fig. 4.1, several responses were excluded because the participants moved to another elementary school district within the study area (449 responses); this eliminates the effect of residence relocation. The relocation of older adult residences represents a process of selective migration. Several responses (2407) were also excluded because of missing data in the questionnaire items regarding the frequency of activity participation. Missing data for individual attributes other than participation in activities were included in the analysis as missing values. Therefore, the analytical sample consisted of 3881 responses. The average number of responses per elementary school was 14.8, and its standard deviation was 6.9. Nagoya comprises 262 elementary school districts. In the case of 200 neighborhoods, the level of 10 samples in each neighborhood was sufficient for model estimation without bias (Clarke and Wheaton 2007). According to this standard, the number of samples was sufficient to estimate a model.

The targets of the JAGES were people aged 65 years and older who did not require long-term nursing care services. The percentage of males within the analytic samples is 50.7%, and the average age and its standard deviation is 71.1 ± 4.66 (years). The characteristics of individuals, such as education, income, household type, walking ability, relationships with neighbors, and frequency of meeting friends, as well as their gender and age, were used as the control variables. These are assumed to be time-invariant variables, and the model uses the covariates measured in 2010.

4.2.2 Neighborhood Environment

The area where older adults spend their time in daily life has two levels of area (Nishino and Omori 2014): (1) the elementary school district (a scale of areas in which older adults can easily walk) and (2) the junior high school district (an area composed of two or three elementary school districts). The elementary school district in which individuals lived was used as a neighborhood unit to investigate the relationship between activity participation and neighborhood environment. Figure 4.2 illustrates the 262 elementary school districts in Nagoya.

The neighborhood environment was assessed by facility density, population density, and percentage of people aged 65 years and older estimated for each elementary school district (Table 4.1). Population density and the percentage of people aged 65 years and older were calculated based on the 2010 census survey. The high population density and high percentage of older adults indicate that there could be many peers with whom to conduct group activities, as well as the great availability of urban services owing to the large population.

Facility data from a telephone directory with associated location information (Zenrin Co., Ltd., Telepoint Pack!) for 2010 and 2016 were linked to the questionnaire responses. This was used to assess the density of facilities in the neighborhood and its changes. Facilities were grouped into (1) recreational facilities (libraries, movie theaters, adult education classes, karaoke boxes, sports facilities, fitness clubs, Mahjong clubs, community centers, etc.), (2) eating places (cafés, pubs, restaurants), (3) medical and welfare facilities (hospitals, welfare facilities for older

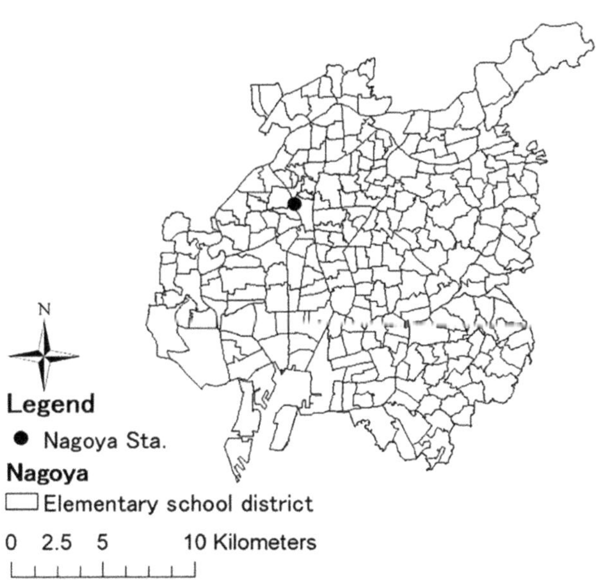

Fig. 4.2 Nagoya and its 262 elementary school districts

Table 4.1 Neighborhood-level variables

	Mean ± SD
Population density (people/ha)	89.1 ± 33.6
Percentage of people who aged 65 years and older	0.21 ± 0.04
Facility density (facilities/ha)	
Recreational facilities	0.12 ± 0.13
Eating places	0.50 ± 0.72
Medical and welfare facilities	0.18 ± 0.16
Food stores	0.13 ± 0.11
Changes in facility density (facilities/ha)	
Recreational facilities	−0.02 ± 0.03
Eating places	−0.11 ± 0.12
Medical and welfare facilities	0.00 ± 0.04
Food stores	−0.02 ± 0.04

Note. The neighborhood is organized into 262 elementary school districts where individuals live. SD: standard deviation

adults), or (4) food stores (grocery stores, convenience stores, and supermarkets). These facilities are places older adults can go to and interact with each other. The number of public recreational facilities was smaller than that of other categories, such as food stores. Recreational facilities include both public and private facilities. Public recreational facilities accounted for less than 10% of the total number of recreational facilities.

4.2.3 Frequency and Changes in Activity Participation

The two dependent variables were the frequency of participation in hobby clubs and sports groups. These variables were assessed using items from the survey regarding the frequency of respondent participation in hobbies or sports groups. The frequency of activity participation was answered on a six-point scale (four times a week or more, two to three times a week, once a week, once to three times a month, several times a year, and not participating).

To secure a certain number of participants, the frequency of activity participation was reclassified into three categories: (1) once a week or more, (2) several times a month or year, and (3) not participating. More precisely, participating in group activities four times a week and more, two to three times a week, and once a week are grouped into "once a week and more," on the other hand, one to three times a month and several times a year are grouped into "several times a month or year." Not participating remains as a category "not participating." These reclassified categories corresponded to high and low frequencies of activity participation and no participation, respectively. This enables the model to assess changes among participation levels in terms of frequency rather than changes within a high frequency of activity participation.

Based on the reclassified categories, changes in the frequency of activity participation of individual i living in area j between 2010 and 2016 were defined as follows:

$$\Delta y_{ij} = \begin{cases} 3, \text{if } y_{ij}^{2016} - y_{ij}^{2010} > 0 \\ 2, \text{if } y_{ij}^{2016} - y_{ij}^{2010} = 0 \\ 1, \text{if } y_{ij}^{2016} - y_{ij}^{2010} < 0 \end{cases} \quad (4.1)$$

where y_{ij}^{2010} and y_{ij}^{2016} indicate the frequency of participation in activities of individual i living in area j in 2010 and 2016, respectively. Consequently, the changes in the frequency of activity participation are defined as "increase" if $\Delta y_{ij} = 3$, "no change" if $\Delta y_{ij} = 2$, and "decrease" if $\Delta y_{ij} = 1$.

4.3 Analytic Method

Multilevel ordinal logistic regression was used because of the non-uniformity of the samples and their hierarchical structure as well as ordinal dependent variables. The regression model was expressed as follows:

$$\log \frac{p\left(\Delta y_{ij} \geq c|X_s, Z_t, u_j\right)}{p\left(\Delta y_{ij} < c|X_s, Z_t, u_j\right)} = \alpha_c + \sum_s \beta_s X_s + \sum_t \gamma_t Z_t + u_j \quad (4.2)$$

where categories for the changes in the frequency of participation are denoted by $c(\in\{1, 2, 3\})$ and α_c indicates the intercept for category c. The random effect (which controls for the effects of other unobserved neighborhood-level variables of area j) is denoted by u_j. The numbers of individual- and neighborhood-level variables are denoted as s and t, respectively. Coefficient β_s indicates s-th individual-level variable X_s, and γ_t denotes t-th neighborhood-level variable Z_t.

The individual-level variables (a set of X_s) comprised both y_{ij}^{2010} and control variables (such as gender, age, education, income, household type, walking ability, relationships with neighbors, and frequency of meeting friends). Owing to the fact that there could possibly be differences in the increases and/or decreases (among the levels of the frequency of activity participation), y_{ij}^{2010} is included in the model. Meanwhile, neighborhood-level variables (a set of Z_t) are composed of demographic neighborhood-level variables (population density and percentage of people aged 65 years and older) and facility density. In addition, the neighborhood-level variables regarding facility density are composed of density of existing (observed in 2010) neighborhood facility of type k in area j (which is denoted by F_{jk}), change in density between 2010 and 2016 (which is denoted by ΔF_{jk}), and interaction term of the two ($F_{jk}\ \Delta F_{jk}$). Given that the interaction term is included, each neighborhood-level variable is mean centered to avoid multicollinearity issues. All facility-related variables were simultaneously included at the same time.

The interaction term enables the model to express the nonlinear relationship between the frequency of activity participation and facility density. The interaction term is positive when both F_{jk} and ΔF_{jk} (mean-centered variables) have positive or negative values. When the interaction term is positively related to the change in the frequency of activity participation, a greater likelihood of participation is shown in neighborhoods where the number of facilities increases in areas with a high density of existing facilities, or their number decreases in areas with low density; this indicates a U-shaped relationship. However, the interaction term has a negative value when either F_{jk} or ΔF_{jk} has a positive value and the other has a negative value. When the interaction term shows a negative coefficient, there is a greater likelihood of participation in the neighborhood where the number of facilities decreases in the area with a high density of existing facilities, or their number increases in the area with a low density, indicating an inverted U-shaped relationship. In this case, there may be a certain facility density with the highest likelihood of increasing the frequency of participation in activities.

The model tests the relationship between the neighborhood environment and changes in the frequency of activity participation. From the model, $p(\Delta y_{ij} = 3)$ indicates that the probability of an increase in the frequency of activity participation can be expressed as a function of F_{jk} and ΔF_{jk}. Given that the model expresses Δy_{ij} using F_{jk} and ΔF_{jk}, the function is denoted as $\Delta y_{ij}(F_{jk}, \Delta F_{jk})$. The amount of change in $p(\Delta y_{ij} = 3)$ when the facility density increases slightly is the marginal effect of the density of neighborhood facilities, and is defined as follows:

$$\text{Marginal effect} = \frac{\partial p\left(\Delta y_{ij}\left(F_{jk}, \Delta F_{jk}\right) = 3\right)}{\partial F_{jk}} \tag{4.3}$$

The marginal effect enables the model to determine the density at which the likelihood of an increase in the frequency of activity participation has the highest/lowest value.

To facilitate the participation of older adults in recreational group activities, a sufficient level of facility density F^*_{jk} can be defined as the density at which the likelihood of an increase in the frequency of activity participation has the highest value. A sufficient level F^*_{jk} is found by the extremum $\hat{F_{jk}}$ where the marginal effect is equal to 0.

$$\frac{\partial p\left(\Delta y_{ij}\left(F_{jk}, \Delta F_{jk}\right) = 3\right)}{\partial F_{jk}} = 0, \quad \text{where } F_{jk} = \hat{F_{jk}} \left(0 \le \hat{F_{jk}} \le \max F_{jk}\right) \tag{4.4}$$

Since it is hard to solve $\hat{F_{jk}}$ analytically, it is numerically solved.

The analysis was conducted in three steps. First, multilevel ordinal logistic regression models were used to test the relationship between older adults' participation in hobbies and sports groups and the density of neighborhood facilities using panel data. It also tested whether a nonlinear relationship existed between the two. Finally, it estimates the density of facilities where the likelihood of an increase in the frequency of activity participation has the highest value based on the model, and visualizes the neighborhoods where a sufficient number of facilities are located.

4.4 Results and Discussion

4.4.1 Change in the Frequency of Activity Participation Between 2010 and 2016

Table 4.2 shows the frequency of participation in hobbies and sports groups in 2010 and 2016. Comparing the frequency of participation between the two time points, more people were found to not participate in hobbies clubs, even though the number of people participating at least once a week increased slightly (once a week or more: 1073 to 1094; several times a month or year: 935 to 843; not participating: 1873 to 1944). On the other hand, in the case of sports groups, the number of participants increased (once a week or more: 1002 to 1129 people; several times a month or year: 399 to 482 people; not participating: 2480 to 2270).

Table 4.3 shows the changes in the frequency of activity participation at each participation frequency level in 2010. Regarding the frequency of participation in hobby clubs, older adults showed an increase (620 responses; 16.0%), no change (2563 responses; 66.0%), and a decrease (698 responses; 18.0%). In terms of the frequency of participation in sports groups, older people showed an increase (552 responses; 14.2%), no change (2968 responses; 76.4%), and a decrease (361 responses; 9.3%). This indicates that older adults who do not need long-term nursing care services are able to continue participating in group activities, and some find and start new activities (Agahi et al. 2006).

Table 4.2 Frequency of participation in recreational group activities

	2010	2016
Hobby clubs		
Once/week and more	1073 (27.6)	1094 (28.2)
Several/month or year	935 (24.1)	843 (21.7)
Not participating	1873 (48.3)	1944 (50.1)
Sports groups		
Once/week and more	1002 (25.8)	1129 (29.1)
Several/month or year	399 (10.3)	482 (12.4)
Not participating	2480 (63.9)	2270 (58.5)

Table 4.3 Changes in the frequency of participation in recreational group activities from 2010 to 2016 ($N = 3881$)

	Increase	No change	Decrease
Hobby clubs			
Once/week and more	–	660 (17.0)	413 (10.6)
Several/month or year	237 (6.1)	413 (10.6)	285 (7.3)
Not participating	383 (9.9)	1490 (38.4)	–
Total	620 (16.0)	2563 (66.0)	698 (18.0)
Sports groups			
Once/week and more	–	756 (19.5)	246 (6.3)
Several/month or year	108 (2.8)	176 (4.5)	115 (3.0)
Not participating	444 (11.4)	2036 (52.5)	–
Total	552 (14.2)	2968 (76.4)	361 (9.3)

4.4.2 Relationships Between Participation and Facility Density

Table 4.4 presents the odds ratios (e^{γ_t}) estimated using this model. The odds ratio for population density (people/ha) indicates the likelihood of a change in the frequency of activity participation when 100 units increase, and the odds ratio for the percentage of older adults indicates the likelihood of change when the percentage increases by 1%. Regarding facility densities (people/ha), the odds ratios are the likelihood of a change in participation frequency when a facility is additionally built in an area comprising 10 ha.

Residents living in neighborhoods with larger populations are more likely to have relationships with those living in the same neighborhood (Boessen et al. 2018). However, population density and percentage of older adults were found to show insignificant relationships with changes in the frequency of participation in hobby clubs and sports groups. In densely populated urban areas (such as Nagoya), differences in population density and percentage of older adults between neighborhoods are not related to the ease of older adults' participation in recreational group activities. In other words, older residents living in urban areas do not suffer from a lack of colleagues or peers to conduct group activities. This suggests that the number of candidates for a group activity does not critically impact the likelihood of an increase in participation frequency once the number reaches a certain level.

The density of existing recreational facilities and their changes were found to show a negative relationship with an increase in participation in hobby clubs or sports groups (for hobby clubs, odds ratios: 0.80, 0.69, respectively; for sports groups, odds ratios: 0.76, 0.68, respectively). Even though public recreational facilities such as community centers function as places for facilitating social interactions among older adults (Hino and Ishii 2014; Makino and Imai 1999), most recreational facilities include private facilities (such as karaoke boxes, fitness clubs, and adult education classes). Although private recreational facilities provide

Table 4.4 Estimates [and 90% Confidence Intervals] for the odds ratio (e^{γ_t}) of the changes in the frequency of participation in hobby clubs and sports groups ($N = 3881$)

	(1) Hobby clubs	(2) Sports groups
Population density[a]	0.94 (0.76–1.17)	0.83 (0.64–1.07)
Percentage of people who aged 65 and older[b]	1.00 (0.99–1.01)	0.99 (0.97–1.01)
Facility density		
Recreational facilities	0.80 (0.69–0.93)*	0.76 (0.63–0.91)*
Eating places	1.05 (1.01–1.08)*	1.10 (1.05–1.14)*
Medical and welfare facilities	1.11 (1.02–1.20)*	0.94 (0.86–1.04)
Food stores	0.89 (0.76–1.03)	0.90 (0.76–1.08)
Changes in facility density		
Recreational facilities	0.69 (0.52–0.92)*	0.68 (0.48–0.96)*
Eating places	1.07 (0.98–1.16)	1.01 (0.91–1.12)
Medical and welfare facilities	1.00 (0.79–1.25)	0.70 (0.54–0.92)*
Food stores	0.95 (0.71–1.27)	1.34 (0.95–1.91)
Facility density × changes in facility density		
Recreational facilities	1.03 (0.94–1.13)	1.01 (0.91–1.13)
Eating places	1.00 (1.00–1.00)	1.01 (1.00–1.01)
Medical and welfare facilities	1.08 (1.00–1.17)*	1.10 (1.00–1.20)*
Food stores	0.91 (0.77–1.08)	0.78 (0.64–0.96)*
AIC	5972.836	4784.692

Note. [a]Odds ratio when 100 units (people/ha) increased. [b]Odds ratio when the percentage increased by 1%. Odds ratio other than a and b occurred when 0.1 units (facility/ha) increased
*$p < 0.1$

activities that meet a variety of needs, this suggests that it is difficult to build social interaction in these facilities and that it could possibly be an obstacle.

On the other hand, areas with a high density of eating places were found to be positively associated with an increase in the frequency of participation in hobby clubs and sports groups (hobby clubs, odds ratio: 1.05; sports groups, odds ratio: 1.10). In the case of densely developed urban areas, there are few places where older people can relax and have conversations; however, eating places (such as cafés, pubs, and restaurants) provide opportunities for social interaction that promote their participation in sports groups, as well as hobby clubs.

However, an increase in the density of eating places was found to have an insignificant relationship with an increase in participation frequency. Although eating places provide places for social interaction, they do not form new social connections; therefore, an increased number of eating places in the neighborhood may not lead to further participation in group activities. Therefore, if the local government aims to encourage activity participation through eating places, it is necessary to create opportunities for collaboration with eating place operators.

In terms of medical and welfare facilities, the relationships between facility density and increase in participation frequency were found to differ between hobby clubs and sports groups. In the case of hobbies clubs, older adults living in areas with a high density of medical and welfare facilities were more likely to participate (odds ratio: 1.11). In areas with a high density of existing medical and welfare facilities, older adults showed a positive relationship with increased participation in hobby clubs when the density increased further (odds ratio: 1.08). Interventions involving collaboration with local health sectors provide opportunities for older adults to participate in hobbies (Haseda et al. 2019; Hosokawa et al. 2019).

On the other hand, in the case of sports groups, increases in the density of medical and welfare facilities are found to show a negative relationship with the increases in the participation frequency (odds ratio: 0.70). However, in areas with a high density of existing medical and welfare facilities, older adults showed a positive relationship with an increase in participation in sports groups when the density increased further (odds ratio: 1.10). These are possible reasons why the density of medical and welfare facilities is related to participation in sports groups, even though the analytic samples included older adults who did not need long-term nursing care services.

In neighborhoods where the number of older adults with limited functional ability is relatively small, people living in the neighborhood, including healthy older adults, may have a negative perception of participation in recreational group activities. Therefore, it is difficult for participation frequency to increase because of fear of their peers' negative judgment. In other words, there could be factors that cause older adults to hesitate to participate. However, as the number of medical and welfare facilities increases, both the presence of older people with limited functional ability and participation in the prevention of functional disability become more common, and older adults may start to have health awareness, which positively influences participation in activities. By contrast, in neighborhoods with a low density of medical and welfare facilities, people are less likely to have a negative image of activity participation. Therefore, they show a higher likelihood of increased participation frequency than neighborhoods where factors that make older adults hesitate to participate in group activities exist. Consequently, there is a U-shaped relationship between the density of medical and welfare facilities and increases in the frequency of participation in sports groups (i.e., a higher likelihood of increases in participation frequency in neighborhoods with both a low and high level of facility density and a relatively lower likelihood in neighborhoods with a middle level of facility density).

In the case of food stores, older adults living in neighborhoods where the number of facilities increased in areas with a high density of existing facilities or decreased in areas with low density were found to be more likely to participate in sports groups (odds ratio: 0.78). This finding suggests that food store density may be sufficient to facilitate older adults' participation in sports. A high food store density means that the average distance to the destination is short when leaving home in daily life. In this case, there are few opportunities for social interaction and conversation. As an extreme example, people living in residential–commercial buildings, which are residential buildings with commercial uses on the ground floor, can easily visit these buildings for grocery shopping; however, going out rarely creates opportunities for chance encounters or obtaining information about sports group activities, owing to the short distance to the ground floor. The same is true at the neighborhood

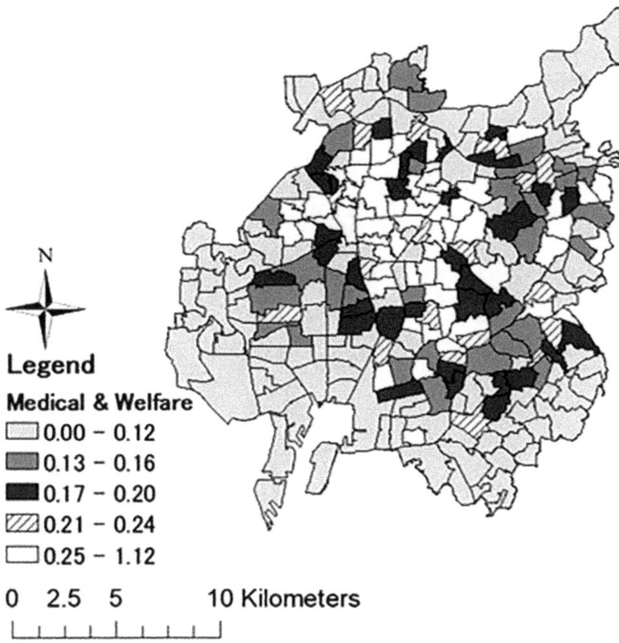

Fig. 4.3 Elementary school districts with the density of medical and welfare facilities where the likelihood of the increase in the frequency of participation in hobby clubs and sports groups is close to the local minimum (black colored areas)

scale. However, if the average distance is too long, willingness to go out decreases; therefore, the likelihood of an increase in participation frequency also decreases.

An extremum in the probability of an increase in participation frequency was found only in the cases of medical and welfare facilities and food stores. Figures 4.3 and 4.4 show neighborhoods (elementary school districts are colored black) with facility density where the likelihood of an increase in activity participation is close to the extremum (i.e., the marginal effect on the increase in participation frequency is close to 0). The density of medical and welfare facilities, where the likelihood of an increase in the frequency of participation in hobby clubs and sports groups has the lowest value, is estimated to be approximately 0.18 facilities/ha. In the case of food stores, the density, where the likelihood of an increase in the frequency of participation in hobby clubs and sports groups has the highest value, is estimated to be approximately 0.22 facilities/ha. In terms of geographical distribution, neighborhoods with facility density (both medical and welfare facilities and food stores) close to the extremum are located more than 2 km from Nagoya Station, which is close to the commercial center of Nagoya.

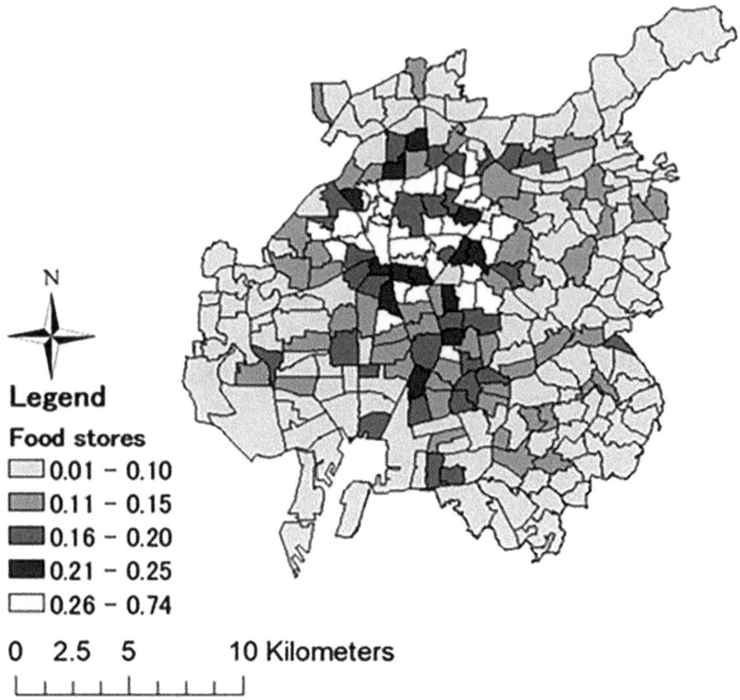

Fig. 4.4 Elementary school districts with a sufficient level of density of food stores in terms of older adults' participation in sports groups. Black colored areas indicate the neighborhoods with facility density where the likelihood of the increase in the frequency of participation in sports groups is close to the local maximum

4.5 Conclusion

This study investigated the relationship between the density of neighborhood facilities and changes in the frequency of older adults' participation in hobby clubs and sports groups. The results of the analysis using panel data to consider self-selection bias indicate that the density of urban facilities is related to participation of older adults in recreational group activities, not just the frequency of their activity participation. In other words, building additional facilities in urban areas at the neighborhood scale can be a solution to promote older adults' participation in recreational group activities, even though its impact on facilitating activity participation could possibly decrease.

In the case of food stores, an inverted U-shaped relationship was found between facility density and increased frequency of participation in sports groups (i.e., a closer distance to food stores is not always better in terms of facilitating older adults' participation in sports groups). In an aging society, it is necessary to consider the opportunities that help form new social connections, as well as proximity to food stores, when policymakers consider the neighborhood environment in which residents can easily shop groceries. For example, in addition to online shopping and home delivery services, it is necessary to focus on the social exchanges involved in shopping among older adults who are unable to shop.

However, there are some limitations because the facility density (the density of unweighted point features) is used to assess each type of facility. When building urban facilities, the scale, attractiveness, and number of facilities are important. In addition, given that it is difficult to obtain detailed information about the geographical location of older adults' homes, elementary school districts where they reside are used as neighborhood units. Therefore, it is unclear exactly how far away urban facilities can still have an effect on older adults' participation. The scale, service quality, and attractiveness of each facility could also be related to the distance that each facility can affect; however, these aspects were not considered in this study. Further studies that consider these issues are warranted.

References

Agahi N, Ahacic K, Parker MG (2006) Continuity of leisure participation from middle age to older age. J Gerontol B Psychol Sci Soc Sci 61(6):S340–S346

Boessen A, Hipp JR, Butts CT, Nagle NN, Smith EJ (2018) The built environment, spatial scale, and social networks: do land uses matter for personal network structure? Environ Plan B Urban Anal City Sci 45(3):400–416

Cabinet Office (2017) Annual report on the ageing Society; 2017. https://www8.cao.go.jp/kourei/english/annualreport/2017/2017pdf_e.html

Clarke P, Wheaton B (2007) Addressing data sparseness in contextual population research: using cluster analysis to create synthetic neighborhoods. Sociol Methods Res 35(3):311–351

Fujita K, Fujiwara Y, Chaves PH, Motohashi Y, Shinkai S (2006) Frequency of going outdoors as a good predictors for incident disability of physical function as well as disability recovery in community-dwelling older adults in rural Japan. J Epidemiol 16(6):261–270

Hand CL, Howrey BT (2019) Associations among neighborhood characteristics, mobility limitation, and social participation in late life. J Gerontol B Psychol Sci Soc Sci 74(3):546–555

Haseda M, Takagi D, Kondo K, Kondo N (2019) Effectiveness of community organizing interventions on social activities among older residents in Japan: a JAGES quasi-experimental study. Soc Sci Med 240:112527

Hino K, Ishii N (2014) Utilization of third places among elderly people and their significance. J Archit Plan 79(705):2471–2477

Hosokawa R, Kondo K, Ito M, Miyaguni Y, Mizutani S, Goto F, Abe Y, Tsuge Y, Handa Y, Ojima T (2019) The effectiveness of Japan's community centers in facilitating social participation and maintaining the functional capacity of older people. Res Aging 41(4):315–335

Jeffres LW, Bracken CC, Jian G, Casey MF (2009) The impact of third places on community quality of life. Appl Res Qual Life 4:333–345

Kanamori S, Kai Y, Aida J, Kondo K, Kawachi I, Hirai H, Shirai K, Ishikawa Y, Suzuki K, JAGES Group (2014) Social participation and the prevention of functional disability in older Japanese: the JAGES cohort study. PLoS One 9(6):e99638

Kondo K (2016) Progress in aging epidemiology in Japan: the JAGES project. J Epidemiol 26(7):331–336

Koshizuka T (1985) On the relation between the density of urban facilities and the distance to the nearest facility from a point in a given area. J City Plan Inst Jpn 20:85–90

Levasseur M, Gauvin L, Richard L, Kestens Y, Daniel M, Payette H, NuAge Study Group (2011) Associations between perceived proximity to neighborhood resources, disability, and social participation among community-dwelling older adults: results from the VoisiNuAge study. Arch Phys Med Rehabil 92(12):1979–1986

Makino Y, Imai N (1999) The place where the elderly relaxing and pleasuring viewed from the residential style of the elderly. J Archit Plan 64(522):131–138

Ministry of Land, Infrastructure, Transport, and Tourism (2015) National survey on urban transportation characteristics. https://www.mlit.go.jp/toshi/tosiko/toshi_tosiko_fr_000024.html

Mouratidis K (2018) Built environment and social well-being: how does urban form affect social life and personal relationships? Cities 74:7–20

Nishino T, Omori K (2014) A case study of senior care services area settings based on junior high school districts: a comparative study of official service area settings and actual living spheres of the elderly in a regional core city. J Archit Plan 79(699):1109–1118

Oldenburg R, Brissett D (1982) The third place. Qual Soc 5(4):265–284

Richard L, Gauvin L, Gosselin C, Laforest S (2009) Staying connected: neighbourhood correlates of social participation among older adults living in an urban environment in Montreal, Quebec. Health Promot Int 24(1):46–57

Richard L, Gauvin L, Kestens Y, Shatenstein B, Payette H, Daniel M, Moore S, Levasseur M, Mercille G (2013) Neighborhood resources and social participation among older adults: results from the VoisiNuage study. J Aging Health 25(2):296–318

Van den Berg P, Kemperman A, De Kleijn B, Borgers A (2015) Locations that support social activity participation of the aging population. Int J Environ Res Public Health 12(9):10432–10449

Vaughan M, LaValley MP, AlHeresh R, Keysor JJ (2016) Which features of the environment impact community participation of older adults? A systematic review and meta-analysis. J Aging Health 28(6):957–978

World Health Organization (2007) Global age-friendly cities: a guide. Retrieved from https://apps.who.int/iris/handle/10665/43755

Chapter 5
Environmental Factors Causing Inconvenience of Store Accessibility for Older Adults in Tokyo: Objective Indicators of Road Environments for Estimating People's Inconvenience

Tatsuya Sekiguchi

Abstract The problem of disadvantaged shoppers has become more serious in Japan. People feel difficulties for grocery shopping. Inconvenience of store accessibility is a factor of the difficulties. To address this problem, it is necessary to specify the elements in the concepts. However, few studies have focused on factors other than distance.

Focusing the older adults, this study aims to (1) clarify the characteristics of dissatisfaction with various road environments, (2) clarify what kinds of road environment dissatisfaction affect inconvenience, and (3) to develop decision tree models that estimate people's dissatisfaction with the objective condition.

A series of analyses were conducted based on data from an online questionnaire survey. First, the results indicated that the degree of tolerance and the amounts of each road elements were related to dissatisfaction. Elderlies tend to avoid several elements. Second, several road element can lead to inconveniences in accessing stores. Additionally, we created decision tree models by combining these data with indicators of objective road conditions. These models provided the threshold values of the objective conditions that significantly changed the subjective evaluation. These results allow us to compare the conditions of regions more efficiently or to identify regions requiring a more detailed survey.

The contents of this paper are based on the following paper originally published in a Japanese journal: Sekiguchi, T. (2023) Factors of inconvenience of store accessibility for older adults in Tokyo: Objective indicators and thresholds for estimating dissatisfaction of road environments that cause the inconvenience. MERA Journal 46(3): 1–10 (in Japanese).

T. Sekiguchi (✉)
Graduate School of Life and Environmental Sciences, Kyoto Prefectural University, Kyoto, Japan
e-mail: ta-sekiguchi@kpu.ac.jp

Y. Asami et al. (eds.), *Studies in Housing and Urban Analysis in Japan*,
New Frontiers in Regional Science: Asian Perspectives 75,
https://doi.org/10.1007/978-981-99-8027-7_5

Keywords Disadvantaged shoppers · Road environments · Questionnaire survey · Subjective evaluation · Objective indicator · Inconvenience of store accessibility

5.1 Introduction

5.1.1 Background and Purpose of Study

The "disadvantaged shopper" problem, where an increasing number of people in a region experience difficulty in daily grocery shopping, has recently become a social issue in Japan. In the administrative context, the term "disadvantaged shopper" has varied ambiguous meanings, such as "elderly people who feel inconvenienced in purchasing groceries" or "people who feel dissatisfied with the shopping environment" (Ministry of Internal Affairs and Communications 2017; Hachioji City 2013). Therefore, the disadvantaged shopper problem is often investigated and evaluated by using subjective indices such as "inconvenience of the shopping" or "the dissatisfaction of the store" (Hirai and Minami 2013; Yakushiji et al. 2013; Sekiguchi and Hino 2021). In addition, studies have pointed out that negative subjective evaluations of the food shopping environment lead to less frequent shopping and inadequate food intake, causing health problems (Sekiguchi and Hino 2021; Hino 2002; Nakamura and Asami 2019).

In Japan, a practical approach for improving this problem is "ensuring accessibility to grocery stores" or "decreasing the inconvenience." However, treating "accessibility" and "inconvenience" as fuzzy concepts prevents the development of concrete countermeasures. It is, therefore, necessary to specify the elements in these concepts. Previously, "distance to stores" has been regarded as the main factor of inconvenience of accessibility globally (Nakamura and Asami 2019; Larsen and Gilliland 2008; Russell and Heidkamp 2011). However, several studies have indicated out the influence of factors other than distance (Sekiguchi and Hino 2021; Takeshima 2007), which must be analyzed when considering these concepts.

This study aimed to answer the following research questions (RQs) based on the results of an online questionnaire survey and the spatial data of road elements in the Tokyo metropolitan area.

(RQ1) Clarify conditions that cause dissatisfaction with each road element
(RQ2) Clarify road elements that cause inconvenience of accessibility to the store
(RQ3) Develop statistical models to describe the objective conditions that cause dissatisfaction with the important road elements using specific objective indicators

We focused on multiple road elements that could become obstacles to traveling to stores, including distance. Additionally, negative evaluations of each element are considered as "dissatisfaction." Our analysis is based on the subjective evaluation structure that dissatisfaction with multiple or single elements of the road environment causes "inconvenience" of accessibility.

Clarifying these RQs will enrich our understanding of store accessibility inconvenience, which is conventionally treated as an abstract concept. Moreover, it helps to create appropriate and concrete countermeasures to overcome disadvantaged shoppers' difficulties. This study will provide a new perspective in the field of research dealing with the issue of disadvantaged shoppers.

5.1.2 Literature Review

Previously, "distance to stores" was regarded as a main factor contributing to the inconvenience of accessibility globally (Nakamura and Asami 2019; Larsen and Gilliland 2008; Russell and Heidkamp 2011). However, in addition to distance, elderly people with declined mobility skills due to aging find other road elements as problems (Takeshima 2007; Shibata et al. 2017; Matsumura et al. 2021).

Several studies have attempted to clarify the specific conditions that cause negative evaluation. Kato et al. (1995) studied the width and slope of sidewalks, whereas Nitta et al. (2004) studied the evaluations of elderly people on slopes and clarified the objective condition of these road elements in which they feel burdened. However, in travel situations such as shopping, people set the travel routes between the starting point and destination. Therefore, the amount and distribution of each element along the route are important; however, existing studies did not consider these points.

This study quantitatively analyzed the relationship between "inconvenience" in visiting stores and "dissatisfaction" with various road elements. In addition, we extract threshold values of each road elements that may lead to dissatisfaction. Clarifying the objective conditions that affect people's negative evaluation of the shopping route will help to set a target value for how much improvement of the road environments is needed and develop an efficient maintenance plan that considers the balance of the entire route, such as focusing on improving the specific sidewalks used by numerous people.

5.2 Materials and Methods

5.2.1 The Survey Area and Overview

The study area was the Tokyo metropolitan area, excluding island areas. This capital area of Japan has the highest population and well-developed public transportation system. People do not depend on cars, and high-density stores are located within walking distance, resulting in a high percentage of people who go shopping on foot. Due to such urban characteristics, road environment factors other than distance are considered problems during shopping travel.

Table 5.1 Demographic attributes of the respondents

Attributes	All respondents	Elderly people
Area of residence		
Tokyo 23 wards	773	381
Suburban Tokyo	227	119
Gender		
Male	544	312
Female	456	188
Age		
Under 65	500	0
65–74	431	431
75–79 (over 75)	69	69
Walking ability		
No difficulty in walking on my own	966	476
Some difficulty in walking on my own, but do not need a supporting device	21	15
Some difficulty in walking on my own, and need a supporting device	13	9
Significant difficulty in walking on my own	0	0
Degree of bulk buying		
Buy just for 1 day	207	102
Buy in bulk for 2 or 3 days	596	317
Buy in bulk for 4 or 5 days	114	53
Buy in bulk for 1 week or more	83	28
Leisure time walking habit (more than 10 min)		
Almost everyday	253	163
4 or 5 days a week	126	77
2 or 3 days a week	159	83
Once a week	115	46
Less than once a week	159	66
No such habit	188	65

An online questionnaire survey was conducted in December 2020 (survey via internet was commissioned by My Voice Communications, Inc.). Thousand responses were received from the respondents living in Tokyo Metropolitan area. The 500 responses were from elderly peoples aged 65–79 years and the rest from non-elderly peoples. All respondents were responsible for their households' shopping and usually went shopping on foot. All of them were selected based on screening questions. Table 5.1 lists the respondents' basic attributes.

5.2.2 *Analysis Flow*

In this study, we mainly focused on the elderly group and performed the following analysis.

In the Sect. 5.3, we analyzed how the tolerance level and the amount of each road element affect the dissatisfaction ratio. In Sect. 5.4.1, we extract dissatisfaction factors in the road environment elements that cause inconvenience in accessibility. In the first half of Sect. 5.4.2, we measured the distribution amount of each road element on each respondent's route to make objective indices for the next analysis. Finally, by analyzing the relationship between the respondent's subjective evaluations and objective indices, we extracted a threshold value using decision tree analysis for the objective situation where the tendency of evaluated satisfaction changed (latter half of Sect. 5.4.2). The method is suitable for extracting the threshold values where the respondents' evaluation trends and the decision tree can visually describe the multi-layered effects explanatory variables. Based on the results, we conclude this study by discussing the application and contribution of the developed models to the disadvantaged shoppers' problem (Sects. 5.4.3 and 5.5).

5.3 Analysis 1: Dissatisfaction of Elderly People with Various Road Elements

5.3.1 *Dissatisfaction Ratio for Each Road Elements*

First, we focused on dissatisfaction with road environment factors. In this section, we deal with the case of the store which respondents most frequently use. Table 5.2 shows the dissatisfaction ratio (the ratio of respondents who evaluated "dissatisfied" and "rather dissatisfied", excluding those who answered "No such element exists") for each item among the elderly respondents.

The dissatisfaction ratio with "Many bicycles run on the sidewalk" is the highest at over 30%, followed by "narrow sidewalks" at over 25%. Therefore, it is important to solve these dissatisfactions to create comfortable walking environment. Other dissatisfaction ratios for the following 15 items were around 15%, though "No rest space," "Steps between road and sidewalk," and "Distance to the store" were below 10%. In particular, the ratio of "Distance to stores" was the lowest. This indicates that elderly people decide the route to stores by prioritizing avoiding long distance. It is presumed that this causes unacceptable road elements in route selection or a large number of dissatisfaction factors on the route, which will leave dissatisfaction with other road elements other than distance.

Table 5.2 Dissatisfaction ratio for each item among the elderly respondents

Item	Ratio (%)	Item	Ratio (%)
Many bicycles run on the sidewalk	31.5	Stairway	14.2
Narrow sidewalk	26.4	Pedestrian bridge	13.8
Hard to walk in summer	18.6	Slope difficult to use	13.4
No sidewalk	17.6	Steep slope	13.3
Crosswalk with short green light	17.0	Many cars run on the road (no sidewalk)	12.9
Hard to walk in bad weather	16.2	Crosswalk with many cars	12.8
Obstacles on the sidewalk	15.7	Many cars run on the road (with sidewalk)	12.2
No nighttime lights	15.5	No rest space	9.1
Many pedestrians on the sidewalk	14.3	Step between road and sidewalk	8.4
Bumpy surface of the sidewalk	14.2	Distance to the store	7.8

5.3.2 Influence of "Tolerance Level" and the "Amount" of the Road Elements on the Respondent's Dissatisfaction

Next, we investigate the influence of "tolerance level" and the "amount" of the road elements on the respondent's dissatisfaction. Table 5.3 summarizes the tolerance level for each factor other than distance and the dissatisfaction ratio by the amount on the route among the elderly respondents.

First, the influence of the tolerance level is significant. When the amount of each element is the same, the dissatisfaction ratio of those who want to avoid such an element ("Unacceptable group") tends to greatly exceed the ratio of the group that accepts the element. In particular, when the amount level of each element is high, the dissatisfaction ratio of the unacceptable group exceeded 50% for 15 out of the 19 items. To solve the dissatisfaction with each road environments, it is important to understand the characteristics of people who do not tolerate each element and estimate the elements likely to be avoided in each region based on residents attributes.

On the other hand, the degree of amount of each element affected the dissatisfaction ratio. Even in the unacceptable group, the dissatisfaction ratio decreases to 20–30% when the amount of each element is "Medium." This indicates that dissatisfaction can be significantly improved by reducing the amount of each element.

However, for several road elements, the dissatisfaction ratio does not decrease significantly even if the amount level of element is reduced to a "Medium" level in the unacceptable group (e.g., narrow sidewalks, steps between sidewalks and roadways, bicycles on sidewalks, crosswalks with short green lights, and no lighting at night). The dissatisfaction with these items cannot be improved unless the amount of their presence becomes "small." In addition, for factors such as narrow sidewalks,

Table 5.3 Dissatisfaction ratio by tolerance/amount for each road element among the elderly respondents

		Distribution amount		
Road elements	Tolerance	Much	Medium	Little
Narrow sidewalk	Acceptable	23.5	22.2	10.4
	Unacceptable	73.2	33.3	16.1
No sidewalk	Acceptable	12.2	0.0	5.1
	Unacceptable	45.1	25.8	9.3
Bumpy surface of the sidewalk	Acceptable	0.0	15.4	4.3
	Unacceptable	71.4	16.1	10.7
Steps between road and sidewalk	Acceptable	7.1	2.8	3.8
	Unacceptable	33.3	31.3	7.1
Obstacles on the sidewalk	Acceptable	0.0	22.2	4.5
	Unacceptable	76.0	22.9	9.8
Many pedestrians on the sidewalk	Acceptable	17.9	0.0	0.0
	Unacceptable	55.6	24.4	5.0
Many bicycles run on the sidewalk	Acceptable	43.8	10.5	7.1
	Unacceptable	73.8	41.9	9.0
Many cars run on road(with sidewalk)	Acceptable	5.5	0.0	1.8
	Unacceptable	52.3	16.7	6.3
Many cars Runn on road(no sidewalk)	Acceptable	14.3	0.0	0.0
	Unacceptable	52.2	24.1	8.0
Crosswalk with many cars	Acceptable	5.6	4.8	2.7
	Unacceptable	50.0	22.9	9.6
Crosswalk with short green light	Acceptable	16.7	4.5	5.1
	Unacceptable	48.0	35.9	10.3
Pedestrian bridge	Acceptable	0.0	0.0	9.1
	Unacceptable	66.7	20.0	15.7
Steep slope	Acceptable	14.3	16.7	0.0
	Unacceptable	57.1	20.0	15.7
Stairway	Acceptable	0.0	0.0	0.0
	Unacceptable	57.1	20.0	15.7
Slope difficult to use	Acceptable	0.0	0.0	0.0
	Unacceptable	100.0	29.4	11.5
No nighttime lights	Acceptable	0.0	0.0	0.0
	Unacceptable	50.0	50.0	8.6
No rest space	Acceptable	1.2	0.0	2.8
	Unacceptable	36.8	40.0	20.0
Hard to walk in summer	Acceptable	21.4	2.9	2.7
	Unacceptable	67.2	16.7	9.8
No rest space	Acceptable	0.0	0.0	0.0
	Unacceptable	65.7	23.3	5.7

Note: The answer of "very unacceptable" and "unacceptable" were integrated into the "Unacceptable" group, while "rather acceptable" and "acceptable" were integrated into the "Acceptable" group. The answer of "very much" and "somewhat much" groups were integrated into the "Much," group, the "neutral" group as "Medium," and the "somewhat little" and "very little" were integrated into the "Little" groups

pedestrian bridges, stairs, and lack of other facilities, the dissatisfaction ratio of the unacceptable group was slightly higher than 10%, even when their level was small. Therefore, these road elements should be eliminated from the route to the store to decrease dissatisfaction dramatically.

5.4 Analysis 2: What Kind of Elements' Dissatisfaction Constitutes the Inconvenience of Accessibility? Objective Conditions that Define Dissatisfaction with Road Elements

5.4.1 Dissatisfaction with Road Elements Related to the Inconvenience of Accessibility

In this section, we attempt to identify the road element dissatisfaction related to the inconvenience of store accessibility.

To ensure a sufficient sample for analysis, we targeted up to three frequently used stores by elderly respondents: 500 respondents answered the first store, 438 answered the second store, and 275 answered the third. The access inconvenience ratio (the percentage of respondents who answered "inconvenient" and "somewhat inconvenient") was 7.0%, 10.5%, and 11.6% for the first, second, and third stores, respectively.

A characteristic of this dataset is that a respondent can have multiple store information. Therefore, an appropriate regression model should be selected based on the data structure. Therefore, we calculated the value of the intraclass correlation coefficients for evaluating the inconvenience of accessibility when all stores used by each individual were considered. The value of intraclass correlation coefficients was 0.545, indicating that it is desirable to assume a structure in which the store evaluation is nested in the individual.

Therefore, ordinal logistic regression was conducted using the data from a total of these 1213 individual evaluations of each store, considering the characteristics of evaluations among respondents as a random effect. The dependent variable was the evaluation of the inconvenience of accessibility (ordered variable on a 5-point scale from 1: inconvenient to 5: not inconvenient), and the explanatory variable was the dissatisfaction level for each road environment element and distance (scored on a 5-point scale from 1: not dissatisfied to 5: dissatisfied, with a cross term of their existing condition on the route: exist (1) and not exist (0)).

The regression results are presented in Table 5.4. It notes that the correlation coefficient of dissatisfaction with "steep slope" and "stairs" exceeded 0.9. In order to avoid multi-collinearity, "stairs" were excluded from the candidate explanatory variables considering of correlation with other variables.

The AIC of the final model is 1915.01, which is smaller than the AIC of 2089.00 when random effects are not considered; model fitting by the likelihood ratio test

Table 5.4 Road elements whose dissatisfaction is statistically related to the inconvenience of store accessibility

Variables	Coefficient	S. E.	Z value	*p*
Narrow sidewalk	−0.234	0.080	−2.905	0.004*
Steep slope	−0.194	0.082	−2.372	0.018**
Distance to the store	−2.200	0.145	−15.172	<0.001*

Note: Significance level *: 1%, **: 5%
Random effect: Deviance 6.266, Standard deviation 2.503

is also significant at the 1% level. The regression coefficients of narrow sidewalks, steep slopes, and distance to the store were significant. The distance had the greatest effect. In addition, the slope and the width of the sidewalk are factors that are often considered in other studies as factors of difficulty for the elderly (Takeshima 2007; Kato et al. 1995; Mizoguchi and Yamakawa 2001; Satoh et al. 2006).

5.4.2 Developing Decision Tree Models Describing Objective Conditions that the Dissatisfaction of the Road Elements Appear

5.4.2.1 Preparation of Quantitative Indicators of Road Elements' Distribution on the Route of Each Respondent

The distance or amount of these road elements extracted in Sect. 5.4.1 is represented by objective indices for the following analysis. To reduce costs, we used a method that does not require a field survey to prepare the objective data.

We randomly selected 203 respondents from the 500 elderly respondents to develop objective data. In addition, only the stores most frequently used by each respondent were selected for ease of interpretation of the model in the next analysis. Figure 5.1 shows an image and outline of data preparation for each respondents. The steps for this preparation are described below in order.

Estimating the Routes Used by Each Respondents
First, we estimated the routes used by respondents. The stores most frequently used, and the bus stops nearest to their residence were identified based on the respondent's answers to the questionnaire. Subsequently, using QGIS (ver. 3.18, one of the GIS (geographical information system) software), each respondent's shortest route and its distance from the bus stop to the store were calculated. For the road network data, we used the road centerline data of the Digital Map (Basic Geospatial Information) published by the Geospatial Information Authority of Japan.

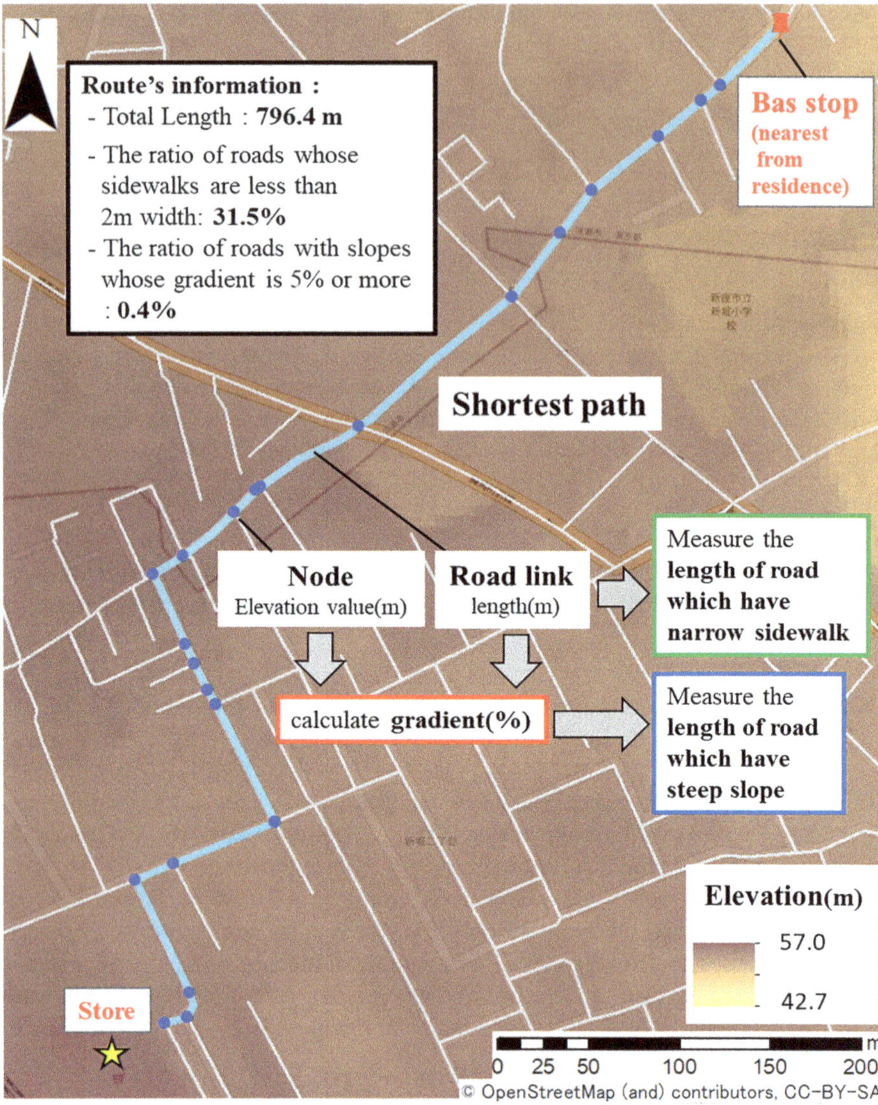

Fig. 5.1 Illustrative example of the data preparation of each respondents

Estimating the Distance to the Store

Using the aforementioned road network data, the lengths of the individual road links were obtained using a GIS tool. The total length of the links comprising the shortest route was used as the total route distance to the store.

Estimating Amount of Steep Slopes

As shown in Fig. 5.1, nodes were created at the endpoints of each link using GIS. Then, the elevation values from a 5 m resolution digital elevation model (The Geospatial Information Authority of Japan) were added to each node using the spatial merging function of GIS. The gradient of each slope was calculated as the percentage of the difference in elevation between the two ends of a road link to the length of the link.

In developing the index of "steep slope," we used the gradient criterion of "5% or more" as steep slope, referring to the results of existing study (Nitta et al. 2004). The ratio of the total length of road links with slopes whose gradient are 5% or more to the total route length (%) (hereafter, "the ratio of roads with slopes whose gradient is 5% or more") was used as an index to evaluate the amount of steep slopes in our analysis.

Estimating Amount of the Roads Which Have Narrow Sidewalks

Commonly available road data do not contain detailed information on the width of the sidewalks on each road. Therefore, we used Google Maps to measure the width of each road link that constituted the shortest path for each individual. The distance measure function was used to measure the width of the walkable parts of each sidewalk. In cases where the width was not constant due to the presence of fences, street trees, and other structures, we divided the walkable part into several sections, and the width of the section with the largest percentage of the road link length was used as the representative value. We used the width of "less than 2 m" as a criterion of "narrow sidewalk" (Doi and Amano 1986). The "ratio of the road links whose sidewalks are less than 2 m width to the total route length (%)" (hereafter, "ratio of roads whose sidewalks are less than 2 m in width") was used as an indicator to evaluate the amount of "narrow" roads.

5.4.2.2 Extraction of Evaluation Bifurcation Points of Dissatisfaction for Each Road Elements by Decision Tree Models

In this section, by combining the indicators prepared, personal attributes and subjective evaluation of the dissatisfaction of each road elements obtained from the questionnaire, decision tree models were developed to identify objective criteria in which the people's evaluation of their satisfaction level for each road elements changes significantly.

Each decision tree model was developed for each road element to describe the respondents' dissatisfaction evaluation structure. The CART (classification and regression tree) method was used to select the model variables. The dependent variable was the dissatisfaction level with each road element, and the explanatory variables were the indices of objective road environments (distance, the ratio of roads with a slope of 5% or more, and the ratio of roads whose sidewalks were less than 2 m in width) of each element on each person's route, the level of acceptance of each road element, and variables of personal attributes (Table 5.5).

Table 5.5 List of variables for developing decision tree models

	Variable name ※D: Dummy variable	Explanation of the variable
Dependent variable	**Dissatisfaction with each road element**	Prepared for each distance to the store, steep slope, and narrow sidewalk. – Three levels(※1): (1)dissatisfied, (2)neutral, and (3) not dissatisfied
Explanatory variables	**Objective road environments (distance to the store, steep slope, and narrow sidewalk)**	(1) Total length of route, (2)the ratio of roads with slopes whose gradient are 5% or more, and (3) ratio of roads whose sidewalks are less than 2 m in width
	Level of tolerance	Prepared for each distance to the store, steep slope and narrow sidewalk. – Three levels(※2): Unacceptable, neutral, and acceptable – this index was not used for the model of distance dissatisfaction)
	Male D	Male: 1, Female: 0
	Age over 75_D	Age over 75: 1, otherwise: 0
	Walking difficulty_D	"No difficulty in walking on my own": 0, otherwise: 1
	Walking regularly_D	If a respondent does leisure time walking (more than 10 min) at least once a week:1, otherwise: 0
	Large baggage_D	If a respondent usually buy groceries in bulk for 1 week or more:1, otherwise: 0

Note: (※1) The answer of "dissatisfied" and "rather dissatisfied" was aggregated into "dissatisfied" category, and the answer of "rather not dissatisfied" and "not dissatisfied" was aggregated into "not dissatisfied" category
(※2) The answer of "very unacceptable" and "unacceptable" was aggregated into "unacceptable" category, and the answer of "rather acceptable" and "acceptable" was aggregated into "acceptable" category

In the analysis, respondents who answered that "I have no such element in their route" were excluded from the 203 candidates. We also calculated the ratio of correct predictions, that is, the percentage of actual responses that matched the predicted response (i.e., predicted responses were estimated as the majority dissatisfaction level in each final node reached by following the branching conditions of the tree model along the respondents' attributes or answers). The results are presented below in order.

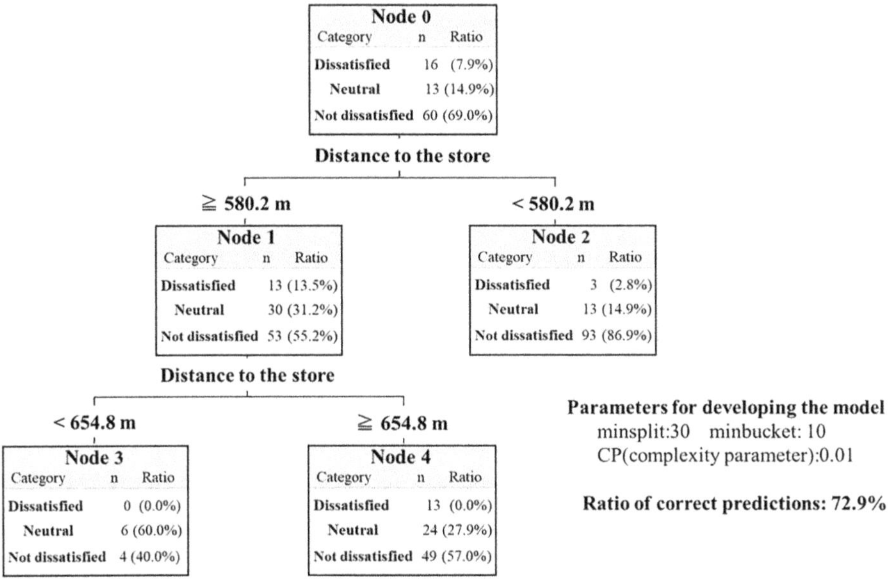

Fig. 5.2 The decision tree model for dissatisfaction of the distance to the store

Decision Tree for Dissatisfaction with the Distance to the Store

Figure 5.2 shows the results. Only the variables about distance value were used as the branching condition, and no personal attributes were included in the model. This finding suggests that the threshold for subjective evaluations does not depend on individual attributes in distance dissatisfaction.

When the respondent's distance was greater than 580.2 m, the overall dissatisfaction level increased. In detail, all dissatisfied respondents at Node 1 have a distance to the store of 654.8 m or more. We assume that the number of "not dissatisfied" respondents decreased when their distance exceeded approximately 580 m, and the number of "dissatisfied" respondents increased after a distance of 650 m.

Decision Tree for Dissatisfaction of the Steep Slope

Figure 5.3 shows the results. First, the nodes were branched according to their tolerance. Almost all respondents answered "not dissatisfied" when they could tolerate the use of steep slopes (Node 2). However, all respondents dissatisfied with the slope did not tolerate its use; thus, it can be said that whether they can tolerance of the steep slope was important in judging their dissatisfaction. Furthermore, the results diverged according to the objective conditions of the slopes on the route. When the "the ratio of roads with a slope of 5% gradient or more" exceeds 2.45%, the dissatisfaction level tends to increase (Node3 vs. Node4). Node 4 was further split according to whether the respondent's age was over 75.

From the above results, the tolerance of steep slopes influences dissatisfaction with steep slopes. Additionally, among those who cannot tolerate steep slopes, the

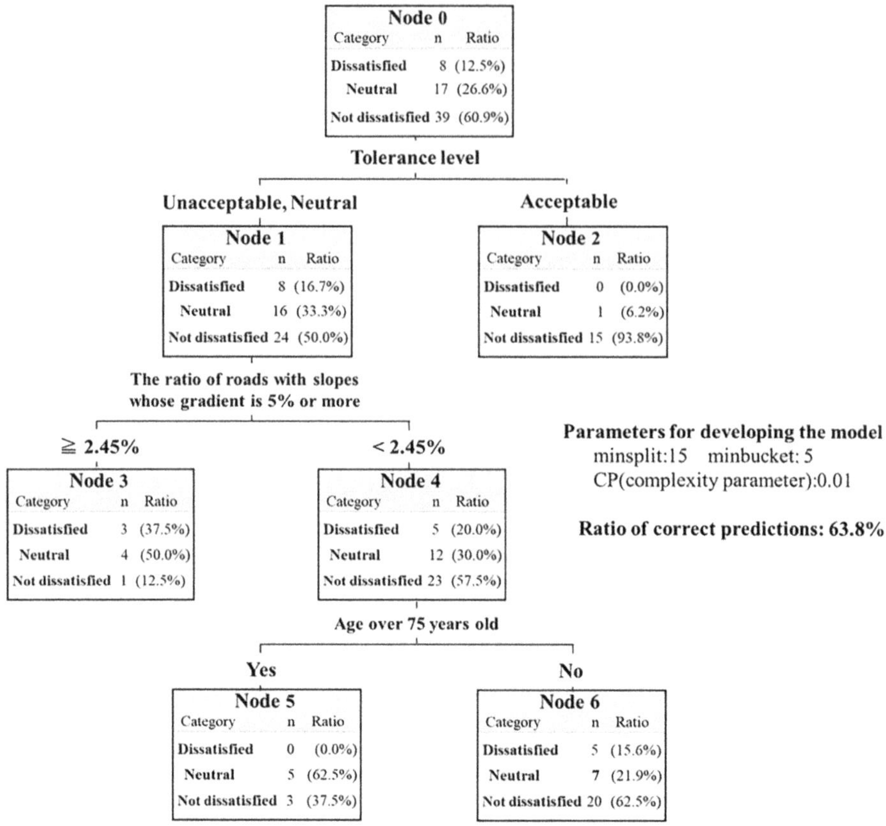

Fig. 5.3 Decision tree model for dissatisfaction with the steep slop

dissatisfaction rate was higher when the road had steep slopes (5% gradient or more) of more than 2.45%.

Decision Tree for Dissatisfaction with the Narrow Sidewalk

Figure 5.4 shows the results. First, the nodes were branched according to their tolerances. If the tolerance is anything other than not acceptable, the respondents were classified as Node 2. The dissatisfaction level is generally low at this node, suggesting that dissatisfaction is unlikely to occur when people are relatively tolerant of roads with narrow sidewalks. However, Node 1 is further divided by the objective condition of the "ratio of roads whose sidewalks are less than 2 m in width." Compared to Node 4, Node 3, with road-to-narrow sidewalk ratios above 22.8%, showed a higher level of dissatisfaction. Additionally, 19 of the 21 respondents who has dissatisfaction in Node 1 were reclassified as belonging to Node 3.

These results indicate that even among those who do not tolerate narrow sidewalks, the ratio of those dissatisfied with the road element tends to be relatively

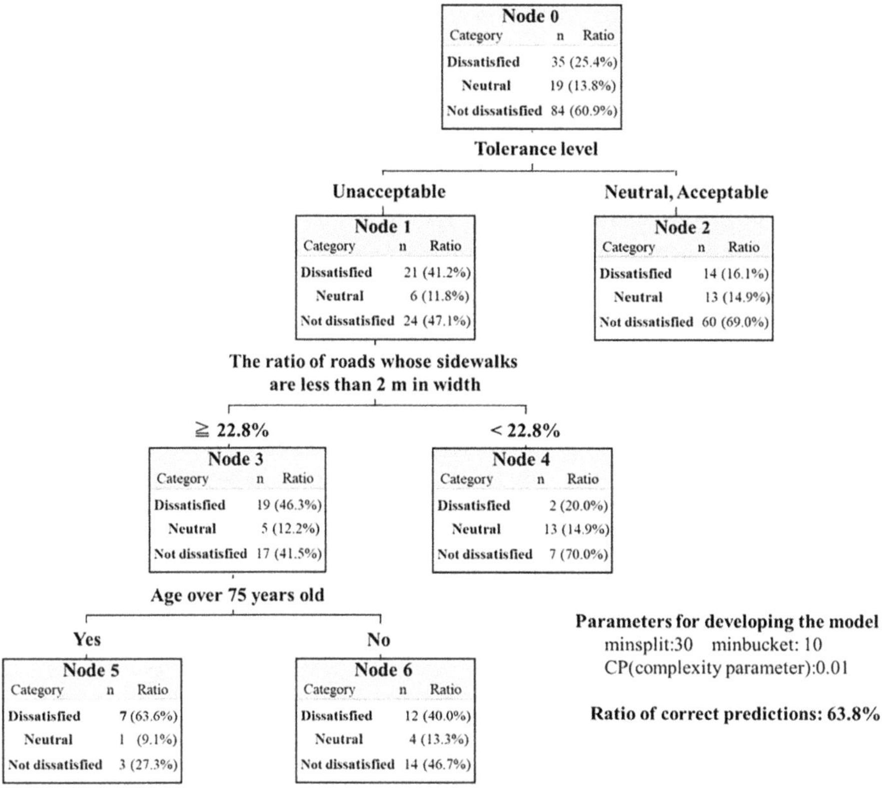

Fig. 5.4 Decision tree model for dissatisfaction with the narrow sidewalk

higher when approximately a quarter of the roads on the route are occupied by narrow sidewalks.

In addition, Node 3, with many roads having narrow sidewalks in the route, was further divided into deeper nodes according to the respondents' age. Comparing Nodes 5 and 6, the ratio of dissatisfied people was higher at Node 5, which was a group of elderly people over 75; in addition, the overall dissatisfaction level was also higher. This suggests that the elderly people over 75 are more likely to be dissatisfied with the roads having many narrow sidewalks.

The above results indicate that tolerance for narrow sidewalks and steep slopes influence dissatisfaction with the road elements. In addition, people tend to feel dissatisfied when narrow sidewalks occupy more than 22.8% of roads. The result also suggests that elderly people over 75 were more likely to feel dissatisfied with narrow sidewalks and steep slopes.

5.4.3 Applications of the Developed Models

In the previous section, we extracted the objective conditions likely to cause dissatisfaction with road elements. The results in Sect. 5.4.2 show that dissatisfaction with these elements affects the inconvenience of accessibility. Therefore, in order to solve disadvantaged shoppers' problems, it is important to improve road conditions of such road elements. Based on the results of our study, the amount of such problematic road elements should be kept below a certain ratio throughout the entire route to prevent dissatisfaction; hence, removing them altogether is unnecessary. This is a useful information for efficient road maintenance planning that prioritizes roads likely to become travel routes of many people.

Furthermore, the results allow easier comparisons of the progress of problems among several regions by estimating the subjective evaluation of people from the objective condition, without time-consuming and expensive questionnaire surveys. Such a simplified estimation can help to narrow problematic regions requiring more detailed field or questionnaire surveys. For estimation, we hypothetically assume an individual with attributes or subjective evaluations representing an area's residents. We set personal attributes with reference to existing local statistics and estimated route used by the person. Once a tentative route is established, we measure its distance or the percentage of roads with steep slopes or narrow sidewalks. We then apply the assumed individual attributes and the objective conditions of road elements to the decision tree models. Based on these settings, we follow the branches of the decision tree. Finally, when reach the final node, we use the majority rating of the classified node as the result of the estimation of the subjective evaluation estimation of the representative resident.

To set a representative individual's characteristics, local statistics can provide statistical information such as age and gender, which can be easily obtained from the national census. In addition, several individual attributes are required to adjust the tolerance level if it is likely to change owing to differences in attributes (see Sekiguchi 2022 to know the individual attributes are used for estimating the tolerance of each road element). For example, walking habits, which is relevant to estimating the tolerance level of steep slopes, can also be set by referring to statistics such as "walking and light exercise" by region, obtained from the Basic Survey of Social Life. In the elderly respondents, the distribution of the degree of tolerance for "narrow sidewalks" and "steep slopes" was as follows: unacceptable: 38.2% and 57.0%; undecided: 18.6% and 17.4%; acceptable: 43.2% and 25.6%, respectively. Based on these percentages, we weighed the aforementioned attributes by multiplying them with a magnification factor to set the tolerance degree of the assumed representative resident. In addition, for estimating the level of tolerance, the following values can be used as a reference: Women are 1.60 times more likely to be "unacceptable" for narrow sidewalks, 2.21 times more likely to be "unacceptable" for steep slope than men, and those whose leisure walking frequency is less than once a week are 1.66 times more likely to be "unacceptable" for steep slopes than those whose frequency is once a week or more (see Sekiguchi 2023 for the details of the derivation).

One of the major strengths of our method for estimating subjective evaluations is the low cost of data collection and the analysis environment. In particular, it can be conducted with a much lower budget than a regional questionnaire survey and can be analyzed for free using applications such as QGIS and R. Objective data, such as elevation data and road width information, can be also obtained at no extra cost. Although obtaining road data (road centerline data) incurs some costs, the cost is low compared with that of distributing and collecting questionnaires in each area. Therefore, our method can be used in many situations requiring efficient surveys with limited budgets, such as by local governments.

5.5 Conclusion

5.5.1 Summary of the Study

This study conducted a series of analyses to clarify people's consciousness structure regarding the inconvenience of store accessibility and dissatisfaction with road elements. These are important concepts because solving the problems associated with these inconvenience and dissatisfaction will help improve the problem of disadvantaged shoppers.

We analyzed road elements contributing to dissatisfaction among the elderly. Dissatisfaction with distance was unexpectedly lower than that of other factors, suggesting that elderly people chose the routes that minimized distance dissatisfaction. However, other environmental factors could not be avoided, leading to dissatisfaction with these road elements.

Furthermore, the results also show that the dissatisfaction ratio for each element was closely related to people's tolerance levels and the amount of each element on the route to the store. Several road elements had high dissatisfaction ratio because of their large amount on the route, even if the use of these element is "acceptable." Conversely, other road elements had high dissatisfaction ratio because they denied to use despite their low occurrence along the route. These results will be useful in deciding which roads or road elements should be prioritized to improve the region.

We also extracted road elements that could cause inconvenience to accessibility. From the ordinal logistic regression analysis, dissatisfaction with narrow sidewalks, steep slopes, and distance to stores are closely related to "inconvenience." Based on the results, we successfully developed decision tree models that describe the objective conditions that can significantly change people's satisfaction levels, considering personal attributes and tolerance levels for each road element. Although the case study was limited to the Tokyo metropolitan area, the significance of this study is that the extracted elements and threshold values can be applied to other road environments. The models developed in this study can help estimate people's subjective evaluations of objective road conditions with high simplicity. Furthermore, the objective data used in the models are generally available at a low cost or can be easily estimated from available data, enabling a more effective estimation of subjective estimation under limited time, human and financial costs.

5.5.2 Future Research Directions

In the future, the applicability of the method to other regions and improvements in prediction accuracy should be considered. If other elements can be indexed in addition to the road elements, and data maintenance can be made more efficient, it will be possible to predict dissatisfaction with various elements. For example, replacing manual acquisition of road widths with image analysis from bird's-eye view images or street views or using machine learning method will allow evaluation of other types of road elements in other regions easily.

In addition, using the proposed concept of "tolerance" to evaluate people's shopping inconvenience and dissatisfaction precisely will improve the prospects for discussing the problems of vulnerable shoppers in the future. These results can help conduct more detailed surveys among regions and take measures to eliminate dissatisfaction, thereby creating a comfortable shopping environment free from inconvenience and dissatisfaction.

Acknowledgments This study was funded and supported by the JSPS KAKENHI (grant number JP20K14901).

This manuscript is an English translation based on the original article "Factors of inconvenience of store accessibility for older adults in Tokyo: Objective indicators and thresholds for estimating dissatisfaction of road environments that cause the inconvenience" in MERA Journal (Vol. 25, No. 2, pp. 1–10). We thank to the Man-Environment Research Association for agreeing to translate the article.

References

Doi M, Amano K (1986) A survey of pedestrian behaviour to determine sidewalk width under desirable service level. Infrastruct Plan Rev 4:237–243

Hachioji City (2013) Report on the survey of shopping environments in Hachioji City. https://www.city.hachioji.tokyo.jp/kurashi/sangyo/sonota/001/p006343_d/fil/hyousi.pdf. Accessed 11 July 2023

Hino K (2002) Connection between inconvenience for shopping and eating habits of old people: analysis of shopping habit of old people in Itabashi area. J Archit Plan (Trans AIJ) 67:235–239

Hirai H, Minami M (2013) Distribution change in grocery store and people with the disadvantage of shopping in Morioka City. J City Plan Inst Jpn 48(3):969–974

Kato F, Wake I, Kawamoto Y (1995) Constructing an easy walking sidewalk for elderly. Infrastruct Plan Rev 17:991–994

Larsen K, Gilliland J (2008) Mapping the evolution of 'food deserts' in a Canadian city: supermarket accessibility in London, Ontario 1961-2005. Int J Health Geogr **7**:16

Matsumura Y, Manabe R, Murayama A (2021) Urban environment that encourages going-out activities of elderly people in the suburban planned residential area—a case study in Koganehara area, Matsudo City. J City Plan Inst Japan 56(1):24–31

Ministry of Internal Affairs and Communications (2017) Report on the survey on measures for disadvantaged shoppers. https://www.soumu.go.jp/main_content/000496982.pdf. Accessed 11 July 2023

Mizoguchi H, Yamakawa H (2001) A study of pedestrian transportation concerning with slopes in hilly residential areas. J City Plan Inst Jpn 36:841–846

Nakamura E, Asami Y (2019) The food desert caused by the economic difficult of access—in Azabu and Takanawa district of Minato-ku, Tokyo. J Archit Plan (Trans AIJ) 84(756):437–445

Nitta Y, Koyama K, Inoi H, Nakamura A (2004) The burden of elderly and disable walker on a slope. Infrast Plann Rev 29:102–105

Russell SE, Heidkamp CP (2011) Food desertification': the loss of a major supermarket in New Haven, Connecticut. Appl Geogr 31:1197–1209

Satoh E, Yoshikawa T, Yamada A (2006) Investigation of converted walking distance considering resistance of topographical features and changes in physical strength by age: model for location planning of regional facilities considering topographical condition and aging society part 1. J Archit Plan (Trans AIJ) 71(610):133–139

Sekiguchi T (2022) Older adults' inconvenience of store accessibility and dissatisfaction caused by the various road environments(Part1): focusing the tolerance and amounts of each road elements. AIJ J Technol Design 28(70):1465–1470

Sekiguchi T (2023) Factors of inconvenience of store accessibility for older adults in Tokyo: objective indicators and thresholds for estimating dissatisfaction of road environments that cause the inconvenience. MERA J 25(2):1–10

Sekiguchi T, Hino K (2021) How individual's various evaluations for the grocery shopping environments affects their shopping behaviors and dietary intake? an empirical study based on a questionnaire survey in a high-aged residential area in an outskirt of Central Tokyo. J City Plan Inst Jpn 55(3):1013–1020

Shibata F, Niwa Y, Oie H, Ito K (2017) The perceived burdens felt by elderly pedestrians in street environments-differences between trips with and without rollators. J Archit Plan (Trans AIJ) 82(732):451–457

Takeshima Y (2007) Research on the walk environment maintenance for the aged -the order of the barrier and how to cope those barriers in the daily going out. J Archit Plan (Trans AIJ) 611:1–6

Yakushiji T, Takahashi K, Tanaka K (2013) Difficulties in accessing food from local Residents' viewpoint: factors of inconvenience and laboriousness involved in shopping for food. J Rural Econ 85(2):45–60

Chapter 6
Relationship Between Crime Rate of Residential Burglary and Local Context

Kimihiro Hino and Takaya Kojima

Abstract This study aimed to assess the risk of residential burglary by examining factors representing the physical and socio-demographic context of the neighborhoods. Structural equation modeling was employed to establish their relationship. The main findings of this study provide evidence-based insights on crime reduction. Particularly, higher density, a larger family size, and daytime population reduced the risk, potentially enhancing natural surveillance. Conversely, neighborhoods with a high rate of rented houses exhibited a higher risk, while those with a high building coverage ratio demonstrated an increased risk, and those with a high floor area ratio demonstrated low risk. In conclusion, considering community safety as an integral aspect of quality of life, a holistic approach is suggested.

Keywords Crime prevention · Residential burglary · Structural equation modeling · Safer places

6.1 Introduction

6.1.1 Background and Purpose

In response to citizens' increasing awareness of crime and the trend toward information disclosure, maps displaying various crimes in each neighborhood have

---

This content is based on the following article originally published in a Japanese journal: Hino, K., and Kojima, T. (2007). Relationship between crime rate of dwelling burglary and local context: On neighbourhoods in 29 wards and cities in Tokyo. *Journal of Architecture and Planning (Transactions of AIJ)*, 616, 107–112 (in Japanese).

K. Hino (✉)
Department of Urban Engineering, The University of Tokyo, Tokyo, Japan
e-mail: hino@ua.t.u-tokyo.ac.jp

T. Kojima
Faculty of Human Sciences, Waseda University, Tokyo, Japan

Y. Asami et al. (eds.), *Studies in Housing and Urban Analysis in Japan*,
New Frontiers in Regional Science: Asian Perspectives 75,
https://doi.org/10.1007/978-981-99-8027-7_6

been published in recent years. These maps aim to inform residents and encourage them to take appropriate measures. However, crime occurrence maps do not offer guidance on promoting community development to prevent crime. The measures proposed mainly focus on individual buildings, sites, and general patrol activities (Schneider and Kitchen 2002).

In July 2003, the Council of Ministries and Agencies compiled research and studies conducted in six model areas across Japan, along with specific measures taken by the relevant Ministries and Agencies, in a report titled "Promotion of Crime-Prevention Community Development." Despite this report providing desirable measures for each type of urban area, a more universal and objective theory must be developed for the widespread implementation of community development to prevent crime.

In this study, a causal model was constructed to assess the risk of residential burglary based on social indicators such as households and population, as well as spatial indicators such as housing type and land use ("neighborhood characteristic indicators"). The premise of this study was the crime opportunity theory, which posits that the environment plays a role in inducing crime. By establishing the causal relationship between crime and the environment through the developed model, this study enables the implementation of community-specific safety measures.

6.1.2 Literature Review

Previous studies have investigated the relationship between community characteristics and crime occurrence using various methodologies. For instance, one study utilized factor analysis to explore the relationship between the number of crimes in each city and socioeconomic indicators (Shimada and Harada 1999a), while another study investigated the relationship between the number of crimes in each neighborhood and spatial indicators (Shimada and Harada 1999b). Recent studies have also investigated the number and rate of residential burglaries in neighborhoods, based on various indicators such as land use area (Hino 2006) and social and spatial indicators based on the crime prevention through environmental design (CPTED) theory (Seo et al. 2006).

Building upon previous studies, this study explains crime occurrence using both social and spatial indicators, selected following a comprehensive crime prevention guideline, described in the next section. Additionally, this study employed structural equation modeling to demonstrate causal relationships among the indicators, contributing to the development of more crime-resistant communities.

6.1.3 "Safer Places"

To select hypothetical neighborhood characteristics indicators, we referred to "Safer Places - The Planning System and Crime Prevention" (ODPM and Home Office 2003), a guideline for crime prevention in the United Kingdom. Safer Places was published in 2003 by the Office of the Deputy Prime Minister and the Home Office. Additionally, it is referenced as a guide in Planning Policy Statement 1 (PPS1) (2005), which outlines the government's national policies on various aspects of land use planning in England. Crime prevention is an integral part of the broader objective of "Delivering Sustainable Development," the title of PPS1. Section 17 of the Crime and Disorder Act requires all planning authorities to exercise their functions with due regard to their likely effect on crime and disorder, and to do all they reasonably can to prevent them. Safer Places was developed in response to requests from professionals such as planners, designers, and architects, which aim to provide practical guidance in line with these requirements.

Safer Places outlines seven attributes derived from a comprehensive review of crime prevention theory, urban design theory, and past practices (Table 6.1). This approach combines traditional theories with insights from multiple researchers, aiming to promote effective crime prevention measures while considering the values that must be emphasized in different practical contexts. Furthermore, it addresses the criticism that traditional theories alone have not consistently yielded positive results in real-world situations.

When comparing the seven attributes of Safer Places with the four from CPTED, which serve as the foundation for various standards in Japan, the following characteristics become evident:

- Broad concept: "Access and movement" consider access for residents, users, and potential offenders, while "Surveillance" focuses on increasing the presence of people and assigning them in appropriate ways.

Table 6.1 Overview of the seven attributes in "Safer Places"

Access and movement: places with well-defined routes, spaces, and entrances that provide for convenient movement without compromising security
Structure: places that are structured so that different uses do not cause conflict
Surveillance: places where all publicly accessible spaces are overlooked
Ownership: places that promote a sense of ownership, respect, territorial responsibility, and community
Physical protection: places that include necessary, well-designed security features
Activity: places where the level of human activity is appropriate to the location and creates a reduced risk of crime and a sense of safety at all times
Management and maintenance: places that are designed with management and maintenance in mind, to discourage crime in the present and the future

Source: ODPM and Home Office (2003)

- Detailed concepts: "Activity" and "Management and maintenance" provide more nuanced perspectives on the human impact on the environment compared to CPTED.
- Existence of higher-level concepts: The higher-level concept of "Structure" addresses housing form, use, and layout planning.

These characteristics arise from Safer Places' emphasis on diversity and resident participation because its goal is not solely crime prevention but also the development of sustainable communities (Hino and Amemiya 2005; Hino and Koide 2006; Hino and Amemiya 2006). Because Safer Places targets the district level, its seven attributes apply to the Japanese neighborhoods examined in this study.

6.2 Methodology

6.2.1 *Target Area*

The research focused on 3048 neighborhoods located within the 23 wards of Tokyo and six adjacent cities (Musashino, Mitaka, Chofu, former Tanashi, former Hoya, and Komae). These neighborhoods were selected based on the availability of necessary data and having a minimum of 300 households. Neighborhoods were chosen as the geographical units for this study owing to their ease of understanding and familiarity to residents. Moreover, they can be readily associated with community development to prevent crime, serving as actionable measures for both residents and local governments.

6.2.2 *Residential Burglary Rate—Dependent Variable*

During the fiscal years 2002–2004, there was a tendency for increased residential burglaries in neighborhoods with a larger number of households that could be potential targets. Figure 6.1 categorizes the neighborhoods into five equally sized groups based on the number of residential burglaries. The neighborhoods highlighted in red are concentrated around Suginami, Nakano, Nerima, and Edogawa. Furthermore, the correlation coefficient between the number of residential burglaries per neighborhood and the number of households was 0.644. To account for the effect of size, this study employed the number of residential burglaries per 1000 households as the dependent variable, which was explained by the neighborhood characteristics index. To address the distributional bias, "crime rate" presented in the equation below was utilized in this study. For logarithmization, the minimum number of residential burglaries in each neighborhood was set at 0.5.

$$\text{Crime rate} = \ln\left(\text{number of residential burglaries}/\text{number of households}^{*}\ 1000\right)$$

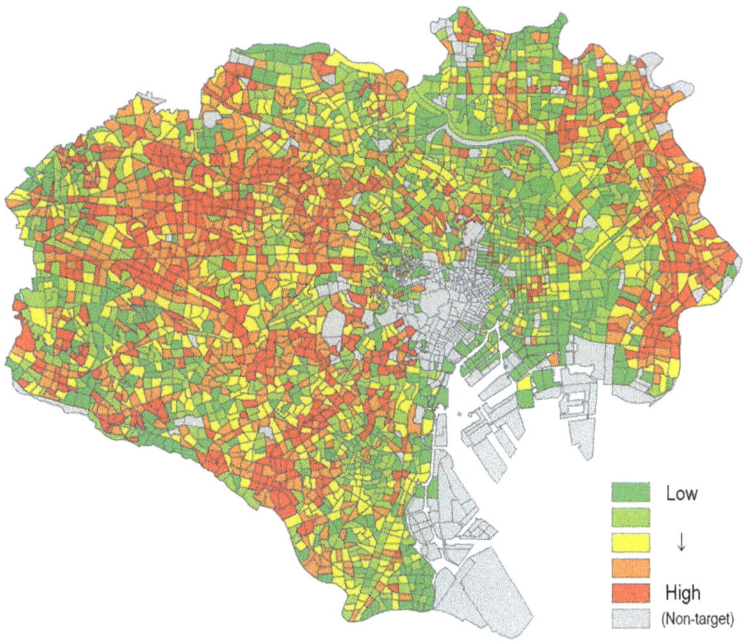

Fig. 6.1 Distribution of neighborhoods divided into five equally sized groups based on crime rate (residential burglary: FY 2002–2004)

The crime rate ranged from a minimum value of −1.60 to a maximum value of 3.95, with a mean value of 2.29 and a standard deviation of 0.72.

6.2.3 Neighborhood Characteristics—Explanatory Variables

Each neighborhood characteristic indicator was developed hypothetically based on the seven attributes of Safer Places (Table 6.2). The housing characteristics indicators were divided into two categories: ownership type (category a) and building type (category b) (apartment buildings were categorized by building floor). Separate models for each category were created to analyze their respective effects.

First, as indicators of neighborhood location, the distance to the nearest station, time distance to central Tokyo (train time from the Yamanote Line to the nearest station in minutes), and road area ratio (which is higher along arterial roads) were utilized. These indicators are related to the "Access and movement" attribute of Safer Places. To capture the attributes of "Activity" and "Surveillance," indicators including the number of persons per household, population density, and daytime population ratio were employed. These indicators influence the amount of natural

Table 6.2 Neighborhood characteristic indicators set with reference to attributes of Safer Places

Categories	Variables	Data sources[a]	Related attributes[b]	Remarks[c]
Population	Number of persons per household Population density Daytime population ratio	A, B	Access, surveillance	= ln (daytime pop/night pop)
Location	Nearest station distance Time distance to central Tokyo	C	(Access)	
Land use	Road area ratio Commercial–residential area ratio	C	(Access), structure	= Road area/nbhd area = ln (com area/res area)
Spatial density	Building coverage ratio Floor area ratio	C	Structure	= $\sum$bldg area/nbhd area = $\sum$(bldg area * no of floors)/nbhd area
Housing characteristics (a)	Owner-occupied household rate Public housing household rate Rented household rate	A	Ownership, management	
Housing characteristics (b)	Detached housing rate Low-rise household rate Mid-rise household rate High-rise household rate	A	Protection	$\leq$2 stories 3–5 stories $\geq$6 stories

[a](A): 2000 Population Census (Tabulation for Small Areas), (B) 2000 Population Census (Report by Ward, City, Town, and Village in Tokyo), (C) Tokyo city planning geographic information system (Wards: FY 2001, Cities: FY 2002)
[b]Access: Access and movement, Protection: Physical protection, Management: Management and maintenance
[c]*pop* Population, *nbhd* Neighborhood, *bldg* Building. For logarithmization, the minimum values of commercial land area divided by neighborhood area and residential land area divided by neighborhood area were set to 0.005

Table 6.3 Summary statistics for neighborhood characteristics indicators

	Min	Max	Mean	S.D.	Corr.
Population					
Number of persons per household	1.16	3.41	2.18	0.30	−0.15
Population density	836	50,986	16,066	6197	−0.11
Daytime population ratio	−1.15	3.77	−0.03	0.66	−0.16
Location					
Nearest station distance	3.39	7.88	6.28	0.66	−0.04
Time distance to central Tokyo	0.00	33.00	10.76	7.07	0.09
Land use					
Road area ratio	0.00	0.52	0.19	0.06	−0.11
Commercial–residential area ratio	−4.93	3.65	−1.55	1.10	−0.21
Spatial density					
Building coverage ratio	0.01	0.57	0.33	0.09	0.09
Floor area ratio	0.08	7.53	1.05	0.53	−0.23
Housing characteristics (a: ownership type)					
Owner-occupied household rate	0.00	0.91	0.42	0.13	0.08
Public housing household rate	0.00	1.00	0.07	0.16	−0.40
Rented household rate	0.00	1.00	0.45	0.14	0.40
Housing characteristics (b: building type)					
Detached housing rate	0.00	0.83	0.32	0.16	0.27
Low-rise household rate	0.00	0.68	0.17	0.11	0.41
Mid-rise household rate	0.00	1.00	0.25	0.13	0.03
High-rise household rate	0.00	1.00	0.21	0.22	−0.40

Max Maximum, *Min* Minimum, *S.D.* Standard deviation, *Corr.* Correlation with crime rate

surveillance within households and neighborhoods. The attribute of "Ownership" is influenced by housing characteristics (category a), including the percentage of owner-occupied households. The level of ownership also impacts the maintenance status. For "Physical protection," housing characteristics (category b) were used because they affect the number of accessible openings and the presence of security facilities. The "Structure" indicator encompasses the commercial-residential area ratio and spatial density. Additionally, it is a significant attribute among the seven attributes, which is connected to several other neighborhood characteristic indicators.

Table 6.3 presents the summary statistics for each variable in the target neighborhoods. To address distributional bias, log transformations were applied to the nearest station distance, daytime population ratio, and commercial–residential area ratio.

6.2.4 Statistical Analysis

First, multiple regression analyses were conducted to examine the relationship between the neighborhood characteristics indicators listed in the previous section (explanatory variables) and the crime rate (dependent variable). Model (a) utilized housing characteristics (a) as the explanatory variable, while model (b) employed housing characteristics (b). To address collinearity among the explanatory variables, a stepwise method ($p < 0.05$) was employed, and "time distance to central Tokyo" was excluded in model (b).

Next, structural equation modeling was employed to determine both the indirect effects on the dependent variable and causal relationships among the explanatory variables. While multiple regression analysis assumes that each explanatory variable is in a parallel relationship and the standardized coefficient of the multiple regression equation reflects the magnitude of the effect of each explanatory variable, structural equation modeling is more appropriate in cases where the explanatory variables do not exhibit a parallel relationship, regarding the neighborhood characteristics indicators in this study.

In this study, each neighborhood characteristic indicator was classified based on the relationship among the seven attributes of Safer Places. Indicators related to neighborhood location were placed at the first level, while indicators associated with the higher-level concept of "Structure" in Safer Places were positioned at the second level. Indicators related to "Ownership" and "Management and maintenance" were situated at the third level, and indicators of "Activity" and "Surveillance" were located at the fourth level. The assumption of the causal model can be summarized as follows: Variables at each level exert an effect on higher-level variables as well as crime rates. Three variables at the first level exhibit correlation. Additionally, there is a correlation between the error variables for each variable at every level.

6.3 Results

6.3.1 Multiple Regression Analysis

The coefficients and significance probabilities of the selected variables for each model are presented in Tables 6.4 and 6.5. The adjusted R-squared values, which indicate goodness of fit, were 0.316 for model (a) and 0.266 for model (b). The correlation coefficients between the estimated and actual number of residential burglaries for each model were 0.792 and 0.766, respectively. These coefficients were higher than the one obtained in the previous section between the number of residential burglaries and the number of households (0.644). Therefore, considering neighborhood characteristics enhanced the explanatory power of the number of residential burglaries.

Table 6.4 Coefficients of the multiple regression equations (Model (a))

	B	S.E.	*β*	*t*	*p*
Constant	1.21	0.29		4.18	0.00
Population					
Number of persons per household	−0.33	0.06	−0.14	−5.25	0.00
Population density	0.00	0.00	−0.27	−9.98	0.00
Daytime population ratio	−0.13	0.03	−0.11	−3.65	0.00
Location					
Nearest station distance	−0.07	0.02	−0.06	−3.30	0.00
Time distance to central Tokyo	0.00	0.00	0.04	2.20	0.03
Land use					
Road area ratio	0.95	0.22	0.08	4.40	0.00
Commercial–residential area ratio	−0.10	0.02	−0.15	−6.01	0.00
Spatial density					
Building coverage ratio	1.72	0.22	0.20	7.89	0.00
Floor area ratio	−0.24	0.04	−0.17	−5.94	0.00
Housing characteristics (a: ownership type)					
Owner-occupied household rate	1.90	0.27	0.34	6.94	0.00
Public housing household rate	0.81	0.26	0.18	3.10	0.00
Rented household rate	2.61	0.26	0.51	10.07	0.00

B Unstandardized coefficient, *S.E.* Standard error, *β* Standardized coefficient, *t* *t*-value, *p* *p*-value

Table 6.5 Coefficients of the multiple regression equations (Model (b))

	B	S.E.	*β*	*t*	*p*
Constant	1.93	0.29		6.67	0.00
Population					
Number of persons per household	−0.47	0.07	−0.19	−7.20	0.00
Population density	0.00	0.00	−0.33	−10.98	0.00
Daytime population ratio	−0.16	0.04	−0.14	−4.34	0.00
Location					
Nearest station distance	−0.10	0.02	−0.10	−4.90	0.00
Land use					
Road area ratio	1.13	0.23	0.10	4.95	0.00
Commercial–residential area ratio	−0.04	0.02	−0.07	−2.46	0.01
Spatial density					
Building coverage ratio	1.84	0.26	0.22	7.13	0.00
Floor area ratio	−0.12	0.05	−0.09	−2.61	0.01
Housing characteristics (b: building type)					
Detached housing rate	2.15	0.28	0.46	7.56	0.00
Low-rise household rate	2.63	0.27	0.40	9.62	0.00
Mid-rise household rate	1.82	0.26	0.34	7.14	0.00
High-rise household rate	1.39	0.26	0.42	5.36	0.00

B Unstandardized coefficient, *S.E.* Standard error, *β* Standardized coefficient, *t* *t*-value, *p* *p*-value

In model (a), which demonstrated higher explanatory power, standardized residuals were calculated based on estimated and measured crime rates. Out of the total, 393 neighborhoods (12.9%) had standardized residuals of −1 or less, suggesting the presence of environmental elements that may contribute to future increases in residential burglaries. Conversely, factors not addressed in this study, such as the prevalence of crime prevention equipment and frequency of crime prevention activities by residents, could potentially suppress burglaries in these neighborhoods. In terms of wards and cities, such neighborhoods were most prevalent in Chiyoda, Chuo, Taito, and Koto wards. Conversely, the 392 neighborhoods (12.9% of the total) with standardized residuals greater than one may attract burglars owing to factors not accounted for in this study, such as a higher average income that criminals expect to be associated with greater gains. In terms of wards and cities, such neighborhoods were more common in Minato, Sumida, and Edogawa wards. Other neighborhoods that exhibited a relatively good fit, despite having high crime rates, may possess the potential to deter burglars through long-term environmental improvements.

6.3.2 *Structural Equation Modeling*

Of the two models analyzed through multiple regression analysis in the previous section, model (a), which exhibited greater explanatory power, was further examined for causal relationships among the neighborhood characteristics indicators using structural equation modeling. The model, obtained by systematically eliminating paths with coefficients less than 0.1 while considering the interrelationships among variables, is depicted in Fig. 6.2, offering insights into the relationship between each variable and the crime rate. The direct and indirect effects of each variable on the

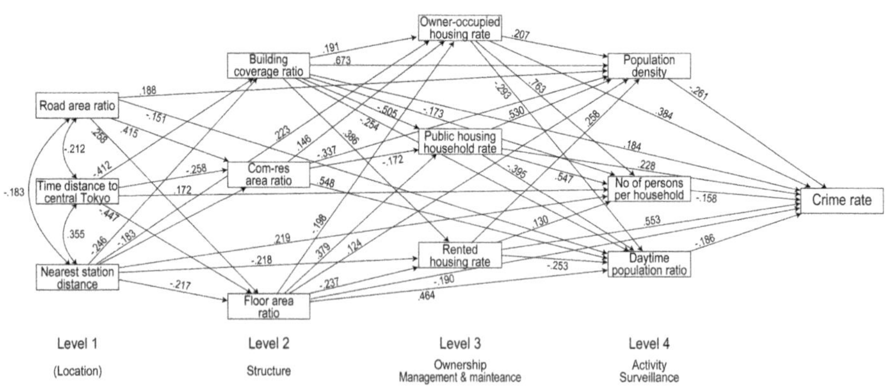

Fig. 6.2 Structural equation model of the residential burglary rate.
The estimated coefficients are standardized. The error terms and error correlations are not presented. Chi-square = 368.5, df = 22, $p = 0.00$

Table 6.6 Direct and indirect effects on residential burglary rate

Neighborhood characteristics indicator	Total effect	Direct effect	Indirect effect
1st level			
Nearest station distance	−0.064	–	−0.064
Time distance to central Tokyo	0.052	–	0.052
Road area ratio	−0.133	–	−0.133
2nd level			
Commercial–residential area ratio	0.012	–	0.012
Building coverage ratio	0.293	0.184	0.109
Floor area ratio	−0.453	−0.190	−0.263
3rd level			
Owner-occupied household rate	0.264	0.384	−0.120
Public housing household rate	0.077	0.228	−0.151
Rented household rate	0.513	0.553	−0.041
4th level			
Number of persons per household	−0.158	−0.158	–
Population density	−0.261	−0.261	–
Daytime population ratio	−0.186	−0.186	–

All the effects are standardized

crime rate are presented in Table 6.6. The goodness of fit indices, with a CFI of 0.987 and an RMSEA of 0.072, indicated a favorable fit for the model.

6.4 Discussion

6.4.1 Effects of Each Variable on Crime Rate

In this section, we discuss the effect of each neighborhood characteristic indicator on the crime rate, starting from the lowest level. To guide our discussion, we will refer to the seven attributes of "Safer Places."

<Level 4>

Among the neighborhood characteristic indicators, population density exhibited the most significant negative direct effect on crime rates. Neighborhoods with high population density have enhanced natural surveillance and a lower risk of victimization. The number of persons per household demonstrated a negative direct effect on crime rates. Neighborhoods with a high number of persons per household have more households with occupants present during the day, reducing the risk of victimization compared to neighborhoods with a larger number of small households, such as single-person households and working couple households, which tend to be unoccupied. The daytime population ratio also showed a negative direct effect on crime rates. Neighborhoods with a higher daytime population ratio have a reduced risk of victimization owing to increased natural surveillance by visitors and commuters, even during the daytime when homes are more likely to be unoccupied and residential burglaries are more probable to occur.

<Level 3>

Owner-occupied households, publicly owned households, and privately rented households each exerted a positive direct effect on crime rates. Based on the seven attributes of Safer Places, privately rented households, primarily consisting of low-rise rental apartments, are more susceptible to victimization owing to the lack of crime prevention measures (physical protection), a low sense of ownership, and insufficient maintenance (Building Research Institute 2006). The rate of public housing households indirectly affected the crime rate through population density and the number of persons per household, partially offsetting the direct effect. While slum-occupied public housing often serves as a crime hotspot in other countries, the risk of residential burglary in public housing is lower in Japan. This is because of the formation of communities owing to many homogeneous residents and the proper management and maintenance of housing by public entities. Additionally, the owner-occupied household rate indirectly affected the crime rate through the number of persons per household, with the total effect (including the direct effect) being approximately half that of the rented household. As aforementioned, these differences can be attributed to variations in physical protection, awareness of ownership, and maintenance.

<Level 2>

Neighborhood building coverage ratio exhibited a positive direct effect on crime rates. Neighborhoods with high building density, characterized by poor visibility and the potential for intrusion via adjacent buildings, are more susceptible to burglaries. The commercial–residential ratio did not have a direct effect on crime rates. Despite some concern that mixed use may increase the anonymity of residential neighborhoods and make them more prone to various types of crime, this study's analysis of residential burglaries did not yield such a result. There was a positive direct effect on the daytime population ratio, indirectly lowering the crime rate. However, the combined effect of all indirect effects almost offset this effect. Neighborhood floor area ratio had a negative direct and indirect effect on crime rates. The negative direct effect can be attributed to the presence of high-rise apartments in neighborhoods with high floor area ratios. These neighborhoods tend to have advanced crime prevention measures (physical protection) and fewer accessible openings (Hino 2005). The negative indirect effect was primarily attributed to higher daytime population ratios in neighborhoods with higher floor area ratios. The total effect, combining both direct and indirect effects, was the largest negative among the neighborhood characteristics variables.

<Level 1>

The road area ratio, time distance to central Tokyo, and nearest station distance did not have a direct effect on the crime rate, indicating that transportation accessibility was not a factor in target selection for residential burglars. However, when considering the total effect of both direct and indirect influences on the crime rate, the road area ratio showed a marginally negative effect.

6.4.2 Main Findings and Implications

Drawing from the crime opportunity theory, this study constructed a model to assess the risk of residential burglary by considering neighborhood characteristics that encompass both social and spatial factors. Furthermore, structural equation modeling was employed to reveal the causal relationships among the neighborhood characteristics indicators. The key findings of this analysis are summarized as follows:

- Neighborhoods characterized by high population density, many persons per household, and a high daytime population benefited from increased natural surveillance and face a lower risk of victimization. In conjunction with measures to revitalize shopping streets, housing policies that promote urban living and enhance the diversity of resident characteristics must be advocated for.
- Compared to neighborhoods with a high proportion of owner-occupied households, those with a high proportion of rented households faced a higher risk of victimization owing to disparities in physical protection, ownership awareness, and maintenance. Conversely, public housing exhibited a lower risk of victimization.
- A positive correlation was observed between the neighborhood building coverage ratio and the neighborhood floor area ratio. However, the former exhibited a positive effect on crime rates, while the latter had a negative effect. Neighborhoods with high building density had a higher risk of victimization, whereas those with a high proportion of high-rise apartments experienced a lower risk.
- The mixture of residential and commercial uses did not increase the risk of residential burglary. To guide community development to prevent crime, urban activities must be enhanced to attract more visitors, thereby increasing surveillance opportunities.
- Despite the presence of highly accessible transportation by railways and automobiles being associated with an increase in the number of crimes, it did not have a direct effect on the crime rate.

These findings are significant for residents and municipalities aiming to foster crime prevention and community development. As emphasized in "Safer Places," a comprehensive perspective that incorporates crime prevention as an integral component of the living environment is essential. Despite the recent decline in the number of residential burglaries, the current level of security is still inadequate for citizens to feel completely safe. As part of our commitment to building a safer and more secure society, our future endeavor involves developing a model that incorporates neighborhood characteristics not examined in this study. This will be accomplished through conducting comprehensive on-site surveys to identify and analyze crucial factors for effective crime prevention.

Acknowledgments This study was supported by a research grant from the Research Foundation for Safe Society in 2005. The authors would like to express their gratitude.

References

Building Research Institute (2006) Results of questionnaire survey on crime prevention in apartment buildings, Building Research Institute, Tsukuba, Japan

Hino K (2005) Evaluation of crime prevention measures for houses: an analysis of the situation in Japan and UK and a residents' attitude survey. Summaries of technical papers of Annual Meeting, Architectural Institute of Japan. pp 1495–1496

Hino K (2006) Relationship between number of crimes and land use. Reports City Plan Inst Japan 5(1):29–32

Hino K, Amemiya M (2005) "Safer Places," a guideline for crime prevention community planning in the U.K. Shintoshi 59(12):82–87

Hino K, Amemiya M (2006) "Structure" and utilization of safer places. Shintoshi 60(5):95–103

Hino K, Koide O (2006) Report on the survey on crime prevention community development in the U.K. Shintoshi 60(4):119–126

ODPM, Home Office (2003) Safer places—the planning system and crime prevention, ICE Publishing, London

Schneider RH, Kitchen T (2002) Planning for crime prevention. Routledge, New York

Seo B, Suzuki T, Hino K (2006) A study on the geographical distribution and environmental of larceny in the Tokyo ward area. J Soc Saf Sci 8:79–82

Shimada T, Harada Y (1999a) Relationship between crime and socioeconomic factors in large cities. Reports Natl Res Inst Police Sci, 39(2)

Shimada T, Harada Y (1999b) Relationship between urban spatial configuration and crime occurrence. Reports Natl Res Inst Police Sci, 40(1)

Chapter 7
A Spatial Analysis of the Effects of Neighborhood Socio-economic Status on Residential Burglaries in Tokyo: Focusing on the Spatial Heterogeneity and the Interactions with Built Environment

Masaya Uesugi and Kimihiro Hino

Abstract This study explores the relationship between neighborhood socio-economic status (SES) and the crime rate of residential burglaries in the wards of Tokyo, focusing on the spatial heterogeneity of this relationship and the moderating role of SES on the relationship between crime rate and the built environment. The results are as follows: (1) While global model shows that the proportion of high-income neighbors is negatively and significantly associated with the neighborhood residential burglary rate, the local model, calibrated by geographically weighed regression estimation, indicates that there are local variations in this relationship. (2) Significant interactions between neighborhood income composition and physical indicators such as road area ratio and distance to the nearest rail station suggest that the links between the residential burglary rate and the built environment are conditioned by neighborhood SES.

Keywords Neighborhood · Residential burglary · Income distribution · Geographically weighed regression · Interaction effect · Tokyo

---

The contents of this paper are based on the following paper originally published in a Japanese journal: Uesugi, M. and Hino, K. (2015) A spatial analysis of the effects of neighborhood socio-economic status on residential burglaries in Tokyo. *Journal of City Planning Institute of Japan* 50(3): 608–615 (in Japanese).

M. Uesugi (✉)
Fukuoka Institute of Technology, Fukuoka, Japan
e-mail: uesugi@fit.ac.jp

K. Hino
Department of Urban Engineering, The University of Tokyo, Tokyo, Japan
e-mail: hino@ua.t.u-tokyo.ac.jp

Y. Asami et al. (eds.), *Studies in Housing and Urban Analysis in Japan*,
New Frontiers in Regional Science: Asian Perspectives 75,
https://doi.org/10.1007/978-981-99-8027-7_7

7.1 Introduction

Spatial patterns of crime occurrences within cities are not necessarily determined only by built environment, such as road connectivity, land use, and building density. In terms of mid- and long-term crime prevention, it is equally important to pay attention to the relationships between crime occurrences and social environment, such as residential composition and industrial structure. Further, recognizing that crime prevention constitutes a constituent of the residential environment, a holistic approach encompassing both spatial and social dimensions becomes imperative (Hino and Kojima 2007).

Neighborhood socio-economic characteristics that reflect the social status and economic bracket of neighborhood residents have long been indicated in research outside of Japan as major explanatory factors of crime occurrence. Logically, a link between neighborhood socio-economic levels and crime is supported from the following perspectives, regardless of whether this relationship is positively or negatively correlated. One perspective stems from the social disorganization theory (Shaw and McKay 1942), which posits that socio-economic difficulties lead to a decrease in social control. This theory is applied in Western countries to explain why neighborhoods with a concentration of disadvantaged residents, including the poor and/or racial minorities, tend to have higher crime rates. Other viewpoints from the routine activity theory (Cohen and Felson 1979) and rational choice theory (Cornish and Clarke 1986) support an inverse relationship between neighborhood socio-economic levels and crime. These theories suggest that affluent neighborhoods are perceived as alluring targets, making crime occurrence within them more likely. Considering these conflicting theories, it is essential to examine which theory supports what kind of case and circumstances.

In Japan, Takagi et al. (2010) found that social capital in a neighborhood has inhibiting effects on residential burglary. With regard to one such element of social capital, Murayama et al. (2013) have demonstrated that social trust is affected by neighborhood social status, as defined by the occupational and educational backgrounds of residents. Of course, it is anticipated that local contexts in Japan differ from those in Western countries, whose circumstances have been emphasized in the development of the above-stated crime theories. Nevertheless, paying close attention to these theories is significant, especially when considering the expansion in economic disparity as well as trends in residential segregation in Tokyo since the 1990s (Uesugi and Asami 2011; Fujita and Hill 2012).

With the above background, this study examines the effects of neighborhood socio-economic status (SES) on crime occurrences in the wards of Tokyo. Residential burglaries in Tokyo have been analyzed in multiple previous studies (Kutsuzawa et al. 2007; Amemiya 2013). The present study contributes to the literature by investigating two characteristics hitherto not considered in Japanese neighborhood-level crime research. The first characteristic involves spatial variability (spatial heterogeneity) in the relationship between neighborhood SES and crime occurrences. Malczewski and Poetz (2005) demonstrated that there is a relationship between crime risk and average house prices, which are an indication of the neighborhood SES, and neighborhoods in which the relationship becomes positive or negative are geographically mixed in a Canadian city. The same point has been indicated for the United States (Cahill and Mulligan 2007; Arnio and Baumer 2012; Zhang and Song 2014). However, residential segregation is not as distinct in Japanese cities as it is in the cities of Western countries. Therefore, to verify this characteristic for Japanese cities, a variety of residential neighborhood types, such as those found in Tokyo, needs to be studied and various built environment need to be controlled for as well.

The second characteristic examined in the present study is interaction effect, such that the relationship between built environment and crime occurrence changes according to neighborhood SES. Predictors previously used to explain crime rates have considered the built environment as independent from the social environment. However, in arguing for crime prevention through environmental design (CPTED), Taylor (2002) stressed that the relationship between environmental design and crime may rely upon neighborhood context. Even studies outside of Japan have conducted little verification work regarding this relationship. An exception is Ward et al. (2014) who found that in neighborhoods with a high concentration of the disadvantaged, the higher the road connectivity, the lower the crime rate, with the converse occurring in districts without such high road connectivity.

Previous studies on Tokyo have predominantly focused on the latter half of the 2000s decade and earlier. However, a significant reduction in the number of confirmed crimes has been observed subsequently, coinciding with geographical leveling (Amemiya and Shimada 2013). Given the contexts within large cities in Japan, understanding the spatial heterogeneity of the association between neighborhood SES and the crime rate of residential burglaries, as well as the moderating role of SES on the relationship between crime rate and the built environment, would contribute to enhancing the explanatory power concerning crime rates.

7.2 Methods

7.2.1 Data

We use data on crime numbers and neighborhood characteristics for *cho* districts for Tokyo. *Cho* districts are address/block divisions in urban areas of Japan. The numbers of crime were obtained from the geographical information database of Amemiya and Iwakura (2012), which provides the number of penal code offenses for *cho* districts by time series. These data are sourced from the confirmed penal code offenses records of the Tokyo Metropolitan Police Department (MPD), which classify these offenses by crime type and method. This study uses "residential burglaries," which comprise the following three MPD classifications: breaking and entering (1) in the absence of residents, (2) in the presence of sleeping residents, and (3) in the presence of residents who are home and awake, but are unaware of a crime occurring elsewhere in their residence. Regarding the statistical years, the three-year period spanning 2009, 2010, and 2011 was used.

For neighborhood characteristics, we use data on land use and building structures as well as population. Population characteristics were obtained from Population Census 2010 (Ministry of Internal Affairs and Communications) and Tokyo Metropolitan Government. Household income distribution, a neighborhood SES measure, was estimated using the method proposed by Uesugi and Asami (2011), with data from both the Comprehensive Survey of Housing Life (Ministry of Land, Infrastructure, Transport and Tourism) and Population Census. "High income" is defined as the upper 20% of household income, and "low income" is defined as the lower 20% of the household income within Tokyo. For land use and building characteristics, GIS data of the Basic City Planning Survey (Tokyo Metropolitan Government) and Zenrin Residential Maps (Zenrin Co., Ltd.) were used. Additionally, rail station location was obtained from the Digital National Land Information (Ministry of Land, Infrastructure, Transport and Tourism).

7.2.2 Variables

Dependent variable: Using the totals for residential burglaries during the three-year period, crime rate was considered to be the number of cases per 10,000 households; to manage zero cases, logarithmic transformation was performed after adding one (Amemiya 2013).

Independent variables: Referencing prior research (Hino and Kojima 2007; Amemiya 2013), the variables were shown in Table 7.1. First, social environment variables include population, housing, and neighborhood SES. For population variables, population density, daytime population ratio, average household size (the number of persons per household), and elderly people rate were used. Considering that the presence of visitors, work and school commuters, and work-at-home persons increase natural surveillance, crime prevention effects are anticipated. Homeownership rate, detached house rate, low-rise apartment rate (the proportion living in apartments 1 or 2 floor), and residential stability (the proportion living in the same residence 5 years previously) were set for housing variables. Homeownership rate and residential stability show the flow rate of residents. Higher flow rates are expected to indicate the weakening of informal social control, as demonstrated by lower awareness and concern of residents for neighborhood problems and unconventional behavior. Further, as it is easier to break into residence types that are detached dwellings and low-rise apartment buildings, these are expected to be easily linked to an increase in crime. Regarding neighborhood SES variables, economic strata are defined by the rates of high-income households and low-income households per resident, respectively.

Built environmental variables are those that represent accessibility, land use, and spatial density. Accessibility is represented by distance to nearest station (distance

Table 7.1 Descriptive statistics

Variables	Mean	SD
Residential burglary		
Count (2009–2011)	3.91	4.19
Social environment		
Population		
Population density (/km^2)	23,613	25,513
Daytime population ratio	2.86	10.57
Average household size	2.00	0.32
Elderly people rate (%)	19.96	5.62
Housing		
Homeownership rate (%)	45.03	14.51
Detached house rate (%)	28.05	16.56
Low-rise apartment rate (%)	13.58	10.34
Residential stability (%)	71.42	9.82
Neighborhood SES		
High-income household rate (%)	18.91	7.30
Low-income household rate (%)	18.83	7.26
Build environment		
Accessibility		
Distance to nearest station (m)	558.41	367.74
Distance to city center (m)	9347.8	4190.2
Land use		
Commercial-residential area ratio	1.65	7.98
Road area ratio	0.02	0.04
Spatial density		
District building coverage ratio	33.67	8.98
District floor area ratio	131.66	78.91

$N = 2999$

from residence to the nearest rail station) and city center (distance from residence to Tokyo station). Land use includes commercial-residential area ratio and road area ratio (the ratio of the road space to land area). Spatial density, such as district building coverage ratio (the ratio of the building footprint area to the land area) and district floor area ratio (the ratio of total of all the floor space to land area), shows how close buildings are built to each other, indicating poor visibility. All of these are considered elements that can attract or trigger crime. However, some research has indicated that accessibility and mixed land use may not necessarily be related to increased crime risk (Amemiya 2013).

Table 7.1 shows the basic statistics of the variables described above. It is noted that when creating the model, we used log transformations for some variables in consideration of their distribution characteristics.

7.2.3 Analysis Method

The neighborhoods studied comprised 2999 *cho* districts with 50 or more households (Population Census 2010) in the wards of Tokyo. The statistical analysis was divided into three major portions.

First, the dependent variable and independent variables mentioned in Sect. 7.2.2 were used to perform multiple regression analysis with an ordinary least squares (OLS) regression model. At this time, independent variables with strong correlations were first excluded to prevent multicollinearity. Then, to extract valid variables as predictive factors for crime occurrence, variables were selected using cross-validation. Here, we examine effects of neighborhood SES on the crime rate by variable significance (the significance level was set at 5%, unless otherwise noted) and model goodness-of-fit.

Next, regarding the relationships between neighborhood SES and crime, we employ geographically weighted regression (GWR) to investigate the spatial variations (spatial heterogeneity). Although the OLS model provides the uniform regression coefficients, t value, and coefficient of determination (R^2), these values were derived for each neighborhood location (in this study, this corresponds to *cho* district), by applying the GWR model of Fotheringham et al. (2002), as shown in Eq. (7.1).

$$y_i = \beta_{0i} + \sum_k \beta_{ki} x_{ik} + \varepsilon_i \tag{7.1}$$

where y_i is a dependent variable at location i, x_{ik} is an independent variable at number k at location i, and ε_i is an error term at location i. β_{ki} is the regression coefficient β_k of location i, and its estimated value $\hat{\boldsymbol{\beta}}$ is determined using Eq. (7.2), as follows:

$$\hat{\boldsymbol{\beta}}(i) = \left(\boldsymbol{X}^T \boldsymbol{W}(i)\boldsymbol{X}\right)^{-1} \boldsymbol{X}^T \boldsymbol{W}(i)\boldsymbol{Y} \tag{7.2}$$

where $\boldsymbol{W}(i)$ is the $n \times n$ spatial weights matrix whose off-diagonal elements are zero and whose diagonal elements denote w_{ij} for i (n is the number of observed values within regression point i). It should be noted that weight w_{ij} of j vis-à-vis regression location i is shown by Eq. (7.3), which uses a Gaussian distance decay function.

$$w_{ij} = \exp\left(-\frac{d_{ij}^2}{b^2}\right) \tag{7.3}$$

where d_{ij} is the direct distance from i to j, and b is the kernel bandwidth, which shows the extent of distance decay. The GWR result is dependent on this bandwidth.

Here, employing the cross-validation (CV) technique, Eq. (7.4) is used to find the optimal bandwidth of b for minimizing the CV score shown in the formula.

$$CV = \sum_i \left[y_i - \hat{y}_{\neq i}(b) \right]^2 \tag{7.4}$$

where $\hat{y}_{\neq i}(b)$ shows the predicted value of y (excluding location i) when using bandwidth b. The result estimated via GWR and OLS model can be examined using indices such as the adjusted degree of freedom R^2, Akaike's Information Criteria (AIC), and Moran's I of the residual. At the same time, we can also use spatial pattern mapping to examine the t value of independent valuables and coefficient of the neighborhood SES variable for each neighborhood.

Finally, we investigate the moderating effect of SES on the relationship between crime rate and the built environment. For the built environment variables not selected as significant variables in the OLS analysis, the interaction terms of the variables with the neighborhood SES variables were introduced. Here, to avoid multicollinearity with interaction terms, mean centering was employed (Cohen et al. 2003). For statistically significant interaction terms, we conducted simple slope analysis to confirm that neighborhood SES significantly changes the effects of the built environment on crime. It is noted that statistical environment R (ver. 3.0.3) was used for all the statistical analyses in this study.

7.3 Relationship Between Crime Rate and Neighborhood SES

In this section, the relationship between crime rate and neighborhood SES is examined using OLS model. First, independent valuables with high mutual correlation were excluded. Specifically, population density, which correlated with daytime population ratio ($r = 0.58$, $p < 0.001$) and district floor area ratio, which correlated with district building coverage ratio ($r = 0.57$, $p < 0.001$), were excluded. Moreover, there were strong correlations between the rates of low-income and high-income household ($r = 0.81$, $p < 0.001$), which were used to represent neighborhood SES, therefore their simultaneous inclusion was avoided.

The estimated results are shown in Table 7.2. Model 0 does not include the neighborhood SES variable. Here, daytime population ratio and average household size were adopted as variables with positive significance with regard to crime rates. Conversely, detached house rate, low-rise apartment rate, and distance to city center were adopted as variables with negative significance.

Model 1 includes high-income household rate as the neighborhood SES variable. Here, daytime population ratio was excluded, and elderly people rate and commercial-residential area ratio were substituted as variables with negative significance. Then, even with consideration of these variables, high-income household

Table 7.2 OLS results for residential burglary

	Model 0			Model 1		
	Coefficient	*t* value		Coefficient	*t* value	
Constant	−1.576	−3.74	***	0.333	0.68	
Daytime population ratio (logged)	−0.270	−9.21	***			
Average household size	−0.634	−7.37	***	−0.594	−6.71	***
Elderly people rate				−0.030	−6.55	***
Detached house rate	0.019	12.67	***	0.018	11.06	***
Low-rise apartment rate	0.025	9.98	***	0.021	7.89	***
Distance to city center (logged)	0.462	8.56	***	0.358	6.39	***
Commercial-residential area ratio (logged)				−0.197	−8.37	***
High-income household rate				−0.030	−8.37	***
Adjusted R^2	0.334			0.345		
AIC	9451			9390		

***: $p < 0.1\%$, **: $p < 1\%$, *: $p < 5\%$

rate was adopted as a variable with negative significance. Here, the adjusted R^2 became 0.345, and the AIC improved from 9451 to 9390.

We can see below that variables with negative effects in relation to crime rates are consistent with prior research; for example, daytime population ratio, average household size, and elderly people rate. As can be imagined, these ratios are of people who are natural "neighborhood watchers," like visitors and residents, and surveillance effects function effectively in these cases.

Conversely, variables that have positive effects with regard to crime rates are detached house rate and low-rise apartment rate. The latter result is especially consistent with prior research (Hino and Kojima 2007; Amemiya 2013). This strongly suggests that the presence of low-rise apartment households is a major predictor that increases crime risk, as such buildings are easier to burglarize.

Even when these variables are controlled for, the fact that the high-income household rate variable was negatively significant shows that higher neighborhood SES reduces crime risk. In other words, considering correlation, the lower the neighborhood SES, the higher the crime risk. Hence, the present study supports the social disorganization theory. In the context of Japan, as highlighted by Kutsuzawa et al. (2007), "persons vulnerable to crime" may exist owing to economic constraints. Moreover, as surmised by Amemiya (2013), it may be possible that neighborhoods with high asset values, which easily became targets of crime in the early 2000s decade, thereafter used their economic leeway to promote anti-crime measures, which may have led to improvements in crime rates.

7.4 Spatial Heterogeneity in the Relationship Between Crime Rate and Neighborhood SES

This section employs GWR to investigate the spatial heterogeneity in the neighborhood SES and crime relationship shown in the previous section. First, Table 7.3 describes the results estimated using GWR. Regarding the goodness-of-fit of the model, comparison with the OLS showed improvements in the adjusted R^2 from 0.417 to 0.345 and in AIC from 9390 to 9120. Moran's I, which shows the spatial autocorrelation of residuals, also reduced from 0.127 to 0.052. To test whether the coefficients are uniform for all neighborhoods, the $F(3)$ test was used (Leung et al. 2000). This test showed statistical significance at the 5% level for the spatial distribution of all independent variables.

Before we explore the effects of the neighborhood SES variable, we first highlight the characteristic points of the results obtained with the present model. Looking at the estimated coefficients in Table 7.3, we see a range in the plus and minus signs for such coefficients as average household size, distance to city center, and commercial-residential area ratio. Thus, we can glimpse spatial heterogeneity in relationships with crime rates, including for high-income household rate. Figure 7.1 shows the spatial distribution of local R^2 values. We can see major fluctuations by neighborhood: for example, in some neighborhoods (e.g., the southern part of Edogawa Ward), the value exceeds 0.5, and in some agglomeration neighborhoods (e.g., Shinjuku, Bunkyo, and Toshima Wards), it exceeds 0.4. On the contrary, the values of surrounding wards are less than 0.2. For such neighborhoods, predictors not considered in the present model are expected to play an important role. Additionally, for low-rise apartment rate variable, most neighborhoods had a t value with positive significance (Fig. 7.2). There are major differences in coefficient size across the east-west axis; this coefficient becomes larger in so-called *Shitamachi* neighborhoods (i.e., older districts in lower lands).

Table 7.3 GWR result for residential burglary

	Coefficient				
	Min	25%	Median	75%	Max
Constant	−4.253	−2.909	−0.506	2.917	20.140
Average household size	−1.713	−0.964	−0.801	−0.623	0.402
Elderly people rate	−0.069	−0.036	−0.030	−0.020	0.020
Detached house rate	0.005	0.017	0.019	0.023	0.031
Low-rise apartment rate	0.004	0.015	0.020	0.031	0.046
Distance to city center (logged)	−1.503	0.041	0.449	0.743	0.991
Commercial-residential area ratio (logged)	−0.297	−0.225	−0.174	−0.127	0.071
High-income household rate	−0.047	−0.026	−0.021	−0.015	0.018
Adjusted R^2	0.417				
AIC	9120				
Moran's I (residuals)	0.052				

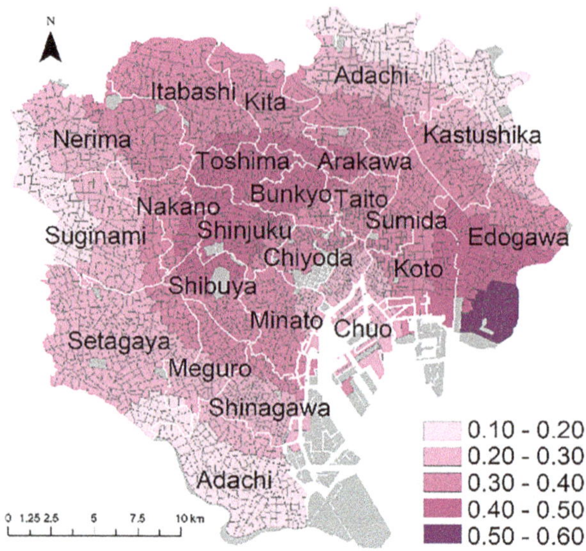

Fig. 7.1 Spatial variation of local R^2

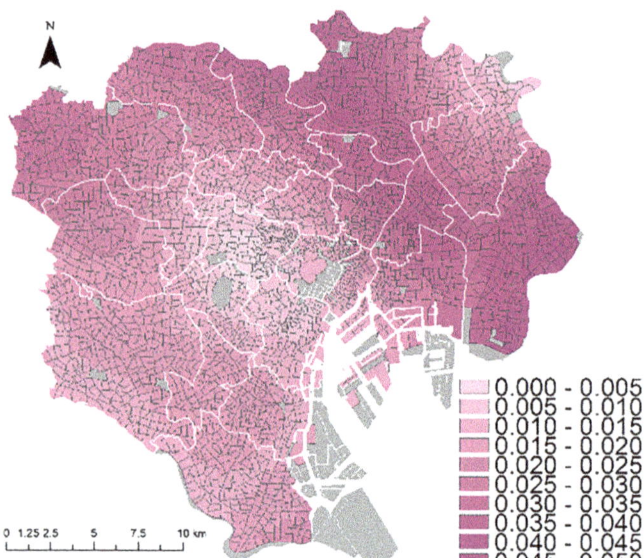

Fig. 7.2 Local coefficients of low-rise apartment rate

Figure 7.3 shows the distribution of the local coefficients of high-income household rate, the particular focus of the present study. In the OLS estimates, this coefficient was negatively significant. However, as shown in Fig. 7.4, neighborhoods with 5% significance ($t > 1.96$) account for only around half of the total (51.3%,

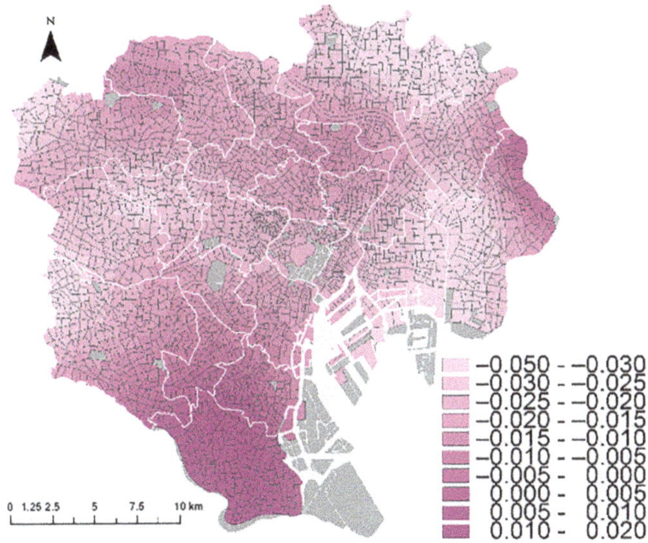

Fig. 7.3 Local coefficients of high-income household rate

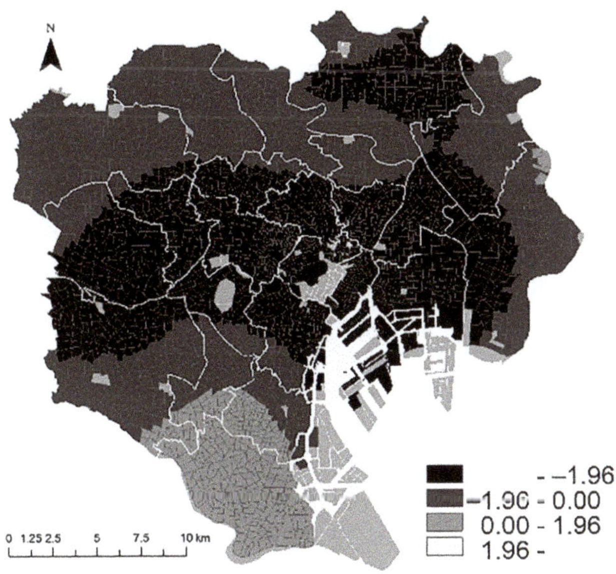

Fig. 7.4 *t* values for high-income household rate

$n = 1537$). Conversely, in some neighborhoods (10.2%, $n = 305$), the coefficient is positive; these neighborhoods are aggregated in the areas surrounding the Ota Ward. This means that, in contrast with other neighborhoods, the higher the high-income household rate is, the higher the crime rate becomes in these neighborhoods.

These geographical differences could be attributed to several potential reasons. For example, although social control could increase with an increase in high-income residents, in affluent neighborhoods where such social control factors were already functioning, there may also be an increase in the allure of concentration of high-income population for persons who are planning to commit crime. Regarding neighborhoods for which the coefficient was negative (e.g., Koto, Edogawa, Adachi, Suginami, and Nakano Wards), there is a tendency of higher income levels corresponding with lower risks of crime. These neighborhoods are among those for which the percentage of high-income strata is relatively low, except for a portion of these neighborhoods, such as the bayside area of Koto Ward, where urban redevelopment has progressed in recent years. The finding that in neighborhoods of low SES, there is a negative correlation between residential burglaries and neighborhood SES is consistent with the findings obtained for foreign cities (Malczewski and Poetz 2005; Zhang and Song 2014).

Conversely, the positive relationship between crime and neighborhood SES, as seen in Ota Ward, shows that in these regions where the percentage of high-income household is high, there is a tendency for increased crime risk. Ota Ward has a mix of relatively high and low economic strata neighborhoods in the south and north. Because neighborhoods where members of the high-income strata are concentrated are more easily targeted for crime, the nearby relatively low-income neighborhoods reduce the incentive to commit crime. However, since statistically significant results were not obtained for these coefficients, more detailed analysis is required.

Through these findings, this study has described geographical differences in crime occurrence factors within Tokyo. However, as confirmed in the beginning, the model goodness-of-fit was not necessarily high for all studied neighborhoods. Markowitz et al. (2001) claimed that trends of crime rates increasing in economically poor neighborhoods are weakened as a result of neighborhood cohesion. This likely suggests the importance of considering not only the demographic attributes of residents, but also dynamic processes (Brisson and Roll 2012). GWR is a superior means of exploring even more effective predictors (Gao and Asami 2002), and in possible future developments, these kinds of neighborhood characteristic variables can also be included within GWR perspectives.

7.5 Neighborhood SES Effects on the Relationships Between Crime Rates and Built Environment

This section analyzes the interaction terms of the neighborhood SES variable concerning the built environmental variables not used in the OLS estimates in Sect. 7.3. Here, we assume that the effects of built environment are dependent on neighborhood SES. The specific built environment variables are distance to nearest station, road area ratio, and district building coverage ratio. Of these, distance to nearest station and road area ratio can be considered variables that are statistically

Table 7.4 OLS results with interaction term (road area ratio)

	Model 1-0			Model 1-1		
	Coefficient	*t* value		Coefficient	*t* value	
Constant	0.346	0.71		0.344	0.71	
Average household size	−0.594	−6.71	***	−0.598	−6.76	***
Elderly people rate	−0.030	−6.56	***	−0.029	−6.34	***
Detached house rate	0.018	11.08	***	0.018	11.10	***
Low-rise apartment rate	0.021	7.91	***	0.021	7.97	***
Distance to city center (logged)	0.355	6.33	***	0.356	6.35	***
Commercial-residential area ratio (logged)	−0.196	−8.33	***	−0.197	−8.40	***
High-income household rate	−0.031	−8.40	***	−0.031	−8.47	***
Road area ratio	0.389	0.67		−0.589	−0.88	
High-income household rate × road area ratio				0.163	2.98	**
Adjusted R^2	0.345			0.346		
AIC	9405			9398		

***: $p < 0.1\%$, **: $p < 1\%$, *: $p < 5\%$

insignificant when taken independently, but become significant when we added their interaction terms with neighborhood SES.

Tables 7.4 and 7.5 show each of their estimated results. Model 1-0 of Table 7.4 does not include an interactive term with neighborhood SES for road area ratio. When the other variables adopted in Model 1 of Sect. 7.3 are controlled for, road area ratio is not a statistically significant variable, and there is almost no improvement in model goodness-of-fit compared with Model 1 of Table 7.2. On the contrary, in Model 1-1, which includes the interaction term, the interaction term becomes a positive significant value at the 1% level. Further, goodness-of-fit is also improved in this model with measures of adjusted R^2 and AIC. Models 2-0 and 2-1 of Table 7.5 show the results for distance to nearest station, and similar results are obtained in that the interaction term becomes a negative significant variable at the 1% level.

Next, to interpret these results, we evaluated the conditional effect (simple slope) of some built environment variables on crime rate at various levels of high-income household rate. Here, "simple slope" is the regression coefficient, with regression of crime rates within built environmental variables under a specific level of neighborhood SES. We can perceive from Figs. 7.5 and 7.6 that the effects of built environment change according to neighborhood SES levels. For example, the effects of road area ratio on crime rate are negative when the high-income household rate in the area is smaller (Fig. 7.5). Conversely, with a higher high-income household rate, this effect becomes positive. Buffer portions show a 95% confidence interval for these effects. When the neighborhood high-income household rate is 10% or less, we observe a significant negative relationship between crime rate and road area ratio, while, conversely, when it exceeds 30%, a positive relationship emerges. Similarly,

Table 7.5 OLS results with interaction term (distance to nearest station)

	Model 2-0			Model 2-1		
	Coefficient	*t* value		Coefficient	*t* value	
Constant	0.351	0.70		0.399	0.79	
Average household size	−0.590	−6.28	***	−0.586	−6.24	***
Elderly people rate	−0.030	−6.54	***	−0.031	−6.74	***
Detached house rate	0.018	10.99	***	0.018	10.78	***
Low-rise apartment rate	0.021	7.89	***	0.022	8.16	***
Distance to city center (logged)	0.359	6.36	***	0.360	6.39	***
Commercial-residential area ratio (logged)	−0.197	−8.31	***	−0.204	−8.57	***
High-income household rate	−0.030	−8.37	***	−0.031	−8.46	***
Distance to nearest station (logged)	−0.006	−0.14		−0.017	−0.43	
High-income household rate × distance to nearest station (logged)				−0.014	−3.08	**
Adjusted R^2	0.344			0.346		
AIC	9406			9398		

***: $p < 0.1\%$, **: $p < 1\%$, *: $p < 5\%$

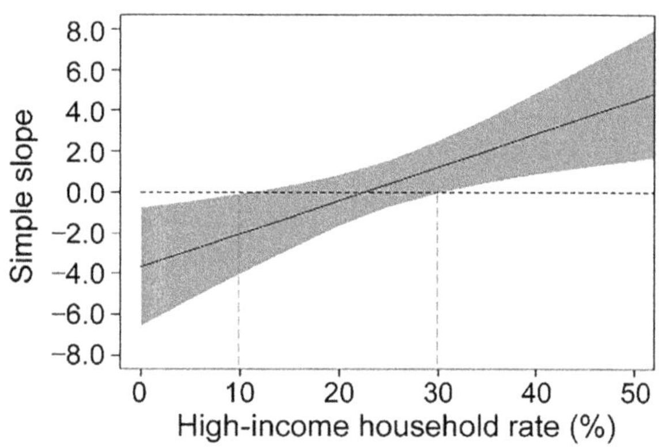

Fig. 7.5 Conditional effect of road area ratio at various levels of high-income household rate. The gray polygon is the 95% interval around the predictions

Fig. 7.6 shows that when the neighborhood high-income household rate is 10% or less, there is a significant positive relationship between crime rate and distance to the nearest rail station, but when it exceeds 26%, this becomes a negative relationship.

These findings indicate that in neighborhoods of high SES, larger road area ratios and smaller distances to the nearest rail station correspond with higher crime rates. Additionally, more convenient access to an arterial roadway and/or a nearby rail station is associated with a higher tendency for crime occurrence. Conversely, in neighborhoods where SES level is low, lower accessibility to roads and rail stations results in higher crime risk.

There are several possible reasons why the effects of accessibility vary according to neighborhood SES levels. Generally, social controls play a role in restraining

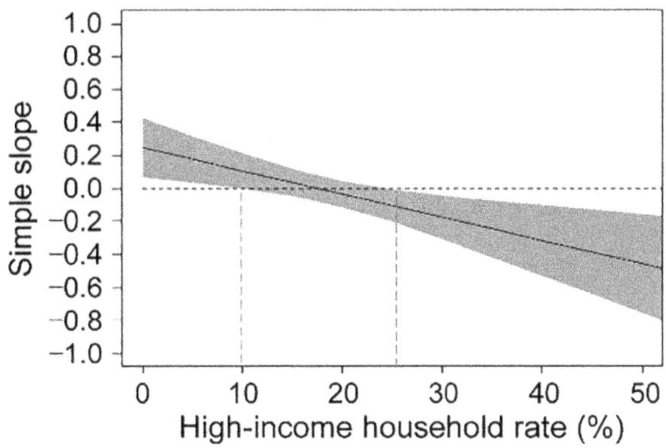

Fig. 7.6 Conditional effect of distance to nearest station at various levels of high-income household rate. The gray polygon is the 95% interval around the predictions

crime in high-income neighborhoods. However, good accessibility also means that unspecified numbers of people can enter and leave the neighborhood, which grants an anonymity that may be an inducement to crime. This may be hypothesized as an increase in crime rate due to the aforementioned anonymity nullifying the original crime restraint effects. In contrast, it is supposed that social controls cannot function in a low-income neighborhood. In such neighborhoods, the natural surveillance effects of residents and passers-by may play a significant role. However, surveillance effects lose their strength in neighborhoods of poor accessibility, where fewer people pass by and visibility is poor. Therefore, more than neighborhoods with good accessibility, these conditions may more readily invite an increase in crime.

7.6 Conclusion

This study focused on neighborhood SES and examined the relationship between neighborhood SES and residential burglary rates. With this approach, we highlighted spatial heterogeneity of the relationship and the interaction effects of neighborhood SES, which have previously not been considered in a Japanese context. The results revealed that neighborhood SES had significant negative effects on crime rates, even when built environment factors were considered. However, the results also revealed that in some neighborhoods, neighborhood SES may have positive effects. Further, in neighborhoods with high neighborhood SES, crime rates were found to be higher in districts with good accessibility in terms of road area ratios and distance to nearest rail station. It was also observed that the converse occurs in neighborhoods with lower district SES. In the past, the fact that there were contextual dependencies for built environment was largely disregarded. These findings suggest that neighborhood SES is useful clue for crime prevention measures in Japan as well, even though the spatial variation of neighborhood SES is smaller in Japan than in Western counties.

However, some aspects of neighborhood SES effects remain to be clarified. As indicated in Sect. 7.4, it will be necessary to pay close attention to how social capital can strengthen or weaken relationships between crime rates and neighborhood SES. Additionally, although the purpose of this study was to explain the current crime tendencies following the progress made in anti-crime measures in the first decade of the 2000s, from the perspectives of mid- and long-term measures to create cities with improved crime prevention, it becomes essential to delve deeper into the associations between trends in residential segregation and changes in the roles of neighborhood SES.

References

Amemiya M (2013) Relationship between time series variations of residential burglary rates and the social and physical environments of local districts in Tokyo's 23 wards: an analysis using latent growth curve modeling. J City Plan Inst Jpn 48(3):351–356 (in Japanese)
Amemiya M, Iwakura N (2012) Construction of a time series micro-scale geographical database of crimes and its potential application for temporal-spatial analysis of crime distribution in Tokyo. Papers and proceedings of the Geographic Information Systems Association 21, CD-ROM (in Japanese)
Amemiya M, Shimada T (2013) Identifying changing patterns in the geographical distribution of residential burglaries in Tokyo: an analysis of crimes that have occurred in the past eleven years. J City Plan Inst Jpn 48(1):60–66 (in Japanese)
Arnio A, Baumer E (2012) Demography, foreclosure, and crime: assessing spatial heterogeneity in contemporary models of neighborhood crime rates. Demogr Res 26:449–488
Brisson D, Roll S (2012) The effect of neighborhood on crime and safety: a review of the evidence. J Evid Based Soc Work 9:333–350
Cahill M, Mulligan G (2007) Using geographically weighted regression to explore local crime patterns. Soc Sci Comput Rev 25(2):174–193
Cohen LE, Felson M (1979) Social change and crime rate trends: a routine activity approach. Am Sociol Rev 44(4):588–608
Cohen J, Cohen P, West S, Aiken L (2003) Applied multiple regression/correlation analysis for the behavioral sciences. Erlbaum Associates, Hillsdale
Cornish D, Clarke R (1986) Reasoning criminal: rational choice perspectives on offending. Springer-Verlag, New York
Fotheringham AS, Brunsdon C, Charlton M (2002) Geographically weighted regression: the analysis of spatially varying relationships. Wiley, Chichester
Fujita K, Hill RC (2012) Residential income inequality in Tokyo and why it does not translate into class-based segregation. In: Maloutas T, Fujita K (eds) Residential segregation in comparative perspective: making sense of contextual diversity. Ashgate, Farnham, pp 54–86
Gao X, Asami Y (2002) Search for spatial influence on price determination for detached houses the quarterly journal of housing and land. Economics 44:10–21 (in Japanese)
Hino K, Kojima T (2007) Relationship between crime rate of dwelling burglary and local context: on neighbourhoods in 29 wards and cities in Tokyo. J Archit Plan 616:107–112 (in Japanese)
Kutsuzawa R, Mizutani N, Yamaga H, Ohtake F (2007) Regional crime rate, land price and housing rent. Q J Hous Land Econ 66:12–21 (in Japanese)
Leung Y, Mei CL, Zhang WX (2000) Statistical tests for spatial nonstationarity based on the geographically weighted regression model. Environ Plan A 32(1):9–32
Malczewski J, Poetz A (2005) Residential burglaries and neighborhood socioeconomic context in London, Ontario: global and local regression analysis. Prof Geogr 57(4):516–529

Markowitz F, Bellair P, Liska A, Liu J (2001) Extending social disorganization theory: modeling the relationships between cohesion, disorder, and fear. Criminology 39:293–320

Murayama H, Arami R, Wakui T, Sugawara I, Yoshie S (2013) Cross-level interaction between individual and neighbourhood socioeconomic status in relation to social trust in a Japanese community. Urban Stud 51(13):2770–2786

Shaw CR, McKay HD (1942) Juvenile delinquency and urban areas. University of Chicago Press, Chicago

Takagi D, Tsuji R, Ikeda K (2010) Crime control by local communities: focusing on social capital and cooperative behaviors in neighborhoods. Japan J Social Psychol 26(1):36–45 (in Japanese)

Taylor R (2002) Crime prevention through environmental design (CPTED): yes, no, maybe, unknowable, and all of the above. In: Bechtel RB, Churchman A (eds) Handbook on environmental psychology. Wiley, New York, pp 413–426

Uesugi M, Asami Y (2011) Neighborhood segregation by household income since the late 1990s: case study of Ota ward, Tokyo. Geogr Rev Japan Ser A 84(4):345–357 (in Japanese)

Ward JT, Nobles MR, Youstin TJ, Cook CL (2014) Placing the neighborhood accessibility—burglary link in social-structural context. Crime Delinquency 60(5):739–763

Zhang H, Song W (2014) Addressing issues of spatial spillover effects and non-stationarity in analysis of residential burglary crime. GeoJournal 79(1):89–102

Chapter 8
Factors Affecting Installation of Residential Photovoltaics in Housing Estates in Kakegawa City, Shizuoka: Focusing on Housing and Surrounding Environment Characteristics

Yuki Okazawa, Kimihiro Hino, and Yasushi Asami

Abstract This study identifies the factors affecting the installation of residential solar photovoltaics (PVs) in housing estates in Kakegawa City, Shizuoka Prefecture. Logistic regression analysis of 1327 houses in housing estates revealed that those that were lower, fully rebuilts, newly builts, or larger as well as newly built were more likely to have PV installed. Furthermore, analysis of 1971 houses revealed that—compared to houses in adjacent blocks—those that were without building agreements, lower, fully rebuilt, newly built, or smaller as well as newly built were more likely to have PV installed. Accordingly, full rebuilds, new builds, building heights (measured in meters), and property areas (measured in square meters) were suggested as factors affecting PV installation.

Keywords Residential solar PV · Residential PV power generation · Housing estate · Building agreement · Logistic regression analysis

The contents of this paper are based on the following paper originally published in a Japanese journal: Okazawa, Y., Hino, K. and Asami, Y. (2022) Factors affecting residential solar photovoltaics installation in housing estates in Kakegawa City, Shizuoka Prefecture. *Journal of City Planning Institute of Japan* 57(3): 1049–1055 (in Japanese).

Y. Okazawa (✉) · K. Hino · Y. Asami
Department of Urban Engineering, The University of Tokyo, Tokyo, Japan
e-mail: okazawa-yuki986@g.ecc.u-tokyo.ac.jp; hino@ua.t.u-tokyo.ac.jp; asami@csis.u-tokyo.ac.jp

Y. Asami et al. (eds.), *Studies in Housing and Urban Analysis in Japan*, New Frontiers in Regional Science: Asian Perspectives 75,
https://doi.org/10.1007/978-981-99-8027-7_8

8.1 Introduction

Residential photovoltaic (PV) power generation has been progressively introduced in Japan since the commencement of an excess electricity purchasing scheme in November 2009 and the transition to a feed-in tariff scheme in July 2012. Moreover, government policy expects further installations of such power generation in the future. The Climate Change Adaptation Plan (Cabinet decision, October 2021) lists "introduction and use of renewable energy, such as self-sufficient PV power generation as a fundamental role of the people." (Ministry of the Environment (MOE) 2021). The Sixth Strategic Energy Plan (Cabinet decision, October 2021) explicitly states "the aim of installing PV power generation facilities on 60% of newly built houses in 2030." (Agency for Natural Resources and Energy (ANRE) n.d.).

According to the following previous research, factors affecting the installation of residential PV power generation can be broadly divided into the characteristics of residents, houses, and the surrounding environment.

The characteristics of residents have been analyzed in aggregation units based on census data. Kosugi et al. (2019) documented that factors affecting installation in census block units in Kyoto, Kyoto Prefecture were lower population density, greater household residents, and shorter residence periods (Kosugi et al. 2019). Schaffer and Brun (2015) argued that the installation of residential PV power generation in Germany is positively related to the home ownership rate and per-capita gross regional product (Schaffer and Brun 2015).

The characteristics of houses focus on the building's height, area, volume, and age; the building method; and changes to the building. Based on residential PV installation permit data from Baltimore, Maryland, USA, Irwin (2021) established that the houses wherein it was installed were relatively larger, newer, and not attached houses (residences that have at least one wall shared with the neighboring residence) (Irwin 2021). Kawaguchi et al. (2020) investigated Nagoya, Aichi and found that compared to buildings wherein PV was not installed, those with PV installed had greater area, height, volume, and solar radiation (Kawaguchi et al. 2020). According to Bollinger and Gillingham (2012), the performance of residential repairs is one factor affecting the installation of residential PV power generation in California, USA (Bryan Bollinger and Kenneth Gillingham 2012).

The third factor comprises characteristics of the surrounding environment. Hachem-Vermette and Singh (2019) created patterns for the arrangement of neighboring houses and buildings as effects from the surrounding buildings and simulated amounts of solar radiation (Caroline Hachem-Vermette and Kuljeet Singh 2019). Narumi and Kim (2017) computed the amount of electricity generated in Yokohama, Kanagawa, accounting for building use, area class, and use districts (Daisuke Narumi and Kim Younghyo 2017). Furthermore, numerous studies have focused on distance, with Müller and Rode (2013) revealing that in Wiesbaden, Germany, installation tends to increase as the installed residential PV capacity in the neighborhood increases (Sven Müller and Johannes Rode 2013). Graziano and Gillingham

(2015) documented that in Connecticut, USA, residential PV installations in the previous 6 months caused an increase in those within 0.5 miles (Marcello Graziano and Kenneth Gillingham 2015). Researchers have noted that distance includes not only physical distance but also peer effects. According to Palm (2017), peer effects refer to "the social influence of peers, such as colleagues, neighbours or friends." Moreover, he concluded that in Sweden, peer effects accompanying direct contact between people were more important than those from simply observing PV panels (Palm 2017). Curtius et al. (2018) revealed a tendency for the number of residential PV installations in a neighborhood and recognition of social expectations for PV installation to increase homeowners' desire for installation (Curtius et al. 2018).

However, these three characteristics were interrelated, and which factor truly had an effect was unclear. Owing to data limitations, controlling all of the characteristics of residents, houses, and surrounding environment was difficult. Therefore, we deliberately focused on housing estates[1] as a space wherein the three aforementioned characteristics were highly homogeneous. Specifically, we assumed that residents would have lived there from the time of development and would be of similar ages and income classes. The houses were built around the same time and have similar building shapes pursuant to building agreements. The surrounding environment was assessed by average distance between nearest houses (see Table 8.1, interval between adjacent buildings (m)), which was similar. However, whether residential PV was installed differs between houses even within a housing estate.

Revealing the factors affecting installation in housing estates is a finding that is useful when evaluating support for residential PV installation in housing estates and residential areas in Japan, which we believe is this study's significant contribution to the literature.

This study aimed to gather information regarding the types of houses and surrounding environments wherein residential PV power generation was installed in housing estates. Moreover, it aimed to verify the robustness of the factors affecting installation in housing estates by conducting analysis adding houses in neighboring areas outside the housing estates.

As presented below, we established hypotheses regarding the factors affecting residential PVs installation. Concerning housing characteristics, PV power generation was more likely to be installed if the building area is larger, the building height is higher, and the building shape changes, such as through being renovated, rebuilt, or newly built. By contrast, if a building agreement contains particulars about roofs, these must be considered and PVs must not be installed. Regarding the characteristics of the surrounding environment, installation makes less progress if the interval between adjacent buildings is closer, owing to the effects of light pollution from light reflecting off the PVs.

[1] A housing estate is defined as "a residential area on the National List of New Towns or judged to be a 'housing estate' by a local government," with reference to the MLIT "National List of Housing Estates (FY2018)." (Ministry of Land, Infrastructure, Transport, and Tourism (MLIT) n.d.).

Table 8.1 Outline of housing estates and areas for comparison in Kakegawa City

Area		Date residence commenced (year)	Project method at development (developing entity)	Area (ha)	No. of houses	No. of houses with residential PVs	Residential PV installation rate (%)	Area mean (see Sect. 8.3.1 for an explanation of variables)						
								Area of houses (m^2)	Height houses (m)	Interval between adjacent buildings (m)	Building agreement dummy	Renovation dummy	Rebuilding dummy	New build dummy
Housing estates	Katsuragaoka estate	1980	General public (prefecture)	30	604	32	5.3%	87.7	8.2	11.8	0.005	0.015	0.070	0.008
	Asahigaoka estate	1985	Development permit (private)	16	379	12	3.2%	83.2	8.1	11.0	1.000	0.013	0.055	0.021
	Akihamichi	1992	Development permit (private)	10	344	8	2.3%	100.4	8.4	11.2	0.305	0.015	0.026	0.029
Comparison targets	Mizutari	—	—	97	358	38	10.6%	110.1	8.0	14.2	0.000	0.014	0.106	0.131
	Ōtarō	—	—	11	129	11	8.5%	96.1	8.5	11.9	0.000	0.016	0.062	0.140
	Kamiyashiki	—	—	11	157	17	10.8%	97.5	8.2	13.2	0.000	0.019	0.051	0.299

Sources: Housing estates: created with reference to MLIT, "National List of Housing Estates (FY2018)"
Areas for comparison: scope and areas identified based on Statistics Bureau, Ministry of Internal Affairs and Communications, "2015 Census: Border Data by Neighborhood, Etc."
No. of houses: used polygons that had apartment building names or private house names in the Zmap-TOWN II Digital Residential Map
Interval between adjacent buildings: used straight-line distance from the center of the building polygon to that of the nearest building polygon in 2020 in Kakegawa City for each house

8.2 Research Methods

8.2.1 Research Subjects

This study's subjects were PVs installed on the roofs of houses in Kakegawa City, Shizuoka.

We selected Kakegawa City, Shizuoka, for the following reasons: a 3D urban model (Project PLATEAU) provided by the City Bureau of the Ministry of Land, Infrastructure, Transport and Tourism (MLIT) has been developed, making it easier to gain information on building heights. Additionally, it seems that residential PV installations will continue into the future because—as a target under the Third Kakegawa City Climate Change Countermeasure Action Plan (Area Policy Volume)—the city aims to grow from 3994 "installations of general residential PV power generation (cumulative total)" as of FY2017 to 6000 in FY2023 and 8000 in FY2030 (Kakegawa City n.d.-a).

Kakegawa City has implemented "New Energy Device, Etc. Installation Support Project", which offers assistance for up to half the expenses (to a maximum of 30,000 yen) incurred for installing devices in PV power generation facilities installed on new houses (connected to the system within 1 year of completion of construction or from purchase of a ready-built house) and up to half (to a maximum of 60,000 yen) those incurred for installing devices installed on existing houses (connected to the system 1 year or more from completion of construction or from purchase of a ready-built house) (Kakegawa City n.d.-b).

According to Kakegawa City's Environmental Policy Division, the aforementioned project has been used for 2216 installations from April 2013 to February 2022 for an installed capacity totaling 12,668.36 kW. The number of houses using the project each fiscal year—divided into new builds and existing buildings—is illustrated in Fig. 8.1. Its usage in new builds has been greater since FY2018, when eligibility for assistance was classified into new builds and existing buildings. The cost of installing PVs on existing buildings is greater than for new builds because Kakegawa City authorities recognized that it would be more expensive than for new builds.[2].

[2] According to the "Opinion on Procurement Prices for FY2022 Onward, Etc." by the Procurement Price Calculation Committee, ANRE (2022), the mean system cost for residential PV power generation installed in 2021 was 302,000 yen/kW and 280,000 yen/kW for existing and new buildings, respectively (Procurement Price Calculation Committee n.d.).

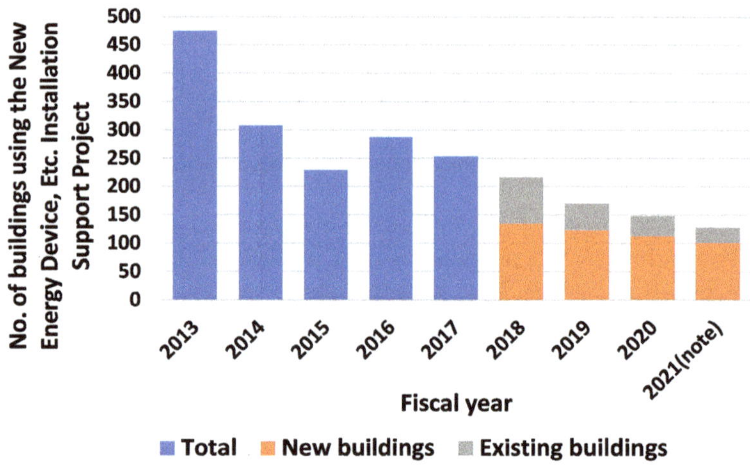

Fig. 8.1 Number of buildings using the Kakegawa City New Energy Device, Etc. Installation Support Project, by new builds and existing buildings. Source: Provided by Environmental Policy Division, Kakegawa City. Note: The figures for FY2021 end in February 2022

8.2.2 Data

8.2.2.1 Building Polygons

We acquired building polygon data on Kakegawa City for 2008/2009 ("2008") and 2020 from the Zmap-TOWN II Digital Residential Maps provided by Zenrin. Only those polygons that had apartment building names or private house names in the Zmap-TOWN II Digital Residential Map were extracted. Consequently, public facilities and business premises were not included in the analysis.

8.2.2.2 Building Height

We acquired building heights from the 3D urban model (Project PLATEAU) provided by MLIT ("Plateau data"). However, because the Plateau data did not match (Sect. 8.2.2.1), we excluded buildings that lacked a height value from the Plateau data and spatially linked data in Sect. 8.2.2.1 with a center of gravity closest to the that in the former on ArcGIS Pro to include building heights in Sect. 8.2.2.1.

8.2.2.3 Presence or Absence of Residential PV Installations

To confirm whether residential PV power generation had been installed, we used the Geospatial Information Authority of Japan's National Up-to-Date Photographs (Seamless) as a background map in ArcGIS Pro. The aerial photographs in

Kakegawa City were based on simple aerial photography (photographed in June to August 2020).

8.2.3 Process of the Study

We analyzed three housing estates to identify the factors affecting residential PV installation in housing estates based on houses and the surrounding environment. To verify the robustness of the factors affecting installation, we examined three neighboring areas outside the housing estates. To confirm the state of residential PV installations, we conducted an on-site survey (January 2022), an interview survey with the Environmental Policy Division of Kakegawa City (March 2022), and an interview survey with the mayor of Akihamichi (April 2022).

8.3 Presence or Absence of Residential PV Installations (Analysis 1)

8.3.1 Analysis Method

Kakegawa City has six housing estates within a 3.5-kilometer radius from the center of Kakegawa Station.[3]

To establish the factors affecting installation of residential PV power generation, we focused on Katsuragaoka, Asahigaoka, and Akihamichi. These housing estates have established building agreements that stipulate particulars relating to roofs pertaining to some or all of the houses in the estate.

An outline of the housing estates and neighboring areas for comparison in 8.4 was presented in Table 8.1, and their positional relationship was illustrated in Fig. 8.2.

In the Third Kakegawa City Climate Change Countermeasure Action Plan (Area Policy Volume), the residential PV installation rate among all households in FY2017 was 9.0% (Kakegawa City n.d.-c), but Akihamichi had a lower rate—only 2.3%. In neighboring areas, the residential PV installation rate was higher than in the housing estates, at a similar level to that for the entire city. The neighboring areas have farmland surrounding houses. Compared to the housing estates, houses in the neighboring areas had a mean area and height of a similar scale but a larger

[3] The housing estates in Kakegawa City were selected based on MLIT, "National List of Housing Estates (FY2018)." Katsuragaoka, Asahigaoka, and Akihamichi estates were identified as the target houses based on polygons in the Statistics Bureau, Ministry of Internal Affairs and Communications, "2015 Census: Border Data by Neighborhood, Etc." because the estate borders corresponded to the neighborhoods.

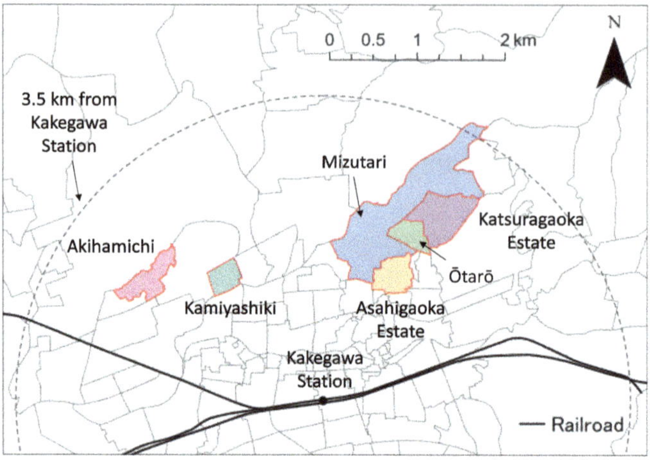

Fig. 8.2 Positional relationship in Kakegawa City. Source: created based on Statistics Bureau, Ministry of Internal Affairs and Communications, "2015 Census: Border Data by Neighborhood, Etc.", National Land Information, "2020 Railway Data"

mean interval between adjacent buildings. Regarding changes to building shape, renovations were about the same level in the housing estates and neighboring areas. While rebuilds occurred predominantly in Mizutari, few occurred in Akihamichi. New builds were observed more in the neighborhoods for comparison than in the housing estates.

We conducted a logistic regression analysis of all 1327 houses in the housing estates[4] using R4.1.2.

The outcome variable was whether residential PV power generation was installed on each house (a dummy variable that was 1 if installed and 0 if not). We adopted the following explanatory variables.

(a) Building area (m^2): We computed the area in ArcGIS Pro from the building polygons in 2020.
(b) Building height (m): We used the building heights acquired from the Plateau data.
(c) Interval between adjacent buildings (m): We measured the straight-line distance from the center of the building polygon to that of the nearest building polygon in 2020 in Kakegawa City in ArcGIS Pro.
(d) Renovation, rebuilding, new build dummies: To clarify the timing when residential PVs were installed on each house, we set three dummy variables (see Fig. 8.3). The renovation dummy represents building polygons that do not accurately match between 2008 and 2020 but, on visual inspection, were

[4] When we confirmed the intra-class correlation coefficient, it was low at 0.005, and therefore, we did not select a multilevel model.

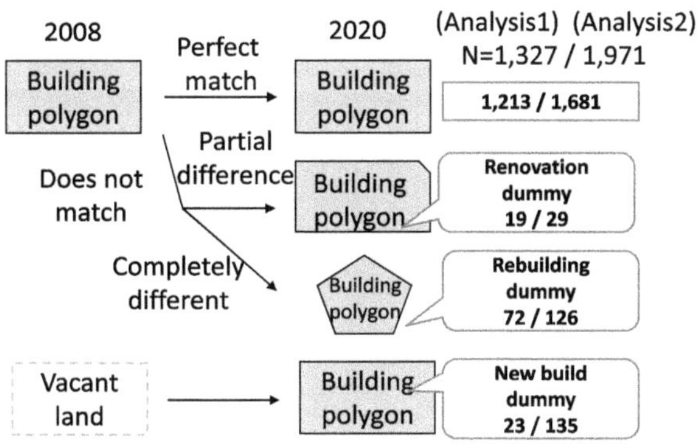

Fig. 8.3 Flowchart for determining renovation, rebuild, or new build

determined to exhibit partial differences. The rebuilding dummy represents building polygons that do not accurately match between 2008 and 2020 but, on visual inspection, were determined to have greatly different shapes. The new build dummy represents building polygons that did not exist in 2008 but do exist in 2020, and these were assigned to building polygons that did not intersect with 2008 building polygons in ArcGIS Pro.

(e) Building agreement dummy: The provisions on roof coloring and materials thought to affect residential PV installation were presented in Table 8.2. Building agreements covered the entirety of the Asahigaoka estate but only some houses in the Katsuragaoka and Akihamichi estates. Houses included in the areas covered were assigned the building agreement dummy (1 if in an area was covered by a building agreement, 0 if not).

Notably, we took the natural logarithm of (a) to (c) calculated the mean of each housing estate where the houses were located, and centered them within the cluster (cwc). Considering the aforementioned relationships between (a) to (c) and (d) and (e) we forced the use of interaction terms.

8.3.2 Analysis results

The analysis results were reported in Table 8.3. The variance inflation factor (VIF) was at most 1.92, and the effects of multicollinearity seemed small.

As main effects, ln(building height (m))$_{\mathrm{cwc}}$, rebuilding dummy, and new build dummy were significant. Lower building heights compared to the mean in each estate tended toward residential PV installation if the houses were rebuilt or newly built. Among the interaction terms, ln(building area (m^2))$_{\mathrm{cwc}}$ × New build dummy was significant. Compared to the mean in each estate, new houses with larger building areas had residential PVs installed more frequently.

Table 8.2 Outline of building agreements with provisions concerning roofs

Area	Building agreement	Provisions concerning roofs	No. of houses affected
Katsuragaoka estate	Kakegawa Oregon village building agreement	Article 11.5. The colors of the roofs and external walls of the buildings will be predominantly white, black, green, or brown, and have sober coloring in harmony with the surroundings.	3 (Katsuragaoka 3-chōme only)
Asahigaoka estate	Green Square Kakegawa building agreement	Article 10 (4). The colors of the roofs of the buildings will be predominantly gray, and mixtures of dark gray and green, blue, or brown may be applied. Furthermore, television antennas and similar items must not be installed on roofs.	379
Akihamichi	Akihamichi building agreement	Article 13.2. In zone B, the roofs of the buildings must be roofed with Japanese tiles, copper roofing, or other similar materials, except for the roofs of eaves, garages, sheds, and similar items.	105
	Akihamichi no. 2 west building agreement	Article 13.2. The colors of the roofs and external walls of the buildings will be predominantly white, black, green, or brown, and have sober coloring in harmony with the surroundings.	

Source: Created with reference to Kakegawa City authorities: "On Building Agreements," https://www.city.kakegawa.shizuoka.jp/gyosei/docs/8135.html (accessed April 2, 2022). The number of affected houses was confirmed visually in ArcGIS Pro

Table 8.3 Logistic regression analysis of whether residential PV power generation is installed

	Analysis 1: Housing estates only ($N = 1327$)				Analysis 2: Housing estates and neighboring areas ($N = 1971$)			
	B	S.E.	*P*-value		B	S.E.	*P*-value	
Constant term	*−3.528*	0.210	0.000	***	*−3.345*	0.155	0.000	***
ln(building area (m^2))$_{cwc}$	0.228	0.931	0.806		0.834	0.488	0.088	
ln(building height (m))$_{cwc}$	*−2.460*	0.901	0.006	**	*−1.811*	0.685	0.008	**
ln(interval between adjacent buildings (m))$_{cwc}$	0.226	0.772	0.770		−0.145	0.531	0.785	
Building agreement D	−0.619	0.389	0.112		*−0.680*	0.327	0.038	*
Renovation D	1.462	0.964	0.129		1.120	0.980	0.253	
Rebuilding D	**2.236**	0.374	0.000	***	**1.909**	0.286	0.000	***
New build D	**2.584**	1.071	0.016	*	**1.835**	0.365	0.000	***
ln(building area (m^2))$_{cwc}$ × building agreement D	−3.237	1.682	0.054		−1.296	1.169	0.267	
ln(building height (m))$_{cwc}$ × building agreement D	0.329	2.320	0.887		−0.404	1.847	0.827	
ln(interval between adjacent buildings (m))$_{cwc}$ × building agreement D	1.794	1.326	0.176		1.762	1.205	0.144	
ln(building area (m^2))$_{cwc}$ × renovation D	1.025	4.836	0.832		−1.933	2.143	0.367	
ln(building height (m))$_{cwc}$ × renovation D	13.371	8.707	0.125		10.600	6.140	0.084	
ln(interval between adjacent buildings (m))$_{cwc}$ × renovation D	−0.928	2.935	0.752		−3.803	2.351	0.106	
ln(building area (m^2))$_{cwc}$ × rebuild D	−0.678	1.635	0.678		−1.610	0.987	0.103	
ln(building height (m))$_{cwc}$ × rebuild D	2.136	2.031	0.293		1.666	1.636	0.308	
ln(interval between adjacent buildings (m))$_{cwc}$ × rebuild D	−0.439	1.289	0.733		0.299	0.980	0.760	
ln(building area (m^2))$_{cwc}$ × new build D	**9.639**	4.523	0.033	*	*−2.530*	0.978	0.010	**
ln(building height (m))$_{cwc}$ × new build D	2.034	6.128	0.740		3.180	1.709	0.063	
ln(interval between adjacent buildings (m))$_{cwc}$ × new build D	−7.370	4.794	0.124		0.469	0.945	0.620	
Nagelkerke R^2	0.15				0.19			
AIC	420.15				796.06			

D Dummy variable ***, **, and * indicate <0.001, <0.01, and <0.05, respectively. Significant variables with a positive and negative coefficient are highlighted in bold and italics, respectively

8.4 Comparison of Housing Estates and Neighboring Areas (Analysis 2)

8.4.1 Analysis Method

Subsequently, we verified the robustness of the factors affecting installation by analyzing the neighboring areas outside the housing estates in addition to the estates themselves. We conducted a logistic regression analysis of 1971 houses, adding three neighborhoods analyzed in 8.3 (Ōtarō, Mizutari, Kamiyashiki)—near to the Katsuragaoka, Asahigaoka, and Akihamichi estates—that comprise housing.[5]

As in 8.3, the outcome variable was whether residential PV power generation was installed on each house. For the explanatory variable, we used the explanatory variables (a) to (e) used in 8.3 and forced use of interaction terms in consideration of the relationships between (a) to (c) and (d) and (e) using R4.1.2 for the analysis.

8.4.2 Analysis Results

The analysis results were reported in Table 8.3. The VIF was at most 5.29—a somewhat greater result—but the effects of multicollinearity seemed slight owing to it being from interaction effects.

The significant main effects were ln(building height (m))$_{cwc}$; and building-agreement, rebuilding, and new build dummies. Having expanded the areas for analysis, the building-agreement dummy was significant. The results indicated that houses covered by a building agreement must show consideration regarding their roofs and tended not to have residential PV power generation installed.

Regarding the interaction terms, ln(building area (m^2))$_{cwc}$ × new build dummy was significant. Compared to the mean in each area, new houses with smaller building areas had residential PVs installed more often. This trend differed from the results of Analysis 1.

Between Analyses 1 and 2, the latter has a larger value for the pseudo-coefficient of determination Nagelkerke R^2, which suggested a good fit. To confirm the divergence between predictions and actual results in Akihamichi, which exhibited the lowest rate among the three housing estates, the installation probability in houses computed from the state of residential PV installation and building changes and the results for Analysis 2 in Table 8.3 were illustrated in Fig. 8.4.

Analysis 2 revealed that the probability of installation was higher in houses that had been rebuilt. However, in Akihamichi, none of the nine such houses had installed residential PVs. In Akihamichi, several new houses have a small area,

[5] When we confirmed the intra-class correlation coefficient, it was low at 0.02, and therefore, we did not select a multilevel model.

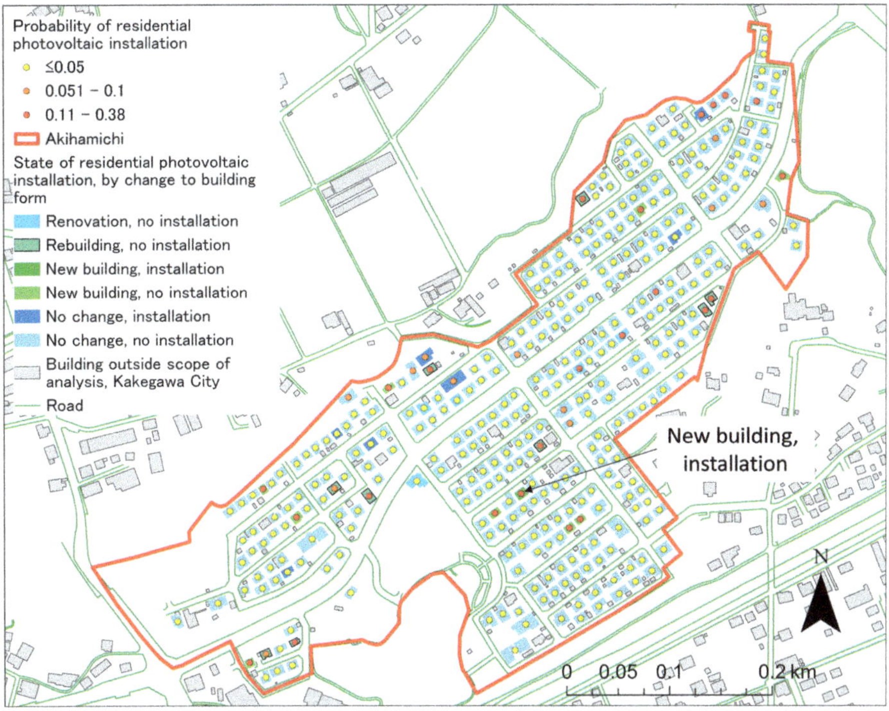

Fig. 8.4 Analysis 2: Probability of residential PV installation in Akihamichi. Sources: Buildings and roads: Zenrin Zmap-TOWN II Digital Residential Map (2020), Akihamichi: Created from Statistics Bureau, Ministry of Internal Affairs and Communications, "2015 Census: Border Data by Neighborhood, Etc."

which was suggested as the cause of the low installation probability computed in Analysis 2, despite these houses being new builds. In fact, of the 10 new houses, only one had installed a residential PV.

8.5 Discussion

8.5.1 House Characteristics

8.5.1.1 Relationship with Building Area and Height

In Analysis 1, contrary to our hypothesis, houses with lower building heights than the mean in each estate had residential PVs installed more often. If there were no tall buildings nearby, installations appeared to spread if houses can obtain sufficient sunlight, even if they were not tall. This trend was also observed in Analysis 2.

In the housing estates, residential PVs were installed in new houses with larger areas compared to the mean in each estate.

Including the neighboring areas as well, it was installed in new houses with areas smaller than the mean in each estate. The mean area of new houses that have residential PV power generation installed was 99.9 and 70.8 m^2 in the housing estates and neighboring areas, respectively. Aerial photographs of the neighboring areas confirmed numerous installations in small-scale detached houses that were newly built on former farmland.

Residential PVs being installed in new houses seem common to Analyses 1 and 2, but the interaction between new houses and house area was affected by characteristics of the area. In housing estates, comparatively large new houses were developed owing to the fixed nature of the lots for each house; however, in the neighboring areas, smaller-scale new houses were developed on former farmland, and residential PVs were installed when each was newly built.

8.5.1.2 Relationship with Changes in Building Form

In both Analyses 1 and 2, rebuilds and new builds exhibited significant results. Larger-scale changes in building form occur than in renovations; hence, seemingly, a greater scope exists for installing PV panels. The rebuilds and new builds were subsidized by new houses under the New Energy Device, etc. Installation Support Project, and the subsidy amount for them was less than that for existing houses. There was a tendency to install residential PVs on the occasion of a change in building form, such as a rebuild or new build, and support for these may expand in the future with the aim of PV installation in more houses on these occasions.

8.5.1.3 Relationship with Building Agreements

Analysis 1 did not reveal any trends unique to houses covered by a building agreement. Analysis 2, which expanded the scope of analysis, established a clear trend against residential PV installation in houses covered by a building agreement. Although building agreements designate colors and materials, they do not directly prohibit the PV installation.

According to the mayor of Akihamichi, areas covered by the Akihamichi Building Agreement were sold in lots in 1992, and areas covered by the Akihamichi No. 2 West Building Agreement on the western side were sold in lots approximately 10 years later. Solar water heaters were popular when the Akihamichi houses were sold, but the building agreements prohibited the installation of these and antennas to preserve the landscape. With PV power generation, the panels can be installed closely attached to the roof; therefore, installation is not prohibited, and it does not seem that the landscape would exhibit any issue. However, according to the mayor of Akihamichi, the estate sales occurred before PV power generation became popular, and as the residents—in their 30 and 40 s when they purchased the houses—have grown older, they may have missed the timing to install such power generation.

8.5.2 Characteristics of the Surrounding Environment (Relationship with the Interval Between Adjacent Buildings)

In Analyses 1 and 2, the interval between adjacent buildings was not a significant result. This may be attributable to area, height, and changes in building form being more important factors affecting installation than the distance between houses.

8.6 Conclusion

Focusing on houses and the surrounding environment, this study determined the factors affecting the installation of residential PV power generation with a focus on houses and the surrounding environment. We adopted housing estates with similar characteristics among residents, houses, and the surrounding environment in Kakegawa City, Shizuoka as a case study.

Common to the housing estates and the neighboring areas, residential PVs were installed more in buildings with heights lower than the mean in each area and as more changes in building form, such as rebuilding and new building, occurred.

In the housing estates, residential PVs were installed more in new houses with larger building areas than the mean in each estate in particular. Irwin (2021) and Kawaguchi et al. (2020) revealed a trend favoring residential PV installation on larger, taller buildings, but the result here differed. In addition to the house repairs noted by Bollinger and Gillingham (2012), rebuilds and new builds were also opportunities to install residential PV power generation. Although the results pertaining to the area of new houses differed outside the housing estates, we hope to conduct a more detailed analysis by expanding the scope in the future.

Regarding the characteristics of the surrounding environment, the interval between adjacent buildings was not a significant result, even when including interactions with dummy variables. This leads to the conclusion that the factors affecting residential PV installation in housing estates were predominantly owing to characteristics of the houses, not those of the surrounding environment.

We may not have been able to accurately gather information regarding residential PV power generation overall because it was determined by visual inspection—a limitation of our study. Currently, the only public data on residential PV power generation (less than 10 kW) are aggregate values by local government area[6]; therefore, owing to information on installation locations being published in the future, we may expect the factors affecting the installation to be revealed in greater

[6] The only business plan approval information for PV power generation projects that is publicly available on ANRE's Website for Publishing Project Plan Approval Information relates to projects for 20 kW or above. The website publishes approved and installed volumes by local government area.

detail. Moreover, owing to data limitations, we could not gather information on resident attributes (e.g., age, household type) or the age of the houses, but we intend to address this as a research question in the future.

Acknowledgments In conducting this research, we interviewed and were provided with information by the Environmental Policy Division of Kakegawa City. The mayor of Akihamichi cooperated for an interview. This study is part of the outcomes of the joint research with CSIS, the University of Tokyo (No. 917) and using the following data (Zenrin, Zmap-TOWN II 2008/09 (Shape version) Shizuoka Prefecture data set and Zmap-TOWN II 2020 (Shape version) Shizuoka Prefecture data set. We sincerely thank the abovementioned individuals and entities.

References

Agency for Natural Resources and Energy (ANRE) (n.d.) "Dai-6-ji enerugī kihon keikaku"[Sixth Strategic Energy Plan], p. 59, Japanese. https://www.meti.go.jp/press/2021/10/20211022005/20211022005-1.pdf (issued 22 Oct 2021; Accessed 2 April 2022)

Bryan Bollinger and Kenneth Gillingham (2012) Peer effects in the diffusion of solar photovoltaic panels. Mark Sci 31(6):900–912

Caroline Hachem-Vermette and Kuljeet Singh (2019) Mixed-use neighborhoods layout patterns: impact on solar access and resilience. Sustain Cities Soc 51:101,771

Curtius HC, Hille SL, Berger C, Hahnel UJJ, Wüstenhagen R (2018) Shotgun or snowball approach? Accelerating the diffusion of rooftop solar photovoltaics through peer effects and social norms. Energy Policy 118:596–602

Daisuke Narumi and Kim Younghyo (2017) Research on the measures to promote introduction of rooftop-type photovoltaic power generation system considering economic efficiency. AIJ J Technol Design 23(55):925–930

Irwin NB (2021) Sunny days: spatial spillovers in photovoltaic system adoptions. Energy Policy 151:112192

Kakegawa City (n.d.-a) Dai-3-ki Kakegawa-shi chikyū ondanka taisaku jikkō keikaku (kuiki shisaku hen) [Third Kakegawa City climate change countermeasure action plan (zone policy volume)], p. 56, Japanese. https://www.city.kakegawa.shizuoka.jp/fs/8/4/9/1/2/_/kuikisesakuhen3rd.pdf (issued March 2019; Accessed 2 April 2022)

Kakegawa City (n.d.-b) Shin enerugī kiki tō no setchi ni tai suru josei [Assistance for the installation of new energy devices, etc.] Japanese. https://www.city.kakegawa.shizuoka.jp/gyosei/docs/101882.html. Accessed 2 April 2022

Kakegawa City (n.d.-c) Dai-3-ki Kakegawa-shi chikyū ondanka taisaku jikkō keikaku (kuiki shisaku hen) [Third Kakegawa City climate change countermeasure action plan (zone policy volume)], p. 22, Japanese. https://www.city.kakegawa.shizuoka.jp/fs/8/4/9/1/2/_/kuikisesakuhen3rd.pdf (issued March 2019; Accessed 2 April 2022)

Kawaguchi N, Hayashi K, Yamada T, Tomino Y (2020) The analysis of the installation indicator for the rooftop-type photovoltaic power generation system of residential districts. Jpn Assoc Human Environ Symbiosis J Human Environ Symbiosis 36(2):151–161

Kosugi T, Shimoda Y, Tashiro T (2019) Neighborhood influences on the diffusion of residential photovoltaic systems in Kyoto City, Japan. Environ Econ Policy Stud 21:477–505

Marcello Graziano and Kenneth Gillingham (2015) Spatial patterns of solar photovoltaic system adoption: the influence of neighbors and the built environment. J Econ Geogr 15(4):815–839

Ministry of Land, Infrastructure, Transport, and Tourism (MLIT) (n.d.) Zenkoku no jūtaku danchi risuto ni tsuite [On the national list of housing estates], Japanese. https://www.mlit.go.jp/jutakukentiku/house/jutakukentiku_house_mn5_000016.html. Accessed 2 April 2022

Ministry of the Environment (MOE) (2021) "Chikyū ondanka keikaku" [Climate Change Adaptation Plan], p. 27, Japanese, http://www.env.go.jp/earth/211022/mat01.pdf (issued 22 Oct 2021; Accessed 2 April 2022)

Palm A (2017) Peer effects in residential solar photovoltaics adoption—a mixed methods study of Swedish users. Energy Res Soc Sci 26:1–10

Procurement Price Calculation Committee (n.d.) ANRE, Reiwa 4-nendo ikō no chōtatsu kakaku tō ni kan suru iken [Opinion on procurement prices for FY2022 onward, etc.], Japanese, p. 22. https://www.meti.go.jp/shingikai/santeii/pdf/20220204_1.pdf (accessed August 10, 2022)

Schaffer AJ, Brun S (2015) Beyond the sun—socioeconomic drivers of the adoption of small-scale photovoltaic installations in Germany. Energy Res Soc Sci 10:220–227

Sven Müller and Johannes Rode (2013) The adoption of photovoltaic systems in Wiesbaden, Germany. Econ Innov New Technol 22(5):519–535

Part II
Urban Analysis

Chapter 9
Statistical Multi-dimensional Scaling with a Geographical Penalty

Development of Bayesian Multi-dimensional Scaling and Its application to Time–Space Mapping

Hayato Nishi and Yasushi Asami

Abstract Understanding the structure of cities and facilities in terms of similarity and proximity when considering wide-area or national land planning is crucial. For this purpose, this chapter proposed a novel method to visualize the similarities or various distances of regions, referred to as Bayesian geographical multi-dimensional scaling. Multi-dimensional scaling (MDS) is the fundamental method that is generally used for this purpose; however, its projections are often substantially different from the geographical locations to facilitate appropriate comparison or understanding. Thus, to overcome this weakness, we introduced a geographical penalty for MDS, and its weight was determined using a statistical criterion. This penalty was introduced because dissimilarities or distances should be small if the two regions are geographically close in many cases. Furthermore, this method was applied to visualize the accessibility of public transit in Japan. The obtained results can be easily compared with geographical locations and facilitate an understanding of the transitions that improve accessibility in each prefecture.

The contents of this chapter are based on the following paper originally published in a Japanese journal: Nishi, H. and Asami, Y. (2019) Statistical multi-dimensional scaling under the geographical constraints: The development of Bayesian multi-dimensional scaling and its application to timespace mapping. *Journal of City Planning Institute of Japan* 54(3):826–832 (in Japanese).

H. Nishi (✉)
Hitotsubashi University, Kunitachi, Tokyo, Japan
e-mail: h.nishi@r.hit-u.ac.jp

Y. Asami
Department of Urban Engineering, The University of Tokyo, Bunkyo-ku, Tokyo, Japan
e-mail: asami@csis.u-tokyo.ac.jp

Y. Asami et al. (eds.), *Studies in Housing and Urban Analysis in Japan*,
New Frontiers in Regional Science: Asian Perspectives 75,
https://doi.org/10.1007/978-981-99-8027-7_9

Keywords Multi-dimensional scaling · Time–space map · Visualization method · Similarity · Bayesian statistics

9.1 Introduction

Understanding the structure of cities and facilities in terms of similarity and proximity when considering wide-area or national land planning is vital, where the relationships among facilities are important. Various indices can be used to define such relationships, such as simple linear distance, time distance, road distance, and similarity indices. The relationships described with these indices are relatively easy to understand when the number of cities and facilities is small. However, it is difficult to understand the overall structure directly from the defined indices when the number of cities and facilities increases.

Multi-dimensional scaling (MDS) is a useful method for visualizing the relationships among cities and facilities located in a geographical space (Shimizu 1992). When the relationship between points such as cities or regions is defined in terms of dissimilarity, MDS arranges these points (typically on a two-dimensional plane) such that the dissimilarity corresponds to the distance between points. This facilitates an understanding of the overall structure as a dissimilarity map.

A typical application is a time–space map. In a time–space map, the time distance between points can be determined on a scale of time units (Shimizu 1992). In particular, several methods of time–space map representation were proposed by Kotoh (2000) based on MDS.

Although MDS is a basic and useful method, its projection can be difficult to understand because the optimal layout solution is not uniquely determined. Moreover, the layout of cities deviates from their actual geographical locations (Kotoh 1997).

To resolve this problem, we focused on the fact that many relationships between cities and facilities are strongly related to actual geographical relationships. For example, time distance has a close relationship with the linear geographical distance between cities and facilities, even if they are not the same. Cities that are close to each other tend to have relatively similar industrial and demographic structures. The method by Tanaka (2018), which is similar to MDS, also provides similarities to the actual geographical location.

However, the comparison of a map constructed using MDS with an actual geographic map can be challenging because MDS ignores rotation, translation, and scaling. A simple solution to this problem involves the application of Procrustes transformation (Dryden and Mardia 1998; Kendall et al. 1999) to the MDS map and then comparing it with an actual map. However, in this two-step method, the expectation that the projected location of the city should be close to the actual geographic location is not reflected in the first-step MDS projection. Therefore, the point-to-point relationships do not reflect this expectation. However, if we introduce a geographical penalty to MDS, the weight that should be assigned to it is not apparent.

Based on the motivations described above, this chapter proposed a method to add a geographical penalty to the relationships among cities and facilities on a projected map by modifying MDS. This method aimed to create maps that can be easily compared to actual geographic maps. To systematically introduce a penalty, this chapter statistically interpreted the MDS. Despite certain existing statistical (probabilistic) MDSs, such as the approach based on maximum likelihood estimation by Zinnes and MacKay (1983), we employed a Bayesian approach to introduce geographical penalties as prior distributions of city locations. Bayesian MDS was introduced for dimension selection in Oh and Raftery (2001) and for variable selection in Lin and Fong (2019). However, to the best of our knowledge, no previous study has introduced this concept from a geographical perspective.

The Bayesian geographical MDS (BGMDS) proposed in this chapter offers the following advantages. First, the geographical penalty introduced into the MDS reflects intuitive expectations in the dissimilarity maps described above and renders its comparison with an actual geographical map easier. Second, the Bayesian approach facilitates an evaluation of the uncertainty of the layouts of cities or facilities in the dissimilarity maps. Third, it provides a statistical criterion for determining the weight of geographical penalties.

The remainder of this chapter is structured as follows. Section 9.2 discusses the standard MDS and its probabilistic interpretation. Section 9.3 proposes BGMDS, which introduces a geographical penalty as a prior distribution to the MDS. In Sect. 9.4, we briefly summarize the inference method employed in the proposed method. In Sect. 9.5, we generate travel time–space maps among prefectures by public transportation using BGMDS as an application of the proposed method to a real-world problem. Section 9.6 presents a comparison of the proposed method with a standard MDS. Finally, Sect. 9.7 provides a discussion and presents future work prospects.

The Python implementation of BGMDS is available at https://github.com/hayato-n/BGMDS.

9.2 Multi-dimensional Scaling

The MDS method creates a dissimilarity map by solving the following optimization problem:

Let $n \in \mathcal{S} = \{1, \ldots, N\}$ denote the index of each observation (city or facility) and the vector $z_n = (z_{n,i} : i \in \{1, 2\})^\top$ denote its location on the dissimilarity map. In this chapter, we assumed a projection onto a two-dimensional map $i \in \{1, 2\}$; however, it is easy to generalize for three or more dimensional maps. Let $d_{n,m}$ be the dissimilarity (or distance) defined between two points $n, m \in \mathcal{S}$[1] and $\delta_{n,m} =$

[1] Here, we assumed that the dissimilarity is symmetric.

$\delta(z_n, z_m)$ is the distance on the MDS map defined in the same manner. Note that $\delta(z_n, z_m)$ is the Euclidean distance function between two points. z_n, z_m[2].

MDS solves the minimization problem

$$\min_{z_n:n\in\mathcal{S}} \sum_{n<m} \left(d_{n,m} - \delta_{n,m}\right)^2. \tag{9.1}$$

This formulation is consistent with the "divergence evaluation model focusing on difference" in Tanaka (2018). As this optimization result is independent of rotation, translation, and scaling, an appropriate retransformation is required for comparison with the actual map.

If we interpret this MDS model statistically, minimization is the maximum likelihood estimation of the statistical model. Specifically, we assume that the observed dissimilarity $d_{n,m}$ contains random errors. Thus, we assume a Gaussian distribution as follows:

$$\mathcal{N}\left(d_{n,m} \mid \delta_{n,m}, \tau^{-1}\right). \tag{9.2}$$

Note that $\mathcal{N}(\cdot \mid \mu, \sigma^2)$ indicates that the random variable $\cdot$ follows a Gaussian distribution with mean μ and variance σ^2. In this model, the potential (unobserved) distance $\delta_{n,m}$ is the mean parameter, and τ is the precision (inverse of variance) parameter.

The maximum likelihood estimation of this statistical model is equivalent to minimization in the MDS. This relationship is similar to that between the maximum likelihood estimation and least squares in linear regression models.

9.3 Bayesian Geographical Multi-dimensional Scaling

The BGMDS method proposed in this chapter introduces a prior distribution that imposes a geographic penalty on the latent position $z_{n,m}$. Specifically, we assumed that it follows a Gaussian distribution.

$$\mathcal{N}\left(z_{n,i} \mid x_{n,i}, \lambda^{-1}\right). \tag{9.3}$$

This implies that the BGMDS projection z_n is distributed around the corresponding observed geographic location $\boldsymbol{x}_n$. Note that λ is the precision (inverse of the variance) parameter of a Gaussian distribution.

Let $p(\cdot)$ be the probability density function of $\cdot$. We define $\mathbf{X} = \{\boldsymbol{x}_1, \ldots, \boldsymbol{x}_N\}$, $\mathbf{Z} = \{z_1, \ldots, z_N\}$, and $\mathbf{D}$ is the matrix whose (n, m) element is $d_{n,m}$. The

[2] The distance function on the dissimilarity map can also be generalized to a symmetric function; however, it should be differentiable for efficient inference.

statistical model $p(\mathbf{D} \mid \mathbf{Z})$ and its prior distribution $p(\mathbf{Z} \mid \mathbf{X})$ are defined by Eqs. (9.2) and (9.3). The potential locations $\mathbf{Z}$ are the parameters (latent variables). Considering two hyperparameters, λ and τ, the posterior distribution of $\mathbf{Z}$ is formally obtained from Bayes' theorem as follows (Suyama 2017):

$$p(\mathbf{Z} \mid \mathbf{X}, \mathbf{D}) = \frac{p(\mathbf{D} \mid \mathbf{Z})\, p(\mathbf{Z} \mid \mathbf{X})}{p(\mathbf{D} \mid \mathbf{X})}. \tag{9.4}$$

The hyperparameters λ and τ control the strength of the geographic penalty. Specifically, the larger λ is relative to τ, the stronger the penalty for the latent location z_n. As the marginal likelihood $p(\mathbf{D} \mid \mathbf{X})$ is the probability of obtaining the observed data $\mathbf{D}$ from the given model and prior distribution, it can be used to evaluate the model. Thus, λ and τ can be chosen to maximize the marginal likelihood $p(\mathbf{D} \mid \mathbf{X}); \lambda, \tau$, which is a function of the hyperparameters[3] (Bishop 2006).

9.4 Inference on BGMDS

In BGMDS, we cannot analytically evaluate the posterior distribution of $\mathbf{Z}$ and the marginal likelihood. Therefore, an approximate method is required. Variational inference is an approach for approximating the posterior distribution of $\mathbf{Z}$ using a parametric distribution $q(\mathbf{Z})$ (Bishop 2006; Suyama 2017). Oh and Raftery (2001) and Lin and Fong (2019) used Markov chain Monte Carlo sampling, which is an alternative approach for approximation. An advantage of variational inference is that it is scalable, even in the presence of many cities or facilities. In variational inference, the "closeness" of the true posterior $p(\mathbf{Z} \mid \mathbf{X}, \mathbf{D})$ and its approximation $q(\mathbf{Z})$ is measured with the Kullback–Leibler (KL) divergence $\mathrm{KL}[q(\mathbf{Z}) \,||\, p(\mathbf{Z} \mid \mathbf{X}, \mathbf{D})]$, which is a non-negative measure between distributes. The relationship between the marginal likelihood and the KL divergence is

$$\log p(\mathbf{D} \mid \mathbf{X}) = \mathcal{L}[q(\mathbf{Z})] + \mathrm{KL}[q(\mathbf{Z}) \,||\, p(\mathbf{Z} \mid \mathbf{X}, \mathbf{D})], \tag{9.5}$$

where $\mathcal{L}[q(\mathbf{Z})]$ is the lower bound of the log-marginal likelihood, referred to as evidence lower bound (ELBO).

In variational inference, the KL divergence is minimized to determine the $q(\mathbf{Z})$ that is "close" to $p(\mathbf{Z} \mid \mathbf{X}, \mathbf{D})$. Upon fixing the hyperparameters, this minimization problem is equivalent to the maximization of ELBO $\mathcal{L}$ because the marginal likelihood is constant. Therefore, we can use ELBO to approximate the log-marginal

---

[3] This procedure can be interpreted as a maximum likelihood estimation wherein the marginal likelihood is the likelihood function to be maximized. Therefore, it is referred to as the "type II maximum likelihood" method.

likelihood. The specific functional form of ELBO[4] is as follows:

$$\mathcal{L}[q(\mathbf{Z})] = \langle \log q(\mathbf{Z}) \rangle_q - \left\langle \sum_{n<m} \log \mathcal{N}\left(d_{n,m} \mid \delta_{n,m}, \tau^{-1}\right) \right\rangle_q - \left\langle \sum_{n=1}^{N} \sum_{i=1}^{2} \log \mathcal{N}\left(z_{n,i} \mid x_{n,i}, \lambda^{-1}\right) \right\rangle_q, \tag{9.6}$$

where $\langle \cdot \rangle_q$ is the expectation with an approximate distribution $q(\mathbf{Z})$. Both q and $\mathcal{N}$ are the probability density functions of the corresponding distributions.

The specific inference procedure is as follows: First, we restricted the approximate distribution $q(\mathbf{Z})$ to a Gaussian distribution.[5] This restriction facilitates the use of Bayes by Backprop algorithm (Blundell et al. 2015) for the inference. This algorithm was proposed to estimate Bayesian neural networks. In this algorithm, the gradient of the ELBO is unbiasedly estimated with a reparameterization trick. Gradient descent methods are used to update the parameters. In this chapter, we used the Adam Optimizer (Kingma and Ba 2015), which is an adaptive gradient descent method.

The hyperparameter settings are important; however, the model is not sensitive to small changes in hyperparameters. In addition, Blundell et al. (2015) reported that simultaneous optimization with variational inference does not work. Therefore, we selected hyperparameters that maximized ELBO using a grid search.

9.5 Time–Space Map Among Prefectures in Japan

As an example of the application of BGMDS, we employed it to visualize travel time with public transportation between prefectures as time–space maps. Such a time–space map expresses accessibility among cities and facilitates an understanding of the relationship between cities and the national land structure.

---

[4] However, it is difficult to compute expectations in ELBO. Therefore, the Bayes by Backprop algorithm approximates them using the Monte Carlo method, which provides an unbiased estimator.

[5] In the Bayes by Backprop algorithm, the approximate distribution is constrained to the following form:

$$q(\mathbf{Z}) = \sum_{n=1}^{N} \sum_{i=1}^{2} \mathcal{N}\left(z_{n,i} \mid \mu_{n,i}, \sigma_{n,i}^2\right),$$

where $\mu_{n,i}, \sigma_{n,i}^2$ are the parameters of the approximate distributions optimized in the variational inference.

Table 9.1 Major railroad station on each prefecture

Prefecture	Major Station	Prefecture	Major Station
Aomori	Aomori	Shiga	Maibara
Iwate	Morioka	Kyoto	Kyoto
Miyagi	Sendai	Osaka	Osaka
Akita	Akita	Hyogo	San-no-miya
Yamagata	Yamagata	Nara	Nara
Fukushima	Fukushima	Wakayama	Wakayama
Ibaraki	Mito	Tottori	Tottori
Tochigi	Utsunomiya	Shimane	Matsue
Gunma	Takasaki	Okayama	Okayama
Saitama	Omiya	Hiroshima	Hiroshima
Chiba	Funabashi	Yamaguchi	Shimonoseki
Tokyo	Tokyo	Tokushima	Tokushima
Kanagawa	Yokohama	Kagawa	Takamatsu
Nigata	Nigata	Ehime	Matsuyama
Toyama	Toyama	Kochi	Kochi
Ishikawa	Kanazawa	Fukuoka	Hakata
Fukui	Fukui	Saga	Saga
Yamanashi	Kofu	Nagasaki	Nagasaki
Nagano	Nagano	Kumamoto	Kumamoto
Gifu	Gifu	Oita	Oita
Shizuoka	Shizuoka	Miyazaki	Miyazaki
Aichi	Nagoya	Kagoshima	Kagochima–Chuo
Mie	Tsu		

Broad railroad networks cover Japan, and they are used by people to move from one city to another. Furthermore, major cities connected by Shinkansen (bullet train) should be placed closer to each other on the time–space map because of the shorter travel time. Moreover, if traveling by airplanes is allowed, cities near popular airports will be placed closer to each other.

In this chapter, an inter-prefectural time–space map was generated using the following procedure. First, we selected a major railroad station (Table 9.1) for each of the 45 prefectures, excluding Hokkaido and Okinawa, which are located far from Honshu Island (the main island of Japan).[6] Next, the travel times among the selected stations were obtained separately in the following two scenarios: (1) people cannot

[6] Hokkaido and Okinawa were excluded for the following reasons. First, travel to Okinawa is more time-consuming than travel to other prefectures, particularly when airplanes are not used owing to available transits being only ships and airplanes. Access routes to Hokkaido from Honshu are also broader than routes to Okinawa, although they are still limited. Thus, the time to Hokkaido is longer than that to the other prefectures.

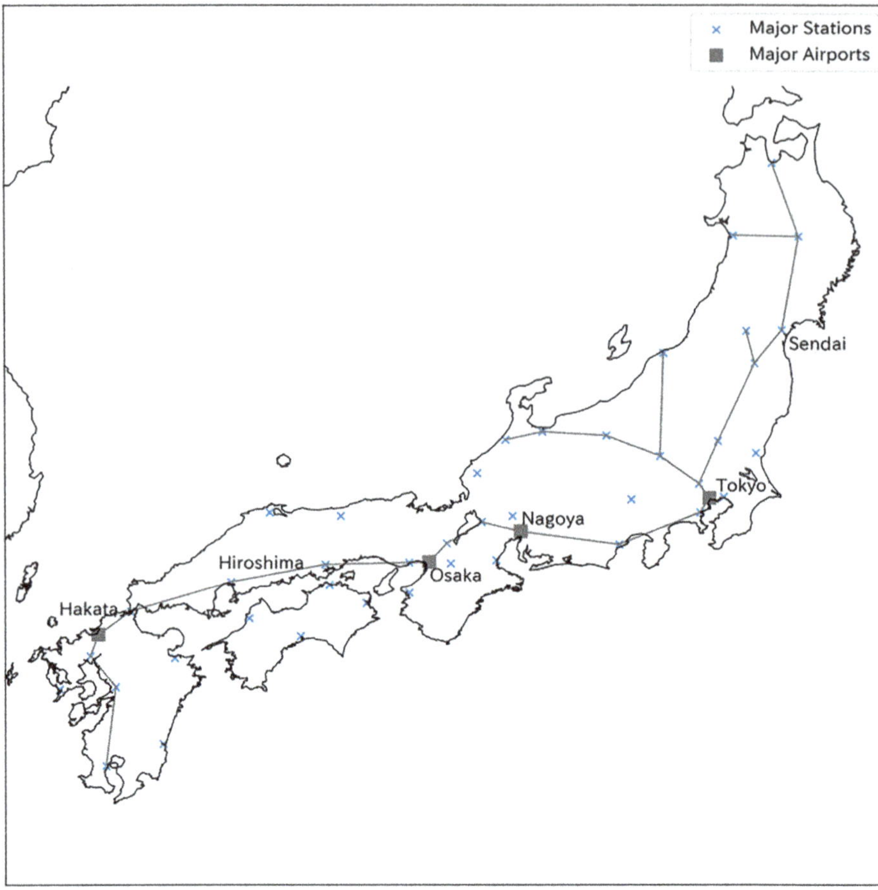

Fig. 9.1 Gray crosses and squares indicate the geographical locations of major stations. Prefectures connected with Shinkansen (bullet train) lines are tied with gray line segments. Gray squares indicate the presence of major airports nearby the stations

use airplanes and (2) people can use airplanes.[7] Thus, we obtained two time distance matrices. Using these matrices and the geographic coordinates of the stations, we generated a time–space map using BGMDS for each matrix.

Figure 9.1 shows the geographical locations of the selected major stations. The prefectures are connected with Shinkansen lines and gray line segments. The Tokyo, Nagoya, Osaka, and Hakata stations are indicated with gray squares because there are major airports near these stations.

[7] We obtained the travel times from https://www.jorudan.co.jp with the departure time set at 12:00 on 2019/4/1. It was assumed that the travel times were symmetrical. In addition, we adjusted the time units such that the average value was the same as that of the geographic Euclidean distance.

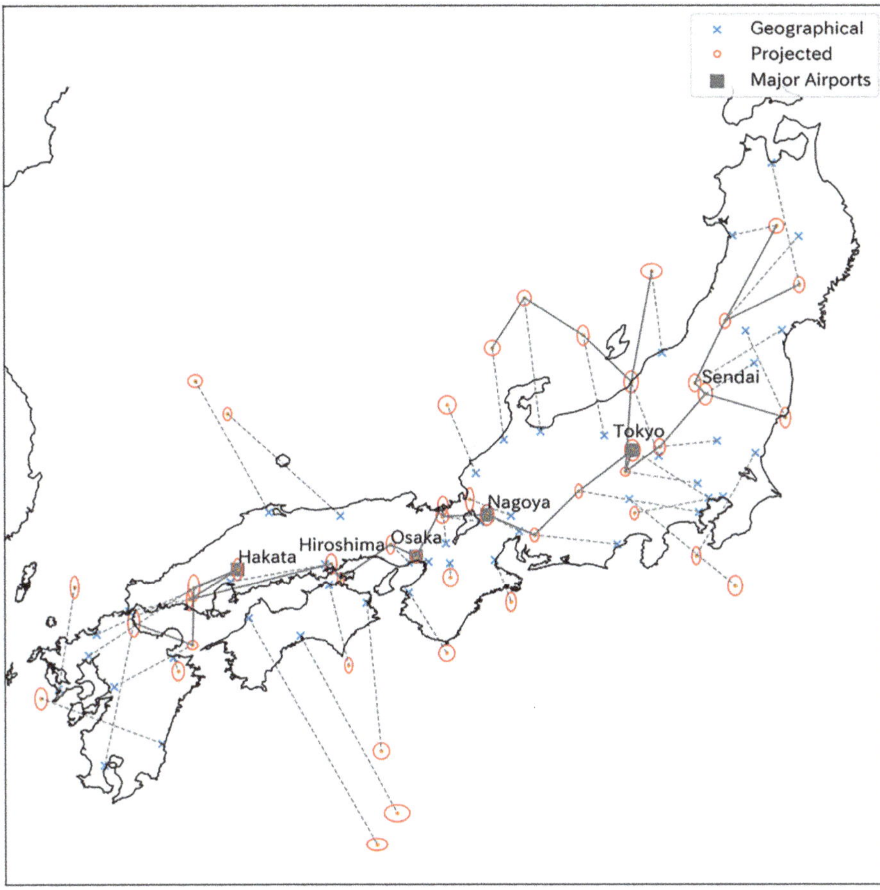

Fig. 9.2 Time–space map provided by BGMDS. Airplanes are not available on this map. Red ellipses show the standard deviations of the approximated posterior distributions. The geographical locations (gray crosses) and the locations in the time-space map (red dots) are connected with gray dashed lines

Before presenting the results of the BGMDS projection, we clarify the results of the hyperparameter selection. We selected the hyperparameters λ, τ to maximize the ELBO from the combinations of $\{0.1, 1, 10, 100\}$ using a grid search. Consequently, $(\lambda, \tau) = (1, 1)$ was selected with and without airplanes.

Figure 9.2 presents the BGMDS results without airplanes. Red ellipses represent the standard deviations of the posterior distributions. The large ellipses indicate the existence of a large amount of locational uncertainty in the BGMDS map. We connected the geographic location and the mean of the posterior distribution using gray dashed lines to clarify the relationship between the geographic and projected locations. The dashed lines indicate that the stations connected by the Shinkansen lines agglomerated toward the center of Japan. However, the prefectures on the Sea

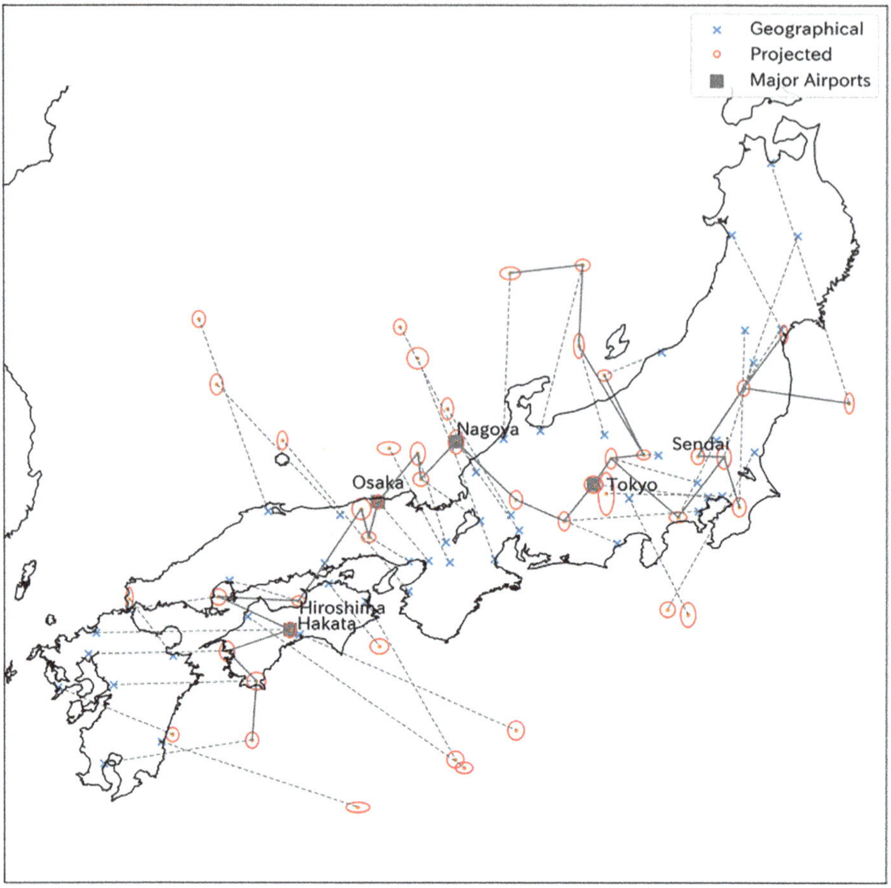

Fig. 9.3 Time–space map provided by BGMDS. Airplanes are available on this map. Red ellipses show the standard deviations of the approximated posterior distributions. The geographical locations (gray crosses) and the locations in the time–space map (red dots) are linked with gray dashed lines

of Japan side (west side of Japan) in the Shikoku and Chugoku regions, which are not connected to Shinkansen lines, are located far from the Shinkansen routes. This result indicates that Shinkansen lines improve accessibility between prefectures in Japan.

Figure 9.3 shows the BGMDS results when airplanes are available in the same format as that shown in Fig. 9.2. In this case, the effect of the Shinkansen is not apparent; however, the entire station layout was distorted such that Hakata Station and Tokyo Station become closer. In addition, Hakata Station dragged the stations in Kyushu (the western island) toward itself. This result indicates that airlines have improved their accessibility from Kyushu to Tokyo. However, prefectures on the

Sea of Japan side of Shikoku and Chugoku tended to move away from the other prefectures.

9.6 Comparison to MDS

To verify the effect of the geographical penalties introduced in the BGMDS, we compared them with the standard MDS. Because BGMDS and standard MDS optimize different evaluation criteria, an alternative criterion was required for a fair comparison. Therefore, in this section, we introduced a measurement criterion to compare time–space maps, which quantified the magnitude of the distortion of the time–space maps from the original geographical maps. We focused on the distortions because we should easily understand the locational relationships among cities if the magnitude of the distortion is small. We only compared the magnitude of the distortion using the time–space map under the scenario without airplanes because the use of airplanes justifies a large distortion of the time–space map.

The comparison procedure is as follows. First, we selected three cities from the time–space map and checked whether the relative positional relationships among the three cities were preserved. The triangles of the three cities were considered here. Without a loss of generality, we can choose an arbitrary city as its origin. Suppose that the left–right relationship of the other two cities from the origin is the same as that in the original map. In this case, the relative position relationships among these cities are preserved.[8] This judgment was made for all city combinations, and we employed the rate of violation of the relative position relationship as a measurement criterion for the magnitude of the distortion. The smaller the violation ratio of the relative position relationship, the closer is the relative position relationship to the geographical map. Thus, the time–space map should easily be compared with the original geographical locations.

Figure 9.4 shows a time–space map generated by MDS. We applied the Procrustes transformation (Dryden and Mardia 1998; Kendall et al. 1999)[9] for the result of MDS before mapping because it is not always appropriate for comparison with the original locations. For BGMDS, such a transformation was not required.

The violation rates of the relative position relationship were 56.0% for MDS and 24.0% for BGMDS. Thus, the relative position relationship was better preserved in the BGMDS time–space map than in MDS.[10] A comparison among certain maps

[8] From a computational perspective, we can examine the left–right relationship with the sign of the outer products of the vectors from the origin to other cities.

[9] The Procrustes transformation is a combination of parallel transition, scaling, and rotation that minimizes the distance between two shapes of points.

[10] The possible reasons for the high violation rate in the MDS are as follows. The prefectures are almost along a straight line because the overall shape of the Japanese islands is long and narrow. Focusing on three cities arranged almost in a straight line, a small locational difference in cities breaks their relative position relationship, for example, when the city in the center moves slightly.

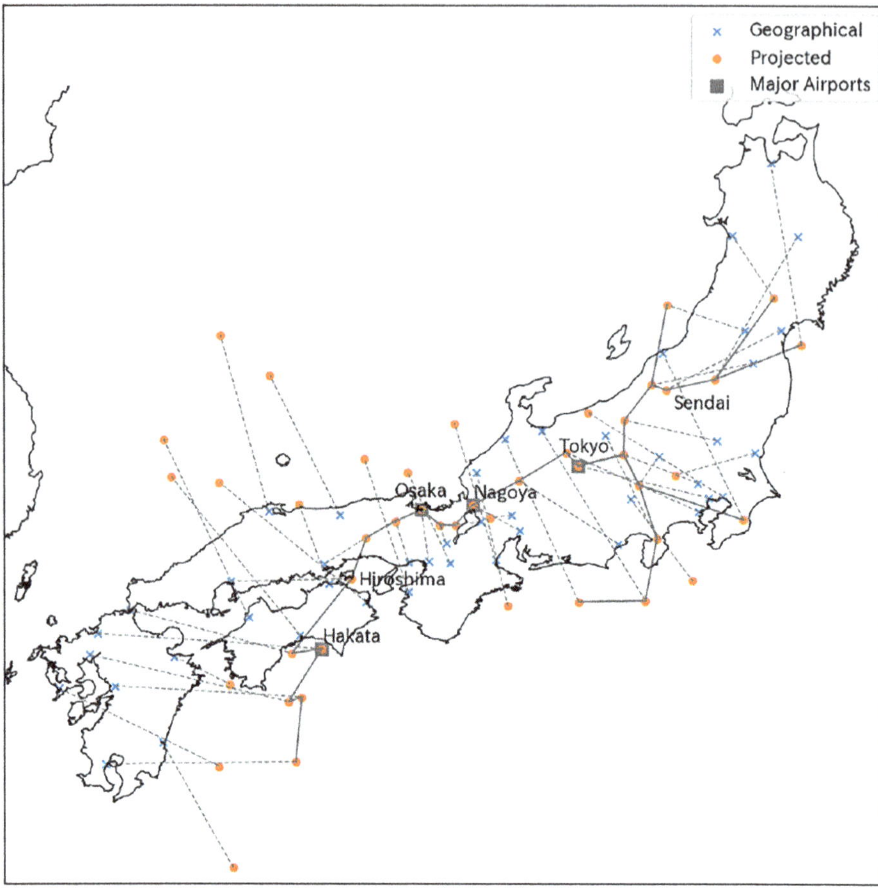

Fig. 9.4 Time–space map provided by MDS + Procrustes transformation. Airplanes are not available on this map

using BGMDS showed that the stronger the geographical penalties, the lower the violation rate. This suggests that geographical penalties are somewhat effective in maintaining relative positional relationships.

The locational characteristics of the time–space maps generated by the MDS are as follows: First, the prefectures in Hokuriku region are located to the south of Tokyo. This provides a large discrepancy with the actual geographical relationships. Prefectures in the Shikoku region are located on the Sea of Japan side. This also contributes to the high violation rate. In the BGMDS, the prefectures in Hokuriku

---

Hence, the violation rate increases if the MDS is applied to Japanese cities. In contrast, it would decrease if we applied MDS for countries such as the United States, where cities are not arranged in a straight line.

region are located to the north of Tokyo and prefectures in the Shikoku region are located on the Pacific Ocean side. Thus, the time–space map using BGMDS is relatively easier to read than that using MDS. However, the locations of prefectures in the Kyushu region were disrupted in both the MDS and BGMDS. This is a common problem in both the methods.

9.7 Discussion and Conclusion

To clarify the features of BGMDS from a different perspective, we interpreted BGMDS as an optimization problem and examined the effect of geographical penalties on the optimization process. As discussed in Sect. 9.4, when we employ variational methods, the approximation of the posterior distribution becomes a maximization problem of ELBO. This is equivalent to the minimization problem of the sign reversal of ELBO, which is the loss function. Because the posterior distributions are proportional to the product of the statistical model $p(\mathbf{D} \mid \mathbf{Z})$ and the prior distribution $p(\mathbf{Z} \mid \mathbf{X})$, the term related to the prior distribution can be isolated from ELBO. Therefore, we can interpret the variational approximation as a penalized optimization with this isolated term as the penalty term.

From this perspective, we obtained the following findings on the optimization process of BGMDS. Figure 9.5 shows how the loss function (the sign reversal of ELBO) changes during the optimization process of the BGMDS model for the scenario without airplanes (blue line). The isolated penalty term related to the prior distribution is indicated by the red line.

First, we focus only on the loss function (blue line). The loss values decreased almost constantly during the iterative optimization procedures (Fig. 9.5, left panel). Next, we focus on the penalty term (red line). It increased temporarily at the beginning and then gradually decreased after reaching a peak (Fig. 9.5, right panel). Thus,

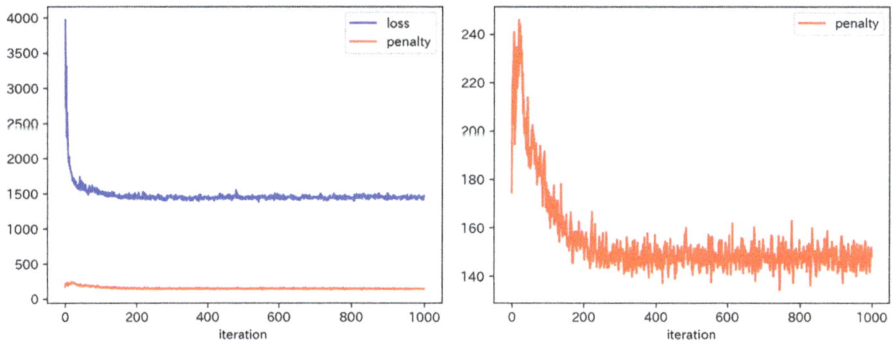

Fig. 9.5 Trajectory of loss values in the optimization procedure of BGMDS for the time–space map in Figure 9.2. The right panel omits the blue line in the left panel to exaggerate the trajectory of the penalty term (red line)

in summary, BGMDS allocated cities without much consideration of geographical penalties at the beginning of the optimization procedure. Subsequently, the locations of the cities were adjusted to be closer to their geographical positions.

Regarding the setting of the hyperparameters λ and τ, the following properties were observed in the time–space maps generated in this chapter. First, if the optimized values of ELBO are similar, we obtained similar city allocations regardless of the hyperparameter settings. However, differences in the size of the standard deviation ellipse were observed in certain cases. Therefore, we should be careful about the setting of the hyperparameters when discussing the uncertainty of the layout from the standard deviation ellipse.

The setting of the hyperparameters also affected the learning of the BGMDS. Specifically, the larger the values of the accuracy parameters λ and τ, the larger the ELBO gradient.[11] Hence, a low learning rate must be selected for the gradient method. Additionally, it is undesirable to set extremely large or small values for the hyperparameters because they make the learning unstable for the reasons mentioned above. Empirically, learning procedures tend to be more stable when the Adam optimizer is used[12] compared with the use of the simple stochastic gradient descent method. In our analyses, the stable values of the learning rate of the Adam optimizer were approximately the reciprocal of the smaller values of the two hyperparameters.

Despite its many advantages, BGMDS has several practical limitations. First, the dissimilarity between cities or facilities $d_{n,m}$ must be symmetric. This constraint condition is reasonable in most scenarios; however, it can limit its applicability to certain situations,[13] such as the road distance, when there are many one-way roads. Second, the units of dissimilarity must be adjusted such that the scales d and δ are comparable because BGMDS provides penalties on the projected location z_n with the distance from the geographical location $\boldsymbol{x}_n$. In the time–space maps presented in Sect. 9.5, the time units were scaled such that the average of time matched the average of the Euclidean distances among geographical locations to visualize the accessibility on the original geographical map. However, there is no optimal scaling method that is applicable to many objectives. An example of a situation wherein our scaling method is unsuitable is as follows. The metaphor "Japan gets smaller" is often used when, for instance, a new transit improves accessibility. The scaling method employed in this chapter does not match this intuitive metaphor. In addition, this method does not allow a direct comparison of, for example, Figs. 9.2 and 9.3 because the distance units of these maps are different. Therefore, when creating two

[11] When the value of the accuracy parameter is large, a small change in the parameter causes a large fluctuation in the loss function. Consequently, the larger the accuracy parameter, the larger the gradient.

[12] The Adam optimizer is an optimization method often used to train deep neural networks, which aims for efficient learning by providing different learning rates and momentums for each parameter.

[13] Another example of a situation that requires asymmetric distances is the travel time of airplanes with long east–west travels. In this case, travel times are asymmetric because of the effects of westerlies and trade winds.

or more time–space maps for comparison, the scaling must be selected such that the distance units are the same.

Despite these limitations, the practical advantages of BGMDS are significant. In particular, the following two facts have notable advantages. First, the coordinates on the dissimilarity map are directly controlled to be close to the geographical coordinates, and second, the strength of the geographical penalties can be determined based on a statistical criterion.

Finally, we discuss the possible future developments in this field. First, certain approaches are required to scale the problem of scaling dissimilarity units, as discussed.[14] In addition, the geographical penalty in BGMDS does not guarantee that relative position relationships, such as top, bottom, left, and right, are preserved (Shimizu 1992). As discussed in Sect. 9.6, BGMDS tends to preserve these relative position relations more often than MDS; however, it is worth explicitly introducing the relative position relations as new penalties or constraints.

Acknowledgments We appreciate the Journal of the City Planning Institute of Japan, where our original Japanese article was published. They have allowed us to publish the English version.

References

Bishop CM (2006) Pattern Recognition and Machine Learning. Springer, New York

Blundell C, Cornebise J, Kavukcuoglu K, Wierstra D (2015) Weight uncertainty in neural network. In: Proceedings of the 32nd International Conference on Machine Learning, vol 37

Dryden I, Mardia K (1998) Statistical shape analysis. Wiley series in probability and statistics. Wiley, Chichester

Kendall D, Barden D, Carne T, Le H (1999) Shape and shape theory. Wiley series in probability and statistics. Wiley, Chichester

Kingma DP, Ba J (2015) Adam: a method for stochastic optimization. In: Proceedings of the 3rd International Conference on Learning Representations. arXiv:1412.6980

Kotoh H (1997) Regional structure and visualized time-distance network. Theory Appl GIS (in Japanese) 5(2):1–10. https://doi.org/10.5638/thagis.5.2_1

Kotoh H (2000) Several time maps and their features. In: The Fall National Conference of Operations Research Society of Japan, abstracts (in Japanese)

Lin L, Fong D (2019) Bayesian multidimensional scaling procedure with variable selection. Comput Stat Data Anal 129:1–13

Oh MS, Raftery AE (2001) Bayesian multidimensional scaling and choice of dimension. J Am Stat Assoc 96(455):1031–1044

Shimizu E (1992) Time-space mapping and its application in regional analysis. Infrastruct Plann Rev (in Japanese) 10:15–29. https://doi.org/10.2208/journalip.10.15

Suyama A (2017) Introduction to machine learning by Bayesian inference (in Japanese). Kodansha, Tokyo

[14] For example, if we wish to reflect the feeling that "Japan gets smaller" due to new transit in a time–space map, a possible scaling approach is the adjustment of the time units by the ratio of travel times before and after the introduction of the new transit. However, the validity of this method depends on the purpose of the time–space map, and it cannot be employed for all purposes.

Tanaka K (2018) Visualization method of spatial flow data based on similarity of destination choice patterns. J City Plann Inst Jpn (in Japanese) 53(3):1464–1471. https://doi.org/10.11361/journalcpij.53.1464

Zinnes JL, MacKay DB (1983) Probabilistic multidimensional scaling: complete and incomplete data. Psychometrika 48(1):27–48

Chapter 10
Factors that Influence Estimation of Building Location in City Blocks in Tokyo Commercial Zones

Masahiro Taima, Yasushi Asami, Kimihiro Hino, and Wataru Morioka

Abstract In the central commercial zones of Tokyo, the importance of city-block restructuring has been strongly emphasized. The predictability of the building development in city blocks is crucial for their planning. This study classifies city blocks by the difference in influential factors of the building location and examines if the city blocks have steady tendencies of building locations. Furthermore, the probability of building coverage for each point in a block is visualized to grasp the spatial image. For our analysis, we used 205 city blocks as reference blocks. Results show influential factors of building location, such as the presence of a wide road and its adjacent direction, the north-south length of city blocks, and the same use of buildings. Further, the study visualizes the probability that each point covered by buildings in a given block and judged for classes has a steady tendency of building location.

Keywords City-block restructuring · Building location · Urgent urban renewal area

---

The contents of this paper are based on the following paper originally published in a Japanese journal: Masahiro Taima, Yasushi Asami, Kimihiro Hino, Wataru Morioka. (2016). Factors that influence estimation of building location in city blocks in Tokyo commercial zones. *Theory and Application of GIS*, 2016, Vol. 24, No.2, pp.85–96. (in Japanese).

M. Taima (✉)
Kyoto University, Kyoto, Japan
e-mail: taima.masahiro.85p@st.kyoto-u.ac.jp

Y. Asami · K. Hino
Department of Urban Engineering, The University of Tokyo, Tokyo, Japan
e-mail: asami@csis.u-tokyo.ac.jp; hino@ua.t.u-tokyo.ac.jp

W. Morioka
University of Illinois at Urbana-Champaign, Champaign, IL, USA
e-mail: watarum2@illinois.edu

Y. Asami et al. (eds.), *Studies in Housing and Urban Analysis in Japan*,
New Frontiers in Regional Science: Asian Perspectives 75,
https://doi.org/10.1007/978-981-99-8027-7_10

10.1 Introduction

10.1.1 Background

Japan is currently entering an era of maturing cities, and the central issue in urban development is the restructuring of existing urban areas through the formation of new ones. There has been recently a strong demand in city centers for the restructuring of the existing urban areas.

The Ministry of Land, Infrastructure, Transport and Tourism formulated the "Guidelines for developing large city blocks" in 2011 and "Work reference materials for utilizing large city blocks to promote urban development" in 2014. These guidelines showed the effectiveness of large city blocks and promoted the efforts of local governments and private urban development businesses (Urban Development and Improvement Division, Ministry of Land, Infrastructure, Transport and Tourism 2011) to restructure them. Restructuring of city blocks and re-organizing them to the desired size enable development by utilizing their regional potentials, such as underutilized land in city centers, motivation for developers, and globally attractive features, such as in business, commerce, and lifestyle.

Based on the above social circumstances, city-block restructuring, particularly in central Tokyo, is thought to be actively implemented in the future. However, empirically predicting how buildings will be located in city blocks following future city-block restructuring is not easy.

10.1.2 Summary of Previous Research

Determining how buildings will be located in city blocks following future city-block restructuring requires a model for estimating building locations in a city block. Existing research that analyzes the relationship between city blocks and building locations includes studies. For example, Saito (2002) has conducted experiments on building locations and showed that rules that maintained road efficiency had formed city blocks. Saito (2004, 2011) has also analyzed how city blocks and road networks differ depending on the adjacency of buildings and measured the state of the surrounding space and used it as an explanatory variable to build a model for estimating the probability that a building is located in an urban area. Vialard (2011) has measured the goodness-of-fit of the buildings already arranged in one city block for another city block. However, no studies estimate building locations from city-block shapes; therefore, in the future, when city blocks are restructured, how buildings will be located within that city block cannot be determined. Asami and Ohtaki (2000) estimate the location of detached houses from the shape of a site; however, this study has a small sample size and a small number of estimated sites. Estimating the location of buildings in a city block requires the use of a larger number of sample sites.

There is research on the utilization of spaces, such as city blocks. Makio et al. (2006) examined changes in the location of buildings owing to condominium conversion at the city-block level for existing residential urban areas. Saito and Kato (2013) targeted central urban areas of provincial cities to clarify changes in land use and current circumstances in terms of city-block units. Nam et al. (2007, 2008) analyzed methods for securing parking spaces in apartment complexes located in existing urban areas as a problem of the balance between securing parking lots and securing green spaces. Nagatomi et al. (2007) determined the current state of land use along major arterial roads for each city block. Nakao and Ito (2012) determined the aspect of urban areas from the density of buildings and building-coverage ratio in a city block. Kawaguchi et al. (2014) quantitatively clarified at the city scale how the size and fluctuations of non-building land were related to the size and fluctuations of green spaces. Matsumiya et al. (2014) measured the distribution of gap rates among buildings in urban areas. However, though these studies have determined spaces to date, they do not estimate future spaces, which requires further analysis.

Various factors determine the location of buildings (henceforth, "influential factors"). Research on the determination of the impact of influential factors on the location of buildings is extensive. Aoki et al. (2010) determined the causal relationship between the regulation of sky-exposure rate and building shape. Kirita et al. (2007) determined the relationship between the location of buildings, such as setbacks, regulation of sky-exposure rate, slopes with front roads, and cut corners. Kawakami and Ohnishi (2013) created a model of usable space for buildings corresponding to building form regulations. Katsuragi and Watanabe (2000) discussed the factors forming building groups lacking continuity from changes in building form regulations. Nakanishi et al. (1996) examined how to introduce wall-line specifications as a method of "placement control" without reducing the feasible volume. Usui and Asami (2011) focused on the densities of buildings and roads to derive the setback distance of buildings. These studies focus on the influence of city-planning regulations on the layout of buildings. However, it is unclear how these urban planning regulations impact which city block characteristics are more or less likely to exhibit building-location trends. Determining building-location trends during future city-block restructuring enables its use in city-block restructuring plans.

10.1.3 Purpose

This study aimed to discuss factors that influence building locations by clarifying the characteristics of city blocks, which are more or less likely to exhibit building-location trends. For city blocks that exhibit these trends, we seek to visualize the probability of building-covered points being within those city blocks and estimate the location of buildings.

In Sect. 10.2, we design a model for estimating the location of buildings in city blocks with an arbitrary shape. In Sect. 10.3, we classify city blocks and use the model for estimation accuracy (magnitude of error) to assess city blocks that are more or less likely to exhibit building-location trends. In Sect. 10.4, we visualize the probability of points being covered by buildings within those city blocks and estimate the location of buildings for city blocks that are likely to exhibit building-location trends. In Sect. 10.5, we discuss the results. Additionally, we discuss factors that influence the building location based on the judgment of city blocks that are likely to exhibit building-location trends within them, as discussed in Sect. 10.3, and the results of estimation of building location outlined in Sect. 10.4. Finally, in Sect. 10.6, we conclude the findings and offer the perspective of this study.

10.2 Research Method

10.2.1 Data Analysis

This study analyzed GIS data provided by Tokyo City (March 2013 edition of Tokyo Metropolitan Planning Geographic Information System).

A city block is an urgent urban renewal area, and a strong demand for city-block restructuring was the target of the study. Different floor-area ratios resulted in big changes in the building location, and detailed analysis becomes difficult. Therefore, targets were limited to city blocks where the floor-area ratio was designated to 600%, which was the largest number for city blocks. We excluded city blocks designated with district plans, districts with special floor-area ratio application, and districts with heights.

10.2.2 Construction of Model for Estimating City-Block Buildings

We adopted the method for estimating the location of detached homes from the site shape (Asami and Ohtaki 2000) in order to build a model for the estimation of the city-block building location.

When two city blocks had similar shapes and environments, the location of the buildings in those city blocks was considered to be similar. This assumption enabled the estimation of building locations from the city-block shape and other environments. We built a model for estimating the location of buildings in a city block based on this assumption.

Building locations in a city block were estimated by superimposing city blocks with the same centroids. The building location was expressed as a probability in the city-block area. For a city block that was referenced by the magnitude of similarity in shape with the city block whose building location we wanted to estimate (henceforth, "reference city block"), we changed the magnitude of the weighting (henceforth, "weight") of the building locations in the reference city block. The probability of the building position estimated from a city block with high similarity is evaluated higher.

The degree of similarity s is defined as following an index representing the degree of similarity between city blocks. A city block is a compact set on a two-dimensional plane. A point on the plane is set as $\mathbf{x}$. A point in the city block X is expressed as $\mathbf{x}$ ($\in X$). The centroid position vector of city block X is expressed as $\mathbf{g}$ (X), and the area of city block X is expressed as A (X). We calculated the degree of similarity s between two city blocks—X and Y—taking the product set when the centroids of the two city blocks were matched and by dividing it with the union set. Therefore, the set obtained by translating the centroid of city block X so that it overlaps the origin was set as $G(X)$. Namely, this is defined in Eq. (10.1):

$$G(X) = \{\mathbf{z} : \mathbf{z} = \mathbf{x} - \mathbf{g}(X), \mathbf{x} \in X\} \tag{10.1}$$

The provisional degree of similarity s^* of city blocks—X and Y—is defined as shown in Eq. (10.2) below based on the concept of Lee and Sallee (1970):

$$s^* (X, Y) = \frac{A\,(G(X) \cap G(Y))}{A\,(G(X) \cup G(Y))} \tag{10.2}$$

In reality, the directions of the roads adjacent to the city block are different, and the buildings in a city block are greatly affected by the direction of roads. Therefore, we considered that city blocks with the changed direction in the range of $\pm\pi/4$ (45°) were almost identical. Therefore, allowing for such directional fluctuations, we defined the final degree of similarity s between city blocks. When a city block is obtained by rotating the city block X by θ with the centroid at its center as $R(X, \theta)$, the degree of similarity s is defined as in Eq. (10.3):

$$s\,(X, Y) = \max_{-\pi/4 \le \theta \le \pi/4} \frac{A\,(G(X) \cap R\,(G(Y), \theta))}{A\,(G(X) \cup R\,(G(Y), \theta))} \tag{10.3}$$

The set of reference city blocks used for estimating the probability of a position of a building in an actual city block is set as reference city block group I. If the city block's shape is similar, the location of the buildings that exist in it is also thought to be similar. Therefore, for a city block X whose building location we wanted to estimate, preferentially using information on the location of buildings in a city block was desirable with a high degree of similarity among those in the reference-city block group I. The degree of similarity value defined above directly could be used as a weight. However, it means that the degree of similarity would never become zero, and no matter how dissimilar the city blocks are, the weight would be positive and would be used in the estimation model of the building location. A better model can be created after city blocks whose shapes are clearly different are excluded. Since city blocks with major differences in shape are excluded, the building location weight needs to be set to 0 for city blocks with clearly different shapes. Therefore, we define the relationship between the degree of similarity s and weight function $f(s)$ as shown in Eq. (10.4):

$$f(s) = \begin{cases} \frac{s-t}{1-t} & s > t \\ 0 & s \leq t \end{cases} \tag{10.4}$$

where t is a parameter between 0 and 1 ($t \in [0, 1]$). When the degree of similarity s is less than or equal to t, the weight becomes 0. The degree of similarity s and weight function $f(s)$ of city block X and city block Y are $s\,(X, Y)$, $f(s\,(X, Y)\,)$, respectively.

Next, we described how to estimate the location of buildings. Estimating the exact shape of a building is difficult; so, we estimated the probability that each point in a city block is covered by buildings. The city block whose building location we wanted to estimate is set as X, and the set of building locations on the set $i(\in I)$ of reference city blocks when the centroid of reference city block $G(i)$ is moved to the origin is set as B_i. When the point $\mathbf{z}$ in $G(X)$ is $\mathbf{z} \in B_i$, it indicates that a building existed at the point $\mathbf{z}$. It is judged that the more B_i covers point $\mathbf{z}$, the more likely the buildings cover the point $\mathbf{x}$ ($\in X$). We define the probability of the estimation of a building's existence as $p\,(\mathbf{x}, X, I)$, which is the probability that a point $\mathbf{x}$ ($\in X$) is covered by a building, and which is weighted by the weight from the degree of similarity of the city block shape defined earlier. First, we defined an indicator function that expressed whether a building B_i of a reference city block existed at an arbitrary point $\mathbf{x}$ on the plane by Eq. (10.5):

$$\chi\,(\mathbf{x}, B_i) = \begin{cases} 1 & \mathrm{x} \in B_i \\ 0 & \mathrm{x} \notin B_i \end{cases} \tag{10.5}$$

Then, the probability of the estimation of a building's existence—$p(\mathbf{x}, X, I)$ at the point $\mathbf{x}(\in X)$ on the city block X—is defined as shown in Eq. (10.6):

$$p\,(\mathbf{x}, X, I) = \frac{\sum_{i \in I} f\,(s\,(X, i))\,\chi\,(\mathbf{z}, B_i)}{\sum_{i \in I} f\,(s\,(X, i))}, \mathbf{x}\,(\in X) \tag{10.6}$$

The parameter t of the weight function $f(s)$ was set to maximize the estimation accuracy of the building. This was done by extracting an arbitrary city block i from the reference city-block group I and estimating the location of buildings with other reference city blocks (the set of remaining reference city blocks after taking i from I is set as I_{-i}), and conducting this operation for each reference city block $i(\in I)$, and finding the value that maximizes the estimation accuracy index ρ defined below. We define the estimation accuracy index ρ as in Eq. (10.7):

$$\rho = \sum_{i \in I} \int_{\mathrm{x} \in B_i} [p\,(\mathrm{x}, i, I_{-i})\,\chi\,(\mathrm{x}, B_i) - p\,(\mathrm{x}, i, I_{-i})\,\,(1 - \chi\,(\mathrm{x}, B_i))\,]\,d\mathrm{x} \tag{10.7}$$

We obtained the estimation accuracy index by subtracting the integral value of the estimation probability of a building's existence at the point where the building was not actually present from the integral value of the probability of the building's existence at the point where the building was actually present. When this sum reached a maximum, the accuracy of a building location's estimation also reached a maximum, and the t at this time was used in the city-block building location estimation model.

10.3 Classification of City Blocks

Next, we classified city blocks to examine whether the error in building-location estimation as per model for estimating city blocks became small. A smaller error after classification than before classification indicated that this classification extracted a city block with a similar building location. Namely, in this classification, a city block was considered where trends were more likely to appear in the city-block building location. Meanwhile, a larger error after classification indicated that the city block of that classification had a large variation in the building location. Namely, in this classification, a city block was considered where trends were less likely to appear in the city-block building location. The above method was used to judge city blocks where trends in city-block building locations were more or less likely to appear.

10.3.1 Method for Calculating Estimation Accuracy (Error Rate) in Model for Estimating City-Block Building Location

First, we calculated the error in estimating the building location in the model for estimating the city-block building location. Research that examined the error of the plane position exists. For example, Teraki (2000a, b, c, d, e) used an error model created under the assumption that the planar position of spatial information was normally distributed around the true position. Chrisman (1987) calculated errors by superimposing spatial information. This study considered it appropriate to calculate the error in estimating the location of the buildings by superimposing the estimated position and true position.

In the present study, the error is calculated as follows. The estimation of the probable existence of the building of a point $\mathbf{x}$ $(\in i)$ in reference city block i $(\in I)$ is expressed as $p\ (\mathbf{x}, i, I_{-i})$. Meanwhile, the function that takes 1 when $\mathbf{x} \in B_i$ and 0 when $\mathbf{x} \notin B_i$ is expressed as $\chi(\mathbf{x}, B_i)$. The error rate of the estimation of the building location as per the model for estimating city-block building location at i is calculated by taking the integral of the absolute value of the difference between $p(\mathbf{x}, i, I_{-i})$ and

$\chi(\mathbf{x}, B_i)$ over the domain of i and dividing by the area $A(i)$ of i. Then this error rate is calculated for each reference city block, and the sum is divided by the total number of reference city-block groups N_I to obtain the error rate of I. If the error rate is set as E, it is defined as shown in Eq. (10.8):

$$E = \frac{\sum_{i \in I} \frac{\int_{\mathbf{x} \in i} |p(\mathbf{x}, i, I_{-i}) - \chi(\mathbf{x}, B_i)| d\mathbf{x}}{A(i)}}{N_I} \tag{10.8}$$

In the present study, E is set as an index of the estimation accuracy of the city-block building location estimation model. Next, city blocks were classified, with city blocks whose error rate E increased or decreased being analyzed, and we clarified city blocks that were more or less likely to exhibit building-location trends.

10.3.2 City-block Classification

Focusing on factors that were thought to impact the location of buildings in city blocks, we classified city blocks.

First, the width of the road in front of the site is thought to be a factor that can affect building locations. Regulations regarding buildings considerably vary depending on the width of the front road. The Building Standards Act sets a width of 12 m as a boundary. In cases where the width of the front road is less than 12 m, the smallest specified floor-area ratio and road width-based floor-area ratio were used as the floor-area ratio of the site. In cases where the width of the front road is 12 m or more, measures, such as the relaxation of setback regulations applied in Category 1 Medium-to-high-rise Exclusive, Residential Districts, Category 2 Medium-to-high-rise Exclusive Residential Districts, Category 1 Residential Districts, Category 2 Residential Districts, or Quasi-residential Districts, were stipulated. Additionally, the Tokyo Metropolitan Comprehensive Design Permission Guidelines set an application requirement in the city center residential comprehensive design system—a comprehensive design system applied to city centers. The requirement is that the width of the front road must be 12 m or more for sites where the floor-area ratio exceeds 600% after the ratio is increased. This shows that the 12-m width of the front road is used as a standard in many circumstances, and the 12-m width of the front road is likely to be a factor that largely impacts the building location.

A width of 12 m is relatively wide for a road. In a few cases in the city center of Tokyo, which is the target area, all roads connected to a city block are wide with a width of 12 m or more. Shape regulations include those that affect the north-south direction of the site, such as regulations on setbacks on adjacent land on the north side, regulations on shade, etc., and the direction is also an important consideration. Also, front roads may impact the building location depending on which direction a wide road connects to the city block.

The shape of a city block can also impact the location of buildings. Kawakami and Ohnishi (2013) show that the vertical shape of the site and the size of the frontage largely impact building-shape regulations and that the magnitude of regulations on shade and on setbacks on the north side differed depending on the north-south width of the city block. When we consider these findings, we can see that the location of buildings is impacted by whether the shape of the city block is long in the east-west direction or in the north-south direction.

Though previous studies have not analyzed the relationship between the land value of a city block with the building location, from the perspective of maximum effective use the land value may impact the building location. The judgment of maximum effective use in real estate appraisal considers the application and shape of the building (Yoshino 2006). The maximum effective use and shape of the building are expected to differ depending on the land value. Therefore, we classify city blocks by their land value.

The building application is thought to be a factor that impacts the building location. The building-design policy differs depending on its application (Naito et al. 2010). Therefore, the building location may differ depending on the building application. City blocks consisting of single-use buildings are expected to exhibit building-location trends more easily. Tokyo commercial zones—the target area of the present study—had city blocks whose building application included only offices, so we decided to consider these city blocks. We also considered limiting the analysis only to other applications (e.g., commercial facilities, housing complexes), but there were few city blocks that consisted of single-use buildings. So we could not conduct such analyses.

We classified city blocks based on the above-mentioned factors—width of the front road, shape of the city block shape, land value, and building application. City blocks whose surrounding roads were classified as less than 12 m or at least 12 m were set as classifications I and II, respectively. We considered the direction in which the front road with a width of at least 12 m is connected to a city block. City blocks connected to a front road with a width of at least 12 m on all four directions were set as classifications III, IV, V, and VI, respectively. City blocks that were long in the east-west and north-south direction were set as classifications VII and VIII, respectively. The average land price per unit area of all city blocks (205 in total) was taken. City blocks with values greater or lower than that average value were set as classifications IX and X, respectively. City blocks where all buildings were used as offices were set as classification XI. For each classification, we applied the model for estimating the city-block building location as explained in Sect. 10.2.2, and we calculated the model-estimation accuracy (error rate E) using the method described in Sect. 10.3.1.

10.3.3 Results

The results of the analysis are summarized in Table 10.1.

Table 10.1 Error rate E of estimations for each classification *E*

Classifications		N_I	t	E
Not classified	All blocks	205	0.79	0.330
Classification I[a]	City blocks whose surrounding roads less than 12 m	127	0.79	0.322
Classification II	City blocks whose surrounding roads at least 12 m	78	0.74	0.350
Classification III[a]	City blocks connected to a front road with a width of at least 12 m on north	30	0.72	0.319
Classification IV[a]	City blocks connected to a front road with a width of at least 12 m on south	21	0.72	0.300
Classification V	City blocks connected to a front road with a width of at least 12 m on east	14	0.64	0.363
Classification VI	City blocks connected to a front road with a width of at least 12 m on west	13	0.68	0.400
Classification VII	City blocks that were long in the east-west direction	101	0.76	0.337
Classification VIII[a]	City blocks that were long in the north-south direction	104	0.79	0.323
Classification IX	City blocks with values greater than that average value	67	0.71	0.356
Classification X	City blocks with values lower than that average value	138	0.74	0.335
Classification XI[a]	City blocks where all buildings were used as offices	34	0.63	0.319

[a]Classification with reduced error rate compared to error rate without classification

City blocks that might exhibit building-location trends were judged as follows. The error rate E in model for estimating city blocks' building location when they were not classified was 0.330. We then classified city blocks, and a lower error rate E of a given classification than 0.330 indicated that this classification extracted city blocks whose building location was similar. This could be thought of as city blocks that were more or less likely to exhibit building-location trends. Meanwhile, a higher error rate E of a given classification than 0.330 indicated that the city block in this classification had a large variation in the building location. It can be said that these city blocks were less likely to exhibit the building-location trends. However, the error rate E increased as the total number of reference city block groups N_I became smaller. So, a definitive conclusion could be made that building-location trends were less likely to appear with these classifications based on our sample size, though its potential could be suggested here. We used this method to judge city blocks whose building-location trends were more or less likely to appear.

The results showed that in classifications I, III, IV, VIII, and XI, the error rate E decreased compared to those that were not classified. Therefore, the building-location trends of these city blocks were more likely to appear. Meanwhile, in classifications II, V, VI, VII, IX, and X, the error rate E increased compared to those that were not classified. So it can be said that building-location trends were less likely to appear in city blocks. We used the above analysis to clarify city blocks whose building-location trends were more or less likely to appear.

10.4 Estimation of City-block Building Locations

In the previous section, we identified city blocks that might exhibit building-location trends. However, how buildings are located in the city block could not be visually understood only through this judgment. When a city block whose building-location trends were more likely to appear was formed by city-block restructuring in the future, knowledge of this building location in advance could provide suggestions for city-block restructuring plans. Therefore, in this section, we used a model for estimating the city-block building location to visualize the building-covered points (estimating the probability of a building's existence) among the city blocks judged to be likely to exhibit building-location trends (classifications I, III, IV, VIII, XI), and we estimated the building location.

The probability of the estimation of a building's existence in the city block was visualized as shown in Figs. 10.1, 10.2, 10.3. A virtual city block was set, and one point per square meter was placed on it. Then, the probability of a building's existence in each point was estimated. This was then converted to raster data, and the probability of the estimation of a building's existence in each point in the city block was visualized. The brightness represents the probability of estimating a building's existence estimation in that part, with brighter areas having a probability closer to 1, and darker areas having a probability closer to 0. The orientation of the figure is north at the top.

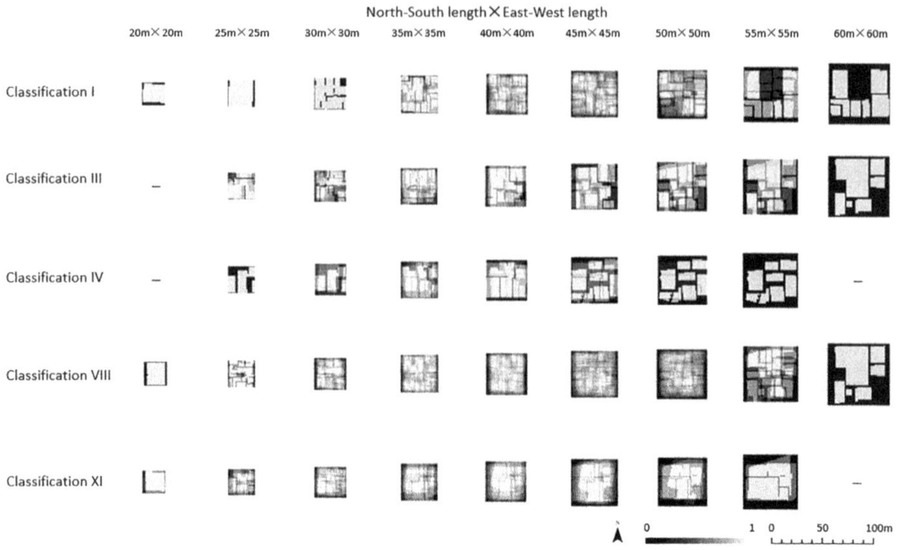

Fig. 10.1 Building existence probability in square city block

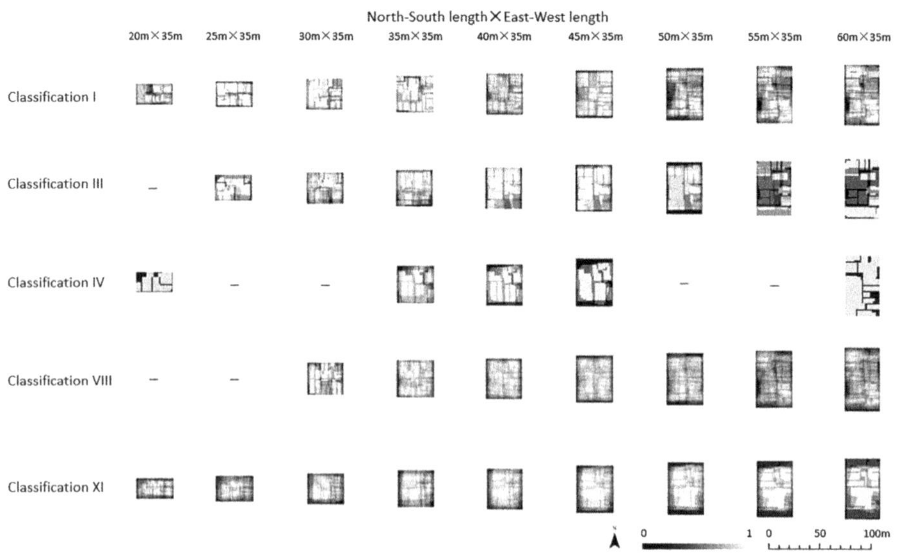

Fig. 10.2 Building existence probability in city block where side extending in east-west direction was unified to 35 m

The shape of the virtual city block was set as a rectangle, and multiple 35 m square city block shapes were set as standards. This shape was set as a standard because the average area and perimeter of all city blocks were 1285.54 m^2 and 147.41 m, respectively. If a city block was set as a rectangle, a city shape with an

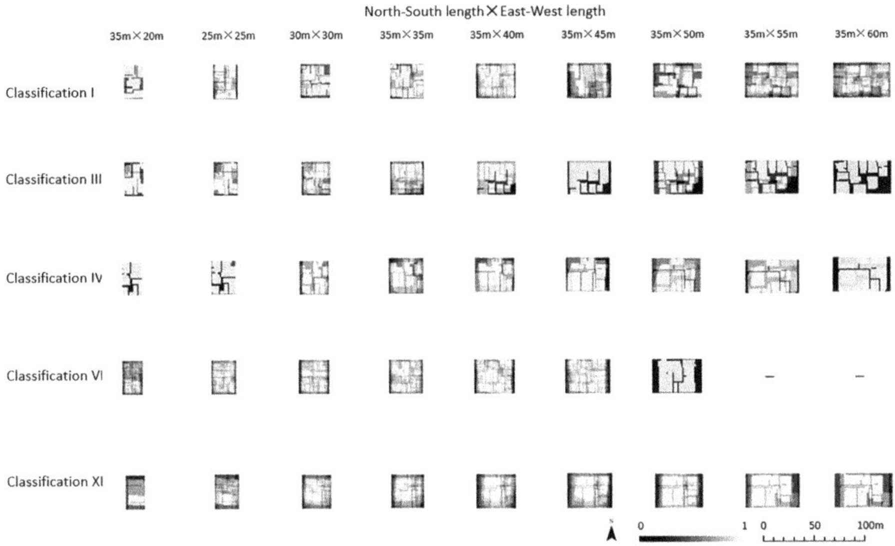

Fig. 10.3 Building existence probability in city block where side extending in north-south direction was unified to 35 m

average shape would be calculated as a city block of roughly 35 m square. A 35 m-square city block was set as a standard for the virtual city block, and the length of one side of the city block was changed at intervals of 5 m between 20 m and 60 m, which is the length in which all city blocks can be roughly accommodated. We then conducted a sensitivity analysis.

Probabilities of estimating a building's existence were visualized when the following city blocks were set: a square virtual city block with a side length of 20–60 m in Fig. 10.1. A virtual city block with a side extending in the east-west direction unified at 35 m with the side extending in the north-south direction to 20–60 m as shown in Fig. 10.2. A virtual city block with a side extending in the north-south direction unified at 35 m with the side extending in the east-west direction to 20–60 m as shown in Fig. 10.3. A virtual city block that does not have a reference city block with similarity greater than, or equal to, t cannot have its building location estimated according to the model for estimating the building location of a city block. So the corresponding virtual city block in the figure is marked with a "–" symbol.

10.5 Discussion

In this section, we discuss the above results.

Among classifications I and II, only classification I resulted in an error rate E, which was lower than that for unclassified cases. So, the presence or absence of a city block connected to a front road with a width of 12 m impacted the location of buildings in that city block. This was an influential factor which occurred because buildings in the city block of classification I were evenly located throughout the block and its location was not biased in the center of the block owing to regulations on setbacks from narrow roads. It can also be seen in Figs. 10.1, 10.2, 10.3. Classification I showed the trend of buildings being more evenly located in a city block compared to other classifications.

Among classifications III, IV, V, and VI, classifications III and IV had an error rate E that was lower than that of unclassified cases. Namely, if the direction where the city block connected the wide road was north or south, it was an influential factor in building location. The north-south direction should be particularly considered for securing sunlight, and it is thought that this affects the location of buildings. Looking at classifications III and IV in Figs. 10.1, 10.2, 10.3, we can see that building locations in classifications III and IV are the north and south sides.

Among classifications VII and VIII, only classification VIII had an error rate E that was lower than the error rate of unclassified cases. The north-south direction is important for securing sunlight, and the buildings in city blocks whose shape is short in the north-south direction needed to be arranged by shifting them to the east and west to secure sunlight, whereas city blocks whose shape is long in the north-south direction can secure sufficient space in the north-south direction, and their buildings do not need to be arranged by shifting them to the east and west. Namely, it is predicted that city block with long shapes in the north-south direction has little building-location variation in the east-west direction, and this becomes an influential factor in the building location. Looking at Fig. 10.2, we can see that city blocks that are longer in the north-south direction (Fig. 10.2: 50 m $\times$ 35 m, 55 m $\times$ 35 m, 60 m $\times$ 35 m) have more buildings concentrated in the north and south sides of the city block.

Both classifications IX and X had an error rate E that was higher than the error rate of unclassified cases. As described in Sect. 10.3.3, a smaller total number of reference city block groups N_I decreases the accuracy of the estimation of the block's building location. Therefore, a higher error rate E in these classifications than that in the non-classified buildings is thought to be simply the absence of building-location trends, rather than the building location variation becoming bigger owing to the classification. If a trend emerges in the building location based on a given land-value boundary, the error rate E of either classification IX or X is expected to be lower than that for unclassified cases. However, since all the target city blocks were city centers, the overall land value was high, and no building-location trends were thought to have appeared. Namely, the land value did not impact the location of the city-block building in the targeted Tokyo commercial zones.

Classification XI had an error rate E lower than that for unclassified cases. Offices are required to have a high rentable ratio and an efficient workplace environment. These characteristics are reflected in the design (Naito et al. 2010). Therefore, city blocks consisting only of offices are thought to have similar building locations. The result of classification XI suggests that, even for non-office applications, constructing a city block with the same building application would be an influential factor for building location. Figures 10.1, 10.2, 10.3 show that classification XI has buildings located in the center of the city block compared to other classifications.

In Sect. 10.4, we used the model for estimating the building location of city blocks judged to be likely to exhibit building-location trends (classifications I, III, IV, VIII, XI). In Sect. 10.3, results showed that the location of buildings could be determined in city blocks of various shapes. In the future, city-block restructuring is expected to be actively implemented in Tokyo commercial zones. However, when a city block formed by city-block restructuring in the future is similar to a virtual city block as shown in Figs. 10.1, 10.2, 10.3 in terms of shape and classification, the location of buildings in the city block after city-block restructuring can be similar to the location of buildings in the corresponding virtual city block. This may not only provide suggestions for city-block restructuring plans, but also be applied to various aspects, such as its utilization in residents' consensus building and rough estimation of future townscapes and wind direction.

10.6 Conclusion

The present study provides an outlook that a future study should analyze in detail—building-location trends and building locations with different floor-area ratios and expand the building-shape estimation from two dimensions on the first-floor level, as was done in this study, to the third dimension.

Acknowledgments The data received in the analysis for this study was provided by the Tokyo Metropolitan Bureau of Urban Development. We would like to express our gratitude here. This paper is translated from the original paper published in Theory and Application of GIS, 2016, Vol. 24, No.2, pp. 85–96.

References

Aoki M, Ohsawa Y, Kirita H, Kobayashi T (2010) Building shape caused by architectural volume maximization under sky exposure criterion regulation. J Archit Plan (Trans AIJ) 75(648):403–410

Asami Y, Ohtaki T (2000) Prediction of the shape of detached houses on residential lots. Environ Plan B Plan Design 27:283–295

Chrisman HC (1987) The accuracy of map overlays: a reassessment. Landsc Urban Plan 14:427–439

Katsuragi K, Watanabe S (2000) An analysis of the process in mixed townscape with strange form buildings under the changes of building controls. J City Plan Inst Jpn 35:763–768

Kawaguchi N, Murayama A, Shimizu H, Takatori C (2014) The state of scale of open space/green covered space in city blocks and their temporal change, and the characteristic feature of their distribution on Nagoya blocks. J City Plan Inst Jpn 49(3):207–212

Kawakami M, Ohnishi H (2013) Study on assessment of regulation on building form using modeling of building space, form and its simulation. J Archit Plan (Trans AIJ) 78(687):1041–1048

Kirita H, Ohsawa Y, Hasuka F, Nakagawa M (2007) Effect of sky exposure rate regulation on building layout. J Archit Plan (Trans AIJ) 78(617):71–78
Lee DR, Sallee GT (1970) A method of measuring shape. Geogr Rev 60(4):555–563
Makio H, Sugiyama S, Tokuono T, Nakaniwa Y (2006) The increases of apartments and the changes in land uses of street-facing open spaces in extent residential areas. J Archit Plan (Trans AIJ) 604:1–7
Matsumiya K, Washizaki M, Oikawa K, Gota M (2014) Quantitative analysis of gap between buildings. J Archit Plan (Trans AIJ) 79(697):693–699
Nagatomi T, Sato S, Kobayashi Y, Himeno Y (2007) Elucidation of the land use condition on the Main road side by building distribution conditions for block units: main roadside of Oita City, Oita prefecture. J City Plan Inst Jpn 42(3):517–522
Naito K, Hashimoto M, Hiiro M, Fujita D (2010) Architectural planning utilized in design. Gakugei Publishing, Kyoto
Nakanishi M, Nakai N, Saito C (1996) A study of building line control and building arrangement. J City Plan Inst Jpn 31:523–528
Nakao N, Ito Y (2012) Study on aspects of city urban space based on building density on blocks. J Archit Plan (Trans AIJ) 77(677):1689–1697
Nam T, Usami M, Sugiyama S, Tokuono T (2007) The parking lot installation methods in the apartment housings located in built-up areas. Journal of Architecture and Planning (Transactions of AIJ) 614:17–24
Nam T, Sugiyama S, Tokuono T (2008) The characteristic of apartment houses located in built-up areas from the parking location and building dispositions. J Archit Plan (Trans AIJ) 73(623):23–30
Saito C (2002) Urban block pattern and efficiency of road network: a study on simulation method of block generation by building arrangement. J City Plan Inst Jpn 37:85–90
Saito C (2004) Effect of lattice on urban block pattern and city shape: a study on influence of neighborhood relations of buildings on urban space. J City Plan Inst Jpn 39:847–852
Saito C (2011) Prediction of buildings emergent position from accessibility, independence and distribution of shortest path between buildings: logistic regression model of buildings arrangement in typical urban block. J City Plan Inst Jpn 46(3):397–402
Saito M, Kato M (2013) An approach to the change and the actual condition of land use in the block unit: case study on Taira central city area in Iwaki City. J City Plan Inst Jpn 48(3):315–320
Teraki A (2000a) Error model of horizontal position on spatial information: estimation of horizontal precision of spatial information: part 1. Theory Appl GIS 8(1):83–90
Teraki A (2000b) Error model of horizontal position on spatial information: estimation of horizontal precision of spatial information: part 2. Theory Appl GIS 8(1):91–97
Teraki A (2000c) Error model of horizontal position on spatial information: estimation of horizontal precision of spatial information: part 3. Theory Appl GIS 8(1):99–105
Teraki A (2000d) Error model of horizontal position on spatial information: estimation of horizontal precision of spatial information: part 4. Theory Appl GIS 8(2):33–41
Teraki A (2000e) Evaluation of accuracy of horizontal position on spatial information. Theory Appl GIS 8(2):43–49
Urban Development and Improvement Division, Ministry of Land, Infrastructure, Transport and Tourism (2011) Formulation of "guidelines for developing large city blocks". Land Readjustment 54(3):12–13
Usui H, Asami Y (2011) Setback distance and density of buildings and roads. J City Plan Inst Jpn 46(3):829–834
Vialard A (2011) Measures of the fit between street network, urban blocks and building footprints Savannah and Atlanta. In Proceedings of the 8th international symposium on space syntax, 8101, pp. 1–17
Yoshino S (2006) Regional analysis and individual analysis. "5th ed: Special areas and appraisals" (Land Appraisal Theory Study Group). Seibunsha, Osaka, pp 199–202

Chapter 11
A Diagnostic Approach to the Multicollinearity Problem for Better Model Selection in the Hedonic Pricing Method

Yuki Hiruta and Yasushi Asami

Abstract The hedonic pricing method is an effective technique for estimating land and building prices. Nonetheless, numerous analysts have expressed concerns regarding the potential for multicollinearity to inflate the variance of the estimated coefficients. Hedonic pricing often involves spatially situated objects, and so strong correlations between two or more variables frequently arise. There are several diagnostics for evaluating multicollinearity, although most—sourced from explanatory variables—only measure the severity of this condition. To assess the risk of multicollinearity in a comprehensive manner, it is imperative to develop diagnostics that consider both the probability of occurrence and the severity of multicollinearity. This paper presents a numerical method for evaluating the influence of multicollinearity from these two perspectives.

Keywords Hedonic regression · Multicollinearity · Variance inflation factor · Numerical analysis · Generalized linear model

---

The contents of this paper are based on the following paper originally published in a Japanese journal: Yuki, H. and Asami, Y. (2018) A diagnostics of multicollinearity problem for better model selection in hedonic approach. Urban housing sciences 102: 113–122 (in Japanese).

Y. Hiruta (✉)
Research Center for Net Zero Carbon Society, Institutes of Innovation for Future Society, Nagoya University, Nagoya, Japan
e-mail: hiruta.yuki.m3@f.mail.nagoya-u.ac.jp

Y. Asami
Department of Urban Engineering, The University of Tokyo, Tokyo, Japan
e-mail: asami@csis.u-tokyo.ac.jp

Y. Asami et al. (eds.), *Studies in Housing and Urban Analysis in Japan*,
New Frontiers in Regional Science: Asian Perspectives 75,
https://doi.org/10.1007/978-981-99-8027-7_11

11.1 Introduction

The hedonic method is frequently used to estimate prices for assets such as land and buildings and to evaluate the economic value associated with environmental factors. Within the hedonic approach, the prices of commodities like real estate are perceived as aggregates of various attributes, encompassing aspects such as convenience and performance. These prices are subsequently modeled through regression analyses, as outlined by Rosen (1974) and Witte et al. (1979). Challenges arise when multiple variables within a regression model exhibit some degree of correlation, inducing multicollinearity effects. In urban contexts, where structures are spatially oriented, it is especially challenging to maintain full independence among the explanatory variables, often because of the spatial locations and underlying, non-explicit common factors (Heikkila 1988; Deaton and Hoehn 2004). The existence of multicollinearity can inflate the variance of the regression coefficients,[1] thereby compromising the model's reliability (e.g., Alin 2010; Kutner et al. 2005, p. 283). Multicollinearity is technically defined in terms of the lack of linear relationships and orthogonality (e.g., Chatterjee and Hadi 2006). Although perfect collinearity is rare in real-world datasets, the more pervasive concern is the near-linear dependence between explanatory variables, leading to a form of imperfect multicollinearity (Asteriou and Hall 2007).

In empirical studies that employ the hedonic approach, the primary objective is often to evaluate the economic value of environmental factors. Inaccurate estimations of the regression coefficients can significantly skew the conclusions. However, excessive cautiousness in variable selection as a result of concerns about multicollinearity can lead to the omission of valuable information. Thus, an accurate diagnosis of the effects of multicollinearity is essential.

Numerous methods for detecting the effects of multicollinearity in advance have been proposed. These include the correlation matrix of explanatory variables, R; the determinant of the correlation matrix, det(R); the variance inflation factor (VIF); the maximum VIF among all explanatory variables, $\mathrm{VIF}_{\max}$; the mean value of VIF_k, $\overline{\mathrm{VIF}}$; and the eigenvalues and eigenvectors of R (Alin 2010). Typically, when constructing hedonic models, these indicators are used to provide a preliminary estimate of the severity of collinearity. If the effects of collinearity are of concern, variable selection or variable consolidation can be applied through principal component analysis or factor analysis (e.g., Yoshida 1987; Chatterjee and Hadi 2006). Among the detection methods for multicollinearity, VIF_k is widely used

[1] The variance inflation of regression coefficients as a result of multicollinearity is explained by a mathematical formula. Regarding Eq. (11.1), the variance of the estimated partial regression coefficient $\hat{a}_k$ is given by $V\left(\hat{a}_k\right) = \sigma^2\left(\mathbf{X}^{\mathrm{t}}\mathbf{X}\right)^{-1}$. Specifically, when $p = 2$, the variance of the estimated partial regression coefficient $\hat{a}_1$, $V\left(\hat{a}_1\right)$ can be written as $V\left(\hat{a}_k\right) = \sigma^2/n\left(1 - r_{12}^2\right)$. Here, r_{12} is the correlation coefficient between variables x_1 and x_2. From this, it can be understood that a larger correlation coefficient r_{12} between the two variables corresponds to a larger variance $V\left(\hat{a}_1\right)$ of the partial regression coefficient $\hat{a}_1$.

as a formal measure (Kutner et al. 2005); when a single metric for a regression model is required, the mean of VIF_k, denoted as $\overline{\mathrm{VIF}}$, is often employed.

When considering the multiple regression Eq. (11.1), the dependent variable is represented by the vector $y = (y_1, \cdots, y_i, \cdots, y_n)'$, which consists of n observations. The explanatory variables are denoted by the matrix $X = [x_1, \cdots, x_k, \cdots, x_p]$, which comprises n observations and p types of variables. The regression coefficients are represented by the vector $a_k = (a_1, \cdots, a_k, \cdots, a_p)'$. VIF_k is expressed by Eq. (11.2). In this context, ${R_k}^2$ is the coefficient of determination when a particular explanatory variable x_k is linearly regressed using explanatory variables other than x_k in Eq. (11.1). VIF_k can be considered as an indicator that represents the extent to which the variance of a particular regression coefficient estimate expands in comparison with the case of no collinearity.[2]

$$\mathbf{y} = \mathbf{Xa} + \boldsymbol{\mu} \quad \boldsymbol{\mu} \sim N\left(\mathbf{0}, \sigma^2\mathbf{I}\right) \tag{11.1}$$

$$\mathrm{VIF}_k = \frac{1}{1 - {R_k}^2} \quad k = 1, 2, \cdots, p \tag{11.2}$$

$\overline{\mathrm{VIF}}$ is defined by Eq. (11.3). This indicator illustrates the extent to which the regression coefficients estimated in the model deviate from the true values (hereafter referred to as the "deviation of the regression coefficients") compared with a model that has no collinearity, $\sum_{k=1}^{p}\mathrm{VIF}_k = p$. This is represented using only explanatory variables and reflects the deviation of the regression coefficients regardless of their variance.[3]

$$\overline{\mathrm{VIF}} = \frac{\sum_{k=1}^{p}\mathrm{VIF}_k}{p} \tag{11.3}$$

However, the deviation of the regression coefficients, as indicated by $\overline{\mathrm{VIF}}$, is merely an expected value. Even for regression models with the same $\overline{\mathrm{VIF}}$, there

[2] In the standardized regression model $\mathbf{y}^* = \mathbf{X}^*\mathbf{a}^* + \boldsymbol{\mu}^*$, with $\boldsymbol{\mu}^* \sim \mathrm{N}(\mathbf{0}, \sigma^2\mathbf{I})$ and using standardized variables from Eq. (11.1), the term $\mathbf{X}^t\mathbf{X}$ in the variance $\mathrm{V}\left(\hat{a}_k\right) = \sigma^2\left(\mathbf{X}^t\mathbf{X}\right)^{-1}$ becomes the correlation matrix $\mathbf{r}_{xx}$. Thus, $\mathrm{V}\left(\hat{a}_k\right)$ is equivalent to $V\left(\hat{a}_k^*\right) = (\sigma^*)^2(\mathbf{r}_{xx})^{-1}$, with its k-th diagonal element being VIF_k. If the determination coefficient ${R_k}^2$ is 0, $\mathrm{VIF}_k = 1$; otherwise, as ${R_k}^2$ increases, VIF_k grows.

[3] The product of the sum of VIF_k and the variance of the error term $(\sigma^*)^2$ become the expected value of the sum of the squared differences between the estimated regression coefficient $\hat{a}_k^*$ and its true value t_k^*, given the standardized regression model $\mathbf{y}^* = \mathbf{X}^*\mathbf{a}^* + \boldsymbol{\mu}^*$, where $\boldsymbol{\mu}^* \sim \mathrm{N}(\mathbf{0}, \sigma^2\mathbf{I})$, which is derived by standardizing all the variables in Eq. (11.1) (see below). When $R_k^2 \equiv 0$, $\mathrm{VIF}_k \equiv 1$, and $\sum_{k=1}^{p}\mathrm{VIF}_k$ equals p. Therefore, the mean value of $\mathrm{VIF_k}$, $\overline{\mathrm{VIF}} = \sum_{k=1}^{p}\mathrm{VIF}_k/p$, can be said to be the expected value that indicates the extent to which the estimated regression coefficient deviates from the true regression coefficient when considering the case where all explanatory variables are not linearly related to other explanatory variables (Kutner et al. 2005).

can be instances where no deviation occurs, or the deviation can be several times larger than the expected value depending on the sample applied. Moreover, there is no objective criterion that indicates the level metrics such as VIF_k or $\overline{\mathrm{VIF}}$ can reach before multicollinearity becomes a concern. This often leads to ambiguous judgments of the form "If VIF_k exceeds 10, the variance might be too large, and its reliability could be compromised" (Minotani 2015). As a result, analysts might consider strategies to address multicollinearity without fully understanding its direct impact on the estimated regression coefficients and values of the dependent variable. When constructing regression models, it is not only important to know the expected deviation of the regression coefficients represented by $\overline{\mathrm{VIF}}$, but also to understand the following:

1. The potential severity of the deviation.
2. The probability of such a deviation occurring (probability of deviation).

It therefore becomes essential to consider not only the expected values, but also the distribution of the deviations of the regression coefficients. This is because, even if $\overline{\mathrm{VIF}}$ (expected value) and the variance are consistent, different distributions of the deviations can change the likelihood of exceeding a certain level of deviation.

The goal of this research is to provide insights for analysts constructing hedonic models in their professional or research endeavors, enabling a better understanding of the implications of multicollinearity. Specifically, this study offers methods and numerical benchmarks for determining the maximum possible deviation of the regression coefficients when a specific $\overline{\mathrm{VIF}}$ is derived, as well as the likelihood of such deviations occurring. Although previous studies have examined the extent of the deviation of the regression coefficients for different explanatory variables through simulations (Yoo et al. 2014), there are no studies with similar objectives and methods to the present research.

11.2 Methods

In this study, we use the generalized linear model (GLM) to simulate the relationship between (1) the deviation of the regression coefficient from its theoretical value, denoted as $\overline{\mathrm{VIF}}$, and (2) the actual deviation of the estimated regression coefficient from its true value. The GLM framework enables the expected value to be estimated and the error structure (residuals) of the dependent variable to be modeled as a probability distribution, allowing for the determination of its parameters. This approach is advantageous for understanding the relationship between the theoretical deviation value $\overline{\mathrm{VIF}}$ and the actual deviation, as it illustrates the variation in the actual deviation for each theoretical value. In contrast, existing methods only evaluate the expected value, without considering the variation.

To elucidate these relationships using the GLM without any dependency on the configuration, we construct regression models across a spectrum of scenarios: from those where explanatory variables have minimal correlation to those with

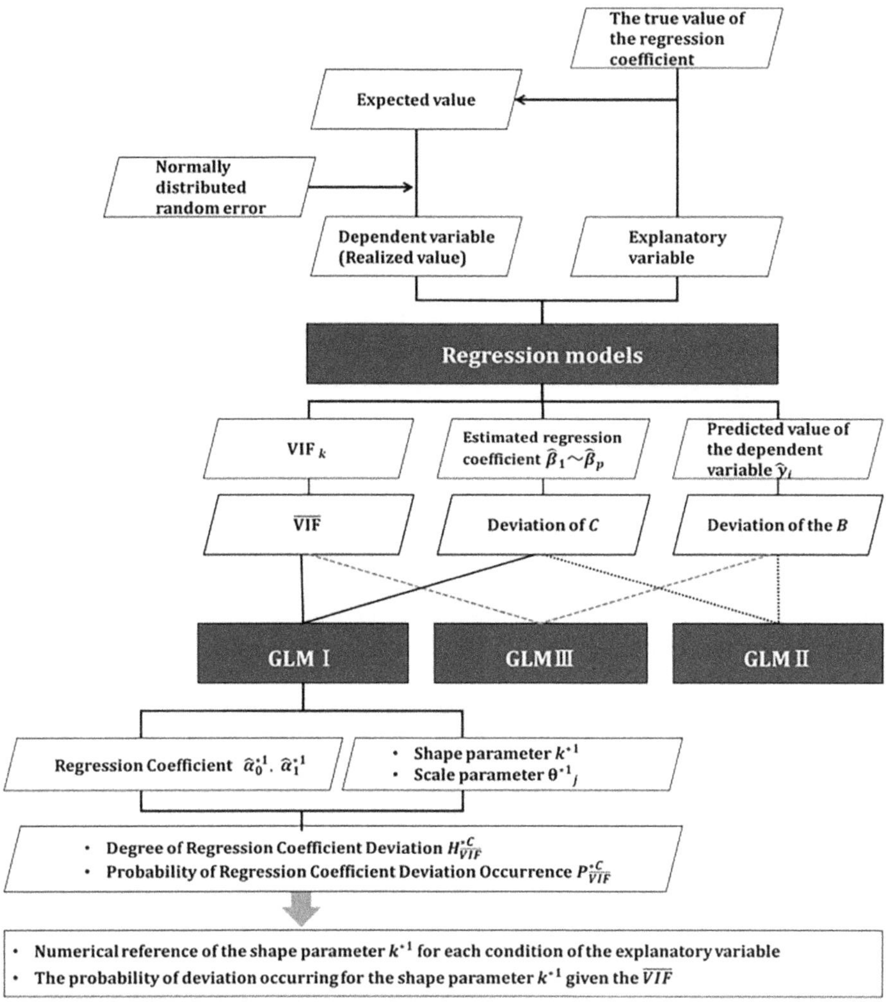

Fig. 11.1 Overview of proposed method

strong correlation. We employ a method that produces diverse variable sets through numerical simulations using random numbers.

An overview of the methodology is depicted in Fig. 11.1. Initially, we create several datasets of explanatory variables with a variety of inter-variable correlations. Using each dataset and pre-defined true regression coefficients, we calculate the true dependent variable (expected value). Subsequently, by incorporating random error terms in the expected value, we derive the actual values for the dependent variable in the regression model. Using the obtained dependent variables (actual values) and the primary explanatory variables, we fashion multiple regression models. Through these models, the deviation between the estimated regression coefficients and their true values is quantified. Furthermore, $\overline{\text{VIF}}$ is directly derived from each explanatory variable dataset. We form a GLM with the metric representing the genuine deviation as the dependent variable and $\overline{\text{VIF}}$ as the explanatory variable.

The proposed method is applied to seven distinct regression model types to illustrate the different impacts of multicollinearity. In addition, the robustness of the constructed GLM and the extracted probability distribution parameters is verified. Finally, considering the influence of multicollinearity on the estimated values of the dependent variable in relation to $\overline{\text{VIF}}$ and the deviation of the regression coefficients, we suggest a diagnostic approach for assessing the effects of multicollinearity.

11.2.1 Construction of Regression Models

11.2.1.1 Generation of Explanatory Variables

The focus of this study is the influence of multicollinearity resulting from the correlation relationships among individual variables in regression models with 2–4 explanatory variables. For regression models with three or more variables, we consider both the presence and absence of independent variables. In this section, we describe the types of datasets and the method for generating variables that exhibit diverse correlations.

1. Types of Datasets

Figure 11.2 presents seven conditions that differ in their number of explanatory variables. The double-lined arrows in the conceptual diagram indicate that the variables were generated using the method described below, ensuring that the 5000 sets of explanatory variables encompass a broad spectrum of correlation relationships, ranging from sets with a correlation coefficient of -1 to those with a correlation coefficient of $+1$. The bold arrows denote correlation relationships that are automatically determined by the correlations of variables not linked by these bold arrows. The dotted arrows symbolize independent relationships between explanatory variables. Variables connected solely by dotted arrows are derived from random numbers that follow a normal distribution with a mean of 0 and a variance of 1. All the generated explanatory variables have been standardized to have a mean of 0 and a variance of 1.

2. Method for Generating Variables Demonstrating Diverse Correlations

The double-lined arrows in Fig. 11.2 imply that the 5000 sets of explanatory variables uniformly include sets exhibiting a variety of correlation relationships, from those with a correlation coefficient of -1 to those with a correlation coefficient of $+1$. These sets of explanatory variables were generated using the method

	Explanation	Number of explanatory variables	Concept Diagram
a.	Two variables showing various correlation coefficients	$p = 2$	
b.	Three variables showing various correlation coefficients	$p = 3$	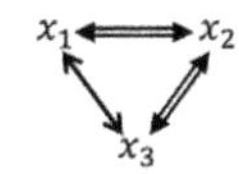
c.	Two variables showing various correlation coefficients and one variable with little to no correlation with the others	$p = 3$	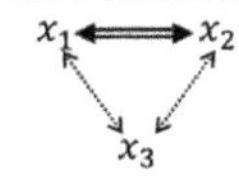
d.	Four variables showing various correlation coefficients	$p = 4$	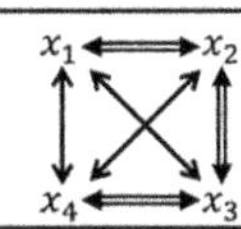
e.	Three variables showing various correlation coefficients and one variable with little to no correlation with the others	$p = 4$	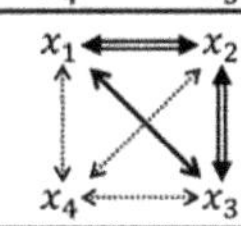
f.	Two variables showing various correlation coefficients and two variables with little to no correlation with the others	$p = 4$	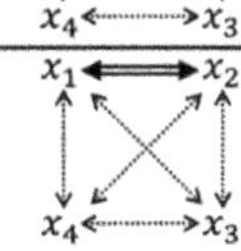
g.	Two sets of two variables each showing various correlation coefficients	$p = 4$	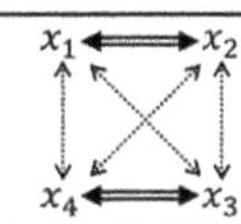

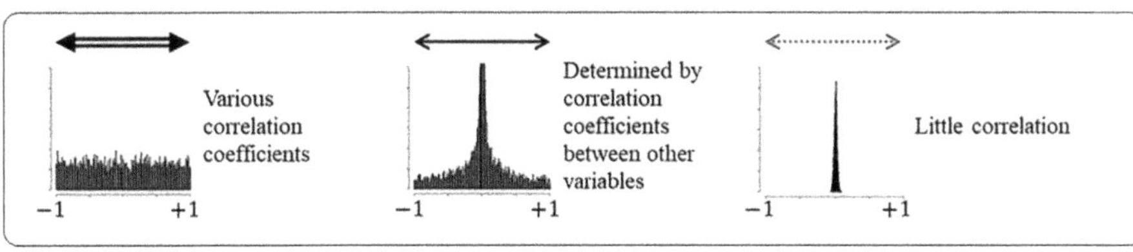

Fig. 11.2 Patterns of explanatory variable generation (seven conditions)

outlined below, leveraging Eq. (11.4), where the variable x_2 predominantly exhibits a correlation coefficient of ρ with x_1:

$$x_2 = \rho x_1 + \sqrt{1 - \rho^2 a} \tag{11.4}$$

Initially, a distinct ρ is determined for each set of explanatory variables. The value of ρ is randomly selected from a uniform distribution ranging from −1 to 1. Subsequently, the explanatory variable x_1 and the variable a are formed by randomly drawing 2000 values from a normal distribution with mean 0 and variance 1. The dependent variable x_2 is computed by applying the determined correlation coefficient ρ along with the independent variables x_1 and a to Eq. (11.4). Repeated iterations of this procedure produce sets of variables (x_1, x_2) with correlation coefficients that span from -1 to $+1$.[4]

11.2.1.2 Generation of the Dependent Variable (Realized Value)

We elucidate the effects of multicollinearity by analyzing the differences between the estimated regression coefficients and the dependent variable in regression models where the true regression coefficients are known. To do this, we set the explanatory variables x_{ik} (where $i = 1, 2, \ldots, n; k = 1, 2, \ldots, p$) and the true regression coefficients ($\beta_0 = 0, \beta_k = 1$) during the variable-generation stage. Using Eq. (11.5), we first calculate the expected value (true value) λ_i. Then, by adding a random error ε_i that is normally distributed with mean 0 and variance σ^2, we generate the dependent variable (realized value) y_i using Eq. (11.6). Note that y_i is not standardized so that the actual degree of discrepancy can be captured.

Expected value:

$$\lambda_i = E[y_i] = \beta_0 + \beta_1 x_{i1} + \beta_2 x_{i2} + \cdots + \beta_p x_{ip} \tag{11.5}$$

Realized value:

$$y_i = \lambda_i + \varepsilon_i \qquad \varepsilon_i \sim N\left(0, \sigma^2\right) \tag{11.6}$$

11.2.1.3 Construction of Regression Models

Using datasets consisting of explanatory and dependent variables, we perform regression according to Eq. (11.7). This regression formula provides the estimated regression coefficients ($\hat{\beta}_0, \ldots, \beta_p$). The estimated value of the dependent variable $\hat{y}_i$ is given in Eq. (11.8). (The estimated value of the dependent variable is simply referred to as the "estimated value").

Regression formula:

$$y_i = \beta_0 + \beta_1 x_{i1} + \beta_2 x_{i2} + \cdots + \beta_p x_{ip} + \varepsilon_i \tag{11.7}$$

[4] Using this method to generate a set of variables introduces some bias. Therefore, we produced a variable set 1.5 times larger than the required sample size. By stratifying and randomly selecting based on the correlation coefficients calculated from the generated variable sets (x_1, x_2) at intervals of 0.005, we ensured that the correlation coefficients of the generated pairs of explanatory variables were as uniform as possible, ranging from -1 to 1.

Estimated value:

$$\hat{y}_i = \hat{\beta}_0 + \hat{\beta}_1 x_{i1} + \hat{\beta}_2 x_{i2} + \cdots + \hat{\beta}_p x_{ip} \tag{11.8}$$

11.2.2 Indicators Representing the Impact of Multicollinearity

11.2.2.1 Calculation of Indicators

To understand the effects of multicollinearity, we calculate the following indicators: $\overline{\mathrm{VIF}}$ using Eq. (11.10); the deviation C of the regression coefficients using Eq. (11.11); and the deviation B of the estimated values using Eq. (11.12). $\overline{\mathrm{VIF}}$ reflects the expected deviation of the regression coefficients, representing how many times greater the deviation is in comparison with a model that has no collinearity. This metric is derived from the explanatory variables. Conversely, C signifies the discrepancy between the estimated regression coefficients ($\hat{\beta}_1, \ldots, \hat{\beta}_p$) obtained from the regression model and the actual regression coefficients ($\beta_1, \ldots, \beta_q$). Both $\overline{\mathrm{VIF}}$ and C are crucial in comprehending the deviation in the regression coefficients because they provide insights into the relationship between the theoretical and actual deviations.

The deviation B of estimated values highlights the difference between the predicted value of the dependent variable $\hat{y}_i$ and its true or expected value, denoted as λ_i. This indicator is especially helpful in gauging the extent to which the deviation of the regression coefficients, induced by multicollinearity, influences the estimated values of the dependent variable.

$$\mathrm{VIF}_k = \frac{1}{1 - {R_k}^2} \quad k = 1, 2, \cdots, p \tag{11.9}$$

$$\text{Mean of } \mathrm{VIF}_k : \overline{\mathrm{VIF}} = \frac{\sum_{k=1}^{p} \mathrm{VIF}_k}{p} \tag{11.10}$$

$$\text{Deviation of regression coefficients} : C = \frac{\sum_{k=1}^{p} \left(\hat{\beta}_k - \beta_k\right)^2}{p} \tag{11.11}$$

$$\text{Deviation of estimated values} : B = \sum_{i=1}^{n} \left(\hat{y}_i - \lambda_i\right)^2 \tag{11.12}$$

11.2.2.2 Basic Statistics of the Indicator Values

For each of the seven conditions illustrated in Fig. 11.2, we construct 5000 independent regression models with a sample size of 2000. From each regression model, we derive $\overline{\text{VIF}}$, C, and B, yielding 5000 values for each indicator. Using these indicator values as variables, we formulate three GLMs with a sample size of 5000:

GLM-I: Explains the deviation C of regression coefficients using $\overline{\text{VIF}}$.
GLM-II: Explains the deviation B of estimated values using $\overline{\text{VIF}}$.
GLM-III: Explains the deviation B of estimated values using the deviation C of regression coefficients.

Each of these GLMs is constructed 500 times using different samples. Table 11.1 presents statistics for all 2,500,000 indicator values employed in the analysis (5000 × 500) for conditions a–g.

11.3 Results

11.3.1 Impact on Estimated Regression Coefficients

In this section, we address the following questions:

- To what extent can the deviation C of regression coefficients arise (severity)?
- What is the likelihood of observing a deviation in the regression coefficients that would not occur in the absence of multicollinearity (probability of deviation)?

These aspects are investigated for each $\overline{\text{VIF}}$, illustrating the risks associated with the deviation of regression coefficients. By constructing GLM-I with the deviation C of regression coefficients as the dependent variable and $\overline{\text{VIF}}$ as the independent variable, we can determine the regression coefficient that describes the relationship between $\overline{\text{VIF}}$ and C and identify the parameters of the probability distribution representing the error structure of C (specifically, k^{*1} and $\theta^{*1}{}_{\overline{\text{VIF}}}$). This study leverages these insights.

11.3.1.1 Relationship Between $\overline{\text{VIF}}$ and Deviation C of Regression Coefficients: GLM-I

For each conditions a–g outlined in Fig. 11.2, we construct GLM-I 500 times with a sample size of $l = 5000$ to determine the relationship between $\overline{\text{VIF}}$ and the actual deviation C of the regression coefficients. After considering the link functions and the error structure of C, we adopt the formula with the smallest value of Akaike's information criterion (AIC). In Eq. (11.13), the subscript $j = 1, 2, \ldots, l$ indicates

Table 11.1 Descriptive statistics for indicator values

		a.	b.	c.	d.	e.	f.	g.
$\overline{\text{VIF}}$	Min.	1.0	1.0	1.0	1.0	1.0	1.0	1.0
	Qu.1st	1.1	1.3	1.0	1.5	1.2	1.0	1.2
	Median	1.3	1.8	1.2	2.2	1.6	1.2	1.6
	Qu.3rd	2.3	3.3	1.9	4.0	2.8	1.6	2.8
	Max.	366,270.1	977,262.1	244,320.9	545,388.1	733,226.5	183,256.6	734,766.7
	Mean	7.1	10.1	5.1	10.9	7.8	4.1	7.8
	sd.	456.1	930.7	304.2	634.9	698.3	228.3	699.0
Deviation C of regression coefficients	Min.	0.000	0.000	0.000	0.000	0.000	0.000	0.000
	Qu.1st	0.000	0.000	0.000	0.000	0.000	0.000	0.000
	Median	0.001	0.001	0.001	0.001	0.001	0.001	0.001
	Qu.3rd	0.001	0.002	0.001	0.002	0.001	0.001	0.001
	Max.	133.202	708.652	306.634	317.798	778.280	151.332	765.021
	Mean	0.003	0.005	0.003	0.005	0.004	0.002	0.004
	sd.	0.193	0.536	0.234	0.325	0.572	0.146	0.564
Deviation B of estimated values	Min.	0.000	0.001	0.002	0.006	0.002	0.014	0.004
	Qu.1st	1.213	1.923	1.923	2.676	2.677	2.675	2.677
	Median	2.366	3.357	3.358	4.351	4.354	4.351	4.355
	Qu.3rd	4.111	5.388	5.386	6.625	6.628	6.626	6.625
	Max.	30.777	33.182	37.556	38.944	35.740	40.588	34.634
	Mean	3.002	4.001	4.001	5.000	5.001	5.000	5.001
	sd.	2.452	2.828	2.830	3.161	3.160	3.164	3.161

the data number, C_j is the dependent variable, and $\overline{\text{VIF}}_j$ is the independent variable. E represents the expected value, α_0^{*1}, α_1^{*1} denote the regression coefficients, η^{*1} is the linear predictor, and f is a logarithmic link function that connects the linear predictor η^{*1} with the expected value $E(C_j)$. Concerning the error structure of C_j, it is assumed that C_j follows a gamma distribution with shape parameter k^{*1} and scale parameter $\theta^{*1}{}_j$.

$$\text{GLM} - \text{I} : f\left[E\left(C_j\right)\right] = \eta^{*1} = \alpha_1^{*1} \log\left(\overline{VIF}_j\right) + \alpha_0^{*1}$$

where C_j follows a gamma distribution:

$$C_j \sim \gamma\left(C_j \mid k^{*1}, \theta^{*1}{}_j\right) \tag{11.13}$$

Table 11.2 presents the estimated regression coefficients ($\hat{\alpha}_0^{*1}$, $\hat{\alpha}_1^{*1}$) and the shape parameter k^{*1} for the gamma distribution. Figure 11.3 (left and center panels) shows the estimated values and confidence intervals for conditions a, b, and d using the mean values of the regression coefficients ($\hat{\alpha}_0^{*1}$, $\hat{\alpha}_1^{*1}$) and the shape parameter k^{*1} derived from 500 regression models. The dots in the diagrams indicate the distribution of all observation values used in constructing the regression models. Figure 11.3 (right panel) shows the probability distribution (gamma distribution) of C for the $\overline{\text{VIF}}$ values marked by gray vertical lines in Fig. 11.3 (left and center panels), i.e., at points $\overline{\text{VIF}} \in \left(2^0, 2^{0.5}, \cdots, 2^{10}\right)$ and for $\overline{\text{VIF}} = 10$.

All models indicate that the regression coefficients ($\hat{\alpha}_0^{*1}$, $\hat{\alpha}_1^{*1}$) are significant at the 1% level. The values of $\hat{\alpha}_0^{*1}$, $\hat{\alpha}_1^{*1}$, and α_{11} are broadly consistent across the seven conditions, with $\hat{\alpha}_0 \approx -7.60$ and $\alpha_1 \approx 1.00$ (as listed in Table 11.2). From Fig. 11.3 (left and center panels), it is evident that the relationship between the expected value of C and $\overline{\text{VIF}}$ for conditions a, b, and d is largely consistent and linear. In contrast, the shape parameter k^{*1} presented in Table 11.2 varies among the seven conditions, and the probability distribution (gamma distribution) of C shown in Fig. 11.3 (right panel) differs in conditions a, b, and d. The probability distribution (gamma distribution) is determined by the shape parameter k^{*1} and the scale parameter $\theta^{*1}{}_j = E(C_j)/k^{*1}$. Therefore, if the expected values $E(C_j)$ estimated for each $\overline{\text{VIF}}$ are identical, the shape parameter k^{*1} determines the form of the distribution.

This implies that even if $\overline{\text{VIF}}$ remains consistent, the expected value of C is stable across explanatory variable conditions a–g. Despite this consistency, the probability distribution of C varies by condition. Even when the expected values align, differences in the probability distributions can lead to variations in the likelihood of observing deviations exceeding a particular threshold.

Table 11.2 GLM-I estimated regression coefficients $\left(\hat{\boldsymbol{\alpha}}_0^{*1}, \hat{\boldsymbol{\alpha}}_1^{*1}\right)$ and shape parameter $\boldsymbol{k}^{*1}$

			Estimated regression coefficients $\hat{\alpha}_0^{*2}$, $\hat{\alpha}_1^{*2}$are all 1% significant						Shape parameter $k^{(*1)}$				
GLM-I			Min	Qu.1st	Mean	Qu.3rd	Max.		Min.	Qu.1st	Mean	Qu.3rd	Max.
2 variables	a.	$\hat{\alpha}_0^{*1}$	−7.562	−7.613	−7.600	−7.587	−7.549	a.	0.794	0.834	0.844	0.853	0.886
		$\hat{\alpha}_1^{*1}$	0.927	0.985	1.001	1.017	1.072						
3 variables	b.	$\hat{\alpha}_0^{*1}$	−7.561	−7.613	−7.600	−7.587	−7.549	b.	1.030	1.069	1.081	1.093	1.166
		$\hat{\alpha}_1^{*1}$	0.934	0.987	0.999	1.011	1.056						
	c.	$\hat{\alpha}_0^{*1}$	−7.545	−7.612	−7.600	−7.588	−7.551	c.	1.143	1.181	1.195	1.208	1.258
		$\hat{\alpha}_1^{*1}$	0.907	0.984	1.000	1.016	1.065						
4 variables	d.	$\hat{\alpha}_0^{*1}$	−7.661	−7.613	−7.600	−7.587	−7.541	d.	1.208	1.250	1.264	1.277	1.332
		$\hat{\alpha}_1^{*1}$	0.953	0.986	0.998	1.010	1.046						
	e.	$\hat{\alpha}_0^{*1}$	−7.646	−7.611	−7.599	−7.587	−7.548	e.	1.279	1.329	1.346	1.364	1.423
		$\hat{\alpha}_1^{*1}$	0.945	0.985	0.999	1.014	1.057						
	f.	$\hat{\alpha}_0^{*1}$	−7.642	−7.610	−7.600	−7.590	−7.547	f.	1.459	1.512	1.532	1.553	1.631
		$\hat{\alpha}_1^{*1}$	0.923	0.982	0.999	1.015	1.083						
	g.	$\hat{\alpha}_0^{*1}$	−7.650	−7.611	−7.599	−7.588	−7.540	g.	1.294	1.341	1.358	1.374	1.460
		$\hat{\alpha}_1^{*1}$	0.940	0.985	0.999	1.014	1.056						

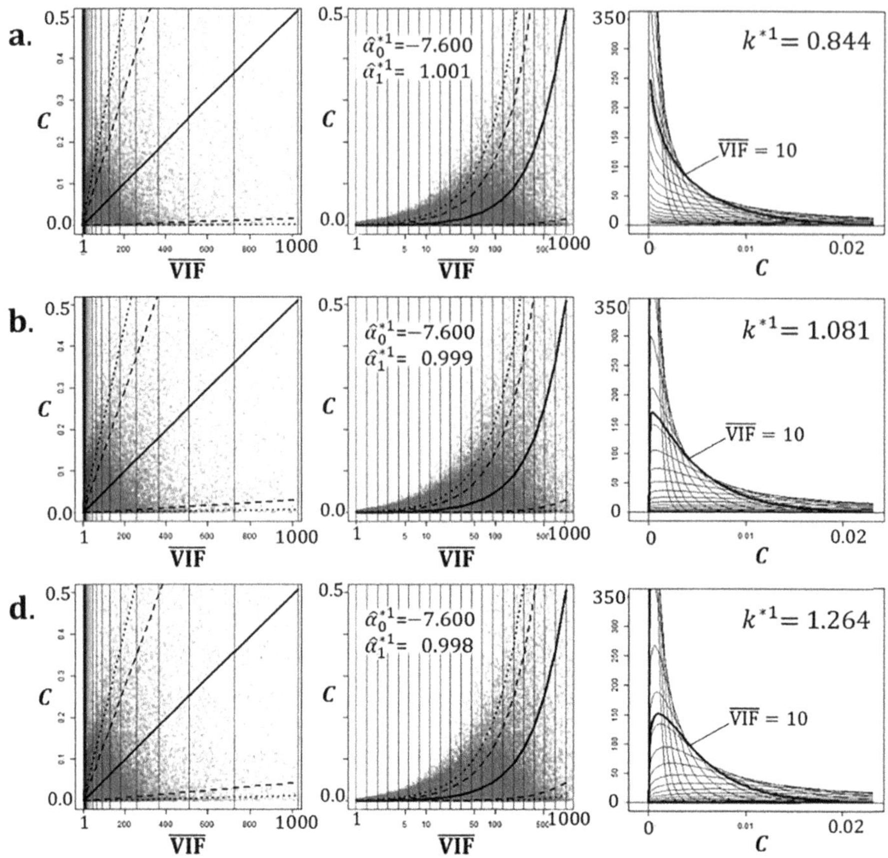

Fig. 11.3 GLM-I for conditions a, b, and d. Left: Relationship between $\overline{\mathbf{VIF}}$ and $\boldsymbol{C}$. Center: Relationship between $\overline{\mathbf{VIF}}$ and $\boldsymbol{C}$ (with a logarithmic scale on the x-axis). Right: Probability distribution for each $\overline{\mathbf{VIF}}$. *For the left and center panels, the solid line represents the mean (estimated value), the dashed line represents the 95% confidence interval, and the dotted line indicates the 99% confidence interval. For clarity, the x-axis ($\overline{\mathbf{VIF}}$) only extends to **1000** and the y-axis ($\boldsymbol{C}$) only extends to **0.5**. For the right panel, for clarity, the x-axis ($\boldsymbol{C}$) is limited to 0.02 and the y-axis (density) only extends to 350

11.3.1.2 Concepts of Severity and Probability of Deviation

We wish to determine the extent to which the deviation C of the regression coefficients can occur (severity) and the likelihood of such a deviation compared with the absence of collinearity (probability of deviation occurrence). However, the maximum deviation that occurs in a regression model showing a certain $\overline{\text{VIF}}$ cannot be specified solely from its probability distribution. In this study, we consider a large deviation that occurs with a 1/1000 probability (the 99.9th percentile) in the gamma distribution as the "maximum deviation." We denote this by $\left(Z^{*C}_{.999}\right)_{\overline{\text{VIF}}}$.

When a particular explanatory variable x_k is regressed linearly using other explanatory variables (i.e., excluding x_k) and the determination coefficient is ${R_k}^2 = 0$, there is no collinearity between x_k and the other explanatory variables and $\mathrm{VIF}_k = 1$. Because $\mathrm{VIF_k}$ takes values in the range $1 \leq \mathrm{VIF}_k < \infty$, if $\overline{\mathrm{VIF}} = \sum_{k=1}^{p} \mathrm{VIF}_k / p = 1$, we can infer that the regression model is not affected by multicollinearity. For $\overline{\mathrm{VIF}} = 1$, we consider a large deviation that occurs with a 1/1000 probability (the 99.9th percentile of the probability distribution) as the maximum deviation in the absence of collinearity, and denote this by $\left(Z^{*C}_{.999}\right)_1$. We define the ratio $\left(Z^{*C}_{.999}\right)_{\overline{\mathrm{VIF}}} / \left(Z^{*C}_{.999}\right)_1$ as an indicator of the extent to which the deviation C of the regression coefficients can occur, and denote this as the degree of regression coefficient deviation, $H^{*C}_{\overline{\mathrm{VIF}}}$. Moreover, in the gamma distribution of C for a regression model with a certain $\overline{\mathrm{VIF}}$, we use the probability $Q\left(\left(Z^{*C}_{.999}\right)_1, k^{*1}, \theta^{*1}{}_{\overline{\mathrm{VIF}}}\right)$ of a larger deviation than $\left(Z^{*C}_{.999}\right)_1$ occurring as an indicator of the likelihood of such a significant deviation in the absence of collinearity. We denote this as the probability of regression coefficient deviation occurrence, $P^{*C}_{\overline{\mathrm{VIF}}}$. The concepts of $H^{*C}_{\overline{\mathrm{VIF}}}$ and $P^{*C}_{\overline{\mathrm{VIF}}}$ are illustrated in Fig. 11.4.

11.3.1.3 Degree of Regression Coefficient Deviation and Probability of Regression Coefficient Deviation Occurrence

The left side of Table 11.3 displays the mean values of the degree of regression coefficient deviation $H^{*C}_{\overline{\mathrm{VIF}}}$ for 500 models with $\overline{\mathrm{VIF}} = (2^0, 2^1, \cdots, 2^{10})$ and $\overline{\mathrm{VIF}} = (5, 10, 100)$ under conditions a–g. Figure 11.5 (upper panel) illustrates the relationship between $\overline{\mathrm{VIF}}$ and $H^{*C}_{\overline{\mathrm{VIF}}}$ for conditions a, b, and d. For conditions a–g, $\overline{\mathrm{VIF}}$ and $H^{*C}_{\overline{\mathrm{VIF}}}$ align almost perfectly. From these results, it is clear that the expected value of C (represented by $\overline{\mathrm{VIF}}$, but also the maximum deviation when using $\overline{\mathrm{VIF}} = 1$, i.e., without the influence of collinearity) can be directly determined by $\overline{\mathrm{VIF}}$.

The right side of Table 11.3 presents the mean values of the probability of regression coefficient deviation occurrence $P^{*C}_{\overline{\mathrm{VIF}}}$ for 500 models with $\overline{\mathrm{VIF}} = (2^0, 2^1, \cdots, 2^{10})$ and $(5, 10, 100)$ under conditions a–g. Figure 11.5 (lower panel) depicts the relationship between $\overline{\mathrm{VIF}}$ and $P^{*C}_{\overline{VIF}}$ for conditions a, b, and d. Whereas $H^{*C}_{\overline{\mathrm{VIF}}}$ aligns with $\overline{\mathrm{VIF}}$ and does not vary across conditions a–g, $P^{*C}_{\overline{\mathrm{VIF}}}$ tends to be higher when there are more explanatory variables. This indicates that even if the expected value of the C is the same, the probability of such a deviation occurring depends on the number of explanatory variables and other factors associated with conditions a–g.

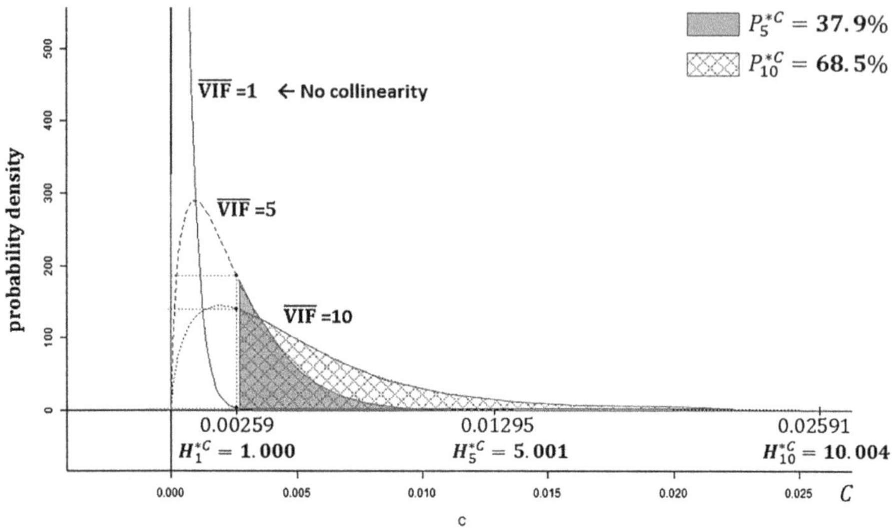

Fig. 11.4 Concepts of the degree of regression coefficient deviation and probability of regression coefficient deviation occurrence. In this figure, the 99.9th percentile of the probability distribution with $\overline{\text{VIF}} = 1$, denoted as $\left(Z^{*C}_{.999}\right)_1$, is 0.00259. We consider this as the "maximum deviation" that occurs when there is no collinearity. Moreover, the 99.9th percentile of the probability distribution with $\overline{\text{VIF}} = 10$, denoted as $\left(\left(Z^{*C}_{.999}\right)_{10}\right)$, is 0.02591, and this represents the "maximum deviation" when $\overline{\text{VIF}}$ is 10. The value $\left(Z^{*C}_{.999}\right)_{10} = 0.02591$ is roughly ten times $\left(Z^{*C}_{.999}\right)_1 = 0.00259$. Therefore, the degree of regression coefficient deviation for the model with $\overline{\text{VIF}} = 10$ $\left(H^{*C}_{\overline{\text{VIF}}}\right)$ is 10. In the probability distribution with $\overline{\text{VIF}} = 10$, the likelihood of a deviation greater than the maximum deviation of a model without collinearity $\left(\left(Z^{*C}_{.999}\right)_1 = 0.00259\right)$ is represented by the area of the shaded region and is given by $Q\left(\left(Z^{*C}_{.999}\right)_1, k^{*1}, \{\theta \mathbin{\hat{}} \{*1\}\} \_ \{10\}\right) = 0.685$, which means that the probability of regression coefficient deviation occurrence for the model with $\overline{\text{VIF}} = 10$, denoted as P^{*C}_{10}, is 68.5%

11.3.1.4 Robustness of GLM-I

Tables 11.2 and 11.3 and Figs. 11.3 and 11.5 summarize the consistency across the 500 models in terms of the GLM-I parameters $H^{*C}_{\overline{\text{VIF}}}$ and $P^{*C}_{\overline{\text{VIF}}}$, indicating that the constructed GLM-I is not heavily dependent on the sample. We further demonstrate the robustness of the model by constructing GLM-I with a constraint that limits $\overline{\text{VIF}}$ to a fixed value of l or below. Table 11.4 presents the results from constructing GLM-I in a similar manner, but using samples that restrict $\overline{\text{VIF}}$ to values of $l = 10, 50, 100, 500, 1000, 5000$, and 10,000. The estimated regression coefficients $(\hat{\alpha}_0^{*1}, \hat{\alpha}_1^{*1})$ and the shape parameter $k*^1$ remain largely consistent, suggesting that GLM-I is a stable model that does not depend on the range of $\overline{\text{VIF}}$.

Table 11.3 Left: Mean value of $\boldsymbol{H^{*C}_{VIF}}$; Right: Mean value of $\boldsymbol{P^{*C}_{VIF}}$

		Degree of regression coefficient deviation H^{*C}_{VIF}							Probability of regression coefficient deviation P^{*C}_{VIF}						
		Mean of 500 simulations							Mean of 500 simulations						
VIF		a.	b.	c.	d.	e.	f.	g.	a.	b.	c.	d.	e.	f.	g.
$2^0=$	1	1.0	1.0	1.0	1.0	1.0	1.0	1.0	0.1	0.1	0.1	0.1	0.1	0.1	0.1
$2^1=$	2	2.0	2.0	2.0	2.0	2.0	2.0	2.0	2.8	3.3	3.6	3.8	4.0	4.4	4.0
$2^2=$	4	4.0	4.0	4.0	4.0	4.0	4.0	4.0	15.4	19.0	20.8	21.8	23.1	25.9	23.3
$2^3=$	8	8.0	3.0	8.0	8.0	8.0	8.0	8.0	37.1	44.7	48.1	50.0	52.4	57.2	52.7
$2^4=$	16	16.1	15.0	16.0	15.9	16.0	16.0	16.0	58.7	67.8	71.5	73.5	75.9	80.3	76.2
$2^5=$	32	32.2	32.0	32.1	31.9	32.0	32.0	32.0	74.9	83.0	85.9	87.4	89.1	92.0	89.3
$2^6=$	64	64.6	64.0	64.2	63.7	64.1	64.0	64.1	85.4	91.5	93.4	94.4	95.4	97.0	95.5
$2^7=$	128	129.5	128.1	128.6	127.4	128.3	128.1	128.2	91.6	95.8	97.0	97.6	98.1	98.9	98.2
$2^8=$	256	259.7	256.5	257.6	254.9	256.9	256.6	256.8	95.3	98.0	98.7	99.0	99.2	99.6	99.3
$2^9=$	512	521.0	513.5	516.2	509.8	514.6	514.2	514.2	97.4	99.0	99.4	99.6	99.7	99.9	99.7
$2^{10}=$	1024	1045.3	1028.2	1034.7	1019.8	1030.8	1030.6	1030.0	98.5	99.5	99.7	99.8	99.9	100.0	99.9
5		5.0	5.0	5.0	5.0	5.0	5.0	5.0	21.8	26.8	29.2	30.5	32.3	36.0	32.5
10		10.0	10.0	10.0	10.0	10.0	10.0	10.0	44.5	52.8	56.5	58.6	61.1	66.0	61.4
100		101.1	100.1	100.4	99.6	100.2	100.1	100.2	89.8	94.6	96.1	96.7	97.4	98.4	97.5

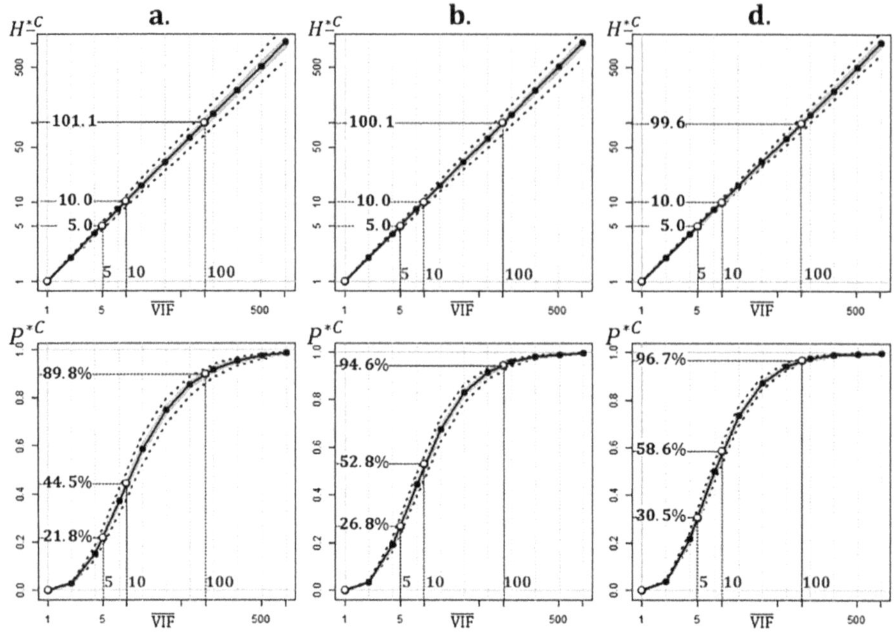

Fig. 11.5 GLM-I for conditions a, b, and d. Upper: Degree of deviation $H^{*C}_{\overline{VIF}}$; Lower: Probability of deviation occurrence $P^{*C}_{\overline{VIF}}$. Solid lines with black dots represent the mean of 500 simulations. The gray area indicates the range from the first quartile to the third quartile, and the dotted lines show the maximum and minimum values

11.3.1.5 Comparison and Test for Differences

Figure 11.6 compares $P^{*C}_{\overline{\mathrm{VIF}}}$ and $H^{*C}_{\overline{\mathrm{VIF}}}$ across models under the seven different conditions for the example cases of $\overline{\mathrm{VIF}} = 5$ and 10. These models differ based on the number of explanatory variables and whether independent variables are present or absent. The kernel density[5] is based on the results of 500 simulations.

For every combination of these seven conditions (21 combinations in total), Welch's t-test[6] is applied to examine the difference in means. For H^{*C}_{5} and H^{*C}_{10}, the null hypothesis that "the means of the two groups are equal" is only rejected at the 1% level (p-value ≥ 0.01) in conditions a and d. For the other conditions, this

---

[5] The bandwidth of a Gaussian kernel density estimator is selected based on a rule-of-thumb (Silverman 1986). The default value is 0.9 times the minimum of the standard deviation and the interquartile range divided by 1.34 times the sample size to the power of −0.2.

[6] Before using Welch's t-test, we tested for normality with the Kolmogorov–Smirnov test and checked for homoscedasticity with the F-test. All data met the 5% significance level for normality. While the variance results varied by group, both the Welch's and Student's t-tests were consistent, so we reported Welch's t-test results.

Table 11.4 Regression coefficient estimates ($\hat{\boldsymbol{\alpha}}_0^{*1}, \hat{\boldsymbol{\alpha}}_1^{*1}$) and shape parameter $\boldsymbol{k}^{*1}$ from GLM-I trained using samples limited to $\overline{\mathbf{VIF}}$ less than or equal to a constant value $\boldsymbol{l}$

GLM-I		Estimated regression coefficients $\hat{\alpha}_0^{*2}, \hat{\alpha}_1^{*2}$ are all 1% significant							Shape parameter $k^{(*1)}$					
			≤10	≤50	≤100	≤500	≤1000	≤5000		≤10	≤50	≤100	≤500	≤1000	≤5000
2 variables	a.	$\hat{\alpha}_0^{*1}$	−7.604	−7.598	−7.597	−7.596	−7.597	−7.600	a.	0.874	0.856	0.852	0.850	0.850	0.850
		$\hat{\alpha}_1^{*1}$	1.016	0.998	0.996	0.994	0.995	1.002							
3 variables	b.	$\hat{\alpha}_0^{*1}$	−7.605	−7.611	−7.615	−7.620	−7.618	−7.616	b.	1.118	1.089	1.080	1.074	1.072	1.071
		$\hat{\alpha}_1^{*1}$	1.025	1.036	1.042	1.050	1.047	1.044							
	c.	$\hat{\alpha}_0^{*1}$	−7.59[illegible]	−7.586	−7.589	−7.592	−7.592	−7.591	c.	1.227	1.205	1.203	1.197	1.197	1.197
		$\hat{\alpha}_1^{*1}$	0.981	0.964	0.971	0.980	0.980	0.978							
4 variables	d.	$\hat{\alpha}_0^{*1}$	−7.589	−7.615	−7.608	−7.605	−7.608	−7.605	d.	1.371	1.302	1.289	1.283	1.281	1.281
		$\hat{\alpha}_1^{*1}$	0.977	1.017	1.008	1.004	1.007	1.005							
	e.	$\hat{\alpha}_0^{*1}$	−7.577	−7.571	−7.565	−7.567	−7.567	−7.565	e.	1.423	1.368	1.355	1.350	1.350	1.350
		$\hat{\alpha}_1^{*1}$	1.004	0.994	0.981	0.985	0.985	0.983							
	f.	$\hat{\alpha}_0^{*1}$	−7.603	−7.613	−7.611	−7.607	−7.607	−7.607	f.	1.567	1.523	1.519	1.506	1.506	1.506
		$\hat{\alpha}_1^{*1}$	1.021	1.040	1.032	1.021	1.021	1.021							
	g.	$\hat{\alpha}_0^{*1}$	−7.571	−7.569	−7.563	−7.566	−7.566	−7.565	g.	1.431	1.378	1.363	1.357	1.357	1.357
		$\hat{\alpha}_1^{*1}$	0.985	0.983	0.972	0.978	0.978	0.976							

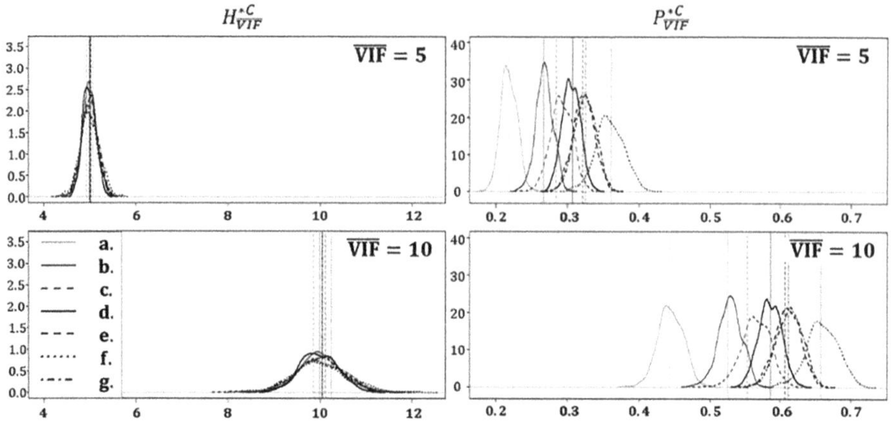

Fig. 11.6 Left: Degree of deviation in regression coefficients $\boldsymbol{H^{*C}_{\overline{VIF}}}$ (no significant difference observed); Right: Probability of deviation in regression coefficients $\boldsymbol{P^{*C}_{\overline{VIF}}}$ (significant difference at the 5% level)

hypothesis is not rejected at the 5% significance level (p-value $\geq$0.05), indicating that there is no substantial difference between the two groups. For P^{*C}_{5} and P^{*C}_{10}, however, all combinations have p-values of less than 0.05, supporting the conclusion that a difference exists between the groups. This suggests that while regression models with the same $\overline{\text{VIF}}$ display little variance in the degree of deviation, the probability of such deviations occurring varies, indicating differing risks of deviations in the regression coefficients.

11.3.2 Impact on the Estimated Values of the Dependent Variable

When multicollinearity influences the estimated regression coefficients, it is postulated that the estimated values of the dependent variable (often referred to as "estimates") will also be affected. In this section, using an approach akin to that applied in Sect. 11.3.1.1, we explore the relationship between $\overline{\text{VIF}}$ and the deviation B of the estimates, as well as the relationship between the deviation C of regression coefficients and the deviation B of estimates.

11.3.2.1 Relationship Between $\overline{\text{VIF}}$ and Deviation B of Estimates: GLM-II

For conditions a–g listed in Fig. 11.2, GLM-II is constructed 500 times with $l = 5000$ to determine the relationship between $\overline{\text{VIF}}$ and the deviation B of estimates. This model is created considering the link function and the error structure of B.

Ultimately, the equation with the smallest AIC is adopted. In Eq. (11.14), the subscript $j = 1, 2, \cdots, l$ represents the data index, B_j is the dependent variable, and $\overline{VIF_j}$ is the explanatory variable. E represents the expected value, α_0^{*2}, α_1^{*2} are the coefficients, η^{*2} is the linear predictor, and g is a logarithmic link function connecting the linear predictor η^{*2} to the expected value $E(B_j)$. Regarding the error structure, it is assumed that B_j obeys a gamma distribution characterized by the shape parameter k^{*2} and the scale parameter $\theta^{*2}{}_j$.

$$\mathrm{GLM-II}: f\left[E\left(B_j\right)\right] = \eta^{*2} = \alpha_1^{*2} \log\left(\overline{VIF}_j\right) + \alpha_0^{*2}$$

where B_j follows a gamma distribution:

$$B_j \sim \gamma\left(B_j \,|k^{*2}, \theta^{*2}{}_j\right) \tag{11.14}$$

For each condition of GLM-II, the estimated regression coefficients $(\hat{\alpha}_0^{*2}, \hat{\alpha}_1^{*2})$ and the shape parameter k^{*2} are presented in Table 11.5. Figure 11.7 (left and center panels) shows the estimated values and confidence intervals for conditions a, b, and d. This representation is derived using the mean values of the regression coefficients $\hat{\alpha}_0^{*2}$ and $\hat{\alpha}_1^{*2}$, and the shape parameter k^{*2} from 500 regression models. The right side of Fig. 11.7 shows the probability distribution (gamma distribution) of the deviation B of estimated values for $\overline{\mathrm{VIF}} \in (2^0, 2^{0.5}, \cdots, 2^{10})$, and specifically for $\overline{\mathrm{VIF}} = 10$, as indicated by the gray vertical lines in the left and center of Fig. 11.7.

The regression coefficient $\hat{\alpha}_1^{*2}$ for GLM-II is not significant in any scenario and is close to zero. Additionally, as illustrated by Fig. 11.7, both the estimated values and confidence intervals of GLM-II are uniform. The probability distribution of B does not change for the various $\overline{\mathrm{VIF}}$ values. This implies that $\overline{\mathrm{VIF}}$, which represents the expected deviation of the regression coefficient, cannot explain the deviation in the estimated values. However, the intercept $\hat{\alpha}_0^{*2}$ is significant at the 1% level for all conditions. This suggests that, although differences in the expected deviation of regression coefficients do not influence the deviation in the estimated values, any variation in factors such as the number of explanatory variables could impact the estimates.

11.3.2.2 Relationship Between Deviation C of Regression Coefficients and Deviation B of Estimates: GLM-III

We now examine the relationship between the deviation of the regression coefficient C and the deviation of the estimated value B using the same method as with GLM-II. For each of conditions a–g presented in Fig. 11.2, we construct GLM-III 500 times with a sample size of $l = 5000$. This demonstrates the relationship between

Table 11.5 GLM-II estimated regression coefficients $\left(\hat{\boldsymbol{\alpha}}_0^{*2}, \hat{\boldsymbol{\alpha}}_1^{*2}\right)$ and shape parameter $\boldsymbol{k}^{*2}$

GLM-II			Estimated regression coefficients $\hat{\boldsymbol{\alpha}}_0^{*2}$ are all 1% significant, $\hat{\boldsymbol{\alpha}}_1^{*2}$ are not significant						Shape parameter $k^{(*1)}$				
			Min.	Qu.1st	Mean	Qu.3rd	Max.		Min.	Qu.1st	Mean	Qu.3rd	Max.
2 variables	a.	$\hat{\alpha}_0^{*2}$	1.065	1.090	1.099	1.110	1.141	a.	1.420	1.480	1.500	1.519	1.577
		$\hat{\alpha}_1^{*2}$	−0.049	−0.010	0.000	0.009	0.037						
3 variables	b.	$\hat{\alpha}_0^{*2}$	1.337	1.378	1.386	1.396	1.429	b.	1.888	1.977	2.003	2.027	2.134
		$\hat{\alpha}_1^{*2}$	−0.031	−0.008	0.000	0.007	0.049						
	c.	$\hat{\alpha}_0^{*2}$	1.351	1.378	1.387	1.395	1.432	c.	1.893	1.977	2.001	2.023	2.107
		$\hat{\alpha}_1^{*2}$	−0.052	−0.009	0.000	0.009	0.039						
4 variables	d.	$\hat{\alpha}_0^{*2}$	1.567	1.601	1.609	1.619	1.657	d.	2.350	2.468	2.501	2.530	2.683
		$\hat{\alpha}_1^{*2}$	−0.029	−0.007	0.000	0.007	0.030						
	e.	$\hat{\alpha}_0^{*2}$	1.572	1.601	1.610	1.619	1.645	e.	2.371	2.476	2.505	2.535	2.667
		$\hat{\alpha}_1^{*2}$	−0.027	−0.008	0.000	0.006	0.032						
	f.	$\hat{\alpha}_0^{*2}$	1.575	1.602	1.610	1.618	1.648	f.	2.359	2.470	2.500	2.530	2.649
		$\hat{\alpha}_1^{*2}$	−0.045	−0.010	−0.001	0.009	0.035						
	g.	$\hat{\alpha}_0^{*2}$	1.565	1.601	1.610	1.618	1.649	g.	2.375	2.478	2.506	2.537	2.667
		$\hat{\alpha}_1^{*2}$	−0.027	−0.007	0.000	0.007	0.038						

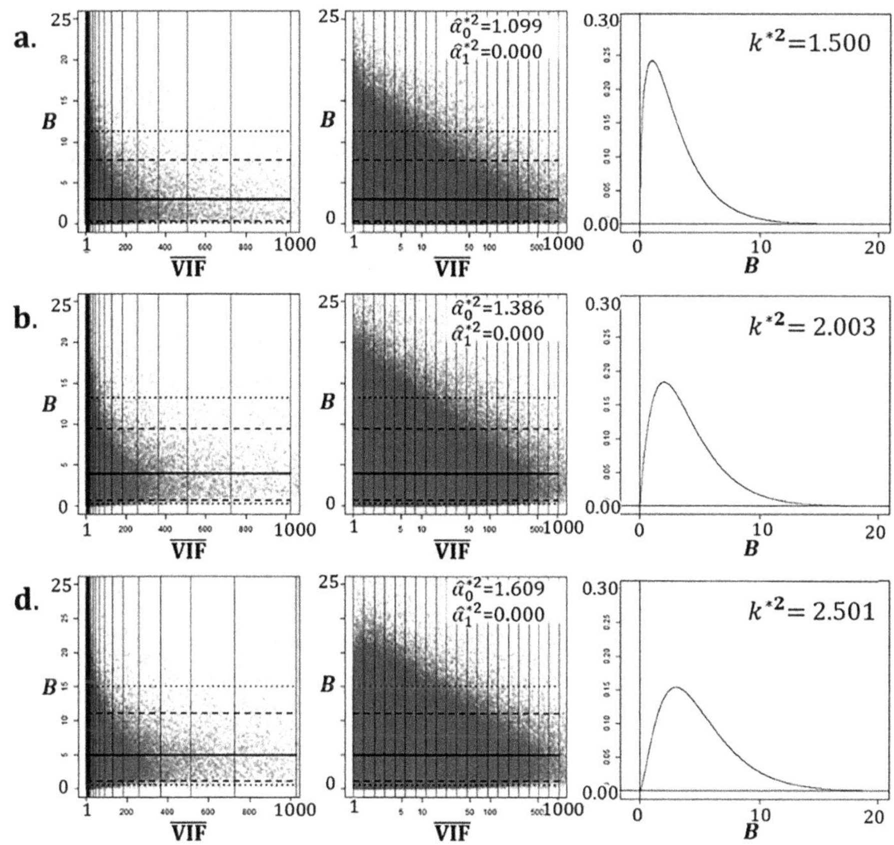

Fig. 11.7 GLM-II for conditions a, b, and d. Left: Relationship between $\overline{\mathbf{VIF}}$ and $\boldsymbol{B}$. Center: Relationship between $\overline{\mathbf{VIF}}$ and $\boldsymbol{B}$ (with a logarithmic scale on the x-axis). Right: Probability distribution for each $\overline{\mathbf{VIF}}$.*For the left and center panels, the solid line represents the mean (estimated value), the dashed line represents the 95% confidence interval, and the dotted line indicates the 99% confidence interval. For clarity, the x-axis ($\overline{\mathbf{VIF}}$) only extends to **1000** and the y-axis ($\boldsymbol{B}$) only extends to **25**. For the right diagram, for clarity, the x-axis ($\boldsymbol{B}$) is limited to 20 and the y-axis (density) only extends to 0.30

the deviation of the regression coefficient C and the deviation in the estimated value B. The model accounts for the link function and error structure of B. Ultimately, the equation with the smallest AIC is adopted. In Eq. (11.15), the subscript $j = 1, 2, \ldots, l$ represents the data number, B_j is the dependent variable, C_j is the explanatory variable, E denotes the expected value, and α_0^{*3} and α_1^{*3} represent coefficients. η^{*3} is a linear predictor. The logarithmic link function h connects the linear predictor η^{*3} to the expected value $E(B_j)$. We assume that the error structure of B_j obeys a gamma distribution with shape parameter k^{*3} and scale parameter $\theta^{*3}{}_j$.

$$\mathrm{GLM}-\mathrm{III}: h\left[E\left(B_j\right)\right] = \eta^{*3} = \alpha_1^{*3}\log\left(C_j\right) + \alpha_0^{*3}$$

where $B_j$follows a gamma distribution:

$$B_j \sim \gamma\left(B_j \,|k^{*3}, \theta^{*3}{}_j\right) \tag{11.15}$$

The estimated regression coefficients of GLM-III for each condition, namely $\hat{\alpha}_0^{*3}$ and $\hat{\alpha}_1^{*3}$ along with the shape parameter k^{*3}, are listed in Table 11.6. Figure 11.8 (left and center panels) is similar to Fig. 11.7, showing the estimated values and confidence intervals for conditions a, b, and d based on the mean values of $\hat{\alpha}_0^{*3}$, $\hat{\alpha}_0^{*3}$, and k^{*3} derived from 500 regression models. Figure 11.8 (right panel) depicts the probability distribution (gamma distribution) of B for points $C \in \left(\left(Z_{.999}^{*C}\right)_{2^0}, \left(Z_{.999}^{*C}\right)_{2^{0.5}}, \cdots, \left(Z_{.999}^{*C}\right)_{2^{10}}\right)$, and $C = \left(Z_{.999}^{*C}\right)_{10}$, marked by gray vertical lines in the left and center panels of Fig. 11.8.

From the estimated regression coefficient $\hat{\alpha}_1^{*3}$ for GLM-III (see Table 11.6), we observe that a larger value of C is correlated with an increased expected value for the deviation in the estimated value $\left(Z_{.999}^{*B}\right)_C$. However, this trend is not as pronounced as for GLM-I. For example, with $\overline{\text{VIF}} = 5$ (i.e., the expected deviation of the regression coefficient C is five times that in a scenario without collinearity), the maximum deviation, $\left(Z_{.999}^{*C}\right)_1$, becomes five times that of a collinearity-free scenario. However, even if C reaches five times its value for a collinearity-free scenario, the resulting maximum deviation of the estimated value $\left(Z_{.999}^{*B}\right)_C$ is only about 1.6 times larger, and the difference is not significant.[7]

Although the expected deviation of the regression coefficient C, represented by $\overline{\text{VIF}}$, does not directly explain the deviation in the estimated value B, models with larger values of C tend to exhibit a larger deviation in B. However, the growth in B is more subtle than the increase in C.

11.4 A Method for Diagnosing the Risk of Multicollinearity

Typically, analysts estimate the degree (severity) of deviation in regression coefficients based on metrics such as $\overline{\text{VIF}}$ derived from explanatory variables and diagnose the regression model accordingly. In this paper, we have proposed a more accurate model diagnosis method by referring to the shape parameter k^{*1} derived from GLM-I in addition to $\overline{\text{VIF}}$.

Table 11.7 presents the shape parameter k^{*1} for each variable condition, as revealed by the numerical experiments. Additionally, Table 11.8 and Fig. 11.9

[7] $\left(Z_{.999}^{*B}\right)_C$ represents the 99.9th percentile (or "maximum deviation") from the probability distribution (gamma distribution) of the deviation B of the estimated values when a deviation C in a certain regression coefficient is indicated. $\left(Z_{.999}^{*B}\right)_1$ denotes the 99.9th percentile (or "maximum deviation") from the probability distribution (gamma distribution) of the deviation B of the estimated values at the maximum deviation of the regression coefficient, $\left(Z_{.999}^{*C}\right)_1$, in the model with $\overline{\text{VIF}} = 1$ (a model without collinearity).

Table 11.6 GLM-III estimated regression coefficients $\left(\hat{\boldsymbol{\alpha}}_0^{*3}, \hat{\boldsymbol{\alpha}}_1^{*3}\right)$ and shape parameter $\boldsymbol{k}^{*3}$

GLM-III			Estimated regression coefficients $\hat{\alpha}_0^{*3}$, $\hat{\alpha}_1^{*3}$ are all 1% significant						Shape parameter $k^{(*1)}$				
			Min	Qu.1st	Mean	Qu.3rd	Max.		Min.	Qu.1st	Mean	Qu.3rd	Max.
2 variables	a.	$\hat{\alpha}_0^{*3}$	−7.662	−7.613	−7.600	−7.587	−7.549	a.	0.794	0.834	0.844	0.853	0.886
		$\hat{\alpha}_1^{*3}$	0.927	0.985	1.001	1.017	1.072						
3 variables	b.	$\hat{\alpha}_0^{*3}$	−7.661	−7.613	−7.600	−7.587	−7.549	b.	1.030	1.069	1.081	1.093	1.166
		$\hat{\alpha}_1^{*3}$	0.934	0.987	0.999	1.011	1.056						
	c.	$\hat{\alpha}_0^{*3}$	−7.645	−7.612	−7.600	−7.588	−7.551	c.	1.143	1.181	1.195	1.208	1.258
		$\hat{\alpha}_1^{*3}$	0.907	0.984	1.000	1.016	1.065						
4 variables	d.	$\hat{\alpha}_0^{*3}$	−7.661	−7.613	−7.600	−7.587	−7.541	d.	1.208	1.250	1.264	1.277	1.332
		$\hat{\alpha}_1^{*3}$	0.953	0.986	0.998	1.010	1.046						
	e.	$\hat{\alpha}_0^{*3}$	−7.646	−7.611	−7.599	−7.587	−7.548	e.	1.279	1.329	1.346	1.364	1.423
		$\hat{\alpha}_1^{*3}$	0.945	0.985	0.999	1.014	1.057						
	f.	$\hat{\alpha}_0^{*3}$	−7.642	−7.610	−7.600	−7.590	−7.547	f.	1.459	1.512	1.532	1.553	1.631
		$\hat{\alpha}_1^{*3}$	0.923	0.982	0.999	1.015	1.083						
	g.	$\hat{\alpha}_0^{*3}$	−7.650	−7.611	−7.599	−7.588	−7.540	g.	1.294	1.341	1.358	1.374	1.460
		$\hat{\alpha}_1^{*3}$	0.940	0.985	0.999	1.014	1.056						

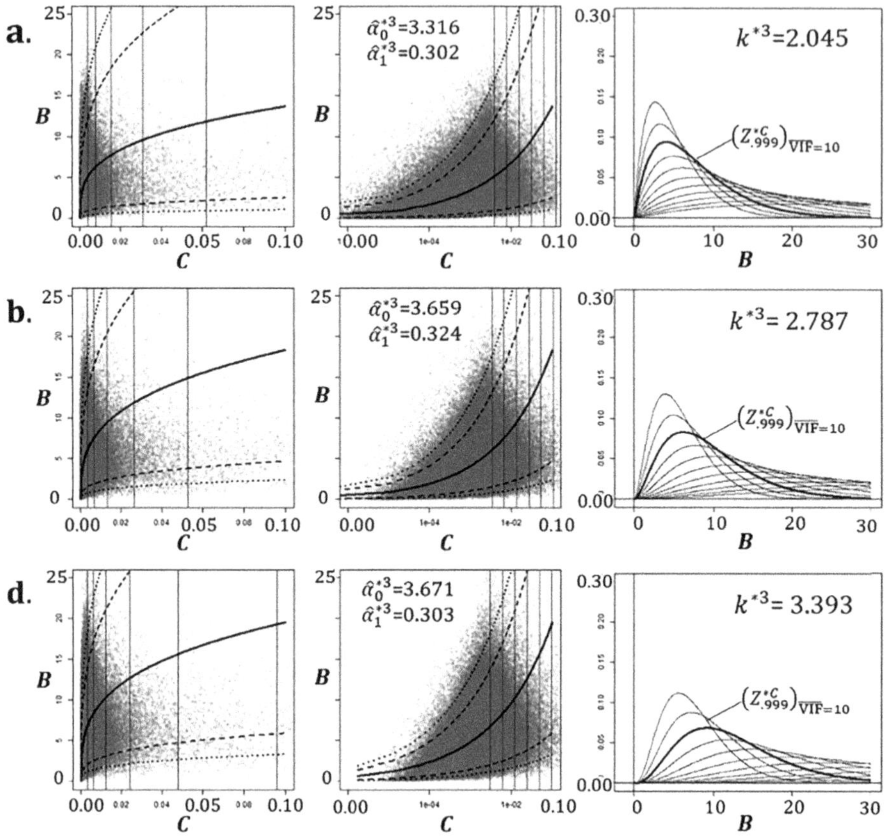

Fig. 11.8 GLM-III for conditions a, b, and d. Left: Relationship between ***C*** and ***B***. Center: Relationship between ***C*** and ***B*** (with a logarithmic scale on the *x*-axis). Right: Probability distribution for each ***C***. For the left and center panels, the solid line represents the mean (estimated value), the dashed line represents the 95% confidence interval, and the dotted line indicates the 99% confidence interval. For clarity, the *x*-axis (***C***) only extends to **0.10** and the *y*-axis (***B***) only extends to **25**. For the right diagram, for clarity, the *x*-axis (***B***) is limited to 20 and the *y*-axis (density) only extends to 0.30

illustrate the probability of a deviation in the regression coefficients in relation to both $\overline{\text{VIF}}$ and the shape parameter k^{*1}. By consulting information similar to that contained in Table 11.8 and Fig. 11.9, the probability P^{*C} of experiencing significant deviations can be determined. Such deviations, in the absence of multicollinearity, are exceptionally rare, occurring only with a 1/1000 probability. This comprehension is based on the $\overline{\text{VIF}}$ derived from the explanatory variables and the shape parameter k^{*1} dictated by the conditions of these variables. Furthermore, as GLM-II and GLM-III have shown, variations in the probability of a deviation in the regression coefficients influence the degree of impact on the estimated values of

Table 11.7 Conditions of variables and shape parameter k_*^{1}

Conditions of the explanatory variables	Shape parameter k_*^{1}
a. 2 variables	0.844
b. 3 variables	1.081
c. 3 variables, one is independent	1.195
d. 4 variables	1.264
e. 4 variables, one is independent	1.346
f. 4 variables, two are independent	1.532
g. 4 variables, two correlated pairs	1.358

Table 11.8 Probability of deviation P_*^{C} for $\overline{\mathbf{VIF}}$ and shape parameter k_*^{1} (in %)

$\overline{\text{VIF}}$ Shape parameter k_*^{1}	1	2	4	8	16	32	64	128	256	512	1024
0.8	0.1	2.7	14.7	35.5	56.7	73.0	83.8	90.5	94.5	96.8	98.2
0.9	0.1	2.9	16.2	38.9	61.0	77.1	87.1	92.9	96.2	97.9	98.9
1	0.1	3.2	17.8	42.2	64.9	80.6	89.8	94.7	97.3	98.7	99.3
1.1	0.1	3.4	19.3	45.3	68.5	83.6	91.9	96.1	98.2	99.1	99.6
1.2	0.1	3.6	20.9	48.3	71.7	86.1	93.6	97.1	98.7	99.4	99.8
1.3	0.1	3.9	22.4	51.2	74.7	88.3	94.9	97.9	99.1	99.6	99.9
1.4	0.1	4.1	23.9	53.9	77.3	90.1	96.0	98.4	99.4	99.8	99.9
1.5	0.1	4.3	25.4	56.6	79.7	91.7	96.8	98.8	99.6	99.9	99.9
1.6	0.1	4.6	26.9	59.1	81.9	93.0	97.5	99.1	99.7	99.9	100.0
1.7	0.1	4.8	28.4	61.4	83.8	94.1	98.0	99.4	99.8	99.9	100.0
1.8	0.1	5.1	29.9	63.7	85.6	95.1	98.5	99.5	99.9	100.0	100.0
1.9	0.1	5.3	31.4	65.9	87.1	95.9	98.8	99.7	99.9	100.0	100.0
2	0.1	5.6	32.9	67.9	88.6	96.6	99.1	99.8	99.9	100.0	100.0

the dependent variables. Employing the proposed method facilitates the comparison of the following aspects:

- The severity of impacts directly inferred from $\overline{\text{VIF}}$
- The probability of a deviation in the regression coefficients
- The magnitude of the impact on the dependent variable across various potential regression models

11.5 Discussion

In empirical studies that use the hedonic model to quantify the economic value of environmental factors, the accurate estimation of regression coefficients is paramount. When constructing such a hedonic model, analysts aim to gauge several potential risks, specifically the maximum extent of the influence of multicollinearity, termed the "severity," and the probability of its occurrence, referred to as the "probability of deviation." This study has provided insights for both practitioners and researchers when building hedonic models, placing particular emphasis on

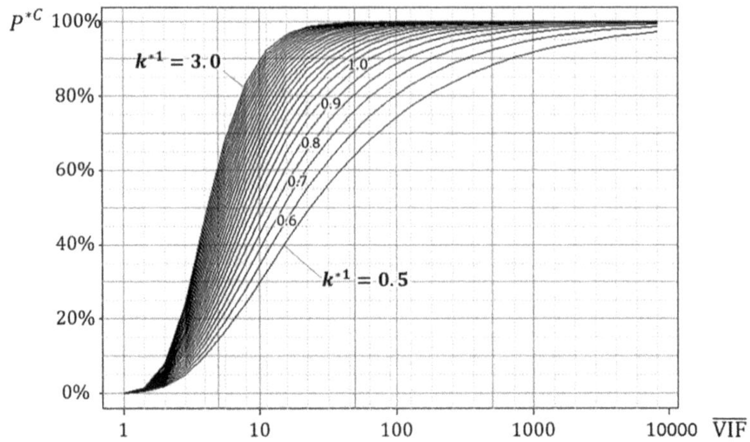

Fig. 11.9 Probability of deviation $P^{*C}_{\overline{\text{VIF}}}$ for $\overline{\text{VIF}}$ and shape parameter k^{*1}

understanding the effects of multicollinearity. To be more precise, we have used GLMs to elucidate the relationships among the mean of VIF_k (denoted as $\overline{\text{VIF}}$), the deviation C in the regression coefficients of the constructed regression model, and the deviation B of the estimated values for dependent variables within the same model.

When using the deviation observed in the absence of collinearity as a baseline, we have confirmed that the actual deviation C of the regression coefficients aligns with the value of $\overline{\text{VIF}}$. For instance, if $\overline{\text{VIF}} = 5$, there is a risk that C might be approximately five times the deviation observed in scenarios without collinearity. This remains consistent irrespective of the number of explanatory variables in the regression model; it holds as long as $\overline{\text{VIF}}$ is the same. Conversely, the probability that C in a specific model surpasses the maximum deviation (which represents the 99.9th percentile) observed without collinearity depends on the number of explanatory variables, even if $\overline{\text{VIF}}$ remains the same. For example, when $\overline{\text{VIF}} = 5$ in a model, the likelihoods with two, three, and four explanatory variables are approximately 21.8%, 26.8%, and 30.5%, respectively.

Although $\overline{\text{VIF}}$ cannot explain the deviation B in the estimated values of the dependent variable, we have observed that as the probability of a deviation in the regression coefficient increases, B also tends to increase. Thus, when comparing the extent of impact on estimated values across multiple models, it is vital to understand the probability of a deviation in the regression coefficients rather than merely relying on $\overline{\text{VIF}}$. Such findings can be attributed to the nature of the ordinary least-squares method, indicating that when a regression coefficient deviates significantly from its true value, other coefficients adjust to minimize the residuals.

Most existing metrics for diagnosing the effects of multicollinearity are based on explanatory variables. However, understanding the risk of multicollinearity solely from these metrics is challenging. Alin (2010) stated that, to genuinely comprehend the real impact, it is essential to consider the variation in the actual deviations. Thus, both analytical and numerical approaches are required.

In this study, through numerical experiments, we have empirically demonstrated that the shape parameter $k*^1$ from the GLM provides valuable information for model selection. More specifically, we showed that the variability in the deviations of regression coefficients can be expressed as differences in the probability distribution for each $\overline{\text{VIF}}$ and is manifested in the gamma distribution. Additionally, the differences in the probability distribution for each $\overline{\text{VIF}}$ are governed by the shape parameter $k*^1$ of the gamma distribution. We have proposed a method for understanding the likelihood of observing deviations which, in the absence of multicollinearity, are almost negligible (occurring with only a 1/1000 probability), by referring to both the $\overline{\text{VIF}}$ derived from explanatory variables and the shape parameter $k*^1$. Furthermore, we have provided numerical guidelines for the shape parameter $k*^1$ across seven different conditions, each distinguished by different numbers of explanatory variables and variations in other parameters.

There are countless combinations of the number of explanatory variables and potential correlations among them. The numerical benchmarks (shape parameter $k*^1$) offered in this study for the seven conditions might not suffice. Moreover, this study has not accounted for the impacts of linear relationships among multiple variables. The methodology used for generating variables might also influence the estimated value of the shape parameter k^1. Therefore, refining and presenting a more precise shape parameter $k*^1$ will be addressed in future research. An additional future challenge is to formulate the relationship between the number of explanatory variables used in the regression model and the shape parameter k^1, with the aim of providing a more widely applicable diagnostic method.

In recent years, available computational resources have expanded rapidly, enhancing our ability to understand the full scope of the impacts of multicollinearity. With these advanced resources, it will be possible to clearly present the probability of deviations and the effects on the estimated values of dependent variables, based on the explanatory variables. This will enable technicians and researchers building hedonic models to choose those that are better aligned with their objectives, supported by clearer reasoning.

References

Alin A (2010) Multicollinearity. Wiley Interdiscip Rev Comput Stat 2(3):370–374. https://doi.org/10.1002/wics.84

Asteriou D, Hall SG (2007) Applied econometrics: a modern approach using EViews and Microfit (Rev. ed). Palgrave Macmillan. http://ci.nii.ac.jp/ncid/BA81825890.bib

Chatterjee S, Hadi AS (2006) Regression analysis by example, 4th edn. Wiley-Interscience. http://ci.nii.ac.jp/ncid/BA78320728

Deaton BJ, Hoehn JP (2004) Hedonic analysis of hazardous waste sites in the presence of other urban disamenities. Environ Sci Pol 7(6):499–508. https://doi.org/10.1016/j.envsci.2004.08.003

Heikkila E (1988) Multicollinearity in regression models with multiple distance measures. J Reg Sci 28(3):345–362. https://doi.org/10.1111/j.1467-9787.1988.tb01087.x
Kutner M, Nachtsheim C, Neter J, Li W (2005) Applied linear statistical models. McGraw-Hill Irwin. https://books.google.co.jp/books?id=OHqBQgAACAAJ
Minotani C (2015) Series statistics library, linear regression analysis (in Japanese). Asakura Publishing Co., Ltd
Rosen S (1974) Hedonic prices and implicit markets: product differentiation in pure competition. J Polit Econ 82(1):34–55. https://doi.org/10.1086/260169
Silverman BW (1986) Density estimation for statistics and data analysis. Chapman and Hall
Witte AD, Sumka HJ, Erekson H (1979) An estimate of a structural hedonic price model of the housing market: an application of Rosen's theory of implicit markets. Econometrica 47(5):1151–1173. https://doi.org/10.2307/1911956
Yoo W, Mayberry R, Bae S, Singh K, He Q, Lillard JW (2014) A study of effects of multicollinearity in the multivariable analysis. Int J Appl Sci Technol 4(5):9–19. http://www.ncbi.nlm.nih.gov/pmc/articles/PMC4318006/
Yoshida M (1987) Multicollinearity and ridge regression in regression analysis. 13:227–242. https://doi.org/10.18910/9749

Chapter 12
Theoretical Relationships Between Building Setback, Plot Frontage, and Plot Depth in Terms of Road and Building Densities

Hiroyuki Usui and Yasushi Asami

Abstract A theoretical investigation into the relationships between building setback, plot frontage, and plot depth in terms of building density (number of buildings per unit area) and road-network density (total length of road networks per unit area) is sparsely documented. Hence, in this study, we constructed a mathematical model to analyze the above relationship and maximize the average plot frontage. We found that (1) the average plot frontage is inversely proportional to the gross building density and is a convex upward quadratic function of the road-network density; (2) the average plot depth and setback allowance do not depend on the building density and are a monotonically decreasing function of the road-network density; and (3) the ratio of average plot depth to frontage is proportional to the gross building density and is a monotonically decreasing function of the road-network density. Based on the above findings, we developed a criterion for determining whether the average plot frontage is longer than the average plot depth. Further, the model was validated via an empirical analysis of the Tokyo metropolitan region. The results revealed that the modes of the average plot frontage, average plot depth, and ratio of plot depth to frontage were 11 m, 16 m, and 1.4, respectively. The mode of setback allowance was 5 m. The ratio of building setback to the setback allowance was in the range 0.4–0.6, which indicated that typically approximately half of the allowance is assigned to building setback. The findings of this study will provide urban planners and practitioners with a firm theoretical foundation for understanding the relationship between building setbacks, plot shape, building density, and road-network density in the district scale.

The contents of this paper are based on the following paper originally published in a Japanese journal: Usui, H. and Asami, Y. (2011) Setback distance and density of buildings and roads. *Journal of City Planning Institute of Japan* 46(3): 829–834 (in Japanese).

H. Usui (✉) · Y. Asami
Department of Urban Engineering, The University of Tokyo, Tokyo, Japan
e-mail: usui@ua.t.u-tokyo.ac.jp; asami@csis.u-tokyo.ac.jp

Y. Asami et al. (eds.), *Studies in Housing and Urban Analysis in Japan*,
New Frontiers in Regional Science: Asian Perspectives 75,
https://doi.org/10.1007/978-981-99-8027-7_12

Keywords Building density · Road-network density · Plot frontage · Plot depth · Ratio of plot depth to frontage · Building setback

12.1 Introduction

Urban spaces primarily consist of buildings and roads (Hillier and Hanson 1989; Kropf 2018; Oliveira 2016). The boundaries between building and road spaces in Western and Japanese cities are discernibly different from each other (Ashihara 1986). Western cities have distinct boundaries between building and road spaces because the buildings along streets have well-defined outlines. However, in Japanese cities, this demarcation tends to be ambiguous because the outlines of the buildings along streets often lack precise definition (Ashihara 1986). The ambiguity in the demarcation between building and road spaces characterizes the physical urban form of Japanese cities (Maki 1980).

The ambiguity between building and road spaces in current Japanese cities can be attributed to the Japanese building-form regulations. In Japan, a building form is governed by land-use zoning regulations, namely the maximum building coverage ratio (BCR) and floor area ratio (FAR) for a given plot. BCR is defined as the maximum ratio of the building area to the plot area, and FAR is defined as the maximum ratio of the total floor area of a building to the area of the plot. The maximum BCR and FAR are legally established in the Japanese building codes and the City Planning Act. Such building-form regulations are called *indirect building-form regulations* because they do not directly determine the building form. Furthermore, building façades are often not aligned with their adjacent front roads, which blurs the boundary between the building and road spaces.

However, building forms in Western cities (for example, cities in France and Germany) are directly controlled by *direct building-form regulations* that mandate guidelines regarding building façade locations and heights according to the proportion of building height, front road width, and building setbacks (Marshall 2011). Although building façades must be aligned with a legally established building line (road perimeter), an adherence to a maximum BCR and FAR is not legally required and is followed only if necessary (Wada 2007). Direct building-form regulations ensure that buildings maintain a uniform height and their facades align with the building lines. However, the close relationship between building façades and roads, as commonly observed in traditional Western cities, has gradually disappeared in modern Western cities since the publication of Buchan's report *Traffic in Towns* (Marshall 2005).

In this study, the ambiguity in the demarcation between building and road spaces in a 2D plane is quantified as the building setback, which is defined as the distance between the façade of a building and the edge of its front road. A comparison between the building-form regulations in Japanese and traditional Western cities indicates that the ambiguous demarcations between building and road spaces in current Japanese cities can be regarded as facilitating an allowance for the building setback. In the literature, this ambiguity has been compared with international urban planning policies and physical urban forms (e.g., building location patterns and road-network patterns) (Kropf 2018; Marshall 2005, 2009; Oliveira 2016) (see a detailed literature review from Japanese journals in the next section). However, the relationship between the setback allowance, average plot area, and plot shape (plot frontage and plot depth) in a district (a collection of urban blocks encircled by roads and large-scale non-built areas) had not been mathematically formulated until the Japanese version of this paper was published in a journal in 2011 (Usui and Asami 2011). Furthermore, qualitatively, the higher the road-network density, the lower the average plot depth and building setback; however, a quantitative and theoretical investigation involving the derivation of a mathematical formulation describing the above relationship had not been undertaken before 2011.

The objective of this study is to investigate the theoretical and empirical relationship between the building setbacks, average plot depth, plot frontage, and plot size within a district in terms of the building density (the number of buildings per unit area) and road-network density (the total length of road-networks represented as road centerlines per unit area) in the district.

The remainder of this chapter is organized as follows. Previous literature is briefly reviewed in the second section. In the third section, the average plot frontage, plot size, and plot depth are mathematically formulated as functions of building density and road-network density. Subsequently, the average allowance for building setback in a plot is mathematically formulated. In the fourth section, we discuss the application of these formulations to approximately 3000 districts in the Tokyo metropolitan region, statistical distributions of the average plot frontage and depth, and ratio of the average plot depth to plot frontage. Next, the relationship between the average setback allowance and road-network density is theoretically investigated. In addition, the ratio of building setbacks to the allowance is empirically analyzed. The final section concludes this chapter.

12.2 Literature Review

In urban planning, the first step to understand the physical form and environment of urban spaces is to compute building density, road-network density, district BCR (the ratio of the total area of building footprints in a district to the area of the district), and district FAR (the ratio of the total floor area of buildings in a district to the area of the district), for which several appropriate criteria have been proposed (Asami 2001; Koshizuka 1988; Koshizuka and Kotoh 1989). Several methods have been proposed to categorize urban forms into different types at the district scale (e.g., *Space Matrix* (Berghauser-Pont and Haupt 2010) and *Morpho* (Oliveira and Medeiros 2016)). The relationship between urban form and density (BCR, road-network density, and FAR) has been theoretically analyzed (e.g., Berghauser-Pont and Haupt 2007, 2010).

In the literature, the ambiguity between building and road spaces has been studied with respect to the physical urban form (the open space around buildings and adjacency of buildings to roads). A previous study investigated the theoretical relationship between the open space area around a building—called the effective vacant area—and building density in a district using a simple mathematical model. The results revealed that the effective vacant area decreased as the average building footprint area increased (Koshizuka 1988; Koshizuka and Kotoh 1989).

In addition, Koshizuka (1985) modelled road networks as a set of random lines uniformly distributed on a 2D plane. He found that the average distance from a location to the nearest road network was inversely proportional to the road-network density. As per theoretical analyses, building density and road-network density must be proportional to each other because all buildings are adjacent to their front roads (Koshizuka 1992). Saito (2004) developed an urban form simulation model on the assumption that the lattice density and building location rule are fixed and investigated the changes in the patterns of urban blocks and road networks with respect to the changes in lattice patterns (e.g., grids, triangles, and hexagons).

Asami and Ohtaki (2000) developed a method for predicting the shape of detached houses from the shape of residential plots by introducing an associative memory model. Empirical analyses conducted in a city in the Kanagawa prefecture revealed that the building setback increased with plot depth. On the basis of the above result, several methods have been developed to assess the adjacency of buildings and roads. The methods involve generating buffer domains from the centerlines of road networks, where the optimal buffer distance can be estimated from the inverse square root of building density (Usui and Asami 2010a, b).

12.3 Mathematical Formulation of the Average Plot Frontage, Depth, and Allowance for Building Setback in a District

12.3.1 *Semi-Gross Building Density*

Consider a district k of area S and perimeter L. Let the number of buildings in the district be n. The gross building density is defined as $\rho_G \equiv n/S$, and the semi-gross building density is defined as

$$\rho_N \equiv n/(S - S_R), \qquad (12.1)$$

where S_R is the area of the roads in the district.[1] In this study, all road networks, including those forming the boundary of the district (denoted by ∂L_k), are represented as road centerlines. Assuming that the maximum degree of road intersections

[1] A district area, S, includes non-residential plots such as urban parks and car parking space. However, the plots containing n number of buildings do not include such non-residential plots. Hence, ρ_G and ρ_N are smaller when non-residential plots are not considered than when they are.

(hereafter called intersections) is four, S_{R} can be formulated as follows:

$$S_{\mathrm{R}}\left(\Lambda, \overline{w}, v_3, v_4\right) \equiv \overline{w}\Lambda - \left(\frac{\overline{w}}{2}L + \overline{w}^2\right) - \left(v_3\frac{\overline{w}^2}{2} + v_4\overline{w}^2\right), \tag{12.2}$$

where Λ denotes the total length of the road networks, v_3 and v_4 denote the number of intersections of degree three and four, respectively, and $\overline{w}$ denotes the average road width. In Eq. (12.2), the second term on the right-hand side is the area of the road network that forms the boundary of the district but is not included in the district. The third term represents the area of the intersections, which is subtracted to avoid counting it twice.

Intersections of degrees three and four are classified into (1) those inside the district and (2) those on the road network forming the boundary of the district. In the former case, the number of intersections of degree three and four are denoted by $v_{3_{\mathrm{in}}}$ and v_4, respectively. Assuming that all roads extending beyond the district are omitted, all intersections of degree four in the latter case can be regarded as those of degree three. The number of intersections of degree three in the latter case is denoted as $v_{3_{\mathrm{peri}}}$.

Using this notation, Eq. (12.2) can be rewritten as follows:

$$\begin{aligned} S_{\mathrm{R}}\left(\Lambda, \overline{w}, v\right) &\equiv \overline{w}\Lambda - \left(\frac{\overline{w}}{2}L + \overline{w}^2 + \frac{\kappa_3 + 2\kappa_4}{2}v\overline{w}^2\right) \\ &= \overline{w}\Lambda - \left\{\frac{\overline{w}}{2}L + (1 + cv)\,\overline{w}^2\right\}, \end{aligned} \tag{12.3}$$

where $v_3 = v_{3_{\mathrm{in}}} + v_{3_{\mathrm{peri}}}$ and $v \equiv v_3 + v_4$. Additionally, $\kappa_3 \equiv v_3/v$ and $\kappa_4 \equiv v_4/v$ denote the ratio of the number of intersections of degree three and four, respectively, to the sum of the number of both intersections, and $c \equiv (\kappa_3 + 2\kappa_4)/2$; $c \in [0.5,1]$. If $v = v_3$, then $c = 0.5$. If $v = v_4$, then $c = 1$. Using Eq. (12.3), Eq. (12.1) can be rewritten as follows:

$$\begin{aligned} \rho_{\mathrm{N}} &= \frac{n}{S - \overline{w}\Lambda + \left\{\frac{\overline{w}}{2}L + (1 + cv)\,\overline{w}^2\right\}} \\ &= \frac{n}{S}\cdot\frac{1}{1 - \overline{w}\left\{\frac{\Lambda}{S} - \frac{1}{2}\cdot\frac{L}{S} \quad \frac{(1+cv)}{S}\overline{w}\right]} \\ &= \rho_{\mathrm{G}}\frac{1}{1 - \overline{w}\left\{\frac{\Lambda}{S} - \frac{1}{2}\cdot\frac{L}{S} - \frac{(1+cv)}{S}\overline{w}\right\}}. \end{aligned} \tag{12.4}$$

The adjacency of a building plot to its front road is not explicitly considered in Eq. (12.4). To address this issue, it is assumed that a building plot has no more than one building, which is termed *the rule of one plot for one building*. This rule has advantages and disadvantages. The rule does not hold true, for example, when the use of two buildings on a lot is inseparably related to one another. Nevertheless, the assumption of the rule can help overcome data limitations regarding plot shapes. Under this assumption, the number of building plots is equal to n.

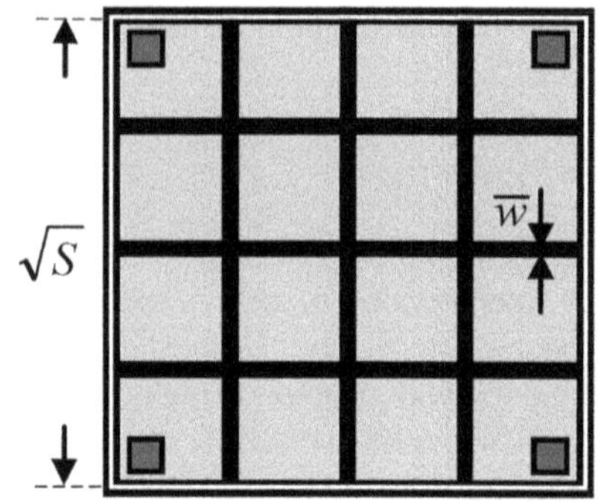

Fig. 12.1 A district modeled as a square of edge length $\sqrt{S}$. It is composed of a square grid of $m \times m$ urban blocks, where m denotes the number of urban blocks along one edge of the square. The white lines show the centerlines of the road networks forming the edges of the square

12.3.2 Average Plot Frontage as the Function of Building Density and Road-network Density

Assuming the rule of one plot per building and that all plots are adjacent to their front roads, the relationship between building density and road-network density was theoretically investigated using the average plot frontage in the district. As illustrated in Fig. 12.1, the district is modeled as a square of edge length $\sqrt{S}$, and it is composed of a square grid of $m \times m$ urban blocks, where m denotes the number of urban blocks along one edge of the square. The white lines show the centerlines of the road networks that form the edges of the square.

The total perimeter of the urban blocks can be formulated using m, $\sqrt{S}$, and $\overline{w}$ as follows:[2]

$$B \equiv 4\frac{\sqrt{S} - m\overline{w}}{m}m^2 = 4m\left(\sqrt{S} - m\overline{w}\right). \tag{12.5}$$

Because the relationship between Λ and m is formulated as $\Lambda = 2(m+1)\sqrt{S}$, Eq. (12.5) can be rewritten as a function of Λ as follows:

$$B = 2\sqrt{S}\left(\frac{\Lambda}{\sqrt{S}} - 2\right) - \left(\frac{\Lambda}{\sqrt{S}} - 2\right)^2 \overline{w}. \tag{12.6}$$

To compute the average plot frontage, the number of plots facing a street in the district, n_F, is expressed in terms of n, v, and c. Inside the district, the numbers of corner plots and plot edges facing an intersection of degree four are four and eight, respectively. Hence, the numbers of corner plots and plot edges facing the total number of intersections of degree four are $4v_4$ and $2 \times 4v_4 = 8v_4$, respectively (see Fig. 12.2a). By contrast, one plot and two corner plots face an intersection of degree three inside the district; therefore, the number of plot edges facing an

[2] B is equivalent to the total plot frontage in the district.

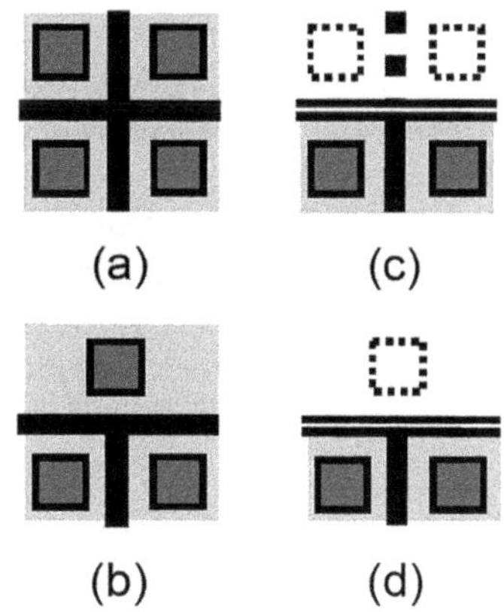

Fig. 12.2 Number of corner plots and the plot frontages: (**a**) corner plots facing an intersection of degree four inside a district, (**b**) corner plots facing an intersection of degree three inside a district, (**c**) corner plots facing an intersection of degree four on the perimeter of a district, (**d**) corner plots facing an intersection of degree three on the perimeter of a district

intersection of degree three is five. Hence, the numbers of plots and plot edges facing the total number of intersections of degree three inside the district are $3v_{3_{\text{in}}}$ and $1 \times v_{3_{\text{in}}} + 2 \times 2v_{3_{\text{in}}} = 5v_{3_{\text{in}}}$, respectively (see Fig. 12.2b). The number of corner plots facing an intersection of degree three on the boundary of the district is two, and the corresponding number of plot edges is four. Hence, the numbers of corner plots and plot edges facing the total number of intersections of degree three on the boundary of the district are $2v_{3_{\text{peri}}}$ and $4v_{3_{\text{peri}}}$, respectively (see Fig. 12.2c,d). Therefore, n_{F} can be formulated as follows:

$$\begin{aligned} n_{\text{F}} &= 8v_4 + 5v_{3_{\text{in}}} + 4v_{3_{\text{peri}}} + \left\{(n+4) - \left(4v_4 + 3v_{3_{\text{in}}} + 2v_{3_{\text{peri}}}\right)\right\} \\ &= (n+4) + 4v_4 + 2v_3 \\ &= (n+4) + (2\kappa_4 + \kappa_3)\, v \\ &= n + 4\,(1 + cv)\,, \end{aligned} \tag{12.7}$$

where $(n+4) - \left(4v_4 + 3v_{3_{\text{in}}} + 2v_{3_{\text{peri}}}\right)$ is the number of plot edges not facing any intersections. Because the corner plots facing the vertices of the square have two edges each facing the vertices, four was added to the right-hand side of Eq. (12.7).

Using Eqs. (12.6) and (12.7), the average plot frontage in the square district is formulated as shown below.

$$\begin{aligned} \overline{F} &\equiv \frac{B}{n_{\text{F}}} = \frac{S}{n} \cdot \frac{n}{n_{\text{F}}} \left\{ 2\left(\frac{\Lambda}{S} - \frac{2}{\sqrt{S}}\right) - \left(\frac{\Lambda}{S} - \frac{2}{\sqrt{S}}\right)^2 \overline{w} \right\} \\ &\approx \frac{1}{\rho_{\text{G}}} \cdot \frac{n}{n_{\text{F}}} \lambda\,(2 - \overline{w}\lambda)\,, \text{ if } \frac{2}{\sqrt{S}} \ll \frac{\Lambda}{S}, \end{aligned} \tag{12.8}$$

where $\lambda \equiv \Lambda/S$ denotes road-network density and n_{F}/n can be regarded as a constant; from empirical analyses in the Tokyo metropolitan region, $n_{\mathrm{F}}/n \simeq 1.2$. Hence, $\overline{F}$ can be regarded as a function of three variables: road-network density, gross building density, and average road width. The average plot frontage is inversely proportional to the gross building density and is a convex upward quadratic function of the road-network density. Therefore, there exists a λ^* that maximizes $\overline{F}$.

$$\overline{F}_{\max} = \frac{n}{n_{\mathrm{F}}} \cdot \frac{S}{n\overline{w}} = \frac{n}{n_{\mathrm{F}}} \cdot \frac{1}{\overline{w}\rho_{\mathrm{G}}}. \tag{12.9}$$

$$\lambda^* = \frac{1}{\overline{w}} + \frac{2}{\sqrt{S}} \approx \frac{1}{\overline{w}}. \tag{12.10}$$

If $\lambda < \lambda^*$, then increasing λ increases the average plot frontage. Otherwise, increasing λ decreases the average plot frontage. In addition, Eqs. (12.9) and (12.10) show that (1) $\overline{F}_{\max}$ is inversely proportional to ρ_{G}, $\overline{w}$, and n_{F}/n, and (2) λ^* is inversely proportional to $\overline{w}$.

$\overline{F}$ in Eq. (12.8) can be regarded as a three-variable function that characterizes the number of buildings and roads in a district. If the infimum of $\overline{F}$, denoted by $\overline{F}_{\mathrm{inf}}$, is a policy variable and the number of roads in a district (λ and $\overline{w}$) is fixed, the supremum of ρ_{G} can be computed using λ, $\overline{w}$, and $n_{\mathrm{F}}/n(\simeq 1.2)$ as follows:

$$\rho_{\mathrm{G}} \leq \frac{1}{\overline{F}_{\mathrm{inf}}} \cdot \frac{n}{n_{\mathrm{F}}} \lambda\,(2 - \overline{w}\lambda) = \sup \rho_{\mathrm{G}}. \tag{12.11}$$

As $\sup \rho_{\mathrm{G}}$ is regarded as a convex upward quadratic function of λ, there exists a λ^{**} that maximizes $\sup\rho_{\mathrm{G}}$ as follows:

$$\max \sup \rho_{\mathrm{G}} = \frac{n}{n_{\mathrm{F}}} \cdot \frac{1}{\overline{w}\overline{F}_{\mathrm{inf}}}, \tag{12.12}$$

$$\lambda^{**} = \lambda^*. \tag{12.13}$$

If $\lambda < \lambda^{**}$, then increasing λ increases $\sup\rho_{\mathrm{G}}$. Otherwise, increasing λ decreases $\sup\rho_{\mathrm{G}}$. Eqs. (12.12) and (12.13) imply that an optimal road-network density exists that maximizes the gross building density, satisfying $\overline{F} \geq \overline{F}_{\mathrm{inf}}$.

12.3.3 Average Plot Depth and Average Plot Depth to Frontage Ratio

The average plot size is given by the inverse of the semi-gross building density, ρ_N.

$$\overline{s} \equiv \frac{1}{\rho_N} = \frac{S}{n}\left[1 - \overline{w}\left\{\frac{\Lambda}{S} - \frac{1}{2}\cdot\frac{L}{S} - \frac{(1+cv)}{S}\overline{w}\right\}\right]$$

$$= \frac{1}{\rho_G}\left[1 - \overline{w}\left\{\lambda - \frac{1}{2}\cdot\frac{L}{S} - \frac{(1+cv)}{S}\overline{w}\right\}\right]. \tag{12.14}$$

Assuming that the plot frontage and depth are independently distributed, the average plot depth can be approximated to $\overline{s}/\overline{F}$ as follows:

$$\overline{D} \approx \frac{\overline{s}}{\overline{F}} = \frac{\frac{S}{n}\left[1 - \overline{w}\left\{\frac{\Lambda}{S} - \frac{1}{2}\cdot\frac{L}{S} - \frac{(1+cv)}{S}\overline{w}\right\}\right]}{\frac{S}{n}\cdot\frac{n}{n_F}\left\{2\left(\frac{\Lambda}{S} - \frac{2}{\sqrt{S}}\right) - \left(\frac{\Lambda}{S} - \frac{2}{\sqrt{S}}\right)^2\overline{w}\right\}}$$

$$= \frac{n_F}{n}\cdot\frac{1 - \overline{w}\left\{\lambda - \frac{1}{2}\cdot\frac{L}{S} - \frac{(1+cv)}{S}\overline{w}\right\}}{\left\{2\left(\lambda - \frac{2}{\sqrt{S}}\right) - \left(\lambda - \frac{2}{\sqrt{S}}\right)^2\overline{w}\right\}}$$

$$\approx \frac{n_F}{n}\cdot\frac{1 - \overline{w}\left\{\lambda - \frac{1}{2}\cdot\frac{L}{S} - \frac{(1+cv)}{S}\overline{w}\right\}}{\lambda\left(2 - \overline{w}\lambda\right)}, \text{ if } \frac{2}{\sqrt{S}} \ll \frac{\Lambda}{S}. \tag{12.15}$$

The validity of this assumption has been confirmed empirically in previous studies (Usui 2019; Usui and Asami 2013). Assuming the rule of one plot for one building and that all plots are adjacent to their front roads, Eq. (12.15) implies that the average plot depth depends not on building density but on λ, intersection density (v/S), $\overline{w}$, and n_F/n($\simeq$1.2).

The ratio of average plot depth to frontage is defined as follows:

$$\gamma \equiv \frac{\overline{D}}{\overline{F}} = \rho_G\left(\frac{n_F}{n}\right)^2\frac{1 - \overline{w}\left\{\frac{\Lambda}{S} - \frac{1}{2}\cdot\frac{L}{S} - \frac{(1+cv)}{S}\overline{w}\right\}}{\left(\frac{\Lambda}{S} - \frac{2}{\sqrt{S}}\right)^2\left\{2 - \left(\frac{\Lambda}{S} - \frac{2}{\sqrt{S}}\right)\overline{w}\right\}^2}$$

$$\approx \rho_G\left(\frac{n_F}{n}\right)^2\frac{1 - \overline{w}\left\{\lambda - \frac{1}{2}\cdot\frac{L}{S} - \frac{(1+cv)}{S}\overline{w}\right\}}{\lambda^2(2 - \overline{w}\lambda)^2} = \rho_G C(\lambda), \text{ if } \frac{2}{\sqrt{S}} \ll \frac{\Lambda}{S}. \tag{12.16}$$

Because $C(\lambda) \equiv \left(\frac{n_{\mathrm{F}}}{n}\right)^2 \frac{1-\overline{w}\left\{\lambda-\frac{1}{2}\cdot\frac{L}{S}-\frac{(1+cv)}{S}\overline{w}\right\}}{\lambda^2(2-\overline{w}\lambda)^2}$ is a monotonically decreasing function of λ, (1) increasing ρ_{G} increases γ, and (2) increasing λ decreases γ. In addition, the dimensional analysis of Eq. (12.16) implies that, because γ is a dimensionless quantity, $C(\lambda)$ must have two dimensions in length; if the other variables are fixed, $C(\lambda)$ can be regarded as the gross average plot size determined by the road-network density. Moreover, if $\gamma < 1$ (i.e., $1/\rho_{\mathrm{G}} < C(\lambda)$), then $\overline{F} > \overline{D}$. Otherwise, $\overline{F} < \overline{D}$. Therefore, $C(\lambda)$ can be used to determine which of $\overline{F}$ and $\overline{D}$ is smaller.

12.3.4 Average Allowance for Building Setback

The average allowance for building setback in a plot is mathematically formulated assuming that (1) the building footprint and its plot are rectangular in shape and similar to each other and (2) the BCR in each plot is the district BCR, denoted by $\mu \equiv S_{\mathrm{B}}/(S - S_{\mathrm{R}})$, where S_{B} is the total area of the building footprints. As illustrated in Fig. 12.3, the building frontage and depth that satisfy Eqs. (12.17) and (12.18) can be defined as the typical building frontage, $\overline{F}_{\mathrm{B}}$, and typical building depth, $\overline{D}_{\mathrm{B}}$, respectively.

$$\mu \overline{F}\,\overline{D} = \overline{F}_{\mathrm{B}}\,\overline{D}_{\mathrm{B}}. \tag{12.17}$$

$$\frac{\overline{D}_{\mathrm{B}}}{\overline{F}_{\mathrm{B}}} = \frac{\overline{D}}{\overline{F}} = \gamma. \tag{12.18}$$

On solving the above two equations, we obtain $\overline{D}_{\mathrm{B}} = \sqrt{\mu}\overline{D}$. The average allowance for building setback in a plot is defined as

$$\overline{D}_{\mathrm{Free}} \equiv \overline{D} - \overline{D}_{\mathrm{B}} = \left(1 - \sqrt{\mu}\right)\overline{D}. \tag{12.19}$$

Equations (12.15) and (12.19) indicate that $\overline{D}_{\mathrm{Free}}$ depends not on the gross building density but on the road-network density when the other variables such as μ, v/S, $\overline{w}$,

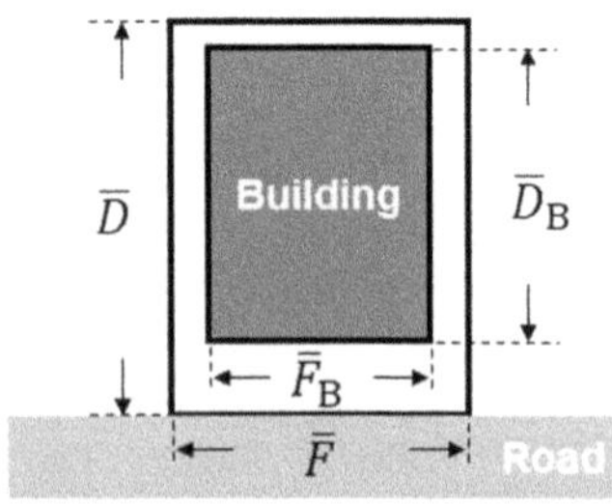

Fig. 12.3 Allowance for building setback in a plot, $\overline{D}_{\mathrm{Free}}$. The building frontage and depth that satisfy Eqs. (12.17) and (12.18) are defined as the typical building frontage, $\overline{F}_{\mathrm{B}}$, and typical building depth, $\overline{D}_{\mathrm{B}}$, respectively

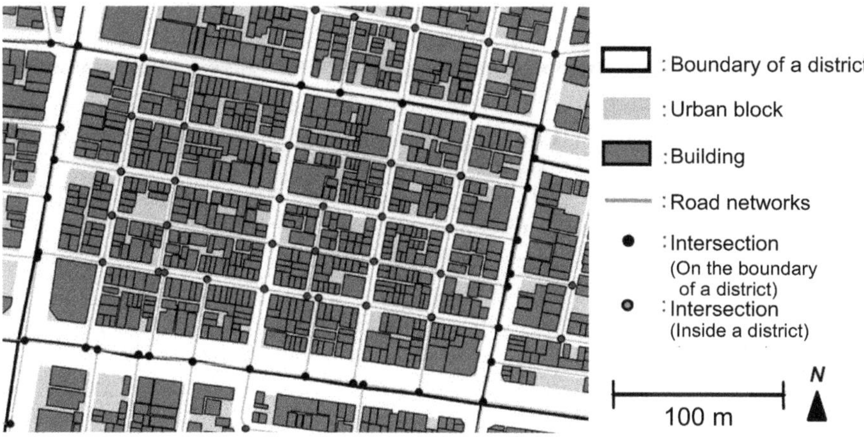

Fig. 12.4 A district in the Tokyo metropolitan region (Torigoe 1 chome). Road networks are modeled as a graph $G = (V, E)$, where $V = \{v_j\}$ denotes a set of nodes representing intersections and $E = \{e_i\}$ denotes a set of edges representing the road segments connecting adjacent intersections

and $n_F/n(\simeq 1.2)$ are regarded as constant. Evidently, $\overline{D}_{\text{Free}}$ is inversely proportional to the square root of μ.

12.4 Empirical Case Analyses in the Tokyo Metropolitan Region

12.4.1 Computing the Empirical Area of the Road Networks

We selected 3115 districts in the Tokyo metropolitan region for the empirical case analysis. Figure 12.4 shows a district in the Tokyo metropolitan region (Torigoe 1 chome: "chome" means the name of district). For each district, n, Λ, v_3, v_4, $\overline{w}$, S_R, and S_B were computed using GIS software (ArcGIS).[3, 4] Hereafter, the software is referred to as GIS. Certain innovative approaches were necessary to compute Λ and $v = v_3 + v_4$ using GIS.

As shown in Fig. 12.4, road networks are modeled as a graph denoted by $G = (V, E)$, where $V = \{v_j\}$ denotes a set of nodes representing the intersections and $E = \{e_i\}$ denotes a set of edges representing the road segments connecting

[3] The spatial data used are *Residential Maps* as of 2005 by Zenrin, Co., LTD. for building footprints and *Mapple 10,000 Digital Data* as of 2007 released by Shobunsha Publications, Inc. for road networks.

[4] Assuming that the maximum degree of intersection is four.

adjacent intersections. A set of polygons representing the urban blocks in the district ($UB = \{UB_k\}$, where UB_k denotes the set of urban blocks in district k) was generated by subtracting the polygons representing roads from the polygon representing the district. In general, ∂L_k and the road networks forming ∂L_k correspond to each other; however, a slight difference between them triggers a significant difference in Λ and v.

To address this issue, the following five-step data processing method must be followed. First, buffer domains are generated from $E = \{e_i\}$. Each buffer domain is denoted by $N(e_i)$, and the buffer distance is computed as the sum of [width of e_i]/2 and $0.35/\sqrt{\rho_G}$ (Usui and Asami 2010a). Second, e_i is considered a part of ∂L_k if $N(e_i)$ includes some parts of ∂L_k, denoted by $N(e_i) \cap \partial L_k \neq \varnothing$. The set of edges satisfying $N(e_i) \cap \partial L_k \neq \varnothing$ is denoted by $E' \equiv \{e_i | \; N(e_i) \cap \partial L_k \neq \varnothing \}$. Third, a set of edges satisfying the condition $N(e_i) \cap \partial L_k = \varnothing$ is denoted by $E'' \equiv \{e_i \mid N(e_i) \cap \partial L_k = \varnothing \} = E - E'$ and decomposed as $E'' = \bigcup_k E''_k$, where E''_k denotes the subset of edges satisfying $N(e_i) \cap \partial L_k = \varnothing$ in district k. Fourth, the number of nodes $v_j \in V$ on E''_k is computed and represented as v. Finally, the total length of edges representing the road segments satisfying the condition $N(e_i) \cap UB_k \neq \varnothing$, including the road networks forming ∂L_k, is computed as Λ. The average road width $\overline{w}$, in a district is then computed as shown below:

$$\overline{w} \equiv \sum\nolimits_h w_h \Lambda_h / \Lambda, \tag{12.20}$$

where w_h and Λ_h denote the width and total length of the edges of the road whose class of road width is h, respectively.

The scatter plot shown in Fig. 12.5 (left) illustrates the relationship between $S_R(\Lambda, \overline{w}, v)$ (given by Eq. (12.3)) and S_R computed using GIS. S_R was estimated by measuring the areas common to the polygon representing the district and those representing the roads. As can be observed from the figure, S_R tends to be lower than $S_R(\Lambda, \overline{w}, v)$. Fig. 12.5 (right) shows the relative frequency distribution of $S_R/S_R(\Lambda, \overline{w}, v) \equiv r$. The average and standard deviation of r are 0.86 and 0.17, respectively. We found that (1) the relative frequency distribution of r is unimodal, and (2) its mean and mode are almost equal.

S_R was lower than $S_R(\Lambda, \overline{w}, v)$ because we treated the width of the roads forming the boundary of the district ∂L_k inappropriately. As can be observed from Fig. 12.4, (1), approximately half of the area of roads forming ∂L_k is not included in district k, and (2) the width of the roads forming ∂L_k tends to be wider than those inside district k. The former was considered for computing $S_R(\Lambda, \overline{w}, v)$ by incorporating the second term on the right-hand side of Eq. (12.3). Because the width of the roads forming ∂L_k tends to be wider than those inside the district, $S_R \simeq 0.86 S_R(\Lambda, \overline{w}, v)$ on average. This property should be considered when estimating S_R from $S_R(\Lambda, \overline{w}, v)$. Hence, it was not $S_R(\Lambda, \overline{w}, v)$ but S_R that we adopted to compute $\overline{F}$, $\overline{D}$, and γ in empirical analyses.

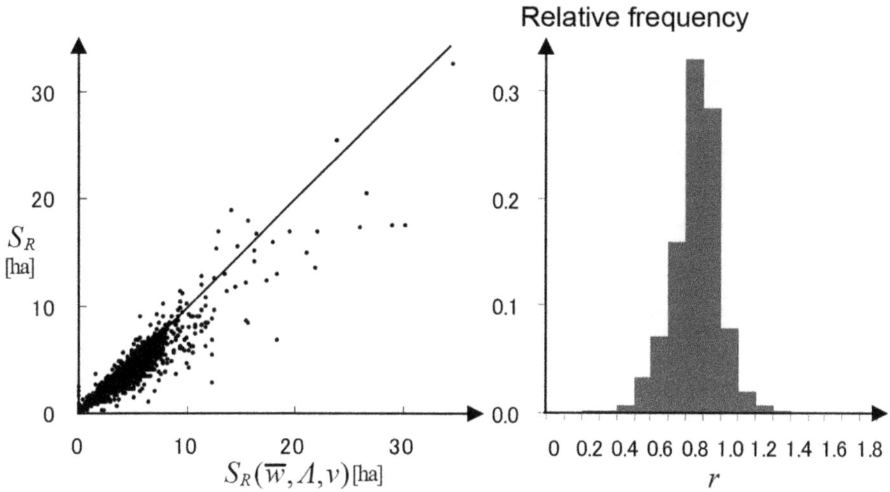

Fig. 12.5 Relationship between $S_R\left(\Lambda, \overline{w}, \nu\right)$ and S_R (left) and the relative frequency distribution of $S_R/S_R\left(\Lambda, \overline{w}, \nu\right) = r$ (right)

12.4.2 Frequency Distribution of the Average Plot Frontage, Depth, and the Average Plot Depth to Frontage Ratio and Their Geographical Distribution in Tokyo

Figure 12.6 (top) shows the frequency distribution of the average plot frontage, $\overline{F}$, and average plot depth, $\overline{D}$, in the districts of the Tokyo metropolitan region. Evidently, (1) the two distributions are unimodal, and (2) the modes of $\overline{F}$ and $\overline{D}$ are 11 m and 16 m, respectively. Figure 12.6 (bottom) shows the frequency distribution of the ratio of average plot depth to frontage, γ, in the districts of the Tokyo metropolitan region. As can be observed, this distribution is also unimodal and the mode of γ is 1.4. This implies that, on average, plot depth tends to be higher than plot frontage in the Tokyo metropolitan region.[5]

The geographical distribution of γ over the districts in the Tokyo metropolitan region is shown in Fig. 12.7. As can be observed, the plot frontage tends to be higher than the plot depth in the districts surrounding the center of the Tokyo metropolitan region (defined as the Tokyo central station). By contrast, the plot depth tends to be higher than the plot frontage in districts that are 5–10 km away from the Tokyo central station.

[5] Considering plots not adjacent to their front roads, $\overline{F}$ computed using Eq. (12.8) is smaller than the observed value. $\overline{D}$ computed using Equation (12.15) is higher than the observed value because $\overline{D} = \overline{s}/\overline{F}$, and γ computed using Equation (12.16) is higher than the observed value because $\gamma = \overline{s}/\overline{F}^2$.

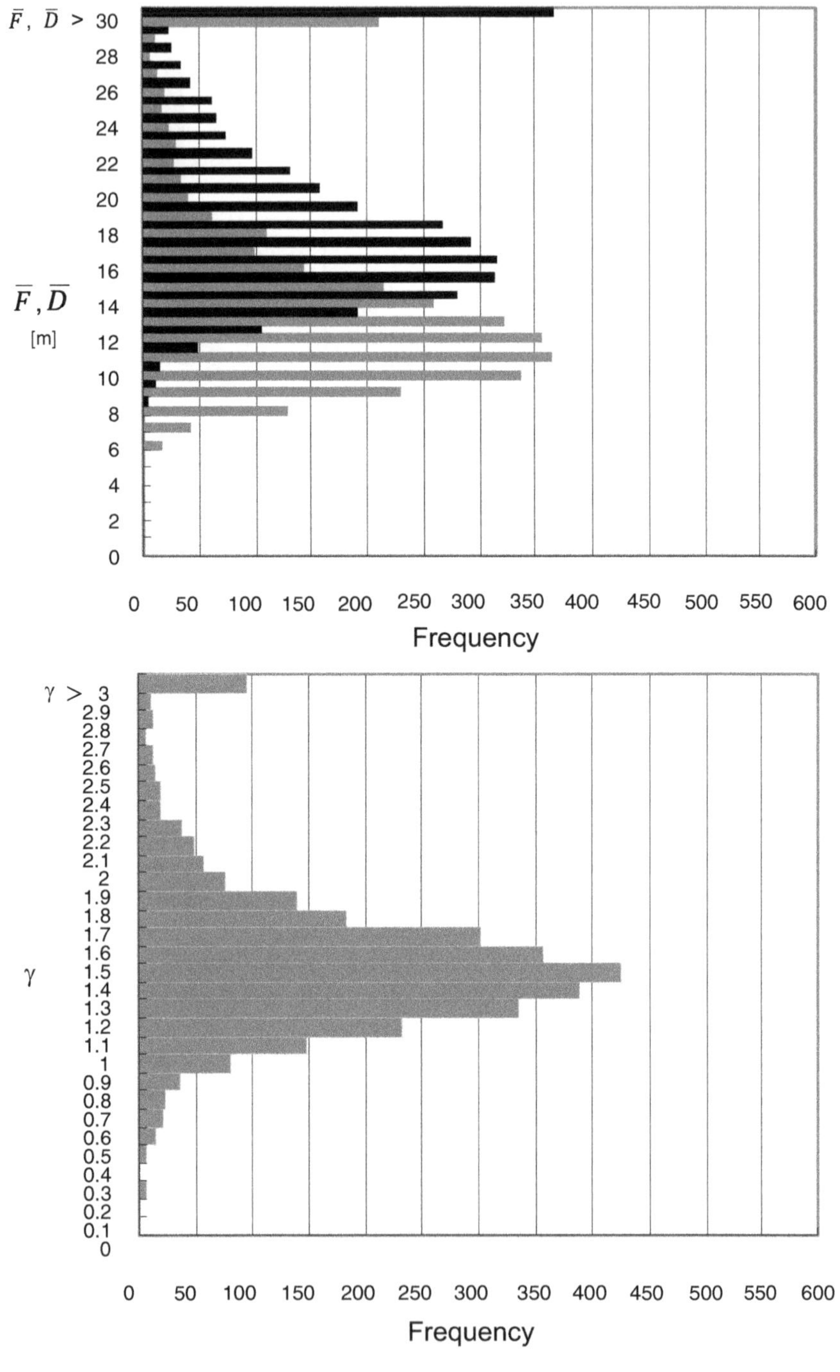

Fig. 12.6 Frequency distribution of the average plot frontage, $\overline{F}$, (gray) and average plot depth, $\overline{D}$, (black) (top) and the frequency distribution of the ratio of average plot depth to frontage, γ, in the district of the Tokyo metropolitan region (bottom)

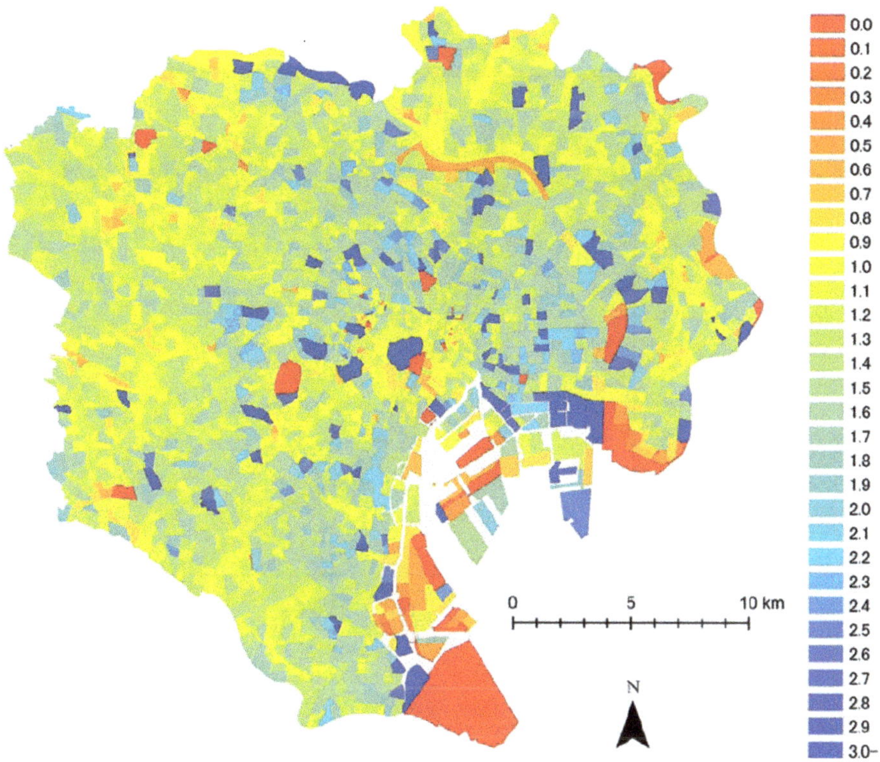

Fig. 12.7 Geographical distribution of γ of each district in the Tokyo metropolitan region

The above empirical findings can be theoretically validated by focusing on the relationship between gross building density and road-network density. Figure 12.8 shows the relationship between road-network density λ and $C(\lambda)$, as defined in Eq. (12.16). The black dots represent the value of $C(\lambda)$ when the values of the other variables in $C(\lambda)$ are the average values over the Tokyo metropolitan region, i.e., $v/S = 45\ \text{ha}^{-1}$, $c= 0.56$, $\overline{w} = 8.7$ m, $n_F/n = 1.38$, and L/S=124 m ha^{-1}. The gray dots represent the values of $C(\lambda)$ when the values of the other variables are those in each district. Compared to the districts of the Tokyo metropolitan region shown in Fig. 12.4, where $c = 0.70$, the average number of intersections of degree three is higher than those of degree four. The gray dots in Fig. 12.8 are distributed along the black dots.

As mentioned previously, $C(\lambda)$ can be used to determine which of $\overline{F}$ and $\overline{D}$ is smaller in each district. In this respect, Fig. 12.8 shows whether the gross building density is higher compared to the road-network density in terms of the comparative relationship between $\overline{F}$ and $\overline{D}$ in each district. The road-network density in the districts with $\gamma > 1$ is less than the building density. By contrast, the road-network density in the districts with $\gamma < 1$ is higher than the building density.

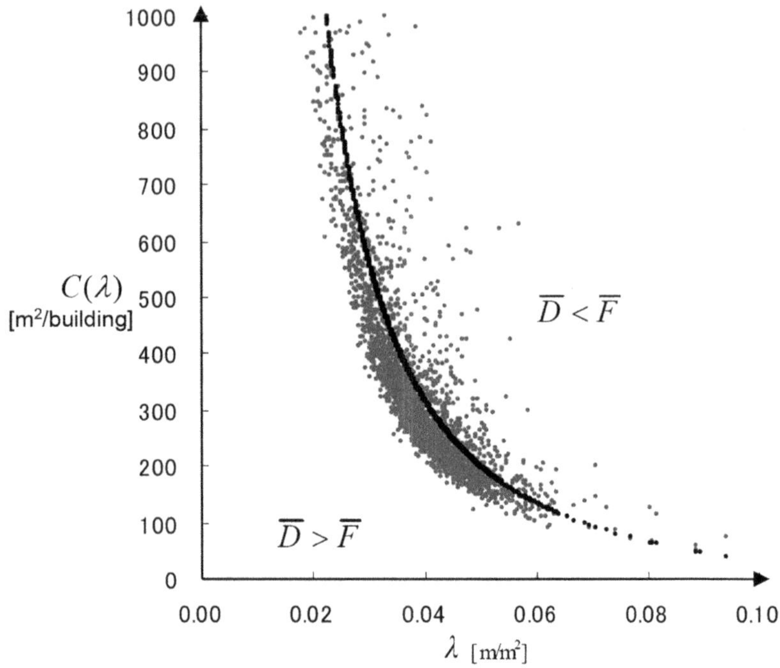

Fig. 12.8 Relationship between the road-network density λ and $C(\lambda)$, as defined in Eq. (12.16). Black dots are the values of $C(\lambda)$ when the values of the other variables in $C(\lambda)$ are their average in the Tokyo metropolitan region, i.e $v/S = 45$ ha^{-1}, $c = 0.56$, $\overline{w} = 8.7$ m, $n_F/n = 1.38$, and $L/S = 124$ m ha^{-1}. Gray dots are the values of $C(\lambda)$ when the values of the other variables are those in each district

12.4.3 Relationship Between Building Setback and Average Plot Depth in a District

Figure 12.9 shows the relationship between road-network density and $\overline{D}_{\text{Free}}$ wherein the values of the variables that constitute $\overline{D}_{\text{Free}}$ are the average values in the Tokyo metropolitan region: $v/S = 45$ ha^{-1}, $c = 0.56$, $\overline{w} = 8.7$ m, $n_F/n = 1.38$, and $L/S = 124$ m ha^{-1}. As can be observed (1) $\overline{D}_{\text{Free}}$ is a monotonically decreasing function of λ, and (2) as λ increases, the ratio $\left|d\overline{D}_{\text{Free}}/d\lambda\right|$ decreases. This implies that (1) the construction of a road network in districts with low road-network density can effectively decrease $\overline{D}_{\text{Free}}$ compared to that in the districts with high road-network density, and (2) relaxing the maximum BCR in each plot can effectively decrease $\overline{D}_{\text{Free}}$ in the districts with high road-network density.

Fig. 12.10 shows the frequency distribution of $\overline{D}_{\text{Free}}$ in the districts of the Tokyo metropolitan region. It is a unimodal distribution with a mode of 5 m. Interestingly, $\overline{D}_{\text{Free}}$ in 33 districts is zero, which implies that the building façades in the districts are aligned with their front roads. The geographical distribution of $\overline{D}_{\text{Free}}$ in each district of the Tokyo metropolitan region is shown in Fig. 12.11. As can be observed,

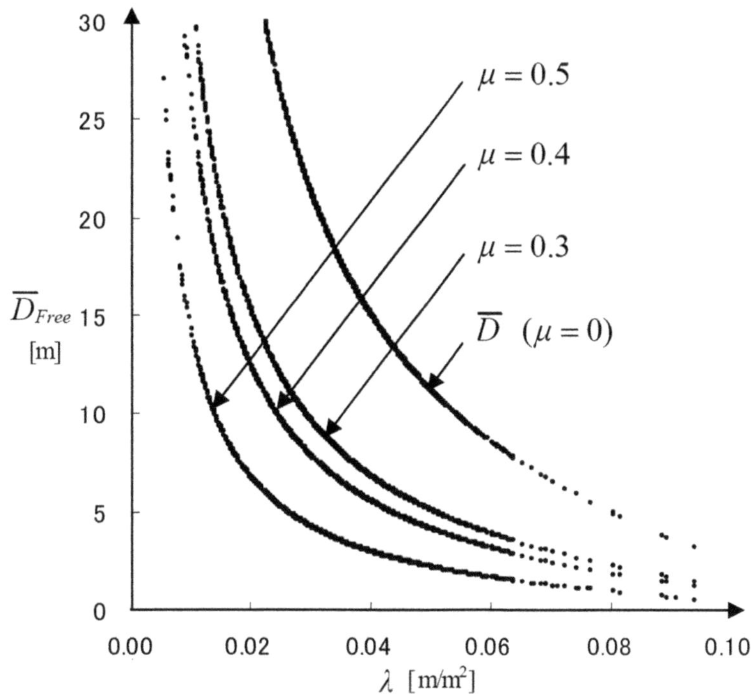

Fig. 12.9 Relationship between the road-network density λ and $\overline{D}_{\text{Free}}$ wherein the values of the other variables in $\overline{D}_{\text{Free}}$ are the average values in the Tokyo metropolitan region: i.e., $v/S = 45\ \text{ha}^{-1}$, $c = 0.56$, $\overline{w} = 8.7$ m, $n_F/n = 1.38$, and $L/S = 124\ \text{m ha}^{-1}$. If $\mu = 0$, $\overline{D}_{\text{Free}} = \overline{D}$

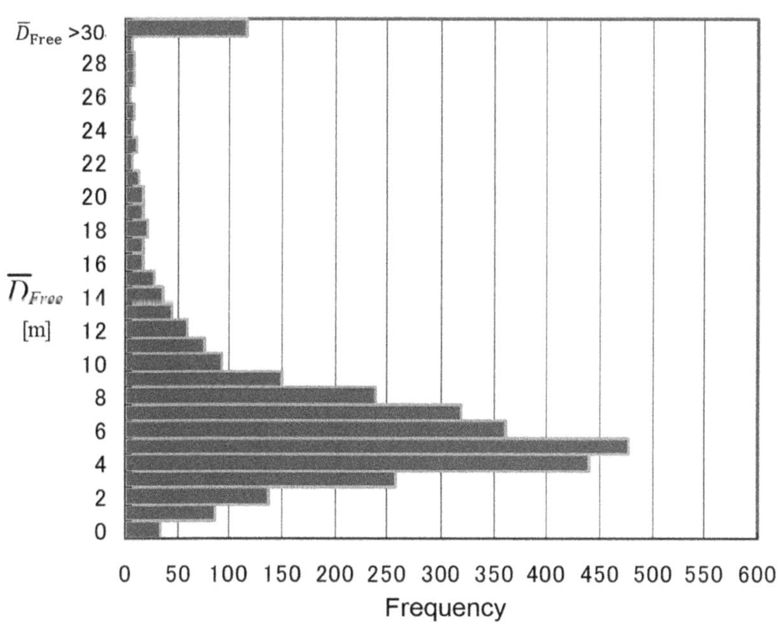

Fig. 12.10 Frequency distribution of $\overline{D}_{\text{Free}}$ in the districts of the Tokyo metropolitan region

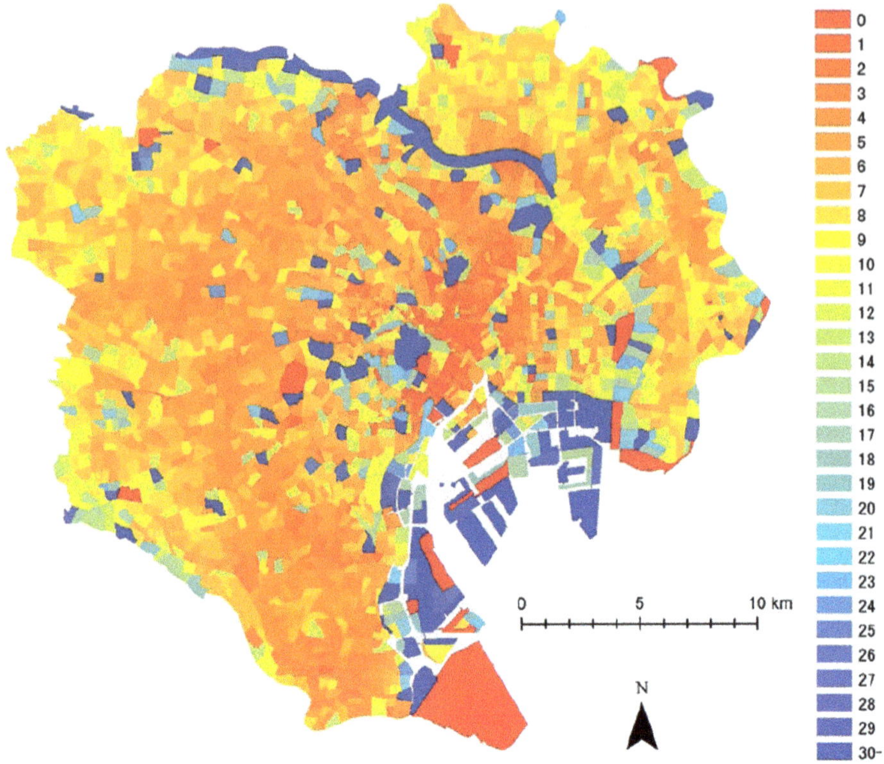

Fig. 12.11 Geographical distribution of $\overline{D}_{\text{Free}}$ over the Tokyo metropolitan region

(1) districts with $\overline{D}_{\text{Free}} = 0$ are geographically clustered around the Tokyo central station and where land readjustment projects were carried out after the Great Kanto earthquake in 1923. By contrast, the districts with $\overline{D}_{\text{Free}} \simeq 5$ m are distributed within 5 km from the Tokyo central station.

Finally, the ratio of building setback to $\overline{D}_{\text{Free}}$ is empirically analyzed for five districts in the Tokyo metropolitan region. On average, $\overline{D}_{\text{Free}}$ in a district can be regarded as the upper limit of the average building setback, denoted by $\overline{D}_{\text{SB}}$.[6] In a real urban space, $\overline{D}_{\text{SB}}$ ranges from zero to $\overline{D}_{\text{Free}}$ according to building-form regulations, namely the BCR and FAR. Table 12.1 shows the ratios $\overline{D}_{\text{SB}}/\overline{D}$ and $\overline{D}_{\text{SB}}/\overline{D}_{\text{Free}}$ in the five districts where the edge length of the square plot computed by $1/\sqrt{\rho_{\text{G}}}$ are 10, 12.5, 15, 17.5, and 20 m. For each district, $\overline{D}_{\text{SB}}$ is computed by measuring the building setbacks using the method developed by Usui and Asami (2010a, b). As shown in the fig. (1) $\overline{D}_{\text{SB}}/\overline{D}$ ranges from 0.1 to 0.2, and

[6] $\overline{D}_{\text{Free}}$ includes the minimum setback distance from the edge of an adjacent building plot as stipulated by the article 236 of Japan Civil Code.

Table 12.1 Computation results of $\overline{D}_{SB}/\overline{D}$ and $\overline{D}_{SB}/\overline{D}_{Free}$

District name	ρ_G	λ	$\overline{D}_{SB}$	$\overline{D}_{Free}$	$\overline{D}$	γ	$\overline{D}_{SB}/\overline{D}$	$\overline{D}_{SB}/\overline{D}_{Free}$
Kyojima 3	98.4	482	1.3	2.8	12.1	1.86	0.11	0.46
Sendagi 2	63.4	503	1.7	3.0	13.1	1.33	0.13	0.57
Yushima 3	44.4	461	1.3	2.8	14.1	1.27	0.09	0.46
Kami-shakujii-Minami	32.3	360	3.2	8.3	18.5	1.32	0.17	0.39
Minami-Shinozakicho 2	25.0	411	3.3	7.5	17.6	1.02	0.19	0.44

Note: ρ_G is in ha^{-1}, λ is in m ha^s, and $\overline{D}_{SB}$, $\overline{D}_{Free}$, and $\overline{D}$ are in m

(2) $\overline{D}_{SB}/\overline{D}_{Free}$ ranges from 0.4 to 0.6. Therefore, $\overline{D}_{SB}/\overline{D}$ and $\overline{D}_{SB}/\overline{D}_{Free}$ tend to remain stable against the changes in $\overline{D}$ and γ. In addition, irrespective of the values of $\overline{D}$ and γ, the fact that $\overline{D}_{SB}$ is approximately half of $\overline{D}_{Free}$is useful in estimating $\overline{D}_{SB}$ from road-network density using Eq. (12.19).

12.5 Conclusion

The objective of this study is to understand the relationship between the allowance for a building setback in its plot, average plot size, and plot shape (plot frontage and plot depth) in a district by theoretically and empirically investigating the relationship in terms of building density and road-network density in a district. We showed that computing the building density and road-network density enables us to compute the average plot size, plot frontage, and plot depth and understand the relationship between the building density, road-network density, and plot shape dimensions. The theoretical and empirical findings of this study will provide urban planners and practitioners with a firm theoretical foundation for understanding the relationship between building setbacks (via allowance for the building setback in its plot), plot shape, building density, and road-network density at the district scale. The study contributes significantly toward understanding the relationship between the part (plot) and whole (district) of physical form of urban space.

Acknowledgment This research was the result of joint research with CSIS, University of Tokyo (No. 785), and used the residential maps provided by Zenrin Co., Ltd.

References

Asami Y (ed) (2001) Residential environment: methods and theory for the evaluation. University of Tokyo Press

Asami Y, Ohtaki T (2000) Prediction of the shape of detached houses on residential lots. Environ Plan B Plan Design 27(2):283–295

Ashihara Y (1986) The hidden order Tokyo through the twentieth century. Chuko Bunko

Berghauser-Pont MY, Haupt P (2007) The relation between urban form and density. Urban Morphol 11:62–64
Berghauser-Pont MY, Haupt P (2010) Spacematrix: space, density and urban form. NAi Publishers
Hillier B, Hanson J (1989) The social logic of space. Cambridge University Press
Koshizuka T (1985) On the relation between the density of urban facilities and the distance to the nearest facility from a point in a given area. J City Plan Inst Japan 20:85–90. (in Japanese)
Koshizuka T (1988) A theoretical study on building density as an indicator of urban environments. J City Plan Inst Japan 23:19–25. (in Japanese)
Koshizuka T (1992) Spatial distribution model. In: Architectural Institute of Japan (ed) Analytic models for architecture and urban planning. Inoueshoin, pp 35–50
Koshizuka T, Kotoh H (1989) Estimations of effective vacant areas by building densities. J City Plan Inst Japan 24:337–342. (in Japanese)
Kropf K (2018) The handbook of urban morphology. John Wiley & Sons
Maki F (1980) City with a hidden past. Kajima Institute Publishing
Marshall S (2005) Street & Patterns. Spon Press
Marshall S (2009) Cities, design and evolution. Routledge, Abingdon
Marshall S (2011) Urban coding and planning. Routledge
Oliveira V (2016) Urban morphology: an introduction to the study of the physical form of cities. Springer
Oliveira V, Medeiros V (2016) *Morpho*: combining morphological measures. Environ Plan Plan Design 43(5):805–825
Saito C (2004) Effect of lattice on urban block pattern and city shape: a study on influence of neighborhood relationships of buildings on urban space. J City Plan Inst Japan 39(3):847–852. (in Japanese)
Usui H (2019) Statistical distribution of building lot depth: theoretical and empirical investigation of downtown districts in Tokyo. Environ Plan B Urban Analyt City Sci 46(8):1499–1516
Usui H, Asami Y (2010a) Method for determining buffer distance for judging adjacency of lots to roads. J Archit Plan 75(651):1175–1180. (in Japanese)
Usui H, Asami Y (2010b) Comparison of accuracy of methods of judging adjacency of buildings to roads. Theory Appl GIS 18(2):53–62. (in Japanese)
Usui H, Asami Y (2011) Setback distance and density of buildings and roads. J City Plan Inst Japan 46(3):829–834. (in Japanese)
Usui H, Asami Y (2013) Estimation of mean lot depth and its accuracy. J City Plan Inst Jpn 48(3):357–362. (in Japanese)
Wada H (2007) France no keikan wo yomu: Hozon to kisei no gendai toshikeikaku (understanding landscape in France: modern urban planning for preservation and restrictions of landscape). Kajima Institute Publishing. (in Japanese)

Chapter 13
A Method of Visual Analytics and Data Visualization in Design Context: Case Study of Spatiotemporal Data Visualization of Urban Retail Agglomeration Growth

Akiyoshi Inasaka

Abstract Urban commercial facilities significantly influence neighborhood residents, with changes like store closures disrupting their routines. Especially for the elderly, proximity to these facilities is crucial. Although introducing well-known brands can rejuvenate areas, they might also lead to overcrowding, potentially diminishing the quality of life for locals. It is imperative in urban planning to understand these dynamics and their implications. This chapter delves into the formation of commercial clusters in urban zones and introduces techniques for analyzing and visualizing these trends. Utilizing data from Shibuya Ward, Tokyo, the study demonstrates the efficacy of these methods. The analysis focuses on the expansion direction of commercial clusters using spatiotemporal point data. By tracking when each store opened, the research identifies how new establishments influence the growth trajectory of existing commercial conglomerates. Using a circular statistical method, the study visualizes the expansion of commercial clusters.

Keywords Visualization · Spatiotemporal data · Circular statistics · Retail agglomeration

This contents of this paper are based on the following papers originally published in a Japanese journal: Inasaka, A. and Sadahiro, Y. (2010): A Method of Analysis and Visualization of Expanding Direction of Retail Distribution, *Transactions of AIJ, Journal of Architecture Planning and Environmental Engineering,* Vol.75, No. 650, pp.889–896 (in Japanese). Inasaka, A. (2013): Visualization of Expanding Direction Pattern of Retail Distribution using High Resolution Spatiotemporal Data, *The 36th Symposium on Computer Technology of Information, Systems and Application of Architectural Institute of Japan*, pp.265–268 (in Japanese).

A. Inasaka (✉)
Department of Design, Faculty of Creative Engineering, Chiba Institute of Technology, Narashino, Chiba, Japan
e-mail: akiyoshi.inasaka@p.chibakoudai.jp

Y. Asami et al. (eds.), *Studies in Housing and Urban Analysis in Japan*,
New Frontiers in Regional Science: Asian Perspectives 75,
https://doi.org/10.1007/978-981-99-8027-7_13

13.1 Introduction

13.1.1 Urban Events as Manifestations of Urban Activities

In urban space, many spatial phenomena manifest people's diverse urban activities. These events have various effects on urban activities, affecting other urban activities. Some of these events affect people's daily lives directly or indirectly. In particular, in urban commercial areas, people's daily lives are intermingled with urban activities that are easily influenced by social conditions, such as commercial activities, and the effects of each of these activities are significant. There are two major types of events in urban commercial areas: visible events that can be seen with the naked eye and invisible events that cannot be seen with the naked eye. The first type of invisible phenomenon is typical of each region's unique atmosphere and image. Although invisible to the eye, they are important indicators for people to choose a good place to live or to verify the affinity between the image of a store and that of the local area. They should be understood when people make important decisions about urban activities.

On the other hand, visible events include, for example, various construction activities such as new building construction and remodeling and the accompanying store closings and vacancies. It is necessary to understand exactly how these activities are manifested and positioned in urban planning to create a harmonious streetscape and a convenient commercial area for residents. Thus, urban events are important in all urban activities, including residents' lives and urban planning, and must be accurately understood.

Many of the urban events mentioned above can be grasped and understood to some extent by reading various spatial data provided by government and public agencies, such as numerical maps, census data, various statistical data, data from basic urban planning surveys, and residential maps and telephone directories provided by private companies. In recent years, with the development of information technology and ubiquitous technology, it has become possible to obtain more detailed spatial data on events in a given location than was previously possible, and data that captures time-series changes at the same location has also been developed. However, displaying a large amount of detailed data does not always accurately capture the essential events. For example, a large amount of data may be graphically close together, densely packed, or overlapping, and the essential event that is being captured may be buried. It is necessary to select, aggregate, or transform the information necessary to capture the essence of the event and perform operations to remove excess information (Monmonier and Watanabe 1995 "How to Lie with a Map"). This study defines visualization as displaying information extracted from such operations.

13.1.2 *Visualization Methods Using Large Amounts of Spatial Data*

Exploratory Spatial Data Analysis (ESA), which utilizes Geographic Information Systems (GIS), has attracted attention in recent years as a method for clarifying phenomena from a large amount of detailed spatial data. Among them, visualization methods are very versatile for two reasons: visualization of information allows for quick and easy verification of more information, and graphical representation allows non-specialists to deepen their understanding of the data intuitively. The graphical representation allows non-specialists to understand the data intuitively (Haining 2003).

Numerous studies have been conducted in various fields on spatial data visualization methods. Here, we will discuss the different types of spatial data and their visualization methods. Specifically, spatial data is divided into low-dimensional spatial data with relatively few attributes and variables ranging from one to three dimensions and high-dimensional spatial data with various variables and attributes of four or more dimensions. The visualization method is organized along two axes: non-spatial visualization, such as graphs and tables of aggregate data, and visualization, in which spatial relationships can be seen on maps. (Table 13.1).

13.1.2.1 Low-dimensional Spatial Data, Non-spatial Visualization

It refers to the visualization of the results of aggregation or statistical processing of data provided in tables, such as census data or various statistical survey data at a certain point in time, by extracting specific attributes or items of the data and visualizing the trends among the data utilizing a numerical line for one-dimensional data, or a two-dimensional or three-dimensional scatter diagram for two-dimensional or three-dimensional data. For one-dimensional data, it is possible to visually capture the trend among the data through a number line. This also applies to histograms, which display data aggregated by section. Methods that belong to this category are used to grasp relationships that do not depend so much on the position of the data as on the attributes of the spatial data or the values of the variables. Since this type of method is generalized and there are abundant research examples, we will not go into detail in this study.

Table 13.1 Organization of existing visualization methods

	Non-spatial	Spatial
Low dimension	Scatterplot, histogram, and	Map, cell, 2D + 3DCG, area cartography Area cartogram
High dimension	Line plot, parallel Coordinate plot)	3DCG, VR, 2D + 3D animation Space syntax

13.1.2.2 High-dimensional Spatial Data, Non-spatial Visualization

This type of data and visualization refers to the visualization of the results of aggregation or statistical processing of comparing various attributes and items and their time-series changes in the census and statistical survey data, as described above. Many of the methods in this category capture the relationships among multiple attributes and variables of data as time-series transitions and changes. For example, in chemistry and physics, data capturing time-series changes of a certain substance, and in statistics, visualization of time-series changes of a certain variable value. In recent years, with the development of information technology, various types of data have become more detailed and voluminous, and the non-spatial visualization methods for low-dimensional spatial data described above are no longer sufficient to handle them. A method for efficiently understanding such multi-attribute and multi-variable data in a unified manner is needed, and many studies have been conducted.

There will be a tool to unify three types of plots: parallel coordinate plots by Gahegan (2005) that capture the variation among variables, scatter plots that look at correlations among multiple attribute variables, and plots that capture distribution trends such as variance, skewness, and kurtosis for each variable's data. kurtosis) of each variable. As an example of its application, an attempt has been made to display the obtained results on a map (e.g., Theus 2005).

13.1.2.3 Low-dimensional Spatial Data and Spatial Visualization

It targets visualization based on the spatial positional relationship of each data, which the above-mentioned non-spatial visualization methods cannot capture. In addition to the census and statistical data described above, this method is used to visualize spatial data that captures spatial and geographical conditions and characteristics, such as topographic elevation data, road maps, and building maps. Many of these methods are based on maps. This type of research is conducted using GIS in cartography and geovisualization.

A GIS visualizes spatial entities (spatial entity) defined by points, lines, and surfaces, which are composed of five properties: position, shape, attribute, function, and time, with two relations: topology and order (Yan 2003, pp. 26). This section will discuss the visualization of spatial data represented by point, line, and surface events, respectively.

13.1.2.4 Point Event(s)

A representative example of point visualization is John Snow's cholera map, commonly used in epidemiology to display the location of disease outbreaks and the like. This is the first example of spatial analysis in epidemiology and has contributed greatly to the development of spatial analysis in epidemiological research. In addition to the visualization of the point of occurrence of an event, there are also methods to represent the point of occurrence of an event and the magnitude of the event in terms of points and their magnitudes. For example, research has been conducted on a visualization method called the circle area cartogram, which represents the population distribution in each area by the location and size of circles (Dykes 1998, Inoue and Shimizu 2005, Dykes 2005, etc.). In recent years, the sophistication and generalization of Global Positioning System (GPS) technology have made it possible to record events at various points, and many studies have been conducted on this most primitive spatial visualization method.

13.1.2.5 Linear Event(line)

Next, visualization of linear events is generally used to represent linear elements in space, such as streets and roads, and to visualize the trajectory of an object's movement. Various navigation tools are a typical example of line data visualization that indicates the optimal route from a starting point to a destination based on the input of time required, distance, and other objectives.

Some methods do not preserve actual locations in real space but visualize their topological relationships. For example, many methods use multidimensional scaling (MDS) and graph theory to visualize the relationships among various locations and data by connecting them with lines (Rodger 2005; Robert 2005, etc.).

The axial map, a part of the space syntax originated by Hillier and Hanson (1984), visualizes the spatial connectivity and accessibility of urban space by dividing the street space into line segments. This is an effective visualization method (Kigawa and Furuyama 2004; Kigawa and Furuyama 2005; Kigawa and Furuyama 2006, etc.).

13.1.2.6 Surface

There are two main types of surface data: those expressed in terms of administrative boundaries such as prefecture, municipality, town, street, etc. Many census and statistical survey data are obtained in this way. In many cases, census and statistical survey data are obtained using this method, and each parcel is assigned a value. For example, a choropleth map is commonly used to visualize population data obtained from a census, in which differences in hue, saturation, and lightness of colors represent values. However, there is not necessarily a relationship between the importance of the value to be displayed and the size of the area of each compartment surrounded by administrative boundaries, etc. For example, when a small value is assigned to a large compartment, the overall distribution of the value appears to be small, which greatly affects the appearance of the information. To eliminate such differences as much as possible, there are many studies on the cartogram method, which expresses each value by the size of the area of each parcel while preserving the spatial positional relationship and the topological shape of each parcel (Inoue and Shimizu 2005, etc.).

In addition, many studies have been conducted using methods that capture events continuously by smoothing point densities and other data using kernel functions. This operation aggregates what cannot be captured by displaying point data alone as a surface. This method is particularly effective for spatiotemporal data and other analyses in which the volume of data is necessarily large (e.g., Sadahiro 2007).

13.1.2.7 High-dimensional Spatial Data and Spatial Visualization

Information has become more abundant in recent years due to advanced information technology, and spatial data is no exception. Therefore, it has become necessary to handle various attributes and variables and their time-series changes simultaneously. Numerous visualization methods have been developed for this purpose. For example, animation methods have been recognized for their importance in visualization research and cartography (Dorling 1992; DiBiase et al. 1992 etc.). As a concrete example, animation is becoming popular as a method to grasp the aspect of change and transition by displaying the numerical change of each position at multiple points in time in a time-series frame-by-frame manner, even if it is not a dynamic object (Inoue et al. 2009, etc.).

With the recent development of remote sensing, GPS, and laser measurement technology in airplanes, wide-area and high-resolution data have become available. This has made it possible to visualize high-resolution 3D urban models (e.g., Yoshikawa 2007a, b). In addition, visualization methods that apply 3D computer graphics and virtual reality technology have become versatile and effective tools for consensus building in urban planning and landscape evaluation (e.g., Amin 2007; Yoshikawa 2007a, b etc.). The methods for capturing physical space have been rapidly improving with the progress of information and computer technology.

On the other hand, research is also being conducted on what is visible but difficult to capture due to the layering and complexity of events. In the aforementioned research related to space syntax, visual simulation methods that consider environmental indicators such as pedestrian distribution, light, wind, and atmospheric flow have been proposed and developed into tools (Takamatsu 2007, Batty 2003 etc.). The method has been used in architectural and urban design (Takamatsu 2007).

As described above, research on visualization has been conducted in various fields, and many of the visualization methods have been developed independently by practitioners and researchers, depending on the events they wish to capture and the attributes and variables they wish to display. The widespread use of GIS has made it easy to analyze data and display the results using a certain set of tools, enabling the capture of a wide range of events. Various general-purpose tools have also been developed by groups such as the GeoVISTA Center at Pennsylvania State University and the Center for Advanced Spatial Analysis (CASA) at University College London. However, both the quality and quantity of spatial data are limited. However, both the quality and quantity of spatial data are increasing daily, and there is still room for further development of visualization methods due to the diversity and complexity of the data, as well as the complexity of spatial events caused by changes in social conditions.

In light of the above, this study proposes a method for analyzing and visualizing representative spatial data for various events in urban commercial areas, which are the most active in urban space. We focus on visualizing invisible events, latent patterns of events that conventional methods cannot grasp, and the visualization of complex and irregularly changing events.

13.2 Analytical Method and Visualization of the Direction of Commercial Concentration Expansion

13.2.1 Introduction

Commercial facilities are an essential part of urban life, and changes in these facilities always have some impact on neighborhood residents. For example, changes such as unexpected store closures in neighborhood shopping areas can disrupt the traditional shopping environment of residents. For the elderly, the commercial facilities they use daily should be located within a short distance from their residences, and this burden may be even greater. While the opening of popular brand-name stores is expected to attract many users and revitalize the surrounding area, there is also concern that the large influx of visitors may cause congestion and other problems that could adversely affect the living environment of residents in the surrounding area. Changes such as the opening and closing of commercial facilities are always occurring in commercial clusters. While such changes can be good for revitalizing the area, they can also significantly impact residents, such as the deterioration of the surrounding residential environment and the decline of the surrounding area due to unjustified development, etc. Therefore, in urban planning and development, it is important to grasp the changes and trends to control them appropriately and accurately. It is necessary to accurately grasp the changes and trends to control them appropriately in urban planning and development.

In light of the above, this chapter focuses on the formation process of commercial clusters in urban areas. It proposes a method for describing, analyzing, and visualizing the trends of changes in these clusters. Section 13.2 proposes a method for describing and analyzing the direction in which commercial clusters expand based on the time when each store in a commercial cluster opens. Section 13.4 describes, analyzes, visualizes, and discusses the results using data from Shibuya Ward, Tokyo, as an empirical analysis to verify the effectiveness of the proposed method. Finally, Section 13.4 summarizes the chapter and discusses future issues.

This chapter was added based on the content of "Analysis Method and Visualization of the Direction of Commercial Agglomeration Expansion" (Inasaka and Sadahiro 2010), which was accepted as a research paper in the Transactions of the Architectural Institute of Japan, Planning Division.

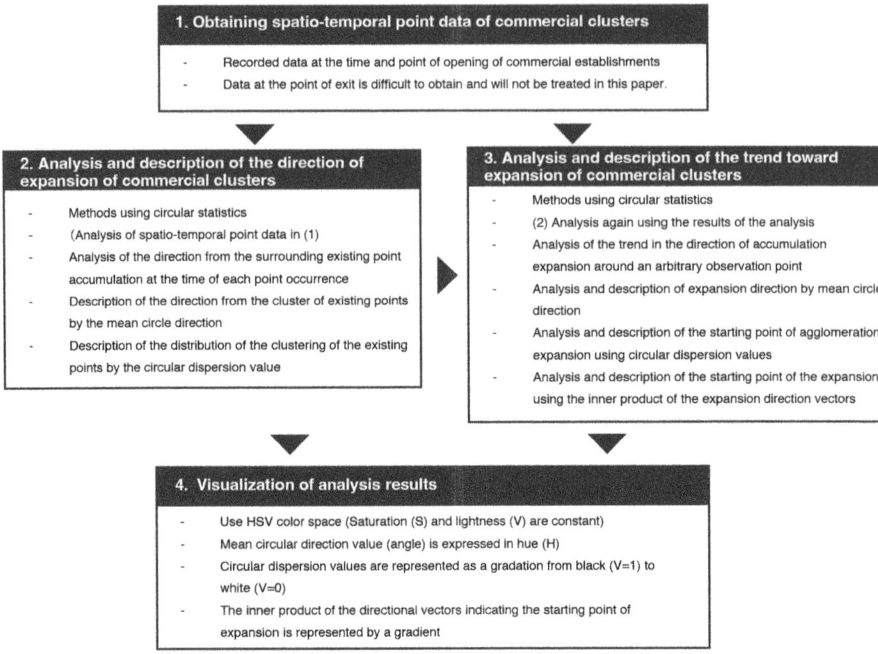

Fig. 13.1 The flow of this section

13.2.2 Analytical Method for Describing the Direction of Expansion of Commercial Clusters

13.2.2.1 Analysis and Description of the Direction of Commercial Agglomeration Expansion Using Spatiotemporal Point Data

Assume that we have spatiotemporal data on existing commercial establishments in a certain region. The location of each commercial facility is represented as a point, and the time when each store opened is recorded. Note that we do not include closed commercial establishments in our analysis since it is often difficult to obtain such information. However, in the "Survey of Establishments and Businesses,"[1] the number of closed and newly opened establishments is almost the same for commercial establishments related to retail, restaurants, and services. New establishments may open in the same location after closing. Therefore, it should be noted that it is necessary to consider the re-filling of commercial clusters, such as new store openings after store closures. In this chapter, we define this data as commercial clusters (Fig. 13.1).

[1] (Note 1) See Statistical Survey of Business Establishments and Enterprises 2004 (Tokyo Metropolitan Government, Table 20, Shibuya Ward). (http://www.stat.go.jp/data/jigyou/2004/)

Next, we propose a method for describing the direction in which commercial clusters expand during their formation process. For the above commercial clusters, we analyze in which direction the individual stores have expanded while joining the existing commercial clusters when they opened new stores. We use the circular statistical method[2] while tracing the past formation process.

From this data, we first determine the current commercial agglomeration. We define a commercial cluster as a set consisting of all stores.

Let us assume that the commercial facilities that comprise the existing commercial clusters are $s_1, s_2, \ldots, s_n$. These facilities are located in the order of s_1, and the aggregate $S = \{s_1, s_2, \ldots, s_n\}$ constitute the current commercial agglomeration.

Next, a circular data is created that describes a certain direction of existing stores in the vicinity, centered on each store to be analyzed. The state of the commercial clusters at each of the intermediate points in time p can be determined by the following equation. s_p the set of existing commercial stores at the point in time when $S_{p-1} = \{s_1, s_2, \ldots, s_{p-1}\}$. The set of points s_p and the set S_{p-1} and the unit vector formed by the points constituting the set $L_p = \{l_1, l_2, \ldots, l_{p-1}\}$ and the angle formed concerning the east on the map is $\Theta_p = \{\theta_1, \theta_2, \ldots, \theta_{p-1}\}$ and the angle formed concerning the east on the map is defined as Here, we assume that s_p on the unit circle centered at Θ_p is defined as the circle data. In this paper, unless otherwise noted, the angles are measured in radian with the east direction set to 0 (rad). This data uses the following equation Θ_p for the mean circular direction $\overline{\theta}$ (Fisher 1993; Brunsdon and Corcoran 2006; Corcoran et al. 2007).

$$C = \sum_{i}^{p-1} \cos\theta_i,\ S = \sum_{i}^{p-1} \sin\theta_i,\ R^2 = C^2 + S^2\ (R \geq 0) \tag{13.1}$$

The mean circular direction $\overline{\theta}$ is a composite vector of a set of unit vectors $L_p = \{l_1, l_2, \ldots, l_{p-1}\}$, and its angle is given by using the value obtained in Eq. (13.1) according to the following:

$$\overline{\theta} = \begin{cases} \tan^{-1}\left(\frac{S}{C}\right) & S > 0, C > 0 \\ \tan^{-1}\left(\frac{S}{C}\right) + \pi & C < 0 \\ \tan^{-1}\left(\frac{S}{C}\right) + 2\pi & S < 0, C > 0 \end{cases} \tag{13.2}$$

R is the composite length of its composite vector, which lies in the range [0, $p-1$]. $\overline{R}$ represents average composite length as follow

$$\overline{R} = R/(p-1) \tag{13.3}$$

[2] (Note 2) Sometimes referred to as directional statistics This method has recently been attracting attention in computational statistics, such as Shimizu (2008)

and are positioned in the range [0, 1]. $\overline{R}$ the maximum and minimum values of $\overline{R} = 1$, indicating that all point data are matched. However, for the average composite length $\overline{R} = 0$ is the case for the average composite length $n(n \geq 3)$. It should be noted, however, that in the case of the average composite length, in addition to the case where the points are evenly distributed in many directions to divide the circumference equally, there is another case where the points are distributed in a straight line up through the center of the unit circle to bisect the circumference.

$$V = 1 - \overline{R} \tag{13.4}$$

V is defined as the circular variance (Note 5). As with the variance of linear data, the smaller the value of the circular variance, the more concentrated the data distribution. The circular variance $V = 1$ property is similar to the above property of the mean composite length $\overline{R} = 0$. Similarly to the above property of the mean composite length, the property of $n(n \geq 3)$ In addition to the case where the point sequences are evenly distributed in many directions to divide the circumference into two equal parts, there is also the case where the point sequences are distributed in the direction above the straight line passing through the center of the unit circle to divide the circumference into two equal parts. However, the cases in which the point sequences are uniformly distributed in multiple directions and the case in which they are distributed in a straight line correspond to the cases in which the accumulation expands in multiple directions around a certain base facility and the case in which it expands in a straight line along a main road, respectively, in urban space, and their meanings are quite different. In the above two cases, it cannot be interpreted that the distribution is uniformly distributed so that the dispersion value is 0 in the linear case. With the above properties in mind, $V < 1$ is considered in this method.

We propose a method for describing the direction of expansion of commercial agglomeration using the above circular statistics method. In this paper, this method is called the "expanded direction description method." The calculation procedure is as follows.

First, the spatiotemporal data of the existing commercial facilities $i(0 < i < t)$ at the point in time $[0, t]$ of the existing commercial facility. p_i The first step is to consider a point that occurred at time $[0, t]$. The point p_i as the base point, and the points that occurred before the point $i(0 < i < t)$. The points that occurred before the point in time $\{p_0, p_1, p_2, \ldots, p_{i-1}\}$ and plot the intersection of the points on the unit circle on the circumference of the circle. The angles formed by the unit vectors connecting the points plotted on the unit circle and the base point p_i. The angle formed by the unit vector connecting the points plotted on the unit circle and the base point is used in Eqs. (13.1) and (13.2) to obtain the average circle direction $\overline{\theta}$. The obtained mean circle direction is calculated using Eqs. (13.1) and (13.2). The

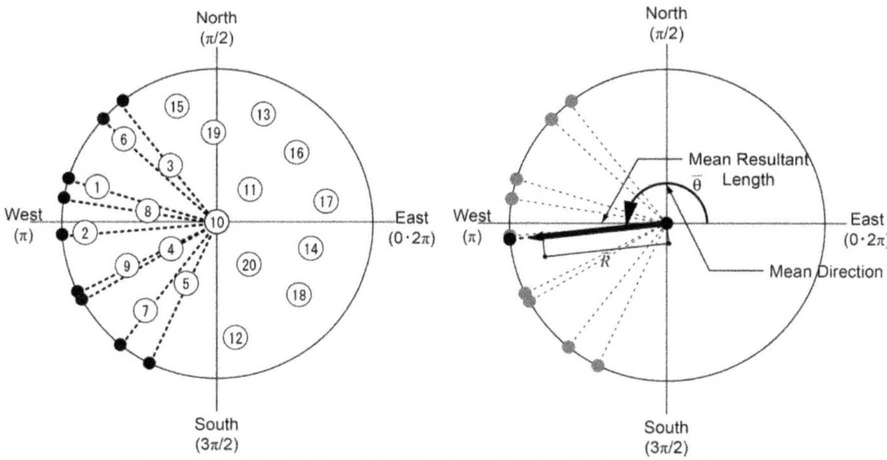

Flow up to the description of the direction using the circular statistical method

(1) The numbers in the circle indicate the order in which the points occur.
(2) In the left figure, the 10th generated point is the base point and connects it with a point that occurred in the past. Plot on the unit circle.
(3) A line connecting a point plotted on the unit circle and the base point. The angle formed by the average circle direction using the circular statistics method. The following equation is obtained.
(4) Invert the obtained result by – π and calculate the existing accumulation or the direction of expansion is the direction of expansion of the
(5) Also, the average of each unit vector in (3) is the average composite vector.

Fig. 13.2 Concept of circular statistics relationship between mean circular direction and circular variance

obtained average circle direction $\overline{\theta}$ by the following equation π. The obtained mean circle direction is inverted by (Note 5). Circular variance is a translation of circular variance.

$$d = \begin{cases} \overline{\theta} + \pi & \left(0 < \overline{\theta} \leq \pi\right) \\ \overline{\theta} - \pi & \left(\pi < \overline{\theta} \leq 2\pi\right) \end{cases}, \quad (0 < d \leq 2\pi) \tag{13.5}$$

Next, the length of the composite vector of each unit vector and the number of points i generated up to that point are substituted into Eq. (13.3) to obtain the average composite length, which is then substituted into Eq. (13.4) to obtain the circle variance V (Fig. 13.2).

13.2.2.2 Analysis and Description of Trends in the Direction of Expansion of Commercial Clusters

Using the results of the expansion direction of each spatiotemporal point obtained through the above expansion direction description method, we analyze the trend of the expansion direction of accumulation at an arbitrary observation point. This paper calls this method the "expansion direction trend description method." The calculation procedure is as follows.

First, assume that there is a spatiotemporal point distribution that includes the magnification direction of each point calculated from the spatiotemporal data of existing commercial facilities at the obtained time using the magnification direction description method proposed earlier. Then, we generate uniform observation points at arbitrary distance intervals on the spatiotemporal point distribution. Generate a unit circle with an arbitrary radius centered on an observation point, and extract the spatiotemporal points encompassed by the circle. The unit circle defines the direction of expansion of each point.

Plot the result on a plot of the mean circle direction. The average circle direction is obtained using Eqs. (13.1) and (13.2). Substitute the length of the composite vector formed by the center of the unit circle and the expansion direction of each point plotted on the circumference of the unit circle and the number of space-time points included in Eq. (13.3) to obtain the average composite length, which is then

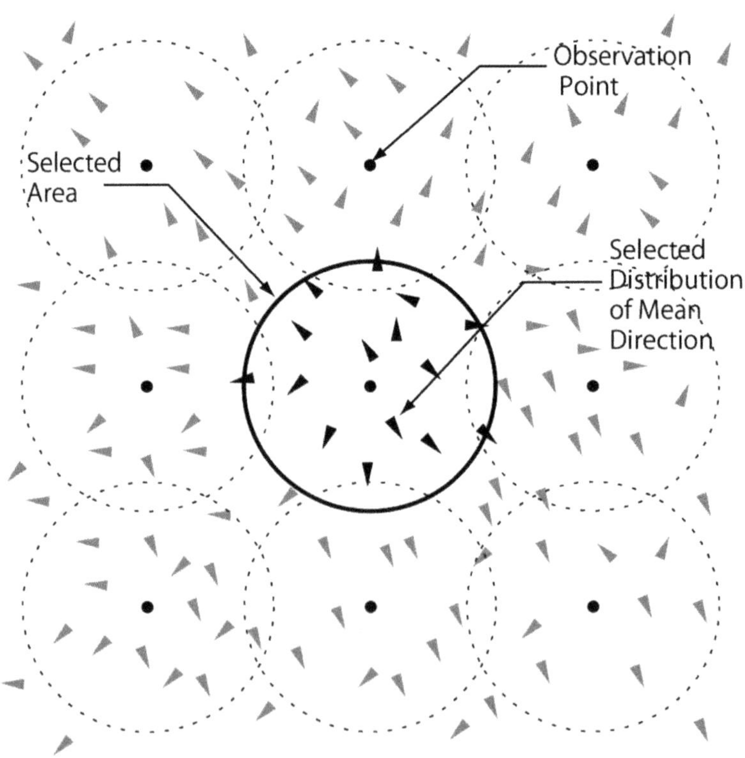

The observation points are generated at arbitrary intervals and encompassed within a buffer range of arbitrary radius. The average direction of each above point is tabulated using circular statistical techniques.

Fig. 13.3 Description of expansion direction trend from an arbitrary observation point

substituted into Eq. (13.4) to obtain the circle variance value (Fig. 13.3). In this case, the observation points generated at arbitrary distance intervals will appear as differences in the resolution of the visualization when the analysis results are displayed, depending on the differences in the intervals. The radius of the circle used to select the spatiotemporal point data around the observation point is the scale that this analysis can read.

The average circle direction obtained here indicates the trend of the expansion direction of the accumulation concerning each observation point. If the circle dispersion is close to 0, the directionality is strongly in the direction of the obtained mean circle. If it is close to 1, the expansion is almost uniform in all directions.

13.2.2.3 Visualization of Analysis Results Using HSV Color Space

The values calculated by the above-mentioned expansion direction description method and the expansion direction trend description method are visualized as follows. The direction of expansion description method calculates the direction of expansion for each spatiotemporal point and the direction of expansion and circular dispersion values obtained. The direction of magnification was calculated for each point by describing the trend of magnification direction, and the mean circle direction and circle variance were obtained. Then, the magnification direction, mean circle direction, and circle dispersion are visualized on a map.

The direction of the mean circle is visualized using ArcGIS 9.3, with the east direction set to 0 (rad) and the angles indicated by the arrows. The HSV color space is defined by hue (H: hue), saturation (S: saturation), and value (V: value) and is suitable for representing the angles of the analysis results. The circular dispersion is visualized by a white-to-black gradient from a dispersion value of 0 to 1 (Fig. 13.4).

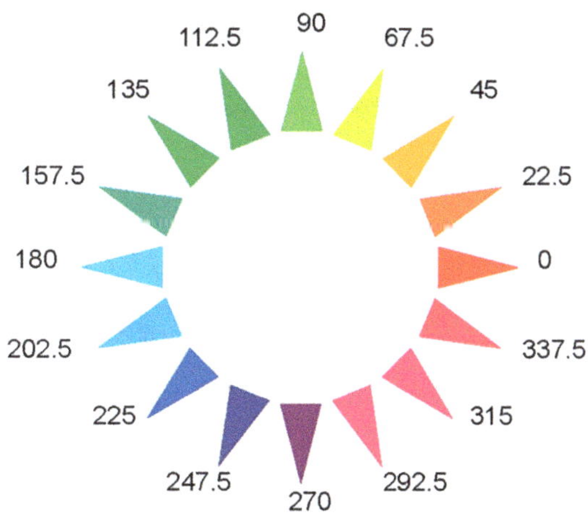

Fig. 13.4 Visualization of mean circle direction using HSV color space. Classify and assign observed angles within ±11.25 degrees

13.2.2.4 A Method for Visualizing the Starting Point of Commercial Agglomeration Expansion Using Vector Analysis

Using directional data describing the angle of expansion of each store in a commercial cluster, we propose a method of searching for the starting point of the expansion of the commercial cluster by applying the method of vector analysis. The computational procedure is as follows.

From the previous method, for a set of commercial stores *SH* for each store in the commercial agglomeration to which they belong, a set of vectors $\overrightarrow{P_n} = \{\overrightarrow{p_1}, \overrightarrow{p_2}, \ldots, \overrightarrow{p_n}\}$ and for a store $\overrightarrow{p}$ for a store in the set, as defined by the vector (x, y) for a store in the set. The data are assumed to be arranged in ascending order by the time of the source.

Vector of stores opened at a given point in time i and the vector of commercial stores that have already opened at a given point in time $\overrightarrow{p_i}$ and the set of vectors of commercial stores that have already opened in the surrounding area is defined by $\overrightarrow{P_{i-1}} = \{\overrightarrow{p_1}, \overrightarrow{p_2}, \ldots, \overrightarrow{p_{i-1}}\}$ where the vectors $\overrightarrow{p_i}$ and the vector $\overrightarrow{p_o} \in \overrightarrow{P_{i-1}}$ the inner product of these two vectors can be calculated using Eq. (13.5) below, and $\cos\theta$ is obtained by Eq. (13.6).

$$\overrightarrow{p_o} = (x_o, y_o), \quad \overrightarrow{p_i} = (x_i, y_i) \text{ than the}$$

$$\overrightarrow{p_o} \bullet \overrightarrow{p_i} = x_o x_i + y_o y_i = |p_o|\,|p_i| \cos\theta \tag{13.5}$$

$$\cos\theta = \frac{\overrightarrow{p_o} \bullet \overrightarrow{p_i}}{|p_o|\,|p_i|} = \frac{x_o x_i + y_o y_i}{\sqrt{x_o^2 + y_o^2}\sqrt{x_i^2 + y_i^2}} \tag{13.6}$$

From the above calculations, it can be seen that the vectors $\overrightarrow{p_i}$ and the set $\overrightarrow{P_{i-1}}$ for all vectors in the set $\cos\theta$ values are obtained for all vectors of using the results, the shape of the distribution of vectors indicating the direction of origin of the expansion of commercial concentration at the location of the store p_i. The following Eq. (13.7) gives the shape of the distribution of vectors that indicate the direction of origin of the expansion of commercial clusters at the location of a store.

$$C_i = \frac{\sum_{k=1}^{i-1} \cos\theta_k}{i-1} \tag{13.7}$$

C_i takes values in the range $[-1, 1]$. Around 1, there is a concentration of the vector distribution, and around -1, there is a divergence of the vector distribution. Furthermore, a random situation is observed near 0. In other words, when the value of C_i around 1 indicates that the area is near the starting point of the expansion of commercial accumulation, while the area around -1 indicates that the area is at the boundary of the expansion.

Using the above method, the values of each store belonging to a commercial cluster are visualized on a map by painting them with a gradient. C_i the above

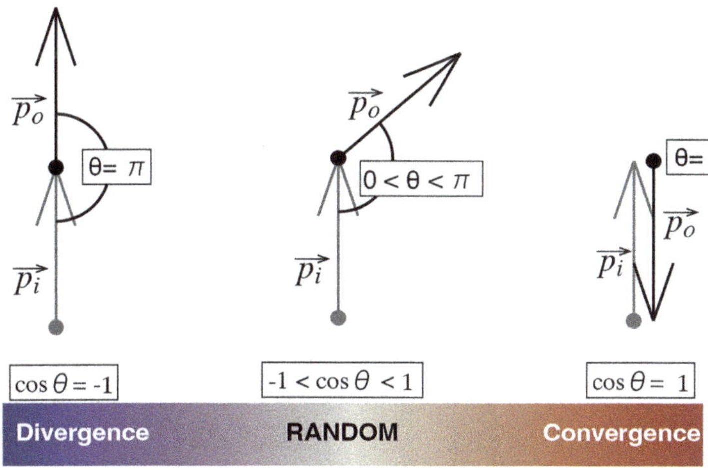

Fig. 13.5 cos θ obtained from the inner product of vectors θ values and directional patterns obtained from the inner product of vectors

method is used to visualize the value of each store belonging to a commercial cluster by painting it in a gradation on a map and to examine the starting points of the expansion of commercial clusters, the boundaries of commercial clusters, and the scale and distribution of commercial clusters (Fig. 13.5).

13.3 Formation Process Analysis of Commercial Agglomeration in Shibuya

13.3.1 About Shibuya Ward

Shibuya Ward (Fig. 13.6) is located in the western part of Tokyo, bordered by Shinjuku and Nakano Wards to the north, Suginami, and Setagaya Wards to the west, Meguro and Shinagawa Wards to the south, and Minato Wards to the east. The area is long and narrow from north to south, with large green areas such as Yoyogi Park and Meiji Shrine in the center. The area is also home to some of Tokyo's most prominent commercial districts, including the areas around Shibuya, Ebisu, and Harajuku stations, as well as a variety of neighborhood shopping areas along the Keio Line, surrounding residential areas, and upscale residential areas such as Shoto. In addition, it is positioned as one of Tokyo's sub-centers, with many major railroad lines and main arterial roads intersecting the city. Therefore, it is considered a suitable site for the empirical analysis in this chapter.

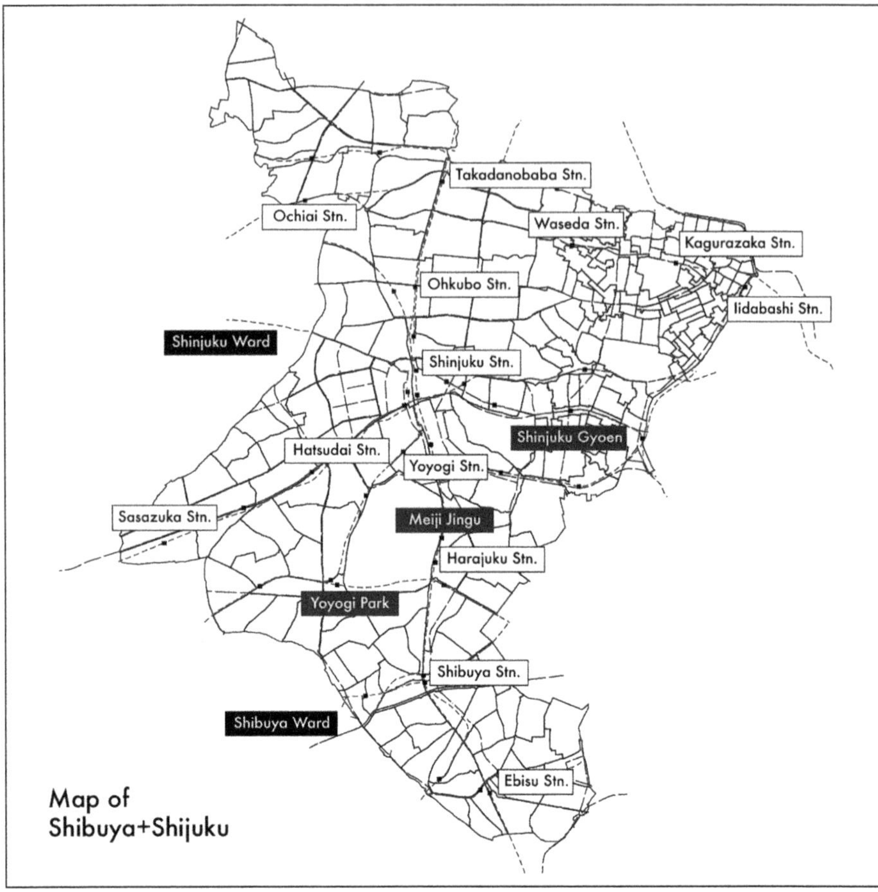

Fig. 13.6 Schematic of Shibuya and Shinjuku Ward

13.3.2 About the Data Set

In this chapter, we use the address-matched "NTT Town Pages Data" for Shibuya Ward, Tokyo, from 1989 to 2007 as an example of applying the above method. This data is suitable for understanding time-series changes because it is published regularly on a yearly or, in recent years, bi-monthly basis. In this study, we use the data published in September of each fiscal year from 1989 to 1994 and published every other month in January, March, May, July, September, and November from 1995 to 2007. A total of 178,859 stores are included in the analysis. The stores covered here appear in the retail, restaurant, and service industries in the NTT Town Pages Data.

The data used in this study is based on the date when each store was first listed on the website, either annually or bi-monthly, and the store is assumed to have opened at that time. However, the data do not show the detailed time of each store opening for each fiscal year or each bi-monthly period. In other words, the order in which stores were opened during each fiscal year or bi-monthly period is unclear. Therefore, in this paper, we assume that multiple store openings at each point in time occurred at the same time.

Since the relationship between stores close to each other is stronger than between stores as far away as possible, the analysis is based on the relationship between stores included within an arbitrary radius of a circle centered on each store to be analyzed.

13.3.3 Analysis of the Direction of Agglomeration Expansion as Seen in the Spatiotemporal Point Data and Discussion of Examples of the Application of Visualization

The method proposed in this chapter makes it difficult to determine its radius during the analysis uniquely. However, since the results depend on the radius distance, this paper analyzes data within a circle of radius 1000 m, 750 m, 500 m, 250 m, 100 m, 50 m, and 30 m, with each store as the base point. The results are as follows. We first discuss the analysis results for the circles with radii of 1000 m, 500 m, and 250 m, where the results are more pronounced.

In this chapter, considering the ease of viewing the analysis results, the angles indicating the direction were divided into 16 azimuths from 0 to 360 degrees, and the observed mean circle angles were classified and visualized for each 22.5-degree width range centered on each angle. The circular dispersion values were expressed as a 10-step gradation.

First, in the analysis of the 1000 m radius buffer (Fig. 13.7), we can observe arrows extending outward from the major railroad stations in Shibuya Ward, namely Shibuya, Ebisu, Yoyogi, and Shinjuku stations. From this, it can be inferred that the stations serve as the base points for the expansion of commercial clusters. Next, focusing on the locations estimated to be the base points of each expansion, we can see that the distribution is not uniform, with large and small distributions in each angular zone. For example, when we focus on the expansion based on the area around Shibuya Station, we observe that the distribution of arrows in the northeast direction is significantly smaller than those in other directions. This indicates that the directionality of expansion in that direction is strong. This is consistent with the location of Meiji-dori Avenue, which suggests that commercial clusters are expanding from Shibuya Station to Harajuku. Similarly, the distribution of arrows along the Keio Line and Koshu Kaido that extend westward from Shinjuku Station is also strongly oriented, suggesting that commercial clusters are expanding westward from the base point of Shinjuku Station.

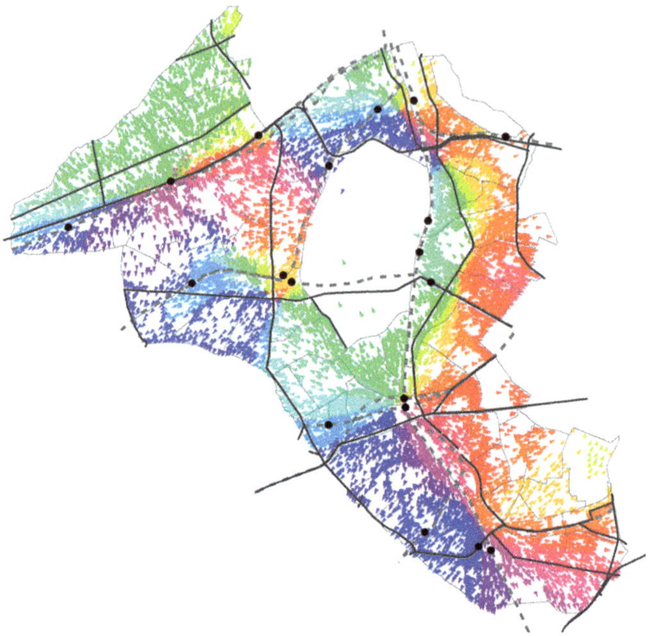

Fig. 13.7 Distribution of expanding direction at 1000 m buffer

In addition, in the circular dispersion distribution chart (Fig. 13.8), we can see results consistent with the above discussion of locations that serve as the base points for expansion. It can be seen that the locations with high circular dispersion are distributed around the major stations of Shibuya, Ebisu, Yoyogi, and Shinjuku. This indicates that relatively old stores are uniformly distributed near the commercial clusters, which is the base point for expanding the commercial clusters, as discussed in the expansion direction distribution map. Similarly, along the main roads, Meiji-dori and Koshu kaido, the locations with high circle dispersion are distributed in a long and thin line, and the commercial clusters are extended outward in parallel with these streets.

In the 500-m buffer (Fig. 13.9), as in the 1000-m buffer, in addition to the extensions centered on major stations, the distribution of arrows in the direction of extensions based on relatively small stations, such as Hatsudai, Hatagaya, and Sasazuka stations on the Keio Line and Yoyogi Uehara station on the Odakyu Line, which was not observed when the buffer radius was large, can be observed. The distribution of arrows in the direction of the road can be observed. We can also observe locations where opposite arrows collide along roads. This is caused by the collision of expansion based on each station and can be assumed to be the boundary line between each expanding commercial cluster.

As in the 1000 m buffer analysis, the circular dispersion distribution (Fig. 13.10) is also higher at the locations that can be inferred as the base points of the

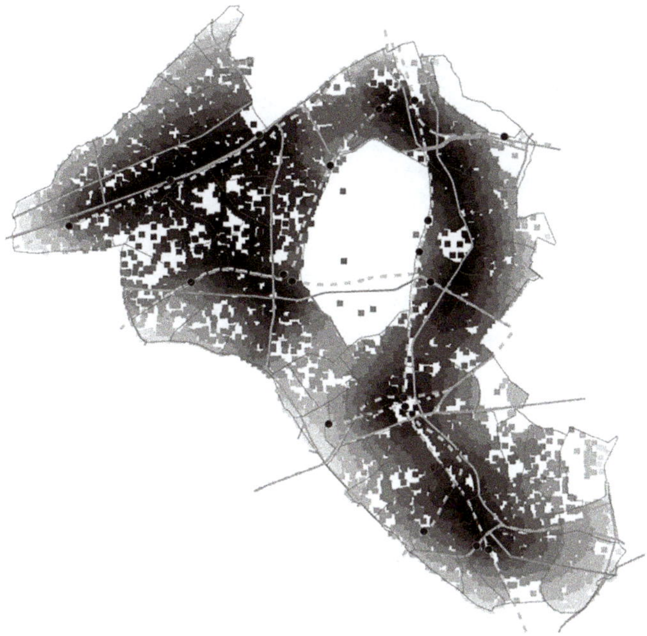

Fig. 13.8 Circular dispersion distribution at 1000 m buffer

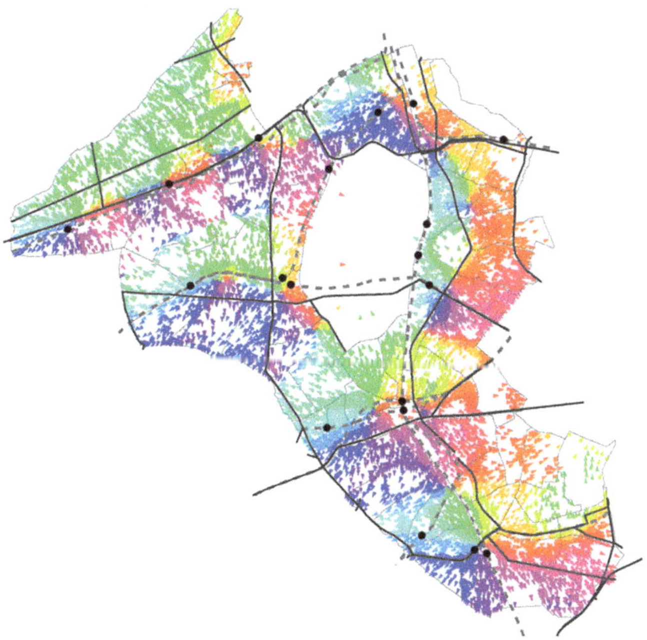

Fig. 13.9 Distribution of expanding direction at 500 m buffer

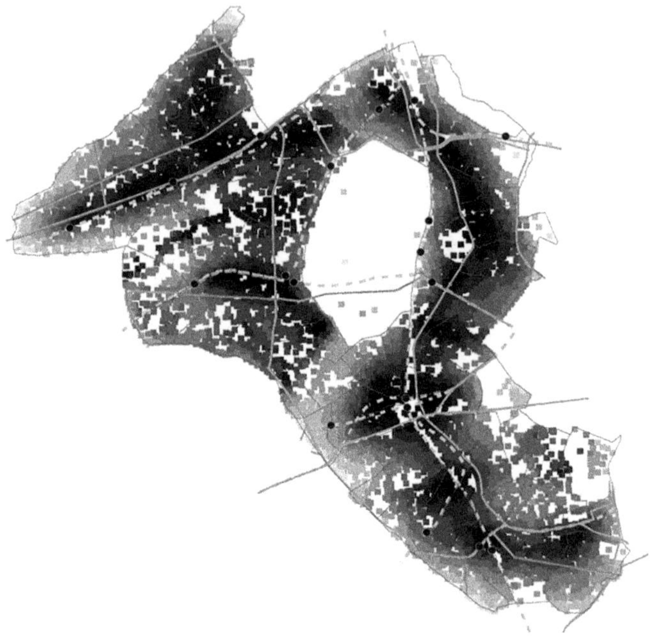

Fig. 13.10 Circular dispersion distribution at 500 m buffer

expansion by visualization of the mean circular direction. In addition, the circular dispersion is high at the locations where arrows pointing in opposite directions collide, which can be inferred as the locations where the expansion collides with the commercial accumulation. In addition, although not shown in the expansion direction distribution map, the Hiroo area on the east side of Ebisu station has a high dispersion value. Although there is no station or other base point, it is considered the base point of the expansion of the area in light of the other trends shown in the circular dispersion.

In the 250-m buffer (Fig. 13.11), we observed more detailed changes than in the analysis using a larger buffer radius. The arrow distribution in the direction of expansion is not observed in the above analysis. Still, it is based on the relatively small station area that does not face an arterial road, such as the Daikanyama station. The arrow distribution along arterial roads also shows a more localized change in the arrow distribution along roads, showing a strong expansion directionality in the analysis results for the large buffer radius. In light of the previous discussion, it can be inferred that the central facility within the smaller regional unit is located at the base of this change. As an example of this being particularly evident, we observed the distribution of arrows extending northward from Hatagaya and Sasazuka stations along the Keio Line. We captured relatively small changes, such as in neighborhood shopping areas.

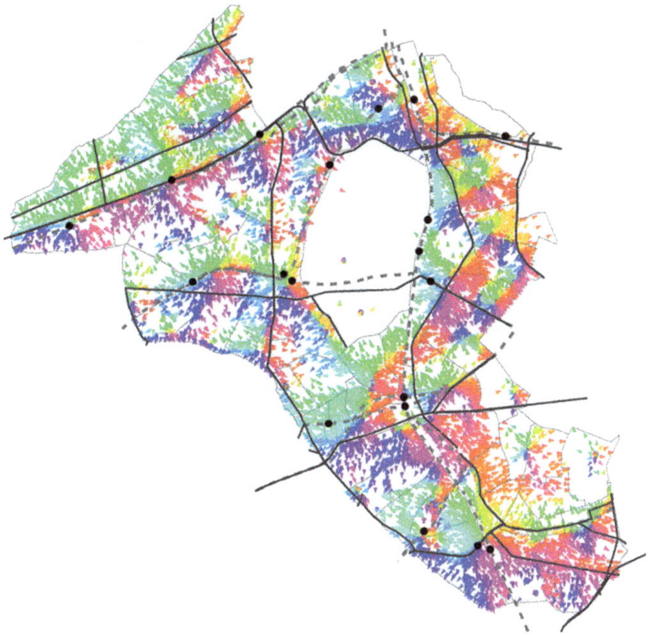

Fig. 13.11 Distribution of expanding direction at 250 m buffer

Similarly, the circular dispersion distribution map (Fig. 13.12) captures minute changes in commercial clusters, which were not observed when analyzing large buffer radii. As mentioned earlier, relatively high circular dispersion is observed for large and small railroad stations and arterial roads. At the same time, high values are also observed for locations where arrows pointing in opposite directions collide. In other words, the locations with high circular dispersion are related to the cases where they are the base points of the expansion of commercial clusters, and the boundary areas where the expansion of commercial clusters collides with each other and where a straight road passes between the clusters and stores are located on the line of the road. It corresponds to the case where the base point is on a straight line with the point sequence bisecting the circumference, which is a property of circular dispersion in this method.

It can be observed that the smaller the buffer radius, the more detailed the analysis results. By decreasing the buffer radius, more detailed changes that were not apparent when the analysis target was selected with a large radius can be captured, such as changes in small stations and areas with characteristics as a region but no facilities that can serve as a base point. On the other hand, if the buffer radius is too small, such as 50 m or 30 m, the arrows will be arranged irregularly, and no trend can be read. This is considered to be a limitation of this method.

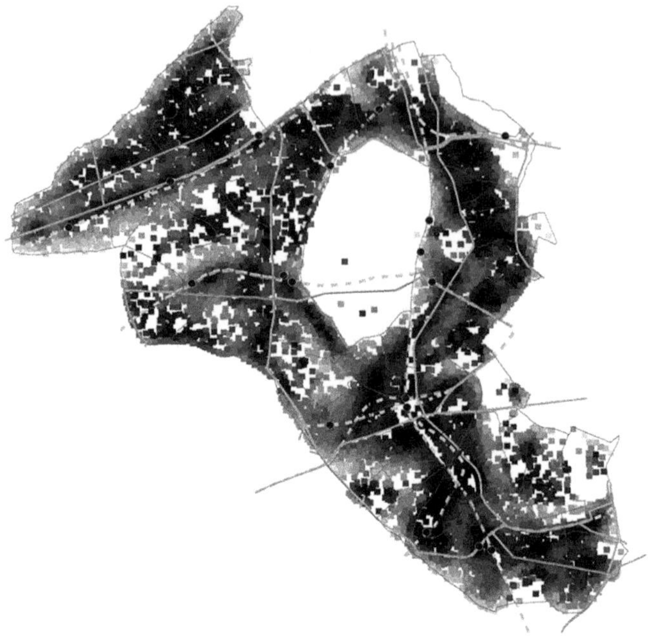

Fig. 13.12 Circular dispersion distribution at 250 m buffer

13.3.4 Analysis of the Trend in the Direction of Agglomeration Expansion and Discussion of Examples of Application of Visualization

Based on the results of the previous analysis, we analyze the trend in the direction of expansion at arbitrary observation points. In this paper, the analysis was conducted using a grid of observation points at intervals of 100 m. The observation area centered on an observation point is affected by the circle's radius encompassing the observation target. The radii of the circles were 1000 m, 750 m, 500 m, 250 m, and 100 m, and the spatiotemporal point data were tabulated for each of these radii. The above analysis found that the 500 m buffer radius captured the relatively wide changes based on major stations and the relatively narrow changes based on arterial roads and small- and medium-scale stations, respectively. The results of this analysis were used to examine the trend in the direction of expansion within the radii of 500 m and 250 m, where the differences in the results from the previously defined observation points were pronounced. The same method was used for the visualization of the results.

As shown in Fig. 13.13, the distribution of arrows indicating the direction of expansion of the concentration is observed near Shibuya and Ebisu stations, which are the main stations in this area, and near the intersection of Meiji-Dori and

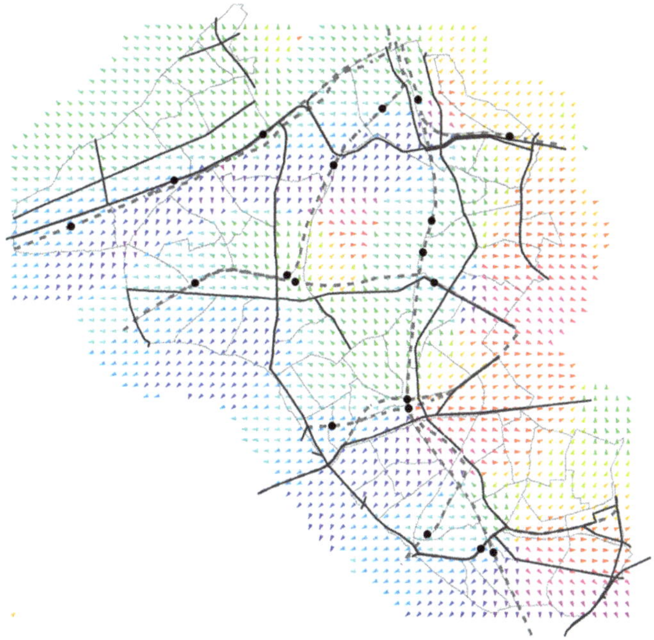

Fig. 13.13 500 m buffer result use of the distribution in the direction of expansion 500 m at buffer Enlarged directional trend distribution

Omotesando, which are the main roads in this area. The same directional arrows can also be observed along Koshu Kaido in a uniform westerly direction, suggesting that the concentration tends to expand along the main road.

The results of this analysis (Fig. 13.14) show that the dispersion values are high near the major stations and trunk roads mentioned above, and these areas can be positioned as the base points for the expansion trend. In particular, the areas with high circular dispersion near trunk roads are long and narrow, indicating that roads are the standard for expansion.

(Expanding trend within a buffer radius of 250 m from the observation point) A more detailed change trend can be read when the observation buffer radius from the observation point in Fig. 13.15 is reduced. In addition to the trends based on major stations and intersections with main roads, small-scale expansion trends can be observed based on stations with residential areas around them, such as Hatsudai, Hatagaya, and Sasazuka stations on the Keio Line and Yoyogi Uehara station on the Odakyu Line. The areas with large values in the circular distribution map (Fig. 13.16) are concentrated near the aforementioned points, and by comparing them with the map showing the trend in the direction of expansion, it can be seen that they are the base points for the expansion of the agglomeration.

(Analysis results for other buffers and overall discussion) Similar to the discussion in the previous Sect. 13.3.3, it can be observed that the analysis results become

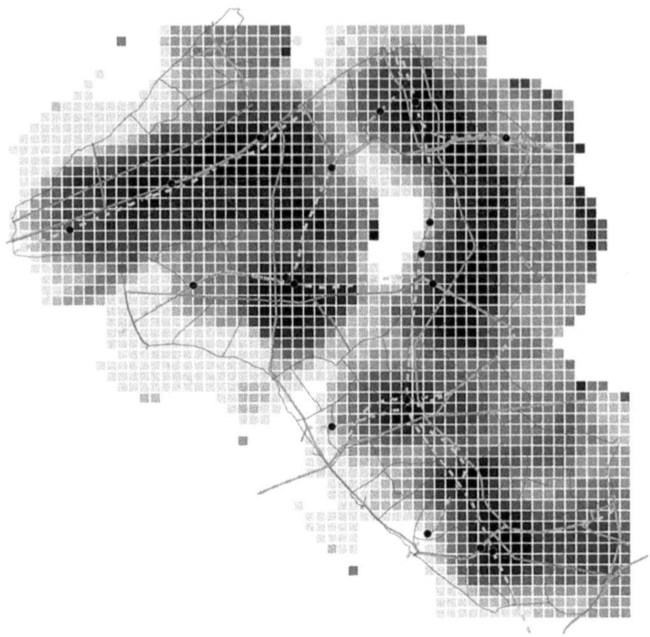

Fig. 13.14 500 m buffer results use of 500 m distribution in the direction of expansion 500 m buffer time Circular dispersion distribution

more detailed as the buffer radius becomes smaller, depending on the size of the buffer radius that determines the area to be analyzed. For the extended directional distribution map results with an irregular array of arrows, such as the results with a buffer radius of 50 m or 30 m, no significant trend can be read in both the extended directional trend distribution map and its circular dispersion distribution map. This is considered to be a limitation of this method.

13.3.5 Analysis and Discussion Using Commercial Agglomeration Expansion Origin Direction Data (Figs. 13.17 and 13.18)

Figures 13.4 and 13.7 show that vectors indicating the direction of origin are concentrated around major stations such as Shinjuku, Shibuya, and Takadanobaba Stations. The 500 m buffer in Fig. 13.7 shows that the vectors are concentrated around Hatsudai, Hatagaya, and Sasazuka stations along the Keio Line, around Kagurazaka, and near the intersection of Meiji-Dori and Omotesando, where many visitors come without depending on nearby shopping areas or major stations.

Fig. 13.15 500 m buffer result use of the distribution in the direction of expansion 250 m at buffer Enlarged directional trend distribution

13.3.6 Visualization of 500 m Buffer Accumulation Expansion Starting Point (Figs. 13.19 and 13.20)

As explained above, the areas around the major stations are painted in red, indicating the starting points of the expansion of commercial clusters. In addition, some locations far from the main station are also indicated as the starting points, indicating that the neighborhood shopping areas and places frequented by many visitors also indicate the starting points of commercial cluster expansion.

Many of the points painted in blue are located between multiple origins, indicating locations where the expansion between each commercial agglomeration converges.

13.3.7 Visualization of 250 m Buffer Accumulation Expansion Starting Point (Figs. 13.21 and 13.22)

Next, when the buffer is set to 250 m to visualize the starting point of the expansion of commercial clusters, the results are almost the same as when the data is aggregated at 500 m. However, when we look at the area around major stations

Fig. 13.16 500 m buffer results use of 500 m distribution in the direction of expansion 250 m at buffer Circular dispersion distribution

such as Shibuya and Shinjuku stations, we can see differences by area divided by stations and railroad lines. In other words, by aggregating the data in 250-m buffers, we can more precisely read the differences in the expansion starting points of the regions.

13.3.8 *Differences in Visualization Results Depending on Buffer Size*

Regarding the effect of buffer size, a relatively large area, such as 500 m, is suitable for reading a more macroscopic situation, such as an expansion centered on a major station. On the other hand, when the buffer size is smaller than 250 m, it is possible to read differences in the area bounded by regional dividing elements such as major stations and railroad tracks, and it is considered to capture more microscopic aspects of the area.

Using detailed spatiotemporal data, we proposed a method to visualize the starting point and convergence point of the expansion of commercial clusters by applying a vector analysis method to the pattern of the expansion of commercial

Fig. 13.17 Commercial agglomeration expansion starting point directional diagram (750 m buffer)

clusters. We also conducted an empirical analysis using actual data. The starting points and convergence points of the expansion were visualized.

In the future, we would like to conduct Monte Carlo simulations and perform statistical tests on the probability distributions obtained from the simulations to investigate the usefulness of this method further.

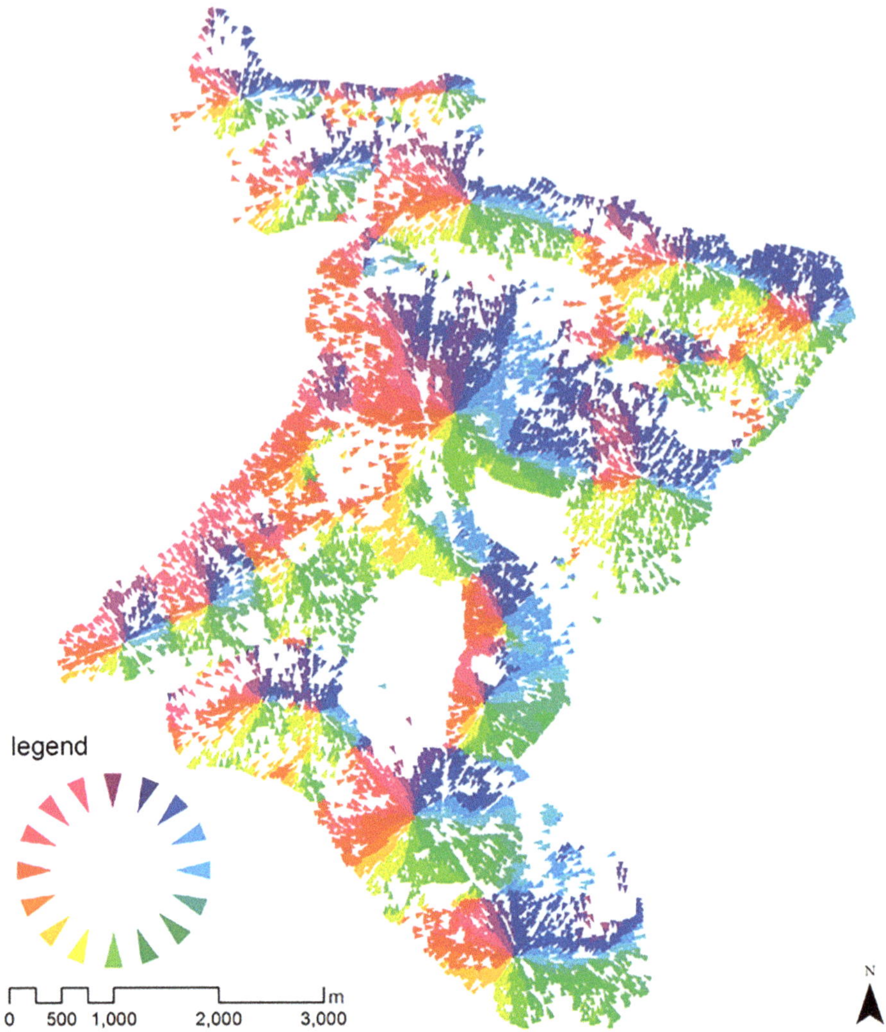

Fig. 13.18 Commercial agglomeration expansion starting point directional diagram (500 m buffer)

13.4 Conclusion

13.4.1 Summary and Future Issues

We proposed a method for describing, analyzing, and visualizing changes accompanying the expansion of complex commercial clusters using a circular statistical method, which is difficult to grasp using conventional methods such as paper maps, analysis based on ordinary statistical data, and animation-like methods that

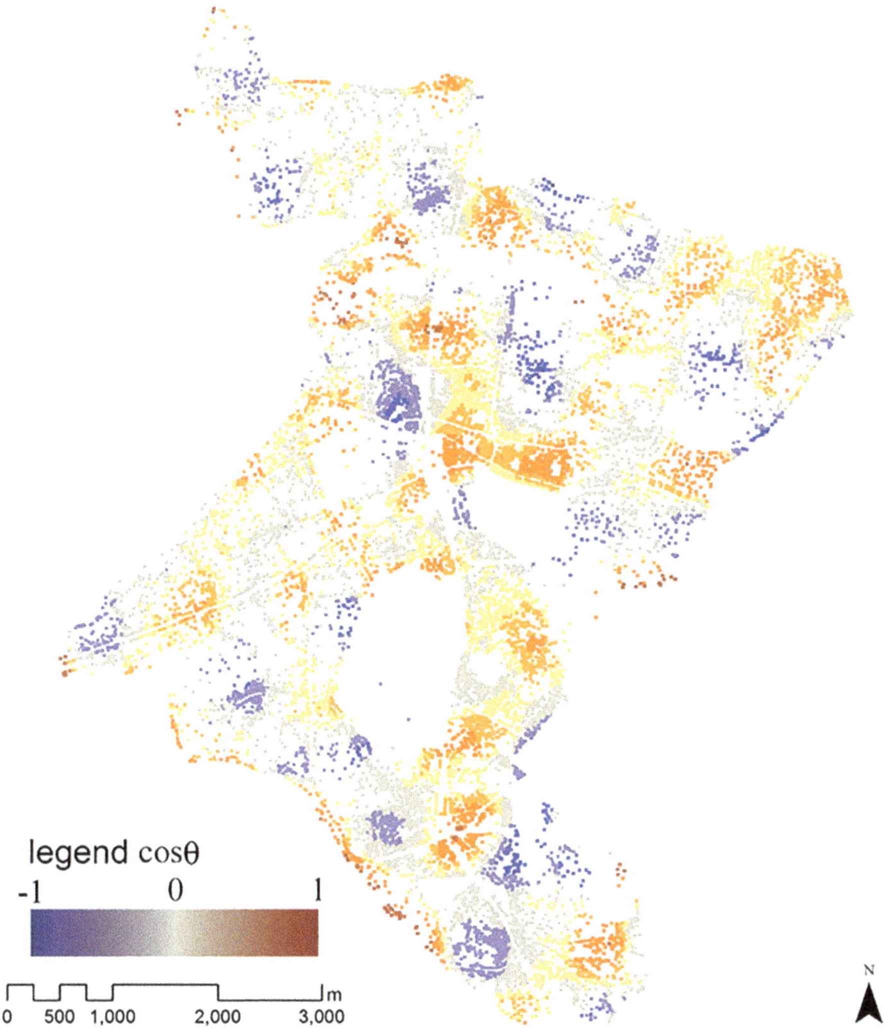

Fig. 13.19 cosθ value distribution (at 500 m buffer) (750 m enlarged starting point directional map)

sequentially display time-series changes. The method is analyzed empirically using spatiotemporal point data of commercial stores in Shibuya Ward, Tokyo, and the results are discussed.

The results obtained are summarized below.

1. First, we analyzed the direction of expansion of commercial clusters by using the circular statistical method to analyze the direction of new store openings relative to existing clusters in each time cross-section and were able to visualize the results. The results show that the arrow points are concentrated and distributed

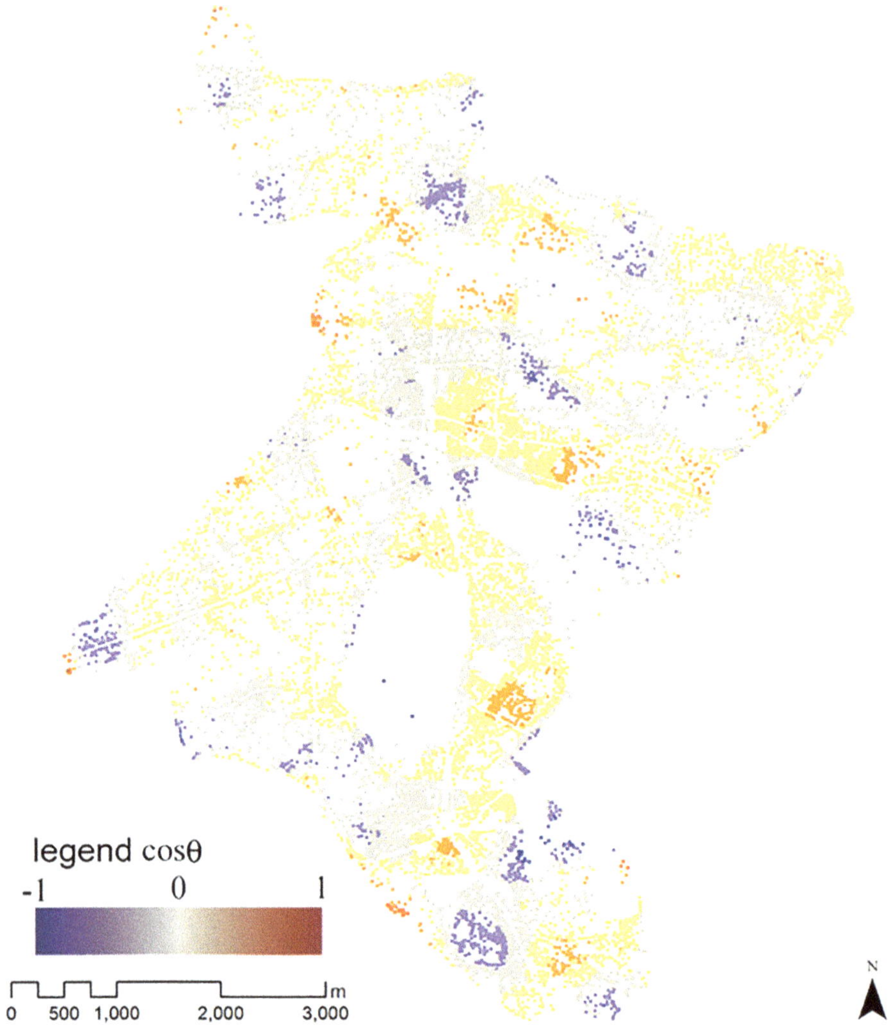

Fig. 13.20 cosθ value distribution (at 500 m buffer) (500 m enlarged starting point directional map)

at the points that are the base points of the expansion of commercial clusters and that they are located near facilities and roads that are generally considered to be the base points of commercial cluster expansion, such as railroad stations and arterial roads, in comparison with the current situation in Shibuya Ward. While conventional mesh data analysis can provide a rough trend of expansion, the analysis using this method can be evaluated to a certain extent in that it can simultaneously identify the base points and directions of expansion.

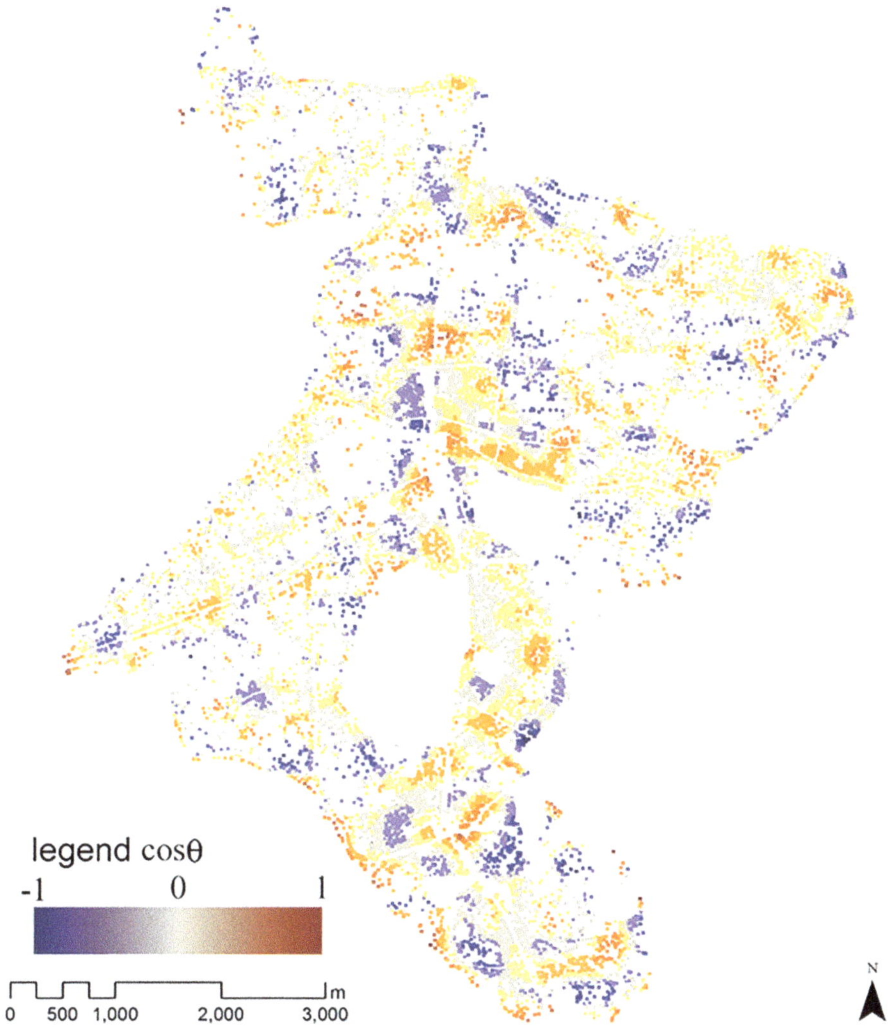

Fig. 13.21 cosθ value distribution (at 250 m buffer) (750 m enlarged starting point directional map)

2. In the map of the distribution of circular dispersion, it was found that the circular dispersion value is decidedly high in the area that can be concluded as the base point in the above map showing the direction of expansion. In other words, high circular dispersion means that the points to be analyzed are uniformly distributed in all directions from the base point. In this case, the area is positioned where relatively old stores are concentrated. The conclusion is generally consistent that the area is the base point of commercial cluster expansion and is being extended outward from there. Although previous studies have captured changes

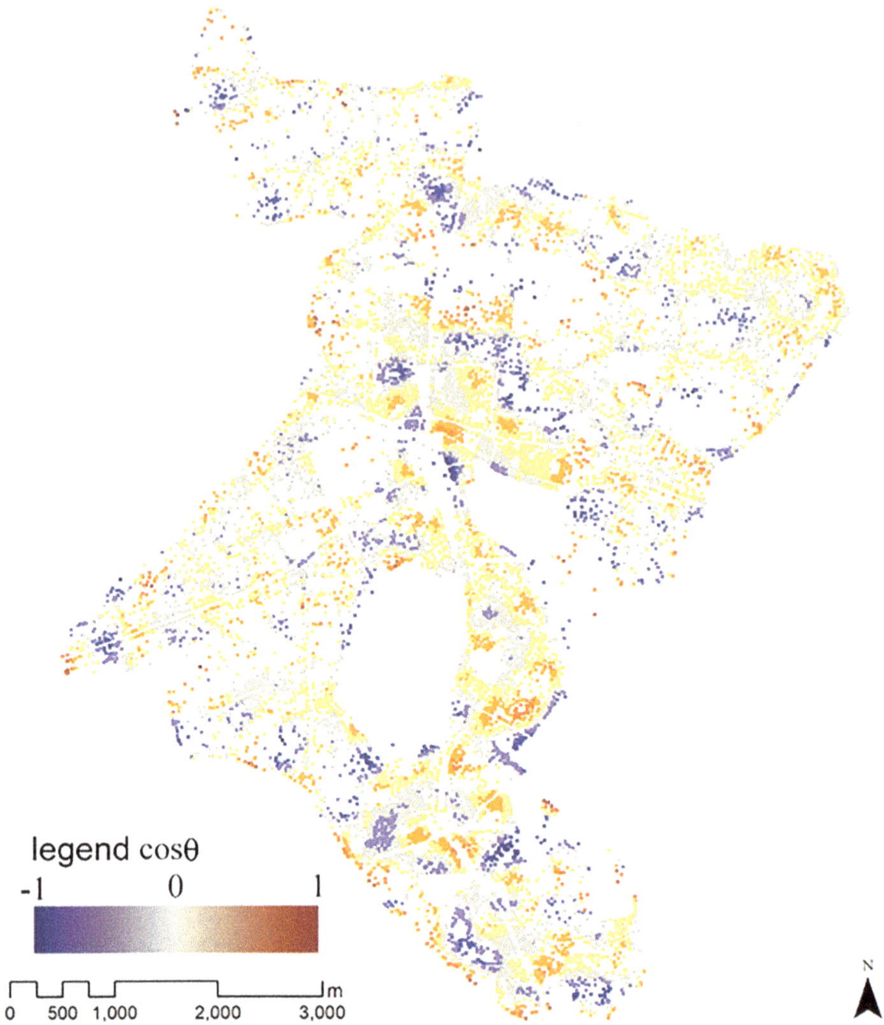

Fig. 13.22 θ value distribution (at 250 m buffer) (500 m enlarged starting point directional map)

in commercial clusters based on train stations, in reality, this method is not limited to such studies but also allows us to understand changes in areas without a clear base point, such as streets or train stations.

3. The conventional method made it difficult to analyze data at different scales, such as township units or mesh data. Still, with this method, it is possible to change the scale of analysis freely. For example, the smaller the buffer radius determining the analysis target area, the more detailed the results become. The smaller the radius, the more detailed the analysis results become. By reducing the buffer radius, we succeeded in capturing more detailed changes, such as changes

in small stations and areas with characteristics as a region but no facilities that can serve as a base point, which were not revealed when the analysis target was selected with a large radius.

4. The analysis from the equally spaced observation points enabled us to capture the expansion trend of the area around each observation point and to show each change more clearly. Similarly, the dispersion diagram showed that the high dispersion values are the points of almost uniform expansion in all directions at the observation points and are the base points of the expansion of commercial clusters.

The following three issues need to be addressed in the future.

1. To simplify the method, we use spatiotemporal data that only considers the location of stores and the time of their opening. In reality, many stores close, and their impact on the formation process is not small. To reproduce the formation process of commercial areas more faithfully, it is necessary to study an analysis method using data that considers store closures.
2. The visualization method of showing the circular dispersion distribution in addition to the expanded directional map and the expanded directional trend distribution sometimes results in a large amount of information, making it difficult to grasp each phenomenon in some cases. Several measures can be considered in the future. A method to automatically detect and highlight the singular points of the vector field from each of the extended directional maps and extended directional trend maps is considered. In addition, since the data display function of a general GIS was used in this study, there are limitations in how the results can be visualized.
3. Although we have only proposed a method to capture the trend of change in commercial clusters based on past changes, we would like to study a future forecasting method to estimate the direction in which commercial clusters are expected to expand in the future, using the results of the analysis based on the method proposed here as test data.

We want to discuss the above issues and others in the future.

13.4.2 Summary and Future Prospects

Urban commercial areas are not only areas with many commercial stores but also places where many people live and are subject to many changes in social and economic conditions, either directly or indirectly. Many of these changes can be seen in the image of the area, the process of formation, and the direction of expansion, which is the subject of this study. The detailed and large amount of spatial data that has become easily available in recent years can be used to capture such urban events, as shown by previous studies and the method proposed in this study.

In this study, we propose an analytical method for extracting correlations among events and the formation process of events and their changes from a large amount of detailed spatial data and a visualization method that enables intuitive interpretation of the data and demonstrates its effectiveness. The methods were empirically validated using actual data from various events in an urban commercial area. On the other hand, the methods are based on a simplified model of events and do not necessarily reproduce real-life events perfectly. Therefore, it should be noted that the method is still insufficient for practical use in urban planning and urban development.

In this study, events such as the exit of a commercial store were treated in a simplified manner to increase the generality of the method. For example, only the opening of a store was treated due to data limitations. In the future, when data that is difficult to obtain, such as at the point of store exit, become available, it will be necessary to consider a method that more closely matches real urban events, including such data.

Next, the analysis of the method proposed in this study uses point data to simplify the method. Many of the events we try to capture in this study are complex and irregular changes and their trends in commercial areas. They are at the basic stage of capturing the nature and aspects of the events themselves. Therefore, to simplify the methodology, we used a point data format to precisely define each attribute, location, and time of occurrence to some extent. However, in an actual urban space, the scale of buildings has not a small influence on each event, and analysis using data such as polygon data, which includes the general area and scale of buildings, as well as the names of tenants, is also conceivable.

In the future, it will be necessary to examine the scale and area of buildings in three dimensions to represent the reality of urban space better. First, we would like to study an analysis method that adds area and other morphological information as attributes to the point data used in this study. Then we would like to develop an analysis method using polygon data that considers shape and scale.

Third, in this study, we could go back to when the latest available data was obtained and capture the changes that occurred during that period and their trends. In urban planning and urban development, it is often discussed in advance what effects a certain plan will have after its implementation and the current situation. The results obtained using the methodology developed in this study capture the trends of each urban event. Therefore, it is thought that using them as test data to predict future changes and transitions will be useful. In the future, it is necessary to consider using the results and findings obtained in each case for methods to clarify the prediction of future changes. The analytical method for this analysis is an issue for the future. The urban commercial areas in this study will be affected by rapid changes in various social conditions, which will accelerate various changes in urban space, such as the exit and exit of commercial stores and the formation of their image. Rapid changes in urban space will burden the lives of residents and various urban activities, and it will continue to be necessary to accurately understand the resulting urban events. In recent years, the development of information technology and ubiquitous technology has made it possible to acquire large amounts of detailed data to ensure a more comfortable life for people. The quality and quantity of spatial data capturing various events will become better and richer, and there will be no shortage of such data in the future. Various spatial analysis and visualization methods have been developed to accurately and efficiently extract various events from these data. As mentioned in Chap. 1, the main objective of this study is to develop an analytical method that can capture trends and structures of events from detailed, diverse, and abundant data resources and derive hypotheses from them, even for exploratory spatial data analysis. As a result, it can be evaluated that each analysis method and visualization method developed in this study can derive hypotheses in an exploratory manner from a discussion of the results of each empirical analysis. We hope this research will serve as a basis for developing analytical methods to capture more advanced spatial phenomena.

References

Akiyoshi I, Yukio S (2010): "A Method of Analysis and Visualization of Expanding Direction of Retail Distribution", Journal of the Architectural Institute of Japan 75(650) 889–895. (in Japanese)

Amin H (2007) Urban visualization with virtual reality. Urban Plan Urban Visual Simul 270:39–42

Batty M (2003) Agent-based pedestrian modelling. In: Longley P, Batty M (eds) Advanced spatial analysis. The CASA book of GIS. ESRI Press, Redland, CA, pp 81–106

Brunsdon C, Corcoran J (2006) Using circular statistics to analyse time patterns in crime incidence. Comput Environ Urban Syst 30(3):300–319

Corcoran J, Chhetri P, Stimson R (2007) Using circular statistics to explore the geography of the journey to work. Reg Sci 88(1):119–132

DiBiase D, MacEachren AM, Krygien JB, Reeves C (1992) Animation and the role of map design in scientific visualization. Cartogr Geogr Inform Syst 19(4):201–214

Dorling D (1992) Stretching space and splicing time: from cartographic animation to interactive visualization. Cartogr Geogr Inform Syst 19(4):β215–β227

Dykes J (1998) Cartographic visualization: exploratory spatial data analysis with local indicators of spatial association using Tcl/Tk and cdv. R Stat Soc Statistician 47(Part 3):485–497

Dykes, Jason (2005): "Facilitating interaction for geovisualization", In: Dykes, J, MacEachren, AM., and Kraak, M-J, (eds.): Exploring geovisualization, Elsevier, Oxford, pp. 143–158

Fisher NI (1993) Statistical analysis of circular data. Cambridge University Press, Cambridge

Gahegan M (2005) Beyond tools: visual support for entire process of GIScience. In: Dykes J, Maceachren AM, Kraak M-J (eds) Exploring geovisualization. Elsevier Ltd, Oxford, UK, pp 83–99

Haining R (2003) Spatial data analysis theory and practice. Cambridge University Press, Cambridge, UK

Hillier B, Hanson J (1984) The social logic of space. Cambridge University Press, Cambridge, UK

Inoue R, Shimizu E (2005) A new method for constructing circle area cartogram. Theory Appl GIS 13(1):43–50

Inoue R, Shimizu E, Yoshida Y, Li Y (2009) Visualization of spatial distribution and temporal transition of land Price in Tokyo 23 wards based on Spatio-temporal kriging interpolation. Theory Appl GIS 17(1):13–24

Kigawa T, Furuyama M (2004) The urban entropy coefficient: a measure describing urban conditions—a morphological analysis on evolutional process of Paris. J City Plan Inst Japan 39(3):823–828

Kigawa T, Furuyama M (2005) Study on a vector in Kyoto's modernization by means of space syntax—a philosophy seen in an urban planning project for installation and broadening streets in Kyoto. J City Plan Inst Japan 40(3):139–144

Kigawa T, Furuyama M (2006) Using space syntax to examine an evolutional process of a local town—a case study on the modernization of Otsu city in Shiga prefecture. J City Plan Inst Japan 41(3):229–234

Monmonier M, Watanabe J (1995) Chizu Ha Usotsukidearu, Shobunsha, Tokyo. Monmonier, Mark (1991): how to lie with maps. University of Chicago Press, Chicago, IL. (Translation).

Roberts, Jonathan, C (2005): "Exploratory Visualization with Multiple Linked Views", In: Dykes, Jason, MacEachren, Alan, M., and Kraak, Menno-Jan,(eds.): Exploring Geovisualization, Elsevier Ltd., Oxford, United Kingdom, pp.159–180

Rodger, Peter (2005): "Graph Drawing Techniques for Geographic Visualization", In: Dykes, Jason, MacEachren, Alan, M., and Kraak, Menno-Jan,(eds.): Exploring Geovisualization, Elsevier Ltd., Oxford, United Kingdom, pp.143–158

Sadahiro, Yukio (2007): "From Spatial Analysis to Spatio-Temporal Analysis: Issues and Future Prospects", Urban Planning <Feature>: Beauty and Reality in Urban Analysis>, 264, pp. 31–35. (in Japanese)

Shimizu K (2008) Recent developments in directional statistics: distributional aspects. Japan Soc Comput Stat 19(2):127–150

Takamatsu S (2007) On visualization methods of form and function: developing techniques in the UK. Urban Plan Urban Visual Simul 270:13–16

Theus M (2005) Statistical data exploration and geographical information visualization. In: Dykes J, MacEachren AM, Kraak M-J (eds) Exploring geovisualization. Elsevier, Oxford, p 127–142

Yan W (2003) GIS no Genri to Ouyou. JUSE Press, Ltd, Tokyo. (in Japanese).

Yoshikawa K (2007a) Application of 3D computer graphics to urban development: assuming its use in municipalities. City Plan Urban Visual Simul 270:43–46

Yoshikawa M (2007b) Digital city and VR. City Plan Urban Visual Simul 270:47–50

Chapter 14
Theoretical Approach for Selection of Public Transport System Considering Urban Density and Travel Distance

Daisuke Hasegawa and Tsutomu Suzuki

Abstract The increasing popularity of ride-sharing services, car sharing, and other transportation options has led to a growing diversification of local public transportation services for short-distance travel. In this study, we aimed to understand how an appropriate public transportation system evolves based on density and travel impedance of passengers using an analytical model. Initially, we developed a model representing five typical transportation systems, each characterized by the presence of fixed routes and varying requirements for access and egress time. Subsequently, we analyzed the influence on expenses that satisfy the service level condition, focusing on the effect of demand density and travel distance. Furthermore, we investigated the current status of the introduction of local transportation in eastern Japan and compared the results with the calculation results of the model. Notably, we observed that increasing the number of buses and demand-responsive transit (DRT) operated by the local government led to enhanced effectiveness of car sharing, particularly in areas where DRT services were introduced.

Keywords Local public transportation · Trip length · Demand density · Operation efficiency · Demand-responsive transit

---

The contents of this paper are based on the following paper originally published in a Japanese journal: Daisuke, H., Tsutomu, S. (2017) Theoretical Approach for the Selection of Public Transport Systems Considering Ur-ban Density and Travel Distance. Journal of City Planning Institute of Japan, 52 (3), pp. 1284–1289 (in Japanese).

D. Hasegawa (✉)
The University of Tokyo, Tokyo, Japan
e-mail: hasega60@e.u-tokyo.ac.jp

T. Suzuki
University of Tsukuba, Ibaraki, Japan
e-mail: tsutomu@risk.tsukuba.ac.jp

Y. Asami et al. (eds.), *Studies in Housing and Urban Analysis in Japan*,
New Frontiers in Regional Science: Asian Perspectives 75,
https://doi.org/10.1007/978-981-99-8027-7_14

14.1 Introduction

Motorization in Japan has made significant strides since the 1960s. The changing urban landscape has resulted in a growing reliance on suburban areas, as evident from the proliferation of roadside shops. Consequently, the number of urban public transportation users has seen a noticeable decline, significantly affecting urban buses with short operating distances. Consequently, an important issue concerning municipalities has been raised: the reduced public transportation threatens the daily lives of mobility-impaired people, such as the elderly, who cannot drive cars. In response to these challenges, there has been a nationwide effort to restructure local public transportation networks, with a primary focus on reorganizing bus route networks. This aims to create more accessible boarding points and ensure regular operation of vehicles to improve the overall public transportation experience.

In recent years, several municipalities have adopted demand-responsive transit (DRT) solutions, DRT systems offer the flexibility to adjust routes dynamically based on user demand taking into account modern modes of transportation like ride-sharing (e.g., Uber) and car-sharing services. These new local public transportation services are distinguished by operational factors including the accessibility of boarding locations and the presence of route/operation schedules. They play a crucial role in bridging transportation gaps (Kato and Takeuchi 1979; Niitani 1993) for individuals who lack access to existing means of transportation and those facing difficulties in utilizing existing means of transportation. However, one key challenge lies in the variable operating route lengths of these new transportation services, which vary based on user demand. Consequently, operational efficiency is heavily influenced by the demand density and the size of the target region. Therefore, the advantages and disadvantages of each means of transportation are determined based on user density and trip length; however, it is difficult to identify these specific conditions quantitatively.

Research has been conducted on the relationship between demand density and the means of transportation. For example, Niitani (1993) categorized public transportation based on running space and transportation capacity and proposed the appropriate coverage area for each mode of transportation based on demand density and trip length. Ishida et al. (1999) examined transportation costs from both supple and user perspectives and identified advantageous regions in the city. Nevertheless, these studies primarily focused on inter-city transport, concentrating on railroads and automobiles, without delving into low demand density and short-distance inter-city transport.

Considering research from the perspective of rationality in delivery, Ieda (1997) and Watanabe and Suzuki (2000) conducted studies focusing on the optimal hierarchical structure of transport transfer facilities (i.e., hubs) in the field of logistics. However, minimal research has addressed the difference in operation methods, such as direct and touring trips in the passenger sector, which exhibit limited hierarchy. Hasegawa and Suzuki (2011) focused on the effectiveness of DRT and quantitatively demonstrated the characteristics of its operational efficiency, including its superiority in low-density cities, rapid decline in operational efficiency due to vehicle shortage, and improvement of operational efficiency through the use of both large and small vehicles. However, certain issues persist, such as the consideration of road density and shape and the continuous representation of the concentration and dispersion of travel demand.

To address these gaps, our study modeled five typical transportation services with varying operational factors, such as the accessibility of boarding locations and the presence of route/operation schedules. We analyzed the effects of demand density and user travel distance on the cost required for each transportation mode to achieve a certain service level. Additionally, we surveyed the current state of their introduction to local public transportation, identified the characteristics of demand density and travel distance, and compared these findings with the model results to theoretically derive the demand density and travel distance where each transportation service becomes advantageous, along with their advantageous regions.

14.2 Comparison of Suitability of Transportation Services Based on City Model

In this section, we provide an overview of the city model and five typical transportation services. The indicators for evaluating the superiority or inferiority of each service method were defined. We then quantitatively clarify the cost required to achieve a certain service level based on the demand density and user-mean travel distance.

14.2.1 City Model Settings

Three main assumptions were considered for our city model:

1. Users with a demand density ρ (people/km^2 h) are uniformly distributed on an infinite plane
2. Users head in various directions from their departure point
3. Travel demand to their destination is attenuated with distance according to an exponential distribution, as shown in Eq. (14.1):

$$q(x) = \lambda e^{-\frac{2}{d}x}. \tag{14.1}$$

Here, if the user mean travel distance d is obtained by averaging Eq. (14.1) over the range of x to infinity, then λ is a parameter related to diminishing distance and is set as s $\lambda = 2\rho/(d\pi)^2$.

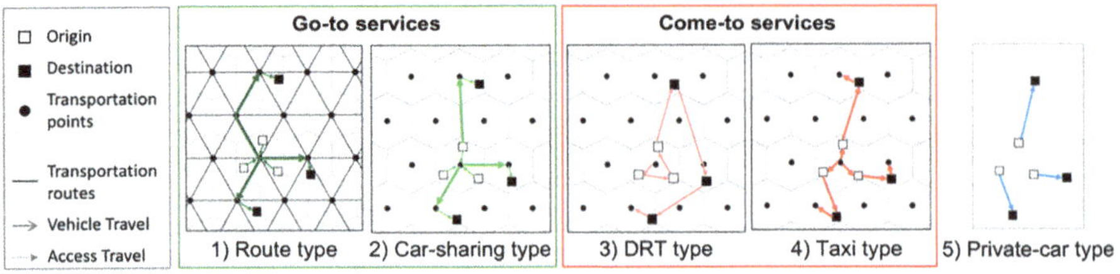

Fig. 14.1 Five types of transportation services

14.2.2 *Modeling Transportation Services*

We defined five transportation services and calculated the cost and service level achieved for each service (mean travel time). Figure 14.1 presents a conceptual diagram of each transportation service.

With respect to transportation points such as train stations, bus stops, and car-sharing parking lots, transportation services can be divided into two categories: go-to services, where users can access or egress the points, and come-to services, where a vehicle picks up the user from the transportation point to the user's departure point.

Five typical transportation services were assumed:

1. Route type: Users go to transportation points and ride vehicles that run at regular intervals on routes connecting the transportation points
2. Car-sharing type: Users access transportation points similarly to the route type but allows for free movement between transportation points
3. DRT type: Shared transportation is used from the user's departure point to the destination
4. Taxi type: These operate on non-fixed schedules and routes, and users are transported from their departure point to their destination on demand without carpooling
5. Private car type: Users go directly from the departure point to the destination by private car, bicycle, or foot

The route and car-sharing types are classified as go-to services, while the demand and taxi types are classified as come-to-services. It is assumed that in a given setting or situation, one of the five types is adopted.

14.2.3 *Provision Cost and User's Mean Travel Time for Each Transportation Service*

In this section, the cost of providing each transportation service and the user travel time achieved by each service are calculated. Each transportation service i has a uniform transportation point, with a travel distance d_{ai} for the access/egress part between the transportation point and departure point or destination, and a travel distance d_{ci} between the departure point and destination.

1) Route Type

A user accesses the nearest transportation point placed at density b_1 (points/km^2) and the access time per unit distance (reciprocal of speed) t_{a1} (h/km) (Fig. 14.1 ┈▸). The user then boards a vehicle that arrives at a fixed operation interval f_1 (h) on a route that connects the transportation points and moves to the transportation point nearest to the destination at the vehicle travel time per unit distance t_{c1} (h/km) (➔). The user then conducts an egress movement from the transportation point to the destination for time t_{a1}. The user selects the shortest route to the transportation point nearest the destination and does not consider the time required for transfers or the acceleration and deceleration times required when the vehicle is stopping. For the user's mean access/egress time d_{a1} (km), the transportation points are assumed to be uniformly distributed in a hexagonal lattice, and because the mean distance to the center of a regular hexagonal region (access travel) is not significantly different from the distance from a disk-shaped region to the center of the circle, this distance is used as an approximation:

$$d_{a1} = \frac{2}{3\sqrt{\pi}} \frac{1}{\sqrt{b_1}}. \tag{14.2}$$

The vehicle travel distance d_{c1} (km) and route density l (km/km^2) are obtained from the user's mean travel distance (from the origin point □ to the destination point ■) $\hat{d}$ (km) and transportation point density b_1 as follows:

$$d_{a1} = \alpha \hat{d}, \tag{14.3}$$

$$l = \beta \sqrt{b_1}, \tag{14.4}$$

where α and β are constant terms in the network travel distance and density, respectively. Although the network distance naturally differs depending on its shape, it is considered proportional to the straight-line distance (Koshizuka and Kobayashi 1983). The α value approaches 1 as the network becomes denser. In this study, the transportation points, which are intersections of routes, are defined as uniformly arranged in a hexagonal lattice (i.e., the route network is connected in a triangular lattice); therefore, α is the area ratio of a sector with a central angle of 60° and an equilateral triangle, whereas β is the total distance of the triangular lattice network clarified by Watanabe (2006)

$$\alpha = \frac{2\pi}{3\sqrt{3}} \approx 1.209, \tag{14.5}$$

$$\beta = 3\sqrt{\frac{2\sqrt{3}}{3}} \approx 3.233. \tag{14.6}$$

The operation frequency f_1 is obtained by dividing the travel time of route density l by the vehicle density n_1 (vehicles/km^2), with half of this value set as the mean waiting time.

$$f_1 = \frac{l t_{c1}}{n_1}. \tag{14.7}$$

The user mean travel time T_1 (h) is set as

$$T_1 = d_{c1} t_{c1} + d_{a1} t_{a1} + f_1/2, \tag{14.8}$$

and the route type provision cost C_1 (yen/h) is set as

$$C_1 = l CL_1 + b_1 CB_1 + n_1 CN_1, \tag{14.9}$$

where CL_1 (yen·km/h), CB_1 (yen·point/h), and CN_1 (yen·vehicle/h) are the cost parameters per hour for the route, transportation point, and vehicle, respectively. The route density l was obtained from the transportation point density using Eq. (14.4), where the total cost is determined using the transportation point and vehicle densities.

2) Car-sharing Type

A user accesses (······▸) a transportation point placed at density b_2 (points/km^2) and unit time t_{a2} (h/km) in the same manner as in the route type, whereas distance d_{a2} (km) is set as follows, similar to Eq. (2):

$$d_{a2} = \frac{2}{3\sqrt{\pi}} \frac{1}{\sqrt{b_2}}. \tag{14.10}$$

It is assumed that the vehicle travels in a straight line (→) to the transportation point nearest to the destination at a unit vehicle travel time t_{c2} (h/km), and that travel distance d_{c2} (km) is set as the user's mean travel distance d (km). In addition, one-way drop-offs were assumed to be possible, and the time required for the vehicle to return to its original transportation point was not considered. Under the assumption that vehicles converge to a given point from multiple directions, the waiting time f_2 (h) until a user is assigned a vehicle is set as the total usage time, which is obtained by dividing the vehicle usage time per person by the demand density divided by the number of vehicles n_2 (vehicles).

$$f_2 = \frac{\rho \hat{d} t_{c2}}{n_2}. \tag{14.11}$$

The car-sharing users' mean travel time T_2 (h) and provision cost C_2 (yen/h) are expressed as follows:

$$T_2 = d_{c2}t_{c2} + d_{a2}t_{a2} + f_2/2, \tag{14.12}$$

$$C_2 = b_2CB_2 + n_2CN_2. \tag{14.13}$$

3) DRT Type

A user has a vehicle operating at regular intervals from a transportation point located at density b_3 (points/km^2) come to them at the departure point at a unit vehicle travel time t_{c3} (h/km), after which the user travels to the destination (→). During this time, the vehicle passes through the departure points and destinations of other users who are riding together, and after completing the rounds of the departure points, the vehicle makes rounds of the destinations. The user requires additional boarding time because of the operation interval and detours to the departure points of other users.

The route length $2d_{c3}$ (km) that connects the user's departure point and destination after the vehicle picks up the user is obtained via the Beardwood–Halton–Hammersley (BHH) theorem (Beardwood et al. 1959). The value obtained by multiplying the region area (reciprocal of transportation point density b_3) of each transportation point by the demand density ρ is set as the number of passengers. For the region where the vehicle makes its rounds, if the departure point is set as a transportation point-specific region and the destination is set as a disk-shaped region where the user mean distance d is set as the radius, then the route distance converges to Eq. (14.14) (where $k = 0.765$):

$$2d_{c3} = \frac{k\sqrt{\rho}}{2}\left(\frac{1}{b_3} + \sqrt{\frac{\pi d^2}{b_3}}\right). \tag{14.14}$$

The mean ride distance d_{c3} (km) to the destination was set to half the value in Eq. (12). The operation interval f_3 (h) of the vehicle is set as the pickup time when a vehicle travels the route length $2d_{c3}$ at vehicle travel time t_{c3} divided by the product of the number of vehicles n_3 (vehicles/km^2) and the transportation point density b_3. Half of the resulting value is set as the mean waiting time of the user.

$$f_3 = \frac{2d_{c3}t_{c3}}{b_3n_3}. \tag{14.15}$$

Therefore, the DRT type user's mean travel time T_3 (h) and provision cost C_3 (yen/h) are as follows.

$$T_3 = d_{c3}t_{c3} + f_3/2, \tag{14.16}$$

$$C_3 = b_3CB_3 + n_3CN_3. \tag{14.17}$$

4) Taxi Type

A user has a vehicle coming to the departure point from a transportation point placed at density b_4 (points/km^2), from which the user goes (→) to the destination at time per distance t_{c4} (h/km). There is no carpooling when a user is directly transported from the departure point to the destination. If ride-sharing services are considered similar to taxis, then ride-sharing can be considered as a case wherein a high density of transportation points for a taxi-type service setup are present, except that instead of professional drivers providing the service, general drivers who are residents pickup and drop-off users. The distance d_{a4} (km) from the transportation point to the departure point and from the destination to the nearest transportation point was calculated based on the transportation point density b_4 using the following equation, which is similar to Eqs. (14.2) and (14.10) as follows:

$$d_{a4} = \frac{2}{3\sqrt{\pi}} \frac{1}{\sqrt{b_4}}. \tag{14.18}$$

The ride distance d_{c4} to the destination is assumed to be equivalent to the user's mean travel distance d. Vehicles are assigned to each user on a first-come-first-served basis; when all n_4 (vehicles/km^2) vehicles are in use, the waiting time f_4 (h) is included. The pickup time when a vehicle travels a distance from the transportation point to the departure point and from the destination to the nearest transportation point at vehicle travel time t_{c4} per unit distance is divided by the product of the number of vehicles n_4 (vehicles/km^2) and transportation point density b_4, with half of the resulting value set as the user waiting time.

$$f_4 = \frac{2d_{c3}t_{c3}}{b_3 n_3}. \tag{14.19}$$

Therefore, the taxi-type user's mean travel time T_4 (h) and provision cost C_4 (yen/h) are as follows:

$$T_4 = d_{c4}t_{c4} + f_4/2, \tag{14.20}$$

$$C_4 = b_4 CB_4 + n_4 CN_4. \tag{14.21}$$

5) Private Car Type

The user owns a vehicle and drives it directly to the destination at a unit vehicle time t_{c5} (h/km) (→). Therefore, the private car type user's mean travel time T_5 (h) is as follows:

$$T_5 = \hat{d}t_{c5}. \tag{14.22}$$

The number of vehicles n_5 (vehicles/h) is defined as the product of demand density and travel time, assuming that each vehicle has a cost associated with the transportation point, which is the vehicle storage location. Therefore, the number of vehicles n_5 and the provision cost C_5 (yen/h) are as follows:

$$n_5 = \rho T_5, \tag{14.23}$$

$$C_5 = (CB_5 + CN_5)\, n_5. \tag{14.24}$$

14.3 Transportation Services that Minimize Costs

Table 14.1 lists the parameters used for the model analysis. For each transportation service type, T_i is given as a parameter from the equations obtained in Sect. 14.2, and the number of vehicles n_i is obtained by solving the equation for f_i. The user travel time T_i was set to 1/2 (h), and the cost to achieve a service level of 30 min on average was calculated. The parameter value per hour was set with reference to automobile transportation business management indicators(Automobile Business Association of Japan 2017). Figure 14.2 illustrates the cost per person c_i (yen/h/person) obtained by dividing the provision cost C_i by the demand density ρ when the transportation point density b_i is changed. Based on these results, the following insights were derived.

When the transportation point density b_i increases, the mean travel time decreases because the access distance d_{ai} and the number of users per transportation point decrease, and there exists an optimal transportation point density that minimizes c_i. However, the number of private vehicles n_5 and transportation density b_5 are uniquely determined by the demand density; thus, these values do not change.

Table 14.1 Parameter settings for each transportation service i

			(1) Route type	(2) Car-sharing type	(3) DRT type	(4) Taxi type	(5) Private car type
T_i	User travel time	(h)	1/2	1/2	1/2	1/2	1/2
t_{ai}	Reciprocal of access speed	(h/km)	1/3	1/3	–	–	–
t_{ci}	Reciprocal of vehicle speed	(h/km)	1/30	1/30	1/30	1/30	1/30
CL_i	Route cost	(Yen·km/h)	90	–	–	–	–
CB_i	Transportation point cost	(Yen·point/h)	50	100	100	100	200
CN_i	Vehicle cost	(Yen·vehicle/h)	5000	2500	2500	2500	2500

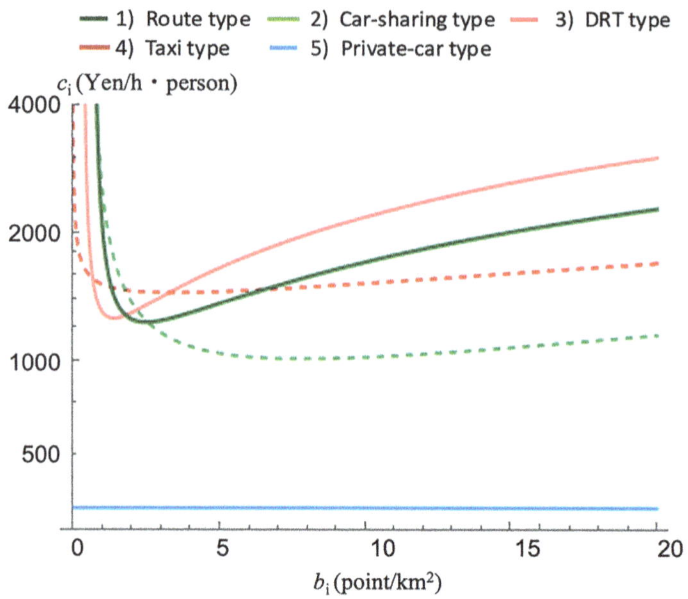

Fig. 14.2 Relationship between transportation point cost b_i and cost c_i per person ($\rho = 5, \hat{d} = 5$)

- The optimal transportation point density is higher for the go-to services category, which experiences a larger increase in mean travel time owing to an increase in access distance.

Here, we verify the change in c_i when the user's mean distance and demand density change when b_i where c_i is at a minimum and b_i* is substituted. Figure 14.3 shows c_i as d changes. From the figure, the following can be observed.

- Route type exhibits increasing costs with travel distance, whereas other means of transportation exhibit diminishing costs. For short-travel distances, DRT is advantageous. As the distance increases, car-sharing becomes more advantageous. However, over time, the DRT type becomes advantageous again, because increases in d have little effect on the waiting time.

Figure 14.4 shows changes in c_i when the demand density ρ is changed, and Fig. 14.5 shows the changes in c_i when both d and ρ are changed. It can be inferred from these figures that:

- The route type is not affected by changes in demand density; thus, when demand is high, the cost per person decreases, and the route type becomes advantageous.
- For the other means of transportation, the cost per person decreases exponentially as demand increases. When demand is high, the most advantageous transportation service is car sharing, followed by private cars.

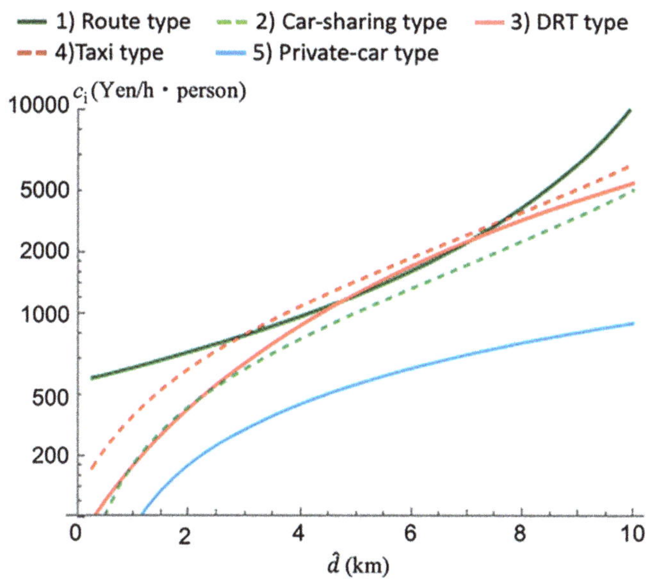

Fig. 14.3 Relationship between user's mean travel distance $\hat{d}$ and cost c_i per person ($\rho = 5$)

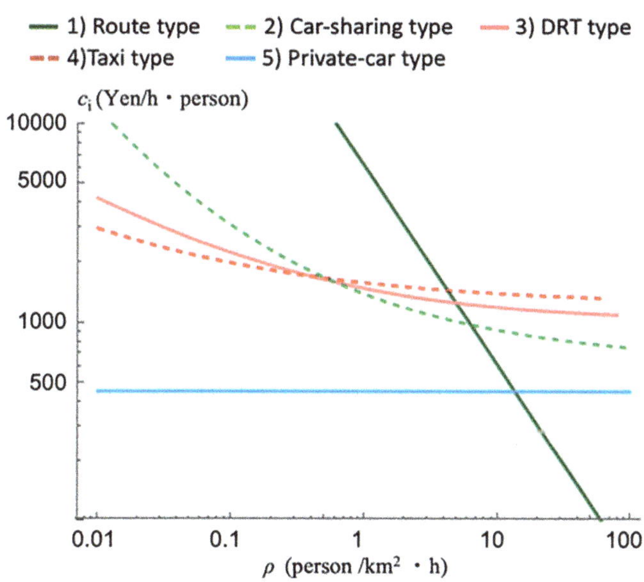

Fig. 14.4 Relationship between demand density ρ and cost c_i per person($\hat{d} = 5$)

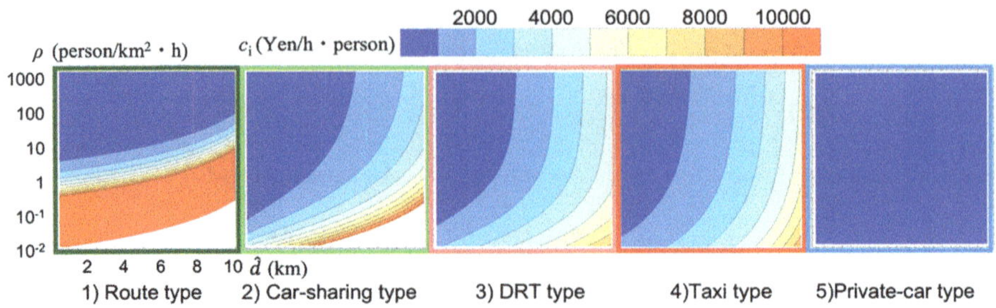

Fig. 14.5 Relationship between user's mean travel distance $\hat{d}$ and demand density ρ and cost c_i per person

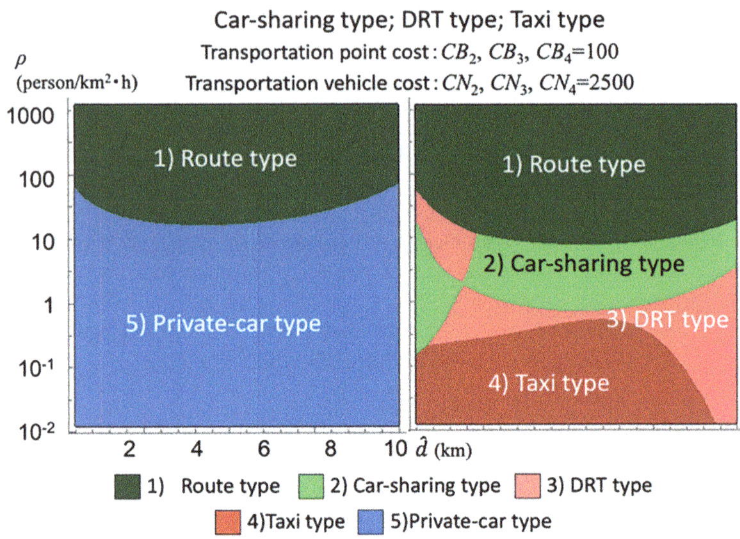

Fig. 14.6 Relationship between user's mean travel distance d and demand density ρ, and means of transportation where C_i is minimized (right: except private car type)

Figure 14.6 shows the means of transportation, where c_i is at a minimum as the ρ and d values are changed, divided between cases where the private car type is considered and where private cars can be used and cases where private cars cannot be used. This result shows that

- The private car type has a large advantageous region, demonstrating the overall superiority of private cars. However, when the demand is high, the route type becomes advantageous.
- When the private car type was not considered, as demand decreased, advantageous regions appeared for the car-sharing type for long distances and for the DRT type for short distances.

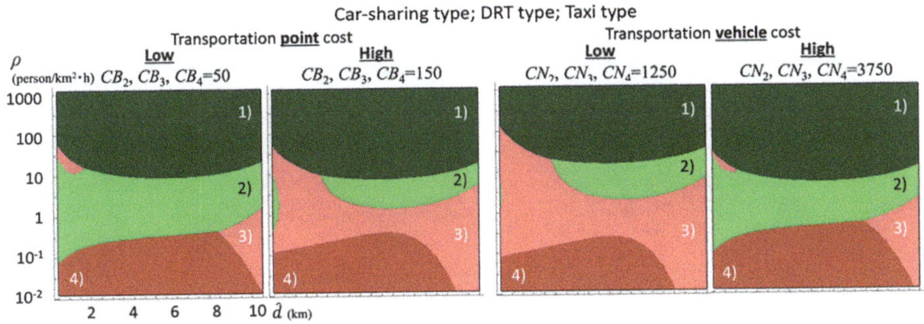

Fig. 14.7 Relationship between changes in car-sharing, demand, and taxi type transportation point and vehicle cost and advantageous regions

Figure 14.7 shows an analysis of the effects on advantageous regions when the car-sharing, demand, and taxi-type transportation point costs CB_2, CB_3, CB_4, and vehicle costs CN_2, CN_3, and CN_4 are changed for unit cost sensitivity analysis. The following conclusions were drawn.

- As shown in Fig. 14.2, fluctuations in transportation-point costs have a significant effect on the car-sharing type, where the number of transportation points is large, whereas decreases in vehicle costs have a considerable effect on the DRT type.

These results show that when demand is low, the difference in costs between the car-sharing and DRT types is small and that the superiority of either type changes with slight changes in cost.

14.4 Comparison Between Actual Circumstances of Introduction of Transportation Services to Local Public Transportation and Results of Theoretical Model

In this section, we examine the circumstances surrounding the introduction of these services to local public transportation, primarily by municipalities in eastern Japan. After comparing the demand densities and travel distance values of the introduced means of transportation, we compared the advantageous regions for these transportation services according to the demand density and travel distance derived in the previous section.

14.4.1 Status of Introduction of Services to Local Public Transportation

Figure 14.8 presents the results of an Internet and literature survey of 734 municipalities across 14 prefectures in eastern Japan, focusing on community buses

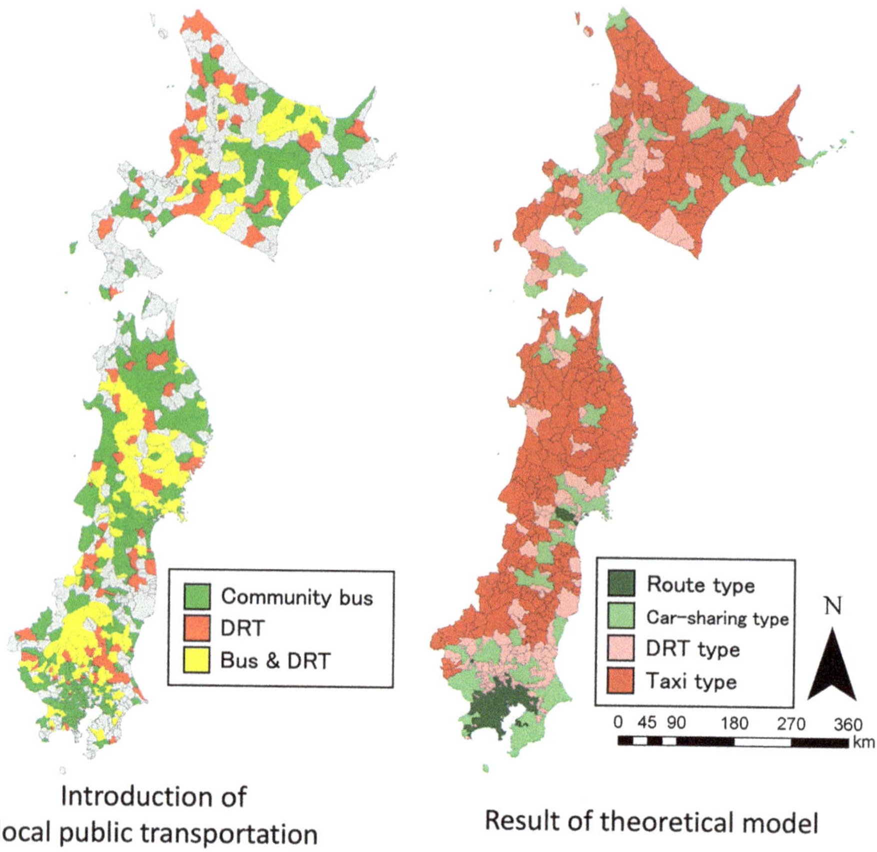

Fig. 14.8 Actual circumstances of local public transportation introduction and model-based classification results

and demand taxis operated mainly by municipalities. The survey identified 371 confirmed cases of introduction of community buses and 193 cases of demand taxis, with 115 cases utilizing both services. Comparing these results with a 2011 survey by Nihon Keizai Shinbunsha (2012), community buses and demand taxis increased by 199 and 117 cases, respectively, indicating the diversification of local public transportation in municipalities.

14.4.2 Comparison with Model Results

We investigated the correspondence between the advantageous regions of each transportation service and the introduction results for each municipality based on the relationship between d/ρ and provision cost. Demand density and travel distance

Table 14.2 Mean travel distance d and demand density ρ values for each local public transportation service

	Count	$\hat{d}$ (km)	ρ (person/km^2·h)
Community bus	303	3.75	40.15
DRT	87	5.29	2.19
Bus and DRT	117	4.51	6.23
Not applicable	258	5.67	9.79

were calculated for each 1-km mesh, aggregated by municipality, and averaged. The demand density ρ_k per hour for a 1-km grid k is calculated using the following equation:

$$\rho_k = P_k\,(Day/7)\,(Hour/24)\,, \tag{14.25}$$

where P_k is the total population (people) of mesh k in 2010, *Day* is the number of outings per week, and *Hour* is the number of outings per day. Assuming that users visit the nearest hospital/supermarket for 1 h per day, we set $Day = 2$ and $Hour = 1$. The mean travel distance d_k of mesh k takes the straight-line distance from the representative point in each mesh to the nearest point among the 8657 hospital points (Wellness Corporation 2011) and 41,351 supermarket points (Shogyo Kai 2013) using GIS, and averages them. Table 14.2 shows the mean travel distance and demand density values by means of transportation, excluding the island areas. These data indicate that municipalities that have introduced community buses have a short mean travel distance and a high demand density of 40 people/h km^2, whereas those that have introduced demand taxis have long travel distances and low demand densities.

In the case illustrated in Fig. 14.6, which does not consider private car types, Fig. 14.9 shows the mean travel distance and demand density values for each municipality. Figure 14.8 (right) shows their geographical distribution. Although this analysis has limitations because of the assumption of uniform demand density by the model and the use of an estimated value for the travel distance, the boundary between the community bus and demand taxi is generally consistent with boundaries (1), (2), (3), and (4) shown in the figure.

Of the 774 municipalities included in the analysis, 154 were classified as route types corresponding to regions with high population densities and destinations within short distances, such as the Tokyo metropolitan area and Sendai City in Miyagi Prefecture. This result reflects the real world, where railroads and bus routes are already densely developed in these regions. In contrast, 148 municipalities were classified as car-sharing type, which included regions such as Chiba, Saitama, Kanagawa, and Miyagi, with demand densities similar to those of route-type settings: some areas of Hokkaido where the demand density is low, but the travel distance is long. However, unlike route-type municipalities, to date, no locations have achieved car-sharing-type local public transportation, with only a mixed presence of community buses and demand taxis. Finally, 188 municipalities were classified as the DRT type, including northern Kanto, the area around Sendai City in Miyagi Prefecture, and central Hokkaido. Among the cities classified as car-sharing

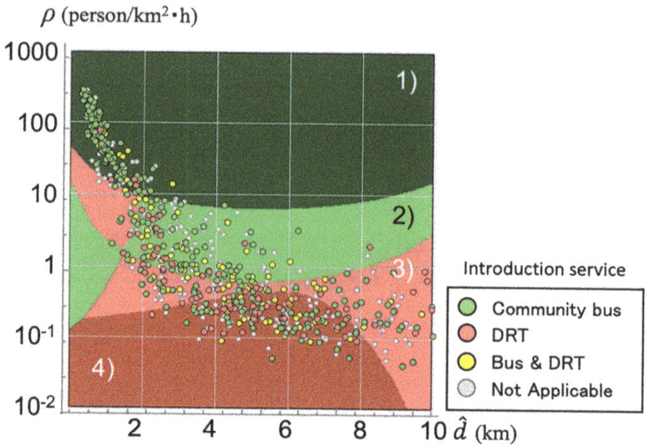

Fig. 14.9 Relationship between advantageous region by each transportation service and travel distance d and demand density ρ

and DRT types, 108 locations already had demand taxis in operation (of which 62 locations used community buses), which is consistent with the analysis results. Meanwhile, 287 municipalities were classified as taxi types, with many distributed in the Tohoku region. However, based on a comparison with Fig. 14.8 (left), many regions introduced community buses instead. Considering the per capita local bus route maintenance costs in Japan Bus Association (2016), Aomori Prefecture ranked third nationwide, whereas Akita Prefecture ranked seventh nationwide, indicating its high costs; thus, it is speculated that introducing a taxi-type service setup could reduce costs. In addition, the explanatory power of the analysis could be increased if cases where habitable land is limited (e.g., hilly and mountainous areas) and where actual demand density is high were considered in the study.

14.5 Summary and Future Issues

This study aims to provide a theoretical discussion on the benefits of increasingly diversifying local public transportation and its advantages over the current transportation situation. We derived the basic conditions for changes in costs for each means of transportation to achieve a certain service level based on the relationship between demand density and user travel distance. These results were then compared with the local transportation services introduced and implemented in specific regions. The following conclusions were drawn:

1. The route type is superior when the demand density is high, while the taxi type is superior when the demand density is low. The car-sharing type and DRT type

fall between these two types, with the DRT-type service setup becoming more advantageous as the travel distance increases.
2. The results of investigating the actual introduction of these transportation services to local public transportation revealed that regions where community buses were introduced tended to have high demand densities and short travel distances, while regions with DRT-type services were introduced showed the opposite tendencies.
3. A comparison of the advantageous regions for each transportation service and the values observed in different municipalities partially confirmed the consistency of the model with regions where route and DRT-type services were introduced.

For future studies, the following topics should be considered: correspondence of parameters to actual conditions, discussions that consider capacity, and addressing imbalanced demand distributions.

Acknowledgments This study was supported in part by JSPS KAKENHI (Grant Numbers JP 26560162 and 16 J02064), the Oobayashi Foundation, and the Division of Policy and Planning Science Commons at the University of Tsukuba.

References

Automobile business Association of Japan (2017) Motor truck transportation business management indicators
Beardwood J, Halton JH, Hammersley JM (1959) The shortest path through many points. Proc Camb Philos Soc 55:299–232
Hasegawa D, Suzuki T (2011) Theoretical consideration of demand-based transportation establishment conditions focusing on urban scale and density. J Urban Plan 46(3):817–822
Ieda H (1997) Formulation and solution of hierarchical transportation systems, such as hub-spokes/point-to-point and aggregated/direct transportation, on a homogeneous infinite plane. J Civil Eng Plan 14:773–782
Ishida T, Taniguchi M, Suzuki T, Furuya H (1999) Analysis of the effectiveness of transportation policies focusing on the feasible range of transportation modes and favorable areas. Transport Policy Res 2(1):14–25
Japan Bus Association (2016) Japanese bus business 2015
Kato A, Takeuchi D (1979) New edition of urban transportation and urban planning. Gijutsu Shoin
Koshizuka T, Kobayashi J (1983) Road distance and straight-line distance. J Urban Plan 18:43–48
Nihon Keizai Shinbunsha (2012) Evaluation of sustainability of national cities
Niitani Y (1993) Urban transportation planning. Gihodo Publishing
Shogyo Kai (2013) Japan supermarket directory
Watanabe D (2006) Mathematical study on network form analysis and transportation system optimization from the perspective of proximity, Doctoral dissertation, graduate School of Tsukuba University, 2006
Watanabe D, Suzuki T (2000) Study on hierarchical collection and distribution transportation systems considering economies of scale. J Urban Plan 35:1027–1032
Wellness Corporation (2011) National hospital directory

GPSR Compliance
The European Union's (EU) General Product Safety Regulation (GPSR) is a set of rules that requires consumer products to be safe and our obligations to ensure this.

If you have any concerns about our products, you can contact us on

ProductSafety@springernature.com

In case Publisher is established outside the EU, the EU authorized representative is:

Springer Nature Customer Service Center GmbH
Europaplatz 3
69115 Heidelberg, Germany

www.ingramcontent.com/pod-product-compliance
Ingram Content Group UK Ltd.
Pitfield, Milton Keynes, MK11 3LW, UK
UKHW022213051026
15624UKWH00001B/140
9789819980260